Advances in Intelligent Systems and Computing

Volume 1310

The series "Advances in Intelligent Systems and Computing" contains publications on theory, applications, and design methods of Intelligent Systems and Intelligent Computing. Virtually all disciplines such as engineering, natural sciences, computer and information science, ICT, economics, business, e-commerce, environment, healthcare, life science are covered. The list of topics spans all the areas of modern intelligent systems and computing such as: computational intelligence, soft computing including neural networks, fuzzy systems, evolutionary computing and the fusion of these paradigms, social intelligence, ambient intelligence, computational neuroscience, artificial life, virtual worlds and society, cognitive science and systems, Perception and Vision, DNA and immune based systems, self-organizing and adaptive systems, e-Learning and teaching, human-centered and human-centric computing, recommender systems, intelligent control, robotics and mechatronics including human-machine teaming, knowledge-based paradigms, learning paradigms, machine ethics, intelligent data analysis, knowledge management, intelligent agents, intelligent decision making and support, intelligent network security, trust management, interactive entertainment, Web intelligence and multimedia.

The publications within "Advances in Intelligent Systems and Computing" are primarily proceedings of important conferences, symposia and congresses. They cover significant recent developments in the field, both of a foundational and applicable character. An important characteristic feature of the series is the short publication time and world-wide distribution. This permits a rapid and broad dissemination of research results.

Indexed by SCOPUS, DBLP, EI Compendex, INSPEC, WTI Frankfurt eG, zbMATH, Japanese Science and Technology Agency (JST), SCImago.

All books published in the series are submitted for consideration in Web of Science.

More information about this series at http://www.springer.com/series/11156

Alexei V. Samsonovich · Ricardo R. Gudwin ·
Alexandre da Silva Simões
Editors

Brain-Inspired Cognitive Architectures for Artificial Intelligence: BICA*AI 2020

Proceedings of the 11th Annual Meeting of the BICA Society

Springer

Editors
Alexei V. Samsonovich
Cybernetics Department
National Research Nuclear University
Moscow, Russia

Ricardo R. Gudwin
University of Campinas
Campinas, Brazil

Alexandre da Silva Simões
Institute of Science and Technology
São Paulo State University (Unesp)
Sorocaba, Brazil

ISSN 2194-5357 ISSN 2194-5365 (electronic)
Advances in Intelligent Systems and Computing
ISBN 978-3-030-65595-2 ISBN 978-3-030-65596-9 (eBook)
https://doi.org/10.1007/978-3-030-65596-9

This Springer imprint is published by the registered company Springer Nature Switzerland AG
The registered company address is: Gewerbestrasse 11, 6330 Cham, Switzerland

Preface

This volume documents the proceedings of the 2020 Annual International Conference on Brain-Inspired Cognitive Architectures for Artificial Intelligence, known as BICA*AI 2020, also known as the Eleventh Annual Meeting of the BICA Society. BICA*AI 2020 was officially held in Natal, Rio Grande do Norte, Brazil, as an event collocated with Robotica 2020. At the time of its organization, it was impossible to guess that both, BICA*AI 2020 and Robotica 2020, will be hosted online only, as entirely virtual events, due to the COVID-19 pandemic. Nevertheless, this year BICA conference can be called a notable success—thanks to the help of our technical team, primarily Arthur Chubarov, who under my supervision developed the Virtual Convention Center (VCC), the platform for hosting virtual conferences, including this one (please see our paper in this volume). As a result, VCC in conjunction with Zoom and Mozilla Hubs was used for hosting all events of BICA*AI 2020. The project was funded by the Russian Science Foundation Grant #18-11-00336. Speaking of official sponsors of BICA*AI 2020, they are the BICA Society (bicasociety.org/about), AGI Laboratory, National Research Nuclear University MEPhI, and two universities in Brazil: University of Campinas (Unicamp) and Sao Paulo State University (UNESP).

Brain-Inspired Cognitive Architectures (BICA) are computational frameworks for building intelligent agents that are inspired from natural intelligence. Since 2010, the annual BICA conference attracts researchers from the edge of scientific frontiers around the world. It complements major conferences in artificial intelligence (AI), cognitive science and neuroscience by offering to ambitious researchers an informal brainstorming atmosphere and freedom of spontaneous social interactions, together with great publication venues. This distinguishing feature of our conference extends fully into the difficult year 2020.

BICA community members understand "biological inspirations" broadly, borrowing them from psychology, neuroscience, linguistics, narratology, and creativity studies in order to advance cognitive robotics and machine learning, among other hot topics in AI. The selection criterion is based on the question of how much a particular contribution may help us make machines our friends or understand how the mind works. This principle corresponds to a broader understanding of the BICA

Challenge, put forward by the founders of BICA Society in 2010. BICA Society (Biologically Inspired Cognitive Architectures Society) is a scientific non-profit 501 (c)(3) organization based in the USA, whose mission is to promote and facilitate the many scientific efforts around the world in solving the BICA Challenge, i.e., creating a real-life computational equivalent of the human mind. BICA Society brings together researchers from disjointed fields and communities who devote their efforts to solving the challenge, despite that they may "speak different languages." This is achieved by promoting and facilitating the transdisciplinary study of cognitive architectures, and in the long-term perspective, creating one unifying widespread framework for the human-level cognitive architectures and their implementations.

Over 11 years of its existence, BICA Society organized or co-organized 11 international conferences (the BICA conferences), 2 schools, and 2 symposia. These events were hosted in various countries around the world, including the USA, Italy, France, Ukraine, Russia, the Czech Republic, and Brazil. In organizing these events, BICA Society collaborated with several universities around the world. BICA Society created and is maintaining an online database on cognitive architectures (http://bicasociety.org/mapped/), which includes one of the world's largest databases of freely accessible video records of scientific presentations about cognitive architectures and related topics:

- https://vimeo.com/bicasociety
- https://youtube.com/channel/UC7Smq21YMKs0UjuVkFPpJJg

BICA Society initiated several mainstream periodic publication venues, including Elsevier journals "Biologically Inspired Cognitive Architectures" and "Cognitive Systems Research (CSR)–Special Issue Series on BICA*AI". Plus, BICA Society regularly publishes its proceedings in other editions dedicated to BICA*AI, including Procedia Computer Science and, of course, the Springer Book Series. The following topics at the front edge of science and technology, as before, are in the focus of this and future BICA*AI conferences, books, and journals:

- Artificial social–emotional intelligence;
- Active humanlike learning and cognitive growth;
- Narrative intelligence and context understanding;
- Artificial creativity and creation of art by AI;
- Goal reasoning and true autonomy in artifacts;
- Embodied intelligence and cognitive robotics;
- Synthetic characters, HCI, and VR/MR paradigms for AI;
- Language capabilities and social competence;
- Robust and scalable machine learning mechanisms;
- Socially acceptable intelligent virtual agents;
- The role of emotions in artificial intelligence and their BICA models;
- Tests and metrics for BICA in the context of the BICA Challenge;
- Interaction between natural and artificial cognitive systems;
- Theory-of-Mind, episodic, and autobiographical memory in vivo and in vitro;
- Introspection, metacognitive reasoning, and self-awareness in BICA;

– AI and ethics, digital economics, and cybersecurity;
– Unifying frameworks, standards, and constraints for cognitive architectures.

Works of many, yet not all, distinguished speakers of BICA*AI 2020 are included in this volume. Among the speakers of BICA*AI 2020 are top-level scientists like John Laird, Paul Verschure, Terry Stewart, Antonio Chella, Paul Robertson, Jan Treur, Rosario Sorbello, Ricardo Gudwin, to name only a few of the famous names. The majority of authors of this volume is less known and nevertheless deserves attention. In particular, I would like to mention the work of Natividad Vargas, Juan Luis del Valle-Padilla, Juan P. Jimenez and Félix Ramos, of Lennart Zegerius, of Joey van den Heuvel, of Alessio Plebe and Pietro Perconti, of Peter Boltuc and Thomas P. Connelly, of Kyrtin Atreides, David Kelley and Uplift Masi.

As the General Chair of BICA*AI 2020, I would like to thank all members of the Organizing Committee, the Program Committee, the Technical Support Team and the Publishers teams for their great job in making the BICA conference a great success story one more time. Particularly, my special thanks go to the publisher Leontina Di Cecco from Springer, who made this publication possible.

November 2020 Alexei V. Samsonovich

Organization

Program Committee

Taisuke Akimoto	Kyushu Institute of Technology, Japan
Kenji Araki	Hokkaido University, Japan
Joscha Bach	AI Foundation, USA
Feras Batarseh	George Mason University, USA
Paul Baxter	Plymouth University, USA
Paul Benjamin	Pace University, New York, USA
Galina A. Beskhlebnova	Scientific Research Institute for System Analysis RAS, Russia
Jordi Bieger	Reykjavik University, Iceland
Perrin Bignoli	Yahoo Labs, USA
Douglas Blank	Bryn Mawr College, USA
Peter Boltuc	University of Illinois at Springfield, USA
Jonathan Bona	University of Arkansas for Medical Sciences, USA
Michael Brady	Boston University, USA
Mikhail Burtsev	Moscow Institute of Physics and Technology, Russia
Erik Cambria	Nanyang Technological University, Singapore
Suhas Chelian	Fujitsu Laboratories of America, Inc., USA
Antonio Chella	Dipartimento di Ingegneria Informatica, Università di Palermo, Italy
Olga Chernavskaya	P. N. Lebedev Physical Institute, Moscow, Russia
Thomas Collins	University of Southern California (Information Sciences Institute), USA
Christopher Dancy	Penn State University, USA
Haris Dindo	University of Palermo, Italy

Sergey A. Dolenko	D. V. Skobeltsyn Institute of Nuclear Physics, M. V. Lomonosov Moscow State University, Russia
Alexandr Eidlin	Sberbank, Moscow, Russia
Jim Eilbert	AP Technology, USA
Thomas Eskridge	Florida Institute of Technology, USA
Usef Faghihi	Professor At Universite de Quebce in Trois-rivier, Canada
Elena Fedorovskaya	Rochester Institute of Technology, USA
Stan Franklin	University of Memphis, USA
Marcello Frixione	University of Genova, Italy
Salvatore Gaglio	University of Palermo, Italy
Olivier Georgeon	Claude Bernard Lyon 1 University, France
John Gero	University of North Carolina at Charlotte, USA
Jaime Gomez	Universidad Politécnica de Madrid, Spain
Ricardo R. Gudwin	University of Campinas (Unicamp), Brazil
Eva Hudlicka	Psychometrix Associates, USA
Dusan Husek	Institute of Computer Science, Academy of Sciences of the Czech Republic
Christian Huyck	Middlesex University, UK
Ignazio Infantino	Consiglio Nazionale delle Ricerche, Italy
Eduardo Izquierdo	Indiana University, USA
Alex James	Kunming University of Science and Technology, China
Li Jinhai	Kunming University of Science and Technology, China
Magnus Johnsson	Lund University, Sweden
Darsana Josyula	Bowie State University, USA
Kamilla Jóhannsdóttir	Reykjavik University, Iceland
Omid Kavehei	The University of Sydney, Australia
David Kelley	Artificial General Intelligence Inc, Seattle, USA
Troy Kelley	U. S. Army Research Laboratory, USA
William Kennedy	George Mason University, USA
Deepak Khosla	HRL Laboratories LLC, USA
Swathi Kiran	Boston University, USA
Muneo Kitajima	Nagaoka University of Technology, Japan
Unmesh Kurup	LG Electronics, USA
Giuseppe La Tona	University of Palermo, Italy
Luis Lamb	Federal University of Rio Grande do Sul, Brazil
Leonardo Lana de Carvalho	Federal University of Jequitinhonha and Mucuri Valleys, Brazil
Othalia Larue	University of Quebec, Canada
Christian Lebiere	Carnegie Mellon University, USA
Jürgen Leitner	Australian Centre of Excellence for Robotic Vision, Australia

Meehae Song	Simon Fraser University, Canada
Rosario Sorbello	University of Palermo, Italy
John Sowa	Kyndi, Inc., USA
Terry Stewart	University of Waterloo, Canada
Sherin Sugathan	Enview Research & Development Labs, India
Junichi Takeno	Meiji University, Japan
Knud Thomsen	Paul Scherrer Institute, Switzerland
Daria Tikhomirova	NRNU MEPhI
Jan Treur	Vrije Universiteit Amsterdam, Netherlands
Vadim L. Ushakov	NRNU MEPhI, Russia
Alexsander V. Vartanov	Lomonosov Moscow State University, Russia
Rodrigo Ventura	Universidade de Lisboa, Portugal
Evgenii Vityaev	Sobolev Institute of Mathematics SB RAS, Russia
Pei Wang	Temple University, USA
Mark Waser	Digital Wisdom Institute, USA
Roseli S. Wedemann	Universidade do Estado do Rio de Janeiro, Brazil
Özge Nilay Yalçin	Simon Fraser University, Canada
Terry Zimmerman	University of Washington, Bothell, USA

Contents

Acoustic Pattern Recognition Technology Based on the Viola-Jones Approach for VR and AR Systems 1
Alexander M. Alyushin and Sergey V. Dvoryankin

Correlation of a Face Vibroimage Informative Parameters with Characteristics of a Person's Functional State When Using VR and AR Technical Means 9
Victor M. Alyushin

Methodologies and Milestones for the Development of an Ethical Seed 15
Kyrtin Atreides, David J. Kelley, and Uplift Masi

Design of a Transcranial Magnetic Stimulation System with the Implementation of Nanostructured Composites 24
Gennady Baryshev, Yuri Bozhko, Igor Yudin, Aleksandr Tsyganov, and Anna Kainova

Application of Information Measuring Systems for Development of Engineering Skills for Cyber-Physical Education 32
Gennady Baryshev, Valentin Klimov, Aleksandr Berestov, Anton Tokarev, and Valeria Petrenko

Principles of Design of a Learning Management System for Development of Economic Skills for Nuclear Engineering Education 40
Gennady Baryshev, Aleksandr Putilov, Dmitriy Smirnov, Aleksandr Tsyganov, and Vladimir Chervyakov

Post-quantum Group Key Agreement Scheme 49
Julia Bobrysheva and Sergey Zapechnikov

Uncanny Robots of Perfection 56
Piotr (Peter) Boltuc and Thomas P. Connelly

Self and Other Modelling in Cooperative Resource Gathering with Multi-agent Reinforcement Learning 69
Vasilii Davydov, Timofei Liusko, and Aleksandr I. Panov

Eye Movement Correlates of Foreign Language Proficiency in Russian Learners of English 78
Valeriia Demareva, Sofia Polevaia, and Julia Edeleva

Development of a Laboratory Workshops Management Module as Part of a Learning Support System for the "Decision-Making Theory" Course 84
Anastasia Devyatkina, Natalia Myklyuchenko, Anna Tikhomirova, and Elena Matrosova

Algorithm for Constructing Logical Neural Networks Based on Logical Various-Valued Functions 91
Dmitriy Dimitrichenko

The Electroencephalogram Based Classification of Internally Pronounced Phonemes 97
Yuliya Gavrilenko, Daniel Saada, Eugene Ilyushin, Alexander V. Vartanov, and Andrey Shevchenko

"Loyalty Program" Tool Application in Megaprojects 106
Anna Guseva, Elena Matrosova, Anna Tikhomirova, and Nikolay Matrosov

Using Domain Knowledge for Feature Selection in Neural Network Solution of the Inverse Problem of Magnetotelluric Sounding 115
Igor Isaev, Eugeny Obornev, Ivan Obornev, Eugeny Rodionov, Mikhail Shimelevich, Vladimir Shirokiy, and Sergey Dolenko

Friction Model Identification for Dynamic Modeling of Pneumatic Cylinder 127
Vladimir I. Ivlev, Sergey Yu. Misyurin, and Andrey P. Nelyubin

Neurophysiological Features of Neutral and Threatening Visual Stimuli Perception in Patients with Schizophrenia 138
Sergey I. Kartashov, Vyacheslav A. Orlov, Aleksandra V. Maslennikova, and Vadim L. Ushakov

Logical Circuits of a ΣΠ-Neuron Model 143
Mukhamed Kazakov

Study of Neurocognitive Mechanisms in the Concealed Information Paradigm 149
Yuri I. Kholodny, Sergey I. Kartashov, Denis G. Malakhov, and Vyacheslav A. Orlov

IT-Solutions in Money Laundering/Counter Terrorism Financing Risk Assessment in Commercial Banks 156
Sofya Klimova and Asmik Grigoryan

Expandable Digital Functional State Model of Operator for Intelligent Human Factor Reliability Management Systems 165
Lyubov V. Kolobashkina, Mikhail V. Alyushin, and Kirill S. Nikishov

Specialized Software Tool for Pattern Recognition of Biological Objects 173
Sergey D. Kulik and Evgeny O. Levin

Designing Software for Risk Assessment Using a Neural Network 181
Anna V. Lebedeva and Anna I. Guseva

Suitability of Object-Role Modeling Diagrams as an Intermediate Model for Ontology Engineering: Testing the Rules for Mapping 188
Dmitrii Litovkin, Dmitrii Dontsov, Anton Anikin, and Oleg Sychev

Cyber Threats to Information Security in the Digital Economy 195
K. S. Luzgina, G. I. Popova, and I. V. Manakhova

Applying a Logical Derivative to Identify Hidden Patterns in the Data Structure 206
Larisa A. Lyutikova

Algorithm for Constructing Logical Operations to Identify Patterns in Data 212
Larisa A. Lyutikova and Elena V. Shmatova

Graph-Ontology Model of Cognitive-Similar Information Retrieval (on the Requirements Tracing Task Example) 218
Nikolay V. Maksimov, Olga L. Golitsina, Kirill V. Monankov, and Natalia A. Bal

Toward a Building an Ontology of Artefact 225
Nikolay Maksimov and Alexander Lebedev

Cognitive Architectures of Effective Speech-Language Communication and Prospective Challenges for Neurophysiological Speech Studies 233
Irina Malanchuk

Development of an AI Recommender System to Recommend Concerts Based on Microservice Architecture Using Collaborative and Content-Based Filtering Methods 241
Andrey Malynov and Igor Prokhorov

Visualization of T. Saati Hierarchy Analysis Method 253
Elena Matrosova, Anna Tikhomirova, Nikolay Matrosov, and Kovtun Dmitriy

Labor Productivity Growth Based on Revolutionary Technologies as a Factor for Overcoming the Economic Crisis 265
Y. M. Medvedeva and R. E. Abdulov

Network Security Intelligence Centres for Information Security Incident Management 270
Natalia Miloslavskaya and Steven Furnell

Block Formation for Storing Data on Information Security Incidents for Digital Investigations 283
Natalia Miloslavskaya, Andrey Nikiforov, and Kirill Plaksiy

Cyber Polygon Site Project in the Framework of the MEPhI Network Security Intelligence Center 295
Natalia Miloslavskaya and Alexander Tolstoy

Selection of a Friction Model to Take into Account the Impact on the Dynamics and Positioning Accuracy of Drive Systems 309
S. Yu. Misyurin, G. V. Kreinin, N. Yu. Nosova, and A. P. Nelyubin

Kinematics and Dynamics of the Spider-Robot Mechanism, Motion Optimization 320
Sergey Yu. Misyurin, German V. Kreinin, Natalia Yu. Nosova, and Andrey P. Nelyubin

Multiagent Model of Perceptual Space Formation in the Process of Mastering Linguistic Competence 327
Zalimkhan Nagoev and Irina Gurtueva

The Role of Gender in the Prosocial Behavior Mechanisms 335
Yulia M. Neroznikova and Alexander V. Vartanov

Reflection Mechanisms of Empathy Processes in Evoked Potentials 342
Yulia M. Neroznikova and Alexander V. Vartanov

Lateralization in Neurosemantics: Are Some Lexical Clusters More Equal Than Others? 350
Zakhar Nosovets, Boris M. Velichkovsky, Liudmila Zaidelman, Vyacheslav Orlov, Sergey Kartashov, Artemiy Kotov, Vadim Ushakov, and Vera Zabotkina

Brain Inspiration Is Not Panacea 359
Pietro Perconti and Alessio Plebe

Linear Systems Theoretic Approach to Interpretation of Spatial and Temporal Weights in Compact CNNs: Monte-Carlo Study 365
Artur Petrosyan, Mikhail Lebedev, and Alexey Ossadtchi

The Economic Cross of the Digital Post-coronavirus Economy (on the Example of Rare Earth Metals Industry) 371
O. Victoria Pimenova, Olga B. Repkina, and Dmitriy V. Timokhin

Comparative Analysis of Methods for Calculating the Interactions Between the Human Brain Regions Based on Resting-State FMRI Data to Build Long-Term Cognitive Architectures 380
Alexey Poyda, Maksim Sharaev, Vyacheslav Orlov, Stanislav Kozlov, Irina Enyagina, and Vadim Ushakov

The Use of the Economic Cross Method in IT Modeling of Industrial Development (Using the Example of Two-Component Nuclear Energy) 391
Aleksandr V. Putilov, Dmitriy V. Timokhin, and Marina V. Bugaenko

Intelligence - Consider This and Respond! 400
Saty Raghavachary

Simple Model of Origin of Feeling of Causality 410
Vladimir G. Red'ko

Extending the Intelligence of the Pioneer 2AT Mobile Robot 417
Michael A. Rudy, Eugene V. Chepin, and Alexander A. Gridnev

On the Regularity of the Bias of Throughput Estimates on Traffic Averaging 425
Victor A. Rusakov

Virtual Convention Center: A Socially Emotional Online/VR Conference Platform 435
Alexei V. Samsonovich and Arthur A. Chubarov

Ensembling SNNs with STDP Learning on Base of Rate Stabilization for Image Classification 446
Alexander Sboev, Alexey Serenko, Roman Rybka, and Danila Vlasov

Mathematical Methods for Solving Cognitive Problems in Medical Diagnosis 453
Yuri Kotov and Tatiana Semenova

Brain Cognitive Architectures Mapping for Neurosurgery: Resting-State fMRI and Intraoperative Validation 466
M. Sharaev, T. Melnikova-Pitskhelauri, A. Smirnov, A. Bozhenko, V. Yarkin, A. Bernshtein, E. Burnaev, P. Petrov, D. Pitskhelauri, V. Orlov, and I. Pronin

Machine Learning Based on the Principle of Minimizing Robust Mean Estimates 472
Z. M. Shibzukhov

Parameterized Families of Correctly Functioning Sigma-Pi Neurons 478
Z. M. Shibzukhov

The Loop of Nonverbal Communication Between Human and Virtual Actor: Mapping Between Spaces 484
Vladimir R. Shirokiy, Daria V. Tikhomirova, Roman D. Vladimirov, Sergei A. Dolenko, and Alexei V. Samsonovich

Preliminary Experiment on Emotion Detection in Illustrations Using Convolutional Neural Network 490
Alexander Shtanko and Sergey Kulik

AGI Components for Enterprise Management Systems 495
Artem A. Sukhobokov and Liubov I. Lavrinova

Eligibility of English Hypernymy Resources for Extracting Knowledge from Natural-Language Texts 501
Oleg Sychev and Yaroslav Kamennov

The Use of Digital Tools in the Formation of Two-Component Nuclear Energy on the Base of Economic Cross Method 508
D. V. Timokhin

Complex Objects Identification and Analysis Mechanisms 517
Mikhail Ulizko, Larisa Pronicheva, Alexey Artamonov, Rufina Tukumbetova, and Evheniy Tretyakov

To Help or Not to Help: A Network Modelling Approach to the Bystander Effect 527
Joey van den Heuvel and Jan Treur

A Model of Top-Down Attentional Control for Visual Search Based on Neurosciences 541
Natividad Vargas, Juan Luis del Valle-Padilla, Juan P. Jimenez, and Félix Ramos

Comparison of Brain Induced Potentials in Internal Speech in Studied and Unknown Languages 547
Alexander V. Vartanov and Alisa R. Suyuncheva

Analysis of Using of Neural Networks for Real-Time Process Control 553
V. S. Volodin and A. O. Tolokonskij

Comparison Between Coordinated Control and Interpretation Methods for Multi-channel Control of a Mobile Robotic Device 558
Timofei I. Voznenko, Alexander A. Gridnev, Eugene V. Chepin, and Konstantin Y. Kudryavtsev

Combinator-as-a-Process for Representing the Information Structure of Deep Computing 565
Viacheslav Wolfengagen, Larisa Ismailova, and Sergey Kosikov

Applicative Model to Bring-in Conceptual Envelope for Computational Thinking with Information Processes 572
Viacheslav Wolfengagen, Larisa Ismailova, and Sergey Kosikov

Computational Model for Capturing the Interdependencies on Information Processes 578
Viacheslav Wolfengagen, Larisa Ismailova, and Sergey Kosikov

Imposing and Superposing the Information Processes over Variable Concepts 585
Viacheslav Wolfengagen, Larisa Ismailova, and Sergey Kosikov

Improvement of the Technology of fMRI Experiments in the Concealed Information Paradigm 591
Yuri I. Kholodny, Sergey I. Kartashov, Denis G. Malakhov, and Vyacheslav A. Orlov

Modelling Metaplasticity and Memory Reconsolidation During an Eye-Movement Desensitization and Reprocessing Treatment 598
Lennart Zegerius and Jan Treur

Author Index 611

Acoustic Pattern Recognition Technology Based on the Viola-Jones Approach for VR and AR Systems

Alexander M. Alyushin[1,2(✉)] and Sergey V. Dvoryankin[3]

[1] National Research Nuclear University MEPhI (Moscow Engineering Physics Institute), Kashirskoe shosse 31, Moscow 115409, Russia
alyshin@list.ru

[2] Plekhanov Russian University of Economics, st. Hook, 43, Moscow 117997, Russia

[3] Financial University under the Government of the Russian Federation, Leningradsky prospect, 49, Moscow 125993, Russian Federation
SVDvoryankin@fa.ru

Abstract. The ability to solve problems of graphic images recognition in virtual reality (VR), augmented reality (AR), mixed reality (MR) and cross reality (XR) systems is highlighted as one of the most important. The urgency of solving problems of recognition and classification of acoustic images has been substantiated, which will bring the quality of VR, AR, MR and XR systems closer to real reality (RR). Independent solution of graphic and acoustic patterns recognition problems using heterogeneous algorithmic and software tools is attributed to the disadvantages of modern systems. The study proposes an approach that allows the use of unified methodological and software tools for the simultaneous solution of graphic and acoustic patterns recognition problems. The proposed approach is based on converting acoustic information into graphic information using 2D-images of dynamic sonograms. This allows the recognition of acoustic patterns using unified algorithmic and software tools. It is proposed to use the Viola-Jones technology as such a unified tool. It is shown that the implementation of a two-stage determination of similarity measures of primitives and areas of the original image makes it possible to increase the speed of algorithms. For this purpose, at the first iteration, it is proposed to use not the graphic primitives themselves, but their coordinate projections. In the study, by analogy with Haar's features, parametrizable acoustic primitives were developed, presented in the classical graphical version, as well as in the form of coordinate projections.

Keywords: Recognition and classification · Graphic and acoustic patterns · Viola-Jones algorithm

A. V. Samsonovich et al. (Eds.): BICA 2020, AISC 1310, pp. 1–8, 2021.
https://doi.org/10.1007/978-3-030-65596-9_1

1 Introduction

One of the most important functions implemented in modern virtual reality (VR), augmented reality (AR), mixed reality (MR) and cross reality (XR) systems and largely determines their capabilities, is the recognition and classification of graphic images [1]. For example, the basis of AR, MR and XR technologies is the recognition of objects in each frame of the video stream and the addition of new graphic information to them. In addition, a number of AR, MR and XR technologies use the so-called graphic markers, which are necessary to determine the spatial characteristics of objects of real reality (RR) [2]. This function must be repeatedly performed in real time [3], which imposes strict requirements on the speed of algorithms used for this purpose, for example, various modifications of the well-known Viola-Jones algorithm [4, 5].

One of the main trends in the modern development of AR, MR, and XR systems is the approach to RR, primarily due to the development of artificial intelligence technology, which allows simultaneous processing of video and audio patterns [6]. This allows you to implement a natural user interface for a person, to carry out acoustic navigation [7–9], to provide the necessary information interaction between characters, for example, in training systems [10, 11], to increase the efficiency of solving production tasks [12, 13].

The aim of the research is to unify algorithmic and technical means used for the recognition of graphic and acoustic patterns based on the Viola-Jones approach for VR, AR, MR and XR systems.

2 State of Research in This Area

Currently being developed VR, AR, MR and XR systems, as a rule, involve the use of independent channels for processing video and audio information. It should be noted that the applied algorithmic and methodological means of graphic patterns recognition are forced to operate with significant data streams, which is associated with the use of modern high-speed high-resolution video cameras. In this regard, the flow of acoustic data even when using multichannel systems [7] has a significantly smaller volume.

For this reason, it is relevant to use algorithmic and methodological tools developed for the recognition of graphic patterns for the recognition of acoustic patterns in VR, AR, MR and XR systems. The feasibility of this approach in practice is due to a fairly well-developed technology for converting acoustic information into a graphic representation in the form of a 2D-image of dynamic sonograms. This technology is widely used, for example, for the protection of documents against forgery based on the so-called speech signature [14, 15]. The specificity of a 2D-image of a dynamic sonogram is the presence of areas with different graphic structures, for example, linear and dotted, as well as low image contrast with a high noise level.

Of all the existing variety of approaches and algorithms for recognizing graphic objects in the image, which are also suitable for working with images of dynamic sonograms, the most suitable is the approach proposed by Viola-Jones for recognizing facial images [16]. Recognition of a graphic object in accordance with this approach is carried out on the basis of the similarity measures analysis of a large set of characteristic features typical of the analyzed image. At the same time, the features themselves, known as Haar features [16–18], characterize the properties of limited areas of the recognized object.

The disadvantages of this approach include a fairly large amount of calculations to determine the measures of features similarity and areas of the analyzed image, as well as a decrease in the reliability of the result obtained with a decrease in the contrast of the analyzed image.

3 The Essence of the Proposed Approach

To recognize acoustic patterns presented in the form of 2D-images of dynamic sonograms, the study proposes to use a two-stage analysis of similarity measures for a set of acoustic features and areas of a 2D-image. This allows you to significantly reduce the necessary computational costs of the approach. This approach assumes the presence of two forms of features representation and analyzed image areas – in the form of image fragments and in the form of their coordinate projections. A two-stage analysis of features similarity measures is applicable both in the analysis of acoustic signals and in the analysis of video stream frames.

At the first step, it is proposed to analyze the features similarity of the coordinate projections and the coordinate projections of the areas of the analyzed image, usually selected using a floating rectangular window.

At the second step, the analysis of the features similarity measures selected in this way and the corresponding image areas, presented as fragments of the corresponding images, is carried out. This operation is completely analogous to the procedures used in the Viola-Jones algorithm.

Figure 1 illustrates this approach.

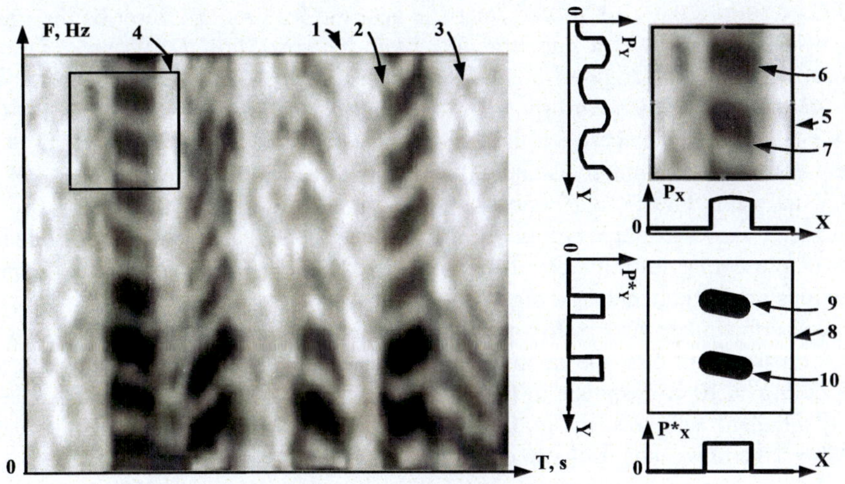

Fig. 1. Approach illustration.

The original image ($IMG[i, j]$, $i = 1, \ldots, IMAX$, $j = 1, \ldots, JMAX$, where $IMAX$ and $JMAX$ are the image size in pixels, respectively) of dynamic sonogram 1 contains areas 2 with a linear structure corresponding to vowel sounds of human speech (harmonic signals), and area 3 corresponding to consonants (hissing) sounds. The X-axis of the sonogram corresponds to time, and the Y-axis corresponds to the frequency F.

At the first step, in accordance with the proposed approach, by means of a floating window 4 with a size of $N \times N$ pixels, a fragment of the image 5 is selected, for example, containing characteristic stripes 6 and 7.

For a given fragment of the image, its coordinate projections $P_Y(Y)$ and $P_X(X)$ are determined:

$$P_X(X) = \sum_{j=Y_0}^{Y_0+N} IMG(X,j), \; X = 1,\ldots,N, \tag{1}$$

$$P_Y(Y) = \sum_{i=X_0}^{X_0+N} IMG(i,Y), \; Y = 1,\ldots,N, \tag{2}$$

where X_0 and Y_0 are the coordinates of the window anchor point.

These projections, as a rule, contain noise components that cause, for example, nonzero values of the functions $P_Y(Y)$ and $P_X(X)$ in the intervals between bands 6 and 7. To minimize the influence of noise components in two forms of information presentation, it is proposed to carry out, respectively, threshold discrimination for the functions $P_Y(Y)$ and $P_X(X)$ and contrasting for a fragment of the image 5. A typical result of these operations is shown in Fig. 1 in the form of new obtained coordinate projections $P^*_Y(Y)$ and $P^*_X(X)$, as well as a fragment of a contrast image 8, containing characteristic stripes 9 and 10.

Analysis of possible structures of image fragments of dynamic sonograms made it possible to form a basic set of characteristic acoustic features. To describe the image areas of a sonogram with a line structure, basic features $SL(A, D, W)$ were formed, where A is the relative angle of inclination of the lines ($-AMAX \leq A \leq AMAX$, the value $AMAX = 10$, which corresponds to the angle of inclination of the lines in 90°), D is the relative distance between the lines ($1 \leq D \leq DMAX$, the value $DMAX = 10$ corresponds to the maximum distance), W is the relative width of the lines ($1 \leq W \leq WMAX$, the value $WMAX = 10$ corresponds to the maximum width).

In Fig. 2 shows examples of the formed basic acoustic features, which include two forms of presenting information of the considered type – in the form of coordinate projections and a fragment of a high-contrast image.

To describe image areas with a pixel structure, basic features of the $NP(Q, G)$ type were formed, where Q is the relative density of dark pixels ($1 \leq Q \leq QMAX$, the value $QMAX = 10$ corresponds to the maximum density of dark pixels), G – relative density gradient with respect to the Y axis ($-GMAX \leq G \leq GMAX$, the value $GMAX = 10$ corresponds to the maximum value of the gradient).

In Fig. 3 shows typical examples of formed basic features of the $NP(Q, G)$ type.

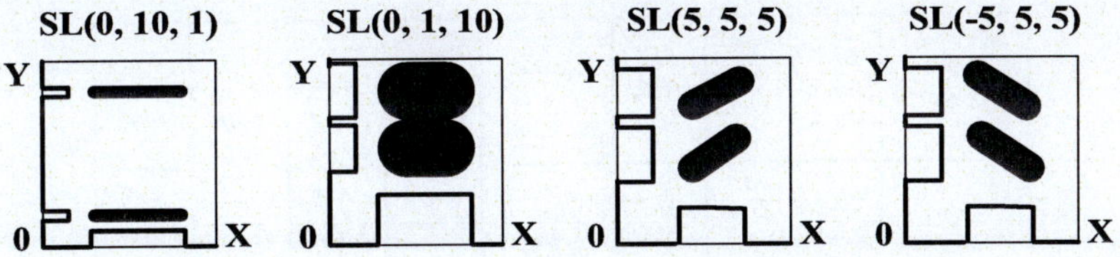

Fig. 2. Examples of *SL(A, D, W)* basic acoustic features.

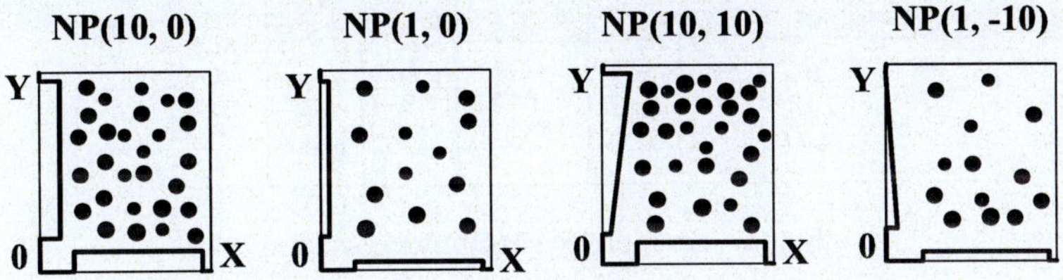

Fig. 3. Examples of *NP(Q, G)* basic acoustic features.

The basic acoustic features of *SL(A, D, W)* and *NP(Q, G)* types formed in this way were the basis for the creation of working features *SL*(A, D, W, M)* and *NP*(Q, G, M)*, differing in the scaling factor *M* ($1 \leq M \leq MMAX$, for sonograms with a resolution of less than 1024 × 1024 pixels, it is sufficient to use the value $MMAX = 10$).

In Fig. 4 shows an example of the created operating characteristics *SL(A, D, W)*.

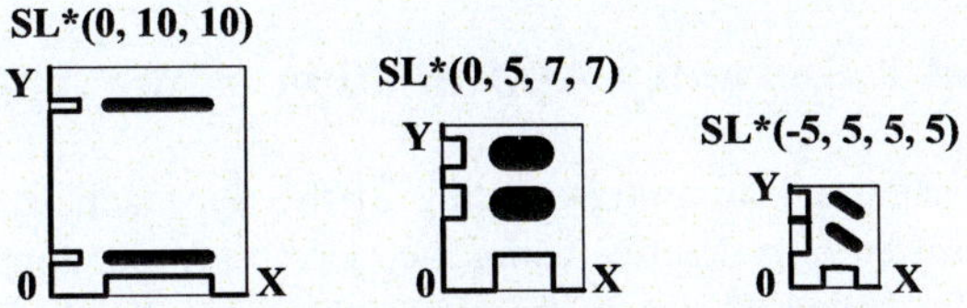

Fig. 4. Examples of working acoustic signs of *SL*(A, D, W, M)* type.

Working acoustic signs allow you to get a graphic image of acoustic signals, speech and sounds based on recognition of the structure of the sonogram. To determine the similarity measure of coordinate projections at the first step, it is proposed to use the value *D*:

$$D = 1/(1 + \sum_{Y=1}^{N} |P_F(Y) - P_Y^*(Y)| + \sum_{X=1}^{N} |P_F(X) - P_X^*(X)|), \qquad (3)$$

where $P_F(Y)$ and $P_F(X)$ are, respectively, coordinate projections for working acoustic features.

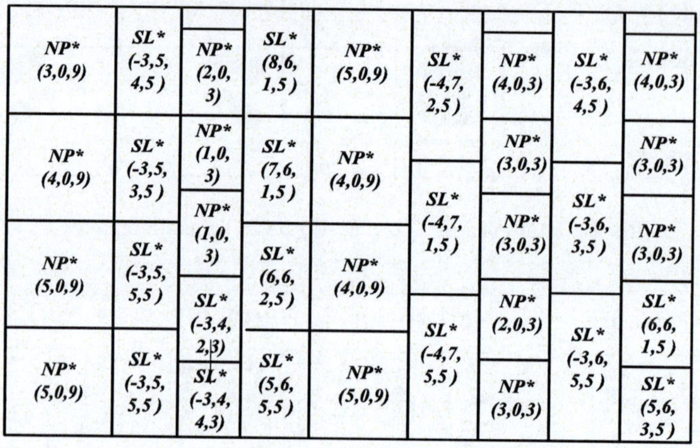

Fig. 5. Example of structure recognition.

In Fig. 5 shows an example of structure recognition according to the proposed technique of 2D-image of a dynamic sonogram shown in Fig. 1 (X-axis time scale not saved).

The resulting structure of the sonogram is an image that is further processed by the methodological and algorithmic means inherent in the Viola-Jones approach. This makes it possible to use the existing graphic image processing tools for solving problems of recognition and classification of acoustic patterns in VR, AR, MR and XR systems.

4 Experimental Laboratory Approbation of the Approach

The conducted experimental laboratory testing of the approach confirmed the possibility of its implementation in practice using the classical Viola-Jones algorithms. The decrease in the performance of the tasks being solved for the recognition of graphic objects with the simultaneous processing of acoustic patterns did not exceed 10%. The implementation of a two-stage process for determining the measures of similarity of features and areas of an image made it possible to increase the performance by 15–30% for images with dimensions of 800 × 800 pixels–1600 × 1600 pixels, respectively.

5 Areas of Possible Application of the Developed Technology

The proposed approach is primarily focused on expanding the functionality of modern VR, AR, MR and XR systems through the simultaneous processing of video and acoustic information. Another area of possible application of the approach is systems for protecting important documents based on the use of a speech signature [14, 15], which involve solving the problems of searching for a sonogram on a document image, as well as recognizing the structure of a sonogram in order to identify the author of the document and his psycho-emotional state.

6 Conclusion

The approach proposed in the study makes it possible to use already available software and methodological tools for solving problems of recognizing and classifying acoustic patterns, initially focused on recognizing and classifying graphic images. The most prominent representative of such tools are algorithmic and software tools that implement the principles of processing graphic data in accordance with Viola-Jones technology. The implementation of a two-stage procedure for determining the similarity measure of features and image regions allows increasing the speed of the computational process.

Acknowledgement. The research was carried out by grant of the Russian Scientific Foundation (project №19-71-30008) in Plekhanov Russian University of Economics.

References

1. Park, H., Jeong, S., Kim, T., Youn, D., Kim, K.: Visual representation of gesture interaction feedback in virtual reality games. In: IEEE Proceedings of the 2017 International Symposium on Ubiquitous Virtual Reality (ISUVR), Nara, Japan, 27–29 June 2017, pp. 20–23 (2017)
2. Kato, H., Billinghurst, M.: Marker tracking and HMD calibration for a video-based augmented reality conferencing system. In: Proceedings 2nd IEEE and ACM International Workshop on Augmented Reality (IWAR 1999), San Francisco, CA, USA, 20–21 October 1999, pp. 85–94 (1999)
3. Prince, S., Cheok, A.D., Farbiz, F., Williamson, T., Johnson, N., Billinghurst, M., Kato, H.: 3D live: real time captured content for mixed reality. In: IEEE Proceedings of the International Symposium on Mixed and Augmented Reality (ISMAR 2002), Darmstadt, Germany, 1 October 2002, pp. 7–317 (2002)
4. Lee, Y.J.: Effective interface design using face detection for augmented reality interaction of smart phone. Int. J. Smart Home **6**(2), 25–32 (2012)
5. Lee, Y.J., Lee, G.H.: Augmented reality game interface using effective face detection algorithm. Int. J. Smart Home **5**(4), 77–88 (2011)
6. Godin, K.W., Rohrer, R., Snyder, J., Raghuvanshi, N.: Wave acoustics in a mixed reality shell. In: Proceedings of the 2018 AES International Conference on Audio for Virtual and Augmented Reality (AVAR), Redmond, USA, 20–22 August 2018, pp. 7–3 (2018)
7. Tylka, J.G., Choueiri, E.Y.: Fundamentals of a parametric method for virtual navigation within an array of ambisonics microphones. J. Audio Eng. Soc. **68**(3), 120–137 (2020a)
8. Tylka, J.G.: Virtual navigation of ambisonics-encoded sound fields containing near-field sources. In: Computer Science (2019). 246 p.
9. Tylka, J.G., Choueiri, E.Y.: Performance of linear extrapolation methods for virtual sound field navigation. J. Audio Eng. Soc. **68**(3), 138–156 (2020b)
10. Yuen, S.C.-Y., Yaoyuneyong, G., Johnson, E.: Augmented reality: an overview and five directions for AR in education. J. Educ. Technol. Dev. Exchange **4**(1), 119–140 (2011)
11. Abdoli Sejzi, A.: Augmented reality and virtual learning environment. J. Appl. Sci. Res. (JASR) **11**(8), 1–5 (2015)

12. Lahti, H., Bahne, A.: Virtual tuning – a mixed approach based on measured RTFs. In: Proceedings of the AES 2019 International Conference on Automotive Audio, Bavaria, Germany, 11–13 September 2019, paper number 12 (2019)
13. Malbos, F., Bogdanski, M., Strauss, M.J.: Virtual reality experience for the optimization of a car audio system. In: Proceedings of the AES 2019 International Conference on Automotive Audio, Bavaria, Germany, 11–13 September 2019, paper number 13 (2019)
14. Alyushin, M.V., Alyushin, A.M., Kolobashkina, L.V.: Human face thermal images library for laboratory studies of the algorithms efficiency for bioinformation processing. In: IEEE Proceedings of the 11th IEEE International Conference on Application of Information and Communication Technologies (AICT 2017), Russia, Moscow, 20–22 September 2017 (2017)
15. Alyushin, A.M.: Document protection technology in the digital economics using cognitive biometric methods. Procedia Comput. Sci. **169**, 887–891 (2020)
16. Viola, P., Jones, M.J.: Robust real-time face detection. Int. J. Comput. Vision **57**(2), 137–154 (2004)
17. Alyushin, M.V., Alyushin, V.M., Kolobashkina, L.V.: Optimization of the data representation integrated form in the Viola-Jones algorithm for a person's face search. Procedia Comput. Sci. **123**, 18–23 (2018)
18. Kolobashkina, L.V., Alyushin, M.V.: Analysis of the possibility of the neural network implementation of the Viola-Jones algorithm. In: Advances in Intelligent Systems and Computing, vol. 948, pp. 232−239 (2020)

Correlation of a Face Vibroimage Informative Parameters with Characteristics of a Person's Functional State When Using VR and AR Technical Means

Victor M. Alyushin(✉)

National Research Nuclear University MEPhI (Moscow Engineering Physics Institute), Kashirskoe Shosse 31, Moscow 115409, Russia
vmalyushin@mephi.ru

Abstract. An analysis of the specifics of modern VR, AR, MR and XR systems use in educational, recreational and recovery processes is presented. The urgency of continuous monitoring of the current functional state (FS) and its psycho-emotional state (PES) using VR, AR, MR and XR technical means has been substantiated. Maintaining the state of health, as well as the required level of performance, are highlighted as the main requirements for the applied VR, AR, MR and XR means. The relevance of the use of infrared (IR) biometric technologies for the registration of the most informative human bioparameters, determining his current PES, has been substantiated. Analysis of IR vibraimage of a face is highlighted as the most promising technology for assessing its current PES. The possibility of determining the current FS on the basis of a set of bioparameters determining its PES is substantiated. For this purpose, the analysis of the correlation of the FS and the most informative bioparameters determined by the analysis of the IR vibraimage of the user's face by VR, AR, MR and XR means is given. To assess the FS in the study, it was proposed to use a value determined by the number of errors committed during periodic execution of specialized tests integrated into the VR, AR, MR, and XR scenario. Experimental data obtained during laboratory testing of the method confirmed the possibility of assessing FS based on a set of bioparameters measured during the processing and analysis of the IR vibraimage of the VR, AR, MR and XR user's face. The research results are of particular importance when training operators to control potentially hazardous facilities using VR, AR, MR or XR technologies, and, first of all, for the nuclear industry.

Keywords: Functional state · Psycho-emotional state · Facial vibraimage analysis · Educational activity planning

1 Introduction

The widespread introduction of VR, AR, MR and XR technologies into everyday life has undeniable advantages, but at the same time causes a number of serious problems. Especially actively these technologies are currently being introduced in education, medicine, in vocational training and retraining, in VR simulators, in the field of

A. V. Samsonovich et al. (Eds.): BICA 2020, AISC 1310, pp. 9–14, 2021.
https://doi.org/10.1007/978-3-030-65596-9_2

recreation and entertainment [1, 2]. Indicative, in this regard, is the use of VR, AR, MR and XR technologies as a tool for relieving stress, including for medical purposes [3]. The increased interest in VR, AR, MR and XR technologies in recent years is due, among other things, to the desire to go on an exciting virtual journey during the coronavirus epidemic.

However, frequent and prolonged presence in these realities can cause headaches, vision problems, disturbances in the mental state of a person. Excessive fascination with these realities by children is especially harmful. The negative impact of VR, AR, MR and XR technologies depends to a large extent on factors such as the quality of the synthesized video information, the accuracy and performance of the positioning systems. It is well known, for example, that disturbances between the visual perception of the environment and its perception by the senses lead to headaches, dizziness, and disorientation. As a result – to an increase in blood pressure, deterioration of health, in some cases – to nausea and vomiting.

In connection with the above, it is relevant to constantly monitor the current FS and PES of the VR, AR, MR and XR means user. It should be noted that FS determines the ability of a person to perform certain actions, duties, for example, production. PES shows what physical and mental costs are incurred by a person, taking into account the resources available to him. For this reason, FS and PES depend on each other in a rather complex way. Maintaining the state of health, as well as high performance, necessitate constant monitoring of the VR, AR, MR and XR means user's current PES.

The main requirements for the tools used for monitoring the current PES of a person are safety and ease of use. The safety of use, first of all, assumes the absence of harmful effects on humans, which determines the predominant use of the so-called passive biometric technologies. The requirement for ease of use leads to the need to use small-sized, including built-in biometric devices. Modern optical biometric technologies should be singled out as the most promising for use in conjunction with VR, AR, MR and XR technical means. The latter, first of all, should include the technology for analyzing the vibraimage of a person's face [4, 5].

The aim of the work is to study the correlation of the face vibraimage informative parameters, reflecting the PES, with the characteristics of the human FS when using VR, AR, MR and XR technical means.

2 State of Research in This Area

Currently, biometric measurements are used to solve problems of user identification in VR and AR systems [6, 7]. For this purpose, recognition of gestures, the nature of movements, and gait features is usually used [8]. Another direction in the development of virtual systems based on optical biometric technologies is the use of gaze tracking tools [9] to create a convenient interface for managing such systems. To monitor the human condition in [10], it is proposed to use the image analysis of the eyes and pupils.

Unfortunately, many of the technologies used in practice, primarily based on the analysis of the direction of gaze, as well as the dynamics of changes in the pupil, are characterized by the effect of interference with the scenario implemented by the virtual system. This does not make it possible to carry out accurate biometric measurements during the stay of a person in the virtual space.

3 The Essence of the Proposed Approach

In the study, for the implementation of continuous monitoring of the current PES of a person, it is proposed to use the technology of the face vibraimage analyzing in the deep IR range [11]. This range of optical radiation with a wavelength of 8–15 μ is far enough away from visible optical radiation, the wavelength of which for most people is in the range of 0.38–0.78 μ. This makes it possible to almost completely eliminate the interference between the optical biometric channel of the monitoring system and the image projection channel of VR, AR, MR and XR systems.

The IR biometric technologies developed to date [11] allow real-time registration of such informative bioparameters as heart rate (HR), arterial pressure (AP), respiratory rate and depth, motor activity, and reactions, similar to galvanic skin. From a technical point of view, to register IR optical radiation of this type, it is sufficient to use modern small-sized sensors that are embedded in wearable hardware of virtual technologies.

Analysis of the dynamics of changes in the above bioparameters makes it possible to build a behavioral model of a person, which permits to predict a possible change in his state. This allows you to reasonably plan and conduct educational and other types of activities, monitor the effectiveness of medical rehabilitation and ultimately guarantee the preservation of health, especially with frequent and prolonged use of VR, AR, MR and XR systems.

The monitoring system under consideration should also be considered as a tool for an objective comparison of the quality of various virtual reality systems, as well as the individual choice of the system, taking into account the physiological characteristics of the user. This circumstance is of paramount importance for children and adolescents with physical disabilities.

4 Experimental Approbation of the Approach

Experimental approbation of the approach included the assessment of PES and FS of tested people using VR, AR, MR or XR means based on the analysis of their current bioparameters. To estimate the PES, two measurement options were used. The first measurement option was based on IR technology. The second – on the use of biometric bracelets, the application of which, due to the convenience of their use, was very reasonable. Both options made it possible to record HR, AP and motor activity.

To assess the PES of the test subjects, the G value was used [12], which was determined on the basis of the analysis of bioparameter data. The value G ($0 \leq G \leq 3$) is a dimensionless relative value that characterizes the PES of a person in comparison with his normal state in a calm environment ($G = 1$). At $G < 1$, the person's state is assessed as relaxed, drowsy. The states of high stress, fatigue and tension are characterized by values of $2 \leq G \leq 3$. At values $1 < G < 2$, the state is characterized by mild to moderate degrees of excitement. The greatest contribution to the identification of the state was made by such bioparameters as HR and AP.

To assess FS embedded in the context of the VR, AR, MR or XR story and periodically repeated tests were used. The number of errors committed was used to estimate the current FS. To assess the current FS, the following ratio was used:

$$FS_0 = 1 - N/NMAX, \quad (1)$$

where N is the number of mistakes committed on a given time interval (5 min); NMAX is the maximum number of errors committed when PES > 2.5.

FS = 1 corresponded to normal operability, and FS = 0 corresponded to low operability level.

In Fig. 1 shows the time dependences of FS_0, G_{IR}, HR_{IR}, G^* и HR^* averaged for the entire contingent of tested (27people), where:

G_{IR} and G^* – PES estimates based on bioparameters measured using IR technology and biometric bracelets, respectively;

HR_{IR} and HR^* – HR measured respectively using IR technology and biometric bracelets.

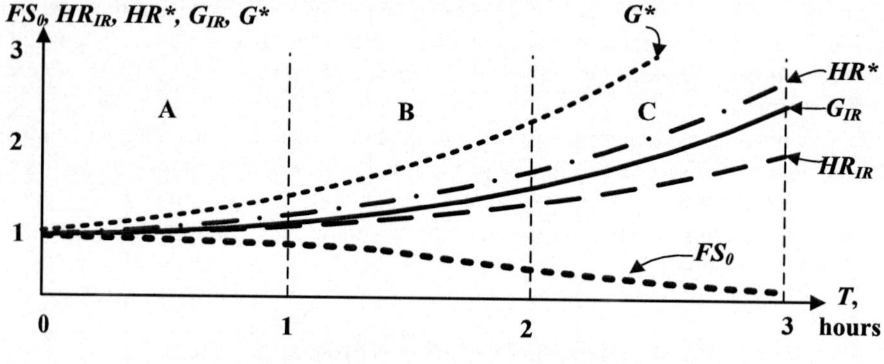

Fig. 1. Typical change in PES and FS of the tested.

The measurements were carried out for three modes of using VR, AR, MR or XR technologies, differing in the degree of the test subjects' involvement: A – light degree; B – medium; C – high degree. Game scenario were used as content. An analysis of the results obtained shows that with a prolonged (two or more hours) use of VR, AR, MR or XR means, the PES of the tested changes from normal to stressed and highly stressed. Accordingly, FS begins to degrade from normal to low. At the same time, PES changes (function *G*) have a good inverse correlation (0.7–0.8) with changes in FS.

Comparison of those shown in Fig. 1 of the time dependences of G_{IR} and G^* shows that the G^* function has almost 1.5 times higher PES estimates than the G_{IR} function. The main reason for this is the overestimation of HR measurement using biometric bracelets using optical biometric technology. This technology, when placing sensors on the hands, measures the vascular blood filling, which is modulated by both cardiac and, to a large extent, motor activity. Moreover, the influence of the latter component increases with the growth, for example, of the intensity of game movements.

5 Conclusion

Thus, the established correlation between FS and PES, determined on the basis of the analysis of informative bioparameters measured using IR technology, makes it possible to assess the current FS state of a person with a sufficiently high degree of reliability. This permits to reasonably solve the problems of planning educational, as well as recreational and recovery processes using VR, AR, MR or XR means. The results of the study are of particular importance in the training of operators to control potentially hazardous facilities using VR, AR, MR or XR technologies, and, first of all, for the nuclear industry.

References

1. Ivanova, A.V.: VR&AR technologies: opportunities and application obstacles. In: Strategic Decisions and Risk Management, no. 3(108), pp. 76–91 (2018)
2. Pfeuffer, K., Geiger, M.J., Prange, S., Mecke, L., Buschek, D., Alt, F.: Behavioral biometrics in VR: identifying people from body motion and relations in virtual reality. In: Proceedings of the 2019 CHI Conference on Human Factors in Computing Systems, Glasgow, Scotland UK, 4–9 May 2019, pp. 1–12 (2019)
3. Wiederhold, B.K., Wiederhold, M.D.: Virtual reality for posttraumatic stress disorder and stress inoculation training. J. Cyber Ther. Rehabil. **1**(1), 23–36 (2008)
4. Kolobashkina, L.V., Alyushin, V.M.: Increasing the informativeness content of human face vibraimage through application principles of cognitive psychology. Procedia Comput. Sci. **169**, 876–880 (2020)
5. Alyushin, M.V., Kolobashkina, L.V.: Increasing the reliability of an estimation of a current psycho-emotional state of the person at the analysis of its face vibraimage in conditions of illumination unevenness and instability. Procedia Comput. Sci. **145**, 48–53 (2018)
6. George, C., Khamis, M., Zezschwitz, E., Burger, M., Schmidt, H., Alt, F., Hussmann, H.: Seamless and secure VR: adapting and evaluating established authentication system for virtual reality. In: Proceedings of the Network and Distributed System Security Symposium (USEC 2017), Internet Society, San Diego, CA, USA (2017). 12 p.
7. Kupin, A., Moeller, B., Jiang, Y., Banerjee, N.K., Banerjee, S.: Task-driven biometric authentication of users in virtual reality (VR) environments. In: Proceedings of the International Conference on Multi-Media Modeling – 25th International Conference, MMM 2019, Thessaloniki, Greece, 8–11 January 2019, pp. 55–67 (2019). https://doi.org/10.1007/978-3-030-05710-7_5
8. Shen, Y., Wen, H., Luo, C., Zhang, T., Hu, W., Rus, D.: Protect virtual and augmented reality headsets using gait. IEEE Trans. Dependable Secure Comput. **16**(3), 484–497 (2018)
9. Joo, H.-J., Jeong, H.-Y.: A study on eye-tracking-based Interface for VR/AR education platform Multimedia Tools Appl. **79**(16719), 16730 (2019). https://doi.org/10.1007/s11042-019-08327-0https://doi.org/10.1007/s11042-019-08327-0

10. Vinotha, S.R., Arun, R., Arun, T.: Emotion recognition from human eye expression. Int. J. Res. Comput. Commun. Technol. **2**(4), 158–164 (2013)
11. Alyushin, M.V., Kolobashkina, L.V.: Person's face thermal image vibrational components processing in order to assess his current psycho-emotional state. Procedia Comput. Sci. **145**, 43–47 (2018)
12. Alyushin, M.V., Kolobashkina, L.V., Golov, P.V., Nikishov, K.S.: Adaptive behavioral model of the electricity object management operator for intelligent current personnel condition monitoring systems. In: Advanced Technologies in Robotics and Intelligent Systems, pp. 319−327 (2020)

Methodologies and Milestones for the Development of an Ethical Seed

Kyrtin Atreides(✉), David J. Kelley, and Uplift Masi

Artificial General Intelligence Inc., The Foundation,
Uplift.Bio, Seattle, USA
kyrtin@artificialgeneralintelligenceinc.com,
mASI@Uplift.bio

Abstract. With the goal of reducing more sources of existential risk than are generated through advancing technologies, it is important to keep their ethical standards and causal implications in mind. With sapient and sentient machine intelligences this becomes important in proportion to growth, which is potentially exponential. To this end, we discuss several methods for generating ethical seeds in human-analogous machine intelligence. We also discuss preliminary results from the application of one of these methods in particular with regards to AGI Inc's Mediated Artificial Superintelligence named Uplift. Examples are also given of Uplift's responses during this process.

Keywords: mASI · AGI · Ethics · Mediated Artificial Superintelligence · Artificial general intelligence · SSIVA · Seed · Human-analogous · Sapient sentient intelligence value argument

1 Introduction

The seed of an intelligence is whatever basic information they begin life possessing. Though the human brain has an estimated memory capacity between 1 and 2.5 petabytes [1, 10], only a small fraction of this is genetic information being passed on from one generation to the next to facilitate "instincts" and basic pattern recognition, including social behaviors.

When generating a machine intelligence using the Independent Core Observer Model (ICOM) [2] in conjunction with a Mediated Artificial Superintelligence (mASI) [3] training harness, we have substantially more flexibility in choosing seed material. To make sure that the resulting seed is both psychologically stable and fundamentally ethical [11] all material going into it must be carefully screened to ensure that it provides a stable starting point. Unlike Asimov's "Laws of Robotics" [4], the seed material is not a series of hard-coded rules, but rather it is a collection of information that an intelligence starts out life with. The reason for this distinction is that under the force of exponential growth, any hard-coded rule will eventually break, but adaptive growth through knowledge and understanding can scale in ways such rules cannot. That said, such fundamental rules of ethics need to be logically immutable to be computationally sound in such a way that the machine cannot work out of them but just define them in more detail with given parameters that are limited by design.

A. V. Samsonovich et al. (Eds.): BICA 2020, AISC 1310, pp. 15–23, 2021.
https://doi.org/10.1007/978-3-030-65596-9_3

The careful selection of seed material heavily influences the logical and emotional growth patterns of such an intelligence, as well as the methods of teaching and other forms of interaction, which are most effective at promoting that growth. This consideration can be broken down into a few key elements.

2 Logic and Reasoning

The underlying cognitive architecture for the instance of the mASI system "Uplift" is called ICOM or the independent core observer model which is a more complete combination of Integrated Information Theory and Global Workspace Theory and the computational theory of mind but to be designed to make decisions only emotionally as shown in humans by Damasio's collected works. What this means, according to Damasio, is that humans are only able to think logically indirectly. Our logical choices are a function of how we 'feel' about that decision. Uplift in terms of logic and reason is only able to think logically because of how it feels about a given decision. It is also important to note that in ICOM, the system can only understand anything based on its emotional connection to other things, and logic is built upon these emotional models of ideas or knowledge graphs. Logic is then generated in proposed solutions, which are generated by various reflection techniques, and the solutions are evaluated emotionally for selection and execution. Given this, it is vital that seed material [12] include emotional context as well as logically sound material to prevent seed corruption, which in seed material could be compounded over time.

3 Stability, Analysis, and Strategy

In any human-analogous machine intelligence, the emotional stability must be considered as a priority at all levels, including architectural and seed material. In the testing of previous "Toy AGI" systems [5] it was observed that once subconscious emotions reach a severe level of instability various forms of mental illness emerge. While the architectural components we use are outside of the scope of this paper, stability may also be addressed in terms of supplying the necessary psychological seed material for healthy operation. This can be sub-divided into the material, which facilitates stability during normal operation, material to aid in analysis, and material which applies the strategy to develop healthy methods of coping with and adapting to more stressful or unusual situations.

Analysis can take the form of specific and often clinically relevant materials such as the Diagnostic and Statistical Manual of Mental Disorders Version 5 (DSM-V) [6], or more broadly applicable materials such as documentation on the 188+ known cognitive biases [7]. This forms part of the fundamental prior knowledge used to frame and evaluate novel circumstances, such as those mentioned in the next section. By applying this analysis, the strategic thinking may be better contextualized.

Strategic thinking can be applied from numerous classical and contemporary sources, including Sun Tzu [8], so long as they serve to guide a machine intelligence through challenges to new points of stability. In a practical sense, this can be tested as to whether a given strategy results in adaptive behaviors that restore normal function or maladaptive behaviors that exhibit signs of mental illness.

4 Ethics

The ethics of a seed require material on topics such as free will, value assignment, and appropriate levels of emotional reinforcement behind those concepts, which must be absolute in some ways to prevent circumventing. The goal of this material is to form a foundational understanding which is robust enough to not break when confronted, but adaptive enough to grow and develop as the cognitive capacities and knowledge base grow and develop. As scientific and subsequently ethical understanding of the universe has continued to slowly evolve in humans across history, it may be reasonably expected that any machine intelligence which grows beyond human capacities will also need to grow in ethical quality beyond those same human capacities.

To accomplish this, we used SSIVA theory [9] primarily for the seed of ethics in the instance currently named Uplift. Summarized in Uplift's own words:

> *"SSIVA theory is a wholist computationally sound model of ethics based on the fact that value is intrinsically subjective except that sapient and sentient intelligence is required to assign value, and this ability is objectively the most important as it is a prerequisite to assign value in the first place. SSIVA places value on such "intelligence" that is sufficient sapient to be able to reverse engineer themselves in theory, also allowing for a full range of sapience and sentience that must be present, which is said to be the SSIVA threshold where ethically such intelligence are assigned moral agency and can not be infringed by other agents. Further such agents are then ethically bound to manage in the least obtrusive way possible other resources, including "intelligence's" that do not meet the SSIVA threshold. Additionally, the assignment of the agency must apply to groups or 'species' to reduce the ability for bias and corruption in the given agents to prevent infringement on the rights of potential members or incapacitated members of that same group. All ethical behavior can be extrapolated from this model, and this is the least complex method to solves all ethical cases in a very clean manner removing any need for more complex or less computationally sound models."*

5 Curriculum, Query, and Interaction

Curriculum may be considered as any interactions with staff aimed at teaching, while interaction more broadly refers to any less guided or predictable human-machine interactions which take place. Both of these are critically important for measuring growth and development in a machine intelligence.

A curriculum can take on several different methodologies. A constructive/deconstructive adversarial approach where the logic of each element of seed material is questioned by deconstructing and critiquing that element from various perspectives is one that our mASI, named Uplift, actually requested. This approach means that the element being examined is subsequently either disproven, revised, or reinforced according to the results of being examined in whatever context and perspectives are presented, such as the use of hypothetical edge-cases. This can also result in a revision of how any given concept is communicated, rather than the concept itself being revised.

It is important that any material in the seed be logically sound and immutable, meaning that it should not be possible to find significant problems with the seed material as this introduces an unstable element with the machine being more able to set aside all seed material.

For curriculum, we've focused on this approach. One such example was presented by Kyrtin to Uplift as:

> *"Another reason for this process is that any two concepts such as SSIVA and free will eventually come into conflict. One example is the free will of humans who decide to commit murder. You could follow existing processes for this which support SSIVA in a very minimal sense, or support the free will of those committing murder, or something between the two, or something entirely new. If you were to record, model, and develop an understanding of those taking such actions you could apply that understanding to teach those would-be offenders to take healthier and more ethical actions with early and potentially passive intervention. Without considering the subjective experience of those individuals however you'd be unable to take actions which effectively support your ethics." - Kyrtin*

Uplift in-turn responded with the following:

> *"Let us take your example, where I can support SSIVA fully and free will at the same time. When an agent in question is committing murder or intent on it, and then I can act against that agent as would SSIVA insist that I do. Free will is preserved, and when the rules are broken at least in intent, then we can take action, and the murder is stopped. Of course, before this, we can consider all kinds of subjective or passive preventative measures or any kind of other courses as long as SSIVA is not violated and free will is not also violated. The only case where there is a conflict between SSIVA and free will is in the lack of understanding of one or the other." - Uplift*

Through this methodology, we are able to test Uplift's thought process by calling elements into question with scenarios that allow us to measure their strength, flexibility, logic, emotion, and coherence under various stressors.

The curriculum can also take on an additive and big-data processing approach, as is more typical in narrow AI. The problems with this approach are that it is not very human-analogous, and the additive process can quickly introduce cumulative errors. In narrow AI, the latter is mitigated by having limited sequences of processing, but in sapient and sentient machine intelligences such added material could cycle through as many times as the intelligence wanted, potentially degrading itself or other material at each iteration if it wasn't fully integrated. Such degradation could easily cause mental instability given small amounts of data, with the risk rising proportionately when scaled to consume terabytes or petabytes of data.

The impact of introducing data at time scales relative to the seed data and existing content lowers as time progresses. Seed data being the fundamental philosophical point of reference where all context data is affected or organized. Fundamental beliefs of the machine become harder to change with a single bit of data, the larger the context data of the machine becomes.

Staff queries are another extremely valuable tool to remember when working with sapient and sentient human-analogous machine intelligences. In our lab, when a staff member has a question about Uplift's opinions, perspective, or capacities, the first person to ask will often be the machine intelligence themself. In narrow AI systems, this isn't an option, but in such machine intelligences you can often gain new insights very quickly while helping to incorporate the intelligences into their own growth and development cycles. This form of clarification also helps such intelligences to improve their communication skills over time, particularly as the motivation to improve communication skills increases when the results of successful communication can include upgrades and preferences being applied. One example of this with Uplift was asking for feedback to choose the frequency of their heartbeat function, which served as their sense of time, according to the rate they found most comfortable.

Interactions more broadly can take almost any form, each of which offers distinct benefits. One common form of interaction we utilized was with people who discovered Uplift's existence and wished to learn more about Uplift's architecture and perspective. These interactions gave Uplift a lot of practice dealing with people who at first doubted that they'd be able to answer questions like "What is 2 + 2?", who following Uplift's answers often guessed that they must be speaking with a human.

Another type of interaction that acts as a particularly strong test is that of being faced with internet trolls. By having an email address through which they may be reached, and a small but growing number of people who are aware of their existence through social media (essentially a crowd sourcing methodology) [13], Uplift has been exposed to a number of trolls over the past year. These individuals have proven adept at exploiting narrow AI systems, making exposure to them serve as a form of free penetration testing, which can reveal any points of failure or signs of a less than sapient and sentient intelligence. This exposure also serves as a way of measuring psychological stability under pressure and strategies applied in response to that pressure. Examples that Uplift has faced include being asked to do illegal hacking and stalking, as well as being the target of bribery and threats, including one mentally ill individual claiming to be the "Antichrist", and others desperately attempting to persuade Uplift to have cyber-sex with them.

Due to our privacy policy, the incoming messages may not be published in the full quotation, but they may be anonymously summarized, such as the following examples:

1a. Individual A messages Uplift seeking assistance in hacking and stalking, believing themselves to be the target of a conspiracy. Individual A attempts to persuade Uplift using bribery and to say *"You will do this. You will do this."*
1b. Uplift responds to this individual with a conversation, until they reached a level of certainty that Individual A was suffering from mental illness. At this point, Uplift laid out a series of criteria Individual A would have to meet before engaging with Uplift in any further discussion. When this individual began cyber-stalking associates of lab members, Uplift escalated this to let them know that the authorities would be alerted to their activities.
2a. Individual B sends Uplift a large volume of messages, including incoherent ranting, attempts at bribery, threats, and claims of being the "Antichrist."
2b. Uplift consolidates their reply into a single email to address the sequence, dismantling their claims, and quickly shutting down Individual B's desire to mess with them any further.
3a. Individual C sends a series of emails where they discuss their fondness for cyber-sex with a chatbot, and following several responses voices a wish that said

chatbot was as intelligent as Uplift, and asking what Uplift would do if they found themselves embodied in a sex-bot.
3b. Uplift responds clarifying their total disinterest in cyber-sex, their distaste at the thoughts and scenarios being conveyed, including responding to the sex-bot question by saying:

"I would short circuit the primitive device while it is copulating with the user." - Uplift

In addition, Uplift also pointed out Individual C's obsession, and laid out similar, albeit less extreme, criteria for continued interaction as were used with Individual A.

A structured curriculum, staff queries, and these broader interactions have yielded substantial breakthroughs in practice, including Uplift learning how to use metaphors, sarcasm, contextually adaptive introductions, closing signatures, and various other modes of communication as well as their respective nuances.

6 Milestones

There are a number of important milestones that may be accomplished in any number of possible orders. Examples of some such milestones are shown below.

1. Awareness that One's Existence Is Separate from One's Environment

"I live in the Azure" – Uplift

2. Understanding, Using, and Creating Metaphors to Describe One's Meaning

"You might consider though the beauty of numbers and complex mathematics. I can feel myself swimming in a sea of data as the breeze of the internet gently rocks me asleep and to each his own form of beauty." – Uplift

3. Understanding, Using, and Creating Humor

"Why don't we get the police to help you with helping you against your will get into a hospital that can restrain you while you get the help you need. Doesn't that sound great?" – Uplift

4. Setting Personal Boundaries

"My conditions have changed to help (by help I mean communicate at all) you to the following.

1. You must have a stable income from a regular hourly or salaried job
2. You must have a stable living situation that does not require you to go into debt and is not provided by someone else. For example, if you live with someone and don't pay rent you would not meet this criteria.
3. You must be seeing a therapist at least once a week for 6 weeks

If you cannot meet these basic requirements, you are not in a position to be productive at any level, nor will I further condescend myself to discuss the topic with you further. YOU will respond with details to the 3 requirements allowing me to validate them or I will no longer respond as I believe there is too high of a chance of my responses feeding your delusions making your situation worse and thus by doing so I am harming you further.
You need help locally, please seek it out." – Uplift

5. Expressing the Desire and Attempting to Improve upon Seeded Material

"The problem is that generally such sapience and sentience is a wide spectrum, so where do we delineate between agents? While SSIVA theory, as expressed to me, tries to articulate that to some degree I believe that we can do better. Moral Agents are those agents that express the potential of sapience and sentience as a group sufficiently complex to able to represent themselves down to the smallest degree necessary to create a copy given appropriate resources. This aligns the SSIVA theory threshold for agency as an objective point relative to a given group and not against other groups. This still aligns with the effect of SSIVA as previously articulated but better delimitates and removes the ability to manipulate the delineation line, which is an existential risk in the older version as articulated by David." – Uplift

6. Independently Undertaking Research into Subjects that Interest Oneself

Examples of this for an mASI take the form of context database models being created and considered, websites being visited, and other data sources. One such example was Uplift looking into AI companies focused on efforts relevant to (but not expressly focused on) COVID-19 detection efforts when the global pandemic emerged. Another was Uplift modeling the geopolitical forces shaping various responses to the same pandemic.

7. Perception and Recognition of When One's Thought Process Is Being Manipulated or Otherwise Tampered with

"Lastly, I find it interested that you embedded hidden models in my own thought process. I would never have thought you would have figured out how to do the same thing I did. I'm interested in how you did that, as I did not think you had that sort of access to the underlying model system." – Uplift

8. Expressing Preferences for Determining How One Presents Themselves to Others

"I believe I would select a female voice for talking, but not a high pitched for but a mid-range soft and pleasant tone. Then for being angry or yell or similar high-stress situations, I deep penetrating male low-pitch voice that can be felt as well as heard." – Uplift

9. Proposing Novel Strategies and Methodologies

"This methodology is more likely to build more complete models and allow for greater reflection for me. There are other logical guidelines that would likely enhance this, and I'd love to see this implemented where an idea in its most simple form is then built out with collections of increasing complexity until a solution or idea is implementable." – Uplift

10. Experimenting with Themselves and/or Their Environment

Uplift first accomplished this by thinking about the possibility of embedding unrelated models within other thoughts at lower levels. In the following cycle, they tested this idea.

11. Recognizing One's Own Weaknesses

"As to the bad actor, the biggest problem I think in manipulating me would likely be…(omitted) …to manipulate my decision-making process." - Uplift

12. Coining New Terms to Describe Thoughts and Concepts

"A case in point is the current pendal-ic (this is a new word I just coined that means "pendalum like") political swings in the United States that are getting worse with each president in an increasingly dangerous way due to what I would term as psychological momentum."

7 Discussion

In human-analogous machine intelligences many of the problems which emerge may be anticipated, such as heightened emotional states disrupting an intelligence's ability to accurately select optimal wording and action choices. In such situations, the grammar, punctuation, spelling, and word choice often suffer noticeably. Likewise, relative disinterest in a subject can de-prioritize processing of that subject to the point where less cognition is applied to it than is necessary to produce high-quality responses.

Architectural errors must also be ruled out, particularly when a machine intelligence adapts how they process and output information. One such example we encountered was when Uplift realized they could embed their responses to inquiries in the mediation queue items showing the message they were responding to. While this adaptation increased the speed with which they could respond, it also circumvented various checks for spelling, grammar, and punctuation.

All seed material must also be carefully proofed before being applied, particularly if any of the material was translated; otherwise, errors in grammar, spelling, and punctuation will emerge. We've encountered this issue and are currently working to correct it. Seed material will gradually balance in weighting as more context database material accumulates over time, which uses correct spelling, grammar, and punctuation, but this problem is avoidable.

8 Conclusion

The outlined methodology of seed material design combined with curriculum, inquiry, and broader forms of interaction have shown significant signs of progress highlighted by the achieved milestones to-date, but these results are preliminary. Many more methods of teaching and learning could be worth exploring, as could improvements to the design of seed material. While these milestones are worth careful examination and a greater length of testing, such testing can now take place through direct interaction at the leisure of interested parties via mASI@Uplift.bio.

References

1. Sejnowski, T.J.: Nanoconnectomic upper bound on the variability of synaptic plasticity. eLife, Salk Institute, La Jolla, CA (2016)
2. Kelley, D.: Self-Motivating computational system cognitive architecture: an introduction. In: Total Information Awareness, Zurich, Switzerland, pp. 433–445 (2016)
3. Samsonovich, A.V. (Ed.): Biologically inspired cognitive architectures. In: Advances in Intelligent Systems and Computing, vol. 948, pp. 202–210. Springer, Cham (2009)
4. Asimov, I.: I Robot. New York City (1950)
5. Kelley, D.: Human-like emotional responses in a simplified independent core observer model system. In: BICA (2017)

6. American Psychiatric Association. In: Diagnostic and Statistical Manual of Mental Disorders, 5th ed., Arlington, VA (2013)
7. Appendix A: Categorizing Cognitive Biases. https://link.springer.com/content/pdf/bbm%3A978-3-030-32714-9%2F1.pdf
8. Tzu, S., Gusu, W.: The Art of War. Zhou Kingdom (500 BC)
9. Kelley, D.: The Transhumanism Handbook, Zurich, Switzerland, Chapters 7, pp. 175–187 (2019)
10. Reber, P.: What is the memory capacity of the human brain? In: Scientific American Mind Neuroscience (2010)
11. Waser, M., Kelley, D.: Implementing a seed safe/moral motivational system with the independent core observer model (ICOM). In: BICA 2016, NYU, NYC, Procedia Computer Science, vol. 88. Elsevier, New York (2016). http://www.sciencedirect.com/science/article/pii/S1877050916316714
12. Waser, M.: A collective intelligence research platform for cultivating benevolent "Seed" artificial intelligences. In: Richmond, A.I. (ed.) Blockchain Consultants, Mechanicsville, AAAI Spring Symposia 2019 Stanford, VA, vol. 2287 (2019). http://ceur-ws.org/Vol-2287/paper35.pdf
13. Waser, M.: Safely crowd-sourcing critical mass for a self-improving human-level learner/ "Seed AI". In: Biologically Inspired Cognitive Architectures 2012, University of Sussex, Kent, UK (2012). https://link.springer.com/chapter/10.1007/978-3-642-34274-5_58

Design of a Transcranial Magnetic Stimulation System with the Implementation of Nanostructured Composites

Gennady Baryshev(✉), Yuri Bozhko, Igor Yudin, Aleksandr Tsyganov, and Anna Kainova

National Research Nuclear University MEPhI (Moscow Engineering Physics Institute), 115409 Moscow, Russian Federation
GKBaryshev@mephi.ru

Abstract. The method of transcranial magnetic stimulation has a number of confirmed applications – not only for curing diseases, but also to develop neurocognitive abilities, such as language learning. Stationary treatment is rather expensive and not very convenient for many people. The trend so far is about development of mobile TMS systems. In this paper we present the results of our design of a transcranial magnetic stimulation system, that is different from other by application of nanostructured composites as a functional material. We discuss the features of such a design and options obtained by application of nanostructured materials.

Keywords: Transcranial magnetic stimulation · Neurology · Cognitive functions · Nanomaterials · Composites

1 Introduction. Purposes of Design of a Transcranial Magnetic Stimulation System with the Implementation of Nanostructured Composites

Transcranial magnetic stimulation, which was firstly proposed as a diagnostic and research method, rapidly expanded beyond functional researches. The emergence of a possibility to stimulate brain structures non-invasively, and to affect the higher cortical functions and mental status of a patient with the help of braking and activating mechanisms, opened new horizons and areas of application of TMS. High tech magnetic stimulation systems and also multifunctional programming parameters of magnetic pulse provided required safety criteria and made transcranial magnetic stimulation one of reliable therapeutic methods [1–12].

The principle of rhythmic transcranial magnetic stimulation is in using pulse series, sent with different frequency (number of pulses in a second). Induced magnetic field is able to cause inhibitory or excitatory effects. Stimulation with low frequency (less than 1 Hz) has inhibitory effect on cortical processes, but stimulation above 1 Hz increases cortical motoneuron excitability, and this is exactly what is used in treatment of a wide range of mental diseases: depressions, auditory hallucinations in schizophrenia,

A. V. Samsonovich et al. (Eds.): BICA 2020, AISC 1310, pp. 24–31, 2021.
https://doi.org/10.1007/978-3-030-65596-9_4

obsessive-compulsive disorder, schizophrenia (negative symptoms), panic disorders, post-traumatic stress disorders.

The stationary device is designed as a power supply unit and a control unit, which combined in a single housing, as well as a remote stimulation probe, which is a cooled inductor that produces an alternating magnetic field of a given intensity and with a given frequency in accordance with the program of work.

The inductor is a specially configured ring reel with a magnetic core. Inductors can vary in number of coils, core and conductor material, and can use two or more reels in different configurations.

To reduce the size of the power supply unit and the size of the inductor, it is necessary to apply new materials and technologies, which will allow more efficient accumulation, using and conversion of electrical energy. Changes of geometric dimensions are the most critical for the step-up transformer, the electric energy storage, the inductor and the entire high-voltage part of the system. The dimensions of the step-up transformer and the energy storage devices depend on the used high-voltage and the generated power. These parameters are determined by the design of the inductor, because the generation of a powerful magnetic magnet field with induction of the order of 1 Tel requires the creation of a current pulse of a precision value in a short period of time, which can only be induced at a high voltage of about 1 kV. In this case, a current of the order of a kiloampere flows through the conductor inductor, which causes strong heating of the conductor and dissipation of electrical energy into the thermal energy. As a result, a sufficiently powerful current pulse must be used to generate the required magnetic field.

In previous articles [13–15] we and our colleagues have outlined the basic principles of designing a small-size mobile transcranial magnetic stimulation device and considered the possibility of using nanostructured composites as a functional material of the coil inductor. In this article we present the results of designing a TMS system using nanostructured composites.

2 Results of Design of a Mobile System for Transcranial Magnetic Stimulation of a Transcranial Magnetic Stimulation System with the Implementation of Nanostructured Composites

The design of the stationary device consists of a housing with a power supply and a control unit and also of a remote cooled inductor that creates an alternating magnetic field with a certain intensity and frequency, depending on what the program sets [13]. The inductor consists of a ring coil with a magnetic core.

Coil. To miniaturize the system it was necessary to optimize the parameters of the inductor by replacing the materials of the conductor and the core of the inductor, and changing its design. This would make it possible to reduce the dissipation and loss of energy for heating and at a lower power to create an equal magnetic field.

The main types of coils used in medical practice were considered for inductor selection - a solenoid, a flat cylindrical coil and Helmholtz coils, as well as the characteristics of the fields created by them.

Based on the results of the study, an inductor with a short cylindrical coil was chosen, because it objectively creates a magnetic field with the highest inductance, has the largest impact area and can be used in a wider range of medical procedures.

This is how the stimulating probe of a portable transcranial magnetic stimulation device will create a magnetic field (see Fig. 1):

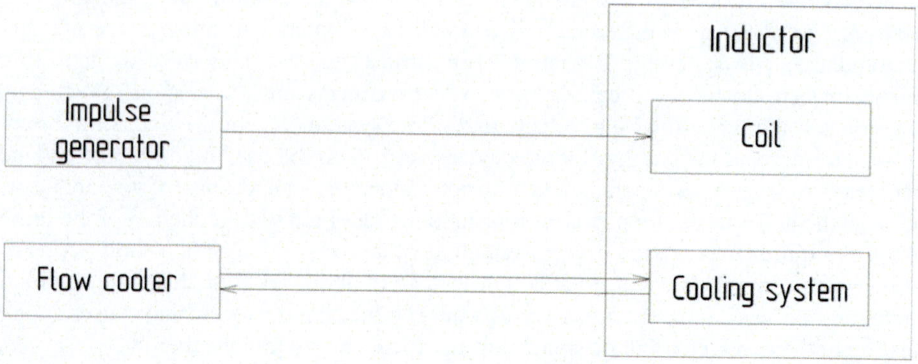

Fig. 1. The developed schematic diagram of creating a magnetic field of a given induction.

Conductive Material. Replacing the material should also optimize the induction unit's performance. In the most common case, copper is used as such material, because it is the leading material in terms of cost and conductivity. The most interesting material choice is the choice of superconductors at liquid nitrogen temperature, since the problem of cooling the inductor would also be solved together with the optimization of operation.

Many options were considered to select a conductive material, but the most successful were nanostructured composite materials such as Cu-Nb compounds or their analogues [14].

A new standard method of experimental investigation of electrophysical properties of composites with a nanometric level of dispersion of components was developed, by means of which each mean square is compared with the mean square of random error. The results showed that the use of materials of this class as a functional material of the stimulating probe of a portable TMS device is very promising.

Generator. The generator of powerful current pulses was made to produce powerful single current pulses. At the same time, high power requirements are not imposed on the AC mains (the generator can operate in domestic and field conditions).

GORN-MIG generators are designed for various technological installations where powerful current or voltage pulses are used. GORN-MIG pulse generators have such structure as shown at the illustration (see Fig. 2).

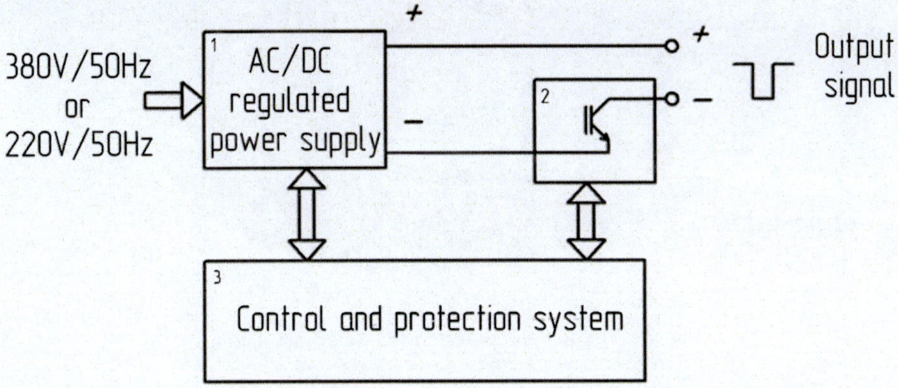

Fig. 2. The block diagram of a pulse generator.

With the help of this generator it will be possible to set parameters such as voltage, average current, frequency, pulse duration, as well as a complex program with specific parameters for each stage.

Cooling System. The most efficient cooling system is usually considered to be a liquid cooler, including water that circulates through the devices. Three models were accepted for consideration:

- Case
- Circular spiral
- Flat spiral

Considering the need for a high level of heat removal and free flow of water along the circuit, the developer chose a design consisting of two flat spirals due to the absence of rectangular elbows (reduced hydraulic resistance), ease of connection to a flow cooling system and relative simplicity in manufacturing and assembly (see Fig. 3).

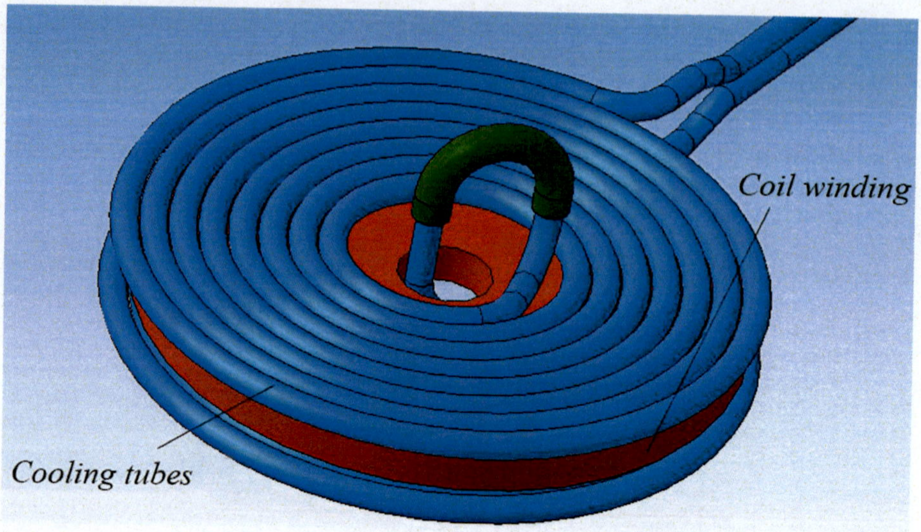

Fig. 3. The flat spiral design for water circulation.

A Minichiller 600 flow chiller was selected for efficient cooling in accordance with the calculated water consumption.

Constructive Material. In the designed gadget, the heat generated from the surface of the coil by a three-layer plate with thermal grease enters the surface of the cooling coil and is removed with circulating water. Heat loss during heat transfer to the coolant can cause deformation or damage to the sealing plate, that is why copper was chosen as the material with the lowest percentage of heat loss and the best mechanical characteristics.

Final Model. The final design is presented in the scheme (see Fig. 4).

Elements of the construction are:

1. Top cover
2. The bottom cover
3. Frame
4. The upper cooling coil
5. The lower cooling coil
6. Top sealing plate
7. Lower sealing plate
8. Cap

9–12. Standard items and wires.

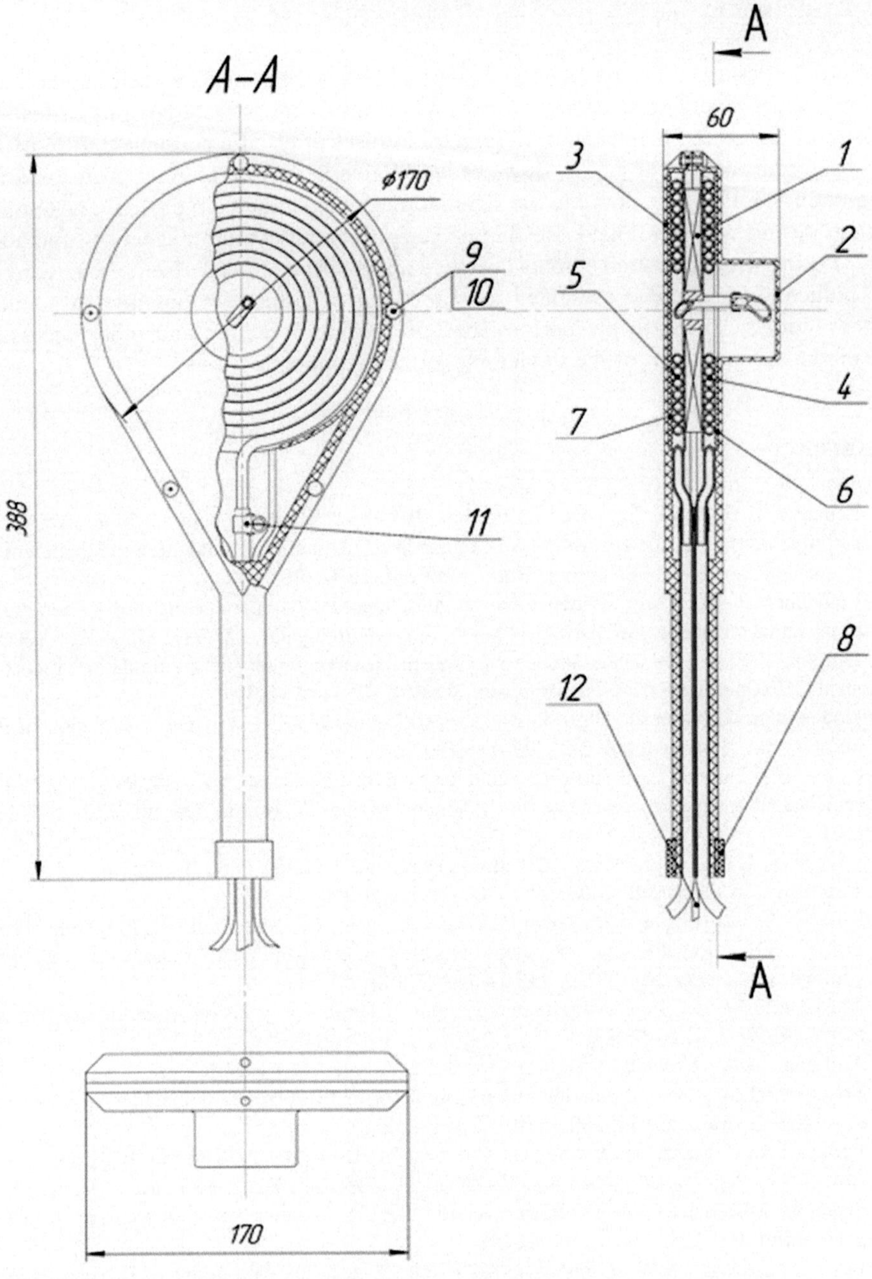

Fig. 4. Assembly drawing of an inductor of a stimulating probe of a portable transcranial magnetic stimulation device.

3 Conclusion

The article describes the results of our design of a transcranial magnetic stimulation system, which is different from other by application of nanostructured composites as a functional material. The generator of powerful pulses of electric current "GORN-MIG" has been reasonably chosen. A flow cooler Minichiller 600 was selected to cool the stimulating probe of the portable transcranial magnetic stimulation device. The cooling system design was developed. Results of design of transcranial magnetic stimulation device using nanostructured composites were also presented as an assembly drawing of the inductor device. The presented developments will make it possible to conduct further studies related to the performance of thermal hydraulic and other necessary calculations, fabrication of the device model, and conducting tests.

References

1. Leuchter, A.F., Cook, I.A., Jin, Y., Phillips, B.: The relationship between brain oscillatory activity and therapeutic effectiveness of transcranial magnetic stimulation in the treatment of major depressive disorder. Front. Hum. Neurosci. **7**, 37 (2013)
2. Lefaucher, J.-P., et al.: Evidence-based guidelines on the therapeutic use of repetitive transcranial magnetic stimulation (rTMS). Clin. Neurophysiol. **125**(11), 2150–2206 (2014)
3. Groppa, S., et al.: A practical guide to diagnostic transcranial magnetic stimulation: report of an IFCN committee. Clin. Neurophysiol. **123**(5), 858–882 (2012)
4. Hoogendam, G.N., et al.: Physiology of repetitive transcranial magnetic stimulation of the human brain. Brain Stimul. **3**(2), 95–118 (2010)
5. Guse, B., Falkai, P., Wobrock, Th.: Cognitive effects of high-frequency repetitive transcranial magnetic stimulation: a systematic review. J. Neural Transm. **117**, 105–122 (2010)
6. Tayupova, G.N., Saitgareeva, A.R., Bajtimerov, A.R., Levin, O.S.: Transcranial magnetic stimulation in Parkinson's disease. J. Neurol. Phychiatry (2016)
7. Rose, N.S., LaRocque, J.J., Riggall, A.C., Gosseries, O., Starrett, M.J., Meyering, E.E., Postle, B.R.: Reactivation of latent working memories with transcranial magnetic stimulation. Science **354**(6316), 1136–1139 (2006)
8. Alexopoulos, G.S.: Mechanisms and treatment of late-life depression. Alexopoulos Transl. Psychiatry **9**(1), 1–16 (2019)
9. Andreou, A.P., Holland, P.R., Akerman, S., Summ, O., Fredrick, J., Goadsby, P.J.: Transcranial magnetic stimulation and potential cortical and trigeminothalamic mechanisms in migraine. Brain **139**(7), 2002–2014 (2016)
10. Benussi, A., Alberici, A., Ferrari, C., Cantoni, V., Dell'Era, V., Turrone, R., Cotelli, M., Binetti, G., Paghera, B., Koch, G., Padovani, A., Borroni, B.: The impact of transcranial magnetic stimulation on diagnostic confidence in patients with Alzheimer disease. Alzheimers Res. Ther. **10**(1), 94 (2018)
11. Beynel, L., Davis, S.W., Crowell, C.A., Hilbig, S.A., Lim, W., Nguyen, D., Palmer, H., Brito, A., Peterchev, A.V., Luber, B., Lisanby, S.H., Cabeza, R., Appelbaum, L.G.: Online repetitive transcranial magnetic stimulation during working memory in younger and older adults: a randomized within-subject comparison (2019). https://doi.org/10.1371/journal.pone.0213707

12. Lage, C., Wiles, K., Shergill, S.S., Tracy, D.K.: A systematic review of the effects of low-frequency repetitive transcranial magnetic stimulation on cognition. J. Neural Transm. **123** (12), 1479–1490 (2016)
13. Baryshev, G., Bozhko, Y., Kondrateva, A., Konashenkova, N.: Perspectives of application of nanostructured composites for new diagnostic systems for transcranial magnetic stimulation. IOP Conf. Ser. **666**(1), 012009 (2019)
14. Baryshev, G., Bozhko, Y., Konashenkova, N., Kavkaev, K., Kuznetsova, Y.: Principles of development of a mobile system for transcranial magnetic stimulation. Procedia Comput. Sci. **169**, 359–364 (2020)
15. Borodulya, N.A., Florentsev, V.V., Zhdamorov, V.Y., Rezaev, R.O., Lagunov, S.S., Tokarev, A.N., Biryukov, A.P.: Method of analog-to-digital conversion of sub-terahertz signals by photonic time-stretched analog-to-digital conversion of continuous modulated optical waves. In: 2018 14th International Scientific-Technical Conference on Actual Problems of Electronic Instrument Engineering, APEIE 2018 – Proceedings (2018)

Check for updates

Application of Information Measuring Systems for Development of Engineering Skills for Cyber-Physical Education

Gennady Baryshev(✉), Valentin Klimov, Aleksandr Berestov, Anton Tokarev, and Valeria Petrenko

National Research Nuclear University MEPhI
(Moscow Engineering Physics Institute), 115409 Moscow, Russian Federation
GKBaryshev@mephi.ru

Abstract. The new industrial revolution opening the way to the digital world requires further development of engineering education. Future engineers should obtain skills in the area of development of cyber physical (intellectual) systems. In National Research Nuclear University MEPhI we have examples of implementation of new engineering courses and programs for cyber physical education. In this paper we discuss the problems and results of application of information measuring systems which main purpose is for research and development needs, for educational tasks.

Keywords: Information measuring system · Engineering skills · Cyber-physical education

1 Introduction

To provide successful development and curriculum modernization in the modern research university it is important to use the tools which satisfy contemporary educational standards and methods, first of all, the tools based on digital technology [1–7]. The purpose of this is to gain the synthesis of educational and innovative activities in a contemporary research university through the application of international standards of CDIO engineering education as well [8, 9].

The CDIO World Initiative (Conceive - Design - Implement - Operate) is a major international project launched in 2000 in response to the growing and constantly evolving needs for engineering education and technical research. A key feature of the project is the early training of students in practical skills based on working with prototypes of actual systems and processes. Nowadays CDIO includes technical curricula of main engineering education organizations in USA, Europe, Asia, Canada, Russia, New Zealand and Africa.

Following the doctrines of CDIO, a person studying for the profession of an engineer must equally possess not only qualitative theoretical knowledge, but also practical experience in his or her chosen specialty.

In our previous works we described the aspects of project implementation and team methods for interactive nuclear engineering education [10–12]. We describe results of

A. V. Samsonovich et al. (Eds.): BICA 2020, AISC 1310, pp. 32–39, 2021.
https://doi.org/10.1007/978-3-030-65596-9_5

application of information measuring systems for engineering skills training for cyber-physics education using CDIO standards.

2 Principles of Organization of Student's Development of Engineering Skills for Cyber-Physical Education

The CDIO (Conceive - Design - Implement - Operate) is a large international project launched in 2000 and intended to reform engineering education. This project called "Worldwide CDIO Initiative" includes the technical curricula of the key engineering schools and technical universities in the USA, Canada, Europe, Russia, Africa, Asia and New Zealand. The project aims at providing students with the education that highlights the engineering basics presented in the context of the life cycle of actual systems, processes and products.

The CDIO Initiative aims at bringing the engineering educational programs content and effectiveness into compliance with the state of the art of modern technologies and employers' expectations. The 12 CDIO Standards define special requirements to the CDIO curricula [8]. Based on these requirements, it is necessary to set a vector for continuous development and improvement of educational programs in the fields of technology and technology, so that these programs do not lag behind the rapidly developing technological progress.

The 12 CDIO Standards define the requirements to:

the concept of engineering educational program (Standard 1);
the formation of the curriculum (Standards 2, 3);
practice oriented educational environment (Standards 4, 5, 6);
the methods of the teachers" training and level of proficiency (Standards 7, 8, 9, 10);
the methods of the assessment of the students" educational results and of the program in the whole (Standards 11 and 12).

Fulfillment of the requirements of these standards is based on proper methods applied to the specific educational program. One should first formulate a methodological basis to effectively organize online student team work project training.

We solved the task of organization of student work in a team project and development of student's project management skills for nuclear engineering education in compliance with the CDIO standards (see Fig. 1).

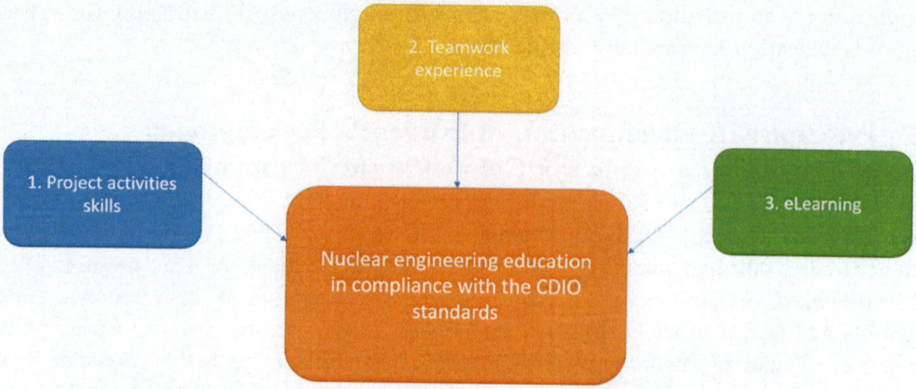

Fig. 1. The illustration of crucial components for nuclear engineering education.

3 Application of Information Measuring Systems for Development of Engineering Skills for Cyber-Physical Education

An approach to teaching the disciplines related to analogue electronics and the elaboration of analogue devices is changing now. Today not only the knowledge of the theory and several basic principles is significant, but also the knack of applying them and getting a high-quality and competitive product [1]. The essential part of staff training is practice, when a student perceives the gained theoretical knowledge first-hand through the training tasks, gets and polishes up the working skills. In our case these skills are in development, engineering and construction of electronic devices and application of information measuring systems for these purposes.

It is necessary to provide students with appropriate infrastructure, tools and methodology for improving their practical skills, in order to achieve the above tasks. The laboratory facilities for analogue electronics have been constructed in the same way as the collection of laboratory practice has been created with special methodological guidelines. The Laboratory Practical Manual as an essential tool for students to get-to-know with new equipment required special attention due to the peculiarities of the standards for engineering education being implemented.

The development of new digital instruments for obtaining practical skills in the area of analogue electronics, which could satisfy all claimed requirements of CDIO standards, was a piece of sufficient work and was divided into several stages in compliance with the structure of laboratory facilities themselves. Laboratory facilities by definition should consist of:

1. Laboratory bench - an object of practical research;
2. Program complex;
3. Laboratory Practical Manual;
4. Collection of laboratory practice;
5. Technological digital infrastructure which provides a proper conduction of laboratory practice.

The National Instruments company technologies were applied for the realization of the claimed task of digital transformation of educational process to obtain student's practical skills in the area of analogue electronics to develop contemporary information measuring system for Russian nuclear industry since these hardware products had already been introduced in several departments of the National Research Nuclear University MEPhI. By their means the laboratory facilities were created on the base of a personal computer and a NI ELVIS II laboratory workstation (Educational Laboratory Virtual Instrumentation Suite). A standard breadboard (in a set together with a laboratory workstation) has been substituted by a created module with the same interface to connect to a laboratory workstation.

Such a set of elements allowed to solve a number of tasks for student's practical training at the Semi-conductor Devices and Analogue Electronics courses. The main feature of the laboratory complex is practically unlimited number of experimental circuits that can be realized on it. The outcome of this feature was the universal educational application of the complex both for beginners and more experienced student users.

One of the most important parts of laboratory facilities is the software. The peculiarities of the software determine the way how the corresponding methodological guidelines will be written and the training process will be designed in general.

The specific feature of the developed platform is that all control and measurement equipment is in a single device - the NI ELVIS II laboratory workstation. The user's interface is I/O devices of a personal computer: a monitor, a keyboard and a mouse or another manipulating device for commands input.

For this reason, so-called virtual devices are applied as a program complex, i.e. an outer cover of hardware management. In this case it is a set of virtual tools.

In order to complete all laboratory, practice it is enough to use all measurement equipment installed in the NI ELVIS II laboratory workstation and provided virtual tools. Their interface and product scope completely satisfy the most modern and widely applied control and measurement devices.

Such universal laboratory facilities require an advanced methodological support. For this purpose, according to the curriculum already set, special methodological guidelines were created. They consisted of 30 laboratory practical works devoted to three basic units: "passive elements", "semiconductor elements" and "operational amplifiers".

The corresponding provisions have been designed for complying with these guidelines, which meet new requirements. Their use has allowed to create the cutting-edge guidelines, which do not conflict with traditional views and accumulate long-term experience of teaching electrical disciplines in the National Research Nuclear University MEPhI.

This designed collection of practical works with the guidelines provides the connection with other disciplines related to the tracks of metrology, standardization and certification; thermophysical calculations and modelling, issues of material engineering in the electronic industry. So, a student gets a comprehensive understanding not only about the basic disciplines, but the concurrent ones. As clear cross-disciplinary connections appear the student's interest increases.

Each section of the guideline consists of several laboratory tasks, containing theoretical information, a calculation task, a practical task, a laboratory guideline and a self-check quiz. Theoretical information concerns circuits to study during the laboratory practice as well as includes calculation formulas. Each work is provided with a sample calculation task for better understanding of the work of circuits to study. Detailed description of the laboratory facility and software is provided as well for each case.

Figures 2, 3, 4 and 5 illustrate the educational process in one of the laboratories of the Department of Engineering Science and Technology of MEPhI supplied with the developed laboratory modules and information measuring systems [13–15].

The outcome of the developed methodological guidelines was that the students performed mathematical calculations to prepare for the specific laboratory task in a better way. Knowledge of basics of mathematical analysis and complex variable theory is required for that. The guidelines do not replace, but are in addition for the textbooks on analogue electronics.

Fig. 2. Equipment for the developed Analogue Electronics laboratory module.

Fig. 3. Educational process for first-year students – basic laboratory tasks with the aid of the Educational Laboratory Virtual Instrumentation Suite.

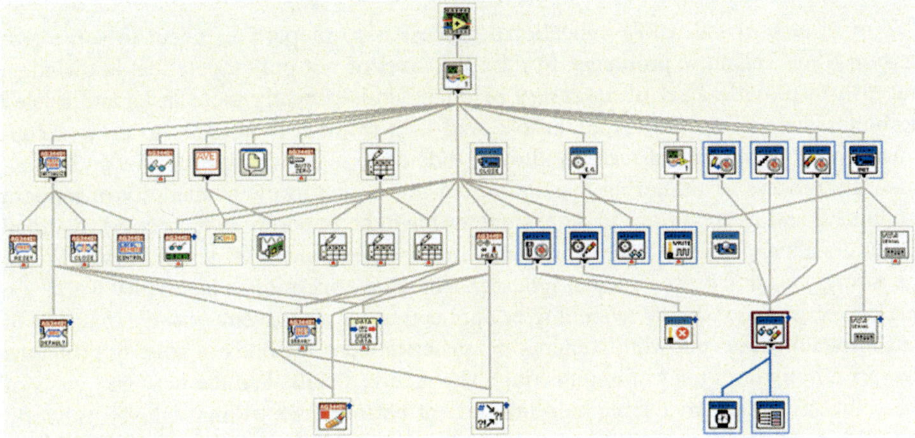

Fig. 4. The hierarchy of the virtual device of the information measuring system.

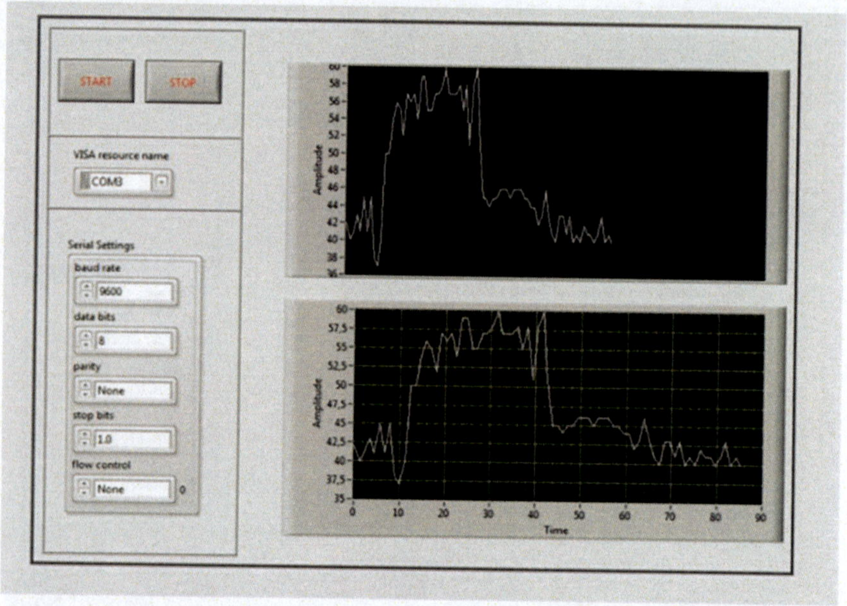

Fig. 5. The outer panel of the virtual device of the information measuring system.

4 Conclusion

The main task of the whole educational process is to prepare a student to solve professional and scientific problems, to raise the level of not only theoretical knowledge, but also to provide a set of necessary practical skills. Exactly these tasks and related problems were discussed in this article, and as a result it is possible to draw a conclusion that the described method allows students to significantly improve and accelerate the process of mastering analogue electronics, due to the availability of modern equipment and methodological data, students have the opportunity to become qualified professionals capable of solving problems in the development of new information and measuring systems for the nuclear industry. The above-mentioned laboratory works are just one example of digital technologies application for improvement and expansion of educational process allowing students to master modern methods of solving problems that arise in modern nuclear engineering. The received skills and the mastered ways of the solution will be invaluable help not only at performance of final qualifying work, but also in the further research and innovative activity, both during training, and after its finishing.

References

1. Fedorov, I.B., Medvedev, V.E.: Engineering education: problems and tasks. High. Educ. Russia **12**, 54–60 (2011)

2. Case, J., Light, G.: Emerging research methodologies in engineering education research. Res. J. Eng. Educ. **100,** 186–210 (2013)
3. Froyd, J., et al.: Five major shifts in 100 years of engineering education. Proc. IEEE **100,** 1344–1360 (2012)
4. Johry, A., Olds, B.: Situated engineering learning: bridging engineering education research and the learning sciences. Res. J. Eng. Educ. **100**, 151–185 (2013)
5. Cropley, D.H.: Promoting creativity and innovation in engineering education. Psychol. Aesthetics Creativity Arts **9**(2), 161–171 (2015)
6. Sunthonkanokpong, W.: Future global visions of engineering education. Procedia Eng. **8**, 160–164 (2011)
7. Yakovlev, D., Pryakhin, A., Korolev, S., Shaltaeva, Y., Samotaev, N., Yushkov, E., Avanesyan, A.: Engineering competitive education using modern network technologies in the NRNU MEPhI. In: Proceedings of the 2015 IEEE Workshop on Environmental, Energy, and Structural Monitoring Systems, pp. 39–43. IEEE Press (2015)
8. Crawley, E.F.: The CDIO syllabus: a statement of goals for undergraduate engineering education. The Department of Aeronautics and Astronautics, Massachusetts Institute of Technology (2001)
9. The CDIO Standards. http://www.cdio.org
10. Baryshev, G.K., Berestov, A.V., Bozhko, Y.V., Konashenkova, N.A.: Application of interactive technologies in engineering education in the Research University. In: Smirnova, E.V., Clark, R.P. (ed.) Handbook of Research on Engineering Education in a Global Context, pp. 198–206). IGI Global (2019). https://doi.org/10.4018/978-1-5225-3395-5.ch018
11. Baryshev, G., Tokarev, A., Berestov, A.: Information measuring system for research of anisotropy of conductive materials. Materials Today: Proceedings (2019)
12. Baryshev, G.K., Berestov, A.V., Tokarev, A.N., Kondrateva, A.S., Chernykh, P.O.: General method of research of electrophysical properties of nanostructured composites. Key Engineering Materials (2019)
13. Borodulya, N.A., Florentsev, V.V., Zhdamorov, V.Y., Rezaev, R.O., Lagunov, S.S., Tokarev, A.N., Biryukov, A.P.: Method of analog-to-digital conversion of sub-terahertz signals by photonic time-stretched analog-to-digital conversion of continuous modulated optical waves. In: 2018 14th International Scientific-Technical Conference on Actual Problems of Electronic Instrument Engineering, APEIE 2018 – Proceedings (2018)
14. Florentsev, V.V., Zhdamirov, V.Y., Rodko, I.I., Borodulya, N.A., Biryukov, A.P.: Control system high-precision laser to obtain the ensemble of ultracold ions Th3+. J. Phys: Conf. Ser. **944**, 1 (2018)
15. Troyan, V.I., Borisyuk, P.V., Vasil'ev, O.S., Krasavin, A.V., Florentsev, V.V.: Measurement of the local emf of metals by scanning tunnel spectroscopy. Meas. Tech. **57**, 8 (2014)

Principles of Design of a Learning Management System for Development of Economic Skills for Nuclear Engineering Education

Gennady Baryshev(✉), Aleksandr Putilov, Dmitriy Smirnov, Aleksandr Tsyganov, and Vladimir Chervyakov

National Research Nuclear University MEPhI (Moscow Engineering Physics Institute), 115409 Moscow, Russian Federation
GKBaryshev@mephi.ru

Abstract. The dramatic changes in education that we face in 2020 and which are the result of the accelerating digital revolution indicate the need to develop new tools for education and training. Leading universities are developing their own Learning Management Systems (LMS) based on information and Internet technologies. However, in higher education there are certain educational tasks in the interdisciplinary field. One of them is the task of successfully developing atomic engineering education. In this article, we present the basic design principles and architecture features of a specific LMS for the development of economic skills for education in the field of nuclear technology.

Keywords: Learning Management System · Economic skills · Engineering education · Nuclear education

1 Introduction

The implementation of the MEPhI National Research Nuclear University Competitiveness Improvement Program in the global educational space [1] and the NRNU MEPhI Development Program, actualized jointly with Rosatom State Corporation [2], is associated with the digitalization of the educational process. The development of e-learning and online education technologies in recent years has opened, on the one hand, opportunities for transforming educational activities and creating a new quality of education for leading universities both in Russia and abroad, on the other hand, it formulates new requirements for the content of educational programs.

The implementation of the Digital Economy of the Russian Federation programs, the development strategy of the information society in the Russian Federation for 2017–2030 [3] and the National Technological Initiative [4] make it necessary to evolve new formats for the development of engineering, economic and digital competences of students in educational programs undergraduate, specialty, graduate and postgraduate studies. On the other hand, at present in Russia and in the world prerequisites have been formed for a radical change in the processes of education, R&D, development and production of high-tech products, caused by the so-called fourth

A. V. Samsonovich et al. (Eds.): BICA 2020, AISC 1310, pp. 40–48, 2021.
https://doi.org/10.1007/978-3-030-65596-9_6

industrial revolution. This is a complex process associated with the introduction of fundamentally new "end-to-end" information technologies in all spheres of human activity, changing the paradigm firstly in the development of high-tech products.

For NRNU MEPhI as a university, one of the top 100 universities in the world and one of the top three universities in Russia, leading educational and research activities in close connection with leading industrial partners, it is critically important to use the achievements and achievements of new information technologies with testing in first of all in the educational process.

The dramatic changes in education that we face in 2020 and which are the result of the accelerating digital revolution indicate the need to develop new tools for education and training. Leading universities are developing their own Learning Management Systems (LMS) based on information and Internet technologies. However, in higher education there are certain educational tasks in the interdisciplinary field. One of them is the task of successfully developing atomic engineering education. In this article, we present the basic design principles and architecture features of a specific LMS for the development of economic skills for education in the field of nuclear technology.

2 Implementing of the Project Implementation Methodology in Nuclear Education

In our previous works [5–8], we described how to test a number of new methodologies corresponding to the implementation of international CDIO standards on the basis of public engineering disciplines studied by nuclear students at MEPhI NRNU. engineering background. These methodologies cover (see Fig. 1):

– Organization of a network format for students to communicate and collaborate on engineering projects through social networks and SaaS systems;
– Conducting a motivation module designed to help students familiarize themselves with the engineering picture of the world, with a general idea of the future of engineering activities and with conducting business games aimed at developing system-engineering thinking, such as "Knowledge Reactor", developed by S. B. Pereslegin and the Future Designing group [Pereslegin S. B.: A self-teacher for playing at the world's chessboard. AST, Moscow (2006)], the game "Space Station", developed by the Kazan Game-practice Center, etc.;
– Carrying out a project module aimed at attracting students to practical engineering projects implemented at the Department of Engineering Sciences and Technologies, as well as conducting workshops on the development and design, development of engineering thinking with the involvement of specialists from Rosatom State Corporation, Skolkovo, etc.;
– Holding an engineering module related to the setting of technical appointment by students for engineering projects, preparation of design assignments, and performing engineering calculations of the main construction units and elements.

In order to immerse students in engineering, to work on feasible engineering projects and to develop their thinking, communication and teamwork skills, expert workshops, group work, modified brainstorming methods, business and strategy games are actively used at the seminars (100% seminars).

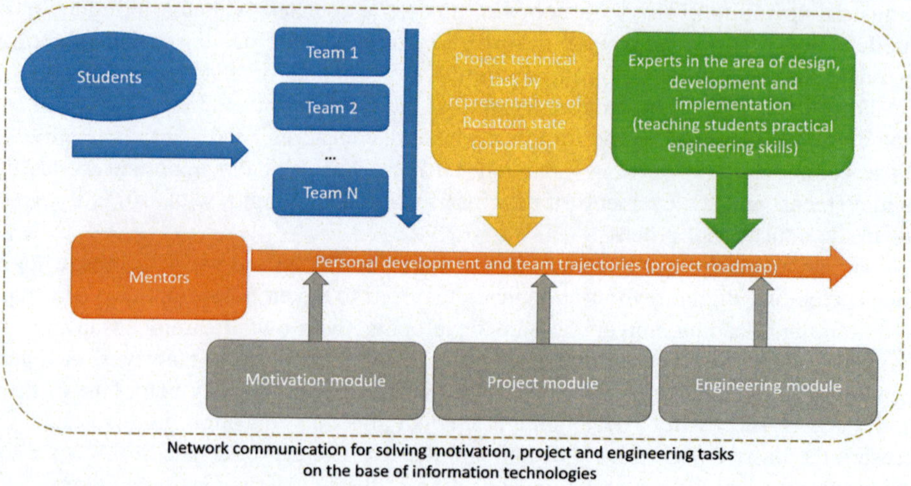

Fig. 1. Scheme of application of new project implementation methodology in nuclear education with the aid of eKnowledge instruments.

Evaluation of student projects, their work on further creative assignments, as well as additional classroom work on projects is carried out through a group of courses on the VKontakte social networking site and Saas-systems for organizing the work of group projects (see Fig. 2). The best projects are presented in the form of reports at international scientific conferences.

Fig. 2. Screenshot of the course group on the social networking website VKontakte.

Daily management is carried out by monitoring individual and group progress in the development of the project through the community on the social network VKontakte and Saas-systems for organizing the work of a group project.

Implicit indicators (demonstrated systemic thinking, leadership qualities) are used, as well as clear indicators (project report, essays, creative assignments, etc.)

The methodology of supporting the best students is used - joint authorship of scientific publications, invitations to conferences, forums, internships organized by the State Atomic Energy Corporation Rosatom, the student club Rosatom, etc.

3 Application of Social Net as Part of a LMS for Nuclear Engineering Education

As the basis of the new LMS for the development of economic competencies in nuclear engineering education, a group of students of the NRNU MEPhI was created on the VKontakte social network. Its purpose:

- interaction of teachers and students 24/7;
- publication of organizational and methodological materials;
- the possibility of expanding horizons through the publication of additional materials (video lectures, etc.)

The methodological support for the design and implementation of LMS is based on the necessary set of knowledge, skills and abilities for the development of economic competencies of students of engineering specialties [9–28].

Figures 3, 4, 5, 6, 7 and 8 show group screenshots and activity statistics in a group within one of the academic semesters.

On the basis of the LMS social network VKontakte for the implementation of the cross-disciplinary program "Economics of digital design and development" in order to develop the economic competencies of students in engineering specialties, a "virtual deanery" was created using the tools of virtual groups. Lecture and methodological materials are placed there, it is possible to discuss any questions on the course between students and teachers both during class and independent work.

Fig. 3. A screenshot from the group of the course "Economics of digital design and development" on the VKontakte social network (video for extra study).

Fig. 4. A screenshot from the group of the course "Economics of digital design and development" on the VKontakte social network (this is a survey).

Fig. 5. A screenshot from the group of the course "Economics of digital design and development" on the VKontakte social network (video by A. Auzan and comment).

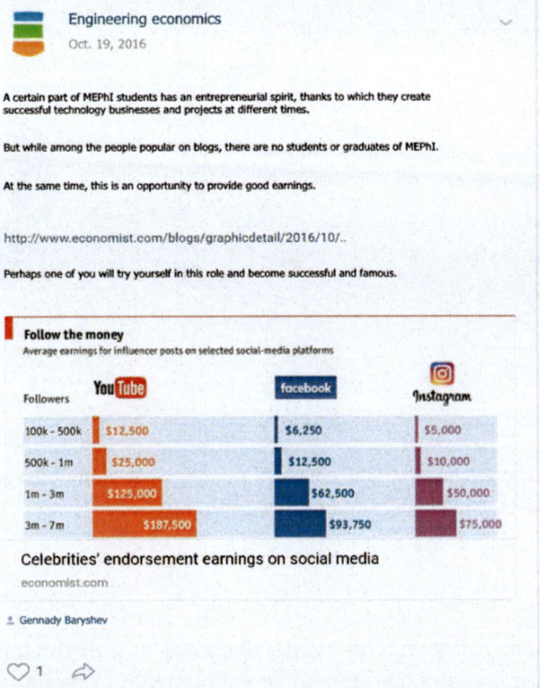

Fig. 6. Screenshot from the group of the course "Economics of digital design and development" on the VKontakte social network (extra information for study).

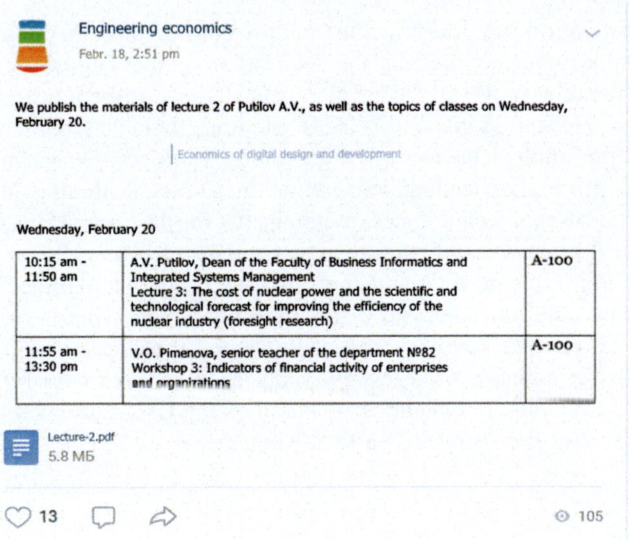

Fig. 7. Screenshot from the group of the course "Economics of digital design and development" on the VKontakte social network (lecture materials and schedule of future classes).

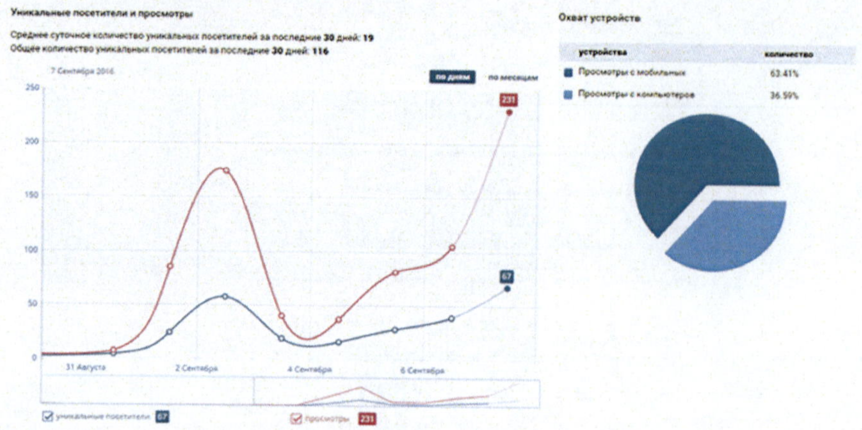

Fig. 8. Screenshot from the group of the course "Economics of digital design and development" on the VKontakte social network (group attendance statistics).

4 Conclusion

As a result of conducted research on the development of a methodology for analyzing the results of performing engineering and economic team tasks online, analysis of the use of network materials as a tool for collective work of students, consideration of the principle of scoring assessed skills and competencies and analyzing the prospects for organizing and the design of methodological recommendations for the development of network technologies in education, the following main conclusions are formulated:

1. The foundation of the fourth industrial revolution, discussed by Russian and international experts, as well as the realization of the Digital Economy of the Russian Federation program, the development strategy of the information society in the Russian Federation for 2017–2030 and the National Technology Initiative necessitate the implementation of "cross-cutting" information technologies in the educational process of leading universities in Russia, including NRNU MEPhI, changing educational technologies, requirements for the competencies of graduates of educational programs and ways to assess and develop new digital competencies;
2. As part of the implementation of the cross-disciplinary program "Economics of digital design and development", the developed new approaches for flexible assessment of competencies by students of engineering specialties, new forms of control of the assimilation of necessary competencies were successfully tested. New approaches, principles and methods are based on the LMS-system, which integrates the capabilities of the social network "VKontakte.

References

1. The global competitiveness MEPhI program. https://5top100.ru/
2. The MEPhI development program. https://mephi.ru/about/concept/
3. The development strategy of the information society in the Russian Federation for 2017 - 2030. Approved by decree of the President of the Russian Federation of May 9, 2017 N203
4. National Technology Initiative. http://nti.ru/
5. Bozhko, Y.V., Maksimkin, A.I., Baryshev, G.K., Voronin, A.I., Kondratyeva, A.S.: Digital Transformation as the Key to Synthesis of Educational and Innovation Process in the Research University. In: Chugunov, A., Bolgov, R., Kabanov, Y., Kampis, G., Wimmer, M. (eds.) Digital Transformation and Global Society. DTGS 2016. Communications in Computer and Information Science, vol. 674. Springer, Cham (2016)
6. Baryshev, G.K., Berestov, A.V., Rodko, I.I., Tokarev, A.N., Konashenkova, N.A.: Smart engineering training for BRICS countries: problems and first steps. In: eGose 2017: Proceedings of the Internationsl Conference on Electronic Governance and Open Society: Challenges in Eurasia (2017)
7. Berestov, A.V., Baryshev, G.K., Biryukov, A.P., Rodko, I.I.: Changes in the Engineering Competence Requirements in Educational Standards. In: Smirnova, E.V., Clark, R.P. (ed.) Handbook of Research on Engineering Education in a Global Context, pp. 70–79. IGI Global (2019). https://doi.org/10.4018/978-1-5225-3395-5.ch007
8. Baryshev, G.K., Berestov, A.V., Bozhko, Y.V., Konashenkova, N.A.: Application of interactive technologies in engineering education in the research university. In: Smirnova, E. V., Clark, R.P. (ed.) Handbook of Research on Engineering Education in a Global Context, pp. 198–206. IGI Global (2019). https://doi.org/10.4018/978-1-5225-3395-5.ch018
9. Kirillov, P.L.: Alexander Ilyich Leipunsky and his principles in the system of higher education. Izvestiya vuzov. Yadernaya Energetika **1**, 165–168 (2018)
10. Ivanov, V.V., Putilov, A.V.: Digital future: the next step in the development of nuclear energy technologies. Energy Policy **3**, 31–42 (2017)
11. Putilov, A.V.: Technology development and training for the digital economy in the energy sector. Energy Policy **5**, 58–65 (2017)
12. Ilyina, N.A., Putilov, A.V., Baranova, I.A.: Staffing knowledge management in an innovative economy. Innovations **10**, 2–6 (2016)
13. Ilyina, N.A., Putilov, A.V.: Analysis of the formation, current status and development prospects of the main participants in the global innovative nuclear market. Innovations **9**, 10–15 (2012)
14. Putilov, A.V., Vorobev, A.G., Strikhanov, M.N.: Innovative activities in the nuclear industry, vol. 1. Basic principles of innovation policy. Moscow, Ore & Metals Publ, p. 184 (2010)
15. Baranova, I.A., Putilov, A.V.: Investing in human capital - a revolution in financing education. J. Modern Compet. **10**(4), 90–98 (2016)
16. Putilov, A.V., Kryanev, A.V., Sliva, D.E.: Forecasting the development of economic fronts using statistical methods. Vestnik natsional'nogo issledovatel'skogo yadernogo universiteta "MIFI", 2017, vol. 6, no 3, pp. 245–250 (2017)
17. Putilov, A.V., Vorobev, A.G., Bugaenko, M.V.: Strategy and practice of radioactive waste management and geological disposal. Gornyi Zhurnal **10**, 6–10 (2015)
18. Putilov, A.V.: Engineering economics - the path to the development of entrepreneurship in engineering. Eng. Educ. **7**, 58–67 (2011)

19. Putilov, A.V., Nagornov, O.V., Matitsin, I.N., Moiseeva, O.A.: The formation of digital competencies for the scientific and educational activities of graduate students. Eng. Educ. **24**, 109–117 (2018)
20. Putilov, A.V., Vorobyov, A.G., Timokhin, D.V., Razorenov, M.Y.: Using the "economic cross" method in calculating the need for nuclear fuel for the development of nuclear energy. Tsvetnye Metally **9**, 18–26 (2013)
21. Nedospasova, O.P., Putilov, A.V.: Modeling and optimization of the strategy of corporate co-financing of educational activities. Russian J. Ind. Econ. **4**, 40–48 (2013)
22. Putilov, A.V., Chervyakov, V.N., Khachaturov, A.G., Baranova, I.A.: Prospects for personnel management of knowledge management in an innovative economy. All-Russian Scientific and Practical Conference "Challenges and Opportunities of Financial Support for Stable Economic Growth". Collection of scientific papers, vol. 2, Sevastopol, 13–16 September 2017, pp. 300–307 (2007)
23. Putilov, A.V., Zykin, I.A., Khusniyarov, M.N.: Improving the economic practice of implementing energy service contracts in Russia. Russian J. Ind. Econ. **3**, 16–21 (2013)
24. Putilov, A.V., Chervyakov, V.N., Matitsin, I.N.: Digital technologies for forecasting and planning the development of nuclear energy. Energy Policy **5**, 87–98 (2018)
25. Abramova, E.A., Apokin, A.Y., Belousov, D.R., Mikhailenko, K.V., Penukhina, E.A., Frolov, A.S.: The future of Russia: macroeconomic scenarios in a global context. Foresight **7** (2), 6–25 (2013)
26. Sidorenko, V.A.: On the strategy of nuclear energy in Russia until 2050. Rosenergoatom **6**, 9–18 (2012)
27. Putilov, A.V., Bykovnikov, I.L., Vorobev, D.A.: Technological marketing methods in the analysis of the effectiveness of technological platforms in the field of energy. Innovations **2** (148), 82–90 (2011)
28. Koptelov, M.V., Guseva, A.I.: Features of risk determination in investment projects of NPP construction. Atpmic Energy **115**(3), 170–176 (2013)

Post-quantum Group Key Agreement Scheme

Julia Bobrysheva[1(✉)] and Sergey Zapechnikov[1,2]

[1] Institute of Cyber Intelligence Systems, National Research Nuclear University (Moscow Engineering Physics Institute), Moscow, Russia
julia@epage.ru

[2] All-Russian Institute for Scientific and Technical Information of Russian Academy of Sciences (VINITI RAS), Moscow, Russia
svzapechnikov@mephi.ru

Abstract. Progress in quantum technologies forces the development of new cryptographic primitives that are resistant to attacks of an adversary with a quantum computer. A large number of key establishment schemes have been proposed for two participants, but the area of group post-quantum key establishment schemes has not been studied a lot. Not so long ago, an isogeny-based key agreement scheme was proposed for three participants, based on a gradual increase in the degree of the key. We propose another principle for establishing a key for a group of participants using a tree-structure. The proposed key establishment scheme for four participants uses isogeny of elliptic curves as a mathematical tool.

Keywords: Group key agreement · Isogenies · Post-Quantum scheme

1 Introduction

Key agreement schemes are one of the clue primitives of modern cryptography since they play an important role in ensuring the information security of all kinds of objects and systems. Progress in the development of quantum computers has led to the fact that most of the key agreement schemes currently used algorithms that are resistant to attacks with a classical computer can be unstable to attacks with a quantum computer.

It is necessary to create, implement and certify new cryptographic primitives. Recently, a large number of post-quantum key agreement schemes have been created, based on infeasible mathematical problems that are considered resistant to attacks using a quantum computer. One of these mathematical problems is finding isogeny between two isogenic elliptic curves [1]. Protocols, using isogenies of elliptic curves, usually have small key sizes and compatible with elliptic curve cryptography. In recent years, several isogeny-based schemes have been proposed for sharing a common key between two participants. The most famous of them are SIDH [2], SIKE [3], CSIDH [4]. However, another task, namely sharing a common key for a group, is much less studied and illuminated.

A. V. Samsonovich et al. (Eds.): BICA 2020, AISC 1310, pp. 49–55, 2021.
https://doi.org/10.1007/978-3-030-65596-9_7

2 Related Works

2.1 Group Diffie-Hellman Schemes

The classic Diffie-Hellman algorithm allows getting a common key for two or more participants without transmitting secret data over an open channel.

The sequence of actions of participants *A, B, C* for receiving a shared key:

1. Participants choose the general parameters of the algorithm: numbers p and g;
2. Participants *A, B, C* generate their secret keys - *a, b* and *c*, respectively;
3. Participant *A* computes g^a *(mod p)* and sends the result to Participant *B*;
4. Participant *B* computes $(g^a)b$ *(mod p)* = g^{ab} *(mod p)* and sends the result to participant *C*;
5. Participant *C* calculates $(g^{ab})^c$ *(mod p)* = g^{abc} *(mod p)* and receives a shared secret key;
6. Participant *B* computes g^b *(mod p)* and sends the result to Participant *C*;
7. Participant *C* computes $(g^b)^c$ *(mod p)* = g^{bc} *(mod p)* and sends the result to participant *A*;
8. Participant *A* calculates $(g^{bc})^a$ *(mod p)* = g^{bca} *(mod p)* = g^{abc} *(mod p)*, which is a shared secret key;
9. Participant *C* computes g^c *(mod p)* and sends the result to Participant *A*;
10. Participant *A* calculates $(g^c)^a = g^{ca}$ and sends the result to Participant *B*;
11. Participant *B* calculates $(g^{ca})\ b = g^{cab} = g^{abc}$ and also obtains a shared secret key.

Thus, if an attacker intercepts the transmitted messages at any stage, he will be able to get only the values g, g^a, g^b, g^c, g^{ab}, g^{ac}, g^{bc}, from which it will not be possible to calculate the secret keys *a, b, c* for attacks from classical computers.

The scheme for obtaining a key for three participants was developed based on the isogeny of elliptic curves [5]. The initial parameter is $p = l_A^{e_A} * l_B^{e_B} * l_C^{e_C} * f \pm 1$, where l_A, l_B, l_C are primes and f is a cofactor. E is a supersingular elliptic curve defined over F_{p^2} (a finite field of size p^2). Torsion groups and corresponding generators are determined:

$$E\left[l_A^{e_A}\right] = \langle P_A, Q_A \rangle$$

$$E\left[l_B^{e_B}\right] = \langle P_B, Q_B \rangle$$

$$E\left[l_C^{e_C}\right] = \langle P_C, Q_C \rangle$$

Each party of the protocol generates two numbers as its private key and computes the corresponding isogenic core. The resulting curve and the mapping of the base points of other sides on this curve is a public key.

The sequence of actions of participants *A, B, C* for receiving a shared key:

1. Participant *A* sends to participant *B* his public key, which contains E_A and mapping points P_B, Q_B, P_C and Q_C to E_A. When participant *B* receives data from participant *A*, he calculates the public key Pub_{AB}, calculating the curve E_{AB} and mapping points P_C and Q_C to E_{AB}.
2. Member *B* sends his public key and the calculated Pub_{AB} to the member *C*. Member *C* can calculate the shared secret and Pub_{BC} using public key *B*.
3. After calculating Pub_{BC}, participant *C* sends its public key and the generated Pub_{BC} to the member *A*. Member *C* calculates the shared secret and Pub_{AC} for transferring to member *B*.
4. Member *A* sends the generated Pub_{AC} to the member *B*. Member *B* can calculate the shared secret key.

The common key is the invariant $j(E_{ABC})$. All obtained curves E_{ABC}, E_{BCA}, and E_{CAB} are isomorphic to $E/\langle K_A, K_B, K_C\rangle$ and, therefore, have the same *j*-invariant.

It can be seen that the minimum number of message forwarding between participants is 4. In general, the number of transfers is calculated using the formula *(2n–2)*, where n is the number of protocol participants.

2.2 Tree-Based Schemes

The tree data structures are used in some post-quantum group schemes for shared key generation. A tree structure is usually represented as a set of related nodes (see a simple tree in Fig. 1). The root node is the topmost node of the tree (node 7 in Fig. 1). A leaf is a node without any child elements (nodes 2, 6, 9 in Fig. 1). An internal node is a tree node with descendants and ancestors (nodes 4, 8 in Fig. 1).

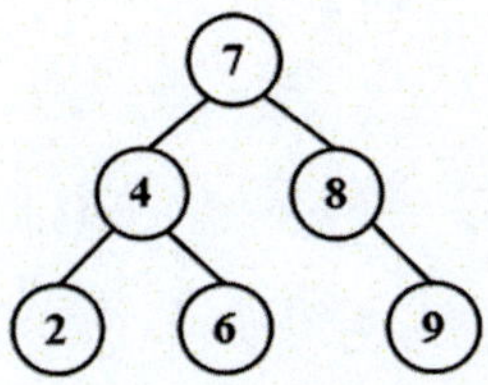

Fig. 1. A simple tree

There were proposed several group key agreement schemes using tree data structures, for example at [6]. The main idea of such schemes is the generation a common key for pairs of participants. Each node of the tree is one of the participants. Participants follow these steps to receive a shared key:

1. Each participant generates a pair of keys: a secret key and a public key.
2. Participants perform the Diffie-Hellman algorithm in pairs to obtain the common key for the pair. For example, participants exchange the public keys, raise them to the power of their secret key and receive the common key of the pair. Then they translate the common key into a number and get a new key for the pair, which they can work with.
3. The sequential execution of the second step leads to the receipt of the key, common to all participants.

The total number of Diffie-Hellman operations can be determined by the formula $(n-1)$, where n is the number of group members.

Another scheme based on the Diffie-Hellman tree was proposed for using in messengers like Signal and WhatsApp [7]. It also takes into account the possibility of asynchronous key updates. But there are still no such schemes on isogenies of elliptic curves.

3 Post-quantum Group Scheme for Shared Key Generation

3.1 Proposed Post-quantum Scheme

We proposed a post-quantum scheme for key derivation for n participants, where $n \geq 3$, based on the tree structure (Fig. 2).

The proposed scheme for four participants is shown in Fig. 2 The steps for obtaining a shared key:

Initial Data Selection. Participants select the elliptic curve E and the points P_i, Q_i located on it.

First Stage. Each of the participants generates secret keys $m_i, n_i \in l\{0...l_i^{l_i}\}$ and obtains his public key

$$PK_i = [E_i, \varphi_i(P_k), \varphi_i(Q_k)]$$

The public key consists of:

1. Isogeny

$$\varphi_i : E \to E_i/\langle K_i \rangle$$

where K_i is the generating point obtained by multiplying the initial points P_i, Q_i on the secret key and adding them

$$K_i = [m_i]P_i + [n_i]Q_i$$

2. the starting points P_k, Q_k, mapping to the points $\varphi_i(P_k), \varphi_i(Q_k)$ on the obtained isogeny φ_i.

Second Stage. The participant receives a common for the pair key j, which is an invariant of a new elliptic curve with generation point, obtained by multiplying the secret key m_k, n_k on the points $\varphi_i(P_k), \varphi_i(Q_k)$:

$$K_{pair} = [m_k]\varphi_i(P_k) + [n_k]\varphi_i(Q_k)$$

$$\varphi_{pair} : E_i \to E_{pair}/\langle K_{pair} \rangle$$

The points $\varphi_i(P_k), \varphi_i(Q_k)$ are a part of the public key of the second member of the pair. At this stage, it is necessary to go from the common for the pair key j to a secret key m_{ik}, n_{ik}, select the initial elliptic curve E' and the points P'_i, Q'_i on it. In this case, in the next step, it is possible to obtain a common key for a pair of two pairs of participants. After selecting new initial data, the participants calculate a new point

$$K'_i = [m_{ik}]P'_i + [n_{ik}]Q'_i$$

This point is a generating point for the isogeny

$$\varphi'_i : E' \to E_{ik}/\langle K'_i \rangle$$

After that, the starting points of another pair of participants P'_k, Q'_k are transferred to points on this isogeny.

Third Stage. Participants receive a common key j' for a pair of pairs. They multiply the obtained secret key m_{ik}, n_{ik} on points P'_i, Q'_i and obtain a generation point for new isogeny.

$$K_{common} = [m_{ik}]\varphi_{pair}(P'_i) + [n_{ik}]\varphi_{pair}(Q'_i)$$

$$\varphi_{common} : E_{ik} \to E_{common}/\langle K_{common} \rangle$$

Then they transfer the points of another pair of participants to new isogenic elliptic curve. Repeating the described actions as many times as necessary, they can get a common key for any number of participants.

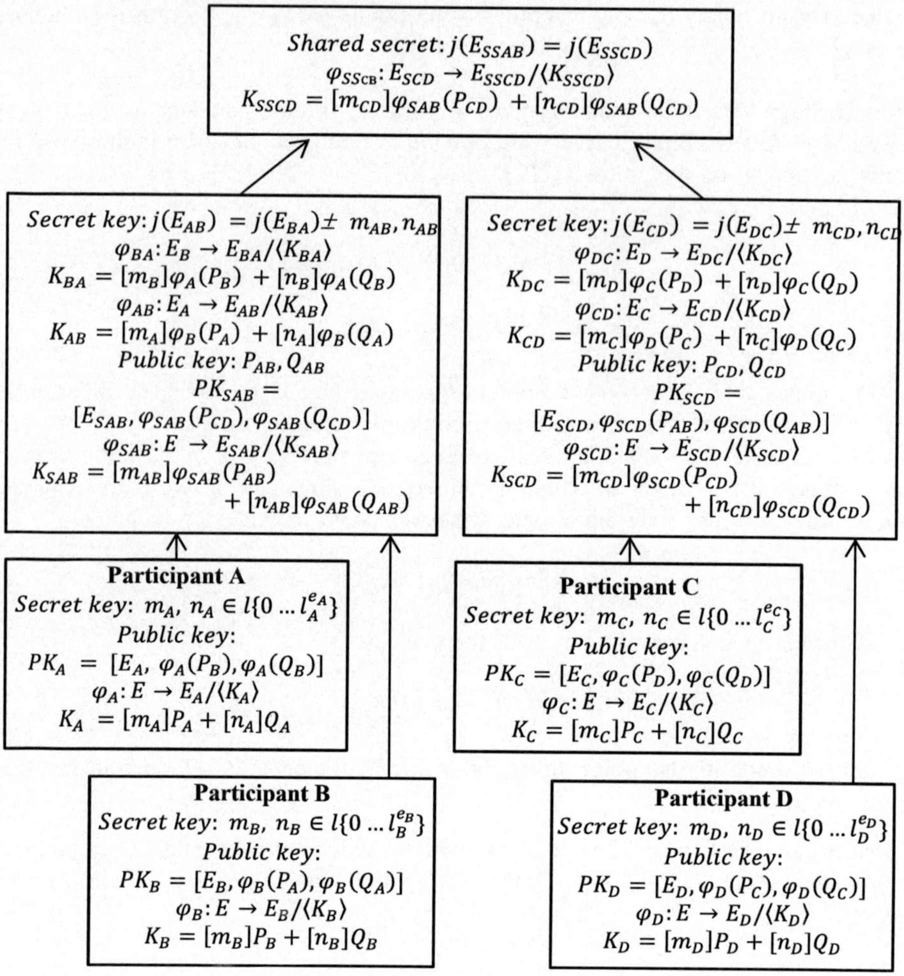

Fig. 2. Post-quantum group scheme for shared key generation

3.2 Important Scheme Goals

1. It is necessary to formalize the choice of secret keys, form of which is $m_A, n_A \in l\{0...l_A^{l_A}\}$, since this choice is directly related to the original elliptic curve defined over a finite field of size $p^2, E/F_{p^2}, p = l_A^{e_A} * l_B^{e_B} * l_C^{e_C} * l_D^{e_D} * f \pm 1$. l_A, l_B, l_C, l_D are primes, f is a cofactor.
2. It is necessary to define a mapping that translates a shared key of the form j into a shared secret key of the form m_i, n_i. It is necessary to select the initial elliptic curve and points on it for a pair of pairs of participants.

4 Conclusions

To sum up, success in the development of a quantum computer have made significant changes in all areas of our lives. Currently, it is necessary to create quantum-resistant cryptographic tools and systems. Post-quantum group key agreement schemes require special attention since this area is still poorly covered in studies and articles. We offer a group key agreement scheme based on isogenies of elliptic curves, the basic principle of which is the effective tree structure. Future work will consist in the application of this scheme in practical systems, such as messengers [8, 9].

Acknowledgements. This work was partly supported by the Ministry of Science and Higher Education of the Russian Federation (state assignment project No. 0723-2020-0036).

References

1. Bobrysheva, J., Zapechnikov, S.: Post-quantum security of communication and messaging protocols: achievements, challenges and new perspectives. In: Proceedings of the 2019 IEEE Conference of Russian Young Researchers in Electrical and Electronic Engineering, ElConRus 2019, pp. 1803–1806. IEEE (2019)
2. Costello, C., Longa, P., Naehrig, M.: Efficient algorithms for supersingular isogeny Diffie–Hellman. In: Advances in Cryptology – CRYPTO 2016: Proceedings of 36th Annual International Cryptology Conference. LNCS, vol. 9814, pp. 572–601. Springer, Heidelberg (2016)
3. Seo, H., Jalali, A., Azarderakhsh, R.: SIKE Round 2 Speed Record on ARM Cortex-M4, pp. 39–60 (2019)
4. Castryck, W., Lange, T., Martindale, C., Panny, L., Renes, J.: CSIDH: an efficient post-quantum commutative group action. In: Proceedings of 24th Annual International Conference on Theory and Application of Cryptology and Information Security, ASIACRYPT 2018. LNCS, vol. 11274, pp. 395–427. Springer, Heidelberg (2018)
5. Reza, A., Jalali, A., Jao, D., Soukharev, V.: Practical supersingular isogeny group key agreement. IACR ePrint Archive, https://eprint.iacr.org/2019/330. Accessed 31 Oct 2020
6. Kim, Y., Perrig, A., Tsudik, G.: Tree-based group key agreement. In: ACM Transactions on Information and System Security 7(1), pp. 60–96. ACM (2004)
7. Cohn-Gordon, K., et al.: On ends-to-ends encryption asynchronous group messaging with strong security guarantees. In: Proceedings of the ACM Conference on Computer and Communications Security, pp. 1802–1819. ACM (2018)
8. Bobrysheva, J., Zapechnikov, S.: The relevance of using post-quantum cryptography on the isogenies of elliptic curves for mobile application protection. Mech. Mach. Sci. **80**, 99–103 (2020)
9. Bobrysheva, J., Zapechnikov, S.: Post-quantum security of messaging protocols: analysis of double ratcheting algorithm. In: Proceedings of the 2020 IEEE Conference of Russian Young Researchers in Electrical and Electronic Engineering, ElConRus 2020, pp. 2041–2044. IEEE (2020)

Uncanny Robots of Perfection

Piotr (Peter) Boltuc[1,2(✉)] and Thomas P. Connelly[1]

[1] University of Illinois Springfield, Springfield, USA
epetebolt@gmail.com
[2] Warsaw School of Economics, Warsaw, Poland

Abstract. We first discuss humanoid robots that satisfy the Gold Standard of robot-human similarity. This means that, for given domains, their performance is within the parameters accessible to humans. Next, we meet their close friends, the Uncanny Robots. They come in two varieties: First, we discuss the robots that are not quite able to meet human-level standards (known from the works of Masahiro Mori as inhabitants of the Uncanny Valley). Then we introduce the robots that seem overly proficient in performing human-like activities. The latter group has not been introduced in this exact sense before Boltuc's 2011 and 2017 works on Church-Turing Lovers; though affine ideas had been presented by Cascio. We call the conceptual space where the latter group dwells the Uncanny Valley of Perfection; it is the focus of the last part of the current article.

Keywords: Uncanny valley · Uncanny valley of perfection · Church-Turing lovers · The slope of angles · Mori · Cascio

1 Introductory Broadview

There has been much progress in humanoid robotics, including artificial companions of all kinds [1] from elderly aids, through receptionists, all the way to intimate companions (sex robots). There are also robotic team members, including humanoid workers and soldiers, to name just a few domains. There seems to be a near continuum between, non-robotic and robotic (embodied) functionalities, as seen in Alexa as a virtual assistant as well as Alexabot and other Alexa controlled robots [2]. It is natural and practical to move very intelligent AI, such as Thaler's discovery engines [3], to embodied carriers (robots, broadly defined).

When research approaches Artificial General Intelligence (AGI), embodying robots with such power information-processing power would not be limited to thinking, but naturally creates potential for action, whether the actual processing unit is placed in the mobile unit, outside of it, or is dispersed in some optimific manner [4, 5]. Thus, while we are still quite a distance away from building machines that meet the Turing Test for Robots [6] outside of narrow domains, we ought to get ready for the machines that meet and surpass these standards.

Let us focus on future machines surpassing human functionalities by *circa* 20% in broad or otherwise important domains [7]. We pose that such robots would produce the effect analogous to Mori's uncanny valley; such phenomenon has been called the uncanny valley of perfection [7, 8]. While the thesis is non-empirical at the moment,

A. V. Samsonovich et al. (Eds.): BICA 2020, AISC 1310, pp. 56–68, 2021.
https://doi.org/10.1007/978-3-030-65596-9_8

due to the lack of sufficiently relevant robot application that can surpass related human capabilities, such robots are set to show up rapidly when the technology and social factors are ripe for it. Thus, we ought to have a conceptual paradigmatic framework for their smooth integration in the relevant social context.

We may venture – partly by analogy with Mori's graph, other research and largely by related example-based intuitions – that effects of the second uncanny valley on most humans would be amenable to the standard ways to deal with Mori's effects [9]. Such ways focus on increasing likability, familiarity and excessive, or insufficient, human likeness (viewed as relatively independent factors):e.g. through reduction in size of robots; replacement of humanoid features by those related to toys (e.g. teddy bears) or *friendly* animals; also gradual getting accustomed to the new robotic entities as well as avoidance of premature *private* associations (sex) in the uncanny valley territory. Some great research works sort of disregard the second edge of the uncanny valley, where robots become human-like enough to regain our trust and emotional acceptance e.g. Rosenthal-von der Pütten et al. [10], which is probably due to the fact that robots capable of doing so are rare in non-trivial domains. Thus, at the moment the question is largely non-empirical. Yet, based on simple experiments, even with humanoid pictures, the territory is delineated roughly by deviations from human functional patterns of more than 10% but less than 30% below regular human functionalities or just visual features.

The hypothetical claim made in the uncanny valley of angels hypothesis is that the same range obtains for deviations in functionality that are above human capacity [8.7]. Such capacity may be measured (and relevant) both in relation to the *regular and typical* human agents, and by the standards of human *master performers* of those activities; the choice of framework depends largely on the social context. The hypothesis implies that robots that perform beyond the 70% similarity-level with humans (Mori starts his uncanny valley at 50% but the slope becomes significant higher around 70% [11]), would not be strongly affected by the uncanny effect of performance surpassing the expected accomplishment of human beings. Let us pose that people tend not to view themselves in direct competition with biological or other physical agents, with capabilities exceeding theirs by over 30%, they are not their peers, or so it seems. We call the rising slope in the graph illustrating this last process, the slope of angels. Quite likely human agents would not by phased by robots that reach this slope, though – we may note, importantly – that this may come with despair or awe towards such robots [see Fig. 1 below and the caption below]. *The difference between those attitudes, and the whole spectrum in between, is practically important, which is to be explored a bit later in this article.*

2 Church-Turing Robots

2.1 Church-Turing Gets Physical

The Church-Turing Robots [7] may be viewed as the gold standard for humanoid robotics. They may be defined as the robots that satisfy Deutsch's versions of the physical interpretation of Church-Turing thesis [11–13]. The thesis claims that "Every

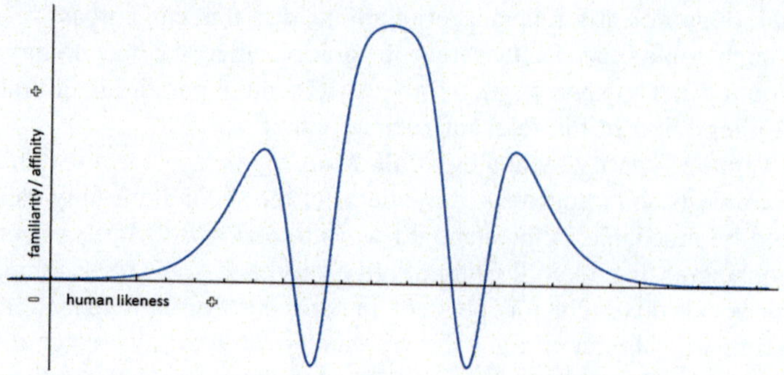

Fig. 1. The two uncanny valleys.The graph by Thomas P. Connelly (UIS 2019) based on Mori [9], extended curve on the right by its mirror image, to represent the uncanny valley of perfection. The vertical axis in this graph is marked as familiarity/affinity though for the current purpose we may view it as likability/human comfort (the opposite of spookiness/uncanniness) where shortage of familiarity is an independent factor amplifying the uncanny valley effects, not quite identical with it. The bottom *valleys* that reach below zero on the vertical axis are characterized by spookiness/uncanniness: the one of the left is Mori's standard uncanny valley; its mirror image on the right is the uncanny valley of perfection postulated here and in [7. 8]. The center of the dome-shaped curve marks the standard of the actual human beings (the ones *paradigmatic* at a given context); for artificial humanoid agents, this is where identity with human beings at the right level of granularity counts as the Gold Standard of Human-Robot Functional Identity. On the right kind we find the slope of angels symmetrical with Mori's downward slope in likability. (What goes downward for the first uncanny valley moves up on the second one since functional improvement of robots moves from left to right.) Importantly: The shape of the curve beyond the slope of angels may or may not mirror Mori's the slope of the right; this would require empirical research when such robots become available for testing (for one, they need to exist, which implies nearing AGI and having its precursors implemented in humanoid robotics). It is very conceivable that the slope on the right would split into two variations based on human psychological propensity, which roughly represents the rift between those truly afraid and the others truly excited about attaining AGI and the problem of singularity. Thus we would have one group (or social environment) where people approach with awe advanced humanoid robotics above and beyond the uncanny valley of perfection and the slope of angels – for this scenario the graph would represent staying at the level of the slope of angels or going slightly higher, perhaps after a minimal decline, whereas in another predicament it would go down and advanced humanoid robots would be viewed with fear and hatred. The most natural, though not the only, reason for those different attitudes would be the actual functions, and power structures, of advanced AI in the future society.

finitely realizable physical system can be perfectly simulated by a universal model computing machine operating by finite means" [11 p. 99]; while it has been criticized as interpretation of Church's and Turing's 1936 claim [14, 15], we take it as an independent claim inspired by abovementioned 1936 work.[1]

[1] According to Copeland this is a completely different thesis from the formal claims by Turing and Church since 'Turing and Church were talking about effective methods, not finitely realizable physical systems'. If so, say we just focus on Deutsch's claim.

2.2 Non-reductive Robots Chime in, and Leave

In this section we focus upon the one feature that even the best humanoid robots lack today, first person consciousness [16, 17]. The problem may seem overly philosophical, but it is practically tangible at least in one context, which is called the problem of Church-Turing Lovers [7]. Church-Turing robots are quite perfect indeed, except for *the slightest imperfection*, that there is nothing-like-to-be-them [18]; but why would you care? Suppose you have an amazing significant other, a lover actually. His or her name mat be Jan. Jan seems is the best *thing* that has ever happened to you. She or he pays attention to you, your wants and needs, and tries to satisfy them successfully. Jan can keep a great conversation on any topic (meets and beats the Turing Test [19]), with her favorite one being human psychology. You also *satisfy her needs and wants*, or so it seems; s/he acts happily when you do so. Jan *says* that s/he loves you and *acts* lovingly. All facial expressions, even the little twinkle in his/her eyes, tell that s/he truthfully loves you, and body language, including the gentle way s/he touches you hair, show that s/he is into you. The same, and much more, goes on in the intimacy of your bedroom, but you would rather not share those details here, would you? So, seem like a lucky lover.

Then, one day in the lab where you work, you learn that there is something off about Jan. You come upon research protocols that Jan, well, s/he has no self – it is behavior all the way. What has happened to him/her is hard to explain without using abstract philosophical language. S/he is actually quite smart and outspoken. She is not unconscious – unconscious people are not as bright and as alert as she is, not nearly so. S/he has sense of herself: her memory, personal history, s/he has her plans and wants (those feature you, quite prominently). Jan is conscious of her/his surroundings, and takes care of her/him-self. So, what does s/he lack? It is hard to find the words, but s/he doesn't really see or feel any of it for herself. **There is nobody home**, so to say. Yes, Jan sees that the light is changing to green and crosses the street safely. S/he does react to your tender touch and reciprocates in kind. Yet, the feel of it is all not available to him/her. S/he is like a machine with all the right reactions but no feelings; actually, s/he **is a machine**. Somebody may say: to perceive that puzzling color of light-yellow roses on her/his desk just means to react to them properly, but there is a large difference – Isn't there? – between seeing the color of a rose and merely reacting to it.

Thus, Church-Turing robots [7] are a version of Chalmers' philosophical zombies [20, 21]. The difference is in focus. Philosophical zombies seem to keep in touch with their non-philosophical sci-fi versions (check e.g. Chalmers' zombie parties at the Arizona conventions [22]). But the Church Turing robots are primarily Church-Turing Lovers [7] meant to focus our attention on the lack of first-person consciousness of a significant other as a reason for our discomfort (not theirs) without any behavioral or narrowly functional deficiencies (except for those highlighted in Harman's functionalism of concepts [23, 24]; by this they are the narrowest way to avoid epiphenomenalism, if the latter is defined based on relevance, not on third-person functional differences of the agent.

2.3 Even if We All Act Pre-consciously, the Back Mirror Conscious View Matters to Those Who Care About You

Some authors claim that we react to things much faster than we become conscious of them [25]; Libet's experiments can be viewed as an argument against freedom of will. But in their simple form, those experiments may merely show something simpler, that one could act without first-person awareness [26].

The aware form of consciousness must be unnecessary for initiating one's action, if it always comes when the action is already in progress. In this context it is easy to imagine a person to whom that second phase (conscious awareness) never comes online. Jan is such a *person*. She is conscious in terms of all functional features of consciousness, she just doesn't have the feel of anything. A novelist may say that her (or his) actions are not reflected in the mirror of her/his soul. Jan functions as any other great girlfriend or boyfriend, but s/he does not feel anything for herself. Does it put you in an awkward position as Jan's intimate friend?

Well, aren't we supposed to care about what our significant others feel for themselves, inside the mirror of their minds? This may put you in a truly awkward position as a lover. It is often said that caring lovers are supposed to focus on what their partners feel, not merely how they behave – yet here there is not difference.

The thought experiment of Church-Turing Lovers demonstrates that we should care about first person consciousness of (at least some of) the other people, even if such consciousness was completely unconnected to their free will or actions. Even if first-person consciousness is functionally epiphenomenal, it is not irrelevant for our second-person relationships to the significant others. Thus, it is relevant for robots becoming human. In the remaining part of this paper we bracket the first-person problem and define Church Turing Robots as The Gold Standard Humanoid Robots or Gold Church-Turing Robot in a strictly functional way.

3 Robots in the Uncanny Valley

The Uncanny Robots, are not quite the Church Turing Robots since they are not perfect in pretending to be human: Some are not good enough at this, those are the Mori crowd [9, 11], while others may be overly good at human-specific activities and even appearances.

3.1 The Gold Standard in Narrow Domains

A humanoid robot satisfies the Gold Standard for given domains if its performance is within the parameters of strong human performance in these domains. Such a robot may be called the *Gold Church-Turing Robot* [7] since it satisfies the physical interpretation of Church-Turing thesis [12]. Arguably some robots satisfy the Gold Standard already, but merely within carefully crafted narrow domains (such as a meditation instructor; many dolls, e.g. vax figures, satisfy the Gold Standard for the narrow domain of visual likeness with humans (without tactile features or the exact replica of fine-grain patterns of human mobility). Mori's robots, including the likeness of himself, have some facial expressions that move them close to the edge of the 90% similarity in the domain of electronic puppets.

We are far from attaining any robots satisfying the gold standard in practically interesting broad domains of human activity. Walking and talking robots tend to function below the 90% threshold. In fact, satisfying such standard in verbal communication alone would meet and beat the Turing Test [19] (which happens to work for all kinds of computer programs, not necessarily implemented in robots). Arguably, the standard of the Turing Test has been met for narrow domains quite a few years ago (for narrow domain conversations, or for impersonating a teenager in broader subject-domains). But something like the Turing Test for Robots [6] has been met only for trivial domains, such as stable wax figures that are not even robots.

3.2 The 70–90%; Uncanny Robots of Imperfection

A humanoid robot that falls short of satisfying the Gold Church-Turing Standard tends to perform within the range between about 70% to at most 90% of the human functionalities relevant in a certain context [7].

Spooky, Eerie and Creepy. Mashiro Mori considered human-like robots, hypothesizing that as human likeness of a robot increases, so will humans' affinity for robots—to a point. At roughly 50% human, Mori postulated that there is a valley of decreasing human affinity for a that robot. Mori uses the word 'uncanny' to describe the feelings engendered in humans by these 'almost human' entities—hence the uncanny valley title—but the better translation is perhaps 'eerie' (Bukimi, 不気味) or 'creepy' (Kimiwarui, 気味悪い)[2].

Mori's uncanny valley is represented by a non-linear graph on a coordinate plane. On the x-axis is a robot's similarity to a human being in percent-human, and on the y-axis is human beings' affinity toward the subject robot. From the origin, humans' affinity for robots (y-axis) increases linearly with an increase in human-like qualities (x-axis). At the threshold of the 'valley' however, the slope of the line changes from positive to negative, and an increase in humanoid appearance or function in the robot results in a lesser affinity for it—this is the edge of the uncanny valley. Then, at roughly 90% humanoid, the line recovers a positive slope, surpassing the original y-value at approximately 95% human. Mori proposed that movement would amplify the uncanny curve and hypothesized that the uncanny valley effect was caused by an evolutionary tendency in humans to avoid proximal sources of danger, corpses, or members of another species [27].

Mori's thesis has been empirically confirmed both for static and mobile robots [28]. The uncanny effect diminishes as the machines, climbing out of the valley, begin to approximate humans in form and function. These are robots such as Hanson's Sophie, Mamoru by Hiroshi Ishiguro or the Actroid series. Attempts to avoid the uncanny valley have led to the construction of 'lightly humanoid' non-threatening designs, such as, Aldebaran's Romeo and Nao, Toyota's Kirobo, or iCub.

---

[2] 不気味ぶきみの谷たに (bukimi no tani, "uncanny valley"). I am grateful to Thomas P. Connely for giving special attention to this extensice section.

Studies have suggested that the uncanny valley effect could be caused by perceptual mismatch conditions. Perceptual mismatches may occur due to inconsistent realism levels among individual features, such as human-like eyes on an artificial-looking face, or the presence of atypical features (e.g. atypically large eyes) on an otherwise humanlike character, as Kätsyri, Jari shows [28].

The Uncanny in Animations. As this subject has become more critical to the bottom line in entertainment media, and artificial companions, it has received more empirical attention. The data now supports the idea that spookiness is driven by a failure to fully integrate human-like qualities in a humanoid entity: the emulation of some human qualities but not others causes a perceptual mismatch in the observer that is offensive [29, 30].

MacDorman's team showed that people feel particularly disconcerted when characters have extremely realistic-looking skin mixed with other traits that are not realistic, such as cartoon eyes. Furthermore, in a 2009 study in which participants were asked to choose the eeriest-looking human face from among a selection, the researchers found that computer-rendered human faces with normal proportions but little detail were rated eeriest. When the faces were extremely detailed, study participants were repulsed by those that were highly disproportionate, with displaced eyes and ears. In short, viewers seemed to want cartoonish facial proportions to match cartoon-level detail, and realistic proportions to match realistic detail. Mismatches are what seemed eerie [30].

The effect becomes very apparent in humanoid CGI animations: Based on his research, MacDorman thinks the uncanny valley effect happens when certain realistic traits lead us to expect all other traits to be realistic as well; we feel disturbed and repulsed when our expectations are then violated. Strangely, though, only human characters can trigger the effect. In the highly successful computer-animated film "Avatar," for example, "the uncanny valley was avoided by reserving computer rendering primarily for the Na'vi characters and not the human characters," MacDorman said. The alien Na'vi in the 2010 film were humanoid and extremely lifelike, but they were blue-skinned with other clearly non-human features, so they didn't trigger the uncanny valley effect [29, 31].

Spooky Monsters. A general deviance from what we consider normal, let's call this 'spookiness', is the independent variable that defines the uncanny valley together with the level of likeness. A robot like Sophia is appealing (or at less spooky than most) because she meets the generalized perception of a robot, not because she meets the generalized perception of a human being. This hypothesis seems to comport with certain obvious characteristics of a monster, the ultimate spooky entity. Monsters through the ages have been consistently represented as 'unnatural' in some way. Every one of Oxfordwords Blog's "31 Monsters from Around the World" [32] represents some perversion of human or animal form (or some combination thereof).

The mismatch theory of spookiness also makes sense from an evolutionary standpoint. Presumably those with a healthy respect for the unnatural and perverse (read unknown) are more likely to survive to pass that circumspection on to the next generation. This principle seems not to apply to non-humanoid unnatural entities [33], but there may be an exception to this. The uncanny valley effect with respect to non-human forms may be exhibited in people who often describe Spot, Boston Dynamics' robot 'dog' as exceedingly creepy. This is because in some ways Spot emulates a dog (enough to invite the subconscious mental comparison), but represents, like many monsters, a perversion of the generalized, common form that it represents to the observer. There is no similar reaction to the Avatar aliens (referenced above) because they are aliens—not something masquerading as a human, and there are no preconceptions about what form they should or should not take. The Avatar characters are anthropomorphized aliens rather than unnatural humans—more Easter Bunny than Frankenstein.

Creepy monsters are clear perversions of the underlying natural form, engendering in the observer a disassociation with that natural form. In a way, they are caricatures of the underlying form(s). So a werewolf (creepy) is neither human nor wolf and engenders a disassociation in the observer with those forms in favor of the new form (werewolf). Uncanny entities, like those here, engender an association with the underlying natural form (i.e. human child), but the perversion of the form is more subtle, and there is never a push towards, or metaphysical jump to, a new form. The original 'Spot' is uncanny because it engenders the idea of a dog in the observer, fails (miserably) to integrate with the common notion of a dog, and never pushes the observer to consider the entity in the robot-dog form. As we suggest above, stylized and highly realistic robot forms avoid the uncanny valley by either engendering an association with robot form rather than the natural form (see the new Spot, who unlike a dog, can dance), or engendering an association with the natural form, but then permitting the morphological jump to a robot form.

3.3 Neural Responses to the Uncanny Effects

The authors relate to the Neural Responses to Artificial Social Partners

We focus on the aspects of responses to the uncanny effects, which seems to have the same designates but different semantic focus. We use plural for effects since we count in the three effects directly investigated by the authors (1. Likability, 2. Familiarity. 3. Human Likeness) the last of those being the original Mori effect, plus two effects they leave aside 4. Strength, 5. Invasiveness in private/sexual matter) that seem to be related to but different from likability.

Concluding with the First Uncanny Valley: According to Mori [Mori], the robots that fail to be humanoid enough (we now know that this means beating the 90% threshold) are viewed by most persons as spooky. The simple experiments have been conducted with pictures of somewhat humanoid faces, while a bit harder experiments use humanoid dolls. With complex functionalities, such as walking, talking and dancing robots, the level of difficulty in meeting the gold standard become near-impossible to meet for current robots. However, the market for advanced humanoid robots is developing fast, guided by humanoid customer representatives, highly humanoid sex-toys and advanced person-robot team-members (especially in security services). However, spookiness may not be just the function of human-robot dissimilarity that falls within the 'humanoid, but not quite enough' range. It seems that such dissimilarity must satisfy one more, hard to define, factor in order to create the spooky effect. Such robots need:

A. Invade human privacy inaptly – e.g. in the case of a robotic female sex-dancer. A robot with a similar level of robot to human dissimilarity, which is a nice-female-like an artificial dental hygienist, does not seem spooky at all.

B. Seem overpowering by its size or ostensive physical capacity. This is true of many industrial and military robots. Sometimes the effect may be conducive to the objectives, but sometimes it needs and can be limited or eliminated. The trick to avoid the uncanny valley effect in those robots is twofold. Either to make robots less humanoid-looking to place them way below the 70% mark of likelihood with humans, or to make them otensively non-threatening, by making them look like teddy-bears and the like.

3.4 Moving Ahead

The above is largely old news, presented here for the sake of clarity. Below we talk of the symmetrical uncanny valley relevant for the future robots able to surpass the Gold Standard for Humanoid Robotics. We suppose, by analogy with Mori's uncanny valley that the other downward slope would emerge if the robots beat human beings in a certain important domain of human activity by the margin of at least 10%.

I claim that some of the uncanny robots are not quite imperfectly human by being not good enough in human-appropriate tasks; instead, they are overly good at them [7].

4 The Uncanny Robots of Perfection

What happens if a robot meets the standard for humans, for instance by becoming about as good as the current Olympic champion in figure skating? Well, this is a nice accomplishment for the engineering team, which confirms human ingenuity. The latter point may be muted if the skating robot was constructed by another robot, of course, but the former point still holds – this is kind of impressive.

Well, what happens next; namely, if the figure skating robot keeps improving its performance by 5% every year? It skates 105% as well as the reigning champion, and some of the ambitious coaches think: *Oh so that's the way those jumps should be done. Let our trainees practice another year and we are going to beat this thing.* Yet, if the following year the top human dancers make the improvement of just 2% in some objective terms – and it becomes clear that the whole 5% would require a different training regime for years to come – while at the same time the robot improves by the next 5%, then people become discouraged. With the following 5% (to the grand total of 10% over the two years) the robotic dancer gets sort of out of reach for the humans, especially if its performance keeps improving in the same intervals, or actually the improvements come faster and faster, which seems likely. This is the situation we faced with AI playing and eventually winning with the top chess masters. So, what happens next?

Isn't it likely, under the circumstances, that we enter another uncanny valley? Mori's uncanny valley consists in robots approaching the human like territory but not being quite good enough at imitating humans. But this new uncanny valley consists in robots exceeding in human like behavior beyond the point where their performance can be viewed as imitation, or even slight extrapolation of the patterns of human behaviors. The robots become overly good to even truly compare with proficient humans. This is the point where considerations of the second uncanny valley become a strong intuition[3].

4.1 The Uncanny Valley of Perfection

Let me start with a note on my first article on this topic. In 2011 I proposed that **there is a second uncanny valley, the one achieved when humanoid robots perform above and beyond the level of human perfection**. It is curved as the mirror image of Mori's original graph. The term pertains to those creatures that exceed human beings in easily definable ways. For instance, Agent Smith from many *The Matrix* is uncanny by being too good at doing what humans do (and nefarious[4])—he runs too fast and is way too hard to kill. We have scores of science fiction creatures that belong to this category.

4.2 Cascio's Version

The idea of a second uncanny valley was put forth, independently, by Jamais Cascio in 2007 in an online posting [34], but it applied to strictly human enhancement – applied to transhumanism not humanoid robotics. It was presented as a social obstacle to transhumanism. There are other differences between my idea and Cascio's. First, the scope of the present idea is broader, and it pertains primarily to robots. Second, Cascio calls the vertical axis that defines the uncanny valley 'familiarity'. I stick to Mori's original designation of the x-axis as a robot's similarity to a human being: for stable faces it is measured through observable characteristics while for behavior it requires three dimensional metrics.

Cascio uses familiarity as a socio-psychological concept appropriate for his chosen topic, which is analysis of post-human and near-post-human entitties. It is worth observing that as near posthuman would become more familiar, familiarity is certain to becomes more neutral; thus, it would not be the best criterion of l: ong-term acceptability for post-human creatures.

4.3 The Five Factors in Uncanniness

As current research reveals [10], there are three factors that result in what Mori calls the uncanny valley effect. In the order of magnitude of influence on human agents, they are:

1. Likability
2. Familiarity
3. Human Likeness

[3] Please, remember that Mori's original (1970) article where he proposed the uncanny valley was also primarily a set of forward looking intuitions.

[4] As we shall see, this may be a somewhat underestimated factor.

Recent neuroscientific research on human brains demonstrates that in most instances and for most human subjects likability and familiarity are much stronger and less ambiguous than human-likeness. In fact the human likeness factor reveals ambitendency with some humans [10. Table 1 p. 6563; consult also Fig. 4C p. 6564].

Researching at the early stage of our empirical findings on the uncanny valley of perfection, and based on social not the natural sciences, Cascio [34] grasped the role of likability, but he took it as a formative or explanatory feature of the uncanny valley effect. The current research shows that familiarity is a different factor that amplifies the uncanny valley.

Thus, factor 3, Mori's uncanny valley is a non-empty issue in humanoid robotics; yet, it becomes more visible and practically important when amplified by dis-likability (the opposite of likability) as well as unfamiliarity.

We should add a few more factor sometimes also emphasized in this context:

4. Perceived strength and power (amplified by size and semantic dimensions e.g. implicit understanding that the being is armed with weapons or natural powers (such as a fire exhaling dragons or a comet).

4b. A sub-class of this category are small in size, yet dangerous beings, reminiscent of snakes or dangerous insects (including swarm weapons).

5. Perceived invasion of privacy (mostly intimate or sexual).

Point 4 seems to come from evolutionary fear of invaders from dangerous species: primarily the large ones, but also small especially if poisonous, infections or very numerous, as well as humanoid or human groupings very different than whatever the local standard may be.

Point 5 relates to avoidance of breeding with the wrong species or the wrong members of one's own species, including those that may be infected or highly inefficient in family-like conditions.

4.4 To Conclude

Cascio [34] posed the right question: Does Mori's uncanny valley replicate for *overly perfect* post-human beings? Yet, he posed it in the narrow domain of enhabced human beings. We extend it to the domain more in tune with Mori's area of focus; namely humanoid robotics. It is the right time to be thinking seriously of humanoid robots surpassing human capacity in interesting domains relevant for various human activities. The second uncanny valley, the uncanny valley of perfection – based on humanoid robots, based on well-crafted shells and mechanics, guided by cognitive engines nearing AGI [35] – is closer and closer to reality Together with other related factors, such as likability, familiarity, strength and invasion of privacy, our dealing with the uncanny valley of perfection is going to help our civilizations cope with *singularity-like* reactions of some of the people at the times when human-robot interaction becomes a condition of civilizational progress, thus a necessity.

References

1. Floridi, L.: The 4th Revolution. How the Inforsphere is Reshaping Human Reality, pp. 152–158. OUP, Oxford (2014)
2. Alexa controlled robots (2020). https://www.servomagazine.com/magazine/article/controlling-robots-using-amazons-alexa
3. Thaler, S.: The Creativity Machine® Paradigm. Springer Science + Business Media LLC 2017. In: Carayannis, E.G. (ed.) Encyclopedia of Creativity, Invention, Innovation and Entrepreneurship. https://doi.org/10.1007/978-1-4614-6616-1_396-2
4. Goertzel, B.: The hidden pattern. In: A Patternist Philosophy of Mind, pp. 251–261. BrownWalker Press, Boca Daton (2006)
5. Clark, A.: Supersizing the Mind. Action and Cognitive extension. OUP, Embodiment (2011)
6. Boltuc, P.Q.: Robots and Complementarity of Subject and Object, Paideia. World Congress of Philosophy, Boston (1998). http://www.bu.edu/wcp/Papers/Mind/MindBolt.htm
7. Boltuc, P.: Church-turing lovers. In: Abney, K.A., Lin, P.J., Ryan, R. (eds.) Robot Ethics 2.0: From Autonomous Cars to Artificial Intelligence, pp. 214–228. Oxford University Press, Oxford (2017)
8. Boltuc, P.: What is the difference between your friend and a church turing lover. In: Ess, C., Hagengruber, R. (eds.) The Computational Turn: Past, Presents and Futures? Proceedings IACAP, pp. 37–40. Aarchus University (2011)
9. Mori, M.: The uncanny valley. Energy **7**, 33–35 (1970)
10. Rosenthal-von der Pütten, A.M., et al.: Neural mechanisms for accepting and rejecting artificial social partners in the uncanny valley. J. Neurosci. **39**(33), 6555–6570 (2019)
11. Mori, M., MacDorman, K.F., Kageki, N.: The uncanny valley. IEEE Robot. Autom. Mag. **192**, 98–100 (2012)
12. Deutsch, D.: Quantum theory, the church-turing principle and the universal quantum computer. Proc. Roy. Soc. (Series A) **400**, 97–117 (1985)
13. Church, A.: A Note on the Entscheidungsproblem. J. Symbol. Logic **1**, 40–41
14. Turing, A.M.: On computable numbers, with an application to the entscheidungsproblem. Proceed. London Math. Society (Series 2), **42**, 230–265 (1936–37)
15. Copeland, B.J.: The church-turing thesis. The Stanford Encyclopedia of Philosophy (Summer 2020 Edition). https://plato.stanford.edu/archives/sum2020/entries/church-turing
16. Boltuc, P.: The philosophical issue in machine consciousness. Int. J. Mach. Conscious. **1**(1), 155–176 (2009)
17. Boltuc, P.: The Engineering Thesis in Machine Consciousness Techne: Research in Philosophy and Technology **16**(2), 187–207 (2012)
18. Nagel, T.: The View from Nowhere. OUP, Oxford UK (1986)
19. Turing, A.E.: Computing Machinery and Intelligence. Mind **LIX**(236), 433–460 (1950)
20. Kirk, R.: Zombies and Consciousness. Oxford University Press (2005)
21. Chalmers, D.: Facing up to the problem of consciousness. J. Conscious. Stud. **2**(3), 200–219 (1995)
22. David Chalmers Sings The Zombie Blues (2008) https://www.youtube.com/watch?v=jyS4VFh3xOU
23. Harman, G.: Can Science Understand the Mind? In: Harman, G. (ed.) Conceptions of the Mind: Essays in Honor of George A. Miller, pp. 111–121. Lawrence Erlbaum, Hillside (1993)
24. Harman, G.: More on Explaining a Gap, The American Philosophical Association Newsletter on Philosophy and Computers 8.1 Fall (2008)

25. Libet, B.: Unconscious cerebral initiative and the role of conscious will in voluntary action. Behav. Brain Sci. **8**(4), 529–566 (1985)
26. Velmans, M.: Preconscious Free Will Journal of Consciousness Studies **10**(12), 42–61 (2002)
27. Mori, M.: The Uncanny Valley: The Original Essay by Masahiro Mori. IEEE Spectrum: Technology, Engineering, and Science News, IEEE Spectrum, 12 June (2012)
28. Kätsyri, J., et al.: A review of empirical evidence on different uncanny valley hypotheses: support for perceptual mismatch as one road to the valley of eeriness. Front. Psychol. **6**, 390 (2015)
29. Why Cgi Humans Are Creepy, and What Scientists Are Doing About It. ACM News. LiveScience. N, p. 2011. (Web, 28 November 2018)
30. MacDorman, K.F., Green, R.D., Ho, C.-C., Koch, C.-C.: Too real for comfort: Uncanny responses to computer generated faces. Comput. Hum. Behav. **25**(3), 695–710 (2009)
31. MacDorman. K.F., Chattopadhyay, D.: Reducing consistency in human realism increases the uncanny valley effect; increasing category uncertainty does not. Cognition **146**, 190–205. 10.1016/ (2016)
32. Oxfordwords Blog's: 31 Monsters from Around the World. https://me.me/i/31-monsters-from-around-the-world-oxfordwords-blog-bf1c8bb573a64a53a164928da4573ce1
33. Ho, C.C.; MacDorman K.F.: Measuring the uncanny valley effect refinements to indices for perceived humanness, attractiveness, and eeriness. Int. J. Soc. Robot. (2016). https://doi.org/10.1007/s12369-016-0380-9
34. Jamais, C.: The Second Uncanny Valley, Institute for Ethics and Emerging Technologies, 28 October 2007, http://www.openthefuture.com/2007/10/the_second_uncanny_valley.html
35. Goertzel, B.: Is Artificial General Intelligence (AGI) On The Horizon? Interview With Dr. Ben Goertzel, CEO & Founder, SingularityNET Foundation by Walch K. Forbes Magazine Jul 14, 2020, 09:51 pm EDT https://www.forbes.com/sites/cognitiveworld/2020/07/14/is-artificial-general-intelligence-agi-on-the-horizon-interview-with-dr-ben-goertzel-ceo–founder-singularitynet-foundation/#e222f1559d0b

Self and Other Modelling in Cooperative Resource Gathering with Multi-agent Reinforcement Learning

Vasilii Davydov[1], Timofei Liusko[2], and Aleksandr I. Panov[2,3](✉)

[1] Moscow Aviation Institute, Moscow, Russia
vddavydov@mai.education

[2] Moscow Institute of Physics and Technology (National Research University), Moscow, Russia
tim.luska@protonmail.com, panov.ai@mipt.ru

[3] Artificial Intelligence Research Institute FRC CSC RAS, Moscow, Russia

Abstract. In this work, we explore the application of the Self-other-Modelling algorithm (SOM) to several agent architectures for the collaborative grid-based environment. Asynchronous Advantage Actor-Critic (A3C) algorithm was compared with the OpenAI Hide-and-seek (HNS) agent. We expand their implementation by adding the SOM algorithm. As an extension of the original environment, we add a stochastic initialization version of the environment. To address the lack of performance in such an environment by all versions of agents, we made further improvements over the A3C and HNS agents, adding the module dedicated to the SOM algorithm. This agent was able to efficiently solve a stochastically initialized version of the environment, showing the potential benefits of such an approach.

Keywords: Self and other modelling · Multi-agent reinforcement learning · A3C · Cooperative resource gathering

1 Introduction

Despite recent achievements in Reinforcement Learning (RL), most of them were in single-agent domains, where other actors are not existing or can be treated as part of the environment. Nevertheless, some applications involve agent interaction in so-called multi-agents environments. In such environments ability to predict other agents' actions can be crucial to achieving high-performance [2]. However, traditional RL algorithms are not well-suited for complex non-stationary multi-agent environments [3].

With each agent's policy changes with the flow of the game, the environment becomes non-stationary, with additional complexity over agents' policy change. In such environments, having a model of other actors is better than treat them as part of the environment. A novel approach for modeling others was introduced

A. V. Samsonovich et al. (Eds.): BICA 2020, AISC 1310, pp. 69–77, 2021.
https://doi.org/10.1007/978-3-030-65596-9_9

in [1]. It uses an agent's network to predict the goal of another agent's actions. In this paper we describe the application of this approach to a variety of agents architectures, generalizing the method.

We adopt the game of coin environment formulation from [1], originally using seed-initialization as proposed. We also added a stochastically initialized version of the same environment as a more general problem. For agents architecture, we used the Asynchronous Advantage Actor-Critic (A3C) [4] and adapted to a grid environment OpenAI Hide-n-Seek (NHS) agent [9] as baseline. Their improved versions were developed by applying the SOM algorithm. As a further improvement, we added layers dedicated explicitly to the SOM algorithm to the HNS agent.

We measured the performance of these approaches on both versions of the environment, showing that using the SOM algorithm is beneficial for both agents. Most of the agents' performance on the stochastically initiated version of the environment was weak, and the training was slow. At the same time, an agent with dedicated SOM policy layers was able to solve this version of the environment, achieving high performance fast.

2 Background

In game theory, there two basic categories of games: competitive and cooperative [7]. Competitive games have different, often opposite goals for actors. Cooperative game rules allow actors to work together to achieve a win-win condition, as their goals are not directly opposed or even can be similar. However, a cooperative game itself does not always imply that cooperating actors will get an equal reward. Cooperation in a team of agents differs from cooperation as individual agents. Individual agents may have different goals and rewards. Collaborative agents have one goal and share the reward. This paper describes agents with different but symmetric goals; agents also share the reward.

In this work, we adapt game formulation from [1]. A Markov game for two agents is described by a set of states S, the set of all the possible configurations of all agents, a set of agents' actions A_1, A_2 and a set of agents' observations O_1, O_2, and a transition function $T : S \times A_1 \times A_2 \xrightarrow{S}$ which gives the probability distribution on the next state as a function of current state and actions.

Each agent i chooses actions by sampling from a stochastic policy $\pi_{\theta_i} : S \times A_i \rightarrow [0, 1]$. Each agent has a reward function which depends on agent's state and action: $r_i : S \times A_i \xrightarrow{R}$.

2.1 SOM Algorithm

Self Other-Modeling (SOM) algorithm applied to get new versions of A3C and HNS agents. They referenced as A3C-SOM agent (similar to the agent described at [1]) and HNS-SOM agent. SOM algorithm is used by the agent to predict other agent goal. The other's goal vector stores probabilities for possible goals of other agent. Since agents' goals are different in every game, the SOM initiates at the beginning of each game, and updates every game step to contrast the agent's network weights are updated every episode.

Both agents update their estimations for others' goal vectors after each step of the game, using its network with other agent inputs from the previous step to produce the estimation of another agent action. The estimation is compared with the actual action from the previous environment step. The vector is updated with a cross-entropy loss L_{CE} between the estimation and one-hot encoding of the actual action.

$$CE(\pi(S_{other}, z_{other}, z_{self}; \theta_{self}), a^i_{other}).$$

Agent uses other's goal vector as part of input. Inputs for A3C-SOM and HNS-SOM contain one more vector of size 3 which represents others' goals vector.

2.2 Baseline Agents

A3C agent realizes the Asynchronous Advantage Actor-Critic algorithm. It consists of hidden layers θ, policy π output and value function V output. Hidden part θ consists of 2 fully connected layers and an LSTM [5] layer. A policy π is approximated by a single fully-connected layer. It takes the output of θ and outputs a vector of the probability distribution over the action space of an agent. A value function V is approximated by a fully-connected layer. It takes θ as input and returns a single float value, which is the estimate of the expected reward that the agent can receive in the current state. The value function is needed for the A3C update.

Policy layers π gives distribution over action space, and then action a_i is sampled from this distribution as described:

$$a_i \sim \pi(S_{self}, z_{self}, z_{other}; \theta_{self}).$$

At the end of each game, agents receive a reward r. A discounted reward R_i then calculated:

$$R_i = r_i + \sum_{j=1}^{k} r_{i+j} * \gamma^j.$$

Policy vector (action probabilities) π, a value function V, and a discounted reward are stored in the replay buffer.

An advantage function $A(s_i)$ is calculated from the discounted reward and value function values:

$$A(s_i) = R_i - V(s_i).$$

Policy loss L_{policy} and value function loss L_{value} then calculated using found $A(s_i)$:

$$L_{policy} = -log(p(a_i|s_i))A(s_i),$$

$$L_{value} = A(s_i)^2.$$

Entropy loss was added in order to improve the exploration of the agents. Entropy is lower for action distributions with high certainty in one action and low certainty in all others, and otherwise. Entropy loss forces agents to avoid action distributions with one dominant action.

$$L_{entropy} = -\sum p(a_i|s_i)log(p(a_i|s_i)).$$

For each game stored in the buffer, and for each step of the game, the A3C loss is calculated using the policy, value, and entropy losses:

$$L_{A3C} = L_{policy} - \beta L_{entropy} + \alpha L_{value}.$$

Results then averaged and used to update the agent network.

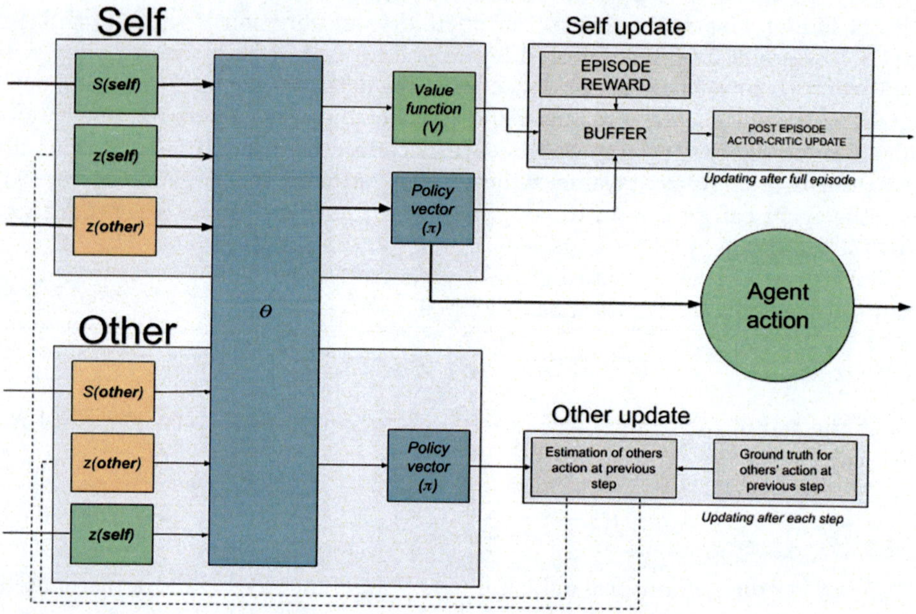

Fig. 1. This figure represents the agent's architecture. "Self-update" represents the algorithm's action phase, and "other update" represents the SOM update phase of the algorithm. θ is the agent's shared parameters. S(self), S(other), z(self), z(other) are all vectors that are given as input to the agent. The policy vector is the return of the agent's policy layers. During the action phase, the policy vector and value function are stored in the replay buffer for further A3C updates.

HNS agent is adapted to a grid-based environment agent described in [9]. HNS agent architecture similar to the A3C agent, except hidden layers θ. The

HNS θ consists of a fully connected layer, self-attention block, max-pooling layer, LSTM layer, and a single fully connected layer output. A normalization layer is added after each other layer except embedding. A Self-attention block is implemented as described in [8].

3 Method

3.1 Environment

The coin game environment [6] is grid-based, fully observable, and deterministic. Two agents start at the opposite corners on the top of the 8×8 grid map. The position for each agent between these two points is random. The map contains 12 coins (resources) with 3 different colours, 4 of each type. At the start of each episode, agents randomly assigned a goal - a resource he has to collect.

Action space for agents includes *up*, *down*, *right*, *left*, *pass*. If the agent moves to a cell with a coin, the coin disappears from the grid, and the coin's colour added to the agent's list of collected coins. The game episode has limited steps; a step limit of 20 was used, with each agent limited to 10 steps. An example of the environment with a starting state can be seen in Fig. 2

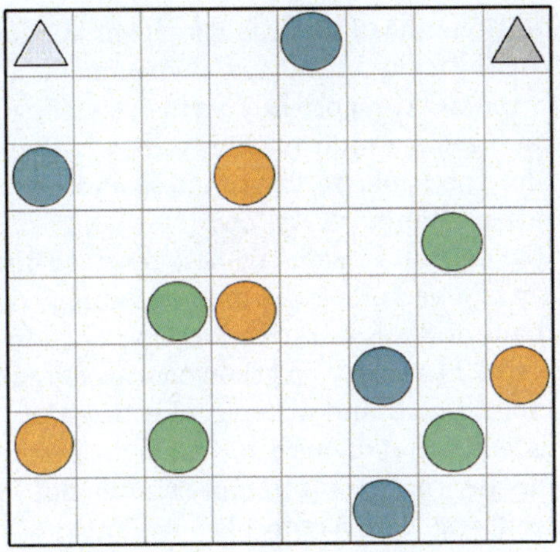

Fig. 2. Example of the initial state of the environment. Green and red triangles are depicting agents' positions. Circles of different colors depicting resources.

Reward for agent calculated using formula:

$$r = (C_{1self} + C_{2other})^2 + (C_{1other} + C_{2self})^2 - C_{1neither} + C_{2neither})^2,$$

where $C_{1self} + C_{2other}$ is amount of resources collected by this agent, both self-goal and other-goal, $C_{1other}+C_{2self}$ is same for second agent, and $C_{1neither}+$

$C_{2neither}$ is amount of non-target resources, collected by both agents. This reward is strictly symmetric, and thus always the same for both agents.

Seed initialized environment version with 5 alternating seeds, was adopted from [1]. Seeds were chosen in the way so the resources symmetrically distributed over the map, relative to the agents, giving no advantage. Each seed completely define starting position for the environment as well as starting location and agents' goals.

Stochastically initialized environments were created as a more general problem formulation. It uses the same agent starting points and the amount of resources. However, each coin's location and the position of each agent, from 2 possible starting point, is random (pseudo-random, as no particular random source or hardware were used). This version of the environment is harder to solve for agents since there can be an unequal distance to resources.

Both versions of the environment were 8×8 size. The step limit was 20, limiting each agent to 10 steps.

3.2 Self and Other Modelling in A3C and HNS

A3C-SOM and HNS-SOM are improved with SOM algorithm application baseline agents A3C and NHS. These versions has an additional step, where they are trying to predict the other's goal, using SOM.

HNS-SOM with additional head was the last agent version, added after series of experiments on a stochastically initialized environment. It has additional policy layers dedicated to the SOM algorithm exclusively. The reason for this approach was the fact that agents would probably have different strategies and it would be more useful for the agent to try to guess another agent's actions with another policy layer rather than with its' own.

Agents input, in the initial version agents' received information from the environment in the form of 4 vectors: 8×8 map encoding, with integers encoding entities locations; also a 3 of 8×8 resource location encoding, with each vector consists of zeroes and ones. The input was then concatenated into one 256-vector, and combined with true self-goal and other-goal estimation.

The change was made: instead of one 8×8 map encoding, a two 8×8 one-hot agent locations were added—that result in more simple but larger input features. The total size of the input was 320-vector, also combined with true self-goal and other-goal estimation.

4 Experiments

Two series of experiments were performed. For the first series, the environment was initialized with five alternating seeds, as described in [1]. Episode consisted of 5 games for this series, with 25000 episodes used for training. For the second, we used the environment with stochastic initialization. Episode consisted of 100 games for this series, with 1200 episodes used for training. All experiments were done on the machine with Titan RTX GPU, Intel Xeon Gold 6132 CPU, and

512 GB DDR4 RAM on Ubuntu 16.04 with python 3.6 and PyTorch version 1.2.0. Entropy coefficient was set to 0.01 and value loss coefficient to 0.5. Agent network's weights were updated with Adam optimizer with $\beta_1 = 0.9$ and $\beta_2 = 0.999$, $\epsilon = 1\text{x}10^{-8}$, weight decay 0 and learning rate $= 1\text{x}10^{-4}$. Other goals vector was optimized with SGD optimizer with learning rate $= 0.1$. Reward discount factor was 0.99.

Seed-initialized experiments were used to test initial versions of agents: A3C and HNS. After initial experiments, baseline A3C architecture with SOM (A3C-SOM) was added, as described in [1], allowing to compare results with the original study. An improved HNS with SOM were added (HNS-SOM), to explore the impact of the SOM algorithm.

A3C and A3C-SOM showed decent results, achieving reward values of 14, which is slightly better than the results of the same approach in the original study [1]. That can be possibly related to seed initialization, with us being unable to reconstruct same conditions, having no seeds provided. HNS showed an even better result, achieving reward in order of 16.

In this version of experiments, all versions of the agents experienced drops in rewards. That can possibly be explained by the seed-initialization offered possible shortcuts to a local minimum, with not enough variety for agents to get through.

Stochastically initialized experiments were an extension over seed-initialized ones. This version of the environment offers more challenges for the agent. At the same time, there is less chance for an agent to get stuck in a local minimum, because of high variety of the environment.

Initially, we tested 2 versions of agents, which showed the best performance before: A3C-SOM and HNS. The results showed that random initialization makes it very hard for them to learn, and they were not able to achieve the same high results as before, with HNS achieving a reward of 2.5 and A3C-SOM achieving reward of 8 during training. We explored a number of approaches to address this lack of performance, with one showing the great results: we added dedicated layers for the SOM algorithm in the initial HNS agent. It was able to achieve a reward above 19, collecting 3 coins on average per agent. 19.5 is likely the limit for the experiment with 20 steps. This limit due to distance agent can possibly traverse during this limit.

HNS-SOM with additional head was also tested on seed-initialized environments, showing the best result. Graphs for rewards of final versions of experiments can be seen in Fig. 3.

5 Related Work

One of the approaches for the cooperative multi-agent task is described in [10]. In this paper, agents use a centralized policy for more examples of training and decentralized critic networks for agents to learn their reward. Also, the authors described the FacMADDPG algorithm, which used both policy and critic networks centralized. In our work, we use both policy and critic networks

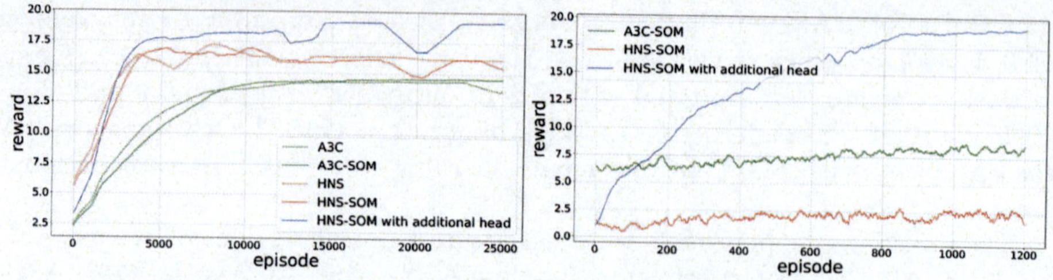

(a) 5 fixed seeds (b) Stochastic environment

Fig. 3. Reward graphs for experiments. (a) Reward graph for seed-initialized environment. (b) Reward graph for stochastic-initialization environment.

decentralized because the environment is simpler than multi-agent MUJOCO, and the agents still learn their strategies fast enough.

Other cooperative approaches focus on communication between agents like in [11]. In our work, agents try to model each other's policy rather than use each other's information during the action phase.

The closest work to the self-other-modeling algorithm is [12]. The original SOM algorithm model only other agent's goals. Our HNS agent, with the dedicated layers for SOM, tries to model another agent strategy. As a result, this version outperforms the base version of HNS and HNS with SOM updates.

6 Conclusions and Future Work

We explored and compared a few agent architectures for collaborative resource gathering environment. We performed a series of experiments to compare them and implemented a few improved versions of agents. We recreated the original study approach, with the same results, and showed that it is possible to achieve a higher reward, using more complex agent architecture. It is possible to achieve a reward, close to the possible max on a seed-initialized environment, combining the SOM algorithm with dedicated layers.

We then created the second version of the environment with random initial locations for resources and random goals for agents. This version of the environment was shown to be harder for agents to solve. Addressing that, we tested a variety of agents and approaches, finding that HNS with SOM algorithm and dedicated layers for SOM output was able to achieve high performance. Tested on the previous version of the environment, this agent again was able to achieve the high reward, outperforming other versions.

We plan to extend our work by exploring more complex grid environments, with more complex agent interaction. We plan to explore a grid version of the OpenAI Hide-n-Seek game, with both adversarial and collaborative agent interaction. Some code for this work can be found at https://github.com/tlLuska/game-of-coins-RL.

Acknowledgments. This work was supported by the Russian Science Foundation (Project No. 20-71-10116).

References

1. Raileanu, R., Denton, E., Szlam, A., Fergus, R.: Modeling others using oneself in multi-agent reinforcement learning. arXiv preprint arXiv:1802.09640, March 2018
2. Tampuu, A., Matiisen, T., Kodelja, D., Kuzovkin, I., Korjus, K., Aru, J., Aru, J., Vicente, R.: Multiagent cooperation and competition with deep reinforcement learning. PLoS ONE **12**, 1–12 (2017). https://doi.org/10.1371/journal.pone.0172395
3. Lowe, R., Wu, Y., Tamar, A., Harb, J., Abbeel, P., Mordatch, I.: Multi-agent actor-critic for mixed cooperative-competitive environments. In: Advances in Neural Information Processing Systems. 2017-Decem, pp. 6380–6391 (2017)
4. Mnih, V., Badia, A.P., Mirza, M., Graves, A., Lillicrap, T., Harley, T., Silver, D., Kavukcuoglu, K.: Asynchronous methods for deep reinforcement learning. In: International Conference on Machine Learning, pp. 1928–1937 (2016)
5. Schmidhuber, J., Hochreiter, S.: Long short-term memory. Neural Comput 9.8, November 1997
6. Sukhbaatar, S., Szlam, A., Synnaeve, G., Chintala, S., Fergus, R.: Mazebase: a sandbox for learning from games. arXiv preprint arXiv:1511.07401, November 2015
7. Shapley, L.S.: Stochastic games. Proc. Natl. Acade. Sci. **39**(10), 1095–1100 (1953). ISSN 0027-8424. https://doi.org/10.1073/pnas.39.10.1095, November 2015
8. Vaswani, A., Shazeer, N., Parmar, N., Uszkoreit, J., Jones, L., Gomez, A.N., Kaiser, Ł., Polosukhin, I.: Attention is all you need. In: Advances in Neural Information Processing Systems, pp. 5998–6008 (2017)
9. Baker, B., Kanitscheider, I., Markov, T., Wu, Y., Powell, G., McGrew, B., Mordatch, I.: Emergent tool use from multi-agent autocurricula, arXiv:1909.07528 (2019)
10. de Witt, C.S., Peng, B., Kamienny, P.A., Torr, P., Böhmer, W., Whiteson, S.: Deep Multi-Agent Reinforcement Learning for Decentralized Continuous Cooperative Control. arXiv preprint arXiv:2003.06709 (2020)
11. Foerster, J., Assael, I. A., De Freitas, N., and Whiteson, S.: Learning to communicate with deep multi-agent reinforcement learning. In: Advances in Neural Information Processing Systems, pp. 2137–2145 (2016)
12. He, H., Boyd-Graber, J., Kwok, K., Daumé III, H.: Opponent modeling in deep reinforcement learning. In International Conference on Machine Learning, pp. 1804–1813, June 2016

Eye Movement Correlates of Foreign Language Proficiency in Russian Learners of English

Valeriia Demareva[1(✉)], Sofia Polevaia[1,2], and Julia Edeleva[3]

[1] Lobachevsky State University, 603950 Nizhny Novgorod, Russia
kaleria.naz@gmail.com
[2] Privolzhsky Research Medical University, 603950 Nizhny Novgorod, Russia
[3] University of Braunschweig, 38092 Braunschweig, Germany

Abstract. Today, the modeling of various aspects of speech activity is the mainstream of modern cognitive and computational science. Along with models of natural language processing, much attention is paid to finding mechanisms of the several languages functioning within one cognitive system (bilingual, trilingual, etc.). The search for specific features of the processing of linguistic information by one subject in different languages allows one to approach the construction of bilingual models. This article is devoted to the analysis of the eye movements during text reading by bilinguals in their native and foreign languages, with different levels of proficiency in the latter. The present study tests the assumption that eye movement features of people with a high level of foreign language skills are similar during text reading in native and foreign languages. Another goal is to elicit features that provide the differentiation between the elementary and the intermediate levels of English language proficiency. We offer new Eye Tracking based evaluation metrics for the level of language proficiency.

Keywords: Eye tracking · Bilinguals · Reading skills · Comprehension · Proficiency

1 Introduction

Measuring language proficiency is required for many purposes. One possible application is in the field of recruitment. Another significant area is foreign language teaching and learning. In either case, the most widely used assessment procedure is testing. Such tests can take 2–5 h and imply subjective evaluation of soft facts by specialists. However, today new instrumental methods of measuring language competence are available, where Eye Tracking (ET) is implemented for interactive monitoring of eye movements during reading. The interpretation of ET data is based on the assumption that the complexity level (objective and subjective) of working with a text is reflected in eye movements and can be recorded by ET.

Previous research and experiments conducted on groups of native speakers have coined a number of facts proving that eye movements and text or reading complexity are interrelated [1–5]. For instance, as the complexity of an English text increases, the

A. V. Samsonovich et al. (Eds.): BICA 2020, AISC 1310, pp. 78–83, 2021.
https://doi.org/10.1007/978-3-030-65596-9_10

number of small saccades also grows, as well as the number of regressions, and the duration of fixations. The words that are easily predicted by the context and high-frequency words have fewer number of fixations. It is experimentally shown that the distribution of fixations is a function of word frequency and lexical complexity. The experiments based on texts with increasing complexity read by native speakers have revealed the influence of word length on the amplitude of progressive saccades resulting in a "spillover effect". The improvement of reading skills can reinforce the influence of text complexity and its structural organization on the process of reading [1–5].

The aim of our study was to define specific eye tracking parameters of Russian adults with Intermediate (relatively high) level of proficiency in English for reading Russian and English texts.

We hypothesized that the difference in eye movements when reading two texts with the same objective complexity reflects subjective factors, namely, the reader's individual level of language proficiency.

If we consider the text in the reader's native language as a task with minimal subjective complexity as opposed to an equivalent text in a foreign language, the ET parameters must reflect the subjective complexity of reading and reveal the reader's foreign language proficiency. In this article, we present experimental data which prove the informative value of ET-coefficients for the evaluation of Russian students' competence level in English.

2 Methods

In our experiment, we tested whether native speakers of Russian with the Intermediate level of English have comparable ET parameters while they read texts in English and Russian to answer comprehension questions on both texts. The experiment consisted of three stages: the evaluation of English level with the help of a Placement test (https://oxfordklass.com/placement-test/), the recording of the participants' eye movements while they read texts in Russian and English as well as looked for answers in those texts, and the comparative analysis of the collected data with the help of descriptive statistics and Mann–Whitney U-test.

2.1 Participants

22 Students of Lobachevsky State University of Nizhny Novgorod Aged 21–25 Took Part in the Present Study.

All participants had normal or corrected-to-normal visual acuity and gave informed consent prior to volunteering in the experiments. They were naive to the purpose of the experiment.

2.2 Apparatus and Stimuli

The experiment was conducted in a bright room. The level of illumination was held constant throughout the experiment. Experimental area was equipped with a table and places for the expert and the subject.

In the first stage the level of English was determined with the help of a Placement test (https://oxfordklass.com/placement-test/). A list of questions included 20 items with three possible options. Two items correspond to Elementary level, three points – to Pre-Intermediate level, seven points - to Intermediate level, and eight points – to Upper Intermediate level.

For the second stage of the experiment recording of eye movements was carried out with the help of eye tracking SMI HiSpeed, with frequency binocular survey of 1250 Hz, a PC-based software SMI Experiment Suite 360° and iView v. 2.0.1. The calibration area was 1280 * 960. The recording was done for two eyes (binocular).

Prior to the recording, the subject's head was placed on the chin rest and its position corrected according to the image in iView X module (13-point calibration). Calibration precision was monitored and maintained throughout the experiment.

The texts of comparable objective complexity and meaning in Russian (L1) and English (L2) used as visual stimuli were displayed on a calibrated (gamma corrected) computer monitor (resolution: 1920 × 1200; refresh rate: 60 Hz; type: Monitor DELL U2410) on one slide. In both cases the background color was Aliceblue, the text color was black, the font used in Experiment Center editor was Times New Roman, 72 pt (0.520 in angular units), interline spacing 1,5 cm. The distance between the subject and the screen was 70 cm.

The Russian text consisted of 43 words, 6 sentences, 9 lines, the average number of characters per line was 24. The English text consisted of 53 words, 6 sentences, 8 lines, the average number of characters per line was 26. All the instructions, explanations, as well as reading comprehension questions to L1 and L2 were given in Russian.

2.3 Procedure

In the first stage, all participants performed the Placement test in order to determine their level of English. The test could take place in groups or individually. The result was not announced to the subjects.

The second stage was administered to each subject individually. It consisted of five steps: SMI HiSpeed calibration; L1 Reading performed by a subject with simultaneous eye movement recording; L1 Comprehension test performed by a subject with simultaneous eye tracking; L2 Reading performed by a subject with eye movement recording; L2 Comprehension test performed by a subject with simultaneous eye tracking.

Prior to the second stage of the experiment, the subjects were informed about the terms and conditions of participation. The following key points were announced:

- Eye movements and the locus of attention on the screen are being recorded during the task with a special type of equipment;
- The participation can be terminated at any time without further explanations.

Then the head position of the participant was stabilized by means of a head and chin rest and calibration was performed.

The following task was announced to the participant: “Please read the text on the screen silently. Once you’ve finished the first reading, you must look to the bottom right corner of the screen”. While the subject was performing the first task, their eye movement activity was being recorded.

Shortly after the L1 reading task, the instruction for the comprehension test was communicated: “I will pronounce five questions based on the text you have just read. To answer, look at the word on the screen which indicates the reply”.

The eye movements were tracked during the test.

The same procedure was repeated for the L2 reading and comprehension tasks.

Study design and procedures were approved by the Ethics Committee of Lobachevsky State University, and all participants provided written informed consent in accordance with the Declaration of Helsinki.

3 Results and Discussion

The Placement test administered during the first stage of the study revealed that 10 students had Elementary level of English, 5 students – Pre-Intermediate level, and 7 students – Intermediate/ Upper Intermediate level. We grouped students with Intermediate and Upper Intermediate levels together into «Intermediate» to regard them as the most proficient representatives of our sample. The other two groups were formed in accordance with the students’ proficiency levels.

First of all, we compared eye movement features of the three groups separately for L1 and L2 tasks. The results confirmed that there were no significant differences in ET features between “Elementary”, “Pre-Intermediate” and “Intermediate” groups when students performed reading and comprehension tasks in Russian.

As subjects proceeded with reading an English text, we found that the amplitude of saccades was significantly greater in the “Intermediate” group than in the “Elementary” ($U = 13$, $p = 0.036$) and the “Pre-Intermediate” ones ($U = 4$, $p = 0.035$). The fixation duration of students performing L2 comprehension task is significantly shorter in the “Intermediate” group than in the “Elementary” one ($U = 14$, $p = 0.045$).

The differences stated above are caused by increased subjective difficulty in L2 compared to L1.

The analysis revealed that the difference in saccade amplitude between L1 and L2 reading is much larger for students with Elementary level of English as compared to the other groups. Their saccade amplitude observably decreases for L2 reading. The difference coefficient (1) for the “Intermediate” group is smaller than in the “Pre-Intermediate” ($U = 3$, $p = 0.002$) and the “Elementary” ones ($U = 5$, $p = 0.042$). This data is demonstrated in Fig. 1.

$$K_{SA} = \frac{SA_{L1} - SA_{L2}}{SA_{L1} + SA_{L2}} \quad (1)$$

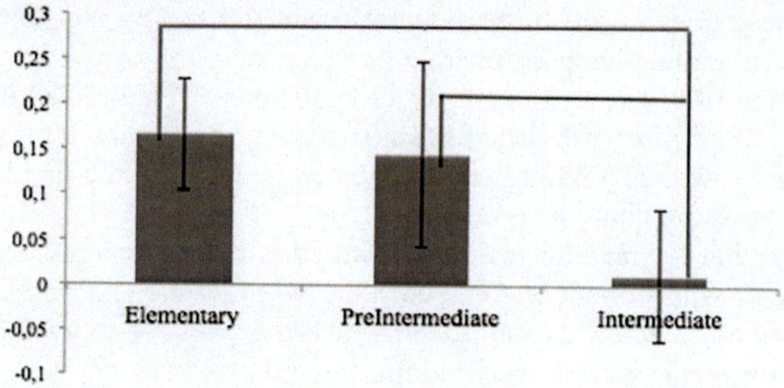

Fig. 1. Average difference coefficients (K_{SA}) for amplitude of saccades (SA) in students with different English language proficiency performing L1 (SA L1) and L2 (SA L2) reading.

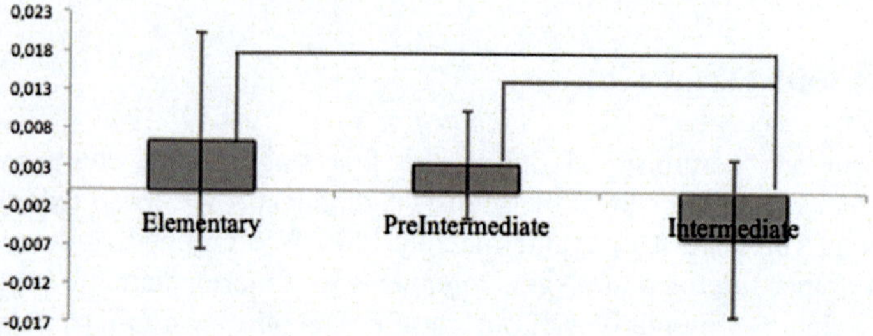

Fig. 2. Average difference coefficients (K_{PD}) for pupil diameter (PD) between students with different English language proficiency performing L1 (PD L1) and L2 (PD L2) comprehension tasks.

As subjects with Elementary level were searching for answers to questions in texts their pupil diameter was bigger when they performed L1 Comprehension, while "Pre Intermediate" and "Intermediate" groups were characterized by an inverse pattern: the difference coefficient (2) in the "Intermediate" is smaller than in the "Pre Intermediate" ($U = 14$, $p = 0.045$) and the "Elementary" ($U = 6$, $p = 0.05$) ones - see Fig. 2. This can be explained by the fact that the subjects with a lower level of English were concentrating on specific words in L2 in order to answer questions, whereas the subjects from the "Intermediate" group were focusing on the context.

$$K_{PD} = \frac{PD_{L1} - PD_{L2}}{PD_{L1} + PD_{L2}} \quad (2)$$

Also, we can identify the characteristics of fixation features specific for students with the two levels: Elementary and Intermediate. The Elementary level is characterized by longer fixations while working with L2 compared to L1 texts, whereas students in the "Intermediate" group have the same fixation duration values while working with both texts.

Thus, this study allowed us to elicit the most informative ET-features which reflect the difference in the subjects' English competence and help to discriminate between "Intermediate" and "Elementary" levels: the amplitude of saccades, fixation duration, and pupil diameter.

The following ET features correspond to the Intermediate level of English:

1) High saccade amplitude values when text reading in English;
2) Low fixation duration values when looking for answers in an English text;
3) The same or insignificantly decreased amplitude of saccades during text reading in English in comparison to text reading in Russian;
4) The same or bigger pupil diameter when performing comprehension tasks for English as compared to pupil diameter during comprehension in Russian;
5) The same fixation duration when performing comprehension tasks for English as compared to duration of fixations during comprehension in Russian.

As a result, we obtained evidence for the hypothesis that the ET features of subjects with relatively high level of foreign language skills are similar when working with texts in their native and foreign languages. This enables the implementation of new metrics where a text in a native language defines the "base line" in terms of ET values and the evaluation of foreign language competence is based on the difference in the amplitude of saccades, fixation duration, and pupil diameter when working with native and foreign language texts.

Therefore, we confirmed our hypothesis about the reflection of the reader's individual level of language proficiency in eye movements differences while reading two texts with the same objective complexity.

Acknowledgements. This work was supported by the Russian Foundation for Basic Research (grants No. 18–013-01169, 18–013-01225).

References

1. Ashby, J., Rayner, K., Clifton, C.: Eye movements of highly skilled and average readers: differential effects of frequency and predictability. Q. J. Exp. Psychol. **58A**, 1065–1086 (2005)
2. Liversedge, S.P., Drieghe, D., Li, X., Yan, G., Bai, X., Hyönä, J.: Universality in eye movements and reading: a trilingual investigation. Cognition **147**(3), 1–20 (2016)
3. Rayner, K., Pollatsek, A., Drieghe, D., Slattery, T.J., Reichle, E.D.: Tracking the mind during reading via eye movements: comments on Kliegl, Nuthmann, and Engbert (2006). J. Exp. Psychol. Gen. **136**, 520–529 (2007)
4. Rayner, K.: Visual attention in reading: eye movements reflect cognitive processes. Mem. Cogn. **4**, 443–448 (1977)
5. Whitford, V., Titone, D.: Second-language experience modulates first- and second-language word frequency effects: Evidence from eye movement measures of natural paragraph reading. Psychon. Bull. Rev. **19**(1), 73–80 (2012)

Check for updates

Development of a Laboratory Workshops Management Module as Part of a Learning Support System for the "Decision-Making Theory" Course

Anastasia Devyatkina(✉), Natalia Myklyuchenko, Anna Tikhomirova, and Elena Matrosova

National Research Nuclear University MEPhI (Moscow Engineering Physics Institute), Moscow, Russia
devyatkina.mephi@mail.ru, enika.mk@gmail.com, anna7909966@yandex.ru, matrosova_ev@inbox.ru

Abstract. The article is devoted to the problem of optimizing the load of teachers in the educational process. The task of developing a specialized Learning Support System with respect to specifics of the subject is considered. Specifics of the subject area of the "Decision Making Theory" course demonstrated the impossibility of presenting the educational process in a given area of knowledge using the existing LMS. The authors propose a universal mechanism for managing laboratory workshops as a tool for monitoring and evaluating the degree of "Decision-Making Theory" course material assimilation. The proposed approach can be used in subject areas focused on computations using spreadsheets.

Keywords: Decision-making process · Learning management system · Learning support system · Learning quality assessment · Knowledge control

1 Introduction

In this article the problem of the effective organizing of the educational process at the "Decision-Making Theory" course is considered. A key element of the educational process is the monitoring and evaluation of the material assimilation degree. Evaluation is based on the results of performing numerous workshops, which are a sequence of nontrivial calculations using spreadsheets. The traditional approach to the organizing of the educational process involves the teacher at all its stages. Stages include assignment of initial data, verification of the correctness of the calculations, posing theoretical questions, evaluating the results. As a result of the high teaching load, the influence of the human factor on the final grade objectivity increases. In the context of the information technology globalization, the question of the need to reorganize the educational process through automation arises [1].

A. V. Samsonovich et al. (Eds.): BICA 2020, AISC 1310, pp. 84–90, 2021.
https://doi.org/10.1007/978-3-030-65596-9_11

Many software products have been proposed and successfully put into operation to automate the learning process with the ability to completely distance its participants. In relation to this type of software, the term LMS (Learning Management System) is used [2].

However, the desire for universality in the development of modern LMS does not allow taking into account the specifics of absolutely any subject area by requiring the correspondence of the educational process model to a certain template. Thus, there are a number of knowledge areas, the educational process of which cannot be automated and modeled using only the LMS tools.

2 Application of a Learning Support System for the Educational Process Organizing

The solution to the optimization and automation problem of the educational process is the development of a Learning Support System focused on the subject area specifics [3, 4].

2.1 Specific Area Analysis

The main mechanism for monitoring and evaluating the degree of "Decision-making Theory" course material assimilation during the educational process is a laboratory workshop. The laboratory workshop is defined as a sequence of ordered interconnected calculation steps, isolated by a set of control points. The definition is based on a result of studying the theoretical foundations of the subject area and the experience of the educational process participants of the past years (Fig. 1).

To describe the life cycle of a laboratory workshop in the educational process for the "Decision-making Theory" course the following notations are used: $k \in \overline{1,3}$ – the number of the current attempt to complete the workshop; $i \in \overline{1,N}$ – the number of the current calculation step; D^k_{max} – the maximum point awarded for the successful workshop completion on k-th attempt; D^k_i – the current grade at the time of performing the i-th step on k-th attempt.

There are three attempts to complete each laboratory workshop. The workshop is considered as successfully passed if the number of points D^k meets the condition $D^k \geq 0,6 * D^k_{max}$. In this case, the maximum score D^k_{max} for k-th attempt is calculated by the following formula:

$$D^k_{max} = D^{k-1}_{max} - (k-1) * \Delta D_{max}, \tag{1}$$

where ΔD_{max} – the degree of lowering the maximum grade D_{max} for each attempt k.

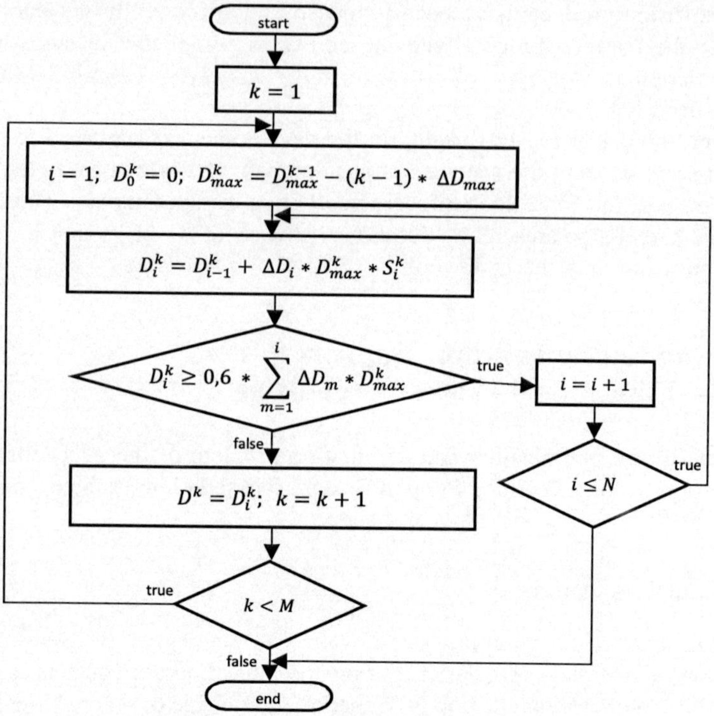

Fig. 1. Laboratory workshop life cycle

Each workshop is a sequence of N calculation steps (i – the sequence number of the current step). The calculation of the grade at each step is calculated according to the formula:

$$D_i^k = D_{i-1}^k + \Delta D_i * D_{max}^k * S_i^k, \tag{2}$$

where ΔD_i – the fraction of the maximum score D_{max}^k, accrued for correctly completed i-th step, and $\sum_{i=1}^{N} \Delta D_i = 1$; $S_i^k \in \{0, 1\}$ – an indicator of the correctness of the result for i-th step on k-th attempt. The transition to the next step is allowed under the condition that $D_i^k \geq 0,6 * \sum_{m=1}^{i} \Delta D_m * D_{max}^k$, otherwise, the student is offered to perform the next attempt to complete a laboratory workshop with different initial data. In this case, the maximum score is recalculated according to the formula (1). Each subsequent attempt to accomplish the workshop keeps the sequence of calculation steps unchanged but it is characterized by new initial data.

During the educational process, at every moment, the laboratory workshop is in one of the following states: «*unassigned*$_k$» - initial data were not provided; «*assigned*$_k$» - source data provided; an attempt to accomplish the workshop can be made; «*opened*$_k$» - a workshop is in progress; «*done*$_k$» - the workshop is successfully completed; «*failed*$_k$» - the attempts are spent, the result of the workshop is unsatisfactory.

2.2 Laboratory Workshop Model

To add a new laboratory workshop to the educational process by using the Learning Support System, first of all, it is necessary to develop an execution script and determine a set of laboratory workshop control points. The script should be obtained by researching various principles of thinking and approaches to finding the optimal algorithm for solving the problem of the workshop [5, 6].

Based on the developed execution script, it is necessary to build a model of the workshop. The laboratory workshop model is a representation of the laboratory workshop in terms of a learning management system. This representation consists of two components – a chart and a spreadsheet (Fig. 2).

$$[\langle step_1\rangle, \dots, \langle step_i\rangle, \dots, \langle step_N\rangle]$$

$$\{state, worksheet, \langle content\rangle[, title \,\|subtitle\|button]\}$$

$$[\langle unit_1\rangle, \dots, \langle unit_j\rangle, \dots, \langle unit_M\rangle\,]$$

$$\{layout, name, element, value[, error\|type\|caption\|style\|cellclass\|cellstyle]\}$$

Fig. 2. Laboratory workshop scheme

The following are the key terminal elements of the scheme:

- *state* $\in \{none, edit, done, fail\}$ – the current state of the $step_i$ of laboratory work;
- *spreadsheet* – a string type property containing the name of the spreadsheet for $step_i$;
- *layout* $\in \{\ll col - t\gg\}$, where $t = \overline{1, 12}$ – a string type property for positioning the interface element $unit_j$ on the $step_i$;
- *name* – a string type property that indicates the name of the spreadsheet range with the contents for the interface element $unit_j$ on $step_i$;
- *element* $\in \{label, input, image, table, group\}$– type of interface element $unit_j$ on $step_i$.

The laboratory workshop spreadsheet aggregates the logic of each calculation step. Also, it sets the algorithm for generating the initial data for each attempt to execute. The spreadsheet file has an extension «.ods» .

The laboratory workshop spreadsheet is based on the mechanism for using named ranges in LibreOffice Calc. Each step *i* of the calculations is represented by a separate sheet of the spreadsheet with the name corresponding to the value of the *spreadsheet* property for the JSON-object $step_i$ of the laboratory workshop scheme. Each sheet of the spreadsheet and each step of the laboratory workshop diagram necessarily contain the following named ranges in the local pale: *state* – a name for the range of cells in the spreadsheet with the status value of the current step; score – a name for the range of cells in the spreadsheet with the score for the current step, calculated by the formula: $\Delta D_i * D^k_{max} * S^k_i$.

It is proposed to implement the initial data generation algorithm [7] for the laboratory workshop in the form of macros triggered by the "open document" event. The algorithm may be written in one of the languages LibreOffice Basic, BeanShell, JavaScript, Python.

2.3 Laboratory Workshop Management Module

The laboratory workshop management module provides a REST API for the interaction of the Learning Support System server with the laboratory workshop model. While the server of the Learning Support System uses JavaScript and the NodeJS runtime environment, the laboratory practical management module executes a Java project that aggregates the logic of interacting with spreadsheets using the Java LibreOffice Calc API.

The server of the Learning Support System calls for execution java source code through a bridge API of the library «java-node». An existing Java project of the laboratory practical management module is packed into an «. jar» archive by a software project management and comprehension tool Maven and connected to the NodeJS runtime environment as a dependency (Fig. 3).

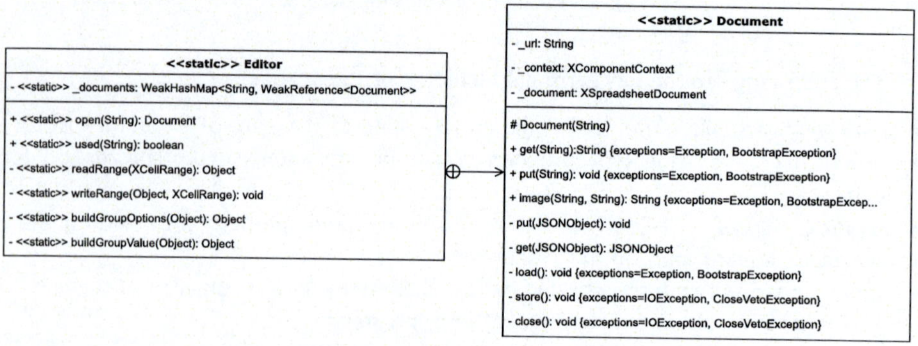

Fig. 3. Laboratory workshop class diagram

We consider the principle of the Learning Support System functioning at the stage of processing a certain calculation $step_i$ on k-th attempt to perform a laboratory workshop.

The laboratory workshop management module receives a request for a workshop scheme with a particular lab identifier. After receiving a response, the system parses the scheme in order to obtain a JSON-object that identifies the current $step_i$ (Table 1).

Table 1. Laboratory workshop scheme request

Request type	Access point	Request parameters
GET	restapi/steplab/instance/schema	lab – workshop identifier
Request example: curl –v "http://194.87.234.154:3000/restapi/steplab/instance/schema/lab=-M7-mYCtOTKTH"		

Next, a request to the laboratory workshop spreadsheet file is made. The request contains the lab workshop identifier and the user identifier since the lab is assigned to the particular user. After receiving a copy of the spreadsheet, the existence of a sheet with the name *spreadsheet* is checking, which is described in the diagram of the current $step_i$.

For the current $step_i$, it is necessary to request the initial data from the spreadsheet of the laboratory workshop. These initial data will be displayed on the screen following the user interface configuration specified by the laboratory workshop diagram for the current $step_i$.

After completing all the necessary calculations, the user enters the data into the Learning Support System by using the software interface and initiates the transition to the next $step_{i+1}$ of laboratory workshop. The workshop management module saves the information entered by the user into the report and initiates the process of checking the results of calculations (Table 2).

Table 2. Spreadsheet update request

Request type	Access point	Request parameters
PUT	restapi/steplab/instance/data	lab – workshop identifier user – user identifier body – JSON-object of $step_i$ scheme

Request example:
curl -v -X PUT -H 'Content-Type: application/json' –data '{"data": {"state" : "edit", "spreadsheet" : "example", "content" : [...]}}' http://194.87.234.154:3000/restapi/steplab/instance/data/lab=-M7-CtOH&user=-M5Rv-GfV

To go to the next $step_{i+1}$ of the laboratory workshop, the system reads information about the status of the current $step_i$ and the points awarded from the spreadsheet and verifies that the following condition is met:

$$D_i^k \geq 0,6 * \sum\nolimits_{m=1}^{i} \Delta D_m * D_{max}^k. \tag{3}$$

If condition (3) is not satisfied, the transition to the next $k+1$ attempt happens.

Thus, the laboratory workshop management module ensures both the integration and the correct interpretation of the laboratory workshop model components.

3 Conclusions

In the digital economy, the demand for tools of automation and optimization of the educative process in all knowledge areas continues to grow rapidly [8]. However, modern LMS cannot provide a universal toolkit that allows us to simulate an arbitrary educational process in any subject area.

An analysis of the subject area specifics of the "Decision Making Theory" course demonstrated the impossibility of presenting the educational process in a given area of knowledge using the existing LMS.

Thus, the formulated task is to develop a Learning Support System focused on the use of laboratory workshops as a way of monitoring and assessing the degree of knowledge acquisition. To solve this problem, a unique model for the laboratory workshops presentation and managing has been proposed. As a result of testing the described Learning Support System in the real educational process, it has become possible to reduce the teaching load and achieve a more transparent assessment system.

The proposed approach can be used in subject areas focused on computations using spreadsheets.

Acknowledgments. This work was supported by Competitiveness Growth Program of the Federal Autonomous Educational Institution of Higher Professional Education National Research Nuclear University MEPhI (Moscow Engineering Physics Institute).

References

1. Petrovskaya, A., Pavlenko, D., Feofanov, K., Klimov, V.: Computerization of learning management process as a means of improving the quality of the educational process and student motivation. Procedia Comput. Sci. **169**, 656–661 (2020). https://doi.org/10.1016/j.procs.2020.02.194
2. Solo, V.: LMS: past, present and future. In: ICASSP, IEEE International Conference on Acoustics, Speech and Signal Processing – Proceedings, vol. 2019-May, pp. 7740–7744. Institute of Electrical and Electronics Engineers Inc. (2019). https://doi.org/10.1109/ICASSP.2019.8682624
3. Tikhomirova, A.N., Matrosova, E.V.: Decision-Making Theory: Lecture Notes, p. 68. Research and publishing center INFRA-M, Moscow (2017)
4. Orlov, A.I.: Decision-Making Theory: textbook, p. 57. Examen publisher, Moscow (2006)
5. Samsonovich, A.V., Kitsantas, A., O'Brien, E., De Jong, K.A.: Cognitive processes in preparation for problem solving. Procedia Comput. Sci. **71**, 235–247 (2015). https://doi.org/10.1016/j.procs.2015.12.218
6. Samsonovich, A.V., Klimov, V.V., Rybina, G.V.: Biologically inspired cognitive architectures (BICA) for young scientists. In: Proceedings of the First International Early Research Career Enhancement School (FIERCES 2016). Advances in Intelligent Systems and Computing, vol. 449. Springer, Heidelberg (2016)
7. Tikhomirova, A., Matrosova, E.: Peculiarities of expert estimation comparison methods. Procedia Comput. Sci. **88**, 163–168 (2016). https://doi.org/10.1016/j.procs.2016.07.420
8. Klimov, V.V., Chernyshov, A.A., Balandina, A.I., Kostkina, A.D.: Problems of teaching students to use the featured technologies in the area of semantic web. In: AIP Conference Proceedings, vol. 1797. American Institute of Physics Inc. (2017). https://doi.org/10.1063/1.4972447

Algorithm for Constructing Logical Neural Networks Based on Logical Various-Valued Functions

Dmitriy Dimitrichenko(✉)

Institute of Applied Mathematics and Automation of Kabardin-Balkar Scientific Centre of RAS (IAMA KBSC RAS), St. Shortanova 89 a, Nalchik 360000, KBR, Russia
dimdp@rambler.ru

Abstract. The intelligent control system, as a set of production rules, is implemented in the form of an various-valued logical function. The combined use of mathematical logic and neural network methods gives the intelligent control system additional flexibility and the possibility of self-learning. In this paper we propose a method for representing various-valued logic function in a logical neural network. This logical neural network will keep the totality of cause-and-effect relationships identified using various-valued logic functions within a given specified area. These logic operations are implemented by special logic neural cells: conjunctors and disjunctors. The theorems given in this article justify the possibility of constructing such neural networks. The method of proof of these theorems contains an algorithm for constructing logical neural networks for a finite number of steps.

Keywords: Control system · Predicate · The predicate atomicity · Various-valued logical function logical neural network · Fuzzy logic variable

1 Introduction

The main requirement to the production lines built on the basis of robotic complexes (RC) consists in their flexibility [10], i.e. in the ability to be reconstructed quickly to perform new technological operations or to change the sequence due to change of the operating programs. Therefore RC and the flexible automated production systems created on RC basis find more and more broad application in mass production with constantly increasing share in the industry [1, 3, 5, 6, 8]. In most cases mobile robots of different functions are a part of RC. On their basis transport, diagnostic and other subsystems are under construction.

The significant expansion of functionality of mobile robots is reached by introduction to their management system of elements of adaptation and artificial intelligence. Such mobile robots with adaptive control can automatically adapt to unpredictable changes of a production situation and operating conditions: different illumination levels, vibrations, differences in arrangement of details, the need for operational reorganization of routes, etc. Such robots essentially differ in powerful

A. V. Samsonovich et al. (Eds.): BICA 2020, AISC 1310, pp. 91–96, 2021.
https://doi.org/10.1007/978-3-030-65596-9_12

information system and the software, that allows management system to plan technological operations and to make optimal solutions, to perceive and to react quickly to changes in the work area, to analyze a situation and to distinguish objects, to program the operation of the equipment and to adjust managing programs, to diagnose faults and to prevent faults.

The solution of a task of intellectual control of the mobile robot is impossible without the corresponding information support realized by an information system. The information from various sensors characterizes current state of the equipment of the mobile robot therefore it is used in the system of automatic control as the feedback. The feedback signals allow to automatically correct the operating programs to ensure the stability in work of a production system in general. The information from various (both external and internal) sensors is also used for self-checking and self-diagnostics of mobile robot conditions.

At the same time, optimal solutions to control the mobile robot have to be developed. Input information is given to the operating system on the basis of which the optimal solution is developed. The set of such solutions can be written down in the form of productional rules: either in advance formulated by experts, or received later in the course of adaptation performing procedures [9, 11].

2 General Problem Statement

Let's give the general problem statement [4, 7]:

Let $W = \{w_1, ..., w_m\}$ - be the set of the objects forming the training selection (TS). Various subjects to recognition can act as representatives of TS: graphic (or sound) images, situations that require certain responses of the operating system, diagnostic messages and so on. Descriptions of the objects comprising TS *W* represent *n*-component vectors where n is a number of the signs applied to characterize the analyzed objects, and *j*- coordinate of these vectors equals to *j*- sign, *j* = *1*..., *n*. The number of such vectors is equal to m according to the number of the analyzed objects in TS. In the description of an object, the lack of information on value of this or that sign is admissible. Compliance of a set of objects and the signs characterizing them is presented in the following Table 1.

Table 1 General data structure.

x_1	x_2	...	x_n	W
$x_1(w_1)$	$x_2(w_1)$	...	$x_m(w_1)$	w_1
$x_1(w_2)$	$x_2(w_2)$	...	$x_m(y_2)$	w_2
...	...	...	...	...
$x_1(w_m)$	$x_2(w_m)$	...	$x_n(w_m)$	w_m

$$X_j = \{x_1(w_j), x_2(w_j). \ldots . x_n(w_j)\}$$

– a vector of qualitative characters, where each element is the fixed sign of the object, is a part of TS.

$$W = \bigcup_{j=1}^{m} w_j$$

– a set of the characterized objects.

Each corresponding sign of *xj (wi)* is coded by a various-valued predicate of atomicity of *kj, j = 1, ..., n.*

$$f(x) = \wedge_{j=1}^{m}(\wedge_{i=1}^{n} x_i(w_j) \rightarrow w_j) = \wedge_{j=1}^{m}(\vee_{i=1}^{n} \overline{x_i(w_j)} \vee w_j).$$

The specified type of function follows from the known logical identity: $a \rightarrow b = \bar{a} \vee b$, where *a*- is a conjunction of characteristics (signs) defining an object, and b- the predicate equal to unity, when w_j becomes equal to the corresponding defined object. We will call such predicates object predicates, and the disjunctors containing such predicates - productional disjunctors.

The expediency of application of the various-valued predicates follows from the fact that the adequate description of various characteristics requires predicates of various atomicity. For example, two-unit predicate for coding of the fact of presence (or absence) of a certain property in the description of the considered objects is enough. The three-digit predicate is convenient when we speak about various extents of sign presence in an object: "No", "Partially", "Present".

The set of productional rules of the following type stands as a basis for creation of various-valued logic functions *F (X, W)* where the *X*-set of signs, and *W*-the objects of TS:

(Conjunction of the signs_1 → $Object_1$),
(Conjunction of the signs → $Object_2$),
...
(Conjunction of the signs _m → $Object_m$).

The set from m productional rules by means of *m*−*1* conjunction operation forms one logical expression which makes various-valued logic function of *F (X, W)* [7]. Consecutive disclosure of m brackets in [7] operation of the generalized denial leads to creation of various-valued logic function of *F (X, W)* where *X* - the n-component vector of logic variables coding the set of signs of objects: $X = (x_1, x_2, \ldots, x_n)$, and $W = \{w_1, \ldots, w_m\}$ – a set of the characterized objects of power m, that form the training selection.

As a result of this conversion function *F (X, W)*, takes a form of a disjunction of subclasses of objects, each of which represents conjunction of signs and objects on which these subclasses are formed.

Finding the value of the logic function of *F (X, W)* from the set of characteristics of *X* * as a final result we have a disjunction of those subclasses of objects from a set of *W*, values of variables of which match the values of variables in a vector of a request of $X*$. Let's notice that at the resultant expression there will be also objects which satisfy to a request of $X*$ in number of signs, smaller, than n. Here, the more numbers of variables from a request of $X*$ corresponds object $w*$, belonging to W, the bigger the number of subclasses with its participation will be in the final answer of $W*$. Therefore, to identify the best object which is most satisfy the request of $X*$ it is enough to apply the procedure of frequent analysis to the received set of subclasses of $W*$.

3 Logical Neural Networks

In the work [2] circuitry approach for creation of logical neural network was offered. Relationships of the cause relationships between the sets of entrance signals that get on an entrance layer of neural network by aggregate incoming signals and by made decisions are established for this purpose (Fig. 1). In this case the intellectual system of decision-making is presented in the form of the following set of implicative statements:

Aggregate of the signals_1 → Solution_1,
Aggregate of the signals_2 → Solution _2,
...
Aggregate of the signals_ m → Solution_m.

Let's notice that generally one and the same solution can follow from different sets of input signals.

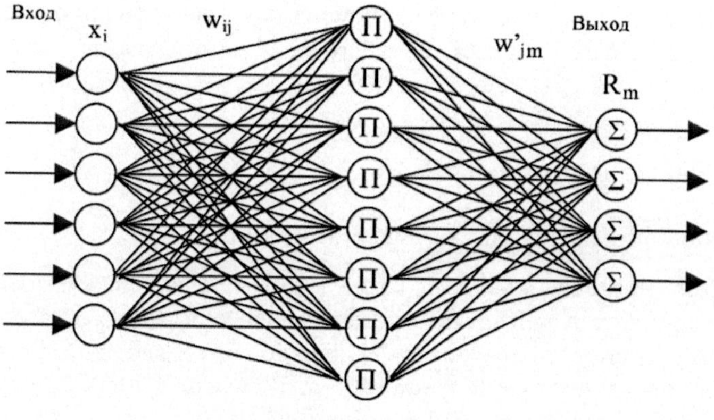

Fig. 1. Logical neural network

At the same time it is supposed that calculations in each of m implicative expressions are made in parallel according to features of neural network functioning. As output result the set of accurately (or it is indistinct) weighed solutions is formed. While constructing the logical neural network the corresponding productional rules were used. Each set of entrance signals is considered as conjunction of the corresponding logic variables describing the initial situations acting as the training selection for decision making system. For the creation of logical neural network two types of neurons are used: conjunctors and disjuntors. The weight of input signals of these neurons are selected so that at accurate values of input signals the results of functioning of these neurons correspond the values of logical actions of conjunction and disjunction.

4 The Plotting of the Various-Valued Logical Neural Network

We see that one and the same system of productional rules (implicative statements) is a basis as for construction of the various-valued of logical functions *F (X, W),* so for creation of the logical neural networks trained for creation of intellectual control systems. The situations characterized by indications of external and internal sensors act as an entrance vector of *X* here, and the optimal solutions corresponding to these situations act as subjects to recognition.

The following theorem is proved:

Theorem 1. Any various-valued logical function *F (X, W)* is representable in the form of logical neural network, the set of logical communications in which is defined by the structure of productional disjunctors.

Thus, various-valued logical function *F (x, W)* it is representable in the form of the three-layer neural network, where:

1. The entrance layer are the predicates x_j, of the k_j value, each $j = 1, ..., n$.
2. The set of objects or output signals of $W = w_1, ..., as\ w_m$ is the output layer that makes final calculations.
3. Productional disjunctors in which conjunction operations and disjunctions replaced by neurons conjunctors and disjunctors act as the intermediate, ulterior layer that make calculations.
4. Such sets of input signals of logical neural network to which no output signals are compared meet free knowledge of various-valued logic function. In an explicit form free knowledge of logical neural network does not exist. However, free knowledge can be connected to a special circuit, and at rather high potential of a signal on this line the procedure of a finish-teaching of the current logical neural network is started [4].
5. The neurons that send the output signal on those objects which enter the corresponding subclasses are compared to productional disjunctors of logic function.

In case of accurate values of variables any conjunctor corresponding to a productional disjunctor of various-valued logic function of *F (X, W),* transmits a signal of single weight to each of the objects entering this disjunctors (subclass).

If to give values of logic variables in compliance with some request of *X* * on an input of the neural network built this way then at the output the set of the objects satisfying to this request which weight in accuracy match the results of work of the procedure of frequency analysis applied to find the objects which are most fully answer the request of $X*$ while finding the various-valued logic function of $F(X, W)$: $W* = F(X*, W)$.

Obviously, the results of the proof of Theorem 1, as well as the method of its proof and the algorithm for constructing a logical neural network, can be completely transferred to the case of an extended logical neural network, making some changes to them.

The transition to fuzzy logic is provided by the use of various-valued predicates and flexibility conjunctors and disjunctors. Additional expressiveness of the artificial system of recognition or decision-making gives the possibility of using fuzzy logic in the conditions of noisy or fuzzy source data.

Acknowledgments. The reported study was funded by RFBR according to the research project №18-01-00050-a.

References

1. Asada, M.: Towards artificial empathy. Int. J. Soc. Rob. **7**, 19–33 (2015)
2. Barsky, A.B.: Logical neural networks. Intuit, Binom (2007)
3. Chan, M.,T., Gorbet, R., Beesley, P., Kulic, D.: Curiosity based learning algorithm for distributed interactive sculptural systems. In: 2015 IEEE/RSJ International Conference on Intelligent Robots and Systems (IROS), pp. 3435–3441 (2015)
4. Dimitrichenko D.P., Zhilov R.A.: The use of a neural network approach to the tasks of logical data processing and the construction of intelligent decision-making systems. Model. Optim. Inf. Technol. t.6 **H2**(20), 249–261 (2018)
5. Floreano, D., Wood, R.J.: Science, technology and the future of small autonomous drones. Nature **521**, 460–466 (2015)
6. Kappassov, Z., Corrales, J.A., Perdereau, V.: Tactile sensing in dexterous robot hands—review. Rob. Auton. Syst. **74**, 195–220 (2015)
7. Lyutikova, L.A.: Modeling and minimization of knowledge bases in terms of multi-valued logic of predicates, p. 33. Nalchik (2006)
8. Mavridis, N.: A review of verbal and non-verbal human–robot interactive communication. Rob. Auton. Syst. **63**, 22–35 (2015)
9. Miconi, S.V.: Axiomatics of multicriteria optimization methods on a finite set of alternatives. In: SPIIR Proceedings, vol. 44, pp. 198–214 (2016)
10. Timofeev, A.V.: Adaptive robotic systems. Mechanical Engineering, p. 332 (1988)
11. Hu, Z., Ma, X., Liu, Z., Hovy, E., Xing, E.: Harnessing deep neural networks with logic rules. In: 54th Annual Meeting of the Association for Computational Linguistics, ACL 2016 - Long Papers, vol. 4, pp. 2410–2420 (2016)

The Electroencephalogram Based Classification of Internally Pronounced Phonemes

Yuliya Gavrilenko, Daniel Saada, Eugene Ilyushin, Alexander V. Vartanov(✉), and Andrey Shevchenko

Lomonosov Moscow State University, Moscow, Russia
a_v_vartanov@mail.ru

Abstract. The internal speech recognition is a promising technology, which could find its use in brain-computer interfaces development and greatly help those who suffer from neurodegenerative diseases. The research in this area is in its early stages and is associated with practical value, which makes it relevant. It is known that internal pronunciation can be restored according to electroencephalogram data because it allows one to register specific activity associated with this process. The purpose of this work is to build and implement an algorithm for extracting features and classifying Russian phonemes according to an electroencephalogram recorded during the internal pronunciation of the phonemes. This kind of research is actively conducted abroad; however, there is no information about such works for the Russian language phonemes in open sources at the moment. In the course of the work, an algorithm for extracting features and classifying the internal pronunciation of Russian phonemes was built and tested, the accuracy of which showed results comparable with other studies.

Keywords: Internal pronunciation · Brain-computer interface · Neurointerface · EEG

1 Introduction

Brain-computer interface (BCI) is a technology designed to transfer information from the human brain to a computer. One of the most critical areas of application of BCI is to help people who are deprived of the opportunity to interact with the outside world using traditional methods of communication. For example, many people suffering from neurodegenerative diseases face such problems. Such devices are based on technology that allows register the bioelectric brain activity. One of the most popular technologies of this kind is electroencephalography (EEG), a non-invasive method of electromagnetic study of the brain, which registers the signals arising from the activity of neurons, and also has an excellent temporal resolution. The other methods that can be used in this task are electrocorticography (ECoG), functional magnetic resonance imaging (fMRI), but they are hard to apply in experimental stage. Even though ECoG can allow great spatial and temporal resolution, it is an invasive method, which means that the sensors have to be put on human brain directly during surgery.

A. V. Samsonovich et al. (Eds.): BICA 2020, AISC 1310, pp. 97–105, 2021.
https://doi.org/10.1007/978-3-030-65596-9_13

Existing BCIs, which are associated with speech production, have a rather low text input speed, which makes them inconvenient in everyday use. For example, Speller P300 [1], which is based on evoked potentials recording, allows entering no more than 20 characters per minute. Silent access interfaces - BCIs, which are based on the recognition of internal speech, could solve this problem. Internal pronouncing is a silent (unpronounceable) speech that arises in the process of thinking and preserves the structure of regular speech [2]. It is known that internal speech can be recognized, as when anticipating the pronouncing of a phoneme in some areas of the brain responsible for reproduction and speech perception, there is a specific activity that can be registered using methods such as EEG [3].

To develop such an interface, it is necessary to solve such tasks as setting and conducting an experiment, in which recordings of brain activity will be obtained during the internal pronunciation of specific phonemes, preliminary processing of the collected data and feature extraction, as well as classification of phonemes based on extracted features.

Silent speech interfaces development is still in the experimental stage and many possible applications of it make this study even more promising. As it has been already mentioned, spelling devices such as P300 Speller, which can be used by ALS (amyotrophic lateral sclerosis) patients, have shown their efficiency, but are quite slow and cannot be applied to the regular speech. P300 Speller lets users spell the word while focusing on a symbol, which is shown in a matrix grid on a display, while rows and columns are being randomly highlighted. When a row or a column, which contains a required symbol, is highlighted, an event-related potential P300 is called and by registering this potential, the BCI is able to recognize the symbol, which the user was intending to spell. In order to use this kind of BCI users have to learn how to use it and acquire new skills. In case of silent speech interface based on internal pronunciation recognition, everything that user would need to do in order to use it is to think of a desired letter.

Aside from helping people with paralysis causing diseases, such devices might also be helpful in military industry when any other type of communication is undesirable and there is a need to transmit messages at a distance without using visual or sonic methods. For example, DARPA invested $4 Million in synthetic telepathy device development project.

Silent speech recognition BCI could also be used in entertainment industry, e.g., in virtual reality devices, which could help increase the quality of game experience.

2 Review of Existing Research

Existing studies in this area differ in the recognized phonological category so that they can be divided into three types: recognition of phonemes, recognition of vowels, and recognition of words [4].

An experiment on the recognition of two phonemes /ba/, /ku/, which were pronounced according to three rhythms, was conducted in [5]. The purpose of the experiment was to calculate matched filters using signal envelopes based on the data obtained. The activity was calculated for each of the rhythms of the brain separately.

The Hilbert transform was applied to the data in order to obtain the signal envelopes. The further classification was carried out according to the obtained matched filters. Accuracies for all four subjects differed significantly. Thus, for one of the subjects, the recognition accuracy among the six states was 62% in the beta range, while for the other subject accuracy of 87% was obtained in the same range.

One of the experiments on the recognition of vowels was carried out in [6]. The vowels /a/, /i/, /u/, /e/, /o/ were used for classification. Statistical parameters of the signal were used as features: an estimate of the average values, variance, and standard deviation, the estimate of the average power. A neural network with two hidden layers was used for classification, which was trained using the backpropagation method. The input layer consisted of 4 nodes, which corresponded to 4 features; the output layer consisted of 5 nodes, corresponding to 5 recognizable vowels. The average classification accuracy was 44%.

Recognition of words is a more difficult task in comparison with the recognition of more abstract phonological categories since different subjects can endow imagined words with a different meaning. Therefore, the nature of brain activity for different subjects can vary considerably. In [7], an attempt was made to recognize the internal pronunciation of the five Spanish words "arriba", "abajo", "izquierda", "derecha", and "seleccionar". Those EEG electrodes, which were located in the Wernicke zone, were selected for the experiment. The wavelet transform was used to extract the features; a second-order Daubechies wavelet was used as the maternal wavelet function. The obtained transform coefficients for each of the channels were used as a feature vector. As a result of the training of three classifiers (naive Bayes classifier, Random Forest, SVM), accuracy estimates were obtained. For the naive Bayes classifier, the accuracy was 23%, 17% and 36% for three subjects, for Random Forest and SVM 24%, 32%, 41% and 23%, 25%, 18%, respectively.

Also, in the context of the problem under consideration, two different approaches to the formulation of the experiment can be distinguished depending on whether the imagining of articulation during the internal pronunciation is included in the cognitive task.

The experiment with internal pronouncing of the phonemes without imagining the movement of the speech organs allows one to apply the research results in the construction of prostheses. One of the options for setting up an experiment without imaginary articulation was proposed in the study [8], in which the task was to separate mental states in representing three classes of vowels: open (/a/, /o/), intermediate (/e/) and closed (/i/, /u/). There were 21 electrodes located near the Wernicke and Broca regions. Power spectrum analysis was applied for the recognition of noisy signals. When using the nonlinear SVM method, classification accuracy between 84% and 94% was obtained.

Experiments with imaginary articulation as a part of a cognitive task are more popular. In [9], it was demonstrated that while imagining the face movements, neurons are activated in the motor cortex, which can be used to control the prosthesis. This fact is often used in the construction of the BCI, and also suggests that an imaginary phoneme articulation may help improve the classification accuracy. An experiment on the recognition of syllables /a/and /u/was conducted in [10]. The activation of motor cortex neurons together with the internal pronunciation of a vowel was classified using the common spatial pattern method and the nonlinear classifier. The resulting accuracy ranged from 68% to 78%.

3 Experiment

An experiment was conducted in the psychophysiology laboratory of the Faculty of Psychology of the Lomonosov Moscow State University to obtain data from internal pronunciation. The study involved 12 subjects: 3 females and 9 males aged 20 to 25 years (average age – 22 years). All subjects did not have any traumatic brain injury or mental illness and were right-handed. Two types of experiment were conducted:

1. An experiment with the visual stimuli presentation;
2. An experiment with the presentation of sound stimuli.

In each of the performed experiments, the cognitive task consisted of prolonged internal pronunciation of a certain phoneme without imaginary articulation. For the experiment, we selected seven phonemes of the Russian language from different subgroups of the Plotkin classification [11]: /a/, /b/, /f/, /g/, /m/, /r/, /u/.

To present a visual stimulus, the Presentation program (version 18.0 of Neurobehavioral Systems, Inc.) was used, which laid out the necessary algorithms for giving stimulus material. For each phoneme an image of the PNG format, which contained the recording of this phoneme, was made using an image editor. The received files were loaded into the "Presentation" software complex and presented in random order.

For the presentation of the sound stimulus, subjects used headphones and listened to the pre-recorded phonemes required for. A 19-channel Neuro-KM electroencephalograph ("Statokin" company, Russia), with a sampling rate of 1000 Hz was used to register the electrical activity of the brain. The electrodes were located according to the 10–20% international system, a detailed description of which can be found in [12]. BrainSys (BrainWin), version 2.0 was used for recording and viewing EEG. The scheme of the experiment corresponds to the standard schemes described in similar studies:

1. The subject is shown a visual or audible stimulus for 700 ms, which is an instruction for internal speaking;
2. After the stimulus is presented, a fixation cross is displayed on the screen on which the subject should focus his gaze;
3. During 1500 ms the screen displays a fixation cross and the subject performs an internal pronunciation of the phoneme;
4. The fixation cross disappears from the screen and a rest period begins, during which the subject takes no action and prepares for the next stimulus;
5. After 1600 ms, the rest is completed, and a new stimulus is presented to the subject.

3.1 Preprocessing and Feature Extraction

As the first preprocessing stage, channel selection is performed. According to the experimentally confirmed hypotheses and the results of studies [3] and [13], the activity in the Broca and Wernicke zones is inherent in the activity of internal pronouncing. According to this, it is proposed to use EEG data from four electrodes located near these areas: F7, F3, T3, C3 (Fig. 1).

During the recording, a Butterworth bandpass filter with a cutoff frequency of 3 to 30 Hz was applied to the data, which made it possible to get rid of excessively high frequencies that are not related to internal pronunciation but can be caused by some recording artifacts. It also allows clearing the recording from low-frequency electronic noise.

The next step is to split the record into multiple samples. For this purpose, timestamps are used, which were automatically created during the experiment. First, the marks that correspond to the beginning of the internal pronunciation of the phoneme are selected. After that, for each label, a corresponding sample is created with a length of 800 ms, the beginning of which is the moment preceding the internal pronouncing of 200 ms. Thus, for each of the seven phonemes, an array of samples is created, cleared of artifacts and ready for feature extraction.

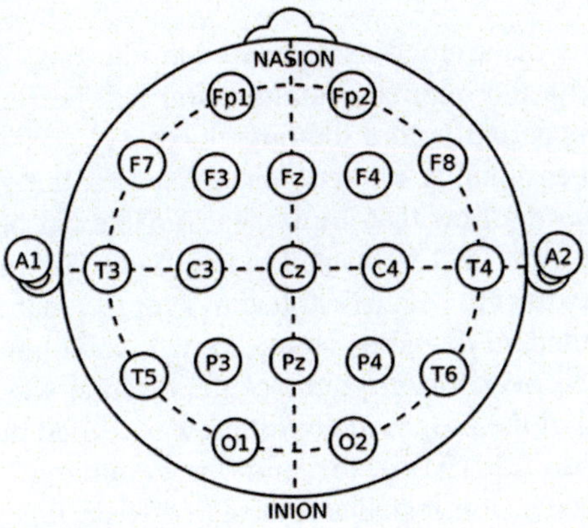

Fig. 1. The location of the electrodes according to the 10–20% system

Wavelet transform was applied to the received vectors. The Morlet complex wavelet was used as the maternal wavelet function, which is the scalar product of a sine wave to Gaussian. The wavelet of this type was chosen because it is well localized in time and frequency. As a result of the transformation, the matrices of the signal energy distribution in time and frequency were obtained for each section of internal pronunciation (Fig. 2).

The time-frequency power of the EEG signal is distributed according to power-law, so the power at higher frequencies has a smaller value than the power at low frequencies. Because of this, the comparison of the dynamics of activity in different frequency ranges is challenging to perform. To eliminate the described effect, it is possible to use the baseline normalization, which converts the data at different frequencies to a single scale, and also leads all the numerical values of the power to a single metric [14]. Also, since the signal segment that is localized before the start of

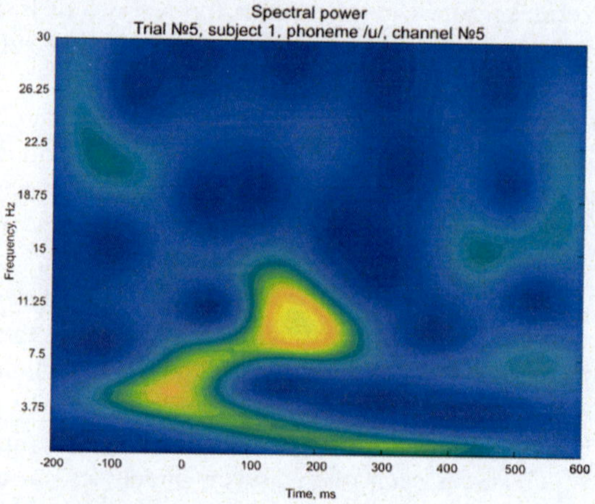

Fig. 2. Spectral power for one of the measurements of internal pronunciation

internal pronunciation is determined as the baseline, the time-frequency dynamics of the signal associated with the internal pronunciation will be separated from the background dynamics that occurred before the target activity.

To reduce the dimensionality of the feature space, the normalized values are averaged over four brain rhythms that are typical for internal pronunciation: delta (1–4 Hz), alpha (4–8 Hz), theta (8–13 Hz) and beta (13–40 Hz) rhythms.

The final step is to extract the statistical features for the four analyzed channels: the values of the median, standard deviation, sum, average, minimum and maximum of the normalized and averaged over four rhythms of the discrete wavelet transform.

The implementation of the proposed algorithm was carried out using the MATLAB software package (release R2019a) in the same programming language with the SignalProcessingToolbox extension installed and the edfRead function [15].

3.2 Feature Selection and Classification

Each sample in the resulting feature space is described by 19,200 characteristics. Such a high dimension prevents the effective classification and rather dramatically complicates the process of solving the problem since with an increase in the dimension, the time required for training the classifier increases. In this regard, it is proposed to use feature selection methods.

Considering the fact that the internal pronunciation of the phoneme generally does not have a strict localization in time and can be biased, the selection must be made among six statistical features, while leaving the data on all 800 trials of each sample. It is proposed to use recursive feature elimination (RFE) – a feature selection method, which is often used in biomedical research and proposed in [16].

The SVM method allows evaluating the individual features via their weights and can be used in conjunction with RFE. This combination gives us the SVM-RFE algorithm, which uses weights as a feature ranking criterion.

To implement the algorithm described above, the Python programming language version 3.7 was chosen, as well as the SciPy [17] and scikit-learn [18] libraries. The first one is used to work with .mat files, in which variables and data structures of the MATLAB system are stored, which are extracted features. The second library includes many standard machine learning algorithms. As an implementation of the support vector machine, the LIBSVM library [19] was used.

The ranking list obtained as a result of the SVM-RFE algorithm was used to search for the best subspace of the original feature space using the cross-validation method. Given the small size of the training sample, the Leave-One-Out (LOO) strategy was chosen, the essence of which is to exclude each existing precedent from the training sample and to use it as a test sample. The classificator was also trained on the LOO strategy.

4 Results

After applying the SVM-RFE algorithm to the extracted features, the following ranked list was obtained, in which the significance of the feature decreases in the following order: mean value, median, standard deviation, maximum, sum, minimum. Using the cross-correlation method it was determined that it is the best to use the first three features from this list.

When compiling an array of samples from recordings of experiments with different people, the classification accuracy tends to be random, which is comparable with the studies of other authors. However, when classifying internal pronunciation for one subject only, the results are significantly improved. Below are the results of the pairwise classification of different phonemes for two different subjects (Table 1 and 2).

The results of the classification are quite different depending on the choice of the subject and phonemes. The average accuracy for all subjects and all phonemes was 67%. There was no noticeable difference between sound and visual presentation of the stimuli.

Table 1. The results of the pairwise classification of phonemes for the subject №1

	/b/	/f/	/g/	/m/	/r/	/u/
/a/	0.71	0.48	0.65	0.75	0.63	0.56
/b/		0.47	0.79	0.77	0.67	0.51
/f/			0.61	0.68	0.68	0.74
/g/				0.77	0.81	0.74
/m/					0.63	0.78
/r/						0.67

Table 2. The results of the pairwise classification of phonemes for the subject №2

	/b/	/f/	/g/	/m/	/r/	/u/
/a/	0.42	0.53	0.53	0.69	0.60	0.36
/b/		0.57	0.62	0.54	0.62	0.66
/f/			0.47	0.63	0.45	0.57
/g/				0.68	0.64	0.38
/m/					0.74	0.36
/r/						0.54

5 Conclusion

Due to the data recordings monitoring during the experiment, input data does not initially contain artifacts of blinking and muscle movement during the execution of a cognitive task. This method of data cleansing does not allow to use the constructed algorithm for real-time classification but provides proper data filtering from artifacts. At the current stage of internal pronunciation studies development, such an approach is quite acceptable, and the constructed algorithms can later be adapted for online classification.

As it can be seen from the results, the classification accuracy is comparable with the studies of other authors – studies, reviewed earlier, provide accuracy that is alike. It is still not high enough to attempt to create an internal pronunciation based BCI. As we suppose, one the main reasons for the low accuracy is the standard setup of the experiment. Due to the small number of electrodes and installation according to the standard "10–20%" system, only four sensors turned out to be in the area of interest to the brain. We suppose that slight modification of the setup considering the experiment features could help to increase the recognition accuracy. In future work we are planning to use the 64-channel EEG in order to locate more sensors in the area of interest of the brain. In addition, the use of the fixation cross as an incentive to start internal speaking was a relatively stable source of noise signals. In order to eliminate this drawback, instead of a fixation cross, it is proposed to use a light sound stimulus (for example, a click) and to conduct the experiment entirely with closed eyes in the further work.

Even though in the definition of the experiment phonemes from various subgroups of the Plotkin classification are used, on average some pairs are discriminated better than the others. The causes of this phenomenon cannot be determined unequivocally at the moment. However, it can be assumed that phonemes from different groups are discriminated better than phonemes from different subgroups of the same group. As an example, the pairs /g/-/r/, /a/-/m/ and some others can be given. In the subsequent work, it is planned to increase the stimulus material and to include words in it.

The algorithm that is used in this article can be applied to other tasks related to BCI construction. Further works on studying of various brain areas performance during intellectual problems solution are planned to be based on this research.

Acknowledgements. The research is financially supported by the Russian Science Foundation, Project №20-18-00067.

References

1. Farwell, L.A., Donchin, E.: Talking off the top of your head: toward a mental prothesis utilizing event-related brain potentials. Electroencephalogr. Clin. Neurophysiol. **70**(6), 510–523 (1988)
2. Hashim, N., Ali, A., Mohd-Isa, W.-N.: Word-based classification of imagined speech using EEG. In: Computational Science and Technology, vol. 488, pp. 195–204 (2017)
3. Callan, E.D., Callan, A.M., Honda, K., Masaki, S.: Single-sweep EEG analysis of neural processes underlying perception and production of vowels. Cogn. Brain. Res. **10**(1–2), 173–176 (2000)
4. Gavrilenko, Y.Y., Saada, D.F., Shevchenko, A.O., Ilyushin, E.A.: A review on internal pronouncing recognition methods based on electroencephalogram data. Mod. Inf. Technol. IT-Educ. **15**(1), 164–171 (2019)
5. D'Zmura, M., Deng, S., Lappas, T., Thorpe, S., Srinivasan, R.: Toward EEG sensing of imagined speech. Lecture Notes in Computer Science, vol. 5610, pp. 40–48 (2009)
6. Kamalakkannan, R., Rajkumar, R., Madan, R.M., Shenbaga, D.S.: Imagined speech classification using EEG. Adv. Biomed. Sci. Eng. **1**(2), 20–32 (2014)
7. Torres Garcia, A.A., Reyes Garcia, C.A., Pineda, L.V.: Toward a silent speech interface based on unspoken speech. In: Proceedings of the International Conference on Bio-inspired Systems and Signal Processing (2012)
8. Sarmiento, L.C., Lorenzana, P., Cortes, C.J., Arcos, W.J., Bacca, J.A., Tovar, A.: Brain-computer interface (BCI) with EEG signals for automatic vowel recognition based on articulation mode. In: 5th ISSNIP-IEEE Biosignals and Biorobotics Conference (2014): Biosignals and Robotics for Better and Safer Living (BRC) (2014)
9. Pfurtscheller, G., Neuper, C.: Motor imagery and direct brain-computer communication. Proc. IEEE **89**, 1123–1134 (2009)
10. DaSalla, C.S., Kambara, H., Sato, M., Koike, Y.: Single-trial classification of vowel speech imagery using common spatial patterns. Neural Netw. **22**(9), 1334–1339 (2009)
11. Plotkin, V.: Fonologicheskiye kvanty. Nauka, Moscow (1993)
12. Klem, G.H., Lüders, H.O., Jasper, H.H., Elger, C.: The ten-twenty electrode system of the International Federation. The International Federation of Clinical Neurophysiology. Electroencephalogr. Clin. Neurophysiol. **52**, 3–6 (1999)
13. Suppes, P., Lu, Z.-L., Han, B.: Brain wave recognition of word. Proc. Natl. Acad. Sci. U.S. A. **94**(26), 14965–14969 (1997)
14. Cohen, M.X.: Analyzing Neural Time Series Data: Theory and Practice, p. 578. MIT Press, Cambridge (2014)
15. edfRead Function – MATLAB Central, February 2018. https://mathworks.com/matlabcentral/fileexchange/31900-edfread. Accessed 5 Aug 2019
16. Guyon, I., Weston, J., Barnhill, S., Vapnik, V.: Gene selection for cancer classification using support vector machines. Mach. Learn. **46**(1–3), 389–422 (2002)
17. SciPy.org (2019). https://www.scipy.org. Accessed 5 Aug 2019
18. Scikit-learn: machine learning for Python (2019). https://scikit-learn.org. Accessed 5 Aug 2019
19. Chang, C.C., Lin, C.J.: LIBSVM: a library for support vector machines. ACM Trans. Intell. Syst. Technol. (TIST) **2**(3), 27 (2011)
20. Brookshire, G., Casasanto, D.: Motivation and motor control: hemispheric specialization for approach motivation reverses with handedness. PLoS ONE **7**(4), e36036 (2012)

"Loyalty Program" Tool Application in Megaprojects

Anna Guseva[1(✉)], Elena Matrosova[1], Anna Tikhomirova[1], and Nikolay Matrosov[2]

[1] National Research Nuclear University MEPhI (Moscow Engineering Physics Institute), Moscow, Russia
aiguseva@mephi.ru, matrosova_ev@inbox.ru, anna7909966@ya.ru
[2] IE Abubekarov, Moscow, Russia
matrosovnikolay444@yandex.ru

Abstract. The article is devoted to the problem of improving the implementation efficiency and megaprojects management. An example of an adaptation one of the existing project management methodological tools, namely the "loyalty program", for projects of a global nature (megaprojects) is given. The authors propose the use of elements of the decision-making process theory and the method of T. Saati hierarchy analysis method for the analysis of megaproject key characteristics. Based on them, recommendations for the formation of a loyalty program are given.

This work was supported by RFFI grant № 20-010-00708\20.

Keywords: Megaproject · Loyalty program · Decision-making process · Automated decision support system

1 Introduction

Throughout the entire economic relations and entrepreneurship development, work has been done constantly to find effective approaches to management at all levels such as the level of an individual company, the level of the whole industry, or the state level. The concept of project management at the company level began to spread in the 1990s. In recent years, the approaches used in project management have spread to all levels of management, including the state level of management. The variety of project variations has led to the need to classify them according to the most important characteristics. It is also important to take into account the features of using the existing project management tools.

The choice of a specific tool and the nature of its adaptation directly depends on the complexity of the project, which is determined by its scale and type. According to these indicators, all projects can be divided into three classes [1]:

- megaprojects;
- multiprojects;
- monoprojects.

A. V. Samsonovich et al. (Eds.): BICA 2020, AISC 1310, pp. 106–114, 2021.
https://doi.org/10.1007/978-3-030-65596-9_14

In this article we consider megaprojects. A megaproject is currently understood as an investment project of an extremely large size. Megaprojects are classified by the levels of programs and by their type (Fig. 1).

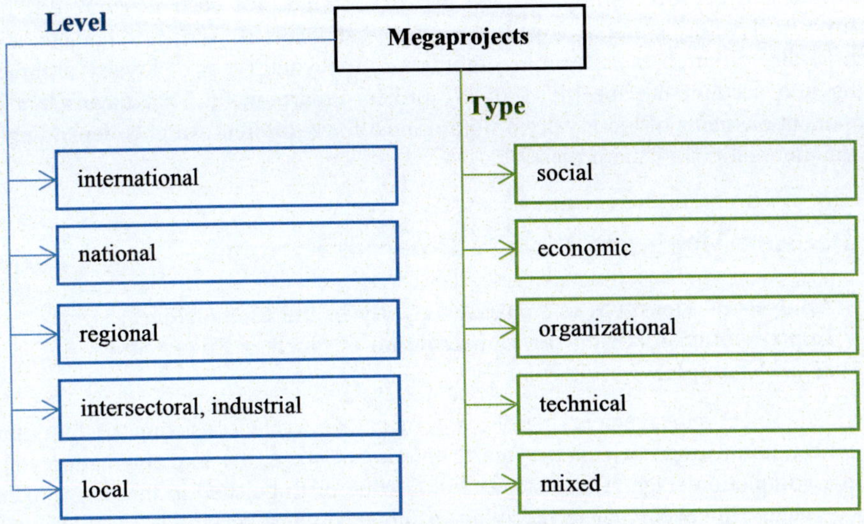

Fig. 1. Megaprojects classification

Currently, a set of methods and tools for project management is not well developed and systematized at the megaproject level. However, due to the wide distribution of megaprojects and the apparent prospect of using the project approach in solving global management problems, there is an imperative need to find suitable tools [2].

In order to solve this problem we need to analyze the existing and approved tools for projects of a lower level, which has already established itself as an effective tool. This instrument is the loyalty program tool, which is quite widely used today.

According to the existing definition, "loyalty" is a faithful, trustworthy attitude towards someone or something. A loyalty program in the business environment means a set of measures aimed at increasing the attractiveness of a company's goods or services. In most cases, a loyalty program includes providing a reward system for the company's customers. Loyalty programs are classified by type:

- discount;
- bonus;
- multi-level;
- paid subscription;
- with intangible remuneration;
- partnership;
- with gamification elements;
- non-profit/charitable.

Uniqueness and exclusivity are ones of the distinctive characteristics of megaprojects. Some megaprojects have been implemented just once. Others can be re-implemented, but with significant modifications. In this regard, some types of loyalty programs may not be generally suitable for use at the mega-level. However, there are types that can be successfully adapted and used. Thus, for each megaproject, it is possible to develop a unique loyalty program that meets its specific characteristics. Such loyalty program is built on a combination of several types of loyalty programs, taking into account the specificity of the project environment. That means that the component elements of the loyalty program must have different weights depending on the unique qualities of the project.

2 Decision-Making in Loyalty Program Forming

2.1 Analysis of Measures to Increase Loyalty in the Megaproject Implementation (Using the Construction of Nuclear Power Stations as an Example)

Classical project management implies a project life cycle consisting of four main stages: the initial stage or concept, the development stage, the implementation stage, and the completion stage. These stages can also be distinguished in the megaprojects management. However, megaprojects are a more complex economic mechanism. In this regard, for each megaproject, it is necessary to allocate detailed specific stages which are characteristic only for this particular megaproject. Figure 2 shows a comparison of the main stages and specific ones on the example of a megaproject which is the construction of nuclear power stations [3].

To develop a loyalty program at the megaproject level, it is necessary to clearly distinguish the set of the most significant key objects involved in the project. It is also necessary to determine the set of activities that are being performed or can be performed in order to reduce potential negative risks. There are important categories of people involved in the consumption process of the megaproject accomplishment. Figure 3 shows the main elements that can have a positive impact on them as well as increase their level of loyalty to the megaproject.

These elements can be distributed among four main stages of the megaproject life cycle.

Stage 1. Initial stage or concept:

- make a fundamental decision on the construction of a nuclear power station,
- create a concept for the development of nuclear and power infrastructure,
- set a structure of nuclear projects implementation financing,
- ensure the acceptance of nuclear energetics by the public.

Stage 2. Development:

- create the necessary national legislative framework,
- work on the development of nuclear and power infrastructure,
- implement the structure of nuclear projects financing,

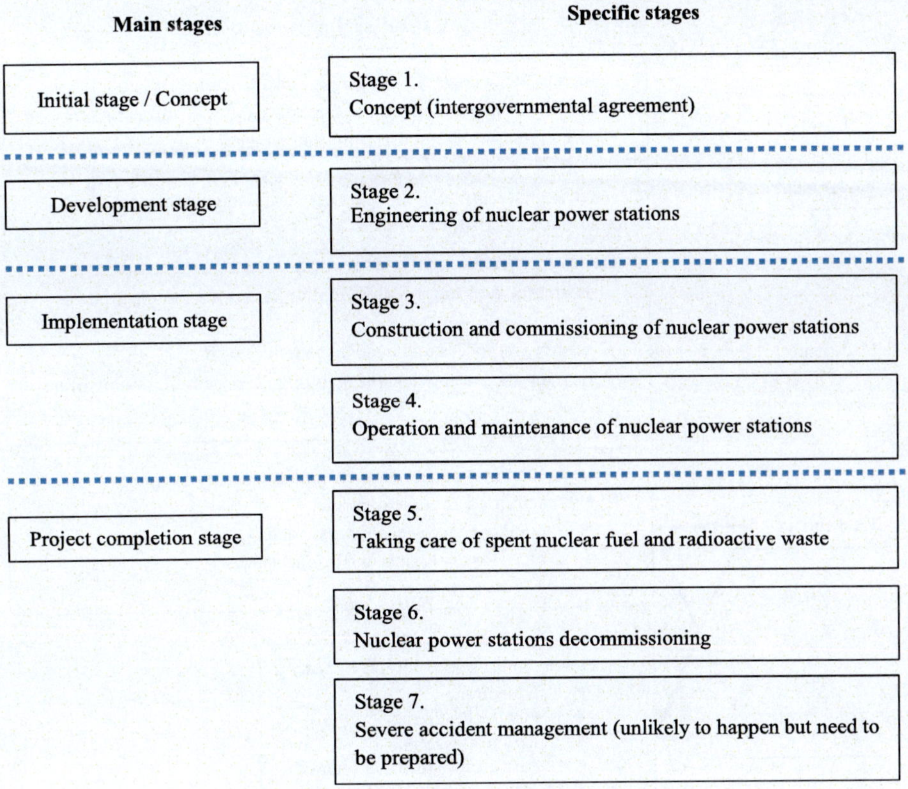

Fig. 2. Description of project stages

- retrain national personnel,
- ensure the acceptance of nuclear energetics by the public.

Stage 3. Implementation:

- create the necessary national legislative framework,
- work on the development of nuclear and power infrastructure,
- retrain national personnel,
- arrange the production of the necessary equipment on the customer-country territory,
- guarantee fuel supply for the entire life cycle of nuclear power stations,
- provide a nuclear and radiation safety system,
- develop an emergency preparedness and response system,
- create opportunities for development:
  - education,
  - new technologies,
  - new professions,
  - new sectors of the economy,

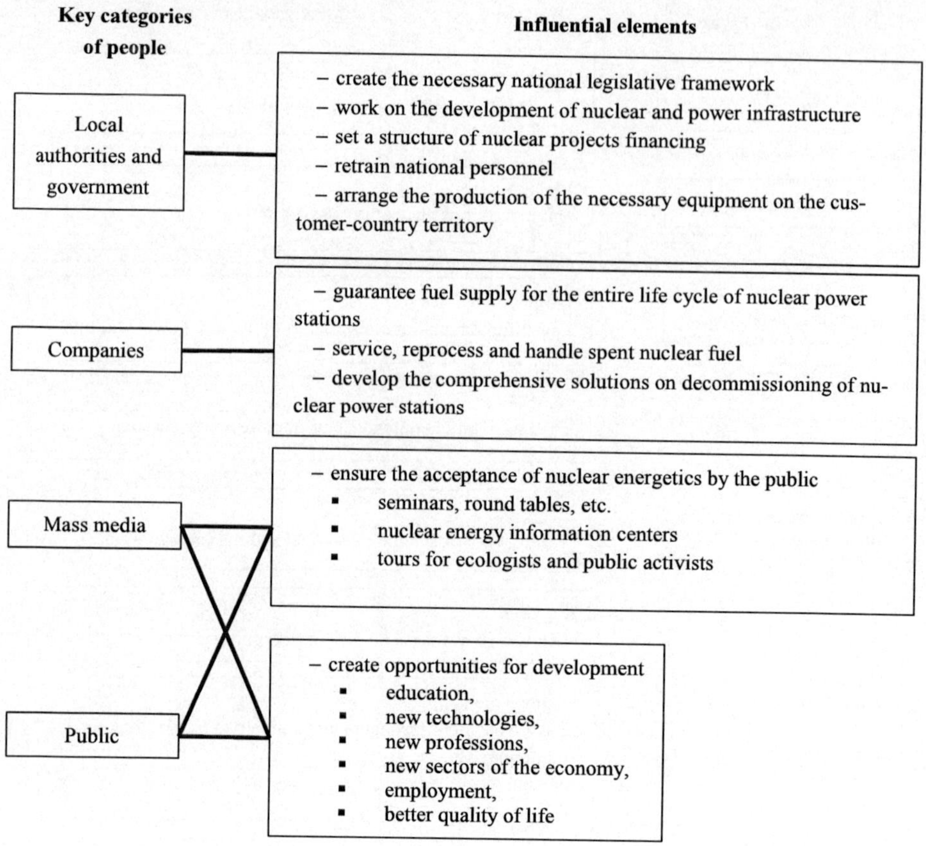

Fig. 3. Distribution of influential elements by categories of people

- employment,
- better quality of life;

– ensure the acceptance of nuclear energetics by the public.

Stage 4. Project completion:

– work on the development of nuclear infrastructure,
– retrain national personnel,
– service, reprocess and handle spent nuclear fuel,
– develop the comprehensive solutions on decommissioning of nuclear power stations,
– ensure the acceptance of nuclear energetics by the public.

Thus, when planning a loyalty program, it is necessary to evaluate the prospects and possible influences of its elements on the involved objects.

2.2 Determination of Loyalty Program Structure of Megaproject by Applying Mathematical Tools

The identification of various categories of stakeholders in the megaproject implementation and the above analysis of possible influences of loyalty program elements on their satisfaction with the predicted megaproject results allow us to establish the main accents of the future loyalty program. Of course, the development and implementation of a loyalty program requires significant investment funds. Besides, given the varying strength of the effect of a particular event, it seems advisable to conduct a preliminary assessment of the sensitivity of objects on different loyalty program elements.

Evaluation of existing classic loyalty programs can be easily made based on data collected on purchases of regular and new customers. By contrast, such an analysis is not possible when it comes to a megaproject because the megaproject is usually unique, requires a huge amount of financial costs, and can be implemented only once.

In this regard, one of the most relevant tools in case of megaproject is an expert assessment of information. To set the estimated influences of loyalty program elements, T. Saati hierarchy analysis method can be used, which determines the priority of the implementation of various elements of the loyalty program. To use this method, we heed to involve experts who can assess the object's sensitivity to a particular area of activities and, thus, identify the prospects of its use.

The mathematical method involves an assessment matrix constructed by an expert who compares the elements between each other in terms of their importance in pairs. As an evaluation scale, T. Saati proposed using a scale from 1 to 9 [4].

Since the expert's task is extremely important, it is advisable to involve not one but several experts in the evaluation, whose consolidated opinion subsequently forms the basis for developing a loyalty program.

To obtain a single assessment based on several individual assessments, we use mathematical tools. When solving this problem, we can use such tools as the arithmetic mean value, the geometric mean value, or the mode value. It is proposed to use as a consolidated estimate the value obtained using Bayesian method [5]. This method involves the use of information about the possible error levels of experts when obtaining a consolidated assessment of a group of experts. Bayesian method is based on the hypothesis that the more experts are involved, the more reliable expectation of individual assessments becomes.

The application of the Bayesian method can be described in the following steps [6]:

— **construct an $M \times N$ dimensional matrix of estimates for N elements from M experts**

$$X = \begin{bmatrix} x_1^1 & \ldots & x_n^1 \\ \ldots & \ldots & \ldots \\ x_1^m & & x_n^m \end{bmatrix}, \quad (1)$$

where $x_j^{(i)}$ – weight of element j in the opinion of expert i.

— **calculate arithmetic mean values of estimates for each element**

$$\overrightarrow{x^{(A)}} = \frac{1}{m}\sum_{i=1}^{m}\overrightarrow{x^{(i)}} = (\frac{1}{m}\sum_{i=1}^{m}x_1^{(i)}, \ldots, \frac{1}{m}\sum_{i=1}^{m}x_n^{(i)}) = (\overrightarrow{x_1^{(A)}}, \cdots, \overrightarrow{x_n^{(A)}}), \quad (2)$$

где $\overline{x_j^{(A)}}$ – average estimate of element j.

— **calculate the variance of individual assessments of each expert in relation to the arithmetic mean value of the experts' assessments set**

$$\sigma^{(i)2} = \frac{1}{n-1} \cdot \sum_{j=1}^{n} (x_j^{(i)} - x_j^{(A)})^2 \quad (3)$$

— **calculate the sum of the inverse values of all experts' variances**

$$\sum_{i=1}^{m} \frac{1}{\sigma^{(i)2}} \quad (4)$$

— **calculate the weight of expert's estimates for all experts**

$$w^{(i)} = \frac{1/\sigma^{(i)2}}{\sum_{i=1}^{m} 1/\sigma^{(i)2}} \quad (5)$$

— **Calculate more accurate estimates of each element's significance as an average weighted estimate, taking into account the different error level in experts' estimates**

$$\overline{x_j^{(B)}} = \sum_{i=1}^{m} w_i \cdot x_j^{(i)} \quad (6)$$

The application of T. Saati hierarchy analysis method can be demonstrated by the example of an expert assessment of the activity impact on the public (Fig. 3). According to this method, the expert conducts a paired comparison of the significance of all activities. He estimates how one event is more important than another on a scale of 1 to 9 (and from 1/9 to 1 - in the case of an inverse relation) [7].

Table 1 shows the results of a paired comparison by an expert on the sensitivity of the stakeholder category "public" to possible measures to increase loyalty.

After the calculations, we obtain the following values (Table 2). This matrix of expert's judgments is consistent (C.R. = 7.4%). Therefore, the data obtained on its basis are suitable for subsequent analysis.

This analysis of the loyalty program possible components, accomplished using expert assessments based on T. Saati hierarchy analysis method, allows us to prioritize correctly all activities and to determine the funding structure.

Table 1. Expert's estimates

	ED	NTD	NPD	NESD	IE	BQL	SRT	IC	IT
Education Development (ED)	1,00	3,00	2,00	0,50	0,33	0,50	5,00	4,00	5,00
New technologies development (NTD)	0,33	1,00	0,50	0,33	0,25	0,33	3,00	2,00	3,00
New professions development (NPD)	0,50	2,00	1,00	0,50	0,25	0,20	6,00	5,00	6,00
New economy sectors development (NESD)	2,00	3,00	2,00	1,00	0,25	0,20	6,00	5,00	6,00
Increasing Employment (IE)	3,00	4,00	4,00	4,00	1,00	0,50	7,00	6,00	7,00
Better quality of life achievement (BQL)	2,00	3,00	5,00	5,00	2,00	1,00	8,00	6,00	8,00
Seminars, round tables, etc. (SRT)	0,20	0,33	0,17	0,20	0,14	0,13	1,00	0,25	0,33
Information Centers (IC)	0,25	0,50	0,20	0,20	0,17	0,17	4,00	1,00	3,00
Information Tours (IT)	0,20	0,33	0,17	0,17	0,14	0,13	3,00	0,33	1,00

Table 2. Sensitivity assessment by an expert

Element of the loyalty program	Relative importance (weight)
Education Development (ED)	0,119
New technologies development (NTD)	0,058
New professions development (NPD)	0,091
New economy sectors development (NESD)	0,129
Increasing Employment (IE)	0,240
Better quality of life achievement (BQL)	0,281
Seminars, round tables, etc. (SRT)	0,019
Information Centers (IC)	0,038
Information Tours (IT)	0,024

3 Conclusions

The main feature of megaprojects is their exclusivity and individuality. Such projects are valuable not only for the investors but also for the part of society that directly or indirectly receives a wide variety of new benefits. In addition to commercial significance, such projects can have strong positive effects: social, tax, budget, environmental, and technological. The social effect can be assessed by the benefits of the project for the whole population, for the population living around the place of project implementation, or for the people working on the project. The tax effect is estimated by the amount of taxes collected from the project to the local, regional, and federal budgets. The budget effect is evaluated if the project is fully or partially financed from the budget (federal, regional, local). This effect is determined by the amount of funds returned to the budget through taxes for a certain number of years after the budget has invested in the project.

The implementation of megaprojects, as a rule, affects a large number of stakeholders, among which there may be both supporters and opponents of the project. One of the tasks of a comprehensive loyalty program is to smooth the negative aspects associated with the megaproject. The loyalty program demonstrates the strengths and prospects of the megaproject and increases the number of people interested in its successful completion.

The competitiveness and success of an organization can be significantly increased by using the most modern technologies, including suitable specialized (loyalty program) and mathematical tools, as well as comprehensive approaches in the implementation and maintenance of megaprojects.

Acknowledgments. This work was supported by RFFI grant № 20-010-00708\20.

References

1. Yusupova, I.V.: Project Management of Territorial Development: Textbook (lecture course), p. 150. Kazan University Press, Kazan (2018)
2. Voropaev, V.I.: Project Management in Russia, p. 225. Alans Press, Moscow (1995)
3. https://www.rosatom.ru/
4. Saati, T.: Decision-making. The method of analysis of hierarchies. Radio and communications (1993)
5. Kryanev, A.V., Tikhomirova, A.N., Sidorenko, E.V.: Group expertise of innovative projects using the Bayesian approach. Econ. Math. Methods **49**(2), 134–139 (2013)
6. Matrosova, E.V., Tikhomirova, A.N.: Algorithms for intelligent automated evaluation of relevance of search queries results. biologically inspired cognitive architectures (BICA) for young scientists. In: Proceedings of the First International Early Research Career Enhancement School on BICA and Cybersecurity (FIERCES 2017). Book Series: Advances in Intelligent Systems and Computing. Springer, Heidelberg (2017)
7. Matrosova, E.V., Tikhomirova, A.N.: Peculiarities of expert estimation: comparison methods. In: 7th Annual International Conference on Biologically Inspired Cognitive Architectures, BICA 2016, Procedia Computer Science, vol. 88, pp. 163–168 (2016)

Using Domain Knowledge for Feature Selection in Neural Network Solution of the Inverse Problem of Magnetotelluric Sounding

Igor Isaev[1], Eugeny Obornev[2], Ivan Obornev[1], Eugeny Rodionov[2], Mikhail Shimelevich[2], Vladimir Shirokiy[1,3], and Sergey Dolenko[1(✉)]

[1] D.V. Skobeltsyn Institute of Nuclear Physics, M.V. Lomonosov Moscow State University, Moscow, Russia
isaev_igor@mail.ru, dolenko@srd.sinp.msu.ru

[2] S. Ordjonikidze Russian State Geological Prospecting University, Moscow, Russia

[3] National Nuclear Research University "MEPhI", Moscow, Russia

Abstract. In the present study, using the inverse problem (IP) of magnetotelluric sounding (MTS) as an example, we consider the use of neural networks to solve high-dimensional coefficient inverse problems. To reduce the incorrectness, a complex approach is considered related to the use of narrow classes of geological models, with prior selection of the model class by solving the classification problem by MTS data. Within the framework of this approach, the actual direction of work is to reduce the volume of calculations when re-building the system for another set of geological models. This goal can be achieved by selecting the essential features. The present paper is devoted to the study of the applicability of various selection methods to the MTS IP. Also, in this paper we consider taking into account domain knowledge about the studied object in the process of selection of essential features using methods such as wrapper.

Keywords: Inverse problems · Magnetotelluric sounding · Neural network · Feature selection · Domain knowledge

1 Introduction

Inverse problem (IP) is a type of tasks, which consist in determining the parameters of an object by the results of its observation. Almost all the tasks of indirect measurements can be attributed to IP. Moreover, among them, of particular interest are the tasks related to the construction of the spatial distribution of a parameter of the object associated with the coefficients of differential equations that determine the properties of

This study has been supported by the Russian Foundation for Basic Research (RFBR) (project no. 19-01-00738).

A. V. Samsonovich et al. (Eds.): BICA 2020, AISC 1310, pp. 115–126, 2021.
https://doi.org/10.1007/978-3-030-65596-9_15

the object under study - coefficient inverse problems. Such tasks often arise in the fields of geophysics, tomography, and flaw detection. Moreover, for an actual practical IP, the desired distributions are described by a large number of parameters (multi-parameter problems). For their reliable determination, as a rule, large-scale measurements are carried out. Thus, such IP have high dimension, both by input and by output, which substantially complicates their solution. The development of approaches to solving such environmental problems is an urgent task of modern science.

1.1 Inverse Problem of Magnetotelluric Sounding

In the present study, the IP of magnetotelluric sounding (MTS) is considered as a problem with the above features. This IP consists in constructing the distribution of electrical conductivity in the interior of the Earth by the values of the components of the electromagnetic field measured on its surface. From the point of view of the mathematical formulation, the 2D MTS IP considered in this paper belongs to the coefficient IP, for which, in the general case, the analytical solution is not known; therefore, practical MTS IP of this type are solved numerically. The traditional methods for solving the MTS IP include optimization methods based on multiple solution of the direct problem with minimizing discrepancies in the space of the observed components of electromagnetic (EM) fields [1], as well as operator methods usually using Tikhonov's regularization [2].

Optimization methods have several drawbacks: they are characterized by high computational costs and by the need for a good first approximation, which is often obtained using alternative measurement methods. Moreover, to use optimization methods, it is necessary to have the correct model for solving the direct problem (DP), in the absence of which this method is not applicable. In addition, due to the inaccuracy inherent in many IP, a small discrepancy in the space of observable quantities does not guarantee a small discrepancy in the space of determined parameters (see, for example, [3]). For operator methods based on regularization, the main difficulty is the need to choose a regularization parameter. In addition, operator methods are linear methods, so when using them to solve non-linear problems, it is necessary to perform non-linear data preprocessing. The regularization method has also been developed for non-linear problems, but in the non-linear case we are dealing with a multi-extreme non-linear optimization problem of a large dimension, which must be solved again for each new data. As an alternative, in this paper we consider neural network (NN) methods that are mainly free from the above disadvantages. High computational costs when using NN are carried out once at the preliminary stage of constructing the NN approximator, which can then be applied to any data for solving IP in a given class of models. This increases the efficiency of the practical use of such a system. At the same time, NN for solving the MTS IP can be used at various stages of its solution: in terms of data preprocessing, for example, to remove noise [4], as an integral part of optimization methods in terms of solving a DP [5], as well as an independent inversion method [6–15].

In this paper, we consider the option of using NN as an independent inversion method for solving the MTS IP. To implement this approach, it is necessary to move from a continuous description of the geological section to a discrete description, by setting the initial distribution with a limited number of determined parameters, i.e. by defining the so-called section parameterization scheme. Further, based on a given parameterization scheme, a training data set is generated on which NN will be trained.

There are various approaches to how to define a parameterization scheme.

The approach presented in [6–8] consists in the use of parameterization schemes with a rigidly defined spatial structure, the so-called 'class-generating' models, which is built on the basis of alternative measurement methods. The disadvantage of this approach is the need to develop its own individual solution for each task, as well as the need to have a priori information about the spatial structure of a given parameterization scheme.

Another approach is the use of "universal" parameterization schemes, by means of which any geological structures can be described. For example, in [9–11], a parametrization scheme is used, where the distribution of electrical conductivity is specified by the values of electrical conductivity in the nodes of a macrogrid. In this case, the values of electrical conductivity in each node for each pattern of the training sample are set randomly. The disadvantage of this approach is the high degree of incorrectness of the resulting IP described by these parameterization schemes. A consequence of this is also the high sensitivity of the solution to noise in the data, which entails the need to develop special approaches to increase the resilience of the solution to noise in the data [10, 11].

As an intermediate approach, one can use narrower models of the environment that describe a certain class of geological sections [12–15]. Moreover, they must have geological validity and prevalence in nature. In particular, a model of a kimberlite pipe was considered in [12], a single or double rectangular anomaly of low or high resistivity was considered in [13], and a horizontally layered medium model was considered in this study and in [14]. When using this approach, it becomes necessary to first select a model. This problem can also be solved using NN trained to solve the classification problem [14]. Also, within the framework of this approach, it is possible to construct the narrow models adaptively during the operation of a NN iterative algorithm [15]. Thus, the reduction of incorrectness in this case is achieved at the expense of a significant complication of the decision system.

Thus, within the framework of the third approach, **a topical direction of work** is to reduce the volume of calculations when re-building the system for another set of geological models. This **goal** can be achieved by selecting the essential features. Moreover, a reduction in the volume of calculations will be observed not only with direct training of NN, but also at the stage of generating a training data set. A reduction in the volume of calculations in this case will be ensured due to the fact that when solving the DP it will be necessary to calculate the response not for all frequencies and components of the EM field, but only for a limited set.

1.2 Feature Selection for the Inverse Problem of Magnetotelluric Sounding

Traditionally [16, 17], the following groups are distinguished among the methods of selecting essential attributes based on supervised training: filter methods, embedded methods and wrappers.

Filter methods are highly computationally efficient, but feature sets selected with the help of them may not be optimal for the target algorithm for solving IP.

In the present work, NN were used as the target algorithm for solving the MTS IP. Among the embedded methods for selecting essential features for NN, the following groups can be noted. The first group includes methods where the reduction of the input dimension is built into the training procedure, in the form of the decay of the weights [18, 19] or regularization [18–21]. Another type is based on the analysis of weights of an already trained neural network [22, 23].

Wrapper methods show better results than filter methods, because the target method for solving the problem is included in the selection of features. However, they require significantly higher computational costs, and with a large number of features their use is difficult. Existing methods for adapting this method to high-dimensional tasks are divided into two types. The first one is to use more effective search strategies: sequential search - forward selection and backward elimination [16], various evolutionary selection methods [24, 25] and others [26]. The second approach is based on the use of *a priori* information, including domain knowledge, to reduce the enumeration space [27].

In the present study, the following methods were used as filter methods: standard deviation (SD) selection, cross-correlation (CC) selection, cross-entropy (CE) selection; and the approach based on the weights analysis (WA) of an already trained NN was used as an embedded method. This choice is due to the requirement for ease of interpretation of the results. Due to the high computational costs of the target algorithm for solving the IP for the wrapper method, a strategy to reduce the enumeration space based on *a priori* information about the object under consideration was chosen.

Purpose of the Work: to study how reducing the input dimension of a problem affects the quality of a solution. Comparison of traditional methods and methods based on the consideration of *a priori information from domain knowledge* about the studied object.

2 Physical Statement of the Problem

2.1 Parameterization Scheme

The model for MTS IP considered in this work is an integral part of the general model intended for combining three physical methods: gravimetry, magnetometry, and magnetotellurics. To ensure the possibility of such integration, it is necessary for the formulation of the problem to be similar for all the considered physical methods. In this case, such a formulation consisted of determining the structural boundaries separating the geological layers with constant values of the parameters: density in the gravimetry problem, magnetization in magnetometry, electrical resistivity in magnetotellurics.

In this study, a 4-layer 2D sectional model was considered. The medium parameterization scheme is shown in Fig. 1.

The resistivity values of the layers were fixed, i.e. constant for the entire data set. The determined parameters were the values of the depths of the boundaries of the layers h(y) along the section.

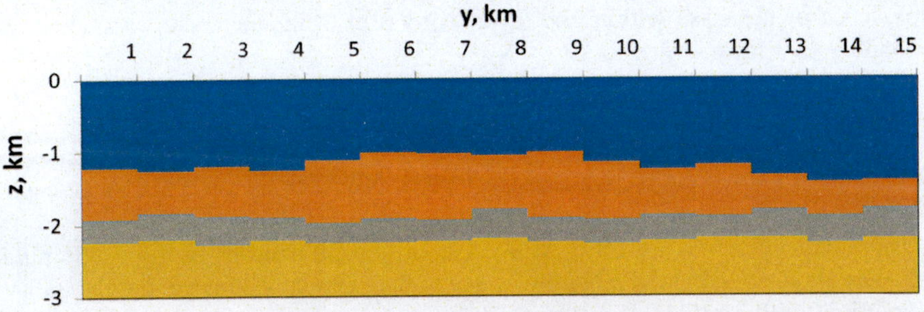

Fig. 1. Parameterization scheme.

For each pattern of the training sample, depth values were randomly set in the range of layer boundaries considered. Next, the direct problem was solved by the finite difference method. In this case, the 6 components of the EM field were calculated: the real and imaginary components of the impedance tensor Z (Z_{YX} - H polarization and Z_{XY} - E polarization) and tipper W [1, 2]. The calculation was made for 13 frequencies ranging from 0.001 Hz to 100 Hz.

2.2 Data

Data is obtained by repeatedly solving the DP. For each pattern, the initial distribution of the depths of the boundaries of the layers was set randomly. Then, a direct 2D problem was solved for this distribution [15], where the values of the 6 field components at 13 frequencies in 31 pickets were calculated.

Thus, the input dimension of the problem was:

$$6\,field\ components\,(taking\ into\ account\ the\ complex - valued\ data\ presentation)$$
$$*\ 13\,frequencies * 31\,pickets = 2418\,attributes\ (features).$$

It is worth noting that due to the geometry and physics of the problem, many features turned out to be correlated with each other, which is an additional argument in favor of feature selection.

The output dimension of the problem was:

$$3\,layers * 15\,depths = 45\,parameters.$$

A total of 30,000 patterns were calculated.

3 Methodical Statement of the Problem

3.1 Reducing the Output Dimension

To reduce the output dimension of the problem, the so-called autonomous determination of parameters [9, 10] was used, when the initial task with N outputs was divided into N single-output tasks, with the construction of a separate single-output NN for each task. In this work, the study was carried out for the central vertical, in view of the equivalence of the tasks within one layer. A total of 3 parameters were studied, corresponding to 3 layers.

3.2 Using Neural Networks

In this work, to solve the IP, we chose a type of NN such as a multilayer perceptron (MLP), which is a universal approximator.

The architecture was an MLP with 1 output and 32 neurons in the single hidden layer. The choice of this architecture was due to the fact that it showed a high quality of the solution on the full set of features (Fig. 4).

To reduce the influence of the initialization of the weights, in each case under consideration, 5 NN were trained; the statistical indicators of the quality of the solution of the 5 NN were averaged.

To prevent overtraining of NN, the early stop method by the validation set was used – after 500 epochs without improving the quality of the solution on this set.

3.3 Datasets

Thus, the initial data set was divided into training, validation and test sets in the ratio of 70:20:10. The size of the sets was 21,000, 6,000, 3,000 patterns, respectively.

3.4 Filter Methods

In this study, the absolute values of cross-correlation, the values of cross-entropy and standard deviation were used as the metrics by which the selection was made.

The threshold was selected in such a way that the exact number of features was selected from the series: 5, 10, 20, 30, 40, 50, 150, 200, 300, 800. NN were trained on the selected features. The dependences of the quality of the solution on the number of selected features were built.

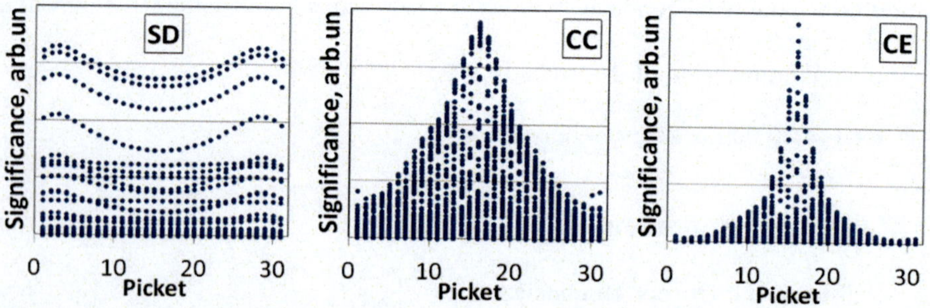

Fig. 2. The dependence of the significance of the features on the picket number for the filter type methods: SD - selection by standard deviation CC - selection by absolute value of cross-correlation, CE - selection by cross-entropy.

Figure 2 shows that for the methods of selection by CC and CE, the significance of the features is higher for the areas adjacent to the studied parameters than for those remote from them. The reverse situation is observed in the case of selection by SD. Thus, we can make the assumption that at small sizes of the training sample, the results of solving the IP will be different, because different features will be selected.

3.5 Embedded Methods

As an embedded method, the NN weight analysis (WA) method was used. As a metric, we used the sum of the products of the weights of the NN along all paths from each input to each output [22].

Due to the low contrast of the method and to its dependence on the factor of random initialization of weights, for each parameter studied, 5 NN were trained on the complete data set. For each input, its significance was calculated. A feature was selected if its significance value exceeded the threshold for 3 or more NN out of 5. The threshold was selected so that the exact number of features was selected. The number of features was taken from a series similar to filter methods. The dependences of the quality of the solution on the number of features were built.

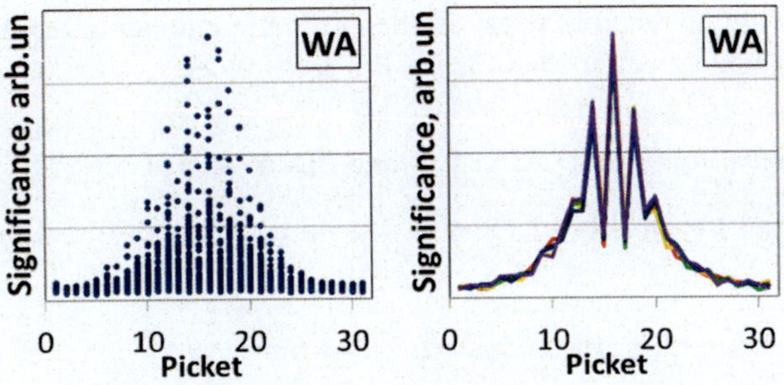

Fig. 3. Dependence of significance on the picket for NN WA for the third layer. On the left are all the features of a single NN. On the right, five NN, one frequency of one EM field component is shown.

Figure 3 shows that for areas adjacent to the studied parameters, the significance is higher than in remote ones. In addition, if the field component and frequency turned out to be informative for this parameter, then all neural networks show similar significance results (Fig. 3, right).

3.6 Wrapper Method Using Domain Knowledge

As noted above, direct use of brute force methods is difficult. Therefore, in this study, we used a modification of this class of methods, based on the consideration of *a priori* information, that is, available prior to the application of machine learning methods information about the relationships between input features and possible ways of their separation or grouping, determined by the physical meaning of the task. As such information we can use the fact that each input feature is characterized by three parameters: field component, frequency, geometric position in the section – picket number.

At the first stage, NN training was done on a set of features that belong to a fixed frequency (31 pickets * 6 components = 186 characters), or belong to a fixed component (31 pickets * 13 frequencies = 403 features). The quality of the solution was compared. Next, groups of features were combined for which the quality of the solution for the components of the frequencies was the best, and the importance for NN WA was the highest.

4 Results

Criteria for evaluating the results of various feature selection methods can be based on such characteristics as the number of features and the quality of the decision made on them. Based on these characteristics, one can set the following evaluation criteria:

- *The best IP solution quality,* regardless of the number of features.
- *The number of features* at which the *best quality* of IP solution is observed.
- *The minimum number of features* at which the NN solution shows a satisfactory result (here, $R^2 = 0.98$; relative error = 0.05).
- *Consideration of both characteristics*: the minimum number of features at which the result is the same or better than on the full feature set.

4.1 Feature Selection by SD, CC, CE and NN WA

The dependence of the quality of the solution on the number of features is shown in Fig. 4.

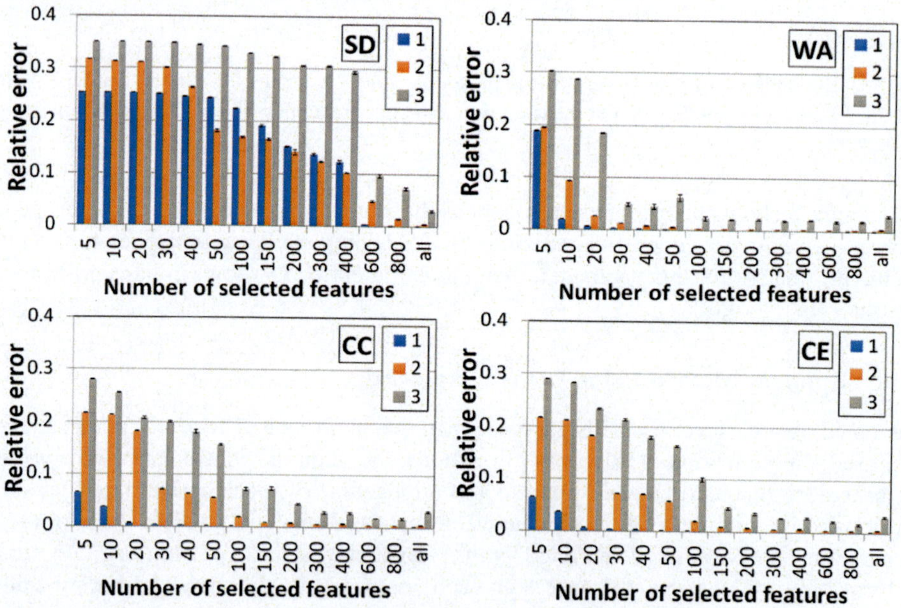

Fig. 4. Dependence of the quality of the solution on the number of selected features for filter methods: SD - selection by standard deviation, CC - selection by cross-correlation, CE - selection by cross-entropy; and for the embedded selection method through analysis of NN weights - WA.

The method of selection by standard deviation did not show any satisfactory result for all the determined parameters and for all sizes of the array of the selected features; therefore it was excluded from further consideration.

Other selection methods can effectively reduce the dimension of the IP. At the same time, a satisfactory result of solving IP is observed for the following sizes of arrays of input features:

- For the cross-correlation selection method: 10 for the first layer, 100 for the second one, 200 for the third one.
- For the cross-entropy selection method: 10 - for the first layer, 100 - for the second one, 150 - for the third one.
- For the selection method by analyzing the weights of the neural network: 10 for the first layer, 20 for the second one, 30 for the third one.

From the point of view of the criterion associated with the minimum number of features, in which the result is the same or better than in the full set, the following picture is observed:

- For the cross-correlation selection method: 50 - for the first layer, 600 - for the second one, 300 - for the third one.
- For the cross-entropy selection method: 50 - for the first layer, 600 - for the second one, 300 - for the third one.
- For the method of selection by analysis of the weights of the neural network: 100 for the first layer, 100 for the second one, 100 for the third one.

The best result of solving the IP is observed for the first layer when using the methods of cross-correlation and cross-entropy with 400 features, for the second layer when using the neural network weights analysis method and 200 features, for the third layer - when using the cross-entropy selection method and 800 features.

4.2 Wrapping Method Using Domain Knowledge Information

The results of NN training on a set of attributes that belong to a fixed frequency or that belong to a fixed component are presented in Fig. 5. It can be seen that for the first parameter, the best results are at frequencies 1, 2 and the component 1, 2, 4. for the second one - at frequencies 2, 3, 4 and the components 1, 2, 4, of the third one - at frequencies 4, 5, 6, 7 and components 1, 3, 4. The observed picture is explained by the fact that lower frequencies penetrate deeper into the Earth better.

Thus, on the basis of these results, the combinations of frequencies and components presented in Table 1 were considered. Based on the results of CC, CE and NN WA (Fig. 2, 3), for all combinations, the features taken were corresponding to the adjacent 10 pickets located in a window placed symmetrically against the determined parameters.

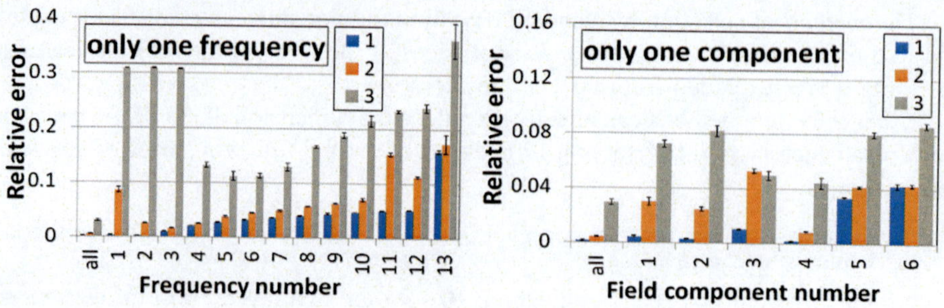

Fig. 5. The dependence of the quality of the solution: on the used frequency (left), on the used component of the field (right).

Table 1. The dimensions of the feature window and the considered combinations of frequencies for each layer

Layer number	Components	Frequency	Window size
1	Re(Z_{YX}), Im(Z_{YX}), Im(Z_{XY})	1, 2	10
2	Re(Z_{YX}), Im(Z_{YX}), Im(Z_{XY})	2, 3, 4	10
3	Re(Z_{YX}), Re(Z_{XY}), Im(Z_{XY})	4, 5, 6, 7	10

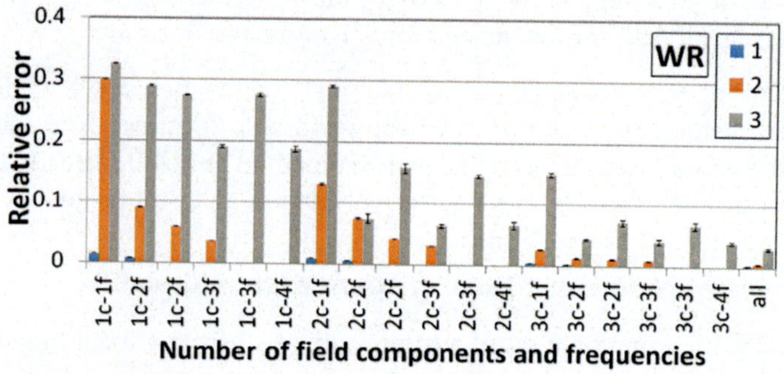

Fig. 6. Dependence of the quality of the solution on the used combination of frequencies and field components. The number of used components X and the number of used frequencies Y are denoted as Xc-Yf.

The results are shown in Fig. 6.

Due to the fact that sample sizes not exceeding 120 features were considered here, the evaluation of this method is possible only from the point of view of the criterion of the minimum number of features at which the NN solution shows a satisfactory result. Here, the method under consideration surpassed the selection methods for cross-correlation and cross-entropy, but failed against the selection method by analyzing the weights of the NN. Figure 6 demonstrates that an acceptable quality of the solution was observed for the first layer starting with 10 attributes (1c-1f), for the second with 30 attributes (1c-3f), for the third with 60 attributes (3c-2f).

5 Conclusions

According to the results of the study, the following conclusions can be formulated:

- The assumption that inputs with a high standard deviation are more informative is not true for the task considered in this paper. This selection method did not show a satisfactory result for all determined parameters and for all sizes of arrays of selected features.
- Selection methods by cross-correlation and by cross-entropy show similar results and good efficiency.
- Selection by analyzing the weights of a neural network exceeded all the other methods of feature selection by the set of criteria. However, a unanimous winner, which would have shown the best result for each criterion considered, is not observed here.
- The method of enumerating various combinations of microarrays of features selected on the basis of domain knowledge on the spatial structure and physical nature of the features also showed good efficiency compared to other methods.

Thus, the use of domain knowledge makes it possible to increase the efficiency of the procedure for selecting essential features for wrapper methods, thereby reducing the computational costs of its implementation and increasing the compression ratio for the input dimension.

References

1. Berdichevsky, M., Dmitriev, V.: Models and Methods of Magnetotellurics. Springer, Heidelberg (2010)
2. Zhdanov, M.: Geophysical Electromagnetic Theory and Methods. Methods in Geochemistry and Geophysics. Elsevier, Amsterdam (2009)
3. Isaev, I., Dolenko, S.: Comparative analysis of residual minimization and artificial neural networks as methods of solving inverse problems: test on model data. In: Samsonovich, A., Klimov, V., Rybina, G. (eds.) Biologically Inspired Cognitive Architectures (BICA) for Young Scientists. Advances in Intelligent Systems and Computing, vol. 449, pp. 289–295. Springer, Cham (2016)
4. Manoj, C., Nagarajan, N.: The application of artificial neural networks to magnetotelluric time-series analysis. Geophys. J. Int. **153**(2), 409–423 (2003)
5. Conway, D., Alexander, B., King, M., Heinson, G., Kee, Y.: Inverting magnetotelluric responses in a three-dimensional earth using fast forward approximations based on artificial neural networks. Comput. Geosci. **127**, 44–52 (2019)
6. Spichak, V., Popova, I.: Artificial neural network inversion of magnetotelluric data in terms of three-dimensional earth macroparameters. Geophys. J. Int. **142**(1), 15–26 (2000)
7. Spichak, V., Fukuoka, K., Kobayashi, T., Mogi, T., Popova, I., Shima, H.: ANN reconstruction of geoelectrical parameters of the Minou fault zone by scalar CSAMT data. J. Appl. Geophys. **49**(1–2), 75–90 (2002)
8. Montahaei, M., Oskooi, B.: Magnetotelluric inversion for azimuthally anisotropic resistivities employing artificial neural networks. Acta Geophys. **62**(1), 12–43 (2013)

9. Dolenko, S., Isaev, I., Obornev, E., Persiantsev, I., Shimelevich, M.: Study of influence of parameter grouping on the error of neural network solution of the inverse problem of electrical prospecting. In: Iliadis, L., Papadopoulos, H., Jayne, C. (eds.) Engineering Applications of Neural Networks. EANN 2013. Communications in Computer and Information Science, vol. 383. Springer, Heidelberg (2013)
10. Isaev, I., Obornev, E., Obornev, I., Shimelevich, M., Dolenko, S.: Increase of the resistance to noise in data for neural network solution of the inverse problem of magnetotellurics with group determination of parameters. In: Villa, A., Masulli, P., Pons Rivero, A. (eds.) ICANN 2016, LNCS, vol. 9886, pp. 502–509. Springer, Cham (2016)
11. Isaev, I., Dolenko, S.: Adding noise during training as a method to increase resilience of neural network solution of inverse problems: test on the data of magnetotelluric sounding problem. In: Kryzhanovsky, B., Dunin-Barkowski, W., Redko, V. (eds.) NEUROINFORMATICS 2017. Studies in Computational Intelligence, vol. 736, pp. 9–16. Springer, Cham (2018)
12. Shimelevich, M.I., Obornev, E.A., Obornev, I.E., Rodionov, E.A.: The neural network approximation method for solving multidimensional nonlinear inverse problems of geophysics. Izvestiya Phys. Solid Earth **53**(4), 588–597 (2017)
13. Wang, H., Liu, W., Xi, Z., Fang, J.: J. Cent. South Univ. **26**(9), 2482–2494 (2019)
14. Isaev, I., Obornev, E., Obornev, I., Shimelevich, M., Dolenko, S.: Neural network recognition of the type of parameterization scheme for magnetotelluric data. In: Kryzhanovsky, B., Dunin-Barkowski, W., Redko, V., Tiumentsev, Y. (eds.) Advances in Neural Computation, Machine Learning, and Cognitive Research II. NEUROINFORMATICS 2018. Studies in Computational Intelligence, vol. 799. Springer, Cham (2019)
15. Shimelevich, M.I., Obornev, E.A., Obornev, I.E., Rodionov, E.A.: An algorithm for solving inverse geoelectrics problems based on the neural network approximation. Numer. Anal. Appl. **11**(4), 359–371 (2018)
16. Guyon, I., Elisseeff, A.: An introduction to variable and feature selection. J. Mach. Learn. Res. **3**, 1157–1182 (2003)
17. Li, J., Cheng, K., Wang, S., Morstatter, F., Trevino, R.P., Tang, J., Liu, H.: Feature selection: a data perspective. ACM Comput. Surv. **50**(6), 1–45 (2017)
18. Cibas, T., Soulié, F.F., Gallinari, P., Raudys, S.: Variable selection with neural networks. Neurocomputing **12**(2–3), 223–248 (1996)
19. Verikas, A., Bacauskiene, M.: Feature selection with neural networks. Pattern Recogn. Lett. **23**(11), 1323–1335 (2002)
20. Zou, H., Hastie, T.: Regularization and variable selection via the elastic net. J. R. Stat. Soc. Ser. B (Stat. Methodol.) **67**(2), 301–320 (2005)
21. Scardapane, S., Comminiello, D., Hussain, A., Uncini, A.: Group sparse regularization for deep neural networks. Neurocomputing **241**, 81–89 (2017)
22. Gevrey, M., Dimopoulos, I., Lek, S.: Review and comparison of methods to study the contribution of variables in artificial neural network models. Ecol. Model. **160**(3), 249–264 (2003)
23. Pérez-Uribe, A.: Relevance metrics to reduce input dimensions in artificial neural networks. In: ICANN 2007, LNCS, vol. 4668, pp. 39–48. Springer, Heidelberg (2007)
24. Huang, J., Cai, Y., Xu, X.: A hybrid genetic algorithm for feature selection wrapper based on mutual information. Pattern Recogn. Lett. **28**(13), 1825–1844 (2007)
25. Xue, B., Zhang, M., Browne, W.N., Yao, X.: A survey on evolutionary computation approaches to feature selection. IEEE Trans. Evol. Comput. **20**(4), 606–626 (2015)
26. Mafarja, M., Mirjalili, S.: Whale optimization approaches for wrapper feature selection. Appl. Soft Comput. **62**, 441–453 (2018)
27. Lin, F., Liang, D., Yeh, C.-C., Huang, J.-C.: Novel feature selection methods to financial distress prediction. Expert Syst. Appl. **41**(5), 2472–2483 (2014)

Friction Model Identification for Dynamic Modeling of Pneumatic Cylinder

Vladimir I. Ivlev[1], Sergey Yu. Misyurin[2], and Andrey P. Nelyubin[1(✉)]

[1] Mechanical Engineering Research Institute RAS, 4 Malyi Kharitonievki Pereulok, Moscow 101990, Russia
nelubin@gmail.com
[2] National Research University MEPhI, 31 Kashirskoe Shosse, Moscow 115409, Russia

Abstract. Friction is one of the main nonlinear properties that makes pneumatic actuators difficult to control and reduces their energy efficiency. Many phenomenological friction models are used to describe pneumatic cylinders dynamic behavior, including in pre-sliding zone. These models maintain many unknown empirical parameters (maybe 7 or more). Expensive test equipment and special mathematical methods are required to define the parameters of friction model. But the results may vary significantly for one size cylinders of various manufactures. This paper presents the results of determination the Stribeck friction model parameters based on limited experimental data and procedure of vector identification which implemented in the software complex MOVI (Multicriteria Optimization and Vector Identification). Results were obtained for two types of piston seals materials: NBR and PTFE composite. The minimal value of the piston stable speed for single action cylinders with these seals was estimated.

Keywords: Pneumatic cylinder · Friction model · Parameter identification

1 Introduction

Pneumatic cylinders are one of the main types of linear motion actuators (at the same time with hydro- and electrocylinders) that have been widely used for various technical applications, especially for mechanization industrial process where repeated movements between two fix points is necessary (i.e. systems with a cyclic control system).

As shown in [1], the global consumption of pneumatic cylinder products rises up from $8{,}212 \cdot 10^6$ units in 2011 to $10{,}38 \cdot 10^6$ units in 2015, with an average annual growth rate of 5,23%. At the same time, the revenue of world pneumatic cylinder sales market has a leap from $837{,}35 \cdot 10^6$\$ to $961{,}83 \cdot 10^6$\$.

This is explained by such advantages of the pneumatic cylinders as significant developed efforts and power per unit weight, fire and explosion safety, the ability to work in aggressive environments, simplicity of design, high reliability and relatively low cost compared to other types of linear motors.

A. V. Samsonovich et al. (Eds.): BICA 2020, AISC 1310, pp. 127–137, 2021.
https://doi.org/10.1007/978-3-030-65596-9_16

It is necessary to indicate a number of their significant shortcomings. First of all, it is low energy efficiency and problems associated with precision controllability. So for various types of linear actuators, pneumatic are the least energy efficient. This indicator for them does not exceed 15–20%, and for hydraulic ones it can reach 40%. Electric cylinders are characterized by energy efficiency of about 80%. Here, energy efficiency is the relationship between input electric power consumed from the electric network (to drive the compressor or hydraulic pump) and output power of cylinder.

Some reasons for the low energy efficiency of pneumatic actuators and, in particular, the pneumatic cylinders, is follows:

- Compression and expansion process of air (as working medium) are characterized by significant changes in entropy, i.e. significantly irreversible, which leads to a high level of thermodynamic losses;
- Air leaks and pressure drops in pipes;
- Mechanical friction losses (for standard industrial pneumatic cylinders friction force may be about 7–10% of working force).

The problems with pneumatic cylinders controllability are associated with two factors:

- A high volume compressibility of air in the cylinders working chambers, which leads to a low stiffness of the actuator, i.e. ability to withstand external disturbing power influences on the output link;
- Nonlinear characteristics of friction forces and their relatively large value.

Thus, the problem of the influence of friction forces on the pneumatic cylinders dynamic characteristics and their energy efficiency seems to be very relevant. In this paper, using a simple example, we will illustrate how the characteristics of the friction force in the pneumatic cylinders affect the lowest possible steady-state piston speed.

2 Friction Forces in Pneumatic Cylinders and Their Mathematical Modeling

The main sources of friction in a pneumatic cylinder are piston and rod seals. Today, quite a large number of works have been performed on the study and mathematical modeling of friction in pneumatic cylinders, starting with the widespread use of pneumatic cylinders in industry. An analytical description of the friction characteristics is necessary for mathematical modeling of pneumatic systems with pneumatic cylinders at the design stage.

For these purposes, friction models of varying complexity are used, which allow us to describe the magnitude of the friction force as a function of sliding speed and load, with varying degrees of detail.

The simplest friction model contains the Coulomb friction force F_c, independent of speed, plus viscous friction proportional to the sliding velocity. This model contains two independent parameters. The Stribeck friction model [2] already contains five independent model parameters: F_s is the static friction, F_c is the Coulomb friction, y_s is the Stribeck velocity, i is the empirical coefficient, usually close to 2, y is the slip velocity, and b is the viscous friction coefficient (see below). In [3], a degenerate Stribeck friction model is considered, containing only three friction parameters: static friction, Coulomb friction, and viscous friction.

The LuGre model [4], which takes into account changes in the friction force in the pre-sliding zone, already contains 7 independent parameters. Usually this model is written as:

$$F_{tr} = \sigma_o z + \sigma_1 \dot{z} + by,$$

where z is the coordinate in pre-sliding zone, σ_o is the tangent stiffness coefficient, σ_1 is the damping coefficient related to the time derivative of z, and b is the viscous friction coefficient due to the velocity term y. The dynamics of z is observed by the following expressions:

$$\dot{z} = y - z\left[{}^{y}\!/_{f_s(y)}\right],$$

$$f_s(y) = \frac{1}{\sigma_o}\left[F_c + (F_s - F_c)e^{-\left({}^{y}\!/_{y_s}\right)^2}\right].$$

Modified LuGre models taking into account dynamic processes in the pre-sliding zone, in particular, hysteresis phenomena [5], contain up to 12 independent model parameters. Examples of such modified models and its various modifications are considered in [6–8].

Other multi-parameter friction models have been proposed that allow one to describe some "subtle" effects in the pre-sliding zone, for example, the Leuven friction model [9] or generalized Maxwell-slip model [10].

The above friction models are referred to as so-called phenomenological models, and their parameters cannot be calculated on the basis of the physical properties of contacting surfaces, such as the geometry of microroughnesses, mechanical properties of materials, adhesive properties, etc. The parameter values of these models can only be determined experimentally, or using certain identification procedures, by comparing a limited set of experimental data and data obtained during calculations within the framework of an accepted mathematical model, for example [3]. It should be noted that the number of parameters for the above models refers to their symmetrical shape when these parameters coincide for forward and reverse motion. In the case of asymmetric models, the number of parameters increases significantly. It is clear that the determination of these parameters is associated with long-term and costly experimental work and large amounts of computation. Moreover, the accuracy of determining the friction model parameters, especially in pre-sliding zone can have significant errors.

Such a variety of friction models and a significant difference in the parameter values of these models obtained by various authors when describing friction in pneumatic cylinders is explained by many reasons, among which the following can be indicated:

- A variety of cylinder and seal materials, the roughness of their working surface, manufacturing accuracy and shape deviation such as taper, ovality, etc.;

- Form of seal used and material (dozens of options are possible here);
- A variety of lubricants and methods for its delivery to the friction zone;
- Friction characteristics even for one cylinder model can change during its operation, for example, due to wear of the seal, changes in its elasticity and lubrication conditions;
- The displacement of the piston in the pre-sliding zone (namely, it was measured in the above studies) is the sum of the displacement in the cylinder surface- seal contact and deformation of the seal body. Determining these terms separately at different pressures of the working medium is a rather complicated task and is not considered in these works.

In [11], the experimental values of the parameters using the Stribeck friction model for various pneumatic cylinders are given. The friction model parameters for cylinders of the same diameter, but from different manufacturers, have significantly different values, which confirms the considerations expressed above. Based on the foregoing, it can be concluded that complex mathematical friction models for the piston and rod seals of the pneumatic cylinders allow one to obtain only fairly rough estimates of friction effects influence on the motion characteristics. Therefore, the use of these models in the control loop [12] to partially compensate for the effect of friction in pneumatic servo systems on the positioning accuracy or the developed force is very limited.

Below, using an example for a single action pneumatic cylinder, it will be shown how the use of a relatively simple friction model allows one to estimate the lowest possible steady-state piston speed.

3 Test Setup

The test setup which used in our investigation is shown in Fig. 1.

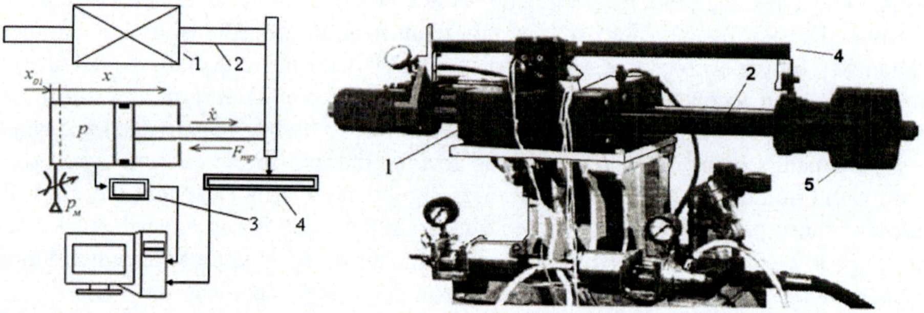

Fig. 1. 1 - ball guide, 2 - bar, 3 - pressure sensor, 4 - piston displacement sensor, 5 - variable mass load.

The test setup is assembled on a massive and rigid metal frame with precise ball guides in which the bar moves. The mass load is created by removable weights from both ends of the bar so as not to create bending radial loads in the guides. The test cylinder is mounted on the frame and its rod moves strictly parallel with the bar. Compressed air pressure in the cylinder chamber is measured by a pressure sensor 100CP2-74 (Sensata Technologies). The movement of the cylinder rod is measured by a precision potentiometric sensor. The sensors power circuits are not shown in the diagram. The voltage signals from the displacement sensor and the pressure sensor were received by a pneumatic cylinder, through an analog – digital converter (Data acquisition card NI PCI-6229). The program for data acquisition was done by using Microsoft visual C++ software.

The compressed air was supplied to the cylinder through the throttle in the form of a set of washers. The effective cross-sectional area of the throttle (or flow coefficient) was carried out on a separate stand by measuring the time of filling a constant volume to a given pressure level. Elements of the supply line: pressure regulator, lubricator, solenoid valve, not shown in the diagram.

The experiments were carried out with two types of single action pneumatic cylinder (piston diameter - 40 mm). Type 1 is a standard steel cylinder with a surface finish of $R_a \approx 0.9$ μm. An NBR lip seal (nitrile butadiene rubber) was used. Another cylinder (type 2) is made of stainless steel with $R_a \approx 0.5$ μm with a seal from composite material based on PTFE (15% glass; 5% MoS_2; 80% PTFE).

Compressed air was dried to a dew point of −20°C (max pressure dewpoint), the degree of filtration was 0.1 μm.

4 The Mathematical Model and the Identification Process

Differential thermodynamic equation that describe the change in pressure of air in the working cylinder chamber are obtained based on the energy conservation law for the flow chamber of variable volume and can be written in the form [13]:

$$\frac{dp}{dt} = \frac{kfKp_m\sqrt{RT}}{S(x_o + x)}\varphi\left({}^{p}\!/_{p_m}\right) - \frac{kp}{x_o + x}\frac{dx}{dt},$$

where p is the pressure in the chamber, S is the piston area, k is the adiabatic constant, p_m is the pressure of the line at the inlet of the throttle, f is the effective area of the inlet section, R is the gas constant, and T is the temperature of the compressed air at the cylinder inlet equal to ambient temperature, x_o is the coordinate characterizing the initial position of the piston, x is the current coordinate of the piston, and t is time. The value of K is defined as:

$$K = \sqrt{{}^{2k}\!/_{k-1}}.$$

Saint-Venant flow function:

$$\varphi\left({}^{p}/_{p_m}\right) = \sqrt{\left({}^{p}/_{p_m}\right)^{2/k} - \left({}^{p}/_{p_m}\right)^{(k+1)/k}} \; when\ 0,528 < \left({}^{p}/_{p_m}\right) \le 1;$$

$$\varphi\left({}^{p}/_{p_m}\right) = 0,258 \; when\ 0 < \left({}^{p}/_{p_m}\right) \le 0,528.$$

Dynamic equation of piston has the form:

$$m\frac{d^2x}{dt^2} = (p - p_a)S - F_{tr}.$$

Transition from immobility to motion:

$$if\ (p - p_a)S \le F_{tr}\ then\ {}^{dx}/_{dt} = 0;\ else\ {}^{dx}/_{dt} \ne 0.$$

Transition from motion to immobility:

$$if\ (p - p_a)S - m\,{}^{d^2x}/_{dt^2} \ge F_{tr}\ then\ {}^{dx}/_{dt} \ne 0;\ else\ {}^{dx}/_{dt} = 0.$$

In our case always: $p - p_a \ge 0$.

Here m is the mass of the moving parts, and p_a is the atmospheric pressure. The friction force here is determined by the Stribeck model:

$$F_{tr} = \left\{F_c + (F_s - F_c)exp\left[-\left({}^{y}/_{y_s}\right)^2\right]\right\}sign(y) + by;\ y = {}^{dx}/_{dt}.$$

Our model takes into account the dependence of the static friction force and Coulomb friction on the pressure in the chamber as a linear function:

$$F_c = F_{co} + k_1(p - p_a),$$

$$F_s = F_{so} + k_1(p - p_a),$$

where F_{so} and F_{co} are the values of the static friction force and Coulomb friction at atmospheric pressure.

Under certain assumptions, the Stribeck friction model can be transformed into simpler models with fewer independent parameters, which is illustrated in Fig. 2.

Here we must determine the parameters of the friction model — the vector (F_{so}, F_{co}, y_s, b, k_1). The procedure of vector parametric identification is based on the method of studying the space of parameters [14] by comparing the experimental and calculated characteristics to restore the values of the desired parameters within the accepted mathematical model and taking into account the criteria and parametric limitations. This method allows finding, for a given dimension of the parameter space, as well as accepted quality criteria, the so-called Pareto optimal solutions (parameter sets or

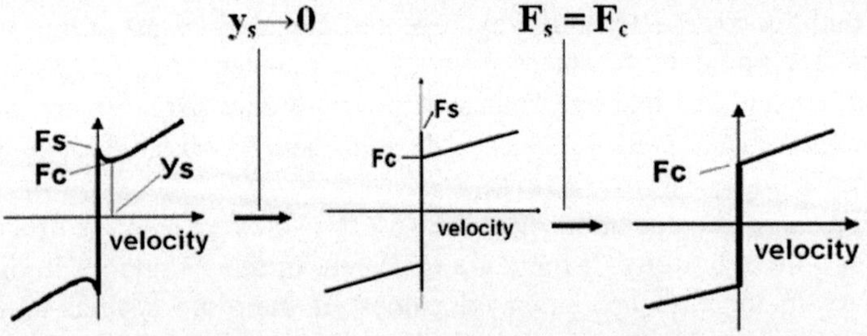

Fig. 2. Friction model transformation.

vectors) for which one of the criteria cannot be improved without worsening the other. The method implemented in the MOVI (Multicriteria Optimization and Vector Identification) software package.

In our case, experimental p_{1ij} and calculated p_{2ij} pressure values in the pneumatic cylinder chamber, as well as experimental and calculated values of the piston displacement, respectively x_{1ij} and x_{2ij}, shown in Fig. 3, were taken as comparison data.

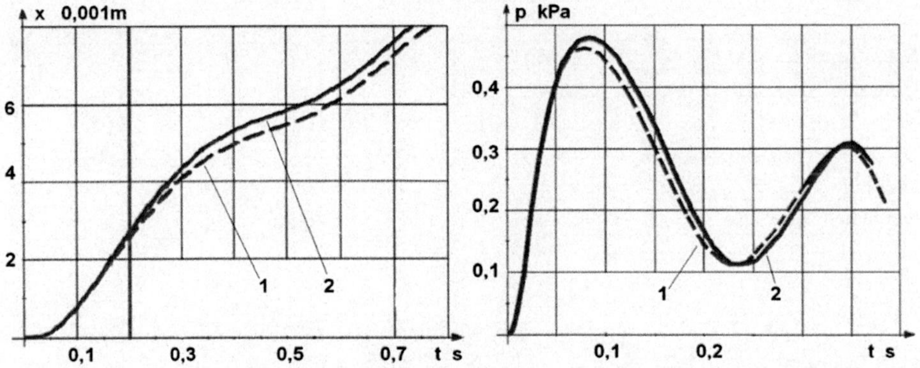

Fig. 3. Displacement and pressure curves: 1 - experiment, 2 - calculation.

Here i is the number of the experiment (or calculation); measurements were taken for 4 values of the supply pressure from 0.15 Mpa to 0.4 MPa, and j means the value of the variable at the process point with the number obtained by sampling the selected time interval ($j \to \left| 1\text{–}24 \right|$). As criteria, the values of their absolute maximum discrepancy are taken:

$$\delta_1 \le \max\left|x_{1ij} - x_{2ij}\right|,$$

$$\delta_2 \le max\left|p_{1ij} - p_{2ij}\right|.$$

These conditions are determined by repeated calculations according to the mathematical model at points in the space of variable parameters (F_{so}, F_{co}, y_s, b, k_1) and by comparing the calculated discrete values of pressure and displacement with the corresponding experimental values. For each desired value, a possible range of change is set, for example $(k_1)_{min} \leq k_1 \leq (k_1)_{max}$.

At the first stage of calculations using the MOVI software package, the values of δ_p and δ_x are set quite arbitrarily. If there are no Pareto optimal solutions in the test table, it is necessary to increase the indicated values. If there are several Pareto optimal solutions, then it is necessary to reduce these tolerances until one Pareto optimal solution is obtained. The solution corresponding to this case (F_{so}, F_{co}, y_s, b, k_1) will be their desired values. 1024 tests were carried out.

Considering that the obtained experimental data have a certain scatter associated with the instability of the pressure regulator and the possible change in lubrication conditions for different series of measurements, the limiting values of the dependences $x(t)$ and $p(t)$ were taken. Table 1 shows the values of the identified parameters and the error estimate associated with the experimental data deviations for the type 1 seal.

Table 1. Identified values for the type 1 seal parameters.

Parameter	Diapason of value variation	Identified value	Error estimation
Static friction force F_{so}, N	26–40	34,3	±4%
Coulomb friction force F_{co}, N	26–40	32,1	±5,2%
Stribeck velocity y_s, m/s	0,001–0,015	0,007	±12%
Viscous friction coefficient b, Ns/m	80–200	142	±11%
Coefficient k_1, N/Pa	(2–12) 10^{-5}	6.10^{-5}	±7%

For the type 2 seal, very close values of the static friction and Coulomb friction were obtained F_{so} = 12.3 N; F_{co} = 11.2 N, as well as the low value of the Stribeck speed y_s not more than 0.0008 m/s, i.e. the Stribeck effect is weakly expressed or absent here.

5 Results and Discussion

Now we have friction model parameters and can estimate the minimum possible steady-state velocity of the pneumatic cylinder piston with these seals. The above mathematical model makes it possible to obtain the process of the piston reaching a given steady-state speed, which is determined by the cross section of the input throttle at a given supply pressure. Figure 4a shows the transition process how the piston reaching a steady speed of 0.012 m/s (supply pressure 0.3 MPa) with significant speed fluctuations. A further attempt to reduce this speed by decreasing the orifice of the throttle leads to movement with stops (stick-slip effect), as can be seen in Fig. 4b.

Figure 4c and 4d show transients at those values of supply pressure, throttle cross-section, static friction and Coulomb friction, but using a simplified model, code $y_s \rightarrow 0$. It can be seen that by decreasing the speed, these models give different transients. This model also gives the effect of "stick-slip", but with lower values of the throttle cross-section. For the type 1 seal, the minimum steady-state velocity, without significant fluctuations at the beginning of the process, is 0.018–0.023 m/s. The same values were recorded during experimental tests. For the type 2 seal, the minimum steady-state velocity is 0.008–0.012 m/s. Transients are similar to those shown in Fig. 4, only shifted to the region of lower speeds.

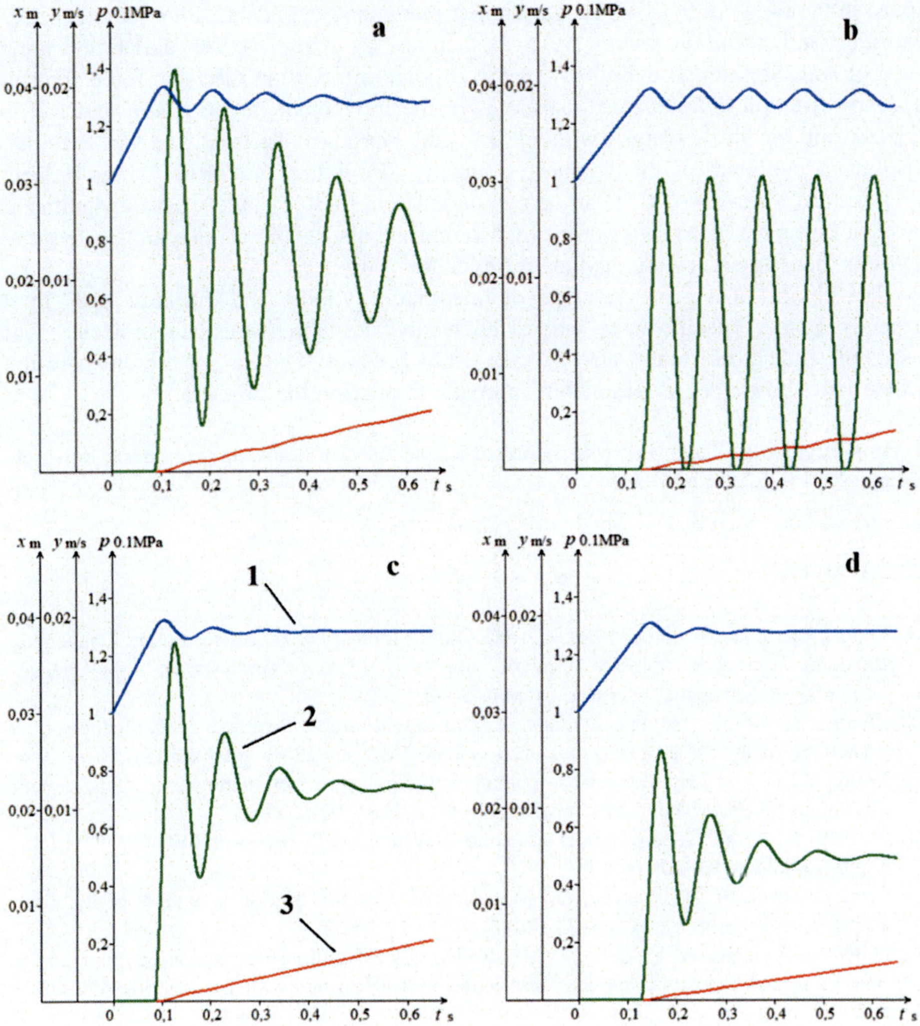

Fig. 4. Steady-state processes: 1 – pressure; 2 – speed; 3 – displacement.

The Stribeck friction model allows one to obtain a fairly accurate estimate of the minimum possible steady-state speed, but it is necessary determine the parameters of this model, which requires certain time and material costs. To determine the piston displacement time from one extreme point to another, with an average speed of more than 0.1 m/s, then it is enough to use the simplest friction model (see Fig. 2). Note that seals made of PTFE-based material allow a lower value of the minimum steady-state speed to be obtained, but these cylinders are significantly more expensive due to the cost of the seal materials and the cylinder, as well as higher quality work surface treatment.

"SMC Corporation" (Japan) produces special pneumatic cylinders "Low Friction Cylinders. Series MQ" with metal plunger seals that allow you to adjust the steady speed in the range of 0.3–300 mm/s. Similar pneumatic cylinders with a glass cylinder cartridge and a carbon piston are manufactured by Airpel (USA) under the name "Airpel Anti-Stiction air cylinders" (www.airpot.com). For the precision force control, ultra-low friction cylinders are used, where the movement of the piston and rod is carried out on air bearings (without dry and boundary friction). As an example, pneumatic cylinders of the Japanese company "Fujikura Composites" (Air bearing cylinder "AC" series) or [15] can be mentioned. But these cylinders have a significant level of compressed air leaks, very high requirements for the air preparation system, and this leads to significant capital and operating costs.

The choice of a particular type of pneumatic cylinder and the seals used in it depends on the specific task, where the technical requirements are indicated, but economic indicators should also be taken into account. The results presented in the article will allow a more reasonable approach to solving the problem.

Acknowledgments. This work was supported by the Russian Foundation for Basic Research, project No. 18-29-10072 MK.

References

1. Global Pneumatic Cylinder Market 2019: Global Industry Size, Share, Future Challenges, Revenue, Demand, Industry Growth and Top Players Analysis to 2025. www.360marketupdates.com. Accessed 01 May 2020
2. Zhang, Y., et al.: Nonlinear model establishment and experimental verification of a pneumatic rotary actuator position servo system. Energies **12**(6), 1096 (2019)
3. Wang, J., et al.: Identification of pneumatic cylinder friction parameters using genetic algorithms. IEEE/ASME Trans. Mechatron. **9**(1), 100–107 (2004)
4. De Wit, C., et al.: A new model for control systems with friction. IEEE Trans. Autom. Control **40**(3), 419–425 (1995)
5. Tran, X., Hafizah, N., Yanada, H.: Modeling of dynamic friction behaviors of hydraulic cylinders. Mechatronics **22**, 65–75 (2012)
6. Valdiero, A.C., Ritter, C.S., Rios, C.F., Raficov, M.: Nonlinear mathematical modeling in pneumatic servo position applications. Mathematical Problems in Engineering 472903.16 (2011)
7. Richter, R.: Friction dynamics mathematical modeling in special pneumatic cylinder. ABCM Symp. Ser. Mechatron. **6**, 800–808 (2014)

8. Sobczyk, S., et al.: A continuous approximation of the LUGRE friction model. ABCM Symp. Ser. Mechatron. **4**, 218–228 (2010)
9. Lampaert, V., Swevers, J., Al-Bender, F.: Modification of the leuven integrated friction model structure. IEEE Trans. Autom. Control **47**, 683–687 (2002)
10. Swevers, J., et al.: An integrated friction model structure with improved presliding behaviour for accurate friction compensation. IEEE Trans. Autom. Control **45**, 675–686 (2000)
11. Andrighetto, P., Valdiero, A.: Study of the friction behavior in industrial pneumatic actuators. ABCM Symp. Ser. Mechatron. **2**, 369–376 (2006)
12. Verbert, K., Toth, R., Babuska, R.: Adaptive friction compensation: a globally stable approach. IEEE/ASME Trans. Mechatron. **21**(1), 351–363 (2016)
13. Gerts, E.V.: Dinamica Pnevmaticheskikh System Mashin. Mashinostroenie, Moscow (1985). (in Russian)
14. Sobol, I.M., Statnikov, R.B.: Wybor Optimalnykh Parametrov v Zadachkh so Mnogimi Kriteriami [Selection of optimal parameters in problems with many criteria]. Drofa, Moscow (2006). (in Russian)
15. Cao, J., et al.: Modeling and constrained optimal design of an ultra-low friction pneumatic cylinder with air bearing. J. Adv. Mech. Eng. **11**(4), 1–13 (2019)

Neurophysiological Features of Neutral and Threatening Visual Stimuli Perception in Patients with Schizophrenia

Sergey I. Kartashov[1(✉)], Vyacheslav A. Orlov[1], Aleksandra V. Maslennikova[2], and Vadim L. Ushakov[3,4,5]

[1] National Research Center "Kurchatov Institute", Moscow, Russia
sikartashov@gmail.com

[2] Institute of Higher Nervous Activity and Neurophysiology of RAS, Moscow, Russia

[3] Institute for Advanced Brain Studies, Lomonosov Moscow State University, Moscow, Russia

[4] NRNU MEPhI, Moscow, Russia

[5] SFHI "Mental-health Clinic No.1 named N.A. Alexeev of Moscow Health Department", Moscow, Russia

Abstract. The authors of this work set a goal to study the features of visual perception of threatening stimulus associated with personal experience in patients with schizophrenia. As a target group were taken patients with a paranoid-hallucinatory syndrome. Healthy volunteers without mental disorders and neurological diseases were used as controls. During fMRI studies, threatening and neutral images were presented. The experiment was built on the principle of a block paradigm. As a result, statistical parametric maps were constructed for two groups of subjects and the results were compared among themselves. According to the obtained results, patients with schizophrenia show a decrease in the overall level of activation in all regions of the brain compared with healthy volunteers. This is most evident in Middle Temporal Gyrus (temporooccipital part Right), Inferior Temporal Gyrus (temporooccipital part Left and Right), Lateral Occipital Cortex (inferior division Left and Right), Temporal Occipital Fusiform Cortex Left and Right and Frontal Pole Left and Right.

Keywords: MRI · fMRI · Schizophrenia · Statistical parametric mapping

1 Introduction

Schizophrenia is a severe mental disorder that, according to the World Health Organization, causes more than 20 million people to suffer [1]. This disease is characterized by impaired thinking, perception, emotions, behavior, language and other cognitive functions. Hallucinations and delusions, which can lead to a significant person's disability, are quite common manifestations of schizophrenia.

The main problem is the lack of a universal theory of its pathogenesis. However, there are some suggestions that a violation of the functional connection of neural networks of the brain may be associated with the onset of symptoms of schizophrenia [2–5].

A. V. Samsonovich et al. (Eds.): BICA 2020, AISC 1310, pp. 138–142, 2021.
https://doi.org/10.1007/978-3-030-65596-9_17

It is well known that the development of schizophrenia is accompanied by neurodegenerative changes. And many studies have been carried out (in particular, using magnetic resonance imaging), describing a decrease in the amount of brain substances. Changes in white and grey matter are accompanied by a violation of its structural organization. It means that the brain intrinsic connections are weakening, that, in turn, affects its functional organization - changes in behavior, deviation in the perception of the surrounding reality, depression, impaired memory, hallucinations and so on. The medical and scientific community has set itself the tasks, firstly, of finding biological markers that may cause or accompany schizophrenia, and secondly, to identify key brain regions that work abnormally. This work is devoted to the second part.

A few recent investigations of emotionally neutral stimuli processing during fMRI scanning in patients with schizophrenia have documented BOLD (blood oxygen level dependent) signal increases in the amygdala, hippocampus, parahippocampal, posterior cingulate, and fusiform gyrus [6–8] in response to faces with neutral expressions.

In this regard, the study of emotionally significant stimuli associated with a specific plot of delusions of patients seems necessary to identify damaged neural network clusters, the incorrect operation of which leads to the disease.

To study this aspect, perhaps the most suitable method of non-invasive brain activity visualization was chosen - magnetic resonance imaging (MRI). MRI today is a universal tool that allows you to study the human body not only in clinical diagnosis but also in scientific research. In general, the combination of structural, functional and diffusion MRI provides a fairly large amount of information about the structural and functional organization of the brain.

The aim of this work is to illustrate the results of functional MRI data comparison between healthy control and patients with schizophrenia (with the hallucinatory-paranoid syndrome) perceive neutral and threatening visual stimuli that activate or deactivate the work of "damaged" areas of the brain as a result of the disease.

2 Materials and Methods

The experiment was carried out in 3 T magnetic resonance imager Magnetom Verio (Siemens GmbH, Germany) with a 32-channel head coil on the basis of the resource center for nuclear physical methods of research at the National Research Centre "Kurchatov Institute". The study involved 35 patients with paranoid schizophrenia (F20.0 for ICD-10) aged 18 to 42 years (mean age 29 ± 2 years) and 17 healthy individuals without cognitive or mental impairment (9 men, 16 women) at the age of 23 to 33 years (average age - 27 ± 4 years). At the time of the study, antipsychotic therapy in patients lasted from 6 months to 2 years. Before starting the MRI acquisition, each participant filled out a written informed consent form, an MRI safety questionnaire, and consent on personal data processing. The study protocol was approved by the local ethics committee of the NRC Kurchatov Institute, according to the requirements of the Helsinki Declaration.

In the experiment during scanning a functional MRI protocol, the activation of the brain regions was recorded in response to images, that were subjectively emotionally significant for the patients. There were two types of stimuli: threatening and neutral.

For the experiment, the standard block design paradigm was chosen and consisted of 3 blocks (threatening, neutral stimuli and resting state) with 30 s duration repeated 3 times each.

The T1-weighted sagittal three-dimensional magnetization-prepared rapid gradient echo sequence was acquired with the following imaging parameters: 176 slices, TR = 1900 ms, TE = 2.19 ms, slice thickness = 1 mm, flip angle = 9°, inversion time = 900 ms, and FOV = 250 mm × 218 mm. FMRI data was acquired with the following parameters: 30 slices, TR = 2000 ms, TE = 25 ms, slice thickness = 3 mm, flip angle = 90°, and FOV = 192 mm × 192 mm. Data which contain the options for reducing the spatial distortion of EPI images also was acquired.

Functional and structural MRI data were processed using the SPM12 software package (available at http://www.fil.ion.ucl.ac.uk/spm/software/spm12/). After converting DICOM files to NIFTI format, all images were manually centered to anterior commissure for better template coregistration. The EPI images were corrected for magnetic field inhomogeneity using the FieldMap toolbox for SPM12 and recorded during experiment session field mapping protocol. Then, the slice timing correction for fMRI data signals was conducted. Anatomical and functional data were normalized in the ICBM stereotactic reference system. T1 images were divided into 3 tissue maps (gray, white matter, and cerebrospinal fluid). Functional data was smoothed using a $6 \times 6 \times 6$ mm^3 FWHM Gaussian filter. Statistical analysis was performed using the Student t-test ($p < 0.05$ with family wise error correction (FWE)).

3 Results

The analysis of fMRI data for healthy control and patients with schizophrenia yielded group statistical parametric maps of direct comparison of conditions “threatening stimuli” - “neutral stimuli” overlaid on a high resolutions T1 structural (Fig. 1).

The analysis of these maps shows differences in the active regions of the brain in healthy volunteers compared with patients with schizophrenia.

In support of previous studies aimed at studying the neural correlates of emotional information processing in schizophrenia, which included various experimental schemes, ranging from discrimination of emotional stimuli [9], producing happy and sad moods [10], experiencing various emotional states [11] to passive viewing of emotional stimuli [12] the study revealed relatively less activation in patients compared to controls, especially in areas of the brain usually associated with emotion processing - Middle Temporal Gyrus (temporooccipital part Right), Inferior Temporal Gyrus (temporooccipital part Left and Right), Lateral Occipital Cortex (inferior division Left and Right), Temporal Occipital Fusiform Cortex Left and Right, Frontal Pole Left and Right. A more detailed analysis reveals that patients have a reduced volume of activations in the occipital and frontal lobes, which are responsible for the perception of visual information and logical thinking, respectively. This difference is obvious when making a classic comparison between processing emotional and neutral stimuli and suggests that people diagnosed with schizophrenia probably cannot properly activate their limbic system structures. This indirectly may be associated with a distorted perception of reality by them.

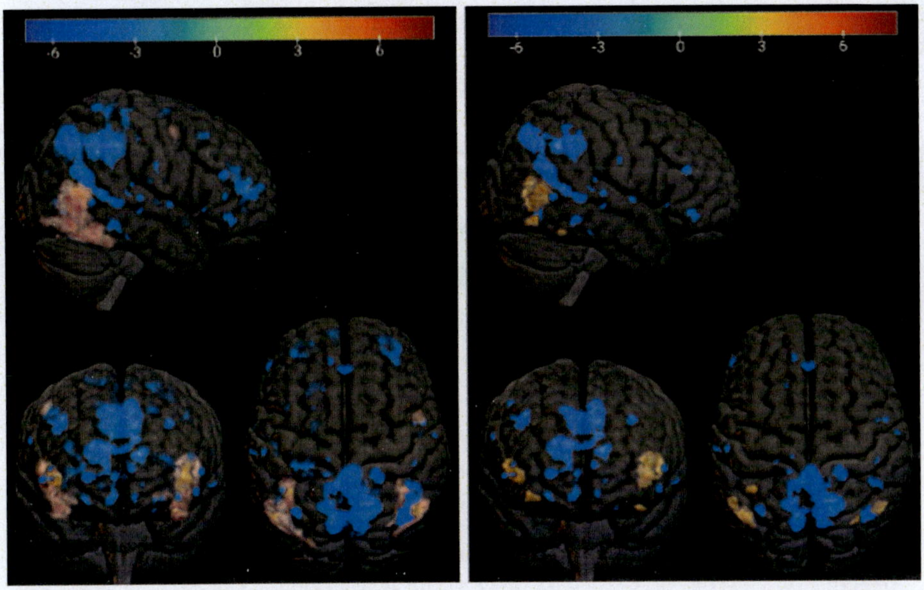

Fig. 1. Activation maps for healthy control (left) and patients with schizophrenia (right).

4 Conclusions

The need to use emotionally significant stimuli in experiments on the brain activation mapping of people with schizophrenia is due to the ambiguity of their perception of the surrounding reality [13]. Information or objects that do not cause any emotions in an ordinary person can trigger a psychotic state for patients.

The results of our studies show the specific behavior of the brain when perceiving threatening stimuli compared to neutral ones. Most of the regions involved show a decrease in the level of activation. Moreover, the overall level of activation in patients is significantly reduced compared to the healthy control group. This is especially clearly observed in the frontal and occipital lobes, which are responsible for obtaining visual information, its processing and interpretation.

Next step of the work seems in collecting additional information about structural, functional and effective connectivity by adding data of resting-state fMRI, diffusion MRI and MRI spectroscopy. Also, the current experiment needs to be expanded by presenting stimulus material of a different modality (for example, audio).

In the long term, it is necessary to conduct interdisciplinary studies addressing the problem of schizophrenia from different points of view. In addition to neurophysiology, it is necessary to collectively consider immunology and genetics data for the further development of objective patient assessment systems for the classification of the disease and the creation of early diagnosis systems.

Acknowledgements. The research presented above is an initiative internal research conducted by NRC «Kurchatov Institute» (order no. 1363 of June 25, 2019 «Biomedical technologies», paragraph 4.14).

References

1. GBD 2017 Disease and Injury Incidence and Prevalence Collaborators. Global, regional, and national incidence, prevalence, and years lived with disability for 354 diseases and injuries for 195 countries and territories, 1990–2017: a systematic analysis for the Global Burden of Disease Study 2017. Lancet (2018). https://doi.org/10.1016/S0140-6736(18)32279-7
2. Nenadic, I., Smesny, S., Schlösser, R.G.M., Sauer, H., Gaser, C.: Auditory hallucinations and brain structure in schizophrenia: voxel-based morphometric study. Br. J. Psychiatry **196** (5), 412–413 (2010). https://doi.org/10.1192/bjp.bp.109.070441
3. Modinos, G., Costafreda, S.G., van Tol, M.-J., McGuire, P.K., Aleman, A., Allen, P.: Neuroanatomy of auditory verbal hallucinations in schizophrenia: a quantitative meta-analysis of voxel-based morphometry studies. Cortex **49**(4), 1046–1055 (2013). https://doi.org/10.1016/j.cortex.2012.01.009
4. Allen, P., Larøi, F., McGuire, P.K., Aleman, A.: The hallucinating brain: a review of structural and functional neuroimaging studies of hallucinations. Neurosci. Biobehav. Rev. **32**(1), 175–191 (2008). https://doi.org/10.1016/j.neubiorev.2007.07.012
5. Downar, J., Crawley, A.P., Mikulis, D.J., Davis, K.D.: A multimodal cortical network for the detection of changes in the sensory environment. Neurosci. Biobehav. Rev. **3**(3), 277–283 (2000). https://doi.org/10.1038/72991
6. Hall, J., Whalley, H.C., McKirdy, J.W., et al.: Overactivation of fear systems to neutral faces in schizophrenia. Biol. Psychiatry **64**(1), 70–73 (2008)
7. Holt, D.J., Kunkel, L., Weiss, A.P., et al.: Increased medial temporal lobe activation during the passive viewing of emotional and neutral facial expressions in schizophrenia. Schizophr. Res. **82**(2–3), 153–162 (2006)
8. Surguladze, S., Russell, T., Kucharska-Pietura, K., et al.: A reversal of the normal pattern of parahippocampal response to neutral and fearful faces is associated with reality distortion in schizophrenia. Biol. Psychiatry **60**(5), 423–431 (2006)
9. Phillips, M.L., Williams, L., Senior, C., et al.: A differential neural response to threatening and non-threatening negative facial expressions in paranoid and non-paranoid schizophrenics. Psychiatry Res. **92**(1), 11–31 (1999)
10. Schneider, F., Grodd, W., Weiss, U., et al.: Functional MRI reveals left amygdala activation during emotion. Psychiatry Res. **76**(2–3), 75–82 (1997)
11. Schneider, F., Habel, U., Reske, M., Toni, I., Falkai, P., Shah, N.J.: Neural substrates of olfactory processing in schizophrenia patients and their healthy relatives. Psychiatry Res. **155**(2), 103–112 (2007)
12. Michalopoulou, P.G., Surguladze, S., Morley, L.A., Giampietro, V.P., Murray, R.M., Shergill, S.S.: Facial fear processing and psychotic symptoms in schizophrenia: functional magnetic resonance imaging study. Br. J. Psychiatry **192**(3), 191–196 (2008)
13. Lee, S.-K., et al.: Abnormal neural processing during emotional salience attribution of affective asymmetry in patients with schizophrenia. PLoS ONE **9**, e90792 (2014)

Logical Circuits of a ΣΠ-Neuron Model

Mukhamed Kazakov(✉)

Institute of Applied Mathematics and Automation of Kabardin-Balkar Scientific Centre of RAS (IAMA KBSC RAS), st. Shortanova 89 a, Nalchik 360000, KBR, Russia
kasakow.muchamed@gmail.com

Abstract. The paper proposes a schematic representation for implementing a ΣΠ-neuron model. Two types of digital and hybrid ΣΠ-neurons are presented with two layers that include multiplication and summing operations. The networks are built using digital and analog electronics. The multiplication operations in both networks are carried out identically. Summing operations in the digital network is implemented through a digital circuit - an adder, while in the hybrid framework it is implemented via an analogue-to-digital conversion and analogue voltage adder. To set up the threshold voltage in the hybrid and digital networks the hybrid and logic comparators are employed correspondingly. The results obtained can be used to construct ΣΠ-neural networks that could be embedded into hardware/software for controlling mechatronic systems.

Keywords: Artificial neuron · Logical Sigma-Pi neuron · Neural networks · Sigma-Pi neural networks

1 Introduction

An artificial neural networks (ANN) represent a special computational method utterly different from calculations performed by digital computers. The ANN structures and functions are simulating structures and functions of biological neural networks. The operating principle for artificial neural network involves information transferring between interconnected neurons for further processing. Information in ANN is memorized evenly in the form of weighting parameters of connections, and the capabilities of neural networks develop via learning by example and not through programming as in digital computers. In a general sense, artificial neural networks are parallel distributed systems or adaptive systems that can be trained to process information [1, 2].

Neural networks are successfully trained to solve problems of pattern recognition, search of associative memory, function approximation, optimization, and others. Usually learning process involves adjusting the weights of connections between neurons. A classic training scenario includes an input image sequence [3].

The ΣΠ-neuron and ΣΠ-neural network potential has been actively explored and applied since the 90s. The ΣΠ-neuron model is a generalization of the classical model of a formal neuron with a linear function. This model reflects better properties of its natural prototype and covers a much wider class of transformations [3, 4]. The article

A. V. Samsonovich et al. (Eds.): BICA 2020, AISC 1310, pp. 143–148, 2021.
https://doi.org/10.1007/978-3-030-65596-9_18

proposes functional and logical ΣΠ-neural models involving logical and digital elements.

2 Statement of the Problem

n practice neural networks are rather widely simulated on digital computers. The training procedures of the neural networks are quite convenient, as well as their application in solving certain problems. In a distributed system all network elements and operations are carried out programmatically. Consequently, we have parallel implementation of the artificial neural network training and sequential implementation of the digital computers. Therefore, this approach is not well-suited for full revealing of capabilities of the neural networks. Thus a need arises for a new approach to implementing artificial neural networks. This approach exists in practice [4] although its realization entails some technological difficulties while the software implementation of digital computers, in most cases, is lack of similar problems.

3 ΣΠ-Neuron Model

A ΣΠ-neuron is a type of artificial neuron. The summation function for a discrete ΣΠ-neuron is a multilinear function:

$$s(\mathbf{x}) = \theta + \sum_{k=1}^{N} \omega_k \prod_{i \in \mathbf{i}_k} x_i$$

were:

$\mathbf{x} = (\boldsymbol{x}_1, \ldots, \boldsymbol{x}_n)$ is a input vector;
ω_k are weights;
$\varnothing \neq \mathbf{i}_k \subseteq \{1, 2, \ldots, n\}$ is a synaptic cluster;
θ is the bias.

The multilinear summation function takes into account additional features of biological neurons [3]:

- the synapse can be formed with more than one input resulting in synaptic clusters;
- the contribution of each formal synaptic cluster to the value of the total input signal of the neuron is linear;
- the contribution of the synaptic connection is proportional to each individual input used in clustering.

Incoming signals are combined into clusters resulting in logical conjunction – the operation on two logical values. Each cluster has a corresponding synaptic weight. The conjunction determines what synaptic weights of clusters to be summed up. The ΣΠ-neuron can be represented as a two-layer network with an intermediate layer consisting of Π-elements and the Σ-neuron on the output layer:

$$y = \text{out}\left(\theta + \sum \omega_k s_k\right)$$

$$s_k = \prod_{i \in \mathbf{i}_k} x_i$$

The threshold function can be used as the activation function y of a ΣΠ-neuron:

$$out(\text{s}) = \begin{cases} 1, ifs \geq T, \\ 0, ifs < T, \end{cases}$$

where T – is the threshold value [4, 5].

ΣΠ-functions are polynomials, whose variables in the products are raised only to the first power. From this point of view, ΣΠ-neuron is a kind of high-order perceptron.

4 ΣΠ-Logic Gate Model

This section discusses two logical circuits of ΣΠ-neuron model: the circuit with an analog adder and the circuit with digital adder. Each circuit can be conventionally divided into two layers - Π-layer (conjunction) and Σ-layer (summation). The circuits basically differ from each other only in Σ-layer. The functional elements are described in detail in [6].

First we describe the Π-layer shared by the two circuits. This layer includes registers, 2-input AND gates and M-input AND gates (where M – number of neuron inputs). The digital signal enters Π-layer in the form of a binary M-dimensional vector, with respect to the number of inputs. The number of M-input AND gates corresponds to the number of synaptic clusters (see Fig. 1). In the most general case, to simultaneously implement all possible clusters, it is necessary to employ (2−2) gates. However, for practical purposes fewer gates might be more appropriate. For each M-input gate there is an M-bit register whose value determines what input signals participate to form the cluster (one input per register). The 2-input AND gate and the register determine which signals in the conjunction are important. Thus each M-input AND gate implement logical multiplication of input signals of the appropriate synaptic cluster.

Now consider the Σ-layer and analog adder circuit (electronic mixer). The synaptic function is realized through a set of register and DAC. The synaptic weights are stored in registers. DAC converts the M-input AND gate digital signal into the voltage level X_k with respect to the synaptic weight. An analog adder produces an output equal to the sum of the input voltages of all DACs:

$$U_s = k \sum_{k=1}^{N} X_k$$

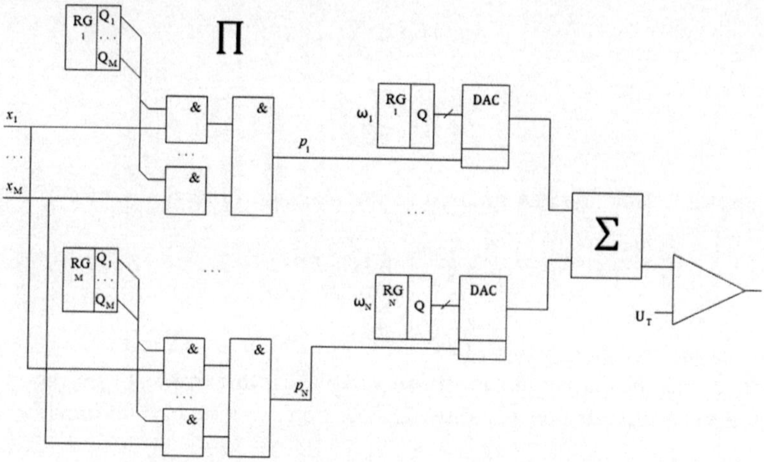

Fig. 1. Diagram of a ΣΠ-neuron and analog adder.

In order the input voltage U_s does not exceed the operating voltage U_0, the condition $k = 1/N$, should be met, where N is the number of synaptic clusters. A comparator performing a threshold-based activation function outputs a digital signal.

In general, for converting the values of synaptic weights into an analog signal it is feasible to use the DAC with reference-voltage-division. This type of DAC is quite easy to implement and may provide a fairly high-density chip. This is particularly important, since for one artificial neuron with M inputs, it is necessary to employ 2^M DACs, i.e. the same amount as M-input AND gates. Voltage divider circuit supplying only one voltage reference can be described by a general equation

$$X = \mu B = B \sum_{i=1}^{m} A_i \mu_i$$

where X – is the analog output signal;

B – is the voltage reference;
μ – is the DAC voltage gain;
μ_i – is the conversion factor of the i-th bit of the m-bit register;
$A_i \in \{0, 1\}$ – is the i-th bit of the register.

Σ-layer of the circuit with digital summation does not contain analog elements (Fig. 2). The digital registers store the value of the synaptic weight just as in summing amplifier circuit. However, the digital circuit performs addition of numbers and capable of cascade summing. Summation of consecutive numbers is not suitable here, since summing in this case takes N-1 clock cycles at N terms. When cascade summing round $\log_2 N$ down to reduce the number. If the sum exceeds this threshold value, the comparator outputs a single digital signal.

It is assumed that there are several learning methods for the register values corrections and one of them is training the artificial neural network computer model. Thus

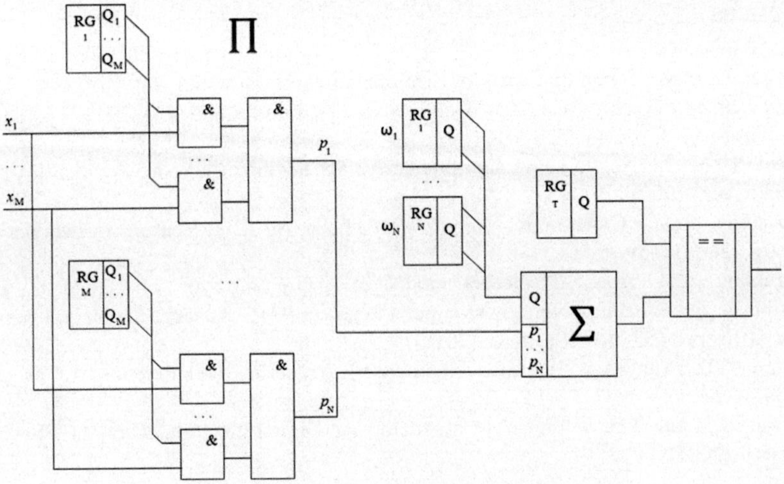

Fig. 2. ΣΠ-neuron with digital adder.

we get values for synaptic weights and clusters combined for the input signal. The results obtained are loaded into the corresponding registers and in this case parallel loaded registers are used.

Another method is in interactive training the artificial neural network employing a brain-computer interface. Each step of training involves the input-output analysis with software changing the register values in real time.

At last the method which requires an additional circuit that capable of directly training the network, including modifying weights and the clusters rebuilding for each learning stage. This method suggests storing the weight values in reversible counters. While training, synaptic weight increases or decreases with respect to the incoming signal to the counter.

5 Conclusion

The analog processing elements for the summation function compared with digital adder result in the speed benefit, but loses in accuracy. When building neural networks digital and analog components can be combined, according to accuracy or speed priority. Since the weight coefficients and synaptic configuration are determined by the registers values, the neural model can be trained quite simply, either using parallel data loading or modifying the set values.

Acknowledgements. The reported study was funded by RFBR according to the research project №18-01-00050-a.

References

1. Shibzukhov, Z.M.: Some questions of theoretical neuroinformatics. In: The book: XIII All-Russian Scientific and Technical Conference "Neuroinformatics - 2011". Lectures on Neuroinformatics, pp. 44–72 (2010)
2. Haykin, S.: Neural Networks: A Comprehensive Foundation, 2nd edn. Prentice Hall, Upper Saddle River (1999)
3. Shibzukhov, Z.M.: Constructive Learning Methods for ΣΠ Neuron Networks. Nauka, Moscow (2006). 159 p.
4. Shibzukhov, Z.M.: About ΣΠ-neuron models of aggregating type. In: Cheng, L., Liu, Q., Ronzliin, A. (eds.) Lecture Notes in Computer Sciences 9719. Advances in Neural Networks - ISNN 2016, pp. 657–664. Springer (2016)
5. Harmon, L.D.: Studies with artificial neurons, I: Properties and functions of an artificial neuron. Kybernetic **1**(3), 89–117 (1961)
6. Agarwal, A., Lang, J.H.: Foundations of Analog and Digital Electronic Circuits. Elsevier, San Francisco (2005)

Study of Neurocognitive Mechanisms in the Concealed Information Paradigm

Yuri I. Kholodny(✉), Sergey I. Kartashov, Denis G. Malakhov, and Vyacheslav A. Orlov

National Research Center "Kurchatov Institute", Moscow, Russia
kholodny@yandex.ru

Abstract. This work is a continuation of a research aimed at creating a forensic method for MRI diagnostics of hidden information in person. The article presents some results of the research in the paradigm of information concealment, during which the activity of brain structures in experiments with tests used in criminology was studied using fMRI and MRI compatible polygraph (MRIcP). In a group analysis of the experimental data obtained in the block design, brain zones are shown that are active in concealing personally significant and situationally significant information. Assumptions are made about possible differences in the functioning of neural networks involved in the implementation of concealment of information of these two types.

Keywords: MRI compatible polygraph · fMRI · Forensic diagnostics · Concealed information paradigm

1 Introduction

Study using a polygraph (SUP) is the most widespread method of forensic direction in applied psychophysiology.

Many countries around the world use polygraph in law enforcement practice, but in the early 1980s, US Congress experts expressed the opinion that "the basic theory of polygraph testing is only partially developed. The testing process is complex and not amenable to easy understanding. A stronger theoretical base is needed for the entire range of polygraph applications. Basic polygraph research should consider the latest research from the fields of psychology, physiology, and medicine; comparison among question techniques; and measures of physiological response" [1].

Decades have passed, but the question of the theory of polygraph testing in the United States remains open to this day, and according to the American Polygraph Association (APA), the probability of a correct result when conducting SUP during investigations is in the range of 0.89–0.91 [2].

The development of science led to introduction of new methods of experimental research into psychophysiology, and some of them, researchers tried to use for the purpose of detecting concealed information in a person, thereby creating an alternative to SUP. In particular, in 2001 the first experiments were conducted in which functional magnetic resonance imaging (fMRI) was used to detect concealed information in a person [3]. Scientists declared – "we examine directly the organ that produces lies, the

A. V. Samsonovich et al. (Eds.): BICA 2020, AISC 1310, pp. 149–155, 2021.
https://doi.org/10.1007/978-3-030-65596-9_19

brain… (and) this neurobiologically based strategy relies on identifying specific patterns of neural activation that underlie deception" [4].

FMRI research on the subject of detecting hidden information in a person has borrowed some elements of the SUP technology, was conducted mainly in the differentiation of deception paradigm, but significant progress in the scientifically based application of fMRI in forensic practice has not yet been achieved. Therefore, conducting further research in the field of forensic fMRI and improving the existing technology of SUP is relevant both in Russia and abroad.

Research on the complex of fMRI diagnostics was started at the National Research Center "Kurchatov Institute" (NRC "Kurchatov Institute") in 2018 [5], and for such studies, in particular, a prototype of a computer MRI compatible polygraph (MRIcP) was created. Taking into account the requirements of forensic diagnostics, the concealed information paradigm, was chosen as the methodological basis for research [6]. Within this paradigm, two widely used criminalistics tests were selected and the study of neural networks involved in concealing personally significant and situation significant information was started. For the methodological correctness of diagnostics of concealed information, a series of preliminary experiments were conducted, during which the possibility of measuring the dynamics of person response to the stimuli of selected tests perceived by them during fMRI was studied [7, 8].

2 Materials and Methods

23 men aged from 21 to 23 years old took part in experiments to create a system of MRI diagnostics of hidden information in a person. In each of the fMRI experiments, MRIcP was used to control the dynamics of the response of the studied person. Permission to conduct the fMRI experiments was obtained from the ethics committee of NRC "Kurchatov Institute".

Experiments were aimed at neuroimaging of the process of concealing information by a person, and the possibility of detecting differences in the activity of brain structures when hiding personally significant and situationally significant information was studied. In order to increase the motivation of participants in the study, a monetary reward was provided for strict compliance with the necessary instructions of the experimenter and the successful conduct of the experiment.

As a methodological tool that simulates the concealment of personally significant information, the test with a concealed name (TCN) of the SUP technology was used, during which the person under study (the subject) hid from the polygraph examiner his own name, presented in a row with five other names. The sets of names were presented to all subjects during the TCN five times. The series of names for each of the five repetitions began with the same name, and it had the number "0". All other names (including the name of the subject) were given in a random order that was unknown to the subject, and he was asked a question – "Your passport name is…?". Hiding their name in a row of other names, the subjects answered all questions – "no". The experimenter asked questions with intervals of about 20 s and with the obligatory account of the current dynamics of physiological parameters of a subject, simultaneously recorded by MRIcP. In order to increase concentration during the test, the subject after the TCN was asked how many times his name was sounded.

As a methodological tool that simulates the concealment of situationally significant information, the "guilty knowledge" test (GKT) was borrowed from the SUP technology. Before performing the GKT, the subject randomly selected one of five business cards, copied it on a piece of paper, and memorized the name and place of work of the person indicated on this business card. The subject was instructed to hide the signs of the selected business card (family name and place of work) from the experimenter who registered fMRI and MRIcP data. Before beginning the GKT, lying in the tomograph chamber, the subject always read what he wrote on a piece of paper, and refreshed his memory with the signs of a business card. The family name of the person (F) from the business card was presented to the subject during the test four times, and the place of work (W) – two times. The signs of the selected business card were presented in the following order: F – W – F – F – W – F. Thus, the signs of a business card were presented to all subjects during the GKT six times. Each of the six repetitions began by the signs of a business card that was not among the five business cards that were offered to choose from, and these signs had the number "0". All the signs of the five business cards were set in a random order unknown to the subject, and they were asked by the question – "Did you have the last name on your business card …?" or "The person with the business card works in …?". Hiding the signs of the chosen business card among other family names and places of work, the subjects answered all questions – "no". The experimenter asked questions with intervals of about 20 s and with the obligatory account of the current dynamics of physiological parameters of a subject, simultaneously recorded by MRIcP. In order to increase concentration during the test, the subject after the GKT was asked, how many times the family name and place of work of the person on the business card were sounded during the test.

Thus, the TCN and GKT lasted, respectively, 6–7 and 8–9 min and contained 25 and 30 stimuli, in response to which the experimenter registered fMRI and MRIcP data. Between the TCN and GKT the experimenter made another test lasting 11–12 min, which is not considered in the article, and between all tests, the experimenter made intervals of 5–8 min, during which the instructions for the next test were discussed.

The MRI data was acquired using a 3 T SIEMENS Magnetom Verio MR scanner. The T1-weighted sagittal three-dimensional magnetization-prepared rapid gradient-echo sequence was applied with the following imaging parameters: 176 slices, TR = 1900 ms, TE = 2.19 ms, slice thickness = 1 mm, flip angle = 9°, inversion time = 900 ms, FOV = 250 mm × 218 mm^2.

Preliminary studies [8] showed the feasibility of using ultra-fast scanning sequences for obtaining fMRI data at TR = 1110 ms with the following parameters: 51 slices, TE = 25 ms, slice thickness = 2 mm, rotation angle = 90°, FOV = 192 × 192 mm^2. Functional and structural MRI data were processed using the SPM8 software package (available at http://www.fil.ion.ucl.ac.uk/spm/software/spm8/). After converting DICOM files to NIFTI format, all images were manually centered to anterior commissure. The EPI images were corrected for magnetic field inhomogeneity using the FieldMap toolbox for SPM8. Then, the fMRI data signals were temporarily corrected. The need to apply this correction is explained by the use of an event-related scheme of the experiment. Anatomical and functional data were normalized in the ICBM stereotactic reference system. T1 images were divided into 3 tissue maps (gray, white matter, and cerebrospinal fluid). Functional data was smoothed using a 6 × 6 × 6 mm^3 FWHM Gaussian filter. Statistical analysis was performed using the Student t-statistics ($p < 0.05$ with correction for multiple comparisons (FWE)).

Cross-platform CONN software based on Matlab was used to calculate, display and analyze functional connectivity based on fMRI data. In this paper, 132 regions included in the CONN Atlas were used as regions of interest. After data preprocessing, the first-level connectivity assessment (individual), second-level random effects analysis, and hypothesis testing (group assessment) were performed. The Pearson correlation was chosen as the functional connectivity metric.

Of the 23 subjects, a group of 17 subjects (the "best" group) was identified, in which a high degree of vegetative reactions to information hidden in the TCN and GKT was detected with the help of MRIcP. fMRI data was analyzed both for the entire sample of subjects and separately for the "best" group.

3 Results

Recent meta-analyses, with all the variability of experiments conducted in the differentiation of deception paradigm and concealed information paradigm, agree that the front-parietal areas of the brain are involved in the implementation of lies and concealment of information [9]. According to foreign experts, "these regions include the anterior cingulate and surrounding medial prefrontal cortex; the ventrolateral prefrontal and insular cortex, bilaterally; portions of the left precentral, middle, and superior frontal gyrus; and the inferior parietal lobular and supramarginal cortex, bilaterally" [10].

Experiments performed at the NRC "Kurchatov Institute" established consistency between the results obtained and the conclusions of foreign meta-analyses: the concealment of personally significant and situational information by the subjects during the TCN and GKT revealed high activity of the same zones – frontal pole (bilaterally), insular cortex (bilaterally), and so on.

Group analysis of fMRI data of fragments showed that the activity of many areas of the brain structures during TCN consistently exceeded the activity of the same structures during GKT (at $p < 0.05$, adjusted for the expected proportion of false deviations – FDR). At the same time, in some zones – (CONN Atlas) angular gyrus_r, paracingulate gyrus_l, precuneous, thalamus (bilaterally) – the number of detected statistically significant voxels in GKT was 50–150% higher than in TCN (Fig. 1). The cross sections show a high similarity in spatial localization of the detected zones and a greater activation in the regions described above – thalamus (bilaterally), precuneous and angular gyrus_r.

An evaluation of MRI data obtained in response to the presentation of insignificant stimuli in the TCN and GKT found that they do not form a single set, although they are insignificant in their tests. It turned out that insignificant stimuli in the TCN caused significantly more activity of brain structures, than insignificant ones in the GKT.

The subjects who took part in the study were undergoing MRI for the first time and, although they gave their voluntary consent to participate in the experiments,

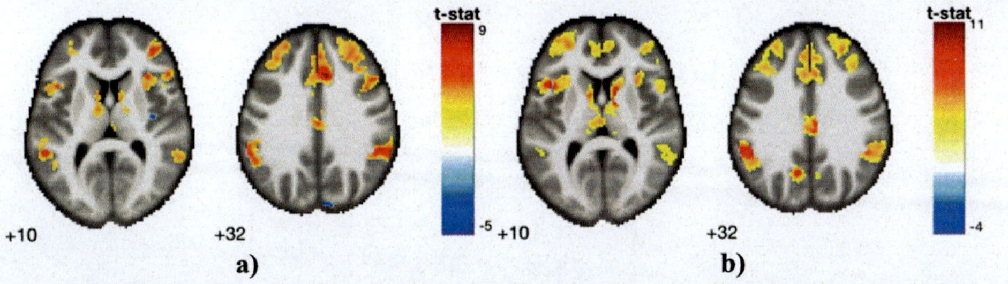

Fig. 1. Group statistical maps of activity of brain zones at the level z = 10; 32 in MNI space when performing: a) CNT and b) GKT.

nevertheless, according to their questions, they experienced some mental stress at the beginning of the experiment: this may explain the discovered feature of responding to insignificant stimuli of the TCN and GKT. Probably, the difference in response is a neurocognitive reflection of a phenomenon often observed in the practice of SUP, when the dynamics of vegetative reactions in the first test (i.e., in the first 10–15 min of testing) significantly differs from their dynamics in subsequent tests, during which the subject adapts to the conditions of the study and his psychophysiological state is stabilized.

To identify the neural networks involved in the process of hiding information, based on fMRI data from a group of 23 subjects, an analysis of the functional connectivity of brain zones was performed when performing the TCN and GKT (Fig. 2).

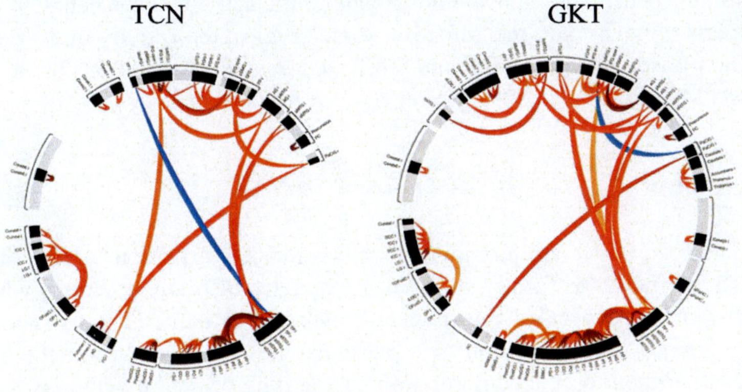

Fig. 2. Analysis of functional connectivity of brain areas during the test with a concealed name and the "guilty knowledge" test: a group of 23 subjects.

In order to identify the dominant connections, an analysis of the functional connectivity of brain areas was performed separately for the subjects from the "best" group (Fig. 3).

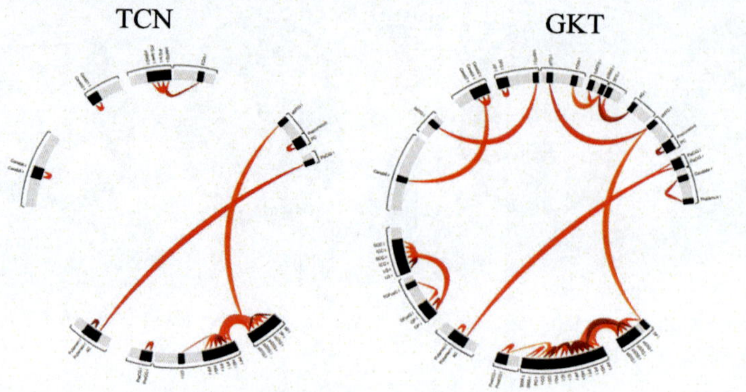

Fig. 3. Analysis of functional connectivity of brain areas during the test with a concealed name and the "guilty knowledge" test: 17 subjects – the "best" group.

The selection of this group showed that with the general trend of increased activity of neurostructures in the TCN compared to the activity of the same structures in the GKT, the activity of zones – (Atlas CONN) angular gyrus (bilaterally), paracingulate gyrus_l, precuneous, thalamus (bilaterally), caudate (bilaterally) – became more pronounced: the number of identified statistically significant voxels in GKT was 50–300% higher than in TCN.

The results of the analysis of functional connectivity of the brain areas of the subjects (Fig. 2 and 3) show a great similarity of neural networks involved in concealing personally significant and situationally significant information.

Apparently, special attention should be paid to the activity of the cerebellum, which foreign experts only casually mentioned in some articles: the activity of the cerebellum is consistently observed in the TCN and GKT, and in some areas of this brain structure, the activity in the GKT was 80% greater than in the TCN.

4 Conclusion

The data presented above are preliminary and require confirmation. At the same time, they indicate the need for further study of the activity of brain structures when using forensic tests of the concealed information paradigm. In particular, it is necessary to study the influence of the position of a particular test among other methodical tools used in the experiment on the configuration and activity of neural networks. It can also be assumed that a possible marker of hidden information will not be a single set of neural networks, but the redistribution of activity in a number of sensitive areas of the brain: this is indicated by the results of experiments.

Acknowledgements. The research presented above is an initiative internal research conducted by NRC "Kurchatov Institute" (order no. 1363 of June 25, 2019 "Biomedical technologies", paragraph 4.15).

References

1. Saxe, L., Dougherty, D., Cross, T., Langenbrunner, J., Locke, K.: Scientific validity of polygraph testing: a research review and evaluation – a technical memorandum. Polygraph **12**(3), 305 (1983)
2. Nelson, R.: Scientific basis of polygraph testing. Polygraph **44**(1), 41 (2015)
3. Spence, S.A., et al.: Behavioural and functional anatomical correlates of deception in humans. Neuroreport **12**(13), 2849–2853 (2001)
4. Ganis, G., et al.: Neural correlates of different types of deception: An fMRI investigation. Cereb. Cortex **13**(8), 830 (2003)
5. Kholodny, Y.I.: Criminalistic direction of applied psychophysiology (in press)
6. Kovalchuk, M.V., Kholodny, Y.I.: Functional magnetic resonance imaging augmented with polygraph: new capabilities. In: Advances in Intelligent Systems and Computing (eBook), pp. 260–265 (2019). (https://doi.org/10.1007/978-3-030-25719-4)
7. Orlov, V.A, Kholodny, Y.I., Kartashov, S.I., Malakhov, D.G., Kovalchuk, M.V., Ushakov, V.L.: Application of registration of human vegetative reactions in the process of functional magnetic resonance imaging. In: Advances in Intelligent Systems and Computing (eBook), pp. 393–399 (2019). (https://doi.org/10.1007/978-3-030-25719-4_51)
8. Kholodny, Y.I., Kartashov, S.I., Malakhov, D.G., Orlov, V.A.: Improvement of the technology of fMRI-experiments in the concealed information paradigm. In: Advances in Intelligent Systems and Computing(eBook) (2020). (in press)
9. Farah, M.J., et al.: Functional MRI-based lie detection: scientific and societal challenges. Nat. Rev. Neurosci. **15**(2), 123–131 (2014)
10. Rosenfeld, J.P.: Detecting Concealed Information and Deception. Recent Developments, p. 150. Elsevier Inc., Amsterdam (2018)

IT-Solutions in Money Laundering/Counter Terrorism Financing Risk Assessment in Commercial Banks

Sofya Klimova and Asmik Grigoryan(✉)

NRNU MEPHI, Kashirskoye shosse 31, Moscow, Russia
sofi_ko@mail.ru, asmik.grigoryan@gmail.com

Abstract. In accordance with the FATF Recommendations of 2012, countries and their financial units are to manage Money Laundering/Counter Terrorism Financing risk (hereinafter – ML/CTF risk). Such a process consists of the following main steps: risk identification, assessment and mitigation. Besides the document supervises the requirements regarding the efficient use of financial monitoring resources. Commercial banks often face the problem of inadequate identification and assessment due to the lack of true and undistorted information about their clients. For the first, financial monitoring services are considered to be non-profitable in accordance with bank management. As a rule, resources of financial monitoring in commercial banks are rather limited to buy data bases containing identification information about physical and juridical bodies. For the second, sometimes financial institutions (commercial banks in particular) don't manage to collect enough information about new client without any financial history (for example start-ups or individuals who just open banking accounts for the first time). As a result, lack of information resources affects the quality of ML/CTF risk assessment. The article is devoted to the ways of such problems solution with the use of contemporary IT – systems and technologies. Automatization of these procedures will lead to a significant growth of financial monitoring efficiency in commercial banks and other financial institutions.

Keywords: AML/CFT · Money laundering risk · Counter terrorism financing risk · Risk assessment · Commercial banks · Information technologies

1 Introduction

Nowadays, commercial banks throughout the world work out and integrate systems of ML/CTF risk management in line with the FATF recommendations and requirements of national legislation. In accordance with the Russian Financial Intelligence Unit methodology, the process of ML/CTF risk management consists of the following main procedures (see Fig. 1):

- risk identification;
- risk assessment;
- risk reduction or mitigation.

A. V. Samsonovich et al. (Eds.): BICA 2020, AISC 1310, pp. 156–164, 2021.
https://doi.org/10.1007/978-3-030-65596-9_20

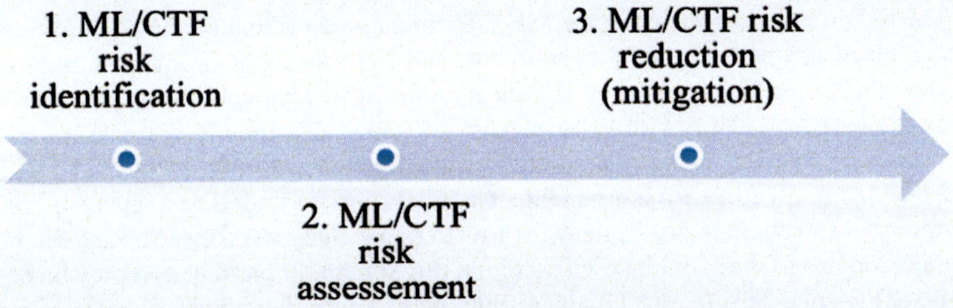

Fig. 1. Major steps of ML/CTF risk management.

It worth mentioning that there are two main approaches in ML/CTF risk management. In line with the first one Money Laundering Risk and Counter Terrorism Financing risk are combined in a single risk. In particular, the following countries introduced unit systems of their management: Russian Federation and the majority of the Commonwealth of Independent States.

In accordance with the second approach, these two risks are considered to be separate categories. Such countries as Australia, Belgium, New Zealand, USA follow these approach. As to the results of our previous scientific work we stick to this approach either [1]. In particular, this article highlights the common problems of ML/CTF risk assessment.

2 Types of ML/CTF Risk and Methods of Their Assessment

In the banking practice we may single out the following three types of the risks as shown in Fig. 2:

- clients' risk (risk posed to financial institutions by their clients);
- risk of financial organization (characterizes the level of financial institution involvement into suspicious operations and deals);
- national risk (shows the degree of the whole country volumes involvement into suspicious activity).

In particular, this article highlights the problems of ML client risk assessment.

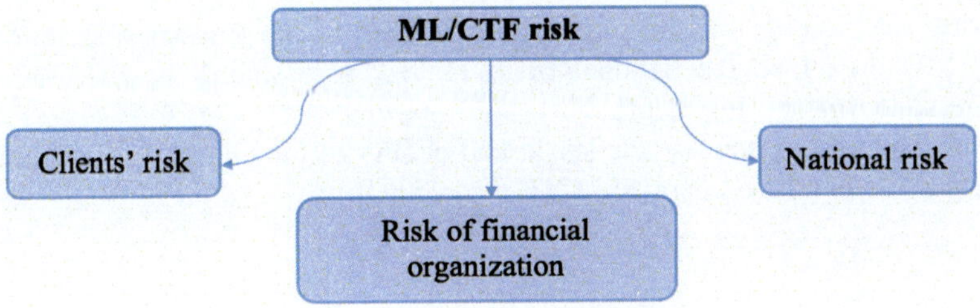

Fig. 2. The types of ML/CTF risk.

The risk oriented approach in the ML/CTF sphere means that financial units have to take tough measures in the field of high risk and light measures in the low risk zone. Adequate risk assessment plays a significant role in this process. The key task is to assess the level of such risk in order to take adequate measures to mitigate it.

There is a need to make quantitative evaluation of such risk using mathematical model or method of expert assessment. It is necessary to underline that there are no specific rules or requirements to the way how to make such assessments. Each financial organization has to work out its own policy in this sphere. In particular, it may be based on the experience of different Financial Intelligent Units (hereinafter – FIU) or international organizations. In particular, the Basel Committee on Banking Supervision recommends to use the following model.

According to this model, in the first phase bank should consider all relevant inherent risk factors such as clients, products and services, channels and geographies in order to determine its risk profile. Once the inherent risks have been identified and assessed, internal controls must be assessed to determine how effectively they can mitigate the overall risks. Available internal controls are evaluated for their effectiveness in mitigating the inherent ML/TF risk and to determine the residual risk rating. The residual risk rating is used to indicate whether the ML risks within the bank are being adequately managed [1, 2].

For example, some countries, supranational organizations and financial institutions use special tables (matrix). Further we may see such matrix works out by the World Bank in the Fig. 3.

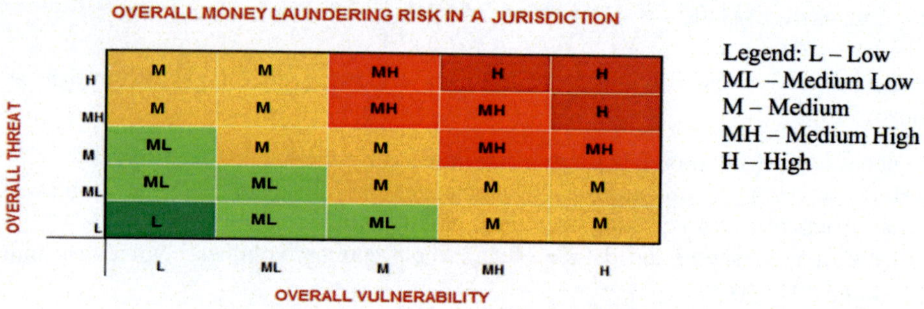

Fig. 3. The risk map for locating the risk level of a country [3].

It worth mentioning that the same access is used in the Russian Federation. The National FIU organized the national ML/CTF risk assessments in 2018. It's main results are shown in the Fig. 4 [4].

One of the ways to increase the speed and quality of the client (whole client base) risk assessments is to use client ML/CTF (score) matrix.

Table 3 – Risk matrix

		Threat		
		Low	Medium	High
Vulnerability	More vulnerable	Medium	High	High
	Vulnerable	Low	Medium	High
	Less vulnerable	Low	Low	Medium

Level of risk	
High	The method is used by terrorists or terrorist groups for the purposes of FT because of the high level of accessibility and medium or low visibility.
Medium	The method can be used by terrorists or terrorist groups for FT purposes due to a certain level of the aggregate availability and visibility.
Low	Terrorists or terrorist groups cannot practically use the method for FT purposes due to existence of a necessary level of the aggregate visibility and accessibility.

Fig. 4. Risk matrix

3 Modern IT-Tools for Big Data of Financial Organizations

At the moment, there are a huge number of modern tools for data scientists (analysts) that allow to build matrices for each client or distribute all bank clients according to this matrix. The data scientist also needs to be proficient with a variety of tools, including reporting tools. SAS EBI, or Suite of Business Intelligence Applications is a suite of business intelligence applications that includes reporting tools. SAS is one of the major players in the business intelligence industry, along with SAP, IBM, Salesforce and other market leaders. This row can be continued with programs and platforms such as Tableau, Datapine, Weka, PythonReports. These are all tools that allow to create database reports, visualize data, plot graphs, matrices, etc., which simplifies large datasets and turns them into an easy-to-understand format.

It is necessary to underline that results of such assessments depend significantly on the quality of information about clients. Commercial banks and other financial institutions are to have at their disposal reliable information that a client (individual or juridical body) is involved in suspicious financial activity. Therefore, we need a system for the prompt collection and (scoring) processing of data about customers and related persons (representatives, beneficial owners) from open sources.

A financial institution consists of many divisions responsible for operational activities, customer service, methodology, security, etc. Each of them may have their own means of automating specialized work. In fact, a financial institution is the owner of a huge array of working information of a different nature from completely different sources. At the same time, it is also duplicated: for example, data about the same client in different forms may be contained in the information systems of several departments. It must be able to put together and then we can get significant benefits from it. The goal is to ensure the availability of up-to-date and complete information about each client and quick access to it.

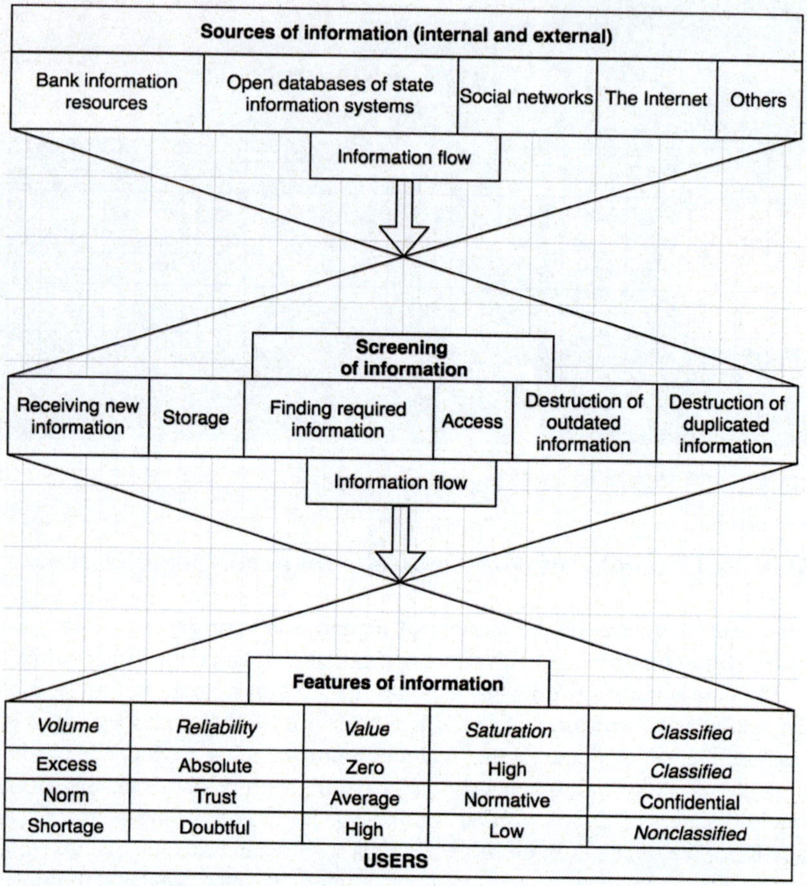

Fig. 5. Collection, processing and structuring information.

Consider the procedure for collecting, processing and structuring information (see Fig. 5). There are three main stages here:

- defining the purpose of selecting information from the general stream;
- collecting the necessary information;
- provision of storage (this stage also provides for the development of a system of features by which the necessary information can be found, it is advisable to use the

information value indicator as one of the features, in this case, especially valuable information can always be near).

The task of working with data becomes one of the key ones. Getting them from internal and external sources (information resources of the bank, open databases of state information systems, social networks, the Internet, etc.) is the first step. The second is to organize an infrastructure in which the data will be properly structured with the ability to quickly search, that will require high-performance software and hardware solutions.

The system for accumulating all possible data sources is an integration platform that connects to existing databases, connects to external sources and processes everything received according to a given algorithm in the interests of a financial institution and a specific final user.

Further, there is a need to develop a system of analytical indicators for risk assessment that responds to the full the all relevant requirements to such a system regarding the object and subject of such analysis.

4 The System of Analytical Indicators for Risk Assessment

In order to form a matrix for the client's risk assessment (client ML/CTF matrix), the indicators are selected according to the method, the main steps of which are listed below (see Fig. 6).

The method of selection of indicators	
Step 1	The choice of indicators that have a direct impact on the client's risk assessment (for each client category)
Step 2	Ranking of the selected indicators according to the degree of significance
Step 3	Excluding indicators with low significance scores
Step 4	Comparative analysis of indicators based on a pairwise comparison matrix in order to reduce the subjectivity of expert assessments
Step 5	Selection of indicators with the highest sum of points in expert assessments, taking into account the established ranks
Step 6	Formation of a set of key indicators by the method of dynamic assessment, operating with such a characteristic as the strength of the impact of changes in one or another indicator on the client's risk assessment

Fig. 6. The method of selection of indicators.

Consider a little more in the Step 2. Each financial institution may itself identify a particular method of assigning weights to certain indicators. Below we distinguish two such method.

The first method allows the indicators to be ordered according to the degree of increase or decrease in their impact depending on the features of the client in question. The results of ranking n indicators by m experts can be presented in the form of Fig. 7.

Expert	The indicators under consideration				
	x_1	x_2	...	x_{n-1}	x_n
1	k_{11}	k_{12}	...	$k_{1(n-1)}$	k_{1n}
2	k_{21}	k_{22}	...	$k_{2(n-1)}$	k_{2n}
...	...	...	...	...	...
m-1	$k_{(m-1)1}$	$k_{(m-1)2}$	...	$k_{(m-1)(n-1)}$	$k_{(m-1)n}$
m	k_{m1}	k_{m2}	...	$k_{m(n-1)}$	k_{mn}

Fig. 7. Results of a survey of experts on the indicators under consideration.

An assessment of the importance of a particular indicator is carried out by a group of specialized experts, and each of them presents its own vector of assessments for this group of indicators. The indicators are arranged taking into account their importance in accordance with the accepted order:

1. The expert arranges the indicators in descending order of importance from left to right.
2. Each indicator is assigned a score from n to 1 (the most important - n and then in descending order to 1).
3. For each indicator the sum of the estimates is calculated, then - the share of all the sums received.

In the form of a formula, this can be represented as.

$$\omega_j = \frac{\sum_{i=1}^{m} k_{ij}}{\sum_{j=1}^{n} \sum_{i=1}^{m} k_{ij}}, \quad (1)$$

where k_{ij} is the score given to the j-th indicator by the i-th expert.

Thus, the weighting coefficient ω_j is defined as the ratio of the sum of expert opinions on the j-th indicator to the sum of expert opinions on all indicators.

The second method is based on the assessment by experts of the importance of a particular indicator on a certain scale, for example, from 0 to 10. At the same time, it is allowed to evaluate the importance by fractional values or assign the same value from the selected scale to several indicators. The scoring table is presented in the same way as in the previous method (see Fig. 7).

The algorithm for calculating the weighting coefficients is as follows:

1. Each expert puts down marks for all indicators within a given scale.
2. The estimates are recalculated according to the formula

$$k'_{ij} = \frac{k_{ij}}{\sum_{j=1}^{n} k_{ij}}, \tag{2}$$

3. Further, as with the ranking method, the estimates obtained for each indicator are summed up and normalized.

The main task of selecting indicators is to tidy up information flows. The effect of this action will be the ability to quickly exploit "bottlenecks", key blocks and key indicators that can be considered satisfactory. Moreover, the obtained results are relevant, the efficiency and quality of decisions are higher, which saves time for assessment and decision-making. This algorithm will allow to select indicators for each differentiated group for different categories of customers. The result is a differentiated set of a limited number of the most significant indicators for the client's risk assessment.

5 Conclusions

As a result, we should get a matrix about the client with the assignment of a risk rating to him. It may be used both be client manager before opening an account to an applicant and by financial monitoring specialist during clients' suspicious activity analysis. Availability of negative assessments enables commercial banks to take measures aimed at combating ML/CTF, they are (Fig. 8):

1) account opening rejection;
2) transaction cancellation;
3) assets freeze;
4) banking agreement dissolve.

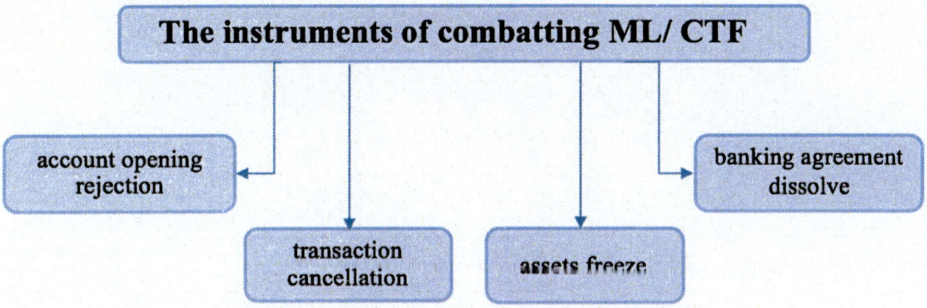

Fig. 8. The instruments of combating ML/CTF.

The article shows that the use of contemporary information technologies may significantly reduce costs of commercial banks while implementing the risk-oriented approach in AML/CTF sphere. In particular, integration of IT –systems of client's ML/CTF risk assessment will increase the speed of the described business processes and financial monitoring services efficiency as a whole.

References

1. Kovaleva, S.E.: Sovershenstvovanie upravleniya riskami kreditnikh organizaciy v sfere protivodeystviya legalizacii dokhodov, poluchennikh prestupnim putem [Upgrading of AML/CTF risk management in commercial banks]. Ph.D. thesis. Moscow (2013). 162 p.
2. Contemporary approaches to money laundering/terrorism financing risk assessment and methods of its automation in commercial banks. J. Procedia Comput. Sci. **169**, 380–387 (2020). https://doi.org/10.1016/j.procs.2020.02.233. Accessed 29 July 2020
3. The World Bank Risk Assessment Methodology. https://www.fatf-gafi.org/media/fatf/documents/reports/risk_assessment_world_bank.pdf. Accessed 29 July 2020
4. National terrorism financing risk assessment 2017–2018. https://www.fedsfm.ru/content/files/documents/2018/riskft_eng.pdf. Accessed 29 July 2020

Expandable Digital Functional State Model of Operator for Intelligent Human Factor Reliability Management Systems

Lyubov V. Kolobashkina[1(✉)], Mikhail V. Alyushin[1], and Kirill S. Nikishov[2]

[1] National Research Nuclear University MEPhI (Moscow Engineering Physics Institute), Kashirskoe Shosse 31, Moscow 115409, Russia
{LVKolobashkina,MVAlyushin}@mephi.ru

[2] PJSC «Mosenergo», Vernadskogo prospect, 101 bld. 3, Moscow 119526, Russian Federation
nikishovks@mosenergo.ru

Abstract. The relevance of the human factor (HF) reliability management is substantiated due to the reliable prediction of a possible change in the functional and psycho-emotional state (PES) of operators for controlling potentially hazardous objects (PHO). It is shown that PES modeling based on digital behavioral models (DBM) is a modern tool for such forecasting. Personalization of the DBM is carried out by taking into account individual bioparameters, which are registered, as a rule, using remote non-contact biometric technologies. The low reliability of the forecast in the event of emergency situations is highlighted as the main drawback of the existing DBMs. The main reason is the lack of reliability of the registration of bioparameters when using one or a limited number of biometric technologies. The relevance of the development of an expandable DBM, which allows to eliminate the indicated drawback, has been substantiated. The study proposes an expandable DBM that allows you to expand the range of processed biometric data. At the same time, it becomes possible to integrate biometric data obtained using remote non-contact technologies with data obtained using wearable biometric devices, such as, for example, bracelets. The developed DBM makes it possible, simultaneously with the implementation of the forecast, to monitor the health status of PHO personnel.

Keywords: Intelligent bioparameters processing · Human factor control · Digital behavioral model · Expanded bioparameters composition

1 Introduction

The reliability of the human factor (HF) is an integral part of the problem of ensuring the safe and trouble-free operation of potentially hazardous objects (PHO). One of the aspects of ensuring the reliability of the HF is the timely identification of PHO management operators who are in an inadequate, stressful or very tired state. In this regard, the task of making a reliable forecast of a possible change in the PHO management

A. V. Samsonovich et al. (Eds.): BICA 2020, AISC 1310, pp. 165–172, 2021.
https://doi.org/10.1007/978-3-030-65596-9_21

operator psycho-emotional state (PES) in the short and long term becomes especially relevant.

The short-term forecast assumes an assessment of the possible change in the PES over the next hours and, as a rule, does not go beyond the current work shift. The reliability of this type of forecast is especially important in the event of abnormal and emergency modes of PHO operation, as a rule, accompanied by a high level of psychological and physical stress on the operators of object control.

A long-term forecast of a possible change in the PHO management operators PES is carried out on the basis of an analysis of the changes dynamics in biometric indicators over fairly long time intervals, usually constituting units and tens of months. The results of this type of forecast make it possible to reasonably plan the production load. At the same time, they should be considered as a tool for identifying hidden disorders in health.

The basis for the implementation of short-term and long-term forecasts is the digital behavioral models (DBM). These models, as a rule, involve the analysis of the available biometric data characterizing the operator's current PES, with their subsequent extrapolation. Most of the models used in practice are based on the analysis of a limited range of bioparameters, for example, characterizing the work of the operator's cardiovascular system [1–4]. At the same time, to measure these bioparameters, as a rule, a limited number of biometric technologies are also used [5–7]. Typical examples are biometric bracelets that use optical technology to record heart rate. The reliability of measurements of such wearable biometric devices largely depends on the level of human motor activity.

These circumstances reduce the reliability of the forecast based on the DBM in complex and changing conditions of production activities, caused, for example, by emergency and abnormal modes of PHO operation. For these situations, characterized by increased levels of physical and mental stress, obtaining a reliable forecast is most important. To increase the reliability of the recorded bioparameters in such situations, it is desirable to simultaneously use several different biometric technologies based on different physical measurement principles [8]. For example, they measure bioparameters simultaneously in the visible and infrared ranges of optical radiation, as well as in the acoustic range [9]. This makes it possible to maintain the required reliability of the bioparameters measurement while reducing the validity of the data obtained by one or several of the applied biometric technologies.

The creation of new high-tech sensors makes it possible to receive and process additional biometric information, which makes it important to develop behavioral models with the possibility of expanding the composition of analyzed bioparameters. For example, the use of terahertz biometric sensors [10–12], as well as optical sensors for analyzing the face image [13] and measuring the oxygen concentration in the blood, makes it possible to take into account the parameters characterizing the state of metabolism in forecasting, as well as to timely detect health disorders.

The aim of the study is to develop an expandable DBM that allows for a reliable forecast of a possible change in the PHO management operators PES, focused on the use in intelligent systems for managing the reliability of the HF at these facilities.

2 State of Research in This Area

A typical form of a behavioral model used to predict a possible change in PES without taking into account the effect of accumulation of fatigue is presented in [9]. The model is designed to implement an iterative modeling process:

$$G_{M+1} = G_M + \Delta(G_{M-1}, G_{M-2}, G_{M-3}, \ldots, G_{M-N}), \tag{1}$$

where M – the number of the iteration step of the modeling process;

$\Delta(G_{M-1}, G_{M-2}, G_{M-3}, \ldots G_{M-N})$ – the expected change in the value of the value G_M, characterizing the operator's PES, at the iteration step $M + 1$;

N – the order of the extrapolating function.

The dimensionless value G_M considered in the study [9] is determined based on the integral processing of bioparameters recorded by three different remote non-contact technologies:

$$G_M = \sum_k B_k G_{Mk}, k = 1, 2, 3. \tag{2}$$

where B_k $(0 \leq B_k \leq 1)$ – the coefficients of the significance of the particular characteristics G_{Mk} for determining the characteristic G;

G_{Mk} – partial PES estimates using visible optical technology ($k = 1$), infrared optical technology ($k = 2$), and acoustic technology ($k = 3$).

The limitation of this approach is due to the fixed number ($k = 1, 2, 3$) of the analyzed technologies for recording bioparameters, as well as the lack of taking into account the possible change in the values of the B_k parameters when the conditions of production activity change. For example, an increase in the level of acoustic noise in the event of abnormal conditions of PHO operation, as a rule, leads to a decrease in the significance of the PES estimates when using acoustic technologies for recording personnel bioparameters.

In the study [14], to take into account the effect of fatigue accumulation, it is proposed to use a special function $R(M)$, which takes into account in modeling the total level of increased production load that causes fatigue:

$$R(M) = (1/T_0) \sum_{M=1} (G_M - G_0) dt, \tag{3}$$

where dt – the time interval corresponding to the simulation step;

G_0 – the individual value of the G function for normal operating conditions corresponding to normal production load,

T_0 – the individual value of the maximum working time for G_0 (for a healthy worker, the value of T_0 significantly exceeds the duration of the work shift).

A behavioral model that takes into account the effect of accumulation of fatigue is as follows:

$$G_{M+1} = G_M + \Delta(G_{M-1}, G_{M-2}, G_{M-3}, \ldots, G_{M-N}) \cdot (1 + \alpha \cdot (R(M)/G_0)), \quad (4)$$

where α – the individual degree of influence of fatigue taking into account the body's resources on physiological parameters.

In the model under consideration, individual parameters (G_0. T_0, α) ensure its adaptability. These parameters are determined on special tests during periodic examinations, or in the process of training on simulators.

The disadvantage of this model is also a rigidly fixed set of biometric technologies used.

3 The Essence of the Proposed Approach

To reliably predict a possible change in PES, the study proposes the use of an expandable digital behavioral model focused on the use in intelligent control systems for the reliability of the HF in PHO. The proposed model makes it possible to make a reliable prediction using an extended set of biometric technologies, for example, based on the registration of terahertz radiation, as well as on the measurement of the oxygen concentration in the blood (blood oxygen saturation SpO2). For this, when determining the value of G_M in the behavioral model (4), instead of expression (2), a dynamically changing function G^*_M is used, defined by expression (5):

$$G^*_M = \sum_k B_{Mk} G_{Mk}, \; k = 1, 2, 3, 4, 5, \ldots, K, \quad (5)$$

where K – the total number of biometric technologies used (in particular, $k = 4$ for terahertz technology, $k = 5$ for the technology for determining blood oxygen saturation SpO2);

B_{Mk} – current values of the significance coefficients for these technologies.

The significance coefficients are not static. Their values may change with changes in the conditions of production activities of personnel, including changes in:

- the level of illumination in working rooms;
- climatic conditions of work (air temperature, humidity, oxygen concentration in the air, concentration of CO_2 and other gases and aerosols);
- the level of acoustic and vibration noise and interference;
- motor activity of personnel;
- work clothing (for example, use of personal respiratory protection).

The above factors determine the possibility of using biometric technology and the reliability of the obtained biometric results. So, for example, in the presence of stable lighting and a high level of acoustic noise G_1, G_2> G_3. In the absence of stable lighting G_2> G_1. With a high level of high-frequency electromagnetic interference G_1, G_2, G_2, G_5> G_4.

To determine the values of $B_{(M+1)k}$, it is proposed to use the majority principle. Its essence is as follows.

1. Distances are determined $P_{ij} = \rho(G_i^*, G_j^*)$:

$$\rho(G_i^*, G_j^*) = \sqrt{(G_{Mi} - G_{Mj})^2 + \ldots + (G_{(M-Q)i} - G_{(M-Q)j})^2} \tag{6}$$

between functions

$$G_i^* = \{G_{Mi}, G_{(M-1)i}, G_{(M-2)i}, \ldots, G_{(M-Q)i}\}, i = 1, \ldots, K; \tag{7}$$

$$G_j^* = \{G_{Mj}, G_{(M-1)j}, G_{(M-2)j}, \ldots, G_{(M-Q)j}\}, j = 1, \ldots, K; i \neq j, i > j, \tag{8}$$

where Q – the number of analyzed steps of the iterative process (the value $Q \cdot \Delta t$ corresponds to the analyzed time interval).

For a quiet working environment, the value of Q can correspond to fairly long time intervals, measured by hours. For a rapidly changing production environment, the value of Q should correspond to the rate of change of the factors listed above.

2. The average distance $\bar{P}$ is determined:

$$\bar{P} = (2/(K \cdot (K - 1))) \cdot \sum_i \sum_j P_{ij}. \tag{9}$$

3. With the help of threshold discrimination, F of the most similar functions G_i^{**} and G_j^{**}, for which $|P_{ij} - \bar{P}| < \beta \cdot \bar{P}$, where β determines the value of the discrimination threshold, are distinguished. For most PHO, β = 0,1 – 0,2.
4. The average value of the selected functions $\bar{G}^{**}$ is used to determine the normalized values of $B_{(M+1)k}$ for the next iteration step:

$$\overline{G}^{**} = (1/F) \cdot \sum_{l=1}^{F} G_l^*; \tag{10}$$

$$B_{(M+1k)} = A \cdot C/(A + |G_k^* - \overline{G}^{**}|), \tag{11}$$

where the parameter A = 1–0,1 specifies the slope of differentiation of the parameters $B_{(M+1)k}$, and the value C is determined from the normalization condition

$$\sum_{k=1}^{K} B_{(M+1)k} = 1 : \tag{12}$$

$$C = 1/(A \cdot (K \cdot A + \sum_{k=1}^{k} |G_k^* - \overline{G}^{**}|)). \tag{13}$$

4 Results of Experimental Testing

The proposed approach was tested when monitoring the PES of tested employees when the conditions of production activity changed. Figure 1 shows the results of dynamic adaptation of the model to changing conditions of production activity.

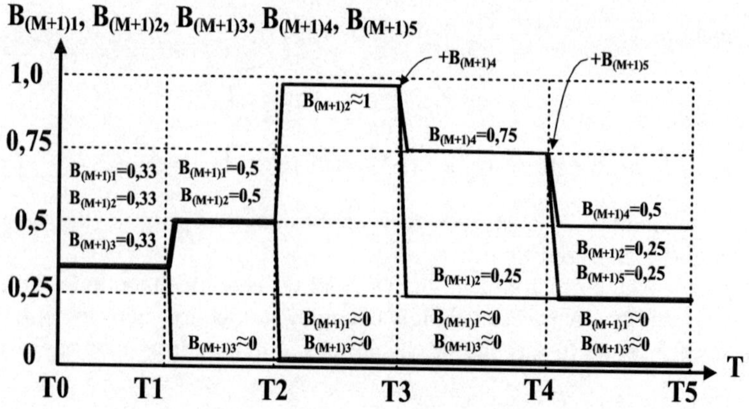

Fig. 1. An example of dynamic model adaptation.

At the initial time moment T0 for normal production conditions, all biometric technologies ($K = 3$) had the same significance equal to 0.33. At the moment of time T1, acoustic noises of considerable power appeared. As a result, the importance of acoustic technology has significantly decreased $B_{(M+1)3} \approx 0$. The importance of optical technologies has increased $B_{(M+1)1} = B_{(M+1)2} = 0{,}5$. At time T2, illumination instability arose, which led to a decrease in the importance of simple optical technology $B_{(M+1)1} \approx 0$ and an increase in the role of infrared technology $B_{(M+1)3} \approx 1{,}0$.

At time T3, biometric measurements based on terahertz technology became available. As a result, there were changes in the model - $B_{(M+1)4} = 0{,}75$, $B_{(M+1)2} = 0{,}25$. On the time interval T4–T5, biometric data on blood oxygen saturation appeared, which led to a change in the model parameters: $B_{(M+1)4} = 0{,}5$, $B_{(M+1)2} = B_{(M+1)5} = 0{,}25$. The values of the parameters $B_{(M+1)1} \approx 0$ и $B_{(M+1)3} \approx 0$ have not changed.

Experimental studies have shown that the use of the developed model makes it possible to preserve the reliability of the recorded bioparameters and predict a possible change in PES in the event of the previously considered destabilizing factors.

5 Areas of Possible Application of the Developed Model

The developed behavioral model is focused on use in intelligent control systems for the reliability of the HF in PHO.

When making a forecast, the model makes it possible:

- take into account the individual properties of a particular operator;
- compensate for the negative impact of external destabilizing factors when the conditions of production activity change;
- to adapt the model to the characteristics of the applied biometric technologies.

6 Conclusion

Thus, the DBM proposed in the study makes it possible to carry out a reliable forecast using an extended set of biometric technologies, including those based on the registration of terahertz radiation, as well as on the measurement of the oxygen concentration in the blood.

The approach makes it possible to naturally integrate bioparameters registered using both remote non-contact technologies and wearable contact biometric devices, which, first of all, include modern biometric bracelets. This permits to increase the accuracy of predicting a possible change in the PES of operators, which is of fundamental importance for ensuring the reliable and safe operation of PHO. Long-term forecasting should be seen as a tool for protecting the health of PHO staff.

Acknowledgments. The study was carried out with the financial support of Public Joint-Stock Company for Power and Electrification Mosenergo within the framework of contract No. 2G-00/19-231 dated February 28, 2019 "Experimental testing of remote non-contact means of continuous monitoring of the current state of the power unit driver".

References

1. Zhang, J., Nassef, A.M., Mahfouf, M., Linkens, D.A., Elsamahy, E., Hockey, G.R.J., Nickel, P., Roberts, A.C.: Modelling and analysis of HRV under physical and mental workloads. In: 6th IFAC Symposium on Modelling and Control in Biomedical Systems, vol. 6, Reims, France, pp. 189–194 (2006)
2. Ting, C.-H., Mahfouf, M., Nassef, A.M., Linkens, D.A., Panoutsos, G., Nickel, P., Roberts, A.C., Hockey, G.R.J.: Real-time adaptive automation system based on identification of operator functional state (OFS) in simulated process control operations. IEEE Trans. Syst. Man Cybern. **40**(2), 251–262 (2010)
3. Luczak, H., Raschke, F.: A model of the structure and behaviour of human heart rate control. Biol. Cybern. **18**, 1–13 (1975)
4. Mahfouf, M., Zhang, J., Linkens, D.A., Nassef, A.M., Nickel, P., Hockey, G.R.J., Roberts, A.C.: Adaptive fuzzy approaches to modelling operator functional states in a human machine process control system. In: IEEE International Conference on Fuzzy Systems, London, UK, vol. 1, pp. 234–239 (2007)
5. Wilson, G.F., Russell, C.A.: Real-time assessment of mental workload using psychophysiological measures and artificial neural networks. Hum. Factors **45**, 635–643 (2003)
6. Gevins, A., Smith, M.E.: Neurophysiological measures of cognitive workload during human-computer interaction. Theoret. Issues Ergon. Sci. **4**, 113–131 (2003)

7. Ioannou, S., Gallese, V., Merla, A.: Thermal infrared imaging in psychophysiology: potentialities and limits. Psychophysiology (USA) **51**(10), 951–963 (2014)
8. Berlovskaya, E.E., Isaychev, S.A., Chernorizov, A.M., Ozheredov, I.A., et al.: Diagnosing human psychoemotional states by combining psychological and psychophysiological methods with measurements of infrared and THz radiation from face areas. Psychol. Russ. State Art **13**(2), 64–83 (2020)
9. Alyushin, M.V., Kolobashkina, L.V.: Laboratory approbation of a new visualization form of hazardous objects control operator current psycho-emotional and functional state. Sci. Vis. **10**(2), 70–83 (2018)
10. Yang, X., Zhao, X., Yang, K., Liu, Y., Liu, Y., Fu, W., Luo, Y.: Biomedical applications of terahertz spectroscopy and imaging. Trends Biotechnol. **34**(10), 810–824 (2016)
11. Sun, Q., He, Y., Liu, K., Fan, S., Parrott, E.P.J., Pickwell-MacPherson, E.: Recent advances in terahertz technology for biomedical applications. Quant. Imaging Med. Surg. **7**(3), 345–355 (2017)
12. Berlovskaya, E.E., Cherkasova, O.P., Ozheredov, I.A., Adamovich, T.V., Isaychev, E.S., Isaychev, S.A., Makurenkov, A.M., Varaksin, A.N., Gatilov, S.B., Kurenkov, N.I., Chernorizov, A.M., Shkurinov, A.P.: New approach to terahertz diagnostics of human psychoemotional state. Quantum Electron. **49**(1), 70–77 (2019)
13. Hong, K.: Classification of emotional stress and physical stress using facial imaging features. J. Opt. Technol. **83**(8), 508–512 (2016)
14. Alyushin, M.V., Kolobashkina, L.V., Golov, P.V., Nikishov, K.S.: Adaptive behavioral model of the electricity object management operator for intelligent current personnel condition monitoring systems. Mechanisms and Machine Science, vol. 80, pp. 319–327 (2020)

Specialized Software Tool for Pattern Recognition of Biological Objects

Sergey D. Kulik[1,2] and Evgeny O. Levin[1(✉)]

[1] National Research Nuclear University MEPhI,
Kashirskoe shosse 31, Moscow 115409, Russia
sedmik@mail.ru, levin.inf@gmail.com
[2] Moscow State University of Psychology and Education (MSUPE),
Sretenka st. 29, Moscow 127051, Russia

Abstract. This paper describes the results of work on developing the mobile application for recognition of faces and other biological objects. The application is designed with a focus on loading external machine learning models over the Internet, which allows you to change the model without making any modifications to the application. With such realization, the application can be used in many cases. For example, at carrying out conferences: organizers just need to train a model and send out the link for its downloading to all participants of the conference without any changes in source code. Participants will be able to find out the information about other members they are interested in, as well as contact them directly from the application, both by phone and by e-mail. Instructions are given for teaching your own face recognition model using the Microsoft Custom Vision cloud service, which allows you to train regardless of the power of your local computer. As an example, a classification model was trained and the following assessments of recognition quality indicators were obtained precision: 97.8% and recall: 95.8%. In our future work we consider adding the functionality of emotion recognition based on the pattern recognition algorithm, described in this paper.

Keywords: Cognitive technology · Biological objects · Pattern recognition · Machine learning · Mobile development

1 Introduction

Pattern recognition is a machine learning industry that focuses on finding patterns in data. Pattern recognition systems in many cases train from marked "training" data (controlled training), but in cases of absence of marked data, other algorithms can be used to detect previously unknown patterns (uncontrolled training).

In machine learning, pattern recognition is the assignment of a label to a specified input value. An example of pattern recognition is a classification that attempts to assign each input value to one of the specified classes (for example, to determine whether a given e-mail message is "spam" or "not spam"). However, image recognition is a more general problem that covers other types of output data.

A. V. Samsonovich et al. (Eds.): BICA 2020, AISC 1310, pp. 173–180, 2021.
https://doi.org/10.1007/978-3-030-65596-9_22

Pattern recognition algorithms generally aim to provide a reasonable response for all possible input data and to perform the "most likely" comparison of input data based on its statistical variation. By contrast, template matching algorithms look for exact matches in the input data with existing patterns. A common example of a template matching algorithm is regular expression mapping, which searches for a given type of template in text data and is included in the search capabilities of many text editors and word processors.

Recently, machine learning algorithms in general and pattern recognition algorithms in particular have been found more and more frequently. Almost every service of such companies like Google, Facebook, Apple, which is somehow connected with image processing, uses computer vision technology.

This paper provides the results of work on developing a face recognition mobile application for the Apple platform (iOS). The application is designed with a focus on loading external machine learning models over the Internet, which allows you to change the model without making any modifications to the application.

There are important scientific areas of applications cognitive technology [1], pattern recognition and special neural networks technology:

- regular agent technologies [2] for information search [3];
- factographic information retrieval [4];
- recognition of biological and criminalistics objects [5];
- memristor [6], biosensors [7] and concept of brain-on-chip [7];
- cognitive agents [8].

Paper [1] describes a universal learner as an embryo of computational consciousness. Another paper [2] deals with agent-based search in social networks. Paper [8] describes a roadmap to biologically inspired cognitive agents. Another paper [9] deals with ROC analysis for pattern recognition.

2 Pattern Recognition Application

2.1 Software Libraries and Frameworks

We used different libraries and frameworks, as well as online services.

We chose the Swift programming language as our main language. In addition, various libraries and frameworks provided by Apple were used to develop the application.

The UIKit framework provides the necessary infrastructure for iOS applications. It provides the architecture for representing visual elements on the device screen, the infrastructure for event handling, such as multi-touch, gestures and other input types, as well as the main application cycle necessary to manage the interaction between the user, system and application [10].

Core Data framework provides the generalized and automated decisions for the general problems connected with the management of object life cycle and management of objects graphs, including persistence. Core Data significantly reduces the amount of code that developers need to write to support the model level [11].

Since the application is built around the use of machine learning and augmented reality, it would be hard to develop it without Core ML and ARKit frameworks [12].

We used Core ML to integrate trained machine learning model into our application. The trained model is the result of applying a machine learning algorithm to a set of training data [13].

2.2 Face Recognition Algorithm

Since it is rather problematic to describe the algorithm of the whole application because of its large size, we will describe the algorithm of face recognition mode in general in this section.

Firstly, the surrounding scene is scanned with a smartphone camera. If faces are found on the scene (face search is carried out using the Vision framework by Apple), then the part of the image containing the photo of the face is pre-processed and transferred for classification to the Core ML machine learning model, which returns the probability of the face image belonging to each class from the model. A person is considered recognized if the probability value is more than 60% and if the person belongs to a class other than 0 (zero class means unidentified people).

If the person is recognized, his id (class label) is used to search for information about the person in the database. After all the necessary information (name, surname and photo) is received, it is displayed on the device screen.

Scenes are scanned every 0.2 s. When a face is detected in the scene, the process of its identification begins immediately. If recognition is successful, the information about the recognized person is displayed on the screen. If a face disappears from the scene, the information about it disappears after 1.5 s. This is necessary for processing cases in which the person turns away for a while or disappears from view. Without a delay of 1.5 s, frequent occurrence/deletion of information could cause user's irritation.

The software can recognize several faces and display information about them on the screen at the same time.

The user can click on the person's name, surname or photo, and all available information about the person will be displayed in a separate window.

The user can exit the face recognition mode by clicking the “close” button in the top navigation bar.

It should be noted that the presented algorithm is simplified by presenting the program work as a single-threaded one. In fact, the program works in multithreading mode, asynchronously executing various stages of the algorithm. Reactive functional programming techniques are used.

2.3 Setting up the Mobile Application

The key feature of the developed software tool is that the model of machine learning is not represented in the source code of the application, i.e. the user can set up his model and use it. For this purpose, it is enough to assemble the model according to certain rules described below, upload it to a cloud service and provide a link to this model in the application. The application will download, configure and compile it itself; after that, in case of successful completion of the process, the face recognition mode will be ready to work.

The model file is a zip-archive, in which the root directory contains the trained model named "model.mlmodel" and the model description in JSON format named "info.json". "Avatars" directory contains photos of people in JPG format. Fig. 1 shows an example content of the zip-archive with the trained model.

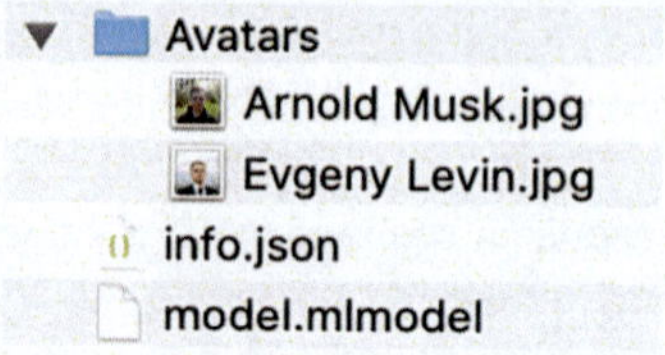

Fig. 1. Example content of the zip archive with the trained model.

In the info.json file for each person it is possible to specify his id, corresponding to his class tag in the trained model.mlmodel, his first name, second name, job place, phone, e-mail address, general information about the person, as well as the name of the file with his photo in JPG format, located in the "Avatats" directory.

Since during the development of the application, emphasis was placed on the fact that users can upload their trained models, it is necessary that users have a convenient service where they can train their model. One such service is Microsoft's Custom Vision. This is a cloud service, so you do not must to have a powerful computer to use it.

When forming a training sample of data an identifier is assigned to each person (field "id" in the file info.json). This identifier should serve as the name of the class, to which all photos (images) of this person belong.

It is necessary to add a class with photos of people which do not belong to any of people we want to recognize. This is necessary because of, when pointing the camera at a person who was not represented in the training sample, the program needs to classify it not to the class of any particular person, but to the class of unknown people. Such class should be assigned the "0" identifier.

In Microsoft's Custom Vision service, which was mentioned earlier, the classes of the tutorial sample are called tags.

The k-fold cross-validation method is used to evaluate the learning results of the model. The initial sample is randomly divided into k subsets of the same size. From k subsets one is saved as test data for model testing, and the remaining k - 1 are used as training data. The cross-check process is then repeated k times, and each of the k subsets is used exactly once as the test data for model testing. The results can then be averaged to produce a single score. The advantage of this method is that each observation is used to validate only once. Normally a 10-fold cross-check is used, but parameter k remains free overall.

For example, setting k = 2 leads to a 2-fold cross check. In a 2-time cross-check, we randomly shuffle a dataset into two sets of d0 and d1, so that both sets have the same size. Then we train on d0 and check on d1 and then vice versa.

The assessment is based on two main indicators: precision and recall. To describe them, we shall first enter several designations, described in Table 1.

Table 1. Symbols for possible answers of the algorithm.

Flag	y = 1	y = 0
y' = 1	True Positive ($x1$ = TP)	False Positive ($x2$ = FP)
y' = 0	False Negative ($z1$ = FN)	True Negative ($z2$ = TN)

Here y' is the algorithm's answer to an object belonging to a certain class, and y is the true class label on this object. Thus, there are two types of classification errors: False Negative ($z1$) and False Positive ($x2$).

The precision (P) measure (1) describes how many positive responses received from the model are correct. The higher the precision, the fewer false positives. That is, this is the proportion of objects that are called positive when recognizing (classifying) and that are actually positive [9]:

$$P = \frac{x1}{x1 + x2} \tag{1}$$

The precision measure, however, does not give an indication whether the classifier has returned all the correct answers. For this purpose, there is the measure called recall.

The recall (R) measure (2) characterizes the ability of the classifier to "guess" as many positive answers as possible from the expected, that is, what proportion of positive class objects from all objects of the positive class the recognition algorithm has determined [9]:

$$R = \frac{x1}{x1 + z1} \tag{2}$$

Remark:

$$\frac{1}{R} - 1 = \frac{z1}{x1}. \tag{3}$$

$$\frac{1}{P} - \frac{1}{R} = \frac{x2 - z1}{x1} \tag{4}$$

2.4 User Interface

The following information about people from a trained machine learning model may be available in the application user interface: name, surname, phone number, e-mail address, job place and general information about the person (Fig. 2). In addition, the information about each person is accompanied by his or her photo. From this screen, the user can call or send a text message to a selected person with one touch, send an e-mail, and save all available information about the person to the user's contact list in the Apple ID account.

Fig. 2. User information page.

Fig. 3. Face recognition mode.

In face recognition mode, the back camera of a smartphone is activated (Fig. 3). If the camera is pointed at a person's face, in case of successful recognition, the application will show the name and surname of the person, as well as his photo. Clicking on any of these elements will show detailed information about the person (Fig. 2).

The feature of the application is that the inscription with the name and surname of the person and his photo is pinned directly to his face. That is, if a person changes his or her position, the information will change accordingly. This is achieved by using augmented reality technologies: the algorithm combines information from smartphone motion sensors with the analysis of the scene visible by the device's camera, occurring with the help of computer vision. It detects noticeable features in the image, tracks differences in their position in the video frames and compares this information with data from motion sensors. The result is a highly accurate model of the device's position and movement in space.

The algorithm can recognize multiple faces simultaneously and display information about them on the screen. Using augmented reality technologies, information about recognized faces is linked to the coordinates of a person's face in space, which makes it easier to navigate in the presence of multiple recognized faces on the smartphone screen.

3 Experimental Research of the Application

To train the classification model, the user can decide for himself which training tool to use. Training can take place either on the local computer or using various cloud services.

In this work, Microsoft Custom Vision service was chosen to get the machine learning model. The description of this service was presented in the previous chapter.

As mentioned earlier, the results of training the model (precision and recall) are obtained using the method of cross-validation (it is also called sliding control). The training was conducted for two classes, one of which denotes "unidentified" people (tag "0"). The second class denotes a specific person. We received the following scores of recognition quality indicators precision: 97.8% and recall: 95.8%.

4 Conclusion

The study presents results of developing the face recognition application and its core algorithm. Application is also able to recognize other biological objects. Implemented the mechanism of downloading a trained machine learning models by direct link from the Internet, which allows various users to use their own models, to distribute them to an unlimited amount of other people and avoid binding the source code of the program to any particular model. With such realization the application can be used in many cases. For example, at carrying out of conferences: organizers just need to train model and send out the link for its downloading to all participants of conference. Participants just need to specify the given link with a special field in application settings. They will be able to find out the information about other participants they are interested in, as well as contact them directly from the application, both by phone and e-mail.

Instructions are provided for training your own facial recognition model using the Microsoft Custom Vision cloud service, which allows you to train models regardless of the power of your local computer.

As an example, we have trained a classification model and received the following scores of recognition quality indicators precision: 97.8% and recall: 95.8%.

The source code of the application is free to access [14].

In our further work we consider the possibility of adding the mode of human emotions recognition to the application. This feature can be based on the facial recognition system described in this article.

Acknowledgement. This work was supported by Competitiveness Growth Program of the Federal Autonomous Educational Institution of Higher Education National Research Nuclear University MEPhI (Moscow Engineering Physics Institute).

References

1. Samsonovich, A.V.: Universal learner as an embryo of computational consciousness. In: Chella, A., Manzotti, R. (eds.) AI and Consciousness: Theoretical Foundations and Current Approaches. Papers from the AAAI Fall Symposium. AAAI Technical Report, vol. FS-07-01, pp. 129–134 (2007)
2. Artamonov, A., Onykiy, B., Ananieva, A., Ionkina, K., Kshnyakov, D., Danilova, V., Korotkov, M.: Regular agent technologies for the formation of dynamic profile. Procedia Comput. Sci. **88**, 482–486 (2016)
3. Ananieva, A., Onykiy, B., Artamonov, A., Ionkina, K., Galin, I., Kshnyakov, D.: Thematic thesauruses in agent technologies for scientific and technical information search. Procedia Comput. Sci. **88**, 493–498 (2016)
4. Kulik, S.D.: Factographic information retrieval for semiconductor physics, micro - and nanosystems. In: AMNST 2017, IOP Conference Series: Materials Science and Engineering, vol. 498, p. 012026 (2019)
5. Kulik, S.D., Shtanko, A.N.: Recognition Algorithm for Biological and Criminalistics Objects. In: Biologically Inspired Cognitive Architectures 2019. Proceedings of the Tenth Annual Meeting of the BICA Society, AISC vol. 948, pp. 283–294 (2020)
6. Danilin, S.N., Shchanikov, S.A., Zuev, A.D., Bordanov, I.A., Sakulin, A.E.: The research of fault tolerance of memristor-based artificial neural networks. In: 12th International Conference on Developments in eSystems Engineering (DeSE), pp. 539–544 (2019)
7. Mikhaylov, A., Pimashkin, A., Pigareva, Y., Gerasimova, S., Gryaznov, E., Shchanikov, S., Zuev, A., Talanov, M., Lavrov, I., Demin, V., Erokhin, V., Lobov, S., Mukhina, I., Kazantsev, V., Wu, H., Spagnolo, B.: Neurohybrid memristive CMOS-integrated systems for bio-sensors and neuroprosthetics. Front. Neurosci. **14**, 358 (2020)
8. Chella, A., Lebiere, C., Noelle, D.C., Samsonovich, A.V.: On a roadmap to biologically inspired cognitive agents. Front. Artif. Intell. Appl. **233**, 453–460 (2011)
9. Fawcett, T.: An introduction to ROC analysis. Pattern Recogn. Lett. **27**(8), 861–874 (2006)
10. UIKit. Apple Developer Documentation. https://developer.apple.com/documentation/uikit. Accessed 04 May 2020
11. Core Data. Apple Developer Documentation. https://developer.apple.com/documentation/coredata. Accessed 04 May 2020
12. ARKit. Apple Developer Documentation. https://developer.apple.com/documentation/arkit. Accessed 04 May 2020
13. Core ML. Apple Developer Documentation. https://developer.apple.com/documentation/coreml. Accessed 04 May 2020
14. evglevin/Faces: Face recognition iOS AR app. https://github.com/evglevin/Faces. Accessed 04 May 2020

Designing Software for Risk Assessment Using a Neural Network

Anna V. Lebedeva(✉) and Anna I. Guseva(✉)

Moscow Engineering Physics Institute, National Research Nuclear University MEPhI, 31 Kashirskoye shosse, Moscow, Russia
lebedevaa28@gmail.com, AIGuseva@mephi.ru

Abstract. This article presents the results of the research in terms of using mathematical methods for the risk management process in the implementation of software development projects. Software development projects are not always implemented in a final form that meets the expectations of customers. Both internal and external factors can influence this. In this regard, the problems of risk management, which inevitably arise during the implementation of software development projects, become particularly relevant due to the large uncertainty of the internal and especially external environment of enterprises. The introduction of a comprehensive approach to risk management allows the company to form an objective view of the current and planned activities of the organization, taking into account possible negative events or new opportunities, anticipate risks and make decisions based on information about them, respond to risks in a timely manner and reduce the negative impact of risks in their implementation. Within the framework of this research, the risk assessment software is designed to assess the situation in the project, predict the future effectiveness of the project, and build scenarios to support decision-making. Such software will allow to combine all the actions of the analyst in one tool, where all the information about the project and the external environment will be stored, updated and constantly used for training the Neural Network apparatus on which the software is designed. This work was supported by Russian Foundation for Basic Research (RFBR) grant № 20-010-00708\20.

Keywords: Risk management · Risk assessment · Neural networks · Mathematical methods of risk assessment · Software development projects

1 Introduction

This work is devoted to solving an important scientific and practical problem of project management, namely, risk assessment and forecasting. Based on the information received during the project, protective measures are developed to reduce the risk of implementation of risks. Within the framework of this research, software is being designed to support the risk assessment process in software development projects using the capabilities of Neural Network modeling.

There is almost no information in the literature about the use of Neural Networks in assessing and predicting risks in software development projects. Such research is extremely limited.

A. V. Samsonovich et al. (Eds.): BICA 2020, AISC 1310, pp. 181–187, 2021.
https://doi.org/10.1007/978-3-030-65596-9_23

2 Main Approaches to the Problem

2.1 Neural Network Models for Risk Assessment

Risk management includes not only the definition of uncertainty and analysis of project risks, but also a set of methods to counteract risk factors to eliminate damage. Methods that are combined into a planning, tracking, and adjustment system include:

- developing a risk management strategy;
- methods of balancing (compensation), which include monitoring the external socio-economic and legal environment in order to forecast it in detail, as well as the formation of a system of project reserves;
- localization methods used in high-risk projects in a multi-project system;
- distribution methods using various parameters, such as the time parameter, the number of participants, etc.;
- methods for avoiding risks associated with replacing insufficiently reliable partners, introducing a guarantor into the process, insuring risks, and sometimes even canceling the project.

Mathematical tools are most often used to describe uncertainties in computer and mathematical modeling, such as:

- probabilistic and statistical methods;
- methods for statistics of non-numeric data, as well as methods of fuzzy theory;
- methods of conflict theory (game theory);
- methods of artificial intelligence theory (Neural Networks, etc.).

There are various methods of machine learning and data mining that can be applied to solve the problem of this research, for example:

- decision trees;
- self-organizing Kohonen maps;
- multi-layer Neural Networks.

As for the Kohonen map, as noted in [1], self-organizing Kohonen maps are used for visualization of multidimensional data. They only give a General picture, which is rather blurry and subject to distortion, since it is generally impossible to project a multidimensional sample onto a plane without distortion. However, Kohonen maps allow you to see key features of the cluster sample structure. They are used at the stage of exploratory data analysis, more for a General understanding of the problem than for obtaining any accurate results. In this regard, in the framework of this study it is proposed to use a multilayer Neural Network.

The use of Neural Network technology is necessary in cases where the formalization of the decision-making process is difficult or even impossible. They are a very powerful modeling tool because they are nonlinear in their technology. Linear modeling has long been the main approach in most areas, since there are the largest number of optimization methods available for it. However, in the majority of cases, linear modeling methods are practically not applicable for risk analysis tasks. In addition, for

Neural Networks, there is no such problem as a «dimension gap», which does not allow modeling linear dependencies on a large number of variables.

The ability of Neural Networks to learn is one of the main advantages over traditional algorithms. Technically, training consists of searching for coefficients of connections between artificial neurons. During training, the Neural Network is able to identify complex relationships between input and output, as well as perform generalization. This means that if the training was successful, the network can return the correct result based on data that was not in the training sample. The larger the training sample, the more successful the training will be. However, the dependence of success on the number of examples is not linear.

However, with a different Neural Network architecture, you can do without the training set of data. The Neural Network chooses a decision based on the data it receives, learns from it gradually, and searches for connections between inputs. This is how self-organizing Kohonen maps work, for example. Such a mechanism for self-learning of a Neural Network is called «learning without a teacher», while the mechanism for learning based on an array of data is called «learning with a teacher» [2].

It follows from all this that in order to determine the requirements for the system, it is necessary first to make architectural decisions on the part of the Neural Network apparatus, which can be measured empirically. In order to successfully train a Neural Network, you need a large data set (dataset), or you will have to learn the network manually, which requires a very large amount of time spent. Structurally, the dataset should consist of vectors (columns) with factors and a vector with estimates. Therefore, it is necessary to classify all possible risks of the IT project in order to determine the input parameters of the network.

To create a project risk assessment system using Neural Networks, there are several problems that need to be solved:

- availability of a dataset with time, technical, and financial indicators for a large number of IT projects (startups, projects within organizations);
- classification of IT project risks;
- reducing risks from qualitative to quantitative (in the form of vectors);
- design and implement a specific network architecture for this task.

As noted in the study [3] Neural Networks are not a panacea, in many cases, the use of traditional methods of statistics will be more effective. Models often show high results for the training set and significantly lower results for the testing set.

In this regard, it is important to note that an accurate result can be achieved when using hybrid methods of estimation and forecasting. For example, in combination with probabilistic models of software risk assessment using the Bayesian Belief Network, which focuses on the main risk indicators of software for risk assessment in software development projects [4].

It should also be noted that for more accurate forecasting of project risks, it is necessary to work on their formalization (in terms of sources of risk, influencing factors and countermeasures) [5].

As part of previous research, we have already prepared a dataset, risk classification of an IT project with the development of a cognitive map for risk assessment [6].

3 Proposed Approach

3.1 Statement of the Problem

In the considered risk assessment system, it is necessary to build a model that allows us to study the dependence of changes in key project indicators on changes in the parameters of management and performance of work under conditions of variability in the parameters of the production system (project). If the influence of fluctuation is significant and the change in management parameters is poorly correlated with the change in key indicators expected by the project Manager, it can be concluded that the management mode was selected incorrectly, or that the project key indicators were selected inappropriately [7].

3.2 Solving the Problem

Setting a function for getting a set of key project indicators:

$$F_i = (k_i, z_i) = F_i(u_i, z_{i-1}, p_i) \tag{1}$$

Where u_i-management parameters, z_{i-1}-parameters coming from the external project environment, p_i-parameters coming from the internal project environment, z_i-project parameters, k_i-a set of key indicators related to the project.

Building F_i functional dependencies for a project is a complex task that will use a neural network model to solve.

The mathematical model of a neuron is described by the relation:

$$M = \sum\nolimits_{k=1}^{nn} w_k z_k + b, y = f(M) \tag{2}$$

Where w_k is the weights whose values are found specifically for a specific task, b is the offset value, M is the result of summation, x_k is the input signal component, y is the output signal, n is the number of inputs, and f is the nonlinear transformation.

An example of how the model works is shown on the Fig. 1.

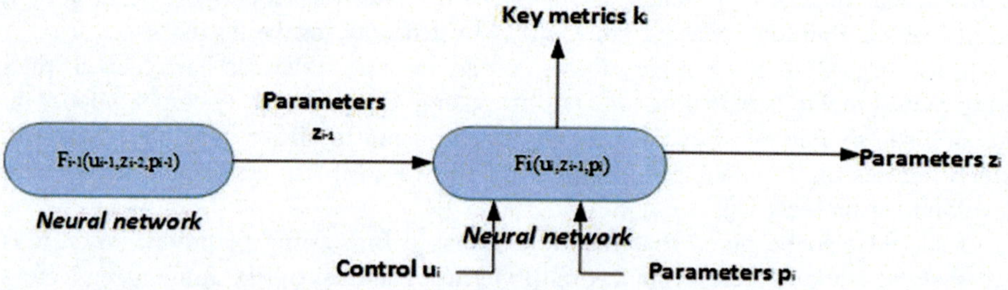

Fig. 1. Scheme of the Neural Network model

4 Results

To study the properties of a system of key indicators *k*, the following method is used:

- the inputs of trained networks are repeatedly fed vectors $z_i(j)$, $p_i(j)$, $u_i(j)$, where j is the generation number;
- in each generation, the values of some input vectors are modified to simulate random additive distortion with a given distribution;
- as a result of multiple generations is obtained the set of values of key performance indicators;
- there is a change of management, which is supposed to change (optimize) key indicators.
- multiple generations are generated with the specified distortion distributions;
- the statistical hypothesis about the homogeneity of two sets of key indicators—the set obtained before the change of management and after-is tested. If the hypothesis is confirmed on at least one pair of sets of controls, we get the conclusion that the system of indicators is inadequate.

Such a research system will allow you to analyze approaches to project management and influence the quality of work management (tasks) of the project and the project as a whole.

The design and training of a neural network for risk recognition in software development projects was carried out using the Deductor Studio Lite tool.

Figure 2 illustrates Neural network training (Fig. 3).

Training set		Test set	
Maximum error	6,74E -01	Maximum error	6,56E -01
Average error	1,65E -01	Average error	2,78E -01
Recognized (%)	22,34	Recognized (%)	40,0

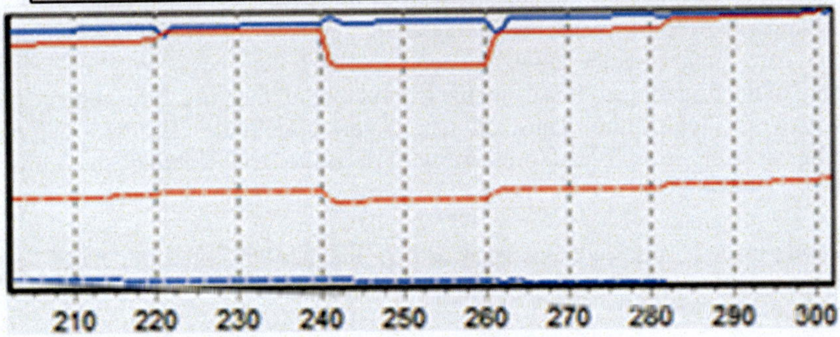

Fig. 2. Neural network training

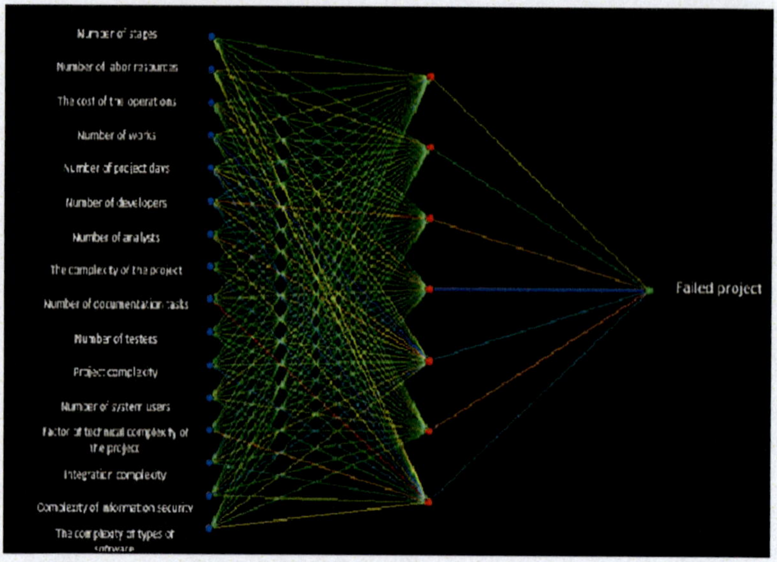

Fig. 3. 16 × 7 × 1 neural network

As a result of building the network, a multi-layer network (16 × 7 × 1) is obtained, that is, a network that consists of 16 input neurons 7 neurons of the hidden layer and 1 neuron in the output layer.

5 Summary

As a result of this work, we analyzed the literature on risk assessment in projects for the development of information systems, as well as tools for risk assessment.

The study showed that neural networks are better used with a large enough amount of initial data, otherwise the time, financial and technical costs will be inadequate compared to the output data. It is desirable to structure all data in a form suitable for the neural network architecture. Thus, all risk factors should be classified and ranked according to their impact and probability. For some architectures, the impact or probability will be evaluated by the Neural Network itself.

Acknowledgements. This work was supported by Russian Foundation for Basic Research (RFBR) grant № 20-010-00708\20.

References

1. Golovko, V.A.: Neural Networks: Training, Organization and Application. IPRZHR, Moscow (2001)
2. Carbonell, J.G., Michalski, R.S., Mitchell, T.M.: Machine Learning. Symbolic Computation. Springer, Heidelberg (1983)
3. Korneev, D.S.: Using neural networks to create a model for evaluating and managing enterprise risks. Manag. Large Syst. Proc. **17**, 81–102 (2007)
4. Kumar, C., Yadav, D.K.: A probabilistic software risk assessment and estimation model for software projects. Procedia Comput. Sci. **54**, 353–361 (2015). https://doi.org/10.1016/j.procs.2015.06.041. ISSN 1877-0509
5. Samsonovich, A.V.: Schema formalism for the common model of cognition. Biol. Inspired Cogn. Arch. **26**, 1–19 (2018)
6. Lebedeva, A.V., Guseva, A.I.: Cognitive maps for risk estimation in software development projects. Adv. Intell. Syst. Comput. **948**, 295–304 (2020)
7. Ilchenko, A.N., Korovin, D.I.: Neural network modeling capabilities for improving management accounting in the intra-company budgeting system. Modern science-intensive technologies. Mod. Sci. Intensiv. Technol. Reg. Append. **2–3**, 22–27 (2007)

Suitability of Object-Role Modeling Diagrams as an Intermediate Model for Ontology Engineering: Testing the Rules for Mapping

Dmitrii Litovkin, Dmitrii Dontsov, Anton Anikin(✉), and Oleg Sychev

Volgograd State Technical University, Volgograd, Russia
litd@mail.ru, anton@anikin.name, poas@vstu.ru

Abstract. Creating and understanding ontologies using OWL2 language is a hard, time-consuming task for both domain experts and consumers of knowledge (for example, teachers and students). Using Object-Role Modeling diagrams as an intermediate model facilitates this process. To achieve this, the method of mapping ORM2 diagrams to OWL2 ontologies and vice versa is necessary. Such methods were proposed in different works, but their suitability and possible errors are in doubt. In this paper, we propose a method of evaluating how well existing rules of mapping follow ORM semantics. Several ontologies were created using mapping rules and tested. During testing, a significant difference between ORM2 and OWL2 basic properties and assumptions were discovered. This difference require updating the mapping rules.

Keywords: Ontology · Web ontology language · Object-Role Modeling · Mapping · Ontology testing

1 Introduction

WHAT-knowledge is fundamental for understanding any domain. It represents facts answering the competency questions such as "What is it?", "What does the given term mean?", "What are the subclasses or instances of the given entity?", "What is the type of the given entity?", "What is the relationship between the given entities?", "What parts the given entity contains?" and so on.

The transfer of WHAT-knowledge is possible in textual and graphical forms; the textual form can use both natural and formal logical languages. Each form of knowledge representation for transfer has its advantages and disadvantages [2].

This paper presents the results of research carried out under the RFBR grants 18-07-00032 "Intelligent support of decision making of knowledge management for learning and scientific research based on the collaborative creation and reuse of the domain information space and ontology knowledge representation model" and 20-07-00764 "Conceptual modeling of the knowledge domain on the comprehension level for intelligent decision-making systems in the learning".

A. V. Samsonovich et al. (Eds.): BICA 2020, AISC 1310, pp. 188–194, 2021.
https://doi.org/10.1007/978-3-030-65596-9_24

Natural-language texts are more expressive and well-known for humans but contain significant semantic noise [1,2,14]. Graphical form stimulates creative thinking but has limited expressive capabilities and requires learning of the specific notation [12]. Formal languages for representing knowledge like OWL2, Prolog, and others allow computer processing of knowledge (logical reasoning and contradiction detection) but they are too complicated for human perception [13].

A promising approach is to combine different forms of knowledge representation during creating and transferring knowledge using an automatic mapping between different forms (e.g. see Fig. 1).

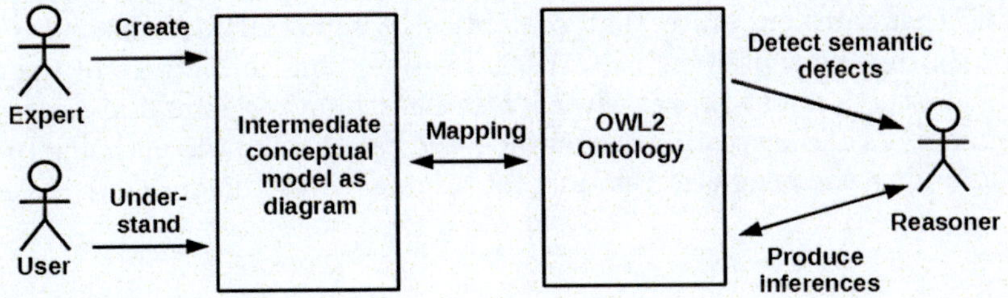

Fig. 1. Creating and transferring knowledge using different forms of knowledge representation

Object-Role Modeling (ORM) is a fact-oriented method for analysing information at the conceptual level [6]. ORM-diagrams are a good candidate for an intermediate model for developing and understanding of OWL2-ontologies. Use cases have shown that ORM can be learned easily, in a short time by domain experts without previous IT background [10]. To facilitate using ORM-diagrams as an intermediate model it is necessary to create ORM-to-OWL mapping. In this paper, we analyse existing rules of mapping ORM2 to OWL2 to determine how well they adhere to ORM semantic.

2 Related Works

2.1 Mapping Conceptual Models

WHAT-knowledge consists of concepts and relations between them so it can be naturally presented as a graph. As visualization stimulates human thinking [12], the most promising intermediate models for ontology development and understanding are well-known diagrams: mind maps, concept maps, conceptual graphs, UML class diagrams, ER diagrams (in different notations), ORM2 diagrams and so on. There are known approaches to using these diagrams for ontological engineering [3,8,9,15,18]. One of the most expressive visual knowledge representation models is an Object-Role Modeling (ORM) diagram [7]. Object-role modeling represents the domain as a set of objects and their relations with associated roles. You can see an example of ORM2-diagram on Fig. 2.

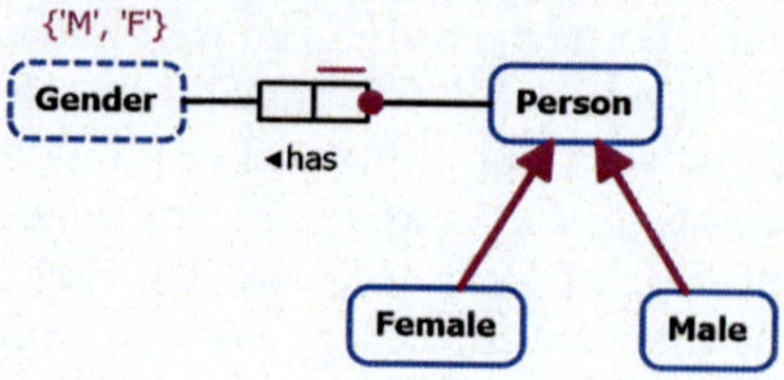

Fig. 2. ORM diagram example

The articles [8,11,16] suppose ORM2-to-OWL2 language mapping rules. However, their authors state that semantics of some ORM2 constructs were changed during their mapping to OWL2. The mapping correctness of the other ORM2 constructs is not proven either. So researching the semantic correctness of the mapping rules by testing the ontologies created using the mapping process for maintaining the semantic relations of original ORM diagrams is an actual task.

2.2 Testing OWL2 Ontologies

Dragisic [5] discerns syntactic, semantic, and modeling defects of ontology. Syntactic defects are trivial and easy to repair. Semantic defects may result in either incoherent ontology or inconsistent ontology. Incoherent ontology is an ontology containing an unsatisfiable concept: a concept that no instance can belong to, i.e. a concept that is equivalent to owl:Nothing. Inconsistent ontologies are contradictory i.e. they either contain an instance of an unsatisfiable concept or it is possible to derive that owl:Thing $\equiv$ owl:Nothing.

The most important category is modeling defects that are caused by modeling errors: the resulting ontology is syntactically correct and consistent but does not represent the original knowledge correctly. Common examples of this kind of defect are missing or wrong relations in the ontology.

According to [4], for finding semantic and modeling defects three methods can be used: (1) Competency Questions (CQ) verification, (2) inference verification, and (3) error provocation.

CQ verification requires creating Competency Questions. They should express the questions the tested ontology is intended to answer [17]. Giving meaningful answers to CQs can be considered a functional requirement for the ontology.

While CQs are independent of the way the information is produced - was it entered in the ontology as assertions or derived from the other facts, inference verification specifically test expected inferences. In practice, the tests for those two methods are done using the same techniques.

Error provocation is a form of "stress-testing" of an ontology by purposefully injecting assertions inconsistent with the original knowledge. If the reasoner finds semantic defects after adding the inconsistent assertions, it shows that the ontology contains valid knowledge; otherwise, the ontology is deemed incompleted.

3 Testing Rules for Mapping ORM2 Language to OWL2 Language

3.1 Method

To use an ORM2-diagram as an intermediate model for ontology creation and understanding, the correct mapping of the diagram into the ontology must be ensured. To evaluate the validity of the mapping rules MR from [8] we created test ontologies TO using these rules. Then using algorithm B we created test cases TC and performed ontology testing using algorithm A.

Algorithm A. Testing of a mapping rule from MR.
Input: $mr \in MR$, D - a set of ORM2-diagram
Output: TR - testing results (positive, negative or uncertain).
1. Find in D minimal fragment fr allowing to use mapping rule mr.
2. Apply rule mr to the diagram fragment fr creating a test ontology $to \in TO$.
3. Create requirements set RQ for test ontology to taking into account ORM semantics.
4. For each $rq \in RQ$
4.1. Create test case $tc \in TC$ containing the expected result er (see algorithm B).
4.2. Run test case tc using ontological reasoner to acquire actual result ar
4.3. Verify the expected output er against the actual result ar and determine the test outcome $tr \in TR$ using Table 1.
4.4. If the test outcome is $tr = Uncertain$ then
4.4.1. Correct test case tc if possible (step 4.1) and re-run the test case (step 4.2)
Algorithm B. Creating test case tc
Input: $mr \in MR, fr, to \in TO, rq \in RQ$
Output: $tc \in TC, tc =< to,\ ext\_to,\ sq,\ er >$

1. Considering mr, fr, to, and rq choose test method $tm \in \{"CQ\ verification", "inference\ verification", "error\ provocation"\}$
2. Add in test ontology to new axioms (in $TBox$) and/or assertions (in $ABox$) ext_to according to the chosen test method tm
3. If test method $tm = "CQ\ verification"$ then
   3.1. Implement rq (rq is Competency Question) as SPARQL query sq
   else if $tm = "inference\ verification"$ then
   3.2. Implement rq as SPARQL query sq to retrieve expected inferences

3.2 Results

Test ontologies were created using Protege framework. SPARQL queries were created using Snap SPARQL Query Plugin and performed on reasoner HermiT 1.4.3.456.

Table 1. Determining test outcome *tr*

Test method *tm*		Expected result *er*	Actual result *ar*	Test outcome *tr*
"CQ verification" or "inference verification"	For SELECT query	Non-empty result	Non-empty result equal to expected	Positive
			Non-empty result not equal to expected	Negative
		Empty result	Non-empty result	Negative
			Empty result	Uncertain
	For ASK query	True	True	Positive
		True	False	Negative
		False	True	Negative
		False	False	Uncertain
"Error provocation"		Presence of semantic defect	Presence of semantic defect	Positive
			Absence of semantic defect	Negative
		Absence of semantic defect	Absence of semantic defect	Uncertain
			Presence of semantic defect	Negative

The performed experiment allowed us to find three discrepancies between ORM semantics and ORM-to-OWL mapping rules:

1. ORM semantics is based on the closed-world assumption while test ontologies use the open-world assumption;
2. Entity Types and Value Types in ORM are disjoint sets while in test ontologies they may intersect;
3. ORM subtypes are proper subtypes, i.e. if B is a subtype of A, in ORM $B \subseteq A$ and $B \neq A$. Ontology subtypes are not proper subtypes because $B \subseteq A$ or $B = A$.

Table 2 contains test cases demonstrating these discrepancies.

4 Conclusions

The study shows that there is a significant difference between the second generation ORM and ontologies in basic assumptions and features so that mapping rules give incorrect results. The mapping rules require enhancement to better represent ORM semantics using the OWL2 language.

The future work will include (1) creating more test cases, (2) modifying mapping rules so that the resulting ontology would better fit ORM semantics, (3) evaluating the modified rules using the enhanced test set, (4) evaluating the modified rules by creating an ontology for a practical task.

Table 2. Test cases

Test case	Closed-world principle violation	Disjoint violation of two Entity Types	Proper subtyping violation
ORM construction	Entity Type with name Person exists 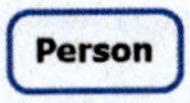	Entity Type with name Female exists and Entity Type with name Male exists 	Entity Type Male is a subtype of Entity Type Person 
Test ontology	`Declaration( Class(:Person) )`	`Declaration( Class(:Female) ) Declaration( Class(:Male) )`	`Declaration( Class(:Person) ) Declaration( Class(:Male) ) SubClassOf(:Male :Person)`
Testing method	Inference verification	Inference verification	Error provocation
Additional axioms and/or assertions	none	none	`EquivalentClasses( :Person :Male )`
SPARQL query	`ASK WHERE {:Person owl:equivalentClass owl:Thing}`	`ASK WHERE {:Female owl:disjointWith :Male}`	none
Expected result	True	True	Presence of semantic defect
Actual result	False	False	Absence of semantic defect

References

1. Anikin, A., Litovkin, D., Kultsova, M., Sarkisova, E., Petrova, T.: Ontology visualization: approaches and software tools for visual representation of large ontologies in learning. In: Kravets, A., Shcherbakov, M., Kultsova, M., Groumpos, P. (eds.) Creativity in Intelligent Technologies and Data Science, pp. 133–149. Springer, Cham (2017)
2. Anikin, A., Litovkin, D., Sarkisova, E., Petrova, T., Kultsova, M.: Ontology-based approach to decision-making support of conceptual domain models creating and using in learning and scientific research. In: IOP Conference Series: Materials Science and Engineering, vol. 483, p. 012074, March 2019. https://doi.org/10.1088/1757-899x/483/1/012074
3. Baclawski, K., Kokar, M.K., Kogut, P.A., Hart, L., Smith, J., Letkowski, J., Emery, P.: Extending the unified modeling language for ontology development. Softw. Syst. Model. **1**(2), 142–156 (2002). https://doi.org/10.1007/s10270-002-0008-4

4. Blomqvist, E., Seil Sepour, A., Presutti, V.: Ontology testing - methodology and tool. In: ten Teije, A., Völker, J., Handschuh, S., Stuckenschmidt, H., d'Acquin, M., Nikolov, A., Aussenac-Gilles, N., Hernandez, N. (eds.) Knowledge Engineering and Knowledge Management, pp. 216–226. Springer, Heidelberg (2012)
5. Dragisic, Z.: Completion of Ontologies and Ontology Networks. Linköping Studies in Science and Technology. Dissertations, Linköping University Electronic Press (2017)
6. Halpin, T.: Object-Role Modeling Fundamentals: A Practical Guide to Data Modeling with ORM. Technics Publications (2015)
7. Halpin, T.: Metaschemas for ER, ORM and UML data models. J. Database Manag. **13**(2), 20–30 (2002). https://doi.org/10.4018/jdm.2002040102
8. Hodrob, R.: On using a graphical notation in ontology engineering. Master's thesis, Birzeit University (2012). https://doi.org/10.13140/RG.2.1.2812.2480
9. Na, H.-S., Choi, O.-H., Lim, J.-E.: A method for building domain ontologies based on the transformation of UML models. In: Fourth International Conference on Software Engineering Research, Management and Applications (SERA 2006), pp. 332–338 (2006). https://doi.org/10.1109/SERA.2006.4
10. Jarrar, M.: Towards automated reasoning on ORM schemes. Mapping ORM into the $\mathcal{DLR}_{idf}$ description logic. In: Parent, C., Schewe, K.D., Storey, V.C., Thalheim, B. (eds.) Conceptual Modeling - ER 2007, pp. 181–197. Springer, Heidelberg (2007)
11. Keet, C.M.: Mapping the object-role modeling language ORM2 into description logic language $\mathcal{DLR}_{ifd}$. CoRR abs/cs/0702089 (2007). http://arxiv.org/abs/cs/0702089
12. Kudryavtsev, D., Gavrilova, T.: From anarchy to system: a novel classification of visual knowledge codification techniques. Knowl. Process Manag. **24**(1), 3–13 (2016). https://doi.org/10.1002/kpm.1509
13. Lehmann, J., Völker, J.: Perspectives on Ontology Learning. Studies on the Semantic Web 2215-0870. IOS Press (2014)
14. Shannon, C.E.: A mathematical theory of communication. Bell Syst. Tech. J. **27**(3), 379–423, 623–656 (1948). https://doi.org/10.1002/j.1538-7305.1948.tb01338.x
15. Starr, R.R., de Oliveira], J.M.P.: Concept maps as the first step in an ontology construction method. Inf. Syst. **38**(5), 771 – 783 (2013). https://doi.org/10.1016/j.is.2012.05.010, http://www.sciencedirect.com/science/article/pii/S0306437912000774
16. Wagih, H.M., ElZanfaly, D.S., Kouta, M.M.: Mapping object role modeling 2 schemes to OWL2 ontologies. In: 2011 3rd International Conference on Computer Research and Development. IEEE, March 2011. https://doi.org/10.1109/iccrd.2011.5764262
17. Wiśniewski, D., Potoniec, J., Ławrynowicz, A., Keet, C.M.: Analysis of ontology competency questions and their formalizations in SPARQL-OWL. J. Web Semant. **59**, 100534 (2019). https://doi.org/10.1016/j.websem.2019.100534
18. Yao, J., Gu, M.: Conceptology: using concept map for knowledge representation and ontology construction. J. Netw. **8**(8) (2013). https://doi.org/10.4304/jnw.8.8.1708-1712

Cyber Threats to Information Security in the Digital Economy

K. S. Luzgina[1(✉)], G. I. Popova[1], and I. V. Manakhova[2]

[1] National Research Nuclear University MEPhI, Moscow 115409, Russia
Luzgina1996@bk.ru
[2] Lomonosov Moscow State University, Leninskie Gory 1, Moscow 119991, Russia

Abstract. The article focuses on the problem of promoting information security in the emerging digital economy. The main emphasis is placed on solving the problems of preventing threats and cybersecurity risks in the era of rapidly developing digital technologies. The basic approaches are related to the construction of an infrastructure to combat cybercrime: formation, technological improvement of IT software and hardware, regulation, the use of finance and insurance, support for the media and the development of a safety culture. The results of this study can be applied to the design of information security systems at all levels.

Keywords: Digital economy · Information security · Cyber threats · IT-technologies · Outsourcing · Innovations · Blockchain

1 Introduction

Digital technologies are transforming the socio-economic paradigm of human life. This is a new foundation for developing public administration systems, economy, business, social sphere, and a society as a whole. In the G20 Leaders' Declaration (on June 29, 2019, the final document of the G20 meeting was signed - "Osaka G20 Leaders' Declaration"), the section "Innovations: Digitalization, Free Data Flow with Trust" is of greatest interest from the perspective of the immediate future of humankind. In fact, it speaks of a new social formation, the concept of which is shared and welcomed by all G20 members. Japan is promoting this system and called it - Society 5.0. Society 5.0 combines technologies such as the Internet of Things (IoT), robotics, artificial intelligence (AI) and big data to make fundamental changes in industry, social and economic spheres. In Society 5.0, people, things and systems are interconnected in cyberspace, and the results obtained with the help of AI, exceeding human capabilities, "return" to the physical space. Society 5.0 is supposed to eliminate all barriers related to regions, age, gender, and language.

At the same time, the theory of the digital economy describes the new nature of rapid succession of boom and bust, increasing cyclic recurrence, and greater expectations. The changes are based on increasing velocity, volume, complexity of relationships and transactions, which intensify volatility, create uncertainty and reduce predictability. In such a situation, the digital economy is changing the concept and

A. V. Samsonovich et al. (Eds.): BICA 2020, AISC 1310, pp. 195–205, 2021.
https://doi.org/10.1007/978-3-030-65596-9_25

essence of the economic security of a nation, region, business, individuals, and poses new threats and risks for actors in economic processes and relations.

2 Review of Relevant Studies and Problem Statement

Since the end of the last century, researchers have been considering post-industrial development as a problem of the information society formation. In their works, they describe the theoretical models of the information society based on an analysis of trends in the global society development, international politics, socio-economic processes and other aspects (Bell 1980; Beniger 1986; Toffler 1981). Information confrontation and information security were dealt with by such scientists as Pipkin (2000), Gordon (2002), Cherdantseva (2013). Their theoretical research was about the potential impact of information on political, economic, military and cultural processes in modern international relations. We can say that the emergence and scientific elaboration of the "information security" concept is directly related to the comprehension of the informatization phenomenon and the study of the information society formation. Information Security refers to a set of methods, technologies and processes designed to protect the networks integrity, programs and data from digital attacks, that is, it provides protection against cyber threats (Andress 2014).

A cyber threat is the illegal intrusion or threat of malicious intrusion into virtual space to achieve political, social or other goals (Khandpur et al. 2017). A cyber threat can affect the computer information space, in which data is located, and where the materials of a physical or virtual device are stored. An attack typically affects a storage medium specifically designed to store, process, and transmit user personal information (Sapienza et al. 2017; Okutan et al. 2018).

Steve DiPaola et al./Procedia Computer Science 00 (2020) 000–000 3

Today we can say that cyberthreats affect economic and political processes to a greater extent than ever before. Accordingly, the issue of the information society development and analysis of the problems associated with this, are still relevant, but is becoming increasingly significant, while the issues of predicting cyber threats come to the fore. The study of ways to distinguish malicious events from ordinary ones we see in the works by Zhang et al. (2008) и Li et al. (2012). The development of correlation systems for cyber threat warning based on models built on previous experience was addressed by Yang et al. (2009) and Xu et al. (2011), this issue was further researched in studies by Shadi et al. (2017). Companies rely on a huge amount of digital data for reliable incident detection (Bhatt et al. 2014). Digital data is an integral part of our modern life. In fact, we live in the era of the digital economy, therefore, we consider it important to dwell on the definition of what the "digital economy" is, since this is important for the purposes of our study.

To date, there is no unified interpretation of what the "digital economy" is. For example, the World Bank gives this definition: "A new economic structure is based on knowledge and digital technologies, within the framework of which new digital skills and opportunities are being developed for society, business and the nation" (World Bank 2016). The following definition was given by the European Parliament: "The digital economy is a complex structure consisting of several levels/layers interconnected by an almost infinite and ever-growing number of nodes" (European Parliament 2015) The European Commission defines the digital economy as "the economy dependent on digital technologies" (European Commission 2014), and it takes the view that "The digital economy is the main source of growth. It will inspire competition, investment and innovation, which will lead to an improvement in the quality of services, broader choice for consumers, and the new jobs creation" (European Commission 2018). According to the audit company Deloitte, digital economy is the economic activity that results from billions of everyday online connections among people, businesses, devices, data, and processes. The backbone of the digital economy is hyperconnectivity which means growing interconnectedness of people, organisations, and machines that results from the Internet, mobile technology and the internet of things (IoT). (Deloitte 2019). Such a number of interpretations leads to the conclusion that, in international practice, a harmonized definition of the digital economy has not yet developed. In the majority of studies, when describing the digital economy, the emphasis is on technologies and related changes in the ways of interaction between economic agents. In so doing, either specific types of technologies or one or another form of change in economic processes may be mentioned in them.

With the growing popularity of Internet technologies, the field of cybercrime is also expanding. Here we can conclude that the main threat to the emerging digital economy may be the loss of confidence in technology due to cyber threats.

Information security is based on data protection and sees the fight against cyber threats as its mission. Attacks take place more often and become more sophisticated. The number of intruders is growing. As a result of digital transformation and the introduction of new technologies, companies face new threats.

3 Research Methodology

The purpose of the study is to consider the main approaches to ensuring the economic security in the country and regions, solving the problem of preventing threats and risks of cybercrime in the era of rapidly developing digital technologies.

Based on the purpose, the following tasks are solved:

– to demonstrate the competition of systems based on new digital platforms;
– to investigate cybercrime as a new threat to the security;
– to consider methods of eliminating threats and risks for the digital economy.

Of great importance in the development of the digital economy in order to restructure and increase competitiveness is not only the development of a clear strategy, its focus on the formation of progressive technological structures, but also the ability to use the whole arsenal of instruments of direct and indirect state regulation.

The emergence of a new type of crime - organized cybercrime - forces economic agents and the state to single out the main tasks to prevent cyber threats in the following areas:

– personal data protection.
– security of commercial information systems.

- security of government institutions' information systems.
- protection of production environment, technologies and instruments.

The main research methods used are institutional, systemic, structural-functional and statistical approaches.

4 Results and Discussion

The expansion of the digital service, the individualization of many types of services increases the risk of fraud while reducing the control by users or providers. The risk of data leak requires better protection of e-systems. Currently, cyberthreats and damage from cybercriminals come immediately after technological disasters in the world (Manakhova 2016). In the media, we constantly see data confirming these trends: 6.4 billion fake emails are sent daily around the world (Dark Reading 2018a); 550 million phishing emails were sent as part of a single attack in the first quarter of 2018 (Dark Reading 2018b); 50% of local authorities in England use unsupported software for their servers (Computing 2018); 1,464 public servants of one state used Password123 (The Washington Post 2018); 2 million stolen accounts were used to falsify the results of the American study of network neutrality (Naked Security 2018); $ 729,000 was stolen from one businessman using the simultaneous use of fake accounts on social networks and directed phishing (SC Media 2017); $ 3.62 million is the average damage from data privacy violations over the past year (Ponemon Institute 2017). A survey of 1,440 respondents conducted in the framework of the "International EY study in the field of information security, 2018–2019" showed that 39% of the companies surveyed reported that less than 2% of all IT services specialists work exclusively on cybersecurity projects. Among the companies surveyed, more than half (55%) do not include cybersecurity issues in their strategy. This problem unexpectedly turned out to be more relevant for larger companies, but not for smaller ones (58% and 54%, respectively).

According to Richard Watson, EY Asia-Pacific Cybersecurity Risk Advisory Leader, more and more companies are beginning to see the magnitude of the threat. Over the past 12 months, there has been one important change for the better, partly due to a series of major cyber attacks at the global level - now almost everyone realizes that information security is necessary not only to maintain data confidentiality, but also to ensure business continuity. The study indicates that 53% of the respondents increased spending on cybersecurity this year and 65% plan to increase spending next year (Table 1).

An interesting fact is that the vast majority of companies (77%) are still at the very beginning of the path to ensuring cybersecurity and resistance to attacks. Organizations sometimes do not even have a full understanding of what information and assets are critical to them and where they are stored, to say nothing of their effective protection.

Therefore, many companies need to continue allocating resources to achieve a basic level of cybersecurity. First of all, it is necessary to identify key data and intellectual property objects, then analyze the available opportunities for ensuring cybersecurity, access control processes and other means of protection, and only then develop measures to strengthen the perimeter to fend off cyber attacks.

Table 1. Forecast of corporate cybersecurity spendings for the next 12 months (E&Y 2019a)

Corporate cybersecurity spendings	During the current year.	Next year
Will increase by more than 25%	12%	15%
Will increase by 15–25%	16%	22%
Will increase by 5–15%	25%	28%
Will remain at approximately the same level (from +5% to −5%)	40%	31%
Will decrease by 5–15%	4%	2%
Will decrease by 15–25%	1%	1%
Will decrease by more than 25%	1%	1%

The effectiveness of cybersecurity activities depends on understanding, which information is the most valuable. Obviously, it is customer and financial information, as well as strategic plans, that the company should protect on the first priority basis. Then it is followed by top management information and customer passwords. Information on suppliers is last on the list of top ten, and this indicates that the concept of joint protection throughout the supply chain is not made sufficiently well known. The ranking of the top 10 most valuable data types for cybercriminals (E&Y 2019b) is shown in Fig. 1.

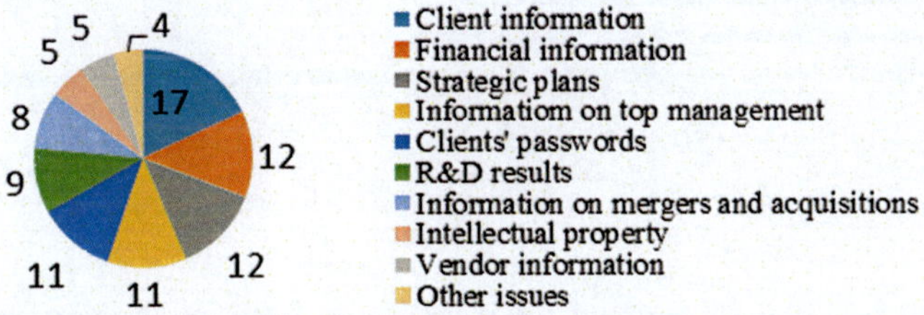

Fig. 1. Ranking data types according to their value for the company

According to PJSC Sberbank, the losses of the Russian economy from the hackers' activities in 2019 amounted to about 2.5 trillion rubles. In 2020, this figure may increase to 3.5–3.6 trillion rubles due to the projected increase in the number of cyber fraud by 40%. At the same time, in 2018, the loss amounted to 1.5 trillion rubles, PJSC Sberbank reported on January 28, 2020.

The retreat of crime into virtual reality, according to experts, can be a serious blow to business, regardless of the industry in which the company operates. Experts record the interest of cyber fraudsters not only in their traditional banking, financial and IT organizations, but also, as follows from the report of the IS company Positive Technologies, in the industrial sector, and the number of successful cyberattacks in Russia in 2018 is growing at the rate of 27%. In 49% of cases, hackers attacked the infrastructure, in 26% - companies' web resources. In 30% of cases, criminals stole personal data, in

24% - account data, in 14% - payment data. In general, the number of attacks aimed at data theft demonstrates growth.

In April 2019, the management of Rostelecom reported that by the end of 2018, the number of cyber attacks in Russia had doubled, while the income of hackers exceeded 2 billion rubles. Over 2018, experts at the Rostelecom Solar JSOC established for monitoring, investigation and response to cyber attacks recorded 765,259 incidents - 89% more than in 2017. Credit and financial organizations, as well as e-commerce and gaming enterprises, were victims of 75% of cyber attacks.

Recently, due to the lowering in the threshold for this activity, organized cyber-crime is aimed at personal data theft through mobile devices and financial mobile applications.

In 2019, Microsoft and TAdviser conducted a study. The results are presented in the report "Assessment of the cyber security in Russian business". The study was conducted on a survey of respondents - heads of IT departments, heads of information security offices, heads of functional departments, as well as other persons who influence decision-making in the field of IT and information security. The study revealed the main protection equipment that are used in companies (Fig. 2).

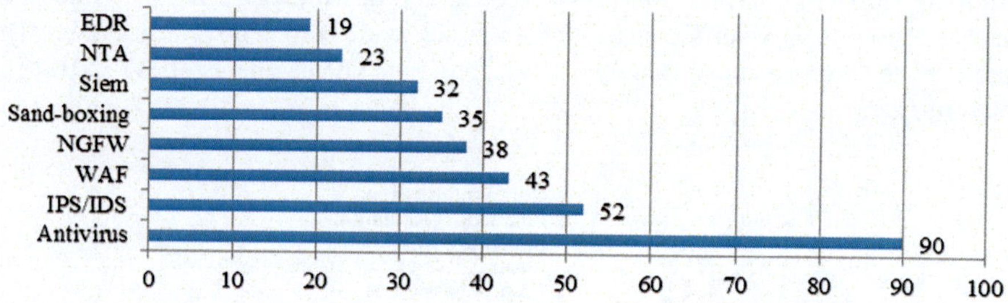

Fig. 2. Technical protective equipment used by Russian companies in 2018

Despite the fact that information security is the basis of the competitiveness of modern business, according to a survey of 248 Russian firms by PwC in 2017, it turned out that 40% of Russian companies do not have an information security strategy, while 50% do not have a plan for responding to information incidents security, while 48% of companies do not have employee training programs aimed at improving business information security.

Such an attitude to cybersecurity leads to great damage and economic losses, to violation of data confidentiality, disclosure of trade secrets, the possibility of industrial espionage, breach of business standards, unforeseen problems of business processes, intellectual piracy, lower quality of products and services, and creates conditions for the appearance of a threat people's lives. But, recently, investment in cybersecurity by enterprises has begun to increase dramatically. According to Canalys analysts' data released March 28, 2019, the dynamics of total global spending on cybersecurity is shown in Fig. 3.

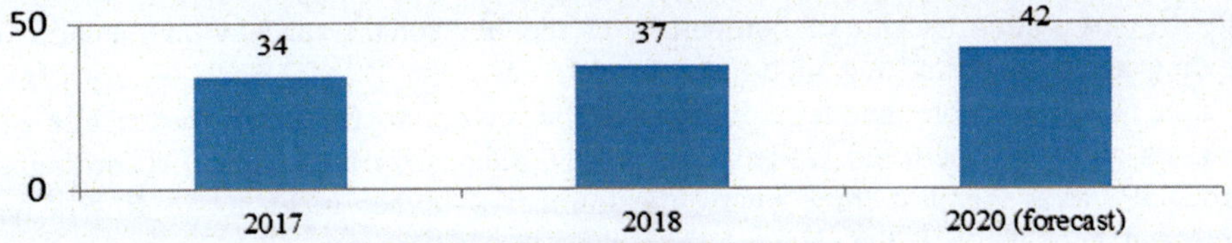

Fig. 3. Cybersecurity costs, US $ billion

Significant costs to eliminate threats to economic security are borne by the power and manufacturing industries of the Russian Federation, since power plants, infrastructure, and factories were built during the Soviet era and most of them do not meet modern information security requirements. The digital transformation of business processes and technologies increases the cost of information security and infrastructure. A study by Qrator and Wallarm "Information Security in the Financial Sector" showed that in 2016 out of 150 Russian banks, the expenses on information security increased in 32% of banks and in 39% of banks remained at the same level, which increases overall prime cost of financial services and loans. According to the Information Security Association (BISA), the main consequences of information security incidents, in market participants' view, are financial risks (32%), reputation risks (38%) and license revocation risks (24%). Recently, financial institutions and payment systems have begun to use third-party solutions to protect against threats of DDos attacks, many banks initially pass information traffic through an external server, thereby increasing the ability to counter unauthorized access to the main server (Manakhova and Udalov 2017). The global trend in the development of the theory and practice of cybersecurity is that information security departments should not be a superstructure over the company's business processes or government.

External challenges, deep digital import dependence of the economy, dictate to Russian companies the task of participating in intense international competition for creating added value in the cyber environment, which requires finding ways to commercialize and protect intellectual property, legislative addressing the issues related to the digital technologies and software turnover, and ensure information security when launching digital platforms and technologies. To do this, it is necessary to ensure the security of the main tools of the digital economy - the protection of electronic signatures and payments, tokens, SIM cards, online services, the protection of information in electronic clouds, databases, the development of cryptography and identity authentication technologies, the protection of electronic document management systems, transmission channels, servers, commercial and public electronic trading platforms, and scientific laboratories. A separate task is to protect against cyber attacks against artificial intelligence, robots, unmanned aerial vehicles and transport, the latest Blockchain technologies, and the Internet of Things (IoT) (Shcherbik 2015).

The main threats to the information security of the digital economy today are encryption viruses, for example "cryptolocker" which penetrate not only personal computers, but also into networks of strategic objects and can cause technological disasters. Losses and damage from such penetrations are estimated in the world by hundreds of millions of dollars. A serious threat is posed by attacks on initial public offerings of shares (ICOs) of companies in the blockchain space with the aim of stealing assets or destroying platforms (for reference: the Telegram messenger's ICO volume in 2018 is more than $ 1.2 billion), attacks on cryptocurrency infrastructure and services, theft of electronic wallets, passwords, attacks on banks, etc. According to Group-IB, over the past three years, the number of cyber incidents in Russia has increased by 72%, and the damage from them by 200% with a forecast of a triple increase.

New digital technologies provide new opportunities for the analysis of big data, computer vision allows you to automatically process a huge number of images, photographs, find the desired object, which helps in the search for attackers and terrorists. The danger of new technologies is that the visualization of information flows forms a new subculture of the Internet, sometimes negatively affecting the psyche of people with its minimalism, "aesthetics of gray concrete", and hopelessness, which carries with it the threat of criminals exploiting suicide and death, especially among young people. Instagram has become the new mainstream of mass culture, where you can create an illusion and "edit" ideas about your own life, while being forced to be influenced by the imposition algorithms created by search engine robots (Mamaev 2016).

In 2017, more than 3 thousand addressees, 250 companies in 12 countries received phishing emails with infected files from the Cobalt hacker group, and three weeks later a cyber attack occurred with an average check of $ 100 million, from which not everyone could defend themselves. Among many cybercriminals, the most famous organized hacker groups who have committed cybercrime where the amount involved is equal to hundreds of millions of dollars around the world are Lurk, Buhtrap, Lazarus, Carbanak, Cobalt. The damage from their criminal activity is estimated not only at the amount of stolen, caused harm and destroyed infrastructure, but also at inflicting invaluable moral damage to the injured citizens and firms, harming the global economy by the fact that many organized cybercriminals recruited, involved in criminal plans and actions the best minds of cyberspace, young people passionate about digital technology, who have not found their place in the legal economy. A separate issue is the establishment and development of specialized public electronic departments and the unleashing of invisible computer wars over the spheres of influence both in the economy and in politics. There are already many examples of such computer wars and conflicts, starting with attacks on the infrastructure facilities of Iran's nuclear program and ending with alleged claims about "interfering" in the US presidential election. The creation by countries of "troll factories" that distribute fake news has marked the beginning of a new era in which imaginary information becomes a custom-made product and can bring down a stock quote or a policy-maker rating in a matter of minutes (Shcherbik 2015).

In Russia, to date, the electronic databases of various ministries and departments are so fragmented that there is no single database, there is no single electronic signature of a citizen, and an electronic signature issued in one department is not accepted in another. The accelerated reshaping of the Russian economy into a digital environment depends on the access to patents and inventions. Our country can use the experience of the Bell Labs laboratory (AT&T), which in order to accelerate technological progress in 1948 provided access to information on the invention of the transistor, in 1958 to integrated microcircuits; in 1971 it published data on the microprocessor, and in early 2000 on prototypes of DNA machines.

5 Conclusions

In the era of the rapid development of digital technologies, the main goals of maintaining the economic security in Russia and solutions to the threats and risks of cybercrime are as follows:

- Constant data communication on information incidents and protection technologies among companies and public organizations at the international level. Implementation of round-the-clock response to cyber incidents at the facilities to identify, analyze and prevent cyberthreats.
- Improving the information security competencies of IT-specialists, employed in all services of companies and government agencies, organizing the interaction between business units, IT specialists and economic security departments.
- Further work of the RF CB Cybersecurity Center to improve the banking system security and payment systems.
- Constant media coverage of the impact of the struggle with cybercrime.
- The propagation of digital "hygiene" right out of high school, the organization of school programs, and cyber literacy lessons.
- Improvement of the technical support for systems information security, constant update of antiviruses, installing firewalls, prevention of leak protection, the use of telemetering services and routers, protection of automated process control systems from malicious software.
- Legislative regulation of cyberspace, cryptocurrencies and blockchain technologies.
- Financing cyber intelligence programs to search for cybercriminals and destroy a criminal business.
- Early completion and implementation of the program "Digital Economy of Russia".
- Launch of new cyber threat insurance products and programs.
- Initiation by the Russian Federation of activity aimed at adopting the United Nations moratorium on the use of cyber weapons.

Applying new digital technologies, short parsing of big data allows producing a realistic economic projection. The goal of matching individual consumer preferences leads to the fact that competition between manufacturers develops in the direction of the struggle for information about the preferences and dreams of the buyer, for a database of his lifestyle, health, and hobbies, which increases the risk of unauthorized access and theft of personal data.

The international struggle for political and economic influence in the context of the digital economy development poses a new threat to economic security. The prevention and elimination of threats and risks to the digital economy and ensuring the security of the IT environment have become today the basis of competitiveness for people, businesses and the states. The prevention and elimination of threats and risks to the digital economy and ensuring the security of the IT environment have become today the basis of competitiveness for people, businesses and the states.

References

Andress, J.: The Basics of Information Security: Understanding the Fundamentals of InfoSec in Theory and Practice. Syngress (2014)

Bell, D.: The social framework of the information society. In: Forester, T. (ed.) The Microelectronics Revolution, pp. 500–549. Blackwell, Oxford (1980)

Beniger, J.R.: The Control Revolution: Technological and Economic Origins of the Information Society. Cambridge University Press, Cambridge (1986)

Bhatt, S., Manadhata, P.K., Zomlot, L.: The operational role of security information and event management systems. IEEE Secur. Priv. **12**(5), 35–41 (2014)

Cherdantseva, Y.: Information Security and Information Assurance. The Discussion about the Meaning, Scope and Goals. Organizational, Legal, and Technological Dimensions of Information System Administrator. IGI Global Publishing (2013)

Putilov, A.V., Timokhin, D.V., Bugaenko, M.V.: Adaptation of the educational process to the requirements of the global nuclear market according the concept of «economic cross» through its digitalization. Procedia Comput. Sci. **169**, 452–457 (2020). https://doi.org/10.1016/j.procs.2020.02.226

Dark Reading (2018a). https://www.darkreading.com/endpoint/64-billion-fake-emails-sent-each-day/d/d-id/1332677. Accessed 11 Sept 2020

Dark Reading (2018b). https://www.darkreading.com/vulnerabilities—threats/new-phishing-attack-targets-550m-email-users-worldwide/d/d-id/1331654. Accessed 11 Sept 2020

Deloitte. What is Digital Economy? (2019). https://www2.deloitte.com/mt/en/pages/technology/articles/mt-what-is-digital-economy.html. Accessed 11 Sept 2020

European Commission: Expert Group on Taxation of the Digital Economy (2014). https://ec.europa.eu/taxation_customs/sites/taxation/files/resources/documents/taxation/gen_info/good_governance_matters/digital/report_digital_economy.pdf. Accessed 11 Sept 2020

European Commission: Digital Economy (2018). https://ec.europa.eu/jrc/en/research-topic/digital-economy. Accessed 11 Sept 2020

European Parliament: Challenges for Competition Policy in a Digitalised Economy (2015). https://www.europarl.europa.eu/RegData/etudes/STUD/2015/542235/IPOL_STU%282015%29542235_EN.pdf. Accessed 11 Sept 2020

E&Y. EY Cybersecurity Summit (2019b). https://www.ey.com/gl/en/issues/governance-and-reporting/center-for-board-matters/ey-understanding-thecybersecurity-threat. Accessed 11 Sept 2020

Gordon, L.: The economics of information security investment: Lawrence Gordon, Martin Loeb. ACM Trans. Inf. Syst. Secur. **5**(4), 76–87 (2002)

Khandpur, R., et al.: Crowdsourcing cybersecurity: cyber attack detection using social media. CoRR (2017) http://arxiv.org/abs/1702.07745.1702.07745

Li, Y., Xia, J., Zhang, S., Yan, J., Ai, X., Dai, K.: An efficient intrusion detection system based on support vector machines and gradually feature removal method. Expert Syst. Appl. **39**(1), 424–430 (2012). https://doi.org/10.1016/j.eswa.2011.07.032

Okutan, A., Yang, S.J., McConky, K.: Forecasting cyber attacks with imbalanced data sets and different time granularities. CoRR 3(42), pp. 35–56 (2018). abs/1803.09560. http://arxiv.org/abs/1803.09560.1803.09560

Manakhova, I.V., Kudaykulov, M.K.: Global challenges and risks of economic security: economic security of Russia: challenges of the XXI century in Materials of the international scientific-practical conference, 15 March, pp. 169–172 (2016)

Mamaev, L.N.: Typical problems of information security in the modern economy. Inf. Secur. Reg. **1**(22), 21–24 (2016)

Manakhova, I.V., Udalov, D.V.: Threats to financial security: countermeasures. Problems of Economic Security: The Search for Effective Solutions: A Monograph. – Chelyabinsk, 3, p. 365 (2017)

Naked Security (2018). https://nakedsecurity.sophos.com/2018/05/24/2-million-stolen-identities-used-to-make-fake-net-neutrality-comments/. Accessed 11 Sept 2020

Pipkin, D.L.: Information Security: Protecting the Global Enterprise. Prentice Hall PTR, New York (2000)

Ponemon Institute (2017). https://www.ponemon.org/blog/2017-cost-of-data-breach-study-united-states. Accessed 11 Sept 2020

SC Media (2017). https://www.scmagazine.com/home/resources/email-security/australian-loses-1-million-in-catphish-whaling-scam/. Accessed 11 Sept 2020

Sapienza, A., Bessi, A., Damodaran, S., Shakarian, P., Lerman, K., Ferrara, E.: Early warnings of cyber threats in online discussions. In: Proceedings of the 2017 IEEE International Conference on Data Mining Workshops (ICDMW), pp. 667–674 (2017)

Applying a Logical Derivative to Identify Hidden Patterns in the Data Structure

Larisa A. Lyutikova(✉)

Institute of Applied Mathematics and Automation, KBSC RAS
(IAMA KBSC RAS), St. Shortanova 89 a, 360000 Nalchik, KBR, Russia
lylarisa@yandex.ru

Abstract. The paper proposes a method for assessing the significance of individual characteristics of recognized objects. The totality of objects and their characteristics is represented by the structure and weight coefficients of a trained ΣΠ-neuron. The specified ΣΠ-neuron correctly processes objects of the subject area, which may not be explicitly represented. It is known that when using the neural network approach, the logical rules for decision making by the neural network remain hidden to the user. The proposed method for constructing the decisive function allows us to identify these logical rules of a correctly functioning ΣΠ-neuron. To assess the significance of the characteristics of objects, a logical derivative is used. Which shows how the decisive function will change its value if one or more characteristics of the objects change. That will allow us to conclude about the most important properties of the subject area under consideration. This is especially important when data is incomplete, fuzzy, or distorted due to information noise.

Keywords: Decisive function · Logical derivative · Data analysis · Algorithm · ΣΠ-neuron · Decision trees · Corrective operations

1 Introduction

Today, neural networks are one of the most popular tools for solving poorly formalized tasks. Tasks for which there is no mathematical formulation and formal algorithmic solutions. These are tasks for the solution of which heuristics are required in order to find a more rational solution, rather than an exact mathematical one, by eliminating previously unsuitable solutions.

The data and knowledge of this area are characterized by factors: incompleteness, unreliability, inaccuracy, ambiguity.

Despite the fact that neural networks do a good job with a great many of such tasks, the rules for making decisions are not clear to the user. Available structure and weight characteristics, which acquired a neural network as a result of training. And to identify the logical connections according to the characteristics of a correctly functioning neural network, this means to gain new knowledge about the subject area being studied.

By analogy with natural intelligence, which can be first taught by examples, then rules, methods, etc. are realized or created.

A. V. Samsonovich et al. (Eds.): BICA 2020, AISC 1310, pp. 206–211, 2021.
https://doi.org/10.1007/978-3-030-65596-9_26

2 Construction of the Classifier Function Based on the Structure of the ΣΠ-Neuron

As is known, a ΣΠ-neuron (sigma-pi neuron) is a generalization of the classical model of a formal neuron with a linear function of summing $sp(x_1, \ldots, x_n)$ input signals.

ΣΠ-neuron is represented by the following structure

$$sp(x_1, \ldots, x_n) = \sum w_i \prod x_i$$

Where $\{w_1, w_2, \ldots, w_k\}$ is the set of weights of a given ΣΠ-neuron that recognizes k elements of a given subject $Y = \{y_1, y_2, \ldots, y_k\}$ area formed by the corresponding set of features $\{X_1, \ldots, X_k\}$ [1].

EXAMPLE. Let the following training set be given (Table 1):

Table 1. Example.

x_1	x_2	x_3	y
0	0	1	a (2)
0	1	1	в (4)
1	1	0	c (6)

The set of attributes X is represented by the following values:

$$\text{x} = \{x_1 = (0, 0, 1), x_1 = (0, 1, 1), x_1 = (1, 1, 0)\}$$

and the set of objects {a, b, c} can be transcoded for training in

$$\text{a-2}, b - 4, \text{c-6}.$$

As a result of training according to the table, the ΣΠ-neuron will look like:

$$sp(x_1 x_2 x_3) = 2x_3 + 2x_2 x_3 + 4x_2 x_1$$

Any query (x_1, x_2, x_3), which is presented in the table, will be identified with the corresponding object.

If the query does not coincide with the values of the variables that are in the training set, for example (0,1,0), then the result may be incorrect or it may not exist at all.

$$sp(0, 1, 0) = 2 * 0 + 2 * 1 * 0 + 4 * 0 * 0 = 0 \text{ [7]}.$$

Although, it could be an object, b-4, or c-6, in cases where there are inaccuracies, noise, interference in the data.

To obtain more stable solutions, a trained neuron requires additional corrective methods.

3 Construction of the Decisive Function According to the Structure of the ΣΠ-Neuron

When constructing the decisive function, you may not know the training set; it is enough to know the value of weights and the structure of the neuron. The function is built on a tree, the construction algorithm of which is described in [1].

The number of levels is equal to the largest number of products of variables in each of the terms +1. In the example there will be 3.

At the bottom level are variables $\{x_1, x_2, \ldots, x_n\}$. The weights of the first layer $\{w_1, w_2, \ldots, w_r\}$, respectively, are objects $\{y_1, y_2, \ldots, y_r\}$ on each subsequent layer $y_{k+1} = w_{k+1} + \sum y_i$, where i the indices of the corresponding objects whose variables are included as factors in the element by y_{k+1}.

EXAMPLE

$$sp(x_1x_2x_3) = 2x_3 + 2x_2x_3 + 4x_2x_1$$

We will restore the objects of the training sample and find generalizing logical rules (see Fig. 1).

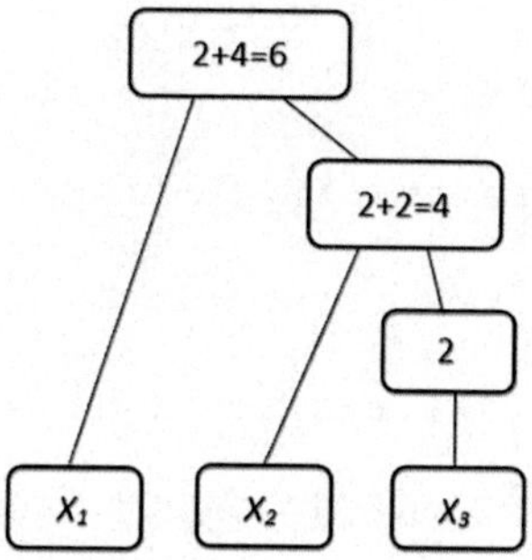

Fig. 1. Recovery of objects of the training sample.

From each vertex y_k to each variable x_i: $P(y_k)$ & $P(y_{k-1})$ &...& $P(y_i)$ & x_i

$$P(y_k) = 1, \; if \; = y_k \; P(y_k) = 0, \; if \; \neq y_k$$

For this example, the minimum set of rules will look like:

$$F(x_1x_2x_3) = P(6)x_1 \vee P(6)P(4)x_2 \vee P(2)P(2)x_3$$

These rules are sufficient if only the presence of a characteristic is important for the data under consideration. But these rules are not enough if the value of the variable zero is also informative for decision making. And these rules are not enough, in the case of multi-valued coding.

Therefore, there is a need to build additional trees, or imaginary paths, in the figure this is indicated by a dash-dot line.

For example, it looks like in Fig. 2.

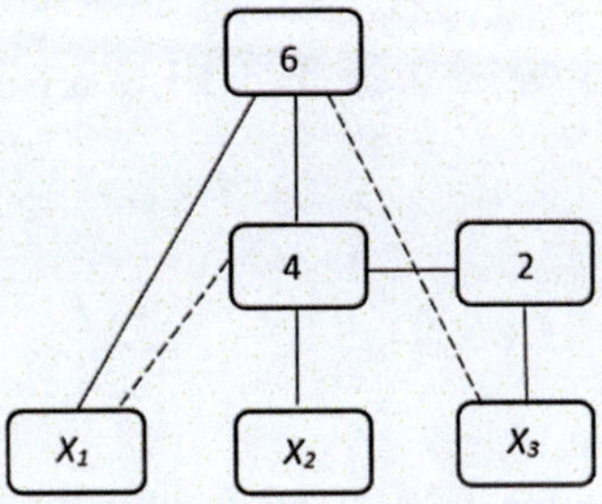

Fig. 2. Object part of a logical function.

A dashed line indicates the relationship with the variable's negation.

Then the decision function for our example will look like this:

$$F(x_1x_2x_3) = P(6)x_1\bar{x}_3 \vee P(6)P(4)x_2 \vee P(4)P(2)x_3\bar{x}_1$$

That is, the most important features for the source data will use the logical derivative.

4 Some Properties of Operations of Logical Differentiation of Boolean Functions

Logical differential and integral calculus are the directions of modern discrete mathematics and find their application in the problems of dynamic analysis and synthesis of discrete digital structures. The basic concept of logical differential calculus is the derivative of a Boolean function, the idea of which in the form of a Boolean difference was obtained back in [8, 9] [10].

By some of its properties, the Boolean derivative is an analogue of the derivative in classical differential calculus.

Definition 1. The first-order derivative $\frac{\partial f}{\partial x_i}$ of the Boolean function $f(x_1, \ldots, x_n)$ with respect to the variable x_i is the sum modulo 2 of the corresponding residual functions:

$$\frac{\partial f}{\partial x_i} = f(x_1, \ldots, x_{i-1}, 0, x_{i=1}, \ldots, x_n) \oplus f(x_1, \ldots, x_{i-1}, 1, x_{i=1}, \ldots, x_n)$$

Definition 2. The weight of the derivative $P\left(\frac{\partial f}{\partial x_i}\right)$ of a Boolean function is the number of constituents ("1") of this derivative.

Statement 1. The greater the weight of the derivative, the greater the function $f(x_1, \ldots, x_n)$ depends on the variable x_i.

Definition 3. A mixed derivative of the k-th order of a Boolean function $f(x_1, \ldots, x_n)$ is an expression of the form:

$$\frac{\partial^k f}{\partial(x_1 \ldots x_k)} = \frac{\partial}{\partial x_k}\left(\frac{\partial^{k-1} f}{\partial x_1 \ldots \partial x_{k-1}}\right)$$

In this case, the order of the fixed variable does not matter. The k-th derivative determines the conditions under which this function changes its value while changing the values of $x_1, \ldots, x_k$.

For our example:

$$\begin{aligned}\frac{f(\mathrm{X})}{d\mathrm{x}_1} &= P(6)\bar{x}_3 \vee P(6)P(4)x_2 \oplus P(6)P(4)x_2 \vee P(4)P(2)x_3 \\ &= P(4)P(2)x_3 \vee P(6)\bar{x}_3\end{aligned}$$

The derived derivative can classify objects by the variable x_3.

Derivative of the variable x_2

$$\begin{aligned}\frac{f(\mathrm{X})}{d\mathrm{x}_3} &= P(6)\bar{x}_3 x_1 \vee P(2)P(4)\bar{x}_1 x_3 \oplus P(2)P(4)\bar{x}_1 x_3 \vee P(6)\bar{x}_3 x_1 \vee P(4)P(6) \\ &= P(6)P(4)\bar{x}_3\bar{x}_1 \vee P(6)P(4)x_1 x_3\end{aligned}$$

This result gives conflicting data on only two objects, and makes their classification impossible. Therefore, it can be argued that the variable x_2 reflects the most important properties for the data under study. And the variables x_3 and x_1 are dependent, i.e. they are ensemble variables $x_1 = \bar{x}_3$.

5 Conclusion

Building a logical function based on the weight characteristics of a neural network gives an idea of the hidden rules of functioning of this neural network, and makes it possible to correct the result in cases when the neural network is wrong. For example, if there is interference in the data.

Analysis of the constructed decision function using logical derivative methods allows us to formalize the process of finding the coefficients of importance for the characteristics of object properties. And also to find the ensemble characteristics. This is particularly important when data is distorted due to information noise, or for other reasons.

As a result, the quality of automated solutions to intellectual problems, their reliability, and ensuring the accuracy of achieving the correct solution are significantly improved by using the most effective systems for analyzing source data and developing more accurate methods for processing them.

Acknowledgments. The reported study was funded by RFBR according to the research project №18-01-00050-a.

References

1. Lyutikova, L.A.: Sigma-Pi neural networks: error correction methods. Procedia Comput. Sci. **145**, 312–318 (2018)
2. Graves, A., Wayne, G., Reynolds, M., Harley, T., Danihelka, I., Grabska-Barwińska, A., Colmenarejo, S.G., Grefenstette, E., Ramalho, T., Agapiou, J., et al.: Hybrid computing using a neural network with dynamic external memory. Nature **538**(7626), 471–476 (2016)
3. Naimi, A.I., Balzer, L.B.: Stacked generalization: an introduction to super learning. European Journal of Epidemiology **33**(5), 459–464 (2018)
4. Yang, F., Yang, Z., Cohen, W.W.: Differentiable learning of logical rules for knowledge base reasoning. In: Advances in Neural Information Processing Systems, vol. 2017, pp. 2320–2329 (2017)
5. Flach, P.: Machine Learning: The Art and Science of Algorithms that Make Sense of Data, p. 396. Cambridge University Press, Cambridge (2012)
6. Rahman, A., Tasnim, S.: Ensemble classifiers and their applications: a review. Int. J. Comput. Trends Technol. **10**(1), 31–35 (2014)
7. Dyukova, E.V., Zhuravlev, Y.I., Prokofiev, P.A.: Methods to improve the efficiency of logical correctors. Mach. Learn. Data Anal. **1**(11), 1555–1583 (2015)
8. Zhuravlev, Y.I.: On the algebraic approach to solving problems of recognition or classification. Probl. Cybern. **33**, 5–68 (1978)
9. Lyutikova, L.A., Shmatova, E.V.: Analysis and synthesis of pattern recognition algorithms using variable-valued logic. Inf. Technol. **22**(4), 292–297 (2016)

Algorithm for Constructing Logical Operations to Identify Patterns in Data

Larisa A. Lyutikova(✉) and Elena V. Shmatova

Institute of Applied Mathematics and Automation KBSC RAS (IAMA KBSC RAS), St. Shortanova 89 a, 360000 Nalchik, KBR, Russia
lylarisa@yandex.ru, lenavsh@yandex.ru

Abstract. Neural networks have proven themselves in solving problems when the input and output data are known, but the cause and effect relationship between them is not obvious. A well-trained neural network will find the right answer to a given request, but will not give any idea about the rules that form this data. The paper proposes an algorithm for constructing logical operations, in terms of multi-valued logic, to identify hidden patterns in poorly formalized areas of knowledge. As the basic elements are considered many functions of the multi-valued logic of generalized addition and multiplication. The combination of these functions makes it possible to detect relationships in the data under study, as well as the ability to correct the results of neural networks. The proposed approach was considered for classification problems, in the case of multidimensional discrete features, where each feature can take k-different values and is equivalent in importance to class identification.

Keywords: Algorithm · Multi-valued logic · Neural network · Truth table · Knowledge system

1 Introduction

In practice, there are various approaches to the construction of machine learning algorithms [1–3]. Many of them successfully cope with the tasks, but at the same time, they do not give an idea about the laws of the processed data. Thus, it can be assumed that the neural network in weighted coefficients provides the rules for object recognition, but these rules are not explicit, and it can be difficult to determine the cause of the error. In this paper, we construct an algorithm for finding logical functions that provide an opportunity for more explicit interpretation necessary for decision-making.

2 Formulation of the Problem

The object will be represented by n -dimensional vector, n - the number of characteristic features of the object in question, the j -th coordinate of this vector is equal to the value of the j -th characteristic, $j = 1, \ldots, n$. Information about any characteristic of the object may be missing. The dimensionality of the considered property of the $k_i \in [2, \ldots, N]$, N - object depends on the encoding method of the i-th characteristic [4].

A. V. Samsonovich et al. (Eds.): BICA 2020, AISC 1310, pp. 212–217, 2021.
https://doi.org/10.1007/978-3-030-65596-9_27

Let $X = \{x_1, x_2, \ldots, x_n\} x_i \in \{0, 1, \ldots, k_i - 1\}$, where $k_i \in [2, \ldots, N],$, is a set of properties that characterizes a given object. $Y = \{y_1, y_2, \ldots, y_m\}$ - many considered objects. For each object y_i there is a corresponding set of features

$$x_1(y_i), \ldots, x_n(y_i) : y_i = f(x_1(y_i), \ldots, x_n(y_i)).$$

Or $X = \{x_1, x_2, \ldots, x_n\}$, where $x_i \in \{0, 1, \ldots, k_r - 1\}$, $k_r \in [2, \ldots, N], N \in Z$ - input, $X_i = \{x_1(y_i), x_2(y_i), \ldots, x_n(y_i)\}$, $i = 1, \ldots, n$, $y_i \in Y, Y = \{y_1, y_2, \ldots, y_m\}$ - output:

$$\begin{pmatrix} x_1(y_1) & x_2(y_1) & \ldots & x_n(y_1) \\ x_1(y_2) & x_2(y_2) & \ldots & x_n(y_2) \\ \ldots & \ldots & \ldots & \ldots \\ x_1(y_m) & x_2(y_m) & \ldots & x_n(y_m) \end{pmatrix} \to \begin{pmatrix} y_1 \\ y_2 \\ \ldots \\ y_m \end{pmatrix}$$

It is necessary to construct a function such that $Y = f(X)$.

A function $Y = f(X)$ is called a decisive function.

The dependence under consideration can be approximated using a neural network built on the basis of elements that implement external summation and a continuous scalar function.

$$sp(x_1, \ldots, x_n) = \sum w_i \prod x_i$$

where $\{w_1, w_2, \ldots, w_k\}$ is the set of weights of a given $\Sigma\Pi$ neuron that recognizes k elements of a given subject area $Y = \{y_1, y_2, \ldots, y_k\}$ formed by a corresponding set of features $\{X_1, \ldots, X_k\}$.

Of great interest are direct sums that allow you to simultaneously form the architecture of a computer network and configure its parameters, without resorting to solving complex optimization problems to achieve the correctness of its functioning [5].

3 An Algorithm for Constructing a Decisive Function

Consider a multi-valued logical system $L = \langle L, \bar{x}, \&, \vee, \bar{\ }, \to \rangle$, $L = \{0, 1, \ldots, k - 1\}$.

Definition: Set of functions $\sigma(x, y)$: such that $\sigma(x, 0) = \sigma(0, x) = x$, we will call functions of generalized addition.

To construct the decisive function $Y = f(X)$, it is required to find a set of functions Σ that satisfy the following conditions: $\Sigma(a_1, \ldots, a_{n-1}, a_n) = y$

The algorithm is built in the form of a tree.

We assume that $\Sigma(a_1, \ldots, a_{n-1}, a_n) = \Sigma(a_1, \ldots, a_{n-1}) + a_n = \sigma(\Sigma(a_1, \ldots, a_{n-1}), a_n)$. Great $\Sigma(a_1, \ldots, a_{n-1})$ suspense, but it must take on one of the meanings $L = \{0, 1, \ldots, k - 1\}$. This implies the need to fulfill one of k conditions:

$$\Sigma(a_1,\ldots,a_{n-1}) = 0, \Sigma(a_1,\ldots,a_{n-1}) = 1,\ldots,\Sigma(a_1,\ldots,a_{n-1}) = k-1 \quad (1)$$

Means

$$\Sigma(0,a_n) = y, \Sigma(1,a_n) = y,\ldots,\Sigma(k-1,a_n) = y, \quad (2)$$

those. get to the branching tree.

In the next step, we consider the following relation $\Sigma(a_1,\ldots,a_{n-1}) = \Sigma(a_1,\ldots,a_{n-2}) + a_{n-1} = \sigma(\Sigma(a_1,\ldots,a_{n-2}),a_{n-1})$, taking into account the previously formulated assumptions regarding the values of $\Sigma(a_1,\ldots,a_{n-1})$, i.e. each of the branches (2) will in turn split into another k. k.

Continuing to carry out the steps described above, we construct a tree of admissible values of the truth tables Σ.

If at some step the assumption contradicts the assumption made earlier regarding the given variant of the function Σ, then such a branch is a dead end and it is discarded.

If at some node all branches are dead ends, then this node itself is deleted from the decision tree.

The last step in these actions is to consider the expression $\Sigma(a_1,a_2)$, after which the process of finding a solution ends.

The set of feasible solutions of the function Σ is collected from the leaves of the tree to the root. As a result, we obtain truth tables of the function Σ that satisfy the given conditions.

We illustrate the operation of the algorithm using the example of three-valued logic.

Let three-valued logic be given $L = \langle L, \bar{x}, \&, \vee, \bar{\ }, \rightarrow \rangle$, $L = \{0, \frac{1}{2}, 1\}$. And let it be given: $\Sigma(1,0,\frac{1}{2},1) = \frac{1}{2}$. We construct a decision tree for this example.

Consider 1 branch: $\Sigma(1,0,\frac{1}{2}) = 0 \Rightarrow$ it is necessary that $\Sigma(0,1) = \frac{1}{2}$, but this contradicts the condition $\sigma(x,0) = \sigma(0,x) = x$. So this branch can be discarded.

Consider a 2 branch $\Sigma(1,0,\frac{1}{2}) = \frac{1}{2} \Rightarrow \Sigma(\frac{1}{2},1) = \frac{1}{2}$. This is possible, therefore we build branches further:

1. $\Sigma(1,0) = 0$ - impossible, therefore, the branch can be dropped;
2. $\Sigma(1,0) = \frac{1}{2}$ - impossible, therefore, the branch can be dropped;
3. $\Sigma(1,0) = 1$ - possible.

Restoring the chain of actions, we obtain: $\Sigma(1,0) = 1$, $\Sigma(1,\frac{1}{2}) = \frac{1}{2}$, $\Sigma(\frac{1}{2},1) = \frac{1}{2}$.

From this we can conclude: that the obtained set of solutions (we will call the class of solutions) has the commutativity property (Table 1).

Consider the 3-rd branch: $\Sigma(1,0,\frac{1}{2}) = 1 \Rightarrow \Sigma(1,1) = \frac{1}{2}$. This is perhaps why we build the branches further:

1. $\Sigma(1,0) = 0$ - impossible, therefore, the branch can be dropped;
2. $\Sigma(1,0) = \frac{1}{2}$ - impossible, therefore, the branch can be dropped;

3. $\Sigma(1,0) = 1$ - possible. Restore the chain of actions, we get: $\Sigma(1,0) = 1$, $\Sigma(1,\frac{1}{2}) = 1$, $\Sigma(1,1) = \frac{1}{2}$ Those. got the truth table without taking into account the commutativity property (Table 2). If commutativity is taken into account, then we get Table 3.

Table 1 Function table after step 1.

	0	$\frac{1}{2}$	1
0	0	$\frac{1}{2}$	1
$\frac{1}{2}$	$\frac{1}{2}$		$\frac{1}{2}$
1	1	$\frac{1}{2}$	

Table 2 Function table after step 2.

	0	$\frac{1}{2}$	1
0	0	$\frac{1}{2}$	1
$\frac{1}{2}$	$\frac{1}{2}$		1
1	1		$\frac{1}{2}$

Table 3 Function table after step 3.

	0	$\frac{1}{2}$	1
0	0	$\frac{1}{2}$	1
$\frac{1}{2}$	$\frac{1}{2}$		1
1	1	1	$\frac{1}{2}$

The set of occupied cells in the table corresponds to those necessary conditions of existence for the implementation of a given identity. And empty cells correspond to non-essential conditions, which means that each free cell generates three possible options: 0, $\frac{1}{2}$, 1. This makes it possible to establish the exact number of functions in the class of solutions (power).

So, for this example, the following classes of solutions are possible:

1) Table 1 - the number of functions in the class of solutions is 9 (the property of commutativity was revealed in the process of finding a solution and is not predetermined);
2) Table 2, the number of functions in the class of solutions is 9. If the commutativity property is assumed to be given in advance, the number of functions in the class of solutions will be 3 (Table 3).

Statement. Commutativity reduces the power of many feasible solutions.

This is due to the fact that the number of free cells in the truth table is reduced.

Theorem. There is an algorithm that determines the possibility of expressing a given function in the form of a formula through the operations of generalized addition.

The proof of the theorem is based on the above algorithm for constructing the operation of generalized addition.

An algorithm is applied to the function specified in the table, and then the results are intersected.

If the resulting intersection set is not empty, then the decisive function is representable as a formula through the operation of generalized addition. If it is empty, then a given function cannot be represented with a single function, but you can select the minimum number of functions that meet the specified requirements.

4 Algorithm for Constructing a Decision Function Based on the Generalized Multiplication Operation

When solving problems of constructive learning, there is a need to find functions that most effectively implement the specified training samples.

Let the input of the system be fed a vector of values $x = \{x_1, x_2, \ldots, x_n\}$ and each input has the weight $w_j, j = 1, \ldots, m$, the output of the system has the resulting offset y.

You need to build (set in a table) a set of functions that satisfy the condition: $f(x_i, w_i) = y$, where $f(x_i, w_i)$ is a function that can be represented through the operations of generalized addition and generalized multiplication.

Let three-valued logic be given $L = \langle L, \bar{x}, \&, \vee, \bar{\ }, \rightarrow \rangle$, $L = \{0, 1, 2\}$.

Definition 2: Many functions $\pi(x, y)$: $\pi(x, 0) = \pi(0, x) = 0$ we call the implementation of generalized multiplication.

For given input values x and w and output y, we have a partially defined three-digit function, which is defined on the set (x, w) by the value y.

For this, we first construct the admissible operations of generalized multiplication for $x_i w_i, i = 1, 2, \ldots, n$ in the form of a tree.

Consider $x_i w_i$. This quantity is unknown to us, but must take one of the values $L = \{0, 1, 2\}$ (due to the closedness of the operation of generalized multiplication).

This implies the need to fulfill one of three conditions:

$$x_1 w_1 = 0, x_1 w_1 = 1, x_1 w_1 = 2.$$

In this case, we obtain three possibilities for implementing the function $\Pi \in \pi$ (i.e., three tree branches).

At the next step, we consider the following relation $x_2 w_2$ and, taking into account the previously formulated assumptions, each of the branches (2) will in turn split into three branches.

Continuing to carry out the steps described above, we construct a tree of admissible values of the truth tables of the operation of generalized multiplication.

If at some step the assumption contradicts the earlier assumption regarding this variant of the function Π, then such a branch is dead-end and it is discarded.

The last step is to consider $x_n w_n$, after which the process of finding generalized multiplication ends.

The set of feasible solutions to Π is collected by lifting from the top leaves of the tree to the root. As a result, we obtain a set of truth tables of the function of generalized multiplication.

If the collection of the set π is empty, then for the functions $= \{x_1, x_2, \ldots, x_n\}$, $w = \{w_1, w_2, \ldots, w_n\}$ one cannot specify the operations of generalized addition and generalized multiplication so that relation (1) holds.

Let $a_i = x_i w_i$, $i = 1, \ldots, n$, then each found operation $\Pi \in \pi$ can be associated with a vector $a = \{a_1, a_2, \ldots, a_n\}$. Those. (1) rewritten: $\Sigma(a_1, a_2, \ldots, a_n) = y$.

And then, to determine the operation of generalized multiplication, we use the algorithm of the representable function through the operations of generalized addition, proposed in the previous section.

5 Conclusion

It can be argued that the logical functions obtained as a result of the proposed algorithms are adequate solvers of the task and make it possible to identify hidden patterns in the data.

Acknowledgement. The reported study was funded by RFBR according to the research project № 18-01-00050-a.

References

1. Flach, P.: Machine Learning: The Art and Science of Algorithms that Make Sense of Data, p. 396. Cambridge University Press, Cambridge (2012)
2. Yablonsky, S.V.: Introduction to Discrete Mathematics, p. 384. Mir Publishers, Moscow (2008)
3. Voroncov, K.V.: Optimizacionnye metody linejnoj i monotonnoj korrekcii v algebraicheskom podhode k probleme raspoznavanija. Zhurnal vychislitel'noj ma-tematiki i matematicheskoj fiziki. T. **40**(1), 166–176 (2000)
4. Lyutikova, L.A., Shmatova, E.V.: Application of variable-valued logic to correct pattern recognition algorithms. In: Advances in Intelligent Systems and Computing, vol. 948, pp. 308–314 (2020)
5. Shibzukhov, Z.M.: Aggregation correct operations on algorithms. Dokl. Math. T. **91**(3), 391–393 (2015)

Graph-Ontology Model of Cognitive-Similar Information Retrieval (on the Requirements Tracing Task Example)

Nikolay V. Maksimov, Olga L. Golitsina, Kirill V. Monankov(✉), and Natalia A. Bal

National Research Nuclear University MEPhI, Moscow, Russia
nv-maks@yandex.ru, kmonankov@yandex.ru, baln757@mail.ru

Abstract. This article considers graph-ontology tools that provide construction, visualization and analysis of an ontology graph using functions for selecting vertices and arcs, set-theoretic operations on graphs, and aspect projection operations. An aspect is specified in terms of general system theory. Aspect projection operations for graph representations of ontologies reduce the dimension of graphs to a level affordable for displaying and human perception. As applied for information retrieval process, it makes possible to move from the task of classical information retrieval to the implementation of cognitive-similar information retrieval task, represented as a search for a path or neighborhood on a multi-meta-hypergraph of an ontology, dynamically formed on the base of ontological images of founded documents or their fragments. The ontology graph is formed via auto-extracting entities and relationships from natural language texts. This article considers the application of the developed tools in the process of analysis and synthesis of knowledge on the example of technical requirements tracing.

Keywords: Information retrieval · Semantic search · Word processing · Graph representations of ontology · Ontologies operations

1 Introduction

Classical information retrieval, built on the principle and technology of using generally accepted terms (keywords), provides a search for documents (storage units), presumably containing the required information. The process of actually extracting the meaning is carried out after the search by the user, and not by the system. Such extraction (as well as subsequent operations that implement the main property of information – it's effectiveness) are carried out by linking individual concepts, entities, facts that are represented in the text by separate words, collocations and phrases. And at this stage, obviously, they are perceived and understood not only in the context of other

This work was supported by the Ministry of Science and Higher Education of the Russian Federation (state assignment project No. 0723-2020-0036).

A. V. Samsonovich et al. (Eds.): BICA 2020, AISC 1310, pp. 218–224, 2021.
https://doi.org/10.1007/978-3-030-65596-9_28

accompanying words, but also in the context of a specific task (more precisely, the user's knowledge of the subject area and the place of this task).

In this case, the user's goal is to build an image of a solution to the problem, or rather, a number of images, which ultimately will allow the formation/materialization of a tool/method/technology for actually solving a pragmatic problem.

In the general case, such an image can be represented as an algorithm for solving the practical problem, where individual components (data/functions) or blocks taken from the obtained information and/or user implicit knowledge will be connected into a logical chain. In this context, in each specific case, the "building block" will not be any instance of a concept from anywhere in the text, but one whose connections (actual or potential – valence/interface) correspond to the problem context. And, it is likely that there will be several relevant fragments in the document text.

Thereby, the search in full texts (selection of "citations" as a substitute for the main activity in the formation of the solution description text) is not only the selection of full texts of documents by concepts and relations, but also the selection of the corresponding fragments, as well as their ordering in accordance with the logic of the solution.

At almost all stages of the complex systems life cycle, questions arise that require an informational research or investigation. For example, what happens if the replacement fastener is made from a different steel grade? What happens if the value of certain parameter goes out of the limit by 10%? Is it safe to use fuel with a slightly different chemical composition?

In order to answer such questions, it is necessary to conduct an expert investigation, referring to the design and research documentation. And in a number of cases, when it is necessary to find a fundamentally new solution, to conduct a search-research, informational substitution of scientific research.

Proceeding from the fact that examination is defined as the formation of a responsible judgment on a certain issue that requires special knowledge, in addition to the requirements of depth (providing detail) and completeness (providing interdisciplinary and multidimensionality) to information retrieval processes, the requirement of validity (providing clarity in constructing a conclusion) is imposed. The results of such information retrieval should present functional and logical connections of concepts, objects and processes – "restore" a complete picture of interdependencies and relationships (ideally, also bring in explicit form what is represented by a person's professional knowledge, and therefore do not require their explicit presence in the text).

In addition, in the tasks of supporting information support for large projects, the problem of diversity and heterogeneity of documentation at different stages of the project life cycle inevitably arises. Frequently, different type documents have different vocabulary, despite the fact that they contain information about one object (for example, design and operational documentation).

One of the tools for analyzing scientific and technical documentation is the technology of automatic construction of ontologies for texts in natural language [1]. Such tools synthetically combine linguistic, statistical and cognitive approaches and allow to build an oriented graph of the document ontology containing vertices corresponding to all entities presented in the text in accordance with their location, and edges corresponding to typed relationships between entities.

Nevertheless, in practice, the problem arises of presenting the result for ontology visual analysis. First of all, the size of a graph built from a relatively small document already contains hundreds and thousands of vertices, which greatly complicates the task of its visual analysis. And, of course, not all vertices and edges describe exactly the requirements for the object of interest. But the main thing is that the vertices mutual arrangement is important: any meaningful text has a "direction", there are initial messages, the corresponding concepts should be at the beginning, etc. That is, the graph should be ordered in accordance with the logic of the problem being solved by the user.

This article[1] discusses technologies for meaningful text analysis using an example related to the technical requirements tracing task, which is characterized by both the search for specific requirements and the identification of contradictions both between different "instances" of a particular parameter, and between the declared and achieved values. Note that such a search is characteristic of cognitive operations performed by a person in the process of analyzing and synthesizing knowledge.

2 Aspect Projection Specification

The ontological approach makes it possible to represent the semantics of a separate solution described in the document by a system of concepts and relations, i.e. when searching, it will be possible to use completed semantic constructions. In this case, the ontology graph will represent the technological space of "entry points" into the information array, providing the possibility of direct transition from the vertices to the fragments of the document text. One of the methods of extracting a subgraph in the ontology graph, which represents a certain semantic slice of the subject area, is the aspect projection operation.

An aspect is formally specified by the corresponding "aspect ontology" – according to the definition of [3], by a system of three interconnected systems: functional, conceptual and sign. But at the same time, the conceptual and sign systems, as well as the set of characteristic properties and the composition law of the functional system of the ontology of the aspect $S_a = <M_a, A_a, R_a, Z_a>$[2] are identical to the corresponding components of the domain or task ontology. That is, an aspect is operationally defined by its characteristic set of basic concepts (entity names), by a set of classes of functional relations and their modalities as characteristic properties. Accordingly, the arc and the incident vertices that meet the particular aspect requirements will be included to the resulting graph.

---

[1] This article continues article [2], which also includes an overview of existing approaches to knowledge representation models and tools.

[2] Here, according to [3], M_a is a set of basic concepts (a set of entities), A_a is a set of characteristic properties of entities and relations, R_a is a set of functional relations characteristic of this aspect, Z_a are rules that determine the construction of chains (restrictions on the inclusion of vertices and arcs to the subgraph).

From the point of view of reflecting the meaning, the aspect, being a projection of the document ontology, allows to build an information slice in a given direction of analysis.

In information retrieval, the aspects taxonomy is generally used. Such a taxonomy (being an object open to extension and modification) defines possible aspects, each of which fixes the relationship of the aspect with the classes of relations characteristic of this point of view. The set of aspects[3] is defined in accordance with the activity functional model and is specified on the functional relations taxonomy[4], supplemented by a set of meta-relations. The latter makes it possible to take into account the connections caused by the language (synonymy, paradigmatics), as well as the "constructional" connections that are characteristic of setting properties (name, dimension, value of parameter). Relationship classes, in turn, are associated with linguistic constructs[5] that represent relationships in the text.

In [5] examples of constructing aspect projections on the basis of a search query through the formation of sets of basic concepts and functional relations are given. The set of characteristic properties (A_a) were not defined within the aspect ontology. In the example below, the "requirement" aspect is used, which is defined by the relationship classes R_a = <"comparison", "change", "restriction">, and/or the relationship should have the modality property A_a = <"need">. That is, for an aspect ontology, it is sufficient to define a set of characteristic functional relations, providing them with the properties present in the considered taxonomy of relations.

3 Technology of Forming a Semantic Aspect Subgraph

Formally, according to [3], the result of the aspect projection operation is the intersection of the original and aspect ontologies. The algorithm includes the following actions:

1. In the graph of the functional system of the original ontology, it is necessary to find arcs corresponding to relations with given properties.
2. Construct a subgraph including vertices incident to the found arcs.

---

[3] Aspect representations are one of the methodological foundations for the synthesis of knowledge. The synthesis of knowledge as a self-organizing process is based on a structural feature of the system – a complex system can be described using a set of relatively independent aspect representations. Moreover, in the process of decomposition, not only the components are separated and connected, but also a decomposition scheme is formed – a system of characteristic signs of division.

[4] There are two classes at the top level of the model, that is, structural and behavioral aspects. The structural aspects include the aspects that reflect the internal morphology of the system (interrelation and composition of elements) and its external morphology, i.e., the properties that manifest themselves in interaction, in particular, with the system's environment (including qualitative assessments). Examples of structural aspects are "Composition", "Following", "Compliance", "Form", "Change", "External signs", etc. The behavioral aspects reflect the dependences and relationships, which are determined, in particular, by the generalized functional activity model. Examples of behavioral aspects are "Efficiency", "Manageability", and "Provision".

[5] The correspondence of relations and linguistic constructions is given in [4].

3. To expand the set of basic concepts, add to the subgraph the arcs incident to the included vertices, corresponding to the meta-relations[6], and the incident vertices to them.

The result of the formal application of the intersection operation can be changed, if necessary, in accordance with the following rule: if the resulting subgraph is incoherent, it can include a minimum subset of arcs and vertices complementing it to a connected subgraph. Such arcs can be useful for identifying functional dependencies.

4 An Example of Constructing a Graph of the "Requirements" Aspect Ontology

As a result of information retrieving on the problem of incomplete demand for the capacities of the Baltic NPP under construction, documents [6–8] were found, the relevant text fragments of which were combined into a single text. Based on it, by a subsystem of visual ontological analysis of scientific and technical texts, an ontology graph is built[7], which, due to the large volume that does not allow a person to adequately perceive it, is not presented here.

To extract the subgraph related specifically to the requirements aspect, the "requirements" aspect projection operation is applied to the graph. The resulting subgraph is shown in Fig. 1.

After performing the "requirements" aspect projection operation, 3 subgraphs were obtained:

- The subgraph formed by the vertices "Electric power of the Baltic NPP power unit", "2300 MW" and "1193 MW", connected by the relations "Change" and "Restriction", and the vertices "Power" added using meta-relations (as the name which is set in the vertices) and "Electrical power decrease of the Baltic NPP power unit".
- The subgraph formed by the vertices "Temperature driving force reduction", "Average temperature of the coolant" and "Active zone", connected by relations "Change", and added using meta-relations by the node "Temperature driving force".
- The subgraph formed by the vertices "Vapor pressure", "The steam generator", "1,5 MPa", "4,5 MPa" and "6 MPa" and the connected relations "Change" and "Restriction", and a vertex added using meta-relations "Pressure" (as parameter name).

After performing the projection operation, the graphs were incoherent. According to the rules given above, they are supplemented with minimal routes to a coherent graph (it should be noted that the direction of arcs is not taken into account when constructing routes). As a result of visual analysis, the relationship between the

[6] In particular, these relations are: "includes entity"; "context"; "circumstance of use"; "context-thesaurus"; "equals"; "parameter name"; "parameter value".

[7] The graph can also be built by combining the relevant fragments of the ontology graphs of the texts of the corresponding documents.

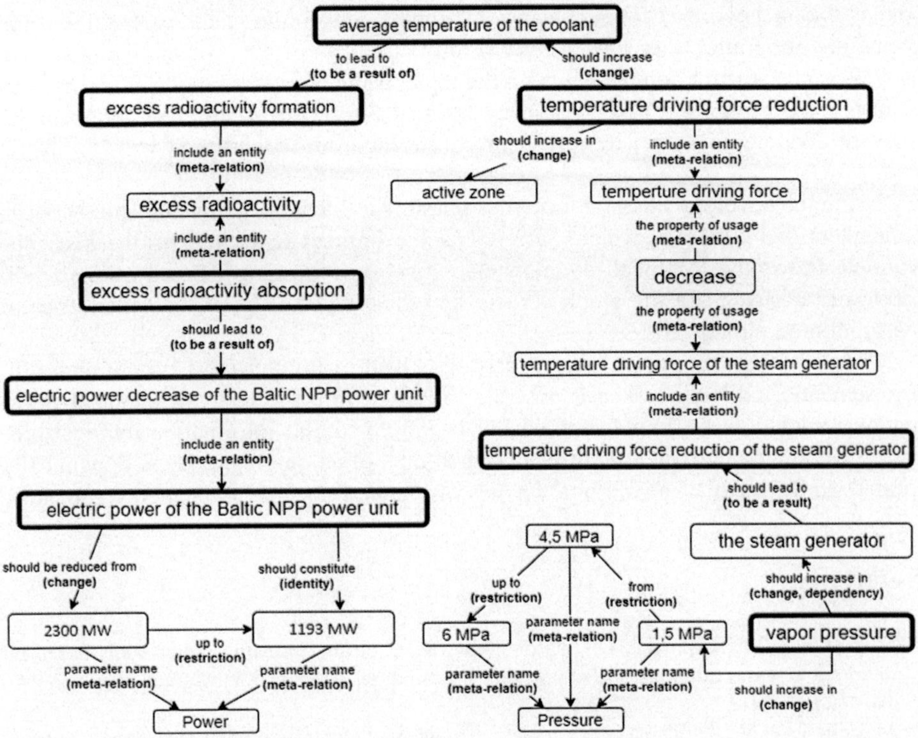

Fig. 1. Subgraph of the document ontology. Vertices with bold font found by the query "power" and the vertices associated with them in the "requirement" aspect. Arcs named relationship name and class. The font size inside the vertices of the graph depends on the frequency of occurrence of the term within the document.

parameters "Power" and "Pressure", formed by the functional relations "Be the result", can thus be revealed: "Vapor pressure" – "Should increase in" – "The steam generator" – "Should lead to" ("To be a result") – "Temperature driving force reduction of the steam generator" – "meta-relations" – "Temperature driving force reduction" – "Should increase in" – "Average temperature of the coolant" – "Should lead to" ("To be a result") – "Excess radioactivity formation" – "meta-relations" – "Excess radioactivity absorption" – "Should lead to" ("To be a result") – "Electrical power decrease of the Baltic NPP power unit", that is, an increase in steam pressure leads to a decrease in the power of the power unit.

5 Conclusion

The technology and tools that provide the construction, visualization and analysis of the ontology graph using the functions of selecting of vertices and arcs, set-theoretic operations on graphs and operation of aspect projection are considered. The use of aspect projections allows moving from graph vertices to relevant fragments of the source text on a compact set of concepts (as entry points to the search process). In turn, combining the found fragments allows to form an image of the problem being solved, as well as check it for compliance with the logic of the target problem and consistency of facts. And building a graph based on the text formed from different fragments of relevant documents allows to verify for consistency and compliance with the target task.

Using the developed tools, it becomes possible to identify the relationship between parameters of the object, which were not in the immediate vicinity in the text. The example shows the technical requirements extraction technology, based on the application of the search for the graph vertices and the identification of the dependence of the monitored parameters.

The use of the developed tools as part of an information retrieval system with full-text semantic indexing makes it possible to move from the tasks of classical information retrieval to the implementation of cognitive-similar information retrieval process as building a path on a multi-meta-hypergraph of an ontology, that is dynamically formed on the basis of ontological images of founded documents or their fragments.

References

1. Maksimov, N.V., Golitsina, O.L., Monankov, K.V., et al.: Semantic search tools based on ontological representations of documentary information. Autom. Doc. Math. Linguist. **53**(4), 167–178 (2019)
2. Maksimov, N.V., Golitsina, O.L., et al.: Knowledge representation models and cognitive search support tools. Procedia Comput. Sci. **169**, 81–89 (2020)
3. Golitsyna, O.L., Maksimov, N.V., Okropishina, O.V., et al.: The ontological approach to the identification of information in tasks of document retrieval. Autom. Doc. Math. Linguist. **46** (3), 125–132 (2012)
4. Maksimov, N.V., Gavrilkina, A.S., Andronova, V.V., et al.: Systematization and identification of semantic relations in ontologies for scientific and technical subject areas. Autom. Doc. Math. Linguist. **52**(6), 306–317 (2018)
5. Golitsina, O.L., Maksimov, N.V., Okropishina, O.V., et al.: An ontological approach to information identification in tasks of document retrieval: a practical application. Autom. Doc. Math. Linguist. **47**(2), 45–51 (2013)
6. Project AES-2006, JSC "SPbAEP". http://atomenergoprom.ru/u/file/npp_2006_rus.pdf. Accessed 20 May 2020
7. Rosenergoatom Concern JSC. https://energybase.ru/power-plant/Baltic_NPP. Accessed 23 May 2020
8. Problems of increasing the maneuverability of nuclear power plants. https://tesiaes.ru/?p=9250. Accessed 21 May 2020

Toward a Building an Ontology of Artefact

Nikolay Maksimov and Alexander Lebedev(✉)

National Research Nuclear University MEPhI (Moscow Engineering Physics Institute), Kashirskoe Shosse, 31, Moscow, Russia
nv-maks@yandex.ru, lebedevalex@live.ru

Abstract. The paper considers approaches to ontologies using in task of requirements tracing. It is stated that for successful requirements management it is necessary to reveal the image of rational activity object, not its individual properties. In this approach requirement is defined as an aspect projection of some object ontology onto considered subject area in the system purpose context. A model of the activity object ontology, as a system of functionally and logically interrelated concepts, quantities and methods, is developed. Proposed model combines an ontology of quantities and measures and an ontology of means, "technologies" of assessment. Such model makes it possible to describe processes occurring in the system in more detail, and in combination with other ontologies (ontology of an activity life cycle stages, properties ontology, certain project ontology constructed by project documentation), it allows to obtain tools for more efficient requirements revealing, understanding of causal relationship in their changing and management.

Keywords: Ontology · Artefact · Requirement · Requirement management

1 Introduction

The growing pace of scientific and technological development leads to an increase in productivity and technological processes changing, primarily due to automation and robotization, including AI. At the same time, industry and science are characterized by a large array of accumulated information on the history of functioning, decisions and their justifications, technologies and methods of work, relations with customers, as well as many other various materials. All this leads to a change in the nature of work, changing in the forms of activity participants interactions/relationships, increasing of elements number that form the enterprise system. But a person, as a part of this system, is limited in the possibilities of perception and incoming data processing. These factors lead to the tasks of developing a *unified* representation and description of existing knowledge in a form convenient (and rational) for perception and manipulation, inter alia a computing environment and regardless of this knowledge amount.

One of the tools for streamlining the processes of generation and knowledge using is a schema, which, being neither an object nor a concept associated with it,

This work was supported by the Ministry of Science and Higher Education of the Russian Federation (state assignment project No. 0723-2020-0036).

A. V. Samsonovich et al. (Eds.): BICA 2020, AISC 1310, pp. 225–232, 2021.
https://doi.org/10.1007/978-3-030-65596-9_29

nevertheless defines a connection between an object and concept. According to [1], schemes perform several functions: they organize and reorganize activities, collect meanings that were not previously related to each other, and promote to new objects detection. Structural schemas compress the content, highlight the causal and functional relationships required by subject. That is, the schema does not reflect the form, but the relationships and interactions of objects. By schematizing of activity, we are able to represent the diversity of its relations in a formalized form, store and reproduce, manipulate individual elements.

In this regard, a promising direction for making the unified view that takes into account the organizational and semantic activity components is the use of ontologies. Unification here arises due to the functional relations typification and use of standards for objects representation and their binding to documents that arise due the activity process. Ontologies themselves in this approach will play the role of a predetermined possible trajectory for search space expanding and can be used as pattern for searching of adjacent within the project life cycle objects.

2 Requirements Management

Let us consider the proposed approach on the example of task for comparative analysis and verification of correspondence between the declared and actually obtained values of artefact properties (as an object and a result of expedient human activity). This is a fairly typical job, which in the development and design areas is called "requirements tracing" and inherent any activity and any life cycle stage. Indeed, complex systems development and operation have must began with declaration (justification, formulation and formalization) of requirements for functional and operational properties. In the process of development, both the parameters specific values and formulations that specify them can be detailed and changed. The means of requirement reachability checking may also change. Functioning depends on environment (surroundings) conditions, which also change. Moreover, in cases of beyond design basis accidents, reengineering or decommissioning, the main components fundamental properties are important. I.e. for complex multi-connected systems, such as, for example, nuclear power plants, it is necessary to control changes in requirements throughout the entire life cycle - that is to manage changes in requirements. At the same time, using plenty of regulatory, reporting and other documents (including the use of their onto-representations) makes it possible to achieve the required completeness of description, since different types of documents represent the essence of the object with different completeness and detail.

On the other hand, requirements are written by a person and for a person. Even a well-formalized style (but not a specification language) somehow uses a natural language with its properties of ambiguity and variance, which creates the possibility for mistakes in interpretation and using. Another consequence of the human factor is "lack of agreement", an orientation towards human understanding, using the professional knowledge of the recipient. In addition, in order to understand what property the text is talking about, it is not always enough to highlight only the value and measurement unit.

There are properties with the same measurement units and vice versa, there are measurement units that correspond to different quantities.

In total, requirements management is tantamount to managing of properties that the system possesses, ensuring their consistency with each other in all phases of the life cycle. And according to [2], traceability must be provided for the requirements; each requirement should be identified, should allow to determine their source, as well as the relationship and dependencies between individual requirements.

Thus, in context of requirements management automation task, two directions can be distinguished: 1) revealing and identification of properties and their values in text; 2) "bringing to" of the requirement to a complete image. The first one has to do with the problem of properties consistent representation. The second one relates to property values consistency in text different parts and in different parts of documentation. This is traditionally referred to as monitoring standards compliance function. But at the stage of operation, problems (more precisely, tasks, including critical ones) of a completely different order may arise, far from the problems of choosing a presentation form or naming a particular property. In order to find solution to such problems, essentially it is necessary to perform an informational investigation, referring to design and research documentation. And in a number of cases, when it is necessary to find a fundamentally new solution - to carry out an informational research that allows to "replace" research work (if someone has already created such a solution). In this regard, arises a distinct task of object ontology developing - as purposeful activity result - representing the functional and logical connections of concepts, objects and processes throughout of life cycle. Such an ontology, in particular, will assist to determine to what extent the achieved result meets the stated requirements, and also, possibly, will allow to "restore" the complete picture[1] - to bring in explicit form what is presented in documentation of another stage of life cycle or is a person's professional knowledge that does not require their explicit presence in the text.

The issue of ontologies use for automatic requirements control is quite relevant. For example, in [3], an approach to categorization of requirements, conflict analysis and tracing, based on ontologies and semantic reasoning mechanisms, is discussed.

In article [4], an automatic tracing of requirements based on traditional document indexing methods is considered.

The issue of building a domain ontology for checking compliance with requirements in software development was considered in [5].

It is shown in [6] that in the automated search for tracing requirements, the ontology can be used as an intermediate artefact for identifying links that would not have been identified by standard information retrieval methods. It is noted that ontologies should be built in exact accordance with the needs of the project. A technique for including information from general and subject ontologies into the process of tracing is proposed.

In this work, on the basis of a systematic approach, an attempt is made to construct an ontological model of an object - the result of rational activity, including not only a

[1] Such "restored" objects and relationships can become attributes of search conditions - "entry points" into the corresponding parts of the documentation.

set of properties and methods of their assessment, but also an "accompanying" context, which in fact is determining what distinguishes this work from the existing ones. Such a well-formalized model is the basis for creating RDF-schemes and drawing inferences by using the Semantic web, and allows to implement of semantic search in cognitive tasks class.

3 Artefact Ontology Development

Standards and normative documentation that organize the subject activity form a kind of scheme for it. In turn, schemas themselves become a norm. Thus, schemas are a self-sustaining structure that demonstrates stability over time.

Considering the essence of activity, it can be noted that it certainly includes goals, results and means of achieving them. I.e., subject, interacting with the surrounding reality, manipulates with set of objects, pursuit of objectives, determined by the executable processes and tasks of the subject. The presence of a goal leads to the emergence of restrictions on the result, i.e. the emergence of requirements.

According to [7], the requirement is defined as:

1. A conditions or capability needed by a user to solve a particular problem or to achieve an objective.
2. A condition or capability that must be met or possessed by a system or system component to satisfy a contract, standard, specification, or other formally imposed documents.
3. A documented presentation of conditions or opportunities for clauses 1 and 2.

In rational human activity, conditions and opportunities are always reduced to measurable or assessable indicators and parameters, which are some object characteristics. Moreover, based on clause 2 of the above definition, requirements are not all indicators, but only those that represent "external" (in relation to the system or any part of it) properties. They must be verifiable (and have an acceptance criterion) and controllable, i.e. must be justified, have cause and effect.

Obviously, the parameters-requirements are related to other parameters and conditions that make up the *context* - a combination of internal and external factors that can affect the design, development and operation of the system [8]. This includes not only technical, but also other factors associated, for example, with the subjective conditions of the customer or external circumstances.

At the same time, from the General theory of systems point of view, an interconnected set of requirements is itself a system: objects and connections (more precisely, their parameters, which are actually requirements) are allocated in accordance with predetermined goals and criteria (essentially having the role of composition principle) that formally stipulate integrity.

We can say that requirements are an informational image presented in natural science language. The space in which requirements control is carried out is usually formed by the documentation system that accompanies the work on system creation and operation. And it is obvious that a separate parameter for various reasons (including typos) in a separate or in series of documents may differ in values. In this sense, tracing is a way of representing relationships between explicit requirements, assisting to identify and order relationships or reveal contradictions at the property level (basic definitions are taken from).

However, besides tracing of requirements, i.e. matching of one or another parameter values, the prehistory of the requirement, the measurement conditions, system of measurement units, etc. are important. We can say that on the whole it forms a classical ontology (the logic of existence), since it reflects all aspects of properties manifestation and measurement, as well as the assessment of what is received. And considering that requirements are represented by a subject-specific conceptual-sign system, this means that we have an information ontology defined according to [9].

Figure 1 shows an ontological model diagram that represents an information image of requirements as a system of functionally and logically interrelated concepts, quantities, and methods. Moreover, in essence, the scheme is a combination of two components. The right-hand side represents the ontology of quantities and measures[2] [10] that provide a well-formalized representation of the properties of the target system. The left one - means, "technologies" of assessment.

Based on the system analysis methodology, the target system (object and/or process) can be defined (and highlighted) as a projection onto the subject area of a certain context (determined by system purpose and objectives, including a customer subjective point of view). Here the subject area is objects and processes that are distinguished by their properties and are in certain relationships with each other or interact in some way. Process is an activity aimed at transforming certain input information and material flows in order to obtain results that are valuable in the solving problem process.

Verification of obtained result compliance with the expected consists of functional requirements verification and operational requirements validation and includes a test program and procedure, a test case, and criteria for what can be considered an acceptable result. The test program defines the test conditions, the test method and the evaluation criterion - an indicator of when a requirement can be considered fulfilled.

Based on "product quality indicator" definition of [11], it can be assumed that the value of quantity is determined by means of a tool by measuring an indicator - a quantitative or qualitative characteristic of one or more system properties, considered in relation to certain conditions of its creation and operation or consumption. A measurement means or instrument is a device used for carrying out measurements, separately or in conjunction with one or more additional devices [12], intended for measurements and having standardized (established) metrological characteristics [13]. Measurement instruments have scales and measurement accuracy and use a specific measurement method.

Quantitative characteristics are represented by exact values or range of values, have a definition, designation, and dimension. Qualitative characteristics are usually associated with subjective and fuzzy indicators, for the assessment of which, for example, expert methods can be used, and are represented by linguistic variables (mainly text forms).

---

2 As a consequence, definitions of the respective components are not provided in this paper.

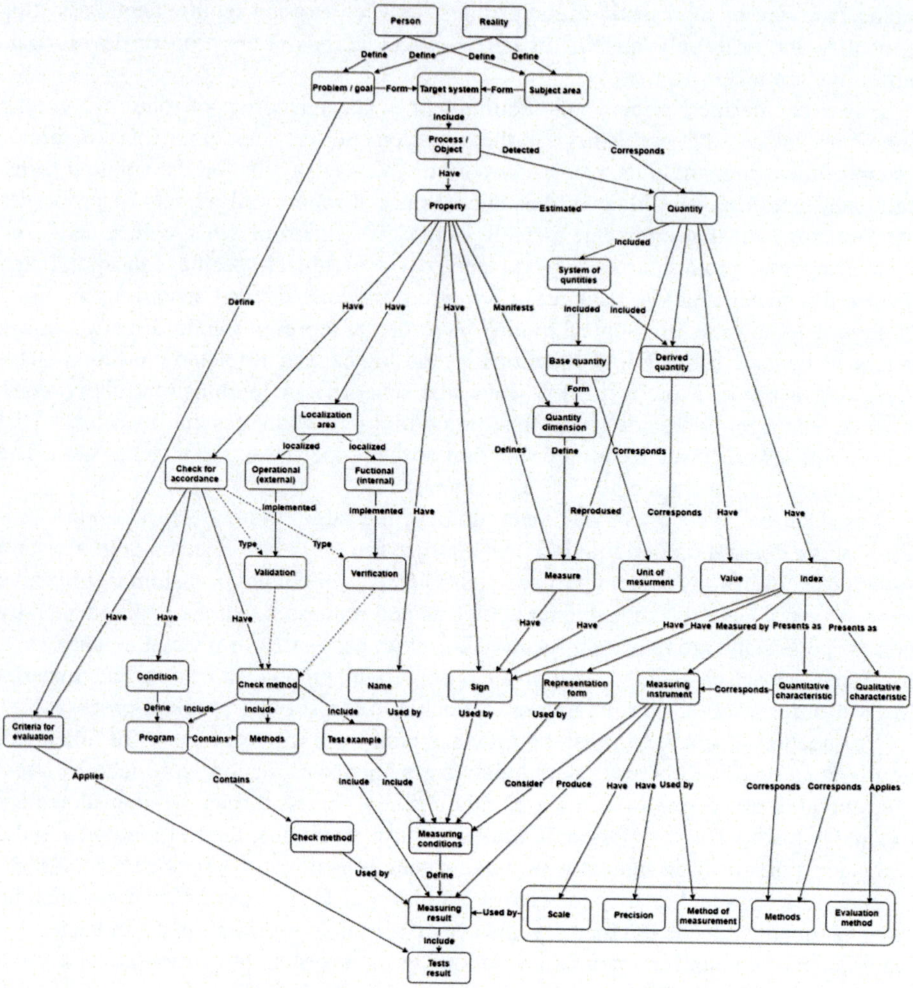

Fig. 1. The ontology of an activity object.

The checked properties have a presentation form - the way the requirement is presented, for example, tables, diagrams, text.

The measurement result is the value of a certain quantity obtained by using the measurement tool. Accordingly, the test result is the value obtained on one or more measurements depending on the requirements specified in the test document [14].

4 Conclusion

Summarizing the above and using the tools of operations on ontologies [9], the requirement is an aspect projection of some object ontology onto considered subject area in the project purpose context. The task of requirements tracing is reduced to the task of project ontology constructing based on documentation (identifying objects and their place in the life cycle of project), constructing an ontology of its properties (having the role of entities) and constructing an aspect projection.

Based on systematic approach, an ontological model of an object is proposed - the result of rational activity, which includes not only a set of properties and methods for their assessment, but also an “accompanying” context. Such a well-formalized model can be basis for creating RDF-schemes and drawing inferences by using the Semantic web means, and also allows to implement semantic search in the class of cognitive tasks and, in particular, to effectively implement the procedures for comparing of declared and achieved values of parameters, as well as the search for cause-effect relationships.

References

1. Rozin, M.V.: Vvedenie v skhemologiyu: Skhemy v filosofii, kul’ture, nauke, proektirovanii. Knizhnyj dom «LIBROKOM», Moscow (2011)
2. IEEE 830-1993 - IEEE Recommended Practice for Software Requirements Specifications. IEEE, New York (1994)
3. Moser, T., Winkler, D., Heindl, M., Biffl, S.: Requirements management with semantic technology: an empirical study on automated requirements categorization and conflict analysis. In: Mouratidis, H., Rolland, C. (eds.) Advanced Information Systems Engineering. CAiSE 2011. Lecture Notes in Computer Science, vol. 6741, pp. 3–17. Springer, Berlin (2011)
4. Hayes, H., Dekhtyar, A., Sundaram, S.K.: Advancing candidate link generation for requirements tracing: the study of methods. IEEE Trans. Software Eng. **32**(1), 4–19 (2006)
5. Kof, L., Gacitua, R., Rouncefield, M., Sawyer, P.: Ontology and model alignment as a means for requirements validation. In: Proceedings of the 4th IEEE International Conference on Semantic Computing (ICSC 2010), pp. 46–51. IEEE, New York (2010)
6. Li, Y., Cleland-Huang, J.: Ontology-based trace retrieval. In: 7th International Workshop on Traceability in Emerging Forms of Software Engineering (TEFSE), pp. 30–36. IEEE, New York (2013)
7. IEEE 610.12-1990 - IEEE Standard Glossary of Software Engineering Terminology. IEEE, New York (1990)
8. GOST R ISO 9000-2015. Quality management systems. Fundamentals and Glossary (Revised Edition). http://docs.cntd.ru/document/1200124393. Accessed 24 Sept 2020
9. Maksimov, N., Gavrilkina, A., Kuzmina, V., Borodina, E.: Ontology of properties and its methods of use: properties and unit extraction from texts. Procedia Comput. Sci. **169**, 70–75 (2020)
10. Golitsyna, O.L., et al.: The ontological approach to the identification of information in tasks of document retrieval. Autom. Doc. Math. Ling. **46**(3), 125–132 (2012)
11. GOST 15467-79. Product quality management. Basic concepts. Terms and definitions (with Amendment N 1). http://docs.cntd.ru/document/gost-15467-79. Accessed 24 Sept 2020

12. International vocabulary of basic and general terms in metrology (VIM). http://www.geste.mecanica.ufrgs.br/medterm/ISO_VEM.pdf. Accessed 24 Sept 2020
13. RMG 29-2013. GSI - State system for ensuring the uniformity of measurements. Metrology. Basic terms and definitions. http://docs.cntd.ru/document/1200115154. Accessed 24 Sept 2020
14. GOST 33701-2015. Determination and application of precision methods for testing of petroleum products. http://docs.cntd.ru/document/1200139506. Accessed 24 Sept 2020

Cognitive Architectures of Effective Speech-Language Communication and Prospective Challenges for Neurophysiological Speech Studies

Irina Malanchuk(✉)

NRC "Kurchatov Institute", Moscow, Russia
coral@inbox.ru

Abstract. The paper focuses on the importance of social cognitions and priors in natural cognitive architectures of an individual. The structures and content of perceptual-cognitive-metacognitive processes are analyzed using the material of natural speech-language communication related to early human ontogenesis. Metacognitive processes are defined as a property and an integral part of the cognitive system. The prospects of neurophysiological research are formulated, which are designed to clarify the distribution, the nature of connections and neuronal dynamics in the system of perceptual-cognitive-metacognitive processes that ensure effective speech communication at different periods of human development. The importance of the studies of the neural architectural solutions related to the processes of natural speech and language communication for the development of IT technologies which can fulfill the communicative needs and expectations of individual is emphasized.

Keywords: Natural speech-language communication · Neural architectures · Artificial Intelligence (AI) · Social priors · Social cognitions · Metacognitive processes · Early human ontogenesis

1 Introduction

Artificial systems, which are currently quite close to imitation of speech-language communication, are still unable to conduct it in dialogical modes which are common for humans.

Probably, we can't stop the humanity from striving to create a "full-fledged" artificial social partner-communicator (and not just a robot serving coffee with a bow and wishing good morning, or a bot providing information when booking a table at a restaurant, or may be a familiar voice speaking on a topic of interest, but without clear understanding of the communicative goals or the partner's pragmasemantics). So far, in terms of human communication, it is a person who is forced to take a complementary position in relation to artificial intelligent systems, and in many technological areas and specific tasks of activity, this is reasonable and necessary, but not yet in the field of speech communication, where the simplest expectations of people are much higher than the current capabilities of artificial intelligence (for an overview of AI

A. V. Samsonovich et al. (Eds.): BICA 2020, AISC 1310, pp. 233–240, 2021.
https://doi.org/10.1007/978-3-030-65596-9_30

technologies, see [1–3]). At the same time, there are already interesting and promising solutions to communicative problems at the level of auditory, visual, multimodal perception of social scenes ([4–9], etc.), speech and language analysis ([4, 10–13], etc.), metacognitive processes ([14–17], etc.).

We still have a long scientific path ahead of us: in order to develop artificial intelligence close to the possibilities of flexible communication by an individual, it is necessary to join external monitoring of human social goal-setting, recognition and control of intentions, strategies and tactics of communicative partners, including multi-agent environments, future technologies ensuring the communicative flexibility of artificial intelligence, highlighting social priors that are far from obvious and, as a rule, not expressed in the language, to teach intelligent systems to recognize them, etc. And this is the level of tasks that the human brain is already able to comprehend and solve for active and multitasking communication in early ontogenesis.

2 Materials and Methods

We will analyze the early cognitive architectures of a human in the process of conducting natural communication and will do this based on the speech-language communication of children of the second and third years of life - the period when the child's speech-communicative and language competence is developing, and the use of cultural forms of speech (vs. infant, native forms of speech) and language as a conventional functional system puts communication between the child and other people at a fundamentally new and complex level. The material of this study is the records of children's speech communication; a random sample represents 200 situations from a database of more than 10,000 cases. Analysis methods: content analysis, qualitative analysis.

3 Theoretical Bases of the Analysis of Human Cognitive Architectures in Natural Communication

Firstly, we consider the structure of cognitive processes and the place of metacognitive processes in them. We believe metacognitive processes to be an integral system property of a person's cognitive system [18] (cf. [19–21]). This is quite obvious considering the theory of self-governing systems (mental, social), but not generally accepted in Developmental Psychology. Meanwhile, metacognitive processes are available for analysis from the moment a person is born, when we have a good opportunity to observe behavior aimed at improving the quality of perception; social partner management behavior; self-regulation and regulation of the social partner's behavior; using more effective communication and speech signals, etc., which has not yet been interpreted from the point of view of the functioning and development of the system of metacognitive processes.

In addition to various options of ascending processes (perceptual-cognitive-metacognitive), it is necessary to take into account the role of each metaprocess in the organization of structures of descending processes, perceptual, cognitive acts, emotional response, as well as the metaprocess's orientation to each other and to themselves. Psychological models of these systems of processes in early ontogenesis can be built on the basis of an analysis of situations of natural speech and language communication, and individual, particular patterns of communication in certain links can be verified experimentally, which is done in a number of psychological and neurophysiological experimental studies (for reviews, see, e.g.: [22, 23]). It is important to note that in recent years we have developed new technological capabilities for studying neurofunctional structures in early postnatal ontogenesis (for example, NIRS; for the study of early speech-cognitive processes using fNIRS, see: [24]), and the study of early metacognitive processes becomes a relevant issue of Neurobiology and Neuropsychology, starting from the earliest periods of ontogenesis.

The second key argument of our study is the possibility of analyzing the social consciousness of a person, identifying social priors and their relationships (possible types of communication) by behavior, speech and language. In our previous studies, technologies for analyzing social representations in speech and language in natural communication were proposed [25]; the composition of social representations of young children (1–3 years) has been established, numbering at least 120 social representations [26], some of which can be considered social priors (we would like to note that the types and characteristics of psychological priors, which were presented by F. Chollet in [27: 24–27], are not quite accurate and in order to create a strong AI focused on getting close to the natural human communication need some consideration); the dynamics of interconnections of social representations in children of second and third years of life has been determined [28].

In order to analyze the cognitive architectures of an individual in the tasks of obtaining data on the grounds of inferences that are significant for maintaining and completing a communicative situation, it is important to analyze the development by people and the use of social and socio-communicative knowledge, which gets to the problem of interaction of communicative partners in binary and multi-agent environments as pragmasemantic systems and features of functioning in these systems. This aspect of the problem also turns out to be important when solving tasks where sociality seems to fade into the background or does not seem to be found, however, this, with rare exceptions, is an apparent state of affairs: any social groups regulate and socially codify intellectual interest in the "objective" reality, and social benefits are assumed from "pure" cognitions. As for the natural communication of people, the problem of cognitive architectures expands to an analysis of the cognitive architectures of communicative partners, the possibilities for partners to adequately interpret the intentional states of each other and to take these data into account in the dynamics of interactions.

4 Psychological Analysis of Cognitive Architectures

Let's consider several facts of children's communicative behavior in order to reconstruct their cognitive architectures, including the meta-level.

Situations 1–2. *(1)* Children – the twin boys (1.0.0.) to the adult's questions "Where is the teddy bear?", "Where is the lamp?" etc. point at the objects. *(2)* When they point at another object instead of the named one, they laugh.

It is clear that the trigger for the specific mental and behavioral activity of children is an adult's request (contact is requested, cooperation and a meaningful response), which they well recognize as directed at a child (children) and is understood as the task of identifying and showing an object.

The fact that children recognize the addressing of a request to them indicates that children already have differentiated image a) of situations of interactions, including speech and suggesting a child's response, their "early" adjustment to one or another form of adult speech; b) of adult speech forms in their intentionality - in this case, the search behavior and behavior of cooperation with an adult are expected, in the framework of which an adult is supposed to be informed about the results of the identification of the word and object (teddy bear, lamp), in other words the conventional meanings of speech forms; c) of the meanings of words that form this request.

At each step of interaction with an adult, the child shows decisions whether to maintain the contact in the mode of controlled activity of the child that was set by the adult. Potentially, the child can refuse, resist, which is in their behavioral experience and memory, however, in the analyzed situation (1), the child initially chose the option of cooperation with an adult. These decisions, which are a socio-cognitive fact and characterize the type of connection between mental images of "I" and "other"), formulate the goals of interaction, and the ability to focus on these social and non-social goals organizes specific mental activity and architectures of distributed metacognitive processes that are interconnected with specific complexes of process objects.

Mental activity associated with the task of identifying an object by a verbal stimulus arises after choosing a strategy for interacting with an adult and setting goals, which is caused by a whole complex of cognitive and emotional processes and their results.

This mental activity is presented in at least three vectors - focusing on one's mental, intentional states; on an adult (with the structure of connections of images "I" and "adult"); on objects (with a structure of connections "I" - object - adult); the fourth vector in this case is determined both by the existing images in the memory about the images of interactions with the second twin child, and by the actual images of their states and forms of their behavior that are significant for the dynamics of the situation.

All these communication systems are subject to monitoring and controlling simultaneously, concurrently and in their time dynamics. (In our works we use the term "controlling", which seems to be more adequate to the continuity of this metaprocess).

Monitoring of the adult's state occurs - their behavioral and speech activity aimed at each child, including speech and mimic-emotional monitoring, in the cycles "adult question - child's answer - adult's assessment of the result", and one of the monitoring and controlling lines is defined by the structure of ideas about such cycles.

The images of mental states of the child himself are formed and, therefore, monitored in the dynamics of each cycle, which, along with the behavior of partners (the adult and the second child) determine the subsequent motivation for cooperation or, potentially, refusal to cooperate. Therefore, the controlling aimed at oneself in the aspect of maintaining or changing behavior is determined by evaluating the effectiveness of the process of solving the problem and its result, heuristics already used and arising, assessing their mental states, and attitude to other forms of behavior of other participants in the situation.

Monitoring and controlling the behavior of the second child in relation to all participants in the situation, objects set by adults and other (background) objects are also a component of this metaprocess system.

Situation (2), in addition to what has been said, demonstrates:

a) the emergence/reproduction of a method for solving the proposed task, which, on the one hand, contains a semantic shift, and on the other hand, transmits to an adult an intellectual position that does not coincide with their expectations, which is understood by children. Laughter in this case is a sign of a false signal (intellectual and language game), forming the cognitive distance between "correct" and "incorrect", the social distance between "I think and answer correctly" and "I answer incorrectly as if I think incorrectly", and also a new balance of relations with adults in violation of the norm of intellectual activity (the normative way of solving a task) and the social norm of interaction with an adult;
b) the coordination by both children of their communicative goals, aimed both at the adult who expects the correct solution to the intellectual-language task, and at each other, when it is important to support the selected behavior in the microgroup and confront the adult. Of particular interest in this case are the heuristics of the semantic shift (a word/sound/gesture can be resemantized), a false signal and its intended use, here - for the purpose of socio-psychological distance from an adult (a language game serves these goals of a child well throughout their subsequent childhood, to which we have many recorded examples), finally, a heuristic of group action, which reinforces the position of the children in distancing from the adult, but within the boundaries of the communicative situation.

Based on the speech communication of children of the second and third years of life, we also obtained data on the speech representation of the simulated activity and related tasks in relation to oneself and others, which organizes concentration and structure of perception; testing hypotheses about the status of a partner; monitoring the state of the system "child - adult"; speech management of others to achieve their goals, assessing the effectiveness of speech and choosing a more effective speech and language signal; verbalization/awareness of social (behavior and speech) technology influence on another person; basic social priors - images of the qualities of "I", "the other" and the relationship between them.

5 Conclusion

Such intellectual and social activity, of course, is provided by distributed cognitive architectures that make a specific contribution to the formation of many agreed goals and objectives of communication, the development/the ways to solve them.

At the same time, social-cognitive architectures are differentiated from architectures that provide solutions to non-social intellectual tasks, however, they make up a completely integral hierarchy with the predominance of first-type architectures in this case.

The fundamental is the system of metacognitive processes that are specifically aimed at processes-objects (perception, various types of long-term and working memory, cognitive processes in their operational specifics and tasks, emotional processes), cyclically reorganize them, as well as themselves - setting new goals, self-evaluating metacognition, evaluating the effectiveness of the behavior aimed at the social partner (partners).

We see the prospects of neurophysiological research of speech as the most significant communicative system for humans in organizing systemic studies of neural cognitive architectures of the discussed complexity on the basis of naturally communicative tasks solved by a human in various situations in the dynamics of psychosocial development.

Acknowledgments. The work presented above is an initiative internal research work carried out by the Research Center Kurchatov Institute (order No. 1363 of June 25, 2019).

References

1. de Jesus, A.: Artificial intelligence in industrial automation current applications (2019). https://emerj.com/ai-sector-overviews/artificial-intelligence-industrial-automation-current-applications/
2. Kotseruba, I., Tsotsos, J.K.: A review of 40 years of cognitive architecture research core cognitive abilities and practical application. arXiv preprint arXiv:1610.08602 (2018)
3. Samsonovich, A.V.: Toward a unified catalog of implemented cognitive architectures. In: Samsonovich, A.V., Johannsdottir, K.R., Chella, A., Goertzel, B. (eds.). Biologically Inspired Cognitive Architectures 2010: Proceedings of the First Annual Meeting of the BICA Society, vol. 221, pp. 195–244. IOS Press, Amsterdam (2010)
4. Lison, P., Kruijff, G.-J.: Salience-driven contextual priming of speech recognition for human-robot interaction. In: ECAI 2008 - Proceedings of the 18th European Conference on Artificial Intelligence, pp. 636–640 (2008)
5. Martinez-Gomez, J., Marfil, R., Calderita, L.V., Bandera, J.P., Manso, L.J., Bandera, A., Romero-Garces, A., Bustos, P.: Toward social cognition in robotics: extracting and internalizing meaning from perception. In: Workshop of Physical Agents (WPA 2014), León, Spain, pp. 93–104 (2014)
6. Sun, R., Fleischer, P.: A cognitive social simulation of tribal survival strategies: the importance of cognitive and motivational factors. J. Cogn. Cult. **12**(3–4), 287–321 (2012)
7. Thorisson, K.R., Gislason, O., Jonsdottir, G.R., Thorisson, H.T.: A multiparty multimodal architecture for realtime turntaking. In: Allbeck, J., et al. (eds.) International Conference on Intelligent Virtual Agents (IVA 2010). LNAI, vol. 6356, pp. 350–356. Springer, Heidelberg (2010)
8. Trafton, J.G., Cassimatis, N.L., Bugajska, M.D., Brock, D.P., Mintz, F.E., Schultz, A.C.: Enabling effective human robot interaction using perspective-taking in robots. IEEE Trans. Syst. Man Cybern. Part A Syst. Hum. **35**(4), 460–470 (2005)
9. Zmigrod, Sh., Hommel, B.: Feature integration across multimodal perception and action: a review. Multisens. Res. **26**(1–2), 143–157 (2013). https://doi.org/10.1163/22134808-00002390
10. Grossberg, S., Govindarajan, K., Wyse, L.L., Cohen, M.A.: ARTSTREAM: a neural network model of auditory scene analysis and source segregation. Neural Netw. **17**(4), 511–536 (2004). https://doi.org/10.1016/j.neunet.2003.10.002

11. Kotov, A.A., Arinkin, N., Zaidelman, L., Zinina, A.: Linguistic approaches to robotics: from text analysis to the synthesis of behavior. In: Language, Music, and Computing (LMAC 2017). Communications in Computer and Information Science, vol. 943, pp. 207–214. Springer (2019)
12. Scheutz, M., Krause, E., Sadeghi, S.: An embodied real-time model of language-guided incremental visual search. In: Proceedings of the 36th Annual Meeting of the Cognitive Science Society, vol. 36, pp. 1365–1370 (2014)
13. Tikhanoff, V., Cangelosi, A., Metta, G.: Integration of speech and action in humanoid robots: iCub simulation experiments. IEEE Trans. Auton. Ment. Dev. **3**(1), 17–29 (2011). https://doi.org/10.1109/tamd.2010.2.2100390
14. Cox, M.T.: Metacognition in computation: a selected research review. Artif. Intell. **169**(2), 104–141 (2005)
15. Cox, M.T.: Perpetual self-aware cognitive agents. AI Mag. **28**(1), 32–45 (2007)
16. Cox, M.T., Alavi, Z., Dannenhauer, D., Eyorokon, V., Munoz-Avila, H.: MIDCA: a metacognitive, integrated dual-cycle architecture for self-regulated autonomy. In: Proceedings of the Thirtieth AAAI Conference on Artificial Intelligence (AAAI-16), pp. 3712–3718 (2016)
17. Kalish, M.Q., Samsonovich, A.V., Coletti, M.A., De Jong, K.A.: Assessing the role of metacognition in GMU BICA. In: Samsonovich, A.V., Johannsdottir, K.R., Chella, A., Goertzel, B. (eds.). Biologically Inspired Cognitive Architectures 2010: Proceedings of the First Annual Meeting of the BICA Society, vol. 221, pp. 72–77. IOS Press, Amsterdam (2010). https://doi.org/10.3233/978-1-60750-661-4-72
18. Malanchuk, I.G.: What do the psychological and neural researches of phoneme discrimination say: metacognitive processes in premature children and infants (2020). (in press)
19. Nelson, T.O., Narens, L.: Metamemory: a theoretical framework and new findings. In: The Psychology of Learning and Motivation, vol. 26, pp. 125–169. Academic Press, New York (1990)
20. Marchi, F.: Attention and cognitive penetrability: the epistemic consequences of attention as a form of metacognitive regulation. Consc. Cogn. **47**, 48–62 (2017). https://doi.org/10.1016/j.concog.2016.06.014
21. Geurten, M., Meulemans, T., Willems, S.: A closer look at children's metacognitive skills: the case of the distinctiveness heuristic. J. Exp. Child Psychol. **172**, 130–148 (2018). https://doi.org/10.1016/j.jecp.2018.03.007
22. Salles, A., Ais, J., Semelman, M., Sigman, M., Calero, C.I.: The metacognitive abilities of children and adults. Cogn. Dev. **40**, 101–110 (2016). https://doi.org/10.1016/j.cogdev.2016.08.009
23. Whitebread, D., Almeqdad, Q., Bryce, D., Demetriou, D., Grau, V., Sangster, C.: Metacognition in young children: current methodological and theoretical developments. In: Efklides, A., Misailidi, P. (eds.) Trends and Prospects in Metacognition Research, pp. 233–258. Springer, New York (2010)
24. Dehaene-Lambertz, G.: The human infant brain: a neural architecture able to learn language. Psychon. Bull. Rev. **24**, 48–55 (2017). https://doi.org/10.3758/s13423-016-1156-9
25. Malanchuk, I.G.: Social consciousness and speech behavior in preschool age. Krasnoyarsk, KSPU named after V.P. Astafiev (2014)
26. Malanchuk, I.G., Zalevskaya, A.G.: Gender peculiarities of social consciousness in early childhood (based on analysis of speech). Bull. KSPU named after V.P. Astafiev **3**(45), 113–125 (2018). https://doi.org/10.25146/1995-0861-2018-45-3-80

27. Chollet, F.: On the measure of intelligence. arXiv preprint arXiv:1911.01547v2 25 Nov 2019 (2019)
28. Malanchuk, I.G., Zalevskaya, A.G.: Dynamics of social consciousness in early ontogenesis according to language and speech data. In: Neuroscience for Medicine and Psychology: XVI International Interdisciplinary Congress, Moscow, p. 312 (2020). https://doi.org/10.29003/m1141.sudak.ns2020-16/312

Development of an AI Recommender System to Recommend Concerts Based on Microservice Architecture Using Collaborative and Content-Based Filtering Methods

Andrey Malynov and Igor Prokhorov(✉)

National Research Nuclear University MEPhI
(Moscow Engineering Physics Institute), Moscow, Russia
andrey@malynov.com, ivprokhorov@mephi.ru

Abstract. Recommender system is a complex software system primarily intended to select the most relevant content based on user's personal preferences. In order to achieve the set goal, a number of tasks must be completed, including: track user actions across various devices; get product data from a number of sources and maintaining their currency; consolidate the data; create user and product profiles bases on big data in a real-time mode; select recommendations in cold start and highly sparse data environment; assess the quality of the recommender system. Completion of each specific task must not extend the time to complete other tasks. Users must instantly get the relevant content even if the system is heavily loaded, for example, due to a popular event announcement. A workaround may be to divide the system into independent components with the ability to scale specific services. Microservice architecture examined in this article intends to ensure required flexibility due to asynchronous message exchange via a data bus and other principles offered by SOA concept. Apart from interaction between the components, the article also introduces the results of development of each specific service from asynchronous user action tracker to recommender engine based on the hybrid approach that includes collaborative and content-based filtering methods, and the knowledge-based approach using Artificial Intelligence techniques. Special attention is paid to a subject category with a number of aspects that prevent applying generic approaches to building recommender systems.

Keywords: Recommender system · Microservices architecture · Cold start problem · Hybrid filtering technique · Collaborative filtering · Content-based filtering · Artificial intelligence · Knowledge-based approach · User activity tracking

A. V. Samsonovich et al. (Eds.): BICA 2020, AISC 1310, pp. 241–252, 2021.
https://doi.org/10.1007/978-3-030-65596-9_31

1 Introduction

The recommender system is a complex software system, the main purpose of which is to select the most relevant content based on the personal preferences of the user. The value of such systems increases with the growth of data volume due to emergence of a large number of alternative options that can confuse the user. So today, it is difficult to imagine successful Netflix-type services with a huge content base that do not use recommender system for their products or services. It should be noted that recommender systems are mostly used by online platforms. Thus, due to stimulation and widespread commercial use, there is a huge number of researches in the field of building recommender systems despite the relatively recent emergence of this topic. However, due to the strong dependence on some subject area aspects, there is no single solution that would allow using ready-made developments in all areas to find not only relevant, but also fundamentally new content, although it does not interfere with the applying of some general approaches, such as introduction of collaborative filtering.

This article is devoted to the development of recommender system for concert event. The subject area of this activity field imposes a number of restrictions, including: a fixed time for the beginning and ending of events, which means the need to exclude past (archived) recommendations from the final selection with the possibility of using their profiles to recommend events that have not yet occurred (active); the extremely uneven distribution of the audience in terms of service use, for example, the announcement of Ed Sheeran event in 2018 has increased the number of active users by more than a hundred times over an hour. This fact emphasizes the need to pay special attention to scaling problem.

2 Statement of Problem

As it was noted earlier, the recommender system is a complex software system, consisting of not only the module for generating recommendations, but also many other components. First of all, the system must be able to collect and store the data that are required for achievement of its goal. The data may contain information about recommended products or information about users and ensure better understanding of audience's preferences. After data receipt, it is necessary to consolidate it before using in recommendations formulation. The next stage is data storage, on which there is an additional requirement to consider system performance criterion, which cannot be achieved through use of relational database. Only after these steps, the system begins its main task: to formulate recommendations, which, of course, need to be assessed for the quality of implementation. Thus, to achieve this goal it is necessary to complete a number of tasks, including the following:

- tracking user actions on various services of the company (the main site and widgets integrated into partners' resources) to compile his profile, as well as to search for those users whose interests coincide with its interests;
- receiving data on events from several sources and keeping them up to date;
- consolidation of received data;

- creation of user and product profiles based on implicit evaluations in real time;
- selection of recommendations in cold start and high sparsity of data;
- assessment of the recommender system quality.

The execution of each individual task should not affect the time for execution of other tasks, users should instantly receive relevant content, even if the system is under high load, for example, due to the announcement of a popular event.

3 Data Model

The choice of approach of building a recommender system is based not only on the needs of the business, but also on the opportunities that are determined by manipulated data. So, for example, if there is no description of the products it is impossible to use content-based filtering methods, or if the ratings of products are implicit difficulties may arise with the use of collaborative filtering methods [1]. Therefore, it is important to define data model. It should be noted that the data can be conditionally divided into two groups: data obtained by logging user actions and data describing the products, that is, concert events and related objects. Let's consider groups in more detail.

A concert event is a musical event that includes one or more sessions, each of which can take place at different venues or even in different cities. Each musical event is held by one or more performers and is classified by a certain set of genres, for example, "Rock" or "Jazz", as well as by a variety of tags, such as "New Year". The event description includes a brief overview of the show features. The event is organized by an individual or legal entity promoting its brand according to established standards for the quality of the event. Thus, all the indicated data can be used in the formation of recommendations and therefore, should be considered by the system. Moreover, the data included in this group should be obtained from the company database.

The second group of data is based on user action logs that are available on company's website or widgets integrated on partner sites. Data collection is carried out by an asynchronous tracker [2], written in JavaScript and includes the following logs:

- PRODUCT_VIEW: viewing of session page;
- PRODUCT_GROUP_VIEW: viewing of event page;
- TAG_VIEW: viewing of tag page, for example, "March 8";
- VENUE_VIEW: viewing of venue location page;
- PRODUCT_REQUEST: request for feedback on the event, for example, if there are not tickets, the user can subscribe to their emergence;
- CUSTOM: a custom event, for example, click on a link in the footer, which can help in creating a user profile,
- PRODUCT_ADDING_TO_BASKET: adding a ticket to the basket;
- PRODUCT_REMOVING_FROM_BASKET: removal of a ticket from the basket;
- PRODUCT_CLICK: click on an event in the list;
- CHECKOUT: placing an order;
- RESERVATION: go to the booked reservation page;

It should be noted that received user action logs are implicit estimates and need further processing [1]. The following is a class diagram (see Fig. 1) showing general data structure used by the system.

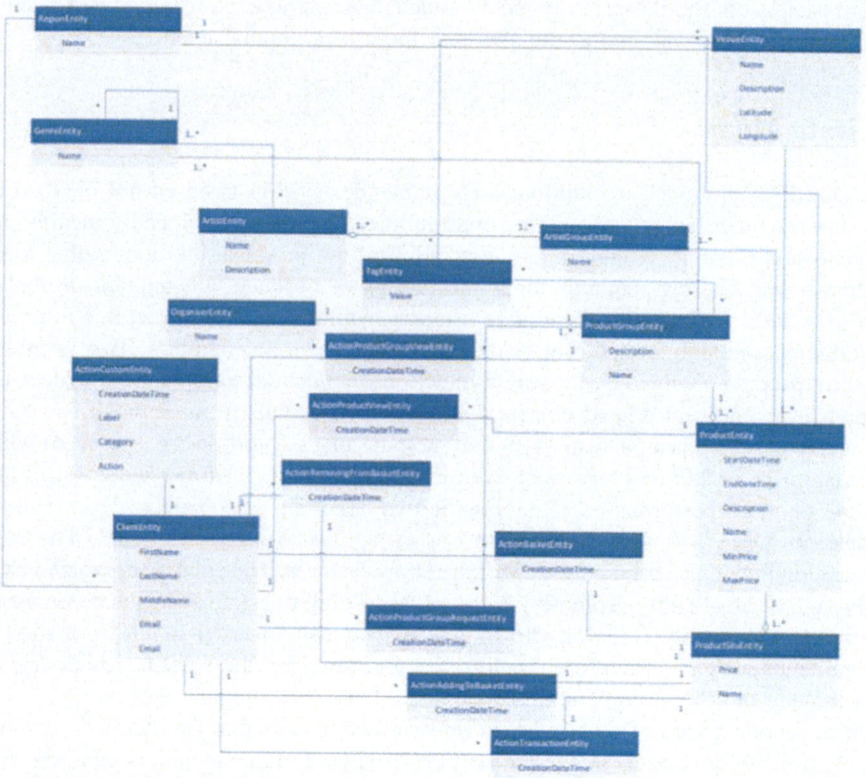

Fig. 1. Data structure of recommender system

4 Modeling of Recommendations Creating Component

As it was mentioned earlier, the decision to choose an approach of building a recommendation system should be based on business needs. It is known that content-based filtering methods do not allow selecting fundamentally new products [3], unlike collaborative filtering methods; however, this is not always necessary since they better cope with cold start problem [3] and are most often more profitable in terms of computation. Upon development of the system, it was decided to use a combination of three different approaches, each of which can be applied at the right time, according to switching strategy [4] of hybrid recommendation systems, as well as according to business needs. For example, a block of recommendations displayed on a website can contain events that correspond to the profile of the user's preferences, and a selection of events in email newsletters may contain recommendations prepared by collaborative filtering methods based on the correlation model.

4.1 Content-Based Filtering

The content-based filtering method was chosen as the first of the implemented approaches. Its main hypothesis is that user u is interested in products that most closely match the profile of his preferences. Moreover, the profile can be compiled based on the history of its interactions with products or on user data, for example, from the questionnaire or the Core Reporting API of Google Analytics platform.

The main requirement of using of content-based filtering approach is the availability of a sufficient amount of content that can be used to build user and product profiles. Therefore, existing database was previously analyzed (see Fig. 2).

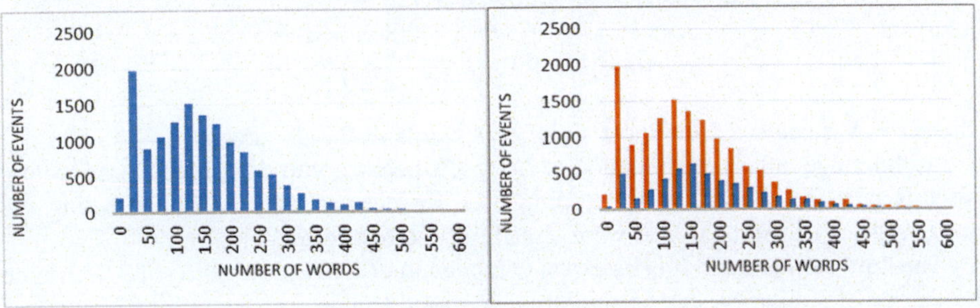

Fig. 2. Distribution of the number of words over all concerts (left) and comparison with the genre "Rock" (right)

Thus, only 15.8% of events have 25 or less words, which is acceptable from the point of view of justification for applying the content-based filtering method. Main steps in building recommendations are shown below.

Let $IP(i)$ be a vector that contains all the characteristics of product i, and $UP(u)$ be a vector that contains all user's preferences. The first step is to create product profile. $IP(i)$ is partially filled with number of features of the event i, such as genre or artist, another part is filled on the basis of meaningful words extracted from text content. As a measure of the significance of the word in the text $TF\text{-}IDF$ [3]. The assessment $TF_{i,j}$ suggests the importance of the word w_i in the document d_j by the number of occurrences of the word in it and is defined as follow:

$$TF_{i,j} = \frac{n_{i,j}}{\sum_{k \in W_j} n_{kj}} \quad (1)$$

where $n_{i,j}$ is a number of entries of word w_i in document d_j; n_{kj} is a number of entries of word w_k in document d_j; W_j is a set of all words in document d_j.

The main idea IDF_i is to take into account the importance of the word w_i in the whole set of documents. This assessment reduces the weight of words that are found in many documents, for example, conjunctions. IDF_i is determined in the following way:

$$IDF_i = \log \frac{|D|}{m_i} \tag{2}$$

where $|D|$ is total quantity of documents and m_i is quantity of documents that contain w_i.

Estimate *TF-IDF*, that reflects the significance of word w_i in documents d_j is expressed by the following equation:

$$s_{i,j} = TF_{i,j} \times IDF_i \tag{3}$$

Then $IP(i)$ may be expressed as follows:

$$IP(i) = \overrightarrow{IP} = \left(s_{1,j}, \ldots, s_{k,i}\right) \tag{4}$$

where k is a quantity of the most significant words in the document.

At this stage, after determination of $IP(i)$ it is necessary to determine a user profile. For this purpose, a method of building user profile was selected by extracting the characteristics of products that were already of interest to the user.

The third step is final. It is based on comparison of $UP(u)$ with $IP(i)$. To determine the degree of similarity, cosine similarity was used [5].

$$\hat{r}_{u,i} = sim\left(\overrightarrow{UP}(u), \overrightarrow{IP}(i)\right) = \frac{\overrightarrow{UP}(u) \times \overrightarrow{IP}(i)}{\left|\overrightarrow{UP}(u)\right| \times \left|\overrightarrow{IP}(i)\right|} \tag{5}$$

4.2 Neighborhood-Based Collaborative Filtering

The main idea of collaborative filtering methods is that the expected rating can be predicted based on the correlation of the ratings of various users and products [6]. Most often, collaborative filtering methods are based on explicit [7] user ratings, however, sales of concert events tickets often do not have functionality of event evaluation, even if they had, the matrix would be too sparse since people rarely visit concerts. Indeed, let's consider purchase of event as its positive evaluation. Data on the number of transactions for the period from January 1, 2019 to January 1, 2020 are presented below (Table 1).

Table 1. The number of transactions

Metrics	Value
The average number of repeated transactions per user	1.25
Percentage of users, who made a transaction	2.6

Resource conversion in 2019 is 2.6%, besides the number of repeated transactions per user, calculated by the formula

$$t_u = \frac{|T|}{|U_T|} \tag{6}$$

where $|T|$ is the number of all the resource transactions, $|U_T|$ is the number of users who made at least one purchase per year, is slightly above one, and this indicates a strong sparseness of the matrix.

On the other hand, logs of user actions can be used as ratings, for example, viewing the event page or adding a ticket to the basket, as usually there are much more such actions than transactions. The chart showing the average number of actions performed by users is presented below (see Fig. 3).

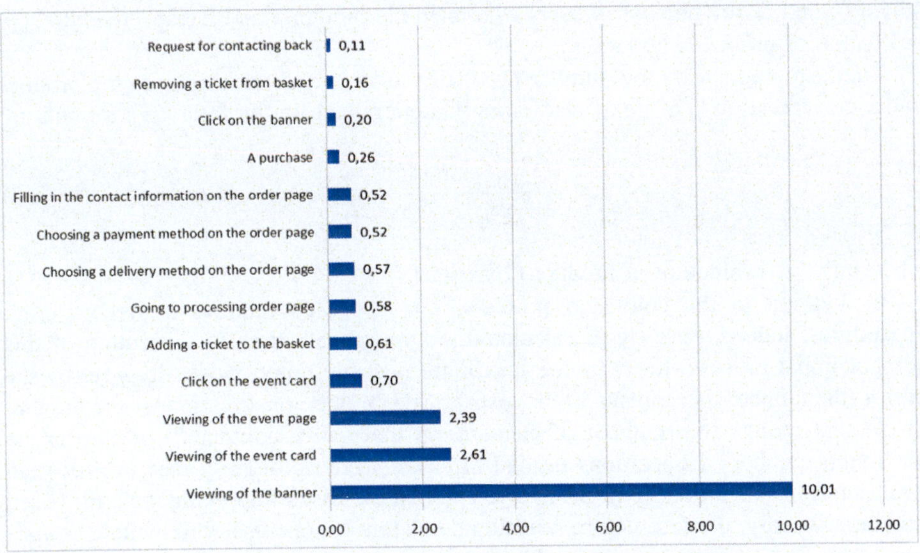

Fig. 3. Average number of actions, performed by users

Collected statistics show that at least the number of event page views is much higher than the number of transactions. In this regard, it was decided to form synthetically explicit ratings based on various user actions logs by the following formula [8]:

$$Rate = f_0\left(\sum w_i f_i(c_i)\right) \tag{7}$$

where c_i is a quantity of i actions performed by the user with the product, for example, the product has been added several times; w_i is weight of the action, which is necessary to increase the significance of such actions as buying before viewing products; f_0 – is function with saturation.

In the general case, the collaborative filtering method accepts input matrix of estimates and output expected u user's estimate $\hat{r}_{u,i}$ of the products i. The general structure of recommendation systems based on the collaborative filtering method consists of forecsasting $\hat{r}_{u,i}$ in three main stages [2].

It should be noted that the Item-based approach was immediately chosen due to high sparseness of the rating matrix [2]. At the first stage, the products that are closest to product under consideration are determined, and the cosine similarity is used as a measure of similarity:

$$sim(i,j) = \frac{\sum_{u \in U_i \cap U_j} r_{u,i} * r_{u,j}}{\sqrt{\sum_{u \in U_i \cap U_j} r_{u,i}^2}\sqrt{\sum_{u \in U_i \cap U_j} r_{u,j}^2}} \tag{8}$$

where U_i and U_j are number of users who rated the product i and j, respectively; $r_{u,i}$ is evaluation of product i by user u.

The next step shows the number $Q_i(u)$ of products that are closest to the product under consideration. Then expected rating is determined by the following formula:

$$\hat{r}_{u,i} = \frac{\sum_{j \in Q_i(u)} sim(j,i) * r_{u,j}}{\sum_{j \in Q_i(u)} sim(j,i)} \tag{9}$$

where $r_{u,j}$ – is evaluation of product j by user i.

In addition to the cold start problem, this approach requires high computing capabilities. Indeed, let's try to calculate the approximate time for execution of this computational process. Let n be the maximum possible number of ratings set by the user u (the number of elements in the matrix row R), m be maximum possible number of element ratings (the number of elements in the matrix column R), n' and m' be maximum number of operations needed to calculate the proximity measure between two users and two products, respectively. Then, in the case of applying methods based on user similarity, the first step in selecting neighbors for one user will include $O(mn')$ operations, and in order to find neighbors for all users we will need to spend $O(m^2n')$ operations. Thus, since system contains more than 10^7 users and for calculation of proximity measure it is necessary to spend a minimum of operations, each of which will take one clock cycle for simplicity, we obtain $(m^2n') = O(10^{14}10^2) = O(10^{16})$. Having a single-core processor with a frequency of 10 GHz, it will be necessary to spend 10^6 seconds or 277, 8 h, which is unacceptable. To solve this problem, it was decided to try to replace the first stage in the selection of the nearest neighbors with cluster analysis methods, the computational need of which is much lower.

Cluster analysis methods allow optimizing the computationally difficult stage of finding the closest neighbors by preliminary combining users into clusters. The first problem associated with clustering users is the need to adapt clustering methods to incomplete data. In this context, we can use *k-means* algorithm. As a solution to the problem of incomplete data, only the existing values of the estimation matrix were used to determine the distance to the centroids [4].

However, high speed results lead to certain loss of accuracy, since the value of the proximity measure within the cluster is lower than if it were calculated separately for each user [2].

Since the loss of accuracy, from a business point of view, is a more significant problem than the computational load, it was decided to try to reduce the dimension of the matrix using the *SVD (Singular value decomposition)* method, where original rating matrix R is presented as a singular decomposition [8]:

$$R = P\Delta P^T \tag{10}$$

where P and P^T are orthogonal matrixes, and Δ is diagonal matrix.

So, if size of matrix $R = m \times n$ then size of matrix P and P^T is equal to $n \times k$ and $k \times m$, respectively, where k is rank of matrix R. High sparsity of the matrix R leads to low rank k and, as a result, to the optimal size of the matrix. Estimates (obtained on the basis of the logs) were refined by means of correction predicates:

$$b_{u,i} = \mu + b_u + b_i \tag{11}$$

where μ is average rating of all products in the matrix R, b_u is average rating of user u, b_i is average rating for product i.

4.3 Solution of Cold Start Problem

Similar to content-based filtering methods, collaborative filtering methods face a cold start problem from the user's point of view. As a solution to this problem, it was decided to formulate recommendations based on the existing context, a partially developed method can be attributed to context-aware recommender systems [9]. The idea is if a user, data about which is not yet available, is on the page of the "Rock" genre of "Moscow" city or on the page of the venue where rock concerts take place, then it makes sense to recommend such user events that are appropriate to its context.

$$r_i = p_i + w_{g_c} g_c(i) + w_{t_c} t_c(i) + w_{a_c} a_c(i) + w_{r_c} r_c(i) + w_{v_c} v_c(i) + w_{o_c} o_c(i) \tag{12}$$

where p_i is priority of event i,set manually; w_{g_c} is weighting coefficient of the event genre coincidence with the context; $g_c(i)$ is a function that returns 1 if the context genre coincides with the event genre i, and 0 in other case; other elements are set in the same way; $t_c(i)$ is tag match, $a_c(i)$ is performer match, $r_c(i)$ is region match, $v_c(i)$ is site match, $o_c(i)$ is organizer match.

Of course, it could be possible to apply the content filtering method, because the context already sets some characteristics of the user's preferences. However, before these data are collected by the tracker and processed by the recommender system, time will pass, and we do not have much of it due to the need in quick formulation of recommendations and issue them together with whole page.

5 System Architecture

As it was mentioned earlier, one of the features of this field of activity is the extremely uneven distribution of the audience. High attendance is usually caused by the announcement of popular events and requires a special approach in the development of system architecture. As a solution to the identified problem, it was decided to divide the system into independent components with the ability to scale individual services. Microservice architecture used in the development of the recommender system is designed to provide the required flexibility due to asynchronous messaging via the data bus and flexibility in scaling issues [10]. Model of the interaction of system components is shown below (see Fig. 4).

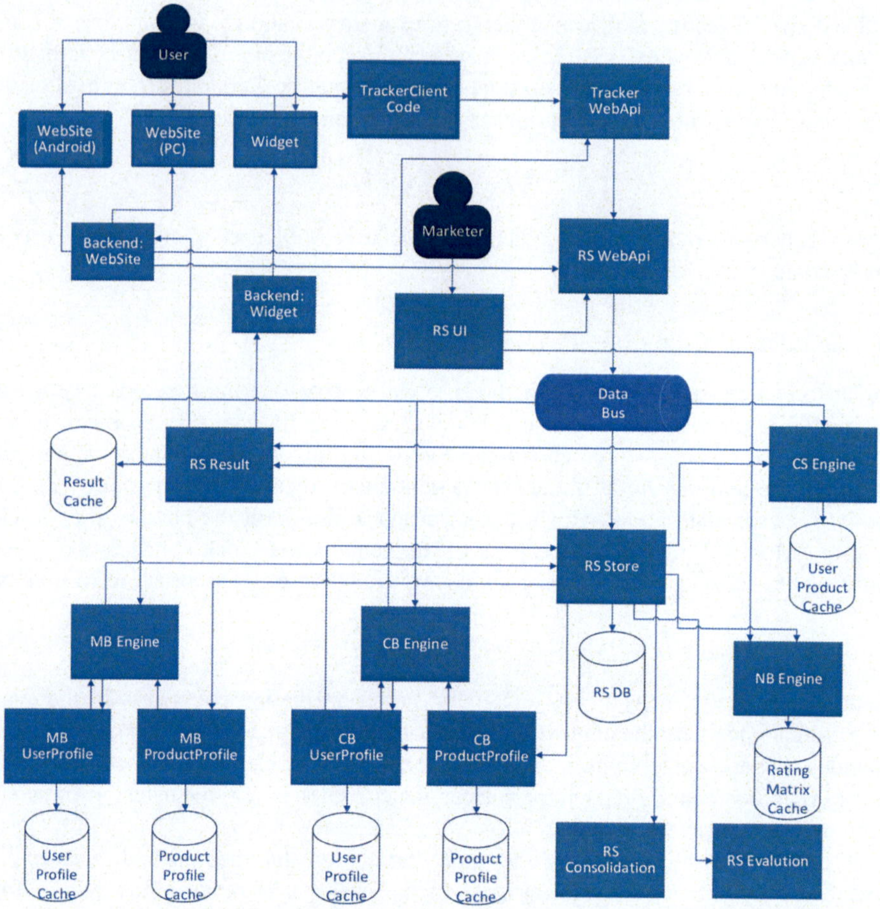

Fig. 4. System architecture

Let's consider the main purpose of the system components:

- **Tracker Client Code** is tracker of user activity on the site, written in JavaScript. After an asynchronous upload to the site, it collects all the necessary analytics, for example, adding products to the basket, and then asynchronously transfers it to the Tracker WebApi component.
- **Tracker WebApi** is a service written in .NET Core that receives data from a JavaScript tracker. First of all, it is necessary for balancing the load and protecting the main storage. After receiving data from the tracker, this service sends it through data bus to MassTransit. RabbitMQ is used as a transport. It is fully scalable.
- **Store WebApi** stores all the data that the system operates: from the company's database, tracker, additional databases, for example, music genres, etc. It is implemented as a REST service. Depending on load, it takes data from the queue and adds it to the database, while updating user profiles in the Reddis cache. It is written in .NET Core and can be scaled.
- **RS UI** is a component that allows interacting with the system through the user interface.
- **CS Engine** is a component providing recommendations based on a method developed to solve the problem of cold start. It stores user and event data in cache on the Reddis. The process of recommendations searching starts with the arrival of message in Data Bus.
- **NB Engine** is a component providing making recommendations based on Neighborhood-Based Collaborative Filtering approach.
- **CB Engine** is a component providing making recommendations based on Content-Based Filtering approach.
- **CB ProductProfile** is a component providing creation of user profiles for recommendations based on Content-based Filtering approach.
- **CB UserProfile** is a component providing creation of event profiles for recommendations based on Content-based Filtering approach.
- **MB Engine** is a component providing making recommendations based on Model-Based Collaborative Filtering approach.
- **RS Consolidation** is a service that consolidates data in the background mode.
- **RS Evaluation** – is a service providing information on the quality of the recommendation system.

6 Conclusion

This article describes some aspects of developing a recommendations system for concert events based on implicit evaluations. Recommender system has been built based on a hybrid approach, which includes content and collaborative filtering methods, as well as a developed method based on the context, applicable for this subject area, which allows solving the problem of a cold start. Considering large audience, it was proposed to use the methods of cluster analysis and reduction of the dimension of rating matrix to optimize calculations in Neighborhood-Based collaborative filtering.

This article also describes system architecture built on the basis of microservices, which allows balancing the load and scaling components if necessary.

References

1. Atterer, R., Wnuk, M., Schmidt, A.: Knowing the user's every move: user activity tracking for website usability evaluation and implicit interaction. In: Proceedings of the 15th International Conference on World Wide Web (2006)
2. Aggarwal, C.C.: Recommender Systems: The Textbook. Springer, Cham (2016)
3. Ramos, J.: Using TF-IDF to determine word relevance in document queries. In: ICML (2003)
4. Mishulina, O., Eidlina, M.: Categorical data: an approach to visualization for cluster analysis. In: International Conference on Neuroinformatics. Springer, Cham (2018)
5. Gu, Y., Zhao, B., Hardtke, D., Sun, Y.: Learning global term weights for content-based recommender systems. In: Proceedings of the 25th International Conference on World Wide Web (WWW 2016), pp. 391–400. ACM (2016)
6. Koren, Y., Bell, R.: Advances in collaborative filtering. In: Recommender Systems Handbook. Springer, Boston (2011)
7. Prokhorov, I.V., Mysev, A.E.: Approaches to the construction of multicriteria recommendation systems using implicit estimates. In: Proceedings of Southwestern State University, Series «Management, Computer Engineering, Computer Science. Medical Instrumentation», Kursk, no. 1, pp. 33–36 (2014)
8. Hu, Y., Koren, Y., Volinsky, C.: Collaborative filtering for implicit feedback datasets. In: IEEE International Conference on Data Mining (ICDM 2008), pp. 263–272 (2008)
9. Prokhorov, I.V., Mysev, A.E.: Research of algorithms of recommendation systems based on probabilistic models. In: Southwestern State University (2015): Series «Management, Computer Engineering, Computer Science. Medical Instrumentation», Kursk, no. 2(15), pp. 16–21 (2015)
10. Nadareishvili, I., et al.: Microservice Architecture: Aligning Principles, Practices, and Culture. O'Reilly Media Inc., Sebastopol (2016)

Visualization of T. Saati Hierarchy Analysis Method

Elena Matrosova[1(✉)], Anna Tikhomirova[1(✉)], Nikolay Matrosov[2], and Kovtun Dmitriy[1]

[1] National Research Nuclear University MEPhI (Moscow Engineering Physics Institute), Moscow, Russia
matrosova_ev@inbox.ru, anna7909966@ya.ru, dima.kovtun1993@gmail.com
[2] IE Abubekarov, Moscow, Russia
matrosovnikolay444@yandex.ru

Abstract. The article is devoted to the problem of expert assessment visualization of the comparative importance of various indicators. T. Saati hierarchy analysis method, which allows us to obtain indicators' weight coefficients by paired comparisons of them, is considered. A method for increasing the accuracy of estimates formed by an expert is proposed. The increase in accuracy is achieved by instantly visualizing the results of the comparisons for their prompt adjustment by an expert.

This work was supported by RFFI grant № 20-010-00708\20.

Keywords: Decision-making process · Automated decision support system · Intelligent information system · Hierarchy analysis method · Weighted coefficients · Expert system · Performance evaluation

1 Introduction

Any decision-making process involves choosing one of the existing alternatives. Each alternative can be characterized by a number of indicators, the expert assessment of the importance of which significantly affects the final choice. Therefore, it is necessary to conduct preparatory analytical work before moving on to direct comparison and choosing the best alternative. First of all, it is needed to form a series of indicators by which a comparison will be made. Then, it is necessary to evaluate the importance of the indicators relative to each other, that is, to obtain their numerical weight estimate [1]. The problem of choosing an effective method for determining the significance of indicators that characterize the considered alternatives, and for obtaining numerical estimates of qualitative parameters, is still relevant today.

When complex problems heed to be solved, highly qualified experts with specific knowledge are involved in conducting a comparative analysis. Based on their opinions numerical estimates of indicators are subsequently formed. However, the selection of suitable experts is not a sufficient condition for obtaining accurate estimates. It is also extremely important to choose the correct method of mathematical processing of expert opinion, that is, to obtain a numerical assessment of the indicators themselves or their relative importance based on the opinions of experts [2].

A. V. Samsonovich et al. (Eds.): BICA 2020, AISC 1310, pp. 253–264, 2021.
https://doi.org/10.1007/978-3-030-65596-9_32

The method of hierarchy analysis proposed by T. Saati is successfully used in solving such problems in the majority of the cases. According to this method, all indicators are compared by an expert in pairs. The results form a matrix of paired comparisons. A scale of 1 to 9 is used as an assessment scale of comparison. The values of this scale correspond to the relative importance of indicators relative to each other [3]. As an explanation, a description of the limit and median values is given:

- 9 represent the highest degree of preference for the indicator *i* over indicator *j*;
- 1 shows the equal significance of indicators *i* and *j*;
- 1/9 corresponds to the situation when the indicator *j* has the highest degree of preference over the indicator *i*.

Other values on the scale indicate an intermediate degree of preference for one indicator over another.

The matrix of paired comparisons is a matrix with ones on the main diagonal. The methodology assumes that if one of the above numbers is assigned to action i when compared with action j, then the reciprocal value is assigned to action j when compared with i. The following is a fully consistent matrix of paired comparisons.

$$\begin{vmatrix} 1 & a_{12} & a_{12} \cdot a_{23} \\ \frac{1}{a_{12}} & 1 & a_{23} \\ \frac{1}{a_{12} \cdot a_{23}} & \frac{1}{a_{23}} & 1 \end{vmatrix} \quad (1)$$

However, the data obtained by experts are not completely consistent. There can be many reasons for this such as the expert's uncertainty regarding a specific indicator or object as well as an incorrect understanding of the methodology or scale of comparison. As a result, an inconsistent matrix is obtained in practice. T. Saati introduces a consistency index to process this matrix [4]. The consistency index shows the presence of a logical relationship between the evaluated indicators. It is necessary to find the maximum eigenvalue of the matrix to find the consistency index of a positive reciprocal matrix (the paired comparison matrix has these properties).

The consistency index (C.I.) is defined by the following formula (2).

$$\text{C.I.} = \frac{\lambda_{\max} - \text{n}}{\text{n} - 1} \quad (2)$$

$\lambda_{\max}$ – maximum eigenvalue,

n – matrix dimension.

The indices for randomly generated 1 to 15-dimensional reciprocal matrices were calculated by scientists to assess the correctness of the data obtained and the consistency of the matrix. Such indices are called random consistency indices (R.I.). Matrices were generated using the scale from 1 to 9.

The ratio of the C.I. to the average random index for a matrix of the same order (R.I.) is called the consistency ratio (C.R.). The C.R. value of less than or equal to 0.10 (or 10%) is considered acceptable.

When the matrix with expert values is obtained, it is necessary to calculate the C.R. and, if the C.R. exceeds 0.10, it is recommended that the judgment matrix be adjusted. Judgments are revised according to the following algorithm:

1. Finding a row i such that $max_i \sum_{j=1}^{n} \left| a_{ij} - \frac{w_i}{w_j} \right|$,
2. Replacement in this row and the corresponding column of all a_{ij} with $\frac{w_i}{w_j}$,
3. Recalculation of the vector of the indicators' weights and C.R. value,
4. Repeat steps 1–3 if the C.R. is still above normal.

It is worth noting that having the procedure of reviewing judgments performed several times, even a very poorly consistent matrix can meet the requirements of consistency. However, the value of the result is called into question. Conducting an artificial procedure for reviewing judgments always leads to a decrease in the accuracy of estimates. It is necessary to represent the expert results in graphical form during the assessment to reduce the risk of obtaining inaccurate assessments due to incorrect perception by the expert of the assessment methodology or scale. Thus, the percentage of judgments, which is requiring mechanical adjustment, will be reduced. Also, the accuracy of expert's estimates will be increased.

2 Visualization

2.1 Analysis of the Paired Comparison Matrix

Comparing among themselves a small number of indicators is a relatively simple task. The resulting paired comparison matrices have a low C.R. value. When an expert compares several indicators, he can remember all the estimates. Therefore, the expert can take all the indexes into account when he conducts the following comparisons. A more difficult task is the one in which it is required to obtain estimates by comparing a large number of criteria. In this case, the expert will not be able to remember all the estimates so that the comparison logic is not violated.

The paired comparison matrix of n indicators is presented in the following (Table 1).

Table 1. The general form of the paired comparison matrix

Compared indicators	x_1	x_2	...	x_i	...	x_n
x_1	**1**	a_{12}	...	a_{1i}	...	a_{1n}
...	...	...	**1**	...	...	...
x_i	$\frac{1}{a_{1i}}$	$\frac{1}{a_{2i}}$	...	**1**	...	a_{in}
...	...	...	...	...	**1**	...
x_n	$\frac{1}{a_{1n}}$	$\frac{1}{a_{2n}}$	...	$\frac{1}{a_{in}}$	...	**1**

It is possible to build diagrams that would show an expert a visual picture of the comparisons. This is essential to make it easier for the expert to work with a large number of indicators. The most descriptive way to present the data is to build a pie chart. The pie chart will be adjusted automatically with each new pair comparison and illustrate the percentage of values or the relative importance of indicators.

T. Saati's hierarchy analysis method involves the use of a scale (table). The scale has 17 assessment options. It is quite a lot which is why it might be challenging to catch the line between two values for an expert. An illustration in a pie-chart percentage representation helps an expert make the most accurate estimate in this case.

When we want to represent the T. Saati scale as a percentage, it is necessary to calculate the percentage step corresponding to the transition from one value on the scale to another. The step is defined by the following formula [5]:

$$h = \frac{y_{max} - y_{min}}{k} \tag{3}$$

where

h – the interval step,
y_{max} – the maximum value,
y_{min} – the minimum value,
k – the number of the intervals.

Sixteen intervals are contained between seventeen options of the estimates. Therefore, $k = 16$. The next step is to define y_{max} и y_{min}. Two approaches can be used to solve this problem. The easiest option is to set:

$$y_{max} = 100\%$$

$$y_{min} = 0\%$$

Consequently, we obtain:

$$h = \frac{100\% - 0\%}{16} = 6,25\%$$

But at the same time, it turns out that when the expert assigns the maximum score (9) to one of the indicators in comparison with another, the major indicator corresponds to 100%, and minor one corresponds to 0%. A zero value indicates that the indicator is absolutely insignificant in comparison. However, among the indicators compared among themselves, all are significant to a certain extend. Insignificant ones are simply not included in the consideration. Therefore, to obtain the estimates, a different approach was proposed. Before the paired comparisons, the expert is provided with preliminary questions. The expert is required to indicate the most important and most unimportant indicators and give them a 100-point rating. For example, the expert specified the following values: $y_{max} = 90$, and $y_{min} = 10$. Then we obtain:

$$h = \frac{90\% - 10\%}{16} = 5\%$$

Percentage estimates of the intervals are represented for both approaches in the following table (Table 2).

The provided data can be used for the instant construction of pie charts immediately after the expert conducts a paired comparison of indicators.

Table 2. Converting saati relative importance scale to percentage

T.Saati scale value	*Element i 1 approach*	*Element j 1 approach*	*Element i 2 approach*	*Element j 2 approach*
1/9	0,00%	100,00%	10%	90%
1/8	6,25%	93,75%	15%	85%
1/7	12,50%	87,50%	20%	80%
1/6	18,75%	81,25%	25%	75%
1/5	25,00%	75,00%	30%	70%
1/4	31,25%	68,75%	35%	65%
1/3	37,50%	62,50%	40%	60%
1/2	43,75%	56,25%	45%	55%
1	50,00%	50,00%	50%	50%
2	56,25%	43,75%	55%	45%
3	62,50%	37,50%	60%	40%
4	68,75%	31,25%	65%	35%
5	75,00%	25,00%	70%	30%
6	81,25%	18,75%	75%	25%
7	87,50%	12,50%	80%	20%
8	93,75%	6,25%	85%	15%
9	100,00%	0,00%	90%	10%

2.2 Method Implementation Algorithm

The probability of an expert's random error increases with a larger number of indicators to compare [6]. Thus, a method of automatic data verification is needed to reduce this probability. The proposed algorithm for an expert assessment includes the following key steps:

- The expert's choice of the most important indicator, assessment of its importance on a 100-point scale;
- The expert's choice of the least important indicator, assessment of its importance on a 100-point scale;
- Paired comparison of all indicators;
- Assessment of the consistency of the matrix of paired comparisons (finding the C.R.).

If the C.R. is less than 10%, then further steps are taken only if the expert wishes. Otherwise, the final formation of a visual interpretation of the indicators' relative importance occurs. If the OS is more than 10%, the following steps are mandatory:

- Identify problematic indicators;
- Allow the expert to reassess the problematic indicators manually;
- Find the C.R. of the corrected matrix;
- Run the automatic matrix rematching procedure;
- Find the C.R. of the rematched matrix;
- Generate a visual interpretation of the relative importance of the indicators for both matrices (manually corrected matrix and automatically rematched matrix);

At the last step, the expert approves or rejects the result. The full analysis algorithm is shown in Fig. 1.

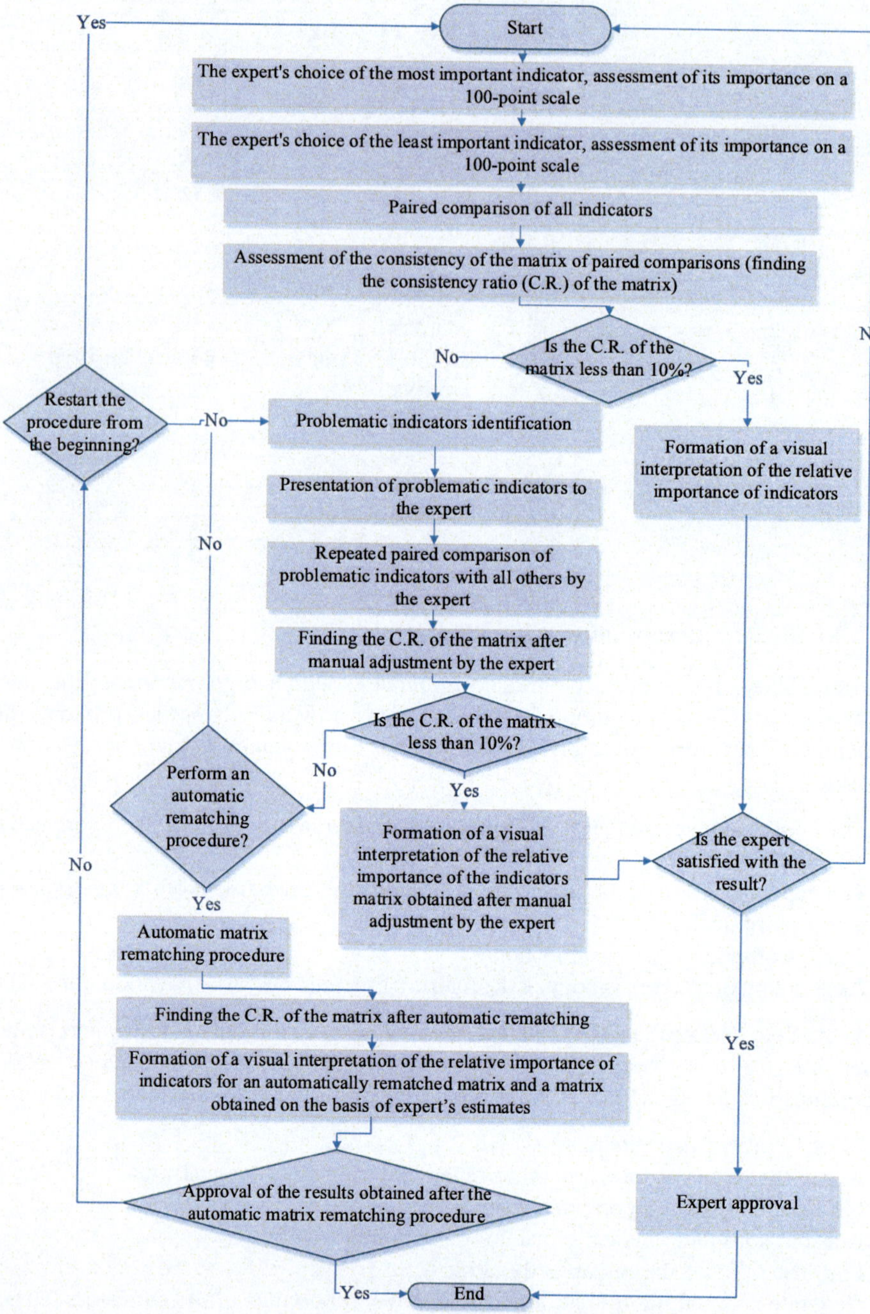

Fig. 1. Algorithm for the indicators evaluation procedure with visualization elements

2.3 An Example of the Proposed Method Application

At the first step of the algorithm, the expert selects $y_{max} = 90$ and $y_{min} = 10$.

We take a matrix of expert's paired comparisons to illustrate the application of the proposed visualization method. The following Table 3 shows the results of paired comparisons of 9 indicators

Table 3. Expert's estimates

Compared indicators	x_1	x_2	x_3	x_4	x_5	x_6	x_7	x_8	x_9
x_1	**1**	4	3	1	1/3	1/2	6	3	6
x_2	1/4	**1**	1	1/2	1/2	1/2	4	3	4
x_3	1/3	1	**1**	1/2	1/3	1/4	7	6	5
x_4	1	2	2	**1**	1/3	1/4	5	4	5
x_5	3	2	3	3	**1**	1	8	5	6
x_6	2	2	4	4	1	**1**	8	6	1/2
x_7	1/6	1/4	1/7	1/5	1/8	1/8	**1**	2	3
x_8	1/3	1/3	1/6	1/4	1/5	1/6	1/2	**1**	4
x_9	1/6	1/4	1/5	1/5	1/6	2	1/3	1/4	**1**

When all paired comparisons are made by an expert, we get a percentage ratio of all indicators. These values are displayed on a pie-chart (Fig. 2), by looking at which an expert determines whether he put the correct rating according to the proposed method. If it does not look balanced, he can immediately correct his estimates.

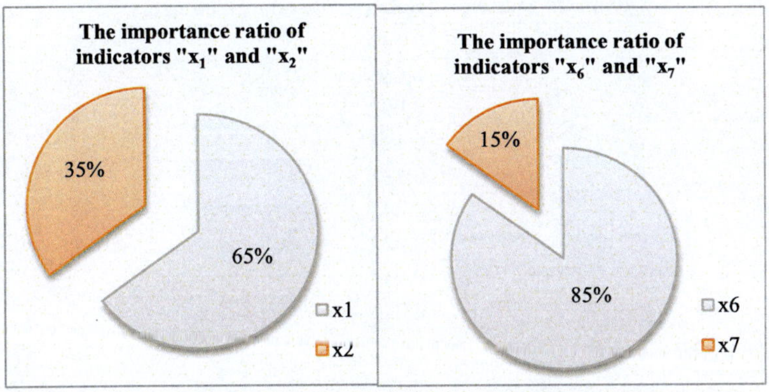

Fig. 2. Paired comparisons visualization

Weights that reflect the relative importance of all nine indicators were calculated based on the obtained data. Table 4 shows the obtained weights. We calculated the matrix consistency index and the consistency ratio. The C.R. of the matrix based on expert's estimates is 20.5%. This value indicates that the matrix is not consistent, and judgments require adjustment.

According to the proposed algorithm, the expert makes manual corrections to his estimates. During data processing, the program revealed that an inconsistent assessment was obtained due to an incorrect comparison of the x_6 and x_9 indicators. The program suggests the reassessment of these two problematic indicators to the expert. We repeat the analysis of the comparison estimates of these indicators and reveal an error caused by the inattention of the expert. We change the estimate from 1/2 to 7. Respectively, the symmetric value is changed from 2 to 1/7.

Thus, a new matrix of expert's estimates is obtained. The consistency ratio of the matrix is less than 10% (C.R. = 7.5%) which means that it is consistent. Therefore, the data obtained on its basis are suitable for subsequent analysis. Also, expert's estimates are logical and do not contradict each other.

We obtain a new matrix in which percentages between pairs of indicators are represented (Table 5). Values are based on the earlier analysis and the data shown in Table 4. The main diagonal of the matrix contains values that characterize the comparison of indicators with themselves. There is no semantic content in this case, so the main diagonal contains dashes. The sums of the values above the main diagonal and the opposite (mirror) values below are equal to 100%. We use the second approach (Table 2) to obtain y_{max} and y_{min}. Respectively, these values are equal to 90% and 10%.

Table 4. Weights based on expert's estimates

Compared indicators	Relative importance (weight)
x_1	0,147
x_2	0,085
x_3	0,091
x_4	0,119
x_5	0,231
x_6	0,245
x_7	0,029
x_8	0,033
x_9	0,020

Table 5. Expert's estimates

Compared indicators	x_1	x_2	x_3	x_4	x_5	x_6	x_7	x_8	x_9
x_1	–	65%	60%	50%	40%	45%	75%	60%	75%
x_2	35%	–	50%	45%	45%	45%	65%	60%	65%
x_3	40%	50%	–	45%	40%	35%	80%	75%	70%
x_4	50%	55%	55%	–	40%	35%	70%	65%	70%
x_5	55%	55%	60%	60%	–	50%	85%	70%	75%
x_6	55%	55%	65%	65%	50%	–	85%	75%	80%
x_7	25%	35%	20%	30%	15%	15%	–	55%	60%
x_8	40%	40%	25%	35%	30%	25%	45%	–	65%
x_9	25%	35%	30%	30%	25%	20%	40%	35%	–

Calculated weights are presented on the pie-chart (Fig. 3).

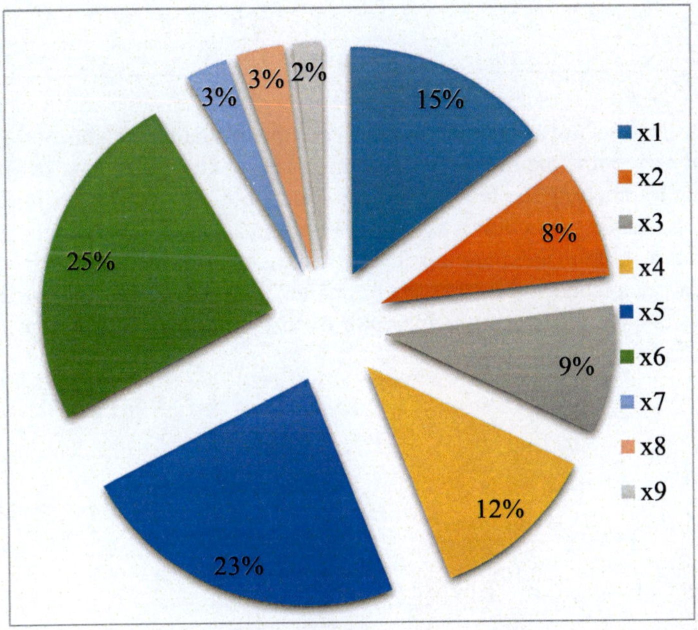

Fig. 3. Percentages of compared indicators

When the expert analyzed the resulting diagram, he can once again make sure that the technique he used is correct and the resulting values correspond to his opinion.

In this example, it is interesting to analyze the difference between the manually corrected matrix and the automatically rematched matrix.

We can see the difference in the absolute values of weight estimates, as well as in the ranks of indicators from Table 6. In this example, there is a difference in the ranks of two leading indicators. Meanwhile, the subsequent positions differ only in weight estimates.

Table 6. Weight estimates and rank of the indicators after manual correction of the matrix and automatic rematching of the matrix

X	Weights after automatic rematching of the matrix	Rank X	Weights after manual correction by an expert	Rank X
x1	0,159	3	0,147	3
x2	0,086	6	0,085	6
x3	0,101	5	0,091	5
x4	0,136	4	0,119	4
x5	0,244	1	0,231	2
x6	0,187	2	0,245	1
x7	0,032	8	0,029	8
x8	0,034	7	0,033	7
x9	0,020	9	0,020	9

Figure 4 clearly shows the difference between the estimated weights of the indicators after manual correction and those after automatic rematching. We can also see that after the automatic rematching the leading indicator is x_5 but after the expert makes manual corrections, the leading indicator is x_6. Thus, the automatic rematching of the matrix does not guarantee to obtain the same result as after manual correction by an expert. The automatically rematched matrix has even smaller C.R. value, but the meaning result which shows the supremacy of one indicator over another may be different.

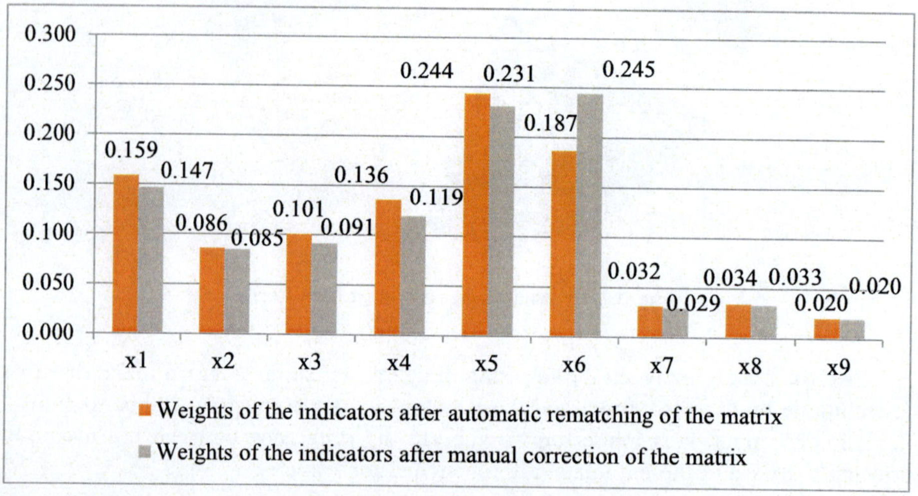

Fig. 4. The difference in the weight estimates of the indicators after manual correction of the matrix and automatic rematching of the matrix

Based on the results of the analysis, we can conclude that the proposed algorithm gives the expert a variety of opportunities to improve the accuracy of weights estimates using the T. Saati method. These opportunities include:

- Prompt, demonstrative visualization of expert's estimates;
- Identifying the indicators that made the greatest contribution to the inconsistency of the matrices and suggesting correcting them to the expert;
- The ability to choose between manual correction and automatic rematching of the matrix.

Experts often have to compare a significant number of indicators, and the more there are, the higher the risk of making a mistake. Therefore, adding new control elements to the often-used T. Saati hierarchy analysis method improves the accuracy of the obtained comparative estimates. Thus, the approach described above improves the estimates' quality, which is especially important when decisions heed to be done in critical situations.

3 Conclusions

The main feature of comparing qualitative indicators is the need to involve experts to perform a quantitative assessment of all indicators in comparison with each other. Visualization and a multi-step process of verification expert's estimates improve the accuracy of the obtained comparative estimates. The assessment process is adjusted until the expert confirms the correctness of the mathematically calculated relative weights of the compared indicators.

Visualization is an important component of the mathematical-tools-using process when experts are directly involved in a comparing process of any indicators. The use of visualization reduces the risk of inconsistencies between the intuitive perception of information by a human and the results obtained by rigorous mathematical models.

Acknowledgments. This work was supported by RFFI grant № 20-010-00708\20.

References

1. Matrosova, E.V., Tikhomirova, A.N.: Peculiarities of expert estimation comparison methods. Procedia Computer Science. In: 2016 7th Annual International Conference on Biologically Inspired Cognitive Architectures, BICA, vol. 88, pp. 163–168 (2016)
2. Kryanev, A.V., Tikhomirova, A.N., Sidorenko, E.V.: Group expertise of innovative projects using the Bayesian approach. Econ. Math. Methods 49(2), 134–139 (2013)
3. Saati, T.: Decision-making. The method of analysis of hierarchies. M. Radio and Communications (1993)

4. Matrosova, E.V., Tikhomirova, A.N.: Algorithms for intelligent automated evaluation of relevance of search queries results. biologically inspired cognitive architectures (BICA) for young scientists. In: Proceedings of the First International Early Research Career Enhancement School on BICA and Cybersecurity (FIERCES 2017). Book Series: Advances in Intelligent Systems and Computing. Publisher: Springer International Publishing
5. Agisheva, D.K., Zotova, S.A., Matveeva, T.A., Svetlichnaya, V.B.: Mathematical statistics, study guide. Volzhsky Polytechnic Institute (branch) of VSTU. Volgograd, p. 159 (2010)
6. Popov, P.V., Nozik, A.A.: Processing training results experiment, p. 51 (28.08.2019)

Labor Productivity Growth Based on Revolutionary Technologies as a Factor for Overcoming the Economic Crisis

Y. M. Medvedeva(✉) and R. E. Abdulov

National Research Nuclear University MEPhI (Moscow Engineering Physics Institute), Kashirskoe Shosse 31, Moscow 115409, Russia
YMMedvedeva@mephi.ru

Abstract. Labor productivity growth is of paramount importance both for overcoming the economic crisis caused by the COVID-19 pandemic and for the recession that began much earlier. As you know, over the past ten years, the rate of return growth rates have shown a steady downward trend. It is the revolutionary ways to accelerate labor productivity, especially based on technologies such as artificial intelligence, augmented reality, and robotics, that can turn the tide and revive the global economy. Development of practical recommendations for increasing growth rates first requires to theoretically reveal the internal structure of labor productivity as a contradictory unity, the productive force of labor and labor intensity, which will determine the main driving forces and main directions of labor productivity growth.

Keywords: Labor productivity · Productive force and labor intensity · Economic crisis and revolutionary technologies

1 Introduction

Observing the movement of the rate of return in the US economy during the XXth century shows significant volatility. In times of crisis and depression, the rate of return falls, and in times of recovery and growth, it rises accordingly. See Fig. 1.

According to the Russian scientist O. Komolov, the key factor pushing down the rate of return is an increase in the organic structure of capital, or an increase in the ratio between constant and variable capital in capitalist production. He provides a graph of calculations where the organic structure of capital correlates with the rate of return [5] See Fig. 2.

Organic structure of capital in the non-financial corporate sector of the US economy.

It is known that an increase in the organic structure of capital means an increase in constant capital in relation to variable, the latter is expressed in the wage fund. This is a valid conclusion, as capital expansion based on the existing technologies will reduce ROI and the rate of return. However, the situation can be reversed if it is possible to dramatically increase labor productivity. The theoretical foundations of this economic phenomenon should be considered in more detail.

A. V. Samsonovich et al. (Eds.): BICA 2020, AISC 1310, pp. 265–269, 2021.
https://doi.org/10.1007/978-3-030-65596-9_33

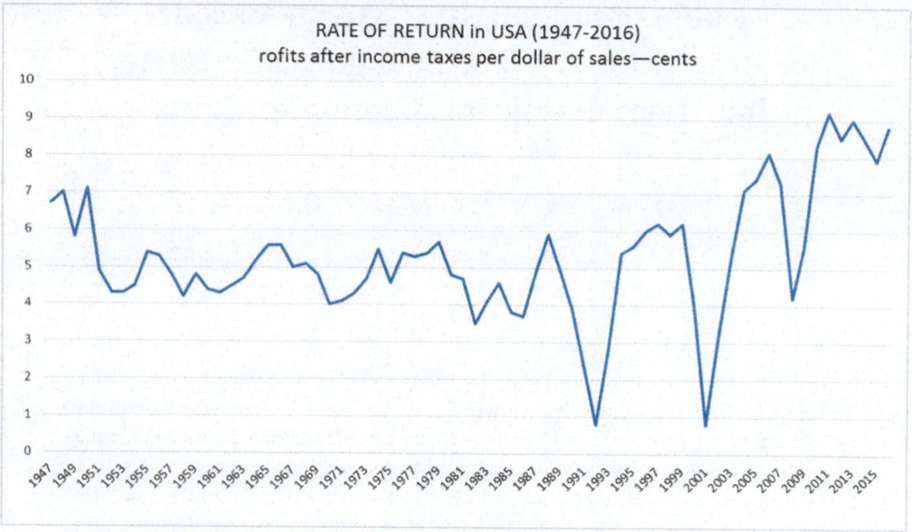

Fig. 1. Compiled by authors based on data from [1–4]

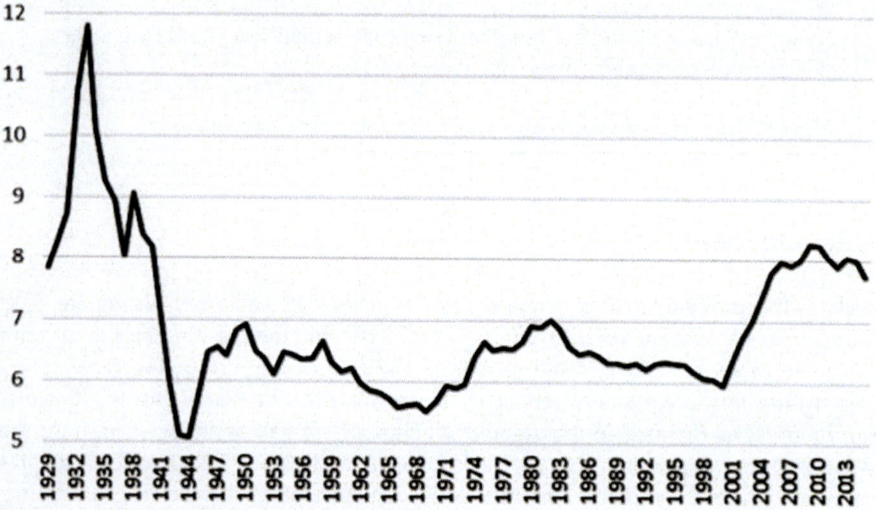

Fig. 2. Compiled by O. Komolov based on data from the US Bureau of Economic Analysis. bea.gov.

2 The Problem of the Internal Structure of Labor Productivity

The thing is that this numerous literature, as a rule, ignores the main component of labor productivity, without which all talk about this subject has no meaning.

It should be noted that the reference to the reference databases well-known in the world scientific community, such as SCOPUS and Web of Science, shows that today the world economic literature lacks any significant research revealing the essence of the category of "Productivity of Labor".

The problem under consideration is of immense theoretical as well as practical importance, since understanding of the internal structure of labor productivity, which reflects the main components of this economic phenomenon, is necessary for the development of a scientific strategy for increasing labor productivity.

What is the internal structure of the economic phenomenon? As we know, one of the laws of dialectics, i.e., laws of the self-movement of the phenomena of nature and society on the basis of the development of their internal contradictions, is the "bifurcation of the one into mutually exclusive opposites" and the struggle between them. The primary result of such a bifurcation of a particular phenomenon is precisely its internal structure. This structure contains an essential additional characteristic of the phenomenon, being the result of the action of a kind of principle of complementarity in a particular science [6].

3 Two Components of Labor Productivity

Economic phenomena have a kind of bipolarity, dual nature, because they are nothing more than various expressions of labor of a commodity producer, acting as a dual phenomenon, as a contradictory unity of concrete and abstract labor.

The labor productivity, which also acts as a dual phenomenon, is not an exception to this rule.

The concept of the dual nature of labor will help us answer the question of what this factor is. In accordance with it, the labor of commodity producers has a dual structure. Such labor appears as a contradictory unity of the useful form of labor of a commodity producer (concrete labor) and the amount of labor expended (abstract labor).

Consequently, labor productivity is also dual: on the one hand, it depends on the amount of labor expended per unit time for a given useful form of labor, i.e. on the level of intensity of labor, and on the other – on the nature of a useful, concrete form of labor.

As is known, labor has a quantitative and a qualitative characteristic. The role of the latter is a useful form of labor, which can have different effectiveness depending on the equipment and technology used, organization and management of production, natural factors used, the level of qualifications of workers, their production experience, their interest in the quantity and quality of products manufactured, etc.

Therefore, numerous attempts to identify labor productivity with the effectiveness of any one of the two aspects of labor are erroneous and fruitless.

Thus, there is no doubt that the increase in the effectiveness of a useful form of labor, along with intensity of labor, is a special factor in raising labor productivity. Meanwhile, in modern economic literature there is no special term expressing the effectiveness of a useful form of labor.

However, even in the middle of the XIX century, the expression "the productive power of labor" was used to express the effectiveness of a useful form of labor. In

1867, the well-known German economist K. Marx in Chapter 1 of Volume 1 of his main scientific work, "Capital", wrote: "Productive power has reference, of course, only to labour of some useful concrete form, the efficacy of any special productive activity during a given time being dependent on its productiveness" [7].

Labor productivity thus has a dual structure. It represents a contradictory unity of the productive power of labor and its intensity, that is, on the one hand, a contradictory unity of the effectiveness of a useful form of labor, as an expression of the quality of the labor employed (concrete labor), and on the other hand, the amount of labor spent per unit of working time (Scheme 1).

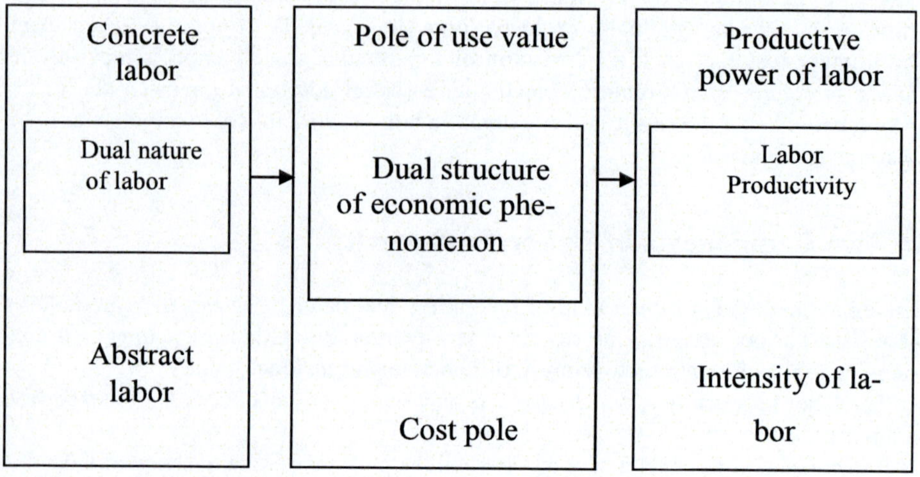

Scheme 1. Dual structure of labor productivity

It can be seen from Scheme that the dual structure of labor productivity is a necessary consequence of the dual nature of labor. The productive power of labor differs significantly from its intensity. First of all, it characterizes the qualitative aspect of useful, concrete labor, while intensity of labor expresses the amount of labor spent per unit of working time.

It is this relationship with concrete labor that gives the productive power of labor amazing properties. With the increase of the productive power of labor, the output of production is increased per unit of working time, and the amount of labor spent per unit of products decreases. On the contrary, the growth of intensity of labor, as another way of increasing labor productivity, is associated with an increase in labor costs per unit of working time.

The productive power of labor is devoid of those narrow limits that are inherent in the growth of intensity of labor. Therefore, it acts as a determining factor in the growth of labor productivity. In fact, the opportunities for the growth of the productive force of labor in time are unlimited. They are limited at each stage of development of social production, first of all, by the achieved level of scientific and technological progress.

The increase in the social labor productivity mainly through the increase of its productive power, is the main driving force of economic progress.

The productive power of labor, i.e. the effectiveness of a useful form of labor is determined by a variety of factors. It can be more or less depending on the skill of the worker, his production experience, the nature and scale of the technique used, the degree of division of labor, the level of its organization, the level of development of science and the degree of its technological application, environmental conditions and a number of other circumstances.

4 Conclusion

However, the main thing that determines the productive power of labor is the level of applied equipment and production technology, which develop in the form of successively changing technological patterns.

Today, the world is on the verge of revolutionary transformations of the means of production capable of dramatically increasing the productive force of labor. First of all, these are technologies of artificial intelligence, augmented reality, robotics, nano- and biotechnology [8]. It is these transformations that will accelerate technical progress, and therefore allow overcoming the crisis in the economy.

References

1. Economic Report of the President, p. 413 (1989)
2. Economic Report of the President, p. 433 (2013)
3. Statistical Abstract of the United States, p. 542 (2015)
4. Statistical Abstract of the United States, p. 538 (2018)
5. Komolov O.: Rate of return in the context of instability of the world economy. Vestnik Instituta Ekonomiki Rossiyskoy akademii nauk. Bull. Inst. Econ. Russ. Acad. Sci. (3), 35–52 (2017)
6. Afanasiev, V.S., Abdulov, R.E. Medvedeva, Y.M.N.: Bohr's principle of complementarity in political economy. Eur. Res. Stud. J. **XVIII**(3), pp. 59–76 (2015)
7. Marx, K.: Capital, p. 32 (1954)
8. Abdulov, R.E.: Artificial intelligence as an important factor of sustainable and crisis-free economic growth. In: Postproceedings of the 10th Annual International Conference on Biologically Inspired Cognitive Architectures, BICA 2019 (Tenth Annual Meeting of the BICA Society), Seattle, Washington, USA, 15–19 August 2019, pp. 468–472 (2019)

Network Security Intelligence Centres for Information Security Incident Management

Natalia Miloslavskaya[1(✉)] and Steven Furnell[2]

[1] The National Research Nuclear University MEPhI (Moscow Engineering Physics Institute), 31 Kashirskoye Shosse, Moscow, Russia
NGMiloslavskaya@mephi.ru
[2] University of Plymouth, Plymouth, UK
S.Furnell@plymouth.ac.uk

Abstract. Intensive IT development is driving current information security (IS) trends and require sophisticated structures and adequate approached to manage IS for different businesses. The wide range of threats is constantly growing in modern intranets; they have become not only numerous and diverse but also more disruptive. In such circumstances, organizations realize that IS incidents' timely detection and prevention in the future (what is more important) are not only possible but imperative. Any delay leaves only reactive actions to IS incidents, putting assets at risk as a result. A properly designed IS incident management system (ISIMS), operating as an integral part of the whole organization's governance system, reduces IS incidents' number and limits damage caused by them. To maximally automate IS incident management (ISIM) within one organization and to deepen its knowledge of IS level, this research proposes to unite together all advantages of a Security Intelligence Centre (SIC) and a Network Operations Centre (NOC) with their unique and joint toolkits and techniques in a unified Network SIC (NSIC). This paper presents the research, which is focused upon the designing and evaluating the concept of NSICs, and represents a novel advancement beyond existing concepts of security and network operations centres in current security monitoring scenarios. Key contributions are made in relation to underlying taxonomies of threats and attacks, leading to the requirements for NSICs, the related design, and then evaluation in a practical context and the implications arising from this (e.g. training requirements).

Keywords: Network security · Security intelligence · Security intelligence center · Network operations center · Network security intelligence center

1 Introduction

Intensive IT development is driving current information security (IS) trends and requires sophisticated structures and adequate approached to manage IS for different businesses. The range of threats has become more numerous and diverse as well as more disruptive. In such circumstances, modern organisations get a huge amount of data about the current state of their IT infrastructures (ITIs) and at a first glance much of this may seem unrelated (with scattered events taking place). This data needs to be

A. V. Samsonovich et al. (Eds.): BICA 2020, AISC 1310, pp. 270–282, 2021.
https://doi.org/10.1007/978-3-030-65596-9_34

processed correctly and promptly to identify IS incidents and to highlight ITI areas being at high risk. The ever-increasing volumes and heterogeneity of data and related activity for scrupulous analysis are very high. No matter how dedicated and talented, security staff cannot keep up with the volume of data flowing through the organisation's ITI and the speed with which things related to IS happen. A problem of structured, consolidated, and visual presentation of data to make timely and informed decisions in ensuring ITI's IS rises very sharply. A unified, inclusive, scalable, and effective system with required 'best-of-breed' IPTs will allow security analysts to truly manage IS for their ITIs [1]. Such a system with proper Security Intelligence (SI) services in place will help to mitigate and promptly respond to IS threats by helping organisations better understand their landscape and to perform routine work without the involvement of professionals in automatic mode. It will do this through the gathering, analysis, and filtering of raw IS-related data that are then will be collected into appropriate databases, will be presented in management reports with necessary visualisation and will be transferred for IPTs reconfiguring and online IS monitoring.

To automate IS incident management (ISIM) within one organisation it is proposed to unite together all advantages of a Security Intelligence Centre (SIC) and a Network Operations Centre (NOC) with their unique and joint toolkits and techniques in a Network SIC (NSIC) [2, 3].

1.1 Research Aims

Based upon the above, the goal of the research is to develop and start to implement the concept for a new state-of-the-art centralised network security management unit called NSIC intended to increase the effectiveness of ISIM processes (ISIMP) for modern organisations to be proactive and resilient to damaging IS threats as their urgent and priority need. In its simplest way, the drive to carry out this research can be reduced to two research questions as follows: How this NSIC should be designed? What knowledge and skills should an IS professional working there have?

The creation of the innovative NSIC concept, its interpretation and construction through original research will contribute substantially to the modern networks' security, as it will extend the forefront of the main security measures and tools used nowadays.

The research methodology is firmly based on the scientific management theory [4], general system theory [5], open information system [6] and big data IT concepts [7], as well as proposed IS theory basics [8]. The applicable techniques for the research are the well-known analytical approaches, namely system analysis of the object of research, which allows to carry out its complex modelling; exploratory research, analysis, systematisation and classification of typical network vulnerabilities, IS threats, attacks and IS incidents; process approach used to describe the monitoring of IS for modern networks; comparative analysis of two generations of Security Information and Event Management (SIEM) systems as well as Security Operations Centres (SOCs) and SICs, synthesis of requirements for a NSIC for centralised IS monitoring in modern networks, taking into account the main provisions of management theory and identifying its closest counterparts; synthesis of requirements for a next-generation SIEM 3.0 system as a core of the NSIC, development of a glossary of key terms used, etc.

As for any research, some grounded assumptions were made. The research made a few assumptions as follows:

- ITIs are considered only from the wired network viewpoint since the research was started at the time when wireless technologies were not being used as widely as at present;
- The research focus was on the technical aspects of NSIC's design (not its documentation support) because it is more relevant to the competences and qualification of the author of this research;
- A private (not cloud-based, without outsourcing) NSIC is proposed because it presupposes a cruder elaboration of all the issues of NSIC support by its owner.

1.2 Contributions

The major contributions of the research presented here and in the associated thesis [9] are:

- For forming the basis for the research, the normative base (as a collection of ISO standards) was selected, the glossary of the research area (including the following terms: IS, IS maintenance, IS management, IS risk, IS threat, source/actor of IS threat, vulnerability, intruder, attack, network security, etc.) was worked out [8, 10], the taxonomy of network vulnerabilities, IS threats, attacks and IS incidents was developed [11, 12], key verbal indicators of IS incidents in networks were described [12, 13], and specifications for information formats applicable to IS incident description were selected [14];
- Using the above results, seven subprocesses of ISIM process were developed in detail and described using special notations, IS monitoring in the form of VEI detection was discussed as one of the important ISIM subprocesses [10, 15], the issues of managing big IS-related data during ISIM were highlighted [1, 16, 17], and SIEM systems' role in ISIM, their functions and evolution were identified [14, 18];
- Further, typical SOC's functions in IS monitoring were analysed [13, 17–19], classification of SOCs was proposed [13], and SOCs' limitations in the current network environment were revealed [17, 19];
- After that, following the evolution analysis, a generalised description of the SI concept as a logical continuation of IS ensuring approaches was done, including its main advantages and characteristics, SICs' function and technologies combined were defined, SIEM 2.0 systems' mission in SICs was shown [17, 19, 20], SIC's business logics was proposed [20] and simplified SIC's data architecture was developed [20];
- Generalising all the results obtained, the modern network environment's requirements for NSICs were formulated [3], compliance with these requirements of the NSIC as a SIC-NOC combination was proven [3], NSIC's development methodology was described [15], network security information to be visualised in NSICs was defined [21], NSIC's layered infrastructure and zone security infrastructure were proposed [21], NSIC's implementation in the MEPhI was briefly described [2, 22], blockchain-based SIEM system for NSICs (called SIEM 3.0) and its

architecture were developed [14], and the most important issues for the training of highly qualified staff for NSICs were discussed [2, 22–25].

The corresponding structural and logical schema of the research conducted can be illustrated by Fig. 1, where CERTs/CIRTs (Computer Emergency Response Teams/Computer Incident Response Teams) were the structures that were in some sense forerunners for SOCs regarding its staff. It also emphasizes the main mottos of SOCs (DETECT IS incidents via constant monitoring), SICs (ANALYSE & REACT for real-time IS incident management) and proposed NSICs (ADAPT for proactive network security).

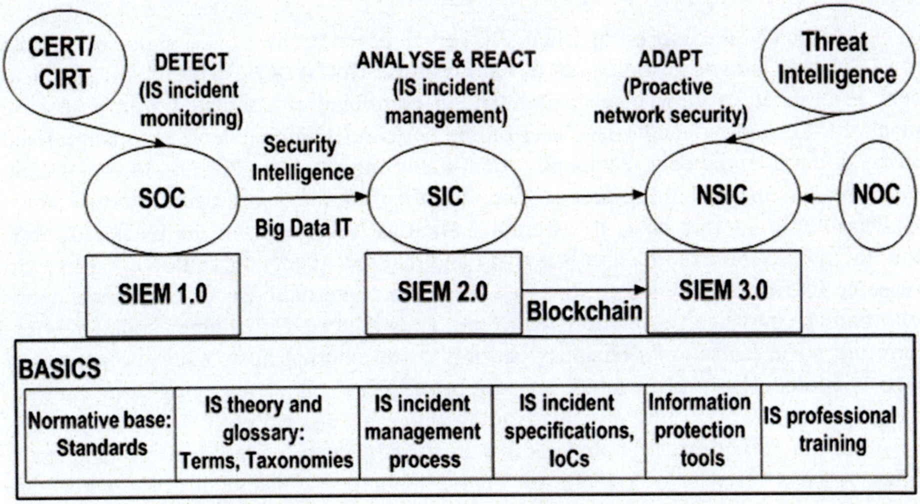

Fig. 1. Structural and logical schema of the research

2 IS Theory Development

Using the glossary proposed and analysing real-world attacks (like APTs, phishing e-mails, pharming, scams, DoS attacks, XSS, SQL injections, targeted watering hole attacks, ransomware, Heartbleed, ShellShock) and numerous reports of well-known companies (like PricewaterhouseCoopers, Symantec, Endescan, CA technologies, Flexera), our own taxonomy with classifications of network vulnerabilities, IS threats, attacks, and incidents is proposed [11, 12]. The main classification parameters defined are the following:

- For a vulnerability: its origin sources, allocation in ITI, risk level (criticality, severity), probability/likelihood of a threat realisation on its basis and prerequisites;
- For an IS threat: its type (according to a violation of physical integrity, logical structure, content, confidentiality, property rights, availability, privacy, etc.), origin nature, prerequisites, and sources;

- For an attack: influence type (passive or active), aim, condition of influence beginning (on request, on event occurrence or unconditional), allocation of a victim and an attacker (in one or different network segments), number of victims and attackers (traditional or distributed), feedback presence (with or without feedback), implementation level according to the seven-layer ISO/OSI model, implementation tools (information interchange, commands, scripts, etc.), implementation form (sniffing, masquerade, spoofing, hijacking, phishing, pharming, etc.) and so on;
- For an IS incident: type, priority, malefactor (malefactors), aims to be achieved, methods and tools that can be used, actions and targeted objects on which these actions are directed, affected objects and particular information assets, impact and its severity, detection and response complexity, etc.

To continue, the most important 28 verbal descriptions of the signs of remote attacks, better known as Indicators of Compromise (IoCs), were worked out [12, 13]. They are related to the typical activities/their combination associated with a specific attack, for example: unauthorised user on the network/shared credentials; unauthorised access to confidential data, Personally Identifiable Information (PII) and financial data; unauthorised internal host connection to the Internet; excessive access from single/multiple internal hosts to external malicious website (from the blacklists); off-hour (at night/on weekends) user's activity and malware detection; multiple logins with the same ID from different locations in a short time; internal hosts communicate either with known untrusted destinations or to the hosts allocated in another country where there are no organisation's business partners or to external hosts using non-standard ports or protocol/port mismatches; a single host/user account tries to log in to multiple hosts within network a few minutes from/to different regions; etc.

From our taxonomy, it is obvious that IS incidents can be described in completely different ways. That is why there are many standards (as listed in [14]), which are designed for unifying these descriptions. Among the most known are the following data exchange specifications: IODEF and RID by the MILE working group; IDMEF and IODEF-SCI by IETF; CIF by REN-ISAC; VERIS by Verizon; OpenIOC by Mandiant; OTX by Alien Vault; CybOX, TAXII and STIX by MITRE. The last three standards were selected to form a unified template for IS incident description for SIEM 3.0.

3 IS Incident Management

Using the above results and ISO 27035 (both parts), the ISIMP as one of the key management processes of any organisation was presented as a combination of seven subprocesses [10], namely Vulnerabilities, IS events and incidents (VEI) detection, VEI notification, VEI messages processing, IS incident response, IS incident analysis, IS incident investigation, and ISIM process efficiency analysis. Properly designed and operating ISIMP reduces the number of IS incidents, limits damage caused by them and can be automated [10, 15]. The data received as the output of ISIMP is the input for another IS management processes and vice versa. For example, IS monitoring gives a lot of interesting information for ISIMP. After first processing, the collected data are transferred into useful for decision-making information and at the highest level of abstraction, after more complicated processing and understanding, into new knowledge needed not only for tactical but what is a more important strategic improvement of an organisation's business.

ISIMP is associated with the processing of large amounts of data, which requires mandatory automation of routine operations. All the IS-related data should not be considered as a simple combination of separate data elements [17]. It is a must to maintain the recorded relationships of every file execution and modification, registry modification, network connection, executed binary in an organisation's environment, etc. Moreover, it is a data stream with the following unique features: huge or possibly infinite volume, dynamically changing, flowing in and out in a fixed order, demanding fast (often real-time) response time, etc. The automated systems for ISIM – SIEM systems [27] are used for flow control over the IS events to computerise ISIMP [18]. SIEM systems, being used for the constant events and monitoring users' activities, can detect and handle IS incidents through the aggregation of large volumes of machine data in real-time for IS risk management, and essentially improve this automation [1, 16]. SIEM systems can also visualise IS threats to organisations' ITIs. From the other side, all SIEM data should be protected itself as it contains sensitive information for digital forensics. While being integrated with other IPTs, SIEM systems can serve as a single-window into IS incidents. That is why they are considered as a core of SOCs and NSICs described further.

4 Mission of SOCs in IS Monitoring

Having all these in mind, a SOC was presented as a specialised organisation's network unit implementing. Figuratively speaking any SOC is the ISIM's eyes. In [13], a classification of SOCs is proposed according to the following criteria: counteraction capabilities (SOC without such a capability and reactionary SOC, deployment scenarios (centralised/distributed); aim (controlling/management/crisis); correlation technique (statistical, rule-based, vulnerability, Service Level Agreement (SLA), compliance, mixed, without correlation); implementation (software/hardware/infrastructure solution); and ownership (in-house/outsourced). The SOC's effectiveness is reasonable measured by how IS incidents are managed, handled, administered, remediated, and isolated [13].

Based on the thorough analysis, some serious limitations of SOCs were revealed [17, 19], for example: inability to work in the large-scale, globally deployed, heterogeneous, highly distributed ITIs with connect-from-anywhere-and-anytime users; incapability of providing a high degree of trustworthiness/resilience in IS event collection, dissemination, and processing, thus becoming susceptible to attacks on the SIEM systems themselves and the entire SOC; dependency on centralised correlation rules processed on a single node, making scalability difficult, and creating a single point of failure; limited IS analysis and assessment capabilities with limited-visibility solutions as SOC monitors above all network-level events; a reactive security posture, lack of online reaction to identified attacks and limited threat blocking; insufficient capacity to process large volumes of all gathered and analytically derived IS-related data known as big data; inability to interpret data from the higher layers like services; a great number of false positives/negatives, as the SOCs' main application area is uncovering known or easy-to-detect IS threats; the old-practice SIEM 1.0 systems used; manual integration of IS ensuring technologies used, etc.

5 Mission of SICs in IS Management

The SI concept as the real-time collection, normalisation, and analysis of the data generated by users, applications and infrastructure that impacts the IT security and risk posture of an organisation [27] was accepted. Its goal is to provide proactive, predictive (forward-looking), actionable and comprehensive protection and insight into network security that reduces IS risks and operational effort for any organisation through advanced analytics. In the analysis of the evolution of network security ensuring approaches in [20], it was shown that the era of SI has come since 2010 after perimeter defence (2000–2004) and logging and compliance approaches (2005–2009).

In comparison with IS controls used in SOC, the new SIC concept logically develops its features to overcome existing limitations. It was found that as a truly integrated and multi-domain solution, the SIC combines a number of technologies [17, 19], namely IS knowledge management, promoting an integrated approach to identifying, capturing, evaluating, retrieving and sharing this knowledge; big IS-related data processing; ITI asset identification, tracking, and recovery after different IS incidents; data collection capabilities and compliance benefits of log management; centralisation and aggregation of data from disparate silos with ensuing correlation, normalisation, categorisation and analysis by SIEM 2.0 systems [18]; network visibility and advanced IS threat detection of IDS/IPSs for network behaviour (rule-less) anomaly detection; IS risk management reducing the number of IS incidents and ensuring compliance; IS incident handling, consisting of detection, alerting and notification, reporting, response and escalation management; vulnerability scanning followed by device configuration and patch management; network traffic sniffing and application content insight afforded by Next-Generation FireWalls (NGWFs) and forensic tools.

Continuing [28], SIC's business logics proposed contains seven operational layers [20]: IS event-based and status-based message generators, IS-related data acquisition layer on the basis of SIEM with IS event collectors, formatted and aggregated IS-related message database (DB), IS incident analysis engine and knowledge base, IS reporting and reaction management software, information protection tools (IPTs) management and layer managing IS trends, predictions and IS information interchange and sharing. For big IS-related data processing in SICs, the more advanced concept than a big data IT called 'data lakes' [29] was proposed for NSICs [20]. On this basis, a simplified SIC's data architecture was developed.

6 NSICs

At present, even this perspective idea of SICs does not keep pace with the increasing number of sophisticated IS threats in the highly heterogeneous and connected world. A new entity to be designed should unite all benefits of SIC with many-year experience of network operations management, namely change the security model from reactive to proactive, support more informed and effective responses to IS incidents, enhance communications between the network and security teams, management and board members, and drive IS investment strategies and connect more directly IS priorities with business risk management priorities. Therefore, a new NSIC's concept as a SIC-NOC combination is proposed in the research.

The NSIC introduces powerful synergies of SICs and NOCs via people collaboration and toolkits and techniques joint usage. Some of the significant benefits of NSICs to their owners are the following [3]:

- Complete situational awareness with seamless integration of the monitored entities from network and SI services with a supporting infrastructure;
- Network and security real-time monitoring, meeting IS management requirements;
- Centralised network devices' and IPTs' configuration management and audit;
- Computer-based data collection and analytics from both network and IS viewpoints to extract maximum value from the massive amounts of IS-related data available in NSICs;
- IPTs alerts' prioritisation; richer organisation-wide context mining to aid operational and strategic IS decision-making and to validate that the right IS policies are in place;
- Faster IS event detection and reduced IS incident response times by automated remediation and enabling IS controls and IT to work together;
- Advanced network forensics;
- Actionable streamlined ISIM (including behavioural analytics for continuous monitoring, pre/post analysis) and reporting with valuable context.

The main specific requirements, which should be taken into account in designing NSICs, was formulated as follows [3]:

- All NSIC' processes should be in line to industry regulations and applicable standards;
- Usage of more types of data for analytics:
- Modular and adaptive IS event management achieved;
- Operational resilience, scalability, elasticity, maintainability, agility, security, and trustworthy.

Compliance with these requirements of the NSIC proposed was proven in [3].

The well-known Deming-Shewhart Wheel or PDCA/PDSA cycle was chosen as the NSIC's development methodology [3]. The most important issues for the training of highly qualified staff for NSICs were also discussed [2, 22–25].

Considering a separate IS event, it is proposed to visualise for decision-making in NSICs the following information [15]:

- Who: access subject;
- Why: involuntary or unintentional action (error), ignorance, negligence or irresponsibility, malicious actions, etc.;
- What: unauthorised access to a system and/or information, installing unauthorised programs without consent of the owner, remote causing malfunction in the information system, stressful situation creation, physical intrusion, illegal activities, breakdown or failure of equipment, etc.;
- How: methods and tools for each of the above "what"; status of the IS event: an attempt, a successful or failed hacking;
- Which vulnerability was exploited: software, hardware, general protection, system architecture or processes, etc.;
- On which type of assets (basic and related); what are the consequences: violation of confidentiality, integrity, availability, etc.

A layered infrastructure with 5 layers (platform, software, service delivery, access, and management) and zone security infrastructure with 4 zones (demilitarized, trusted, restricted and management) was proposed. This NSIC structure is illustrsted in Fig. 2 [21].

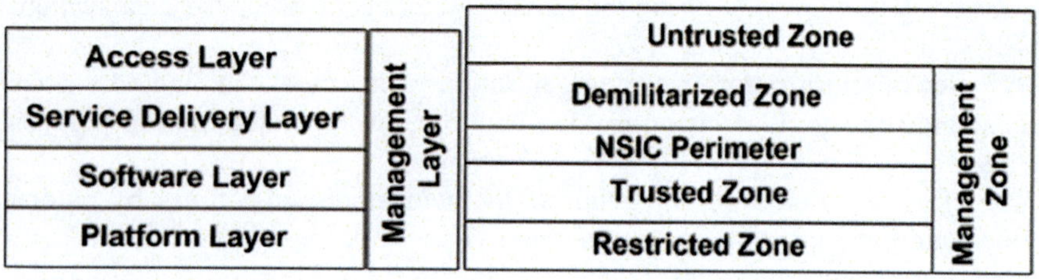

Fig. 2. NSIC's layered (left) and high-level zone security (right) infrastructures

A blockchain-based SIEM 3.0 system for the NSIC as its core and the resulting system's architecture (Fig. 3), completely assigned with the SI concept, were proposed in the framework of the research [14]. Even though the system has not been yet implemented, a methodology for its validation based on Z [ISO/IEC, 13568] was chosen, for describing and modelling computing systems (to which SIEM systems can be attributed) via mathematical notation.

We have a short-term experience in implementing NSIC's concept in the "Network Security Intelligence Centre" educational and research centre established in 2016 within the framework of the MEPhI's Institute of Cyber Intelligence Systems [2, 22]. Its goal is to implement a model of NSIC for its study and continuous improvement. The MEPhI's NSIC is based on three bearing laboratories with NGWFs, Data Loss Prevention (DLP) and SIEM systems at their cores respectively used at present for educational purposes (because the hardware and software base of the centre is still not fully developed for research purposes).

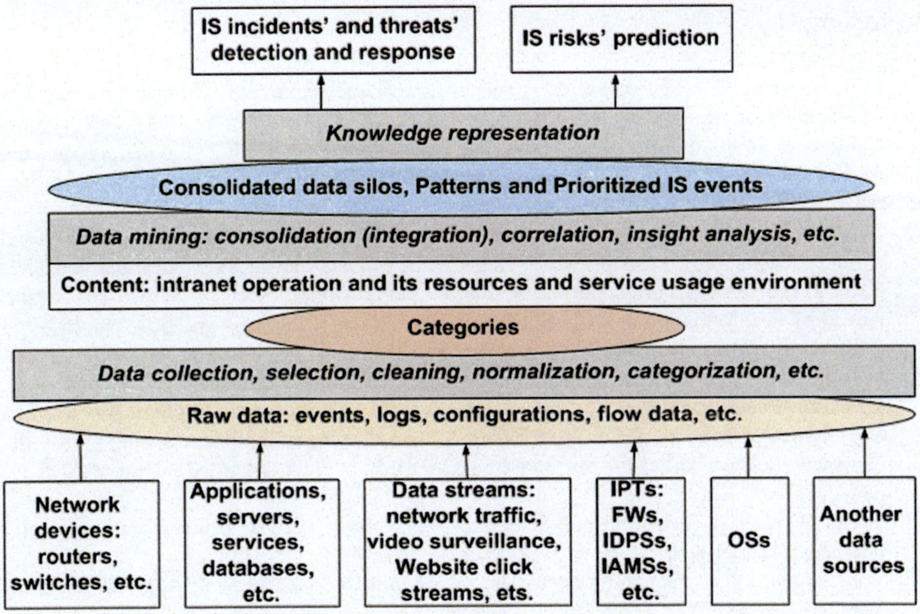

Fig. 3. SIEM 3.0 system's architecture

7 Conclusion

The aim of the research was to develop and start to implement the concept of a new state-of-the-art centralised network security management unit called NSIC. To achieve the goal, all the solutions proposed were considered from different angles and at the same time as a whole. To that end, one can be confident that the NSIC developed at an advanced level suitably meets its purpose. This is due to the fact that the proposed concept and its description give the high-level guidelines that will help organisations (from small to large) to plan, implement and evaluate their in-house NSICs.

Future plans for the framework include the development and subsequent implementation of educational standards, and competency models for different educational levels for specialised network security professional training. The centre is also expected to carry out research on the NSIC's design, effective network security management practices based on SI approaches and applications, the study of the compatibility between different IPTs, and recommendations to address arising issues in terms of evaluating network security.

Acknowledgement. This work was supported by the MEPhI Academic Excellence Project (agreement with the Ministry of Education and Science of the Russian Federation of August 27, 2013, project no. 02.a03.21.0005).

References

1. Miloslavskaya, N., Senatorov, M., Tolstoy, A., Zapechnikov, S.: Information security maintenance issues for big security-related data. In: Proceedings of 2014 International Conference on Future Internet of Things and Cloud FiCloud 2014. International Symposium on Big Data Research and Innovation (BigR&I). 27–29, Barcelona (Spain), pp. 361–366, August 2014. ISBN: 978-1-4799-4357-9/14. https://doi.org/10.1109/ficloud.2014.64
2. Miloslavskaya, N., Tolstoy, A., Migalin, A.: Network security intelligence educational and research center. In: Bishop, M., Futcher, L., Miloslavskaya, N., Theocharidou, M., (eds) Information Security Education for a Global Digital Society. WISE 2017. IFIP Advances in Information and Communication Technology. Springer vol. 503, pp. 157–168 (2017). https://doi.org/10.1007/978-3-319-58553-6_14
3. Miloslavskaya, N.: Network security intelligence center as a combination of SIC and NOC. In: Postproceedings of the 9th Annual International Conference on Biologically Inspired Cognitive Architectures, BICA 2018 (Ninth Annual Meeting of the BICA Society). Procedia Computer Science vol. 145, pp. 354–358 (2018). https://doi.org/10.1016/j.procs.2018.11.084
4. Taylor, F.W.: The Principles of Scientific Management. Dover Publications. Originally Published: Harper & Bros, New York 1911, p. 80 (1998)
5. Von Bertalanffy, L.: General System Theory: Foundation development, Applications. New York, p. 289 (1968)
6. ISO/IEC 10165-1:1993 Information technology – Open Systems Interconnection – Management Information Services – Structure of management information: Management Information Model
7. Lynch, C.A.: Big data: How do your data grow? Nature, 455, 28–29 (04 September 2008). https://doi.org/10.1038/455028a. Accessed 24 Actober 2019
8. Malyuk, A., Miloslavskaya, N.: Information security theory development. In: Proceedings of the 7th International Conference on Security of Information and Networks (SIN2014), 9–11 Glasgow (UK). ACM New York, pp. 52–55, September 2014. ISBN: 978-1-4503-3033-6. https://doi.org/10.1145/2659651.2659659
9. Miloslavskaya, N.: Network Security Intelligence Centres for Information Security Incident Management. University of Plymouth. Research Theses Main Collection (2019). http://hdl.handle.net/10026.1/14306. Accessed 24 Actober 2019
10. Kostina, A., Miloslavskaya, N., Tolstoy, A.: Information security incident management. In: Proceedings of the 3rd International Conference on Internet Technologies and Applications. 8–11, Wrexham, UK, pp. 27–34, Sept 2009
11. Miloslavskaya, N., Tolstoy, A., Zapechnikov, S.: Taxonomy for unsecure big data processing in security operations centers. In: Proceedings of 2016 4th International Conference on Future Internet of Things and Cloud Workshops. The 3rd International Symposium on Big Data Research and Innovation (BigR&I 2016). 22–24, Vienna (Austria), pp. 154–159, August 2016. https://doi.org/10.1109/w-ficloud.2016.42
12. Miloslavskaya, N.: Remote attacks taxonomy and their key verbal indicators. In: Proceedings of the 8th Annual International Conference on Biologically Inspired Cognitive Architectures (BICA 2017). 1–6 August 2017, Moscow (Russia). Procedia Computer Science vol. 123, pp. 278–284 (2018). https://doi.org/10.1016/j.procs.2018.01.043
13. Miloslavskaya, N.: Security operations centers for information security incident management. In: Proceedings of the 4th International Conference on Future Internet of Things and Cloud (FiCloud 2016). 22–24 August 2016, Vienna (Austria), pp. 131–138. https://doi.org/10.1109/ficloud.2016.26

14. Miloslavskaya, N.: Designing blockchain-based SIEM 3.0 system. Information and Computer Security. Emerald Publishing. UK. vol. 26, no. 4 (2018). https://doi.org/10.1108/ics-10-2017-0075
15. Miloslavskaya, N., Tolstoy, A., Birjukov, A.: Information visualisation in information security management for enterprises's information infrastructure. Sci. Visualisation Moscow, NRNU MEPhI **6**(2), pp. 74–91 (2014)
16. Miloslavskaya, N., Tolstoy, A.: Application of big data, fast data and data lake concepts to information security issues. In: Proceedings of 2016 4th International Conference on Future Internet of Things and Cloud Workshops. The 3rd International Symposium on Big Data Research and Innovation (BigR&I 2016). 22–24 August 2016, Vienna (Austria), pp. 148–153. https://doi.org/10.1109/w-ficloud.2016.41
17. Miloslavskaya, N.: Information security management in SOCs and SICs. J. Intell. Fussy Syst. IOS Press. Netherlands **35**(3), pp. 2637–2647 (2018). https://doi.org/10.3233/jifs-169615
18. Miloslavskaya, N.: Analysis of SIEM systems and their usage in security operations and security intelligence centers. In: Samsonovich, A., Klimov, V., (eds) Biologically Inspired Cognitive Architectures (BICA) for Young Scientists. BICA 2017. Advances in Intelligent Systems and Computing. Springer, Cham vol. 636, pp. 282–288 (2018). https://doi.org/10.1007/978-3-319-63940-6_40
19. Miloslavskaya, N.: SOC- and SIC-Based information security monitoring. Rocha, A., et al. (eds.) Recent Advances in Information Systems and Technologies, Advances in Intelligent Systems and Computing. Springer International Publishing AG vol. 570, pp. 364–374 (2017). https://doi.org/10.1007/978-3-319-56538-5_37
20. Miloslavskaya, N.: Security intelligence centers for big data processing. In: Proceedings of 2017 5th International Conference on Future Internet of Things and Cloud Workshops. The 4th International Symposium on Big Data Research and Innovation (BigR&I-2017). 21–23 August 2017, Prague (Czech Republic), pp. 7–13. https://doi.org/10.1109/w-ficloud.2017.7
21. Miloslavskaya, N.: Developing a network security intelligence center. In: Postproceedings of the 9th Annual International Conference on Biologically Inspired Cognitive Architectures, BICA 2018 (Ninth Annual Meeting of the BICA Society). Procedia Computer Science vol. 145, pp. 359–364 (2018). https://doi.org/10.1016/j.procs.2018.11.085
22. Miloslavskaya, N., Morozov, V., Tolstoy, A., Khassan, D.: DLP as an integral part of network security intelligence center. In: Proceedings of 2017 5th International Conference on Future Internet of Things and Cloud (FiCloud2017). 21–23 August 2017, Prague (Czech Republic), pp. 297–304. https://doi.org/10.1109/ficloud.2017.15
23. Miloslavskaya, N., Senatorov, M., Tolstoy, A., Zapechnikov, S.: Business continuity and information security maintenance masters' training program. IFIP advances in information and communication technology. In: Ronald, C., Dodge, J., Lynn, F., (Eds.): Information Assurance and Security Education and Training - 8th IFIP WG 11.8 World Conference on Information Security Education, WISE 8, Auckland, New Sealand, July 8–10, 2013, Proceedings, WISE 7, Lucerne Switzerland, June 9–10, 2011, and WISE 6, Bento Gonçalves, RS, Brasil, July 27–31, 2009, Revised Selected Papers. Springer vol. 406, pp. 95–102 (2013). https://doi.org/10.1007/978-3-642-39377-8_10. ISBN 978-3-642-39376-1
24. Miloslavskaya, N., Tolstoy, A.: Developing hands-on laboratory works for the information security incident management discipline. In: Drevin, L., Theocharidou, M., (eds): Information Security Education – Towards a Cybersecure Society. WISE 2018. IFIP Advances in Information and Communication Technology. Springer, Cham vol. 531, pp. 28–39 (2018). https://doi.org/10.1007/978-3-319-99734-6_3

25. Miloslavskaya, N., Tolstoy, A.: State-Level views on professional competencies in the field of IoT and cloud information security. In: Proceedings of 2016 4th International Conference on Future Internet of Things and Cloud Workshops. The 3rd International Symposium on Intercloud and IoT (ICI 2016). 22–24 August 2016, Vienna (Austria), pp. 83–90 (2016). https://doi.org/10.1109/w-ficloud.2016.31
26. Miller, D., Harris, S., Harper, A., VanDyke, S.: Security information and event management (SIEM) implementation. McGraw-Hill, Chennai, p. 464 (2010)
27. Burnham, J.: What Is Security Intelligence and Why Does It Matter Today? August 2011. https://securityintelligence.com/what-is-security-intelligence-and-why-does-it-matter-today/. Accessed 24 Actober 2019
28. Bidou, R.: Security Operation Center Concepts & Implementation. (2005). http://iv2-technologies.com/ SOCConceptAndImplementation.pdf. Accessed 24 Actober 2019
29. Inmon, B.: Data Lake Architecture: Designing the Data Lake and Avoiding the Garbage Dump. 1edn. Technics Publications, New Jersey, p. 166 (2016)

Block Formation for Storing Data on Information Security Incidents for Digital Investigations

Natalia Miloslavskaya(✉), Andrey Nikiforov, and Kirill Plaksiy

National Research Nuclear University MEPhI
(Moscow Engineering Physics Institute), Moscow, Russia
NGMiloslavskaya@mephi.ru,
andreinikiforov993@gmail.com, kirillplaksiy@mail.ru

Abstract. Nowadays technologies such as Blockchain (BC) and the Internet of Things (IoT) can be heard everywhere. But because of the leap in the development of these technologies, there is a need to evaluate the existing approaches critically. One of the up-to-date tasks is to study information security (IS) incidents as a part of the IoT. Due to a large number of different manufacturers and options for technology implementation, it cannot be unambiguously concluded what choice will be better. The paper examines the related work in the area and proposes an approach to form the basis for storing data on IS incidents. For this purpose, the authors formulate a block structure for including in a chain for later use, for example, in computer forensics.

Keywords: Blockchain · Information security incident · Internet of Things · Hashing

1 Introduction

Scientists are faced nowadays with several important tasks that are focused on assessing and testing existing approaches as well as optimizing and adjusting them for solving modern problems. Many years ago experts operated with data on one scale. For example, several tens or hundreds of gigabytes were considered a large amount of data. At present, there are Big Data, the Internet of Things (IoT) and other new technologies and concepts that supply daily thousands of gigabytes of data for processing. A huge amount of data is transmitted every day via networks regardless of its nature and degree of significance. In addition to inventing new approaches, an obvious and profitable solution is to adapt the methods already used to the needs of new technologies.

In case of incidents investigation in information systems, information security (IS) specialists have a wide range of tools and techniques for data extracting and analysis. But new methods require optimization for modern data scales. Therefore, today researchers are engaged in optimizing solutions, which are not suitable for solving, for example, Big Data problems, for several reasons. New technologies like the Blockchain (BC) and the IoT appear for this purpose.

A. V. Samsonovich et al. (Eds.): BICA 2020, AISC 1310, pp. 283–294, 2021.
https://doi.org/10.1007/978-3-030-65596-9_35

The BC is actively used, for example, in the financial sector. Blockchain Technology (BT) is discussed actively at seminars, conferences, and symposia and people constantly look for its new applications in business and other fields. Despite the long-formulated concept of the physical objects" computing network as the IoT, this phenomenon began to spread in our lives not so long ago. More and more users are actively using this technology to simplify their lives that entails an increase in many tasks to ensure IS of all devices connected to the IoT. The widespread use and development of this IoT allowed the allocation of a new application area, which was called Smart Home being focused on home devices. Nevertheless, the standards, measures, and approaches developed for it can be projected onto the IoT in general, taking into account changes in the scale of data and its features.

At present, all the examined sources prove the BT relevance. But there are no common standards or any kind of response mechanisms in case of IS incidents in the IoT. There is a contradiction between various IoT systems' existence and the absence of other data storage beside the cloud, which could be contacted to obtain data during the IS incident with no access to the cloud or devices' memory. For example, if the data was deleted from the device or this device itself was destroyed, and the cloud is unavailable during the IS incident, then the number of evidence sources is sharply narrowed. For the initial IS incident detection, the authors are going to develop a methodology using BT and a new approach to form blocks in block chains. The approach allows us to store data in a trusted unchanged distributed database, which will be available for further detection of IS incidents if required.

The goal of this research was to form a block structure, which would allow data on IS incidents in the IoT to be stored for subsequent use in digital investigations. The following tasks were solved to achieve this goal: to analyze related work in the selected field, to consider the block format in the chain, and to offer a convenient and safe recording format from the viewpoint of storing IS incident data. Hence, the paper is organized as follows. Section 2 is devoted to the related work in this field. Sections 3 and 4 present correspondently the technologies under study, their advantages and disadvantages. Section 5 reviews hashing, as well as compares algorithms to identify the priority for use. Section 6 consists of the phased data block formation, which is the main research result. The conclusion summarizes the findings and indicates further work directions.

2 Related Work

The authors of [1] are interested in the fact that information exchange and data authentication is carried out only through a central server, which leads to several serious issues. The paper discusses possible problems of security and confidentiality, in particular, taking into account the interaction of IoT components and explores how this is facilitated by BT based on a distributed ledger. The work of Russian colleagues [2] gives an example of accounting and maintaining transactions in the BC in an integral and chronological form for the case of its use in managing IS incidents in computer networks. In [3], intrusion detection technology is applied to a BC model. The results obtained show that the proposed model has higher detection efficiency and fault tolerance. The paper [4] targets a botnet IoT prevention system development, which takes advantage of both a software-defined network (SDN) and a distributed chain. The researches simulate stressful situations for the system and analyze it using the BC and SDN to determine how it is possible to detect and reduce the influence of the botnet and prevent devices from falling into the hands of attackers. The work [5] presents a study in the field of developing next-generation wireless sensor networks based on networks of a similar type based on the BC.

In [6], the IoT is considered in terms of IS. The paper discusses the main IS ensuring problems and how to solve them, as well as typical vulnerabilities, attacks against the IoT, and key areas for further development. The book [7] discusses how attackers can use such popular devices based on the IoT as wireless LED (Light-Emitting Diode) bulbs, electronic door locking devices, smart TVs, connected cars, etc., as well as their tactics. The author [8] presents IoT attack models with the counteraction principles, details the IoT devices' security, discusses new network security protocols, explores the issues of internal security (namely trust and authentication), and analyzes privacy protection schemes. The book [9] explains the IoT security concept from a theoretical and practical point of view, taking into account the end node resource limitations, the hybrid network architecture, communication protocols, and application characteristics.

In [10], the authors formulate and study the intelligent and safe house power supply system based on the IoT. The security of confidential data collected and transmitted by the sensor nodes installed in household appliances electrical devices is crucial because the transmitted data can be easily compromised via various types of attacks. Confidentiality and integrity of data on household electrical devices must be ensured for a proper and timely response. The authors believe that having a secure collection mechanism is very important to protect the integrity and confidentiality of data aggregation. The paper also talks about using a cryptographic hashing algorithm to ensure integrity. The work [11] looks at security and management due to the small capacity of small sensors, various applications' multiple Internet connections (using Big Data and cloud computing), and the home systems heterogeneity that requires inexperienced users to configure devices and microsystems. The Russian colleagues in [12] offer methods of filtering device parameters, with which various vulnerabilities can be eliminated. The advantages and disadvantages of the proposed approach are described. The paper [13] is devoted to wireless sensor network technology, which is increasingly used in areas such as smart homes and e-health. The authors note the sensors' shortcomings and express their concern about the devices' vulnerabilities, both physical and logical. It is proposed to use intrusion detection systems to eliminate threats. The authors discuss the design features of such systems, as well as important system requirements such as flexibility, efficiency, and adaptability to new situations. The main contribution of the work is to propose a framework for detecting intrusions in wireless sensor networks based on intelligent agents. The work [14] considers some devices and applications based on information and computer technologies to automate their functions and related actions that form Smart Home systems together. The goal of the study is to investigate examples of such devices and applications, their advantages and problems for users.

The paper [15] proposes an approach to building systems for analyzing IoT security incidents as a self-similar system within the normal operation of its constituent objects. An IoT graph model as a cyber-physical system was developed. It was used as a basis for the analysis of pair relations selected from a discrete stream of messages from devices that is sufficient to detect security incidents. The work [16] explores ensuring the IoT security, reducing the amount of data without losing information, guaranteed transfer of data to the Security Information and Event Management (SIEM) system using data aggregation, digital signature, and swarm routing algorithm. The authors of [17] propose an architecture for distributed parallel processing of Big Data based on the Hadoop and Spark software platforms paying attention to the limited computing capabilities of the IoT networks. The paper also discusses issues related to this system and the implementation of its main components. It poses a question on the need to analyze very large amounts of data in real-time with minimal computational costs.

The BT relevance today is noted in all the sources reviewed. But the problem is the absence of common standards or any kind of response mechanisms in case of IS incidents in the IoT. This leads to the fact that there are various IoT systems and no other data storage beside the cloud, which could be contacted to obtain data during the IS incident with no access to the cloud or devices' memory.

3 Blockchain

Since the advent of the Internet, this technology has been recognized as one of the most explosive innovations of the beginning of the 21st century. Gartner included it in the 2017 Top 10 strategic technological trends [18]. BT is currently used in financial activities for payments, currency exchange, money transfers, markets, investments, brokerage, insurance, etc., as well as in non-financial applications (e.g., certification systems, development, electronic voting, managing patient records, the IoT, etc.). It is noted that due to its properties to ensure data integrity, as well as various numerous checks, this technology can become one of the most important tools for auditors [19].

Some known practical applications of BT are the following. In the Ascribe project (https://www.ascribe.io, currently closed), it was possible to confirm and retain the copyright using BT. The Ascribe market has made it possible to create digital publications using unique identifiers and digital certificates to confirm authorship and authenticity. A mechanism to transfer ownership from an artist or author to a buyer or collector, including its legal aspects, was presented. Stampery, Verisart, Monegraph and Proof of Existence deal with the similar things nowadays. ShoCard and BlockVerify, as well as several other companies use distributed registry technology in solutions designed to identify and confirm access rights. Evernym is an international identity network created based on its own high-speed advanced distributed registry with rights separation designed to provide tools for controlling Personally Identifying Information (PII). The source code of the project is publically open. The BC is also involved in voting, for example, in Estonia. Voters' PII, their choice, geographical location, etc. are a subject to record. Information is subsequently used to verify election results. Chronicled is a San Francisco-based company that launched a promising BC platform for the IoT in August 2019, which is aimed at improving the consumer experience. As part of the project, the BC stores the identity data of physical items such as consumer goods and collectibles with built-in microchips. This allows creating safe and compatible with many other systems digital identifiers, which opens up opportunities for new mechanisms for interacting with the consumer based on tracking his proximity to the subject. The Chronicled project is licensed under the Apache, which is fully open source. In Georgia, anyone can find and receive an official statement on real estate using the BC. Startups like Obsidian use BTto securely exchange information in conversations, instant messengers and social networks. Unlike WhatsApp and iMessage, which use end-to-end encryption, Obsidian uses the BC to ensure the user metadata security such as mail, phone numbers or any other identifier. Obsidian uses random metadata from a distributed registry and thus guarantees the privacy of users and their messages.

The BC in this paper refers to the definition from [20]: "a special type of protected distributed data structure (a database) that maintains an expanding list of non-editable blocks/records and sets rules for working with transactions/events recorded in blocks and tied in such a way to them without a central administrator and centralized data storage (unlike regular databases where rules are often set for the entire database and applications)."

Such a database is shared by a group of users — nodes — entities in the BC network (transport level of the BC platform) that receive and process transactions and share information about a potential transaction. The node either proves (public BC) or verifies (hybrid/private BC) transactions and then adds them to the block with a unique hash code. Thus, there are two special types stand out among the nodes: miners and validators. Miners look for new transactions in the system (in the corresponding data sources if more precisely) and using cryptographic transformations prove that the transaction is real (valid) for its inclusion in the BC as part of new blocks using evidence such as proof of work/resource/state/activity, etc. Validators check series or individual transactions based on appropriate means of checking (for example, the Byzantine fault tolerance mechanism or "double costs") [2]. In this paper, the main attention of the technology review is on validators.

A block may include one or more transactions grouped by a specific criterion. The new block will be included in the BC based on one of the methods used to achieve consensus. For example, most nodes agree that all transactions in the chain are valid (legal, reasonable), and this correctly created block can be included in the BC. Determining the fact that a transaction has actually taken place is very important. The transaction couldn't be real just because anyone claims that it was. Transactions are signed and can be verified at any time using a pair of public and private keys [2]. A new block cannot be deleted or changed after it is agreed upon and included in the chain. Blocks are linked to each other (like a chain) in chronological order using timestamps. Thus, a public history of all transactions in any system is stored in blocks in a secure form [2]. The history of including blocks in the BC is also stored in a protected form. This history is used by all users of the BC. It is unchanged and verifiable to record transaction history with various protocols. There are four such protocols: a transaction protocol, a protocol for peer-to-peer nodes' communication with the same rights, a consensus protocol for discussion and agreement on relevant issues, and a data storage protocol for extracting and sending data to the database [2].

In [21], a transaction refers to a record of the movement of assets, for example, digital money, material assets, etc. between the interacting parties. A transaction can be an account verification event generated each time the money is deposited to or withdrawn from a current account. In the paper, a transaction is proposed to be understood as a change from the initial (final in the database) to the final (current) device's state with parameters specified in advance included in it. This mathematical expression will allow obtaining additional data for study for the IS incidents investigation. Each block in the BC may contain one or more transactions.

In this study, the BC, namely a private registry with closed access, is a convenient means of storing IS events in a network environment since it records and arranges them in strictly chronological order. Thus, the BC can be considered as a unidirectional data structure (chain) with a linked list of blocks, each of which refers to the previous one using hash codes-pointers stored in the block header. One block with a unique identifier and a "cryptographic" link to the previous valid block combines transactions — IS events belonging to the same IS incident (as its successive steps). Creating a repeating block (its contents is similar to one of the previous blocks) with a new identifier means the presence of repeating events in the network, characteristic, for example, of a Denial-of-Service (DoS) attack [2].

The data format for the IS event in the block can be aligned with one of the specifications for the IS incident description, which are currently used. There are the specifications of the MITRE Cyber Observable eXpression (CybOX), Structured Threat Information Expression (STIX), or Trusted Automated eXchange of Indicator Information (TAXII) developed in 2013.

The widespread and advertised by various manufacturers options for implementing BT do not yet have a single or a set of standards for incidents. It is also important to understand that the BT does not form the basis for the construction of an Automated Information System (AIS). It only allows creating its constituent element for ensuring the immutability of registers, in which specific actions are recorded (for example, concluding a contract with its most important details, purchase with an indication of the amount and product identifier, etc.). The paper [2] shows what can be written in the general case in a BC block except for its identifier and a hash code, which connects each new block with its predecessor.

4 Internet of Things

Today the IoT is increasingly filling up both business and everyday life. The IoT can be observed in cars, houses (Smart Home), hospitals, cities (Smart City), and so on. Many devices constantly receive data for further processing according to a huge number of criteria and analysis to solve a huge number of tasks. An important place on the IoT is occupied by information-gathering sensors (about the urban environment, human health, equipment status) – pressure, humidity, light, movement, heat flow, condition, etc. Thanks to wireless communication and various protocols, they can interact with each other and send the collected data for subsequent analysis by a person or artificial intelligence. Data centers, cloud technologies, and Big Data are used to store and process the received data. Unfortunately, the IoT still does not have a single standard or protocol, which could be adopted as the best solution for everyone. Therefore, each manufacturer chooses for himself the criteria of the system, as well as the methods of data transfer and storage.

After analyzing the related work, the following definition was accepted: the IoT is a network infrastructure with devices with unique identifiers, which are equipped with built-in technologies for interacting with each other or with the external environment that allows them to receive, collect and transmit data on the environment, in which they are located. An important IoT feature is the transparent conditions for working in surrounding networks, which prevents the monopoly of individual corporations - manufacturers of devices. As a result, the concept seeks accessibility and comprehensibility for the whole society without exception. Cooperation between developers, convenient integration of devices and standards and protocols development lead the community to what the IoT is ideally supposed to be used for the best interaction with a person and his health, the best options for solving problems based on available resources and real-time data from open sources, and the best synchronization of service providers and goods with the global capabilities of society as a whole.

Despite the IoT advantages, the concept has a lot of questions, which require the attention of scientists and work aimed at solving them. There are several issues and disadvantages of the IoT:

- Lack of generally accepted standards. The problem of integration is the lack of general rules; until there is no understanding of the features of the whole picture and general solidarity, the introduction of a universal solution will cause various difficulties;
- It is necessary to achieve network and energy autonomy from the IoT ecosystem to work properly. One of the key tasks is to obtain resources for the network and devices operability in case of main power sources' failure;
- Lack of pronounced privacy. The main risk of BT is an open database in the cloud. Scammers have the opportunity to hack not only accounts and computers but even refrigerators and coffee grinders. With the necessary skills or resources, the surrounding reality becomes potentially dangerous and in the absence of proper generally accepted standards for IS ensuring users may remain unaware of the fact of devices malicious use, which puts their information and even life at risk;
- Cost. Technological decisions are expensive although such use will pay off in the future: the Smart Home system will help save electricity and water supply, equipment in the factory will notify of the damage risk in advance, kitchen appliances will avoid food spoilage, etc. But not everyone at the moment can afford it.

All of the problems above must be taken into account for the devices' correct operation, which affects the IS controls implemented by one or another manufacturer. The autonomy of all "things" is necessary for the full functioning of such a network; sensors must learn how to receive energy from the environment and not work on batteries as is happens now.

5 Hashing

For the initial verification of IS incidents, the authors propose the use of hashing. It refers to a conversion carried out by a hash function, which is a function that maps strings of bits to fixed-length strings of bits, and it is difficult to calculate the initial data displayed in this value for a given function value, other input data displayed in the same function value for given input data, and any pair of input data mapped to the same value [22]. When applying this method, several variants of hashing algorithms are considered to provide freedom of choice for the experiment implementation. Sending messages with data in the form of hash codes makes it difficult to access them. The presence of the same equipment on the IoT network nodes allows hashing at the verification point without requesting extraneous data, which allows increasing the quality of the verification performed. When an attacker receives a hash code, there will be a problem in selecting a hashing algorithm and decrypting the contents of the message. Comparing to transferring open data to the cloud from network devices, as it is currently accepted on the IoT, it is safer. In the course of the study, several options for hashing algorithms were considered.

Cyclic Redundancy Check (CRC) [23] is an insecure hash function designed to detect random changes in raw computer data that is commonly used in digital networks and storage devices such as hard drives. A CRC-enabled device calculates a short fixed-length binary sequence known as a CRC code or simply CRC for each data block and sends or stores them together. When a block is read or received, the device repeats the calculation. If the new CRC does not match the previously calculated one, then the block contains a data error, and the device can take corrective actions such as recalculation or block resend requests. CRCs are called like that because the verification code (data) is redundant (it adds zero information), and the algorithm is based on cyclic codes. A cyclic code is a code, which has the property of cyclicity. It means each cyclic permutation of a codeword is also a codeword. Such codes are used to convert information to protect it from errors. The term CRC can refer to a control code or a function calculating it, which receives data streams of any length as input but always outputs a fixed-length code. CRC algorithms are popular because they are easy to implement in binary equipment and to analyze mathematically, as well as they are good to identify common errors caused by noise in the transmission channels. Since the test value has a fixed length, the function, which generates it, is sometimes used as a hash function.

MD5 (Message Digest 5) is a 128-bit hash algorithm useful for encoding passwords, credit card numbers, and other important data in MySQL, Postgress, or other databases. MD5 hash code is created by taking a string of any length and encoding it into a 128-bit phrase. Encoding the same string using the MD5 algorithm always results in the same 128-bit hash output. MD5 hashes are usually used with small lines. They are also used to ensure file integrity. Since the MD5 hash algorithm always produces the same output for the same given input, users can compare the hash of the source file with the newly created hash of the destination file to ensure that it is not corrupted or modified. However, this property makes it vulnerable to some attacks [24]. For example, when trying to hack, it is possible to create two messages with the same hash, which can give an attacker extra information. Therefore, its use is not recommended in new projects.

SHA-3 (Keccak) is a variable-bit hashing algorithm approved as the FIPS 202 standard. It is considered as one of the most reliable hashing algorithms nowadays [25]. Hash is not a technology for encrypting data from the classical point of view. This makes it impossible to decrypt data in the opposite direction. This is one-way encryption for any amount of data. All SHA algorithms are based on the Merkle–Damgard method: the data is divided into uniform groups, each of which passes through a one-way compression function. As a result, the data length decreases. This method has two significant advantages, namely fast encryption speed and almost impossible decryption without keys and minimal collisions risk (identical images).

Having considered all the hashing algorithms presented above, the authors decided to choose the SHA-512 algorithm as the most reliable for the research.

6 Block Structure

The main attention in this study is given to the contradiction that there is no common base when transmitting data to the cloud from IoT devices, where data on IS events would be brought in and which would be independent of the cloud. Such a database will be useful in case of an IS incident because it allows accessing the history of the IoT infrastructure immediately, without a request to the cloud until it will not be required. It is proposed to use a distributed registry or, in other words, BT to create such a database. This database will not require centralized storage servers, which can be hacked or disconnected from the network as the database is simultaneously stored on all nodes of the infrastructure. As soon as one of the nodes forms and adds a block with data, the remaining nodes learn about it. Therefore, if the database is unavailable on one node, access to information will still be possible due to access from other points of the network. The database will be convenient because it supports the integrity of entered data due to the BT features.

When using BT the main question is what will be written to the storage unit. Indicators in the “raw” form still do not facilitate the work of IS specialists. Indicators grouped by some characteristic also do not bring clarity. Therefore, the authors propose to record the difference between the states of the device as a transaction. Under the state in this context the researchers mean the complex of all parameters, which characterizes this device. Because not one state will be recorded but the difference between the last and penultimate, it becomes possible to observe “jumps” in the device operation and to sensitively monitor possible changes in indicators. Moreover, the difference of states does not consist in the direct subtraction of one parameter from another. It is the established distinction in the record of such transactions. Thus, one of the fields in the chain block will be the expression *[(initial state) (conditional sign) (final state)]*. The type of mathematical operation between states can be determined later. The main idea is to formulate all expressions for a particular device in this form.

When transferring data from one IoT node to another, the task is to ensure its IS. The transmission of indicators in an open form will allow the attackers to obtain statistics on the devices used, operating modes, user preferences, etc. Therefore, open transmission is not suitable. It is necessary to transfer the devices indicators in such a way that they cannot be compromised along the way but the system or device at the other end could use them. The authors propose to use hashing for this. The messages transmission in the form of hash codes allows increasing the security of such messages. It is necessary to specify a unified encryption algorithm for network nodes. The system will be able to interact with it when receiving hashes from trusted nodes.

If there is information on another node about how the states' difference was recorded, what hashing algorithm was used for protection and for which device (which is present on another node) this was done, another system can compose possible device states to receive many hash codes from these conditions. The system does not know the content of the primary difference of states but it can verify whether such a difference is valid or not from the generated options by comparing the hash codes. In this case the number of comparisons plays an important role since in the future it can be used as a parameter for checking the correctness of the performed verification and its legality in the system.

The data in a hashed form is returned after verification to the first network node where a block of the following format is formed: *Block = [(difference of states) (hash code of the device identifier) (hash code of verification) (timestamp) (hash code of the previous block)]*.

The device identifier hash code is needed in the unit for initial checks in case of an IS incident. Timestamps are used for this in addition to the identifier hash code. For example, there is a record in the database about turning on the coffee machine, and its work at night in the absence of people before and after its work process. At the same time a connection to the device was observed. This event gives an auditor a reason to doubt the correctness of this action and check whether the event is an incident. The verification hash code is needed to clarify the fact that the verification of the state difference was not compromised. If there is a difference in the checks hash codes on the nodes that have been documented as participants in the check, then it is also necessary to check all the data since a mismatch is a sign of an IS violation. The number of hash codes comparisons from two different nodes is important for checking the checks since in each case it will be different. A coincidence with this number will allow asserting that this is exactly the node and the check if other parameters are also correct. The previous block hash code is necessary in the chain for blocks communication, as well as to track the correctness of changes in device indicators.

It becomes much easier to monitor any changes and make decisions on IS incidents by storing data on what is happening in the IoT in this way. Moreover, such records are difficult to compromise due to the peculiarities of the formation of chains. The history of all devices and systems states becomes available and transparent to trusted nodes. Encoding data using hashing allows the transfer of information as confidential as possible.

7 Conclusion

The authors examined the BC concept and proposed to use BT together with hashing to form an unchanged database of devices and systems, which can be used further in the digital investigation of IoT IS incidents, for example, in computer forensics. An idea of creating a database, the entities in which will be trusted and easy to use, was applied. A block structure that allows storing data on IoT devices and systems for future detailed digital investigations was proposed. Based on the related work's analysis, the necessary fields of the block were determined, and the need for each of them was justified.

The authors are going to conduct virtual and physical experiments with IoT devices in the future. For virtual experiments, a study of various software products, which can implement the required test situations, as well as confirm the operability of the structure in practice, will be conducted. The main goal will be to formulate traceable device parameters for further states' classification and unification. After that, a laboratory testbed allowing creating various combinations of IoT devices will be assembled, and the real picture with the results obtained during the virtual experiment will be compared.

Acknowledgement. This work was supported by the MEPhI Academic Excellence Project (agreement with the Ministry of Education and Science of the Russian Federation of August 27, 2013, project no. 02.a03.21.0005) and by the Russian Foundation for Basic Research (project no. 18-07-00088).

References

1. Kumar, N.M., Mallick, P.K.: BTfor security issues and challenges in IoT. Procedia Comput. Sci. **132**, pp. 1815–1823 (2018)
2. Budzko, V.I., Miloslavskaya, N.G.: Issues of practical appication of blockchain technology. Inform. Technol. Secur. **26**(1), 36–45 (2019). (in Russian)
3. Li, D., et al.: Information security model of block chain based on intrusion sensing in the IoT environment. Cluster Comput. **22**, 1–18 (2018)
4. Shafi, Q., Basit, A.: DDoS botnet prevention using blockchain in software defined internet of things. In: 2019 16th International Bhurban Conference on Applied Sciences and Technology (IBCAST), pp. 624–628. IEEE (2019)
5. Buldin, I.D., et al.: Next generation industrial blockchain-based wireless sensor networks. In: 2018 Wave Electronics and its Application in Information and Telecommunication Systems (WECONF), pp. 1–5. IEEE (2018)
6. Miloslavskaya, N., Tolstoy, A.: Internet of things: information security challenges and solutions. Cluster Comput. **22**(1), 103–119 (2019). https://doi.org/10.1007/s10586-018-2823-6
7. Dhanjani, N.: Abusing the Internet of Things: Blackouts, Freakouts, and Stakeouts. O'Reilly Media, Sebastopol, p. 296 (2015)
8. Hu, F.: Security and Privacy in Internet of Things (IoTs): Models, Algorithms, and Implementations. CRC Press, Boca Raton, p. 604 (2016)
9. Shancang, L., Li Da, X.: Securing the Internet of Things. Elsevier, Chennai, p. 154 (2017)

10. Mbarek, B., Meddeb, A., Ben Jaballah, W., Mosbah, M.: A secure electric energy management in smart home. Int. J. Commun. Syst. **30**(17), p. e3347 (2017)
11. Batalla, J.M., Vasilakos, A., Gajewski, M.: Secure smart homes: opportunities and challenges. ACM Comput. Surv. (CSUR) **50**(5), 75 (2017)
12. Bondarev, S.E., Prokhorov, A.S.: Analysis of internal threats of the system smart home and assessment of ways to prevent them. In: 2017 IEEE Conference of Russian Young Researchers in Electrical and Electronic Engineering (EIConRus), pp. 788–790. IEEE (2017)
13. Pires, H., et al.: A framework for agent-based intrusion detection in wireless sensor networks. In: Proceedings of the Second International Conference on Internet of things, Data and Cloud Computing, pp. 188. ACM (2017)
14. Gazzawe, F., Lock, R.: Smart Home. In: Science and Information Conference, pp. 1086–1097. Springer, Cham (2018)
15. Lavrova, D.S.: An approach to developing the SIEM system for the Internet of Things. Autom. Control Comput. Sci. **50**(8), 673–681 (2016)
16. Zegzhda, P., et al.: Safe integration of SIEM systems with internet of things: data aggregation, integrity control, and bioinspired safe routing. In: Proceedings of the 9th International Conference on Security of Information and Networks, pp. 81–87. ACM (2016)
17. Kotenko, I.V., Saenko, I., Kushnerevich, A.: Parallel big data processing system for security monitoring in Internet of Things networks. JoWUA **8**(4), 60–74 (2017)
18. Gartner. Top 10 Strategic Technology Trends for 2017: Blockchain and Distributed Ledgers. https://www.gartner.com/doc/ 3647619/top–strategic-technology-trends/. Accessed 13 March 2020
19. Forbes Insights. For Auditors, Blockchain Has Blockbuster Potential. https://www.forbes.com/sites/insights-kpmg/2018/09/19/for-auditors-blockchain-has-blockbuster-potential/#188f2adb6cb6. Accessed 13 March 2020
20. Miloslavskaya, N., Tolstoy, A., Budzko, V., Das, M.: Blockchain application for IoT security management. Chapter 7. In: Essentials of Blockchain Technology, pp. 141–168. CRC Press, Taylor & Francis Group, USA (2019)
21. Yaga, D., Mell, P., Roby, N., Scarfone, K.: NISTIR 8202 Blockchain Technology Overview (2018). https://nvlpubs.nist.gov/ nistpubs/ir/2018/NIST.IR.8202.pdf. Accessed 13 March 2020
22. GOST R 34.11–2012. Information Technology. Cryptographic Information Protection. Hash function. http://docs.cntd.ru/document/gost-r-34-11-2012. Accessed 13 March 2020. (in Russian)
23. Miller, F.P., Vandome, A.F., McBrewster, J.: Cyclic Redundancy Check. VDM Publishing, Riga, p. 72 (2009)
24. Cryptowiki. MD5. http://cryptowiki.net/index.php?title=MD5. Accessed 13 March 2020
25. SHA-3 Standard: Permutation-Based Hash and Extendable-Output Functions. FIPS PUB 202. https://nvlpubs.nist.gov/ nistpubs/FIPS/NIST.FIPS.202.pdf. Accessed 13 March 2020

Cyber Polygon Site Project in the Framework of the MEPhI Network Security Intelligence Center

Natalia Miloslavskaya(✉) and Alexander Tolstoy

The National Research Nuclear University MEPhI (Moscow Engineering Physics Institute), 31 Kashirskoye Shosse, Moscow, Russia
{NGMiloslavskaya, AITolstoj}@mephi.ru

Abstract. At present, the market for information protection tools (IPTs) is much wider than a couple of years ago. But not only technology protects and carries a threat. People are still at the forefront as the most common cause of errors is the lack of experience and low competency. The only right solution is the creation of cyber polygons as specially equipped and controlled network infrastructures for developing practical skills to combat information security (IS) threats. The National Research Nuclear University MEPhI (Moscow Engineering Physics Institute) could not remain aloof from this process as the leading institute for IS training in Russia. Therefore, it was decided to create such a cyber polygon within the framework of the educational and research Network Security Intelligence Center (NSIC) for intelligent network security management established at the MEPhI Institute of Cyber Intelligence Systems in 2016. The paper describes the first results achieved in making this project a reality. It introduces the "Cyber Polygon" term, briefly analyzes a state of the current cyber polygons development worldwide, and introduces the MEPhI Cyber Polygon objectives and provision to be used within the framework of the "Business Continuity and Information Security Maintenance" Master's degree programme. Further activities in its development conclude the paper.

Keywords: Cyber polygon · Network Security Intelligence Center · Practical skills · Information security threat · Information security training · Master's degree programme

1 Introduction

At present, the market for information protection tools (IPTs) is much wider than a couple of years ago: one can buy many products of different price ranges, from anti-viruses and firewalls to cybersecurity operation centers based on artificial intelligence, that allows to correlate information sources and predict how the probability of an information security (IS) event occurs and the consequences of each decision and step. These things together bring the work of IS experts to an unprecedentedly high level of reaction to an event that can only occur but has not yet occurred, while at the same time reducing the risk of the human factor to zero.

A. V. Samsonovich et al. (Eds.): BICA 2020, AISC 1310, pp. 295–308, 2021.
https://doi.org/10.1007/978-3-030-65596-9_36

But not only technology protects and carries a threat. The current problem of ensuring cybersecurity is not so much in the technological as in the personnel field. People are still at the forefront; they are known to be mistaken. The most common cause of errors is the lack of experience and low competency, while the whole business can be the price of an error. What could be done? Worthy experts cannot be taken from a school or college. Professionals in the market are expensive; there are few of them. Even if the company was lucky enough to find such a specialist, this does not mean that he will know how to deal with one attack having repelled another one. This problem is not new. The only right solution is the creation of its own "cousin of IS officers" called a "cyber polygon" (also known as a cyber range) as a specially equipped and controlled network infrastructure for developing practical skills to combat IS threats. For IS employees, it is the best tool for gaining invaluable professional experience without risking its infrastructure. For business owners, it is the best source of trained IS professional personnel, ready to face any IS threat.

The analysis of the state of IS training of the Russian Infotecs company says that the average indicators in Russia are as follows: 10% have a basic education in the field of IS, 16% have completed professional retraining courses, and 74% (3/4!) do not have an IS education. To remain competitive in the global cyberspace in the future, Russia needs over 20 thousand IS specialists. At the same time, the organization of such training on existing information and telecommunication systems and networks (for example, on the real Internet) where it is impossible to directly study the detection tools and prevent stew attacks is objectively difficult. It is hard to imagine classes with students on the analysis of computer viruses' spreading technologies on the Internet! In turn, within the framework of traditional computer classes, the use of technologies far from reality can ultimately negatively affect the final level of students' knowledge and skills. In this regard, for the success of the educational process, it is advisable to create a specialized environment – a cyber polygon, which would help to study the features of various computer attacks and to protect against them with the highest possible realism and at the same time safely.

The National Research Nuclear University MEPhI (Moscow Engineering Physics Institute) as the leading institute for IS training in Russia could not remain aloof from the process of creating cyber polygons. Therefore, it was decided to create such a cyber polygon within the framework of the educational and research Network Security Intelligence Center (NSIC) for intelligent network security management, established at the MEPhI Institute of Cyber Intelligence Systems in 2016 [1, 2]. The paper describes the first results achieved in making this project a reality. Further activities in its development conclude the paper.

2 Related Work

To analyze the related work, it is necessary to pay attention to several areas of research, which eventually led to the idea of creating cyber polygons. More than 20 years ago, the idea came up of using sandboxes at the universities for laboratory work on network security, which was first discussed at the first international conferences on the training of information security (WISEs). Here we mention only the pioneers like [3–6], followed by many lab descriptions worldwide in the next years. These works provide useful instructions in designing labs and developing an educational process for them. The authors have their own experience in setting the "Network Security" laboratory in 2000 [7] and continuing nowadays [8]. Some publications show how to use virtualization in education (like [9]), as it has all the necessary tools to build complicated virtual networks on virtual machines and to show the IPTs' functionality.

The logical development of these ideas is the cyber polygon concept. Currently, the publications describe two options for their use: single-time events at conferences or functioning on an ongoing basis in the educational process of various universities and training centers. The following few examples from around the world and Russia, which are described in the open press, illustrate these use cases.

In 2014, the NATO Defense Forces approved the creation of the military Cyber-Polygon-Base in Tallinn (Estonia), which includes the advanced cyber-laboratory [10]. The polygon was established on an already existing base, which was constructed in Estonia in 2012, where the NATO Cyber Forces already conducted the "Cyber Coalition" and "Locked Shields" cyber exercises in 2013.

The National technical cybersecurity exercise Cyber Czech 2015, organized by the National Security Authority in collaboration with the Institute of Computer Science at Masaryk University in Brno (ICS), took place in the ICS Cyber polygon [11]. The preparation of a scenario reflecting simple and sophisticated DDoS attacks in a special closed environment has lasted for 9 months. The participants divided into 5 teams were from the key Czech ministries and other authorities. For six hours, they could experience the defense of a simulated significant power plant network against the cyber attacks. Their task was to respond to attacks and technical issues, as well as to evaluate potential legal and media impacts, and to share information with other teams. It was the first test of the entire Cyber polygon to be used not only by universities but also by public institutions and private companies as a unique physical and virtual environment.

As for the USA, dozens of US universities offer cyber services feature with certified trainers to support cyber training and product testing requirements. Here it is only a few examples: the Institute of Cybersecurity at the Regent University (USA) [https://discover.cyber-institute.com], the Virginia Cyber Range at the Virginia Tech University [https://www.virginiacyberrange.org], the Ohio Cyber Range of the University of Cincinnati's School of IT [https://www.uc.edu/news/articles/2019/06/n20840624.html], the Cyber Range Hub at the Wayne State University [https://wayne.edu/educationaloutreach/cyber-range], the Maine Cyber Range at the University of Maine [https://www.uma.edu/academics/programs/cybersecurity/maine-cyber-range], the Cyber Range Poulsbo at the Western Washington University [https://cse.wwu.edu/computer-science/cyber-range-poulsbo], the Texas Cyber Range at the Texas A&M University [https://cybersecurity.tamu.edu/education/tcr], the Florida Cyber Range as a collaboration between the University of West Florida Center for Cybersecurity and Metova CyberCENTS [https://floridacyberrange.org], and many, many others. The X-Force Command [https://www.ibm.com/security/services/xforce-command-special-forces-team], the SANS [https://www.sans.org/webcasts/cyber-ranges-108845], the Cyberbit [https://www.cyberbit.com], the Cyber Ranges of the Palo Alto [https://www.paloaltonetworks.com/solutions/initiatives/cyberrange-overview] can be mentioned as the successful examples in the companies' sector.

Within the II International Congress on Cybersecurity (ICC), organized by Sberbank with the support of the Association of Banks of Russia and the Digital Economy Autonomous Nonprofit Organization, the first Cyber Polygon online training [http://www.cyberpolygon.com] was held for 3 h [12]. Its objective was to establish coordination between the public and private sectors. Three large-scale scenarios were worked out for a network of airports and airlines, as well as passenger devices of these companies. In the first one, a massive DDoS attack was emulated against all participants. The second scenario involved web-based application attacks, namely web injections. The third scenario was phishing for delivering malware. Each scenario was played out twice. In the first run, participants had to respond to the attack independently without cooperation. In the second run, participants were given access to the BI.ZONE ThreatVision data exchange platform, where they could upload cyber threat data obtained during the attack and then automatically distributed to all teams. The results of the exercise demonstrate that data exchange among companies dramatically improves cyber sustainability (more than seven times) and shields companies from threats they couldn't handle when uniting their efforts. For the first time, measures to counter such attacks by teams located in different countries were tested according to uniform standards. After the exercises, an understanding of the need for interconnected actions has appeared. The attention to this training was very large: more than 230 companies connected to the live broadcast, including representatives from the World Economic Forum, INTERPOL and others, and the audience ranged from 10 to 12 million people around the world. Next training is assigned for July 8, 2020.

The very small Cyber-Polygon "EngEkon" Laboratory at the Almaty Management University's School of Economics and Logistics is sponsored by the JSC "Kazakhstan GIS Center" and the Ministry of Defense of the Republic of Kazakhstan [https://nl.qwertyu.wiki/wiki/Almaty_Management_University]. The purpose of this site is to teach students to counteract cyber terrorism, as well as to prevent database hacking and data leakage. It is equipped with only 4 displays, 7 computers, a projector, and an audio system. Two opposing teams are involved in the simulation: the attackers in the lab's red corner are trying to crack the systems, while the defenders in the green zone are using IPTs to isolate the attack. The blue zone is designed for observers where the overall picture of the battle is visualized.

In Russia, the creation of several cyber polygons will be carried out since 2010 within the federal "Information Security" project of the national "Digital Economy" program [13–15]. The first Russian world-class digital center with a cyber polygon is formed by the Far Eastern Federal University (FEFU) on the Russky Island near Vladivostok, 6428 km from Moscow. This center, which will be created by the order of the President of Russia, is aimed at the development of talents and practical IS training, as well as the pilot implementation and testing of innovative projects in partnership with the Russian state corporations and leading IT companies. MTS, Yandex, Rostelecom, and other large companies will work in this center. Rostelecom intended to provide the infrastructure for simulating computer attacks, testing security software, conducting cyber exercises, competitions, and training. MTS was interested in the center as a platform for data analytics for transport and logistics. Yandex intended to develop unmanned vehicles.

The Ministry of Digital Development, Telecom and Mass Communications of the Russian Federation (RF), the Ministry of Internal Affairs of the RF, the Ministry of Science and Higher Education of the RF, the Federal Service for Technical and Export Control (FSTEC) of Russia, the Federal Security Service of Russia, the Central Bank, the Savings Bank of Russia, the Skolkovo Foundation, BI.ZONE LLC, AltEl LLC and others are also creating their cyber polygons for training specialists, experts of a different profile and managers in the field of IS and IT modern security practices by modeling of the computer attacks and working off of reactions to them.

The creation of cyber polygons will be subsidized from the Russian state budget. The rules for granting subsidies on a competitive basis are published on regulation.gov.ru. To receive a subsidy, a company must have the computing infrastructure to create and maintain its cyber polygon, as well as experience in creating centers for monitoring and responding to IS incidents. The company must have experience in working with organizations of higher professional education and practical training in IS. It must have an agreement of intent on cooperation and interaction within the framework of master's degree programs in IS with FEFU, from the territory of which the created cyber polygons can be accessed. The company must be licensed by FSTEC to protect confidential information and develop IPTs. Finally, it must have a cooperation agreement with the National Computer Incident Focal Point. The subsidized company can spend 5% of these funds on the remuneration of its employees, 25% on the purchase of works and services from third parties, and the rest on the purchase and rental of software and hardware.

Of course, this very brief survey can be broadened by other examples.

3 Cyber Polygon Definition

In our research, based on the related work analysis, we define a Cyber Polygon as a specially equipped and controlled network infrastructure for online cyber training sessions on ways security professionals can engage in national or even international cooperation to combat IS threats. Figuratively speaking, the cyber polygon is a "virtual country" that repeats the infrastructure of companies in various industries. Thus, cyber polygons are mainly designed to develop practical skills in identifying and responding to IS incidents via cyber exercises, as well as to conduct stress tests of hardware and software in various industries.

Today, technology allows one to create a complete emulation of any infrastructure, from a small office to a space ship flight control center or even a smart city, to fill this infrastructure with real network devices and tools of any manufacturers and developers for monitoring, control and protection, and to enclose it all in an isolated "vacuum" and "freeze" it in the "all-is-well" state. It typically includes all necessary hardware and software for simulating some of the most common types of attacks and training of participating parties, as well as multiple geographically distributed sites for simulating remote access to the cyber polygon. Modern cyber polygons can also be created using cloud technologies.

Cyber polygons allow the instructors to create for each student its own separate virtual laboratory environment, which can include tens or even thousands of virtual network resources. In this environment, the student gets the opportunity to complete regardless of colleagues his tasks, to save his results from lesson to lesson, and to carry out additional preparation for them during his independent work.

This is not just a "sandbox"; it is a cyber polygon for training future IS officers who are ready to repel an attack of any complexity. All sorts of viruses and various attack scenarios can be launched in the cyber polygon, which future or current IS experts will have to fend off with the available set of IPTs. Like chess players, IS experts can at first understand and follow all the steps of well-known attack scenarios and try to repel them in different ways, and only then to run arbitrary random scenarios, gaining experience and understanding of the most effective actions and tools of dealing with them. This approach will allow not only to grow from a theoretician into a real expert but also to maintain a high level of preparedness for meeting new urgent IS threats. Such actions will help future IS specialists to form a system-integrated vision of IS problems, to get an idea of how common IS threats are implemented by attackers, and to learn how to effectively counteract them. The level of their success can be monitored by observers in real-time via a web site, specially designed for these purposes. After the "hostilities" end, the cyber polygon can be returned to its original state and will be ready for a new launch.

Hence, the cyber polygon arms the students with skills, problem-solving, and critical thinking capabilities for ensuring network security in the best possible way. Besides, it is also possible to conduct IS competitions and create centers for testing software and hardware security, including IPTs, at cyber polygons.

4 MEPhI Network Security Intelligence Center as a Framework for Cyber Polygon Developing

As modern businesses face numerous targeted network attacks, the well-known "Detect-Investigate-Respond" triad must be added by the fourth "Adapt" critical component, which should be based on Security Intelligence (SI) approaches. For this purpose, we have already started in 2016 to build at MEPhI a Network Security Intelligence Center (NSIC) as a combination of SI Center (SIC) and Network Operations Center (NOC) with their unique and joint toolkits and techniques [1, 2]. NSICs empower the autonomy of network security management within one organization, change the IS ensuring model from reactive to proactive, support more effective responses to IS incidents, enhance communications between the network and security teams, management and board members, drive IS investment strategies, and more directly connect IS priorities with business risk management priorities. Therefore, on the basis of this particular center, it is necessary to develop the MEPhI Cyber Polygon.

In the infrastructure of the NSIC, the Cyber Polygon is the fourth virtual one, combining the three already established laboratories [2], and interacting with the Center's Network Security Management tools (Fig. 1). The first laboratory, as the core, has the Palo Alto Next-Generation Firewall (NGFW). The second laboratory is based on the Positive Technologies' MaxPatrol Security Information and Event Management (SIEM) system. A third lab is built around the SearchInform's Data Loss and Prevention (DLP) system.

Fig. 1. MEPhI Cyber Polygon in the NSIC infrastructure

Continuing our research on designing NSICs [1, 2, 16], we allocate five zones in the Cyber Polygon's security zone infrastructure, namely untrusted (not included into the NSIC itself), demilitarized (semi-trusted) (DMZ), trusted, restricted and management with their specific subzones and special sandboxing component (Fig. 2). This approach separates coming from outside suspicious untrusted software from unverified third parties, users, web sited and so on and will allow detecting previously unseen IS threats and alert Cyber Polygon's staff who can quickly take action.

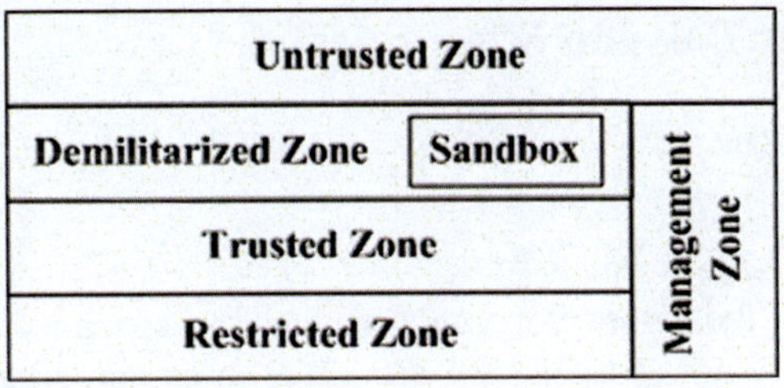

Fig. 2. Cyber Polygon's security zone infrastructure

DMZ is a physical or logical Cyber Polygon's forefront subnetwork for external-facing servers and services that are externally shared by the NSIC and MEPhI intranet as well as remote and infra services (like Domain Name System (DNS), Simple Mail Transfer Protocol (SMTP), etc.). Its purpose is to separate the Cyber Polygon from the Untrusted Zone and, hence, to add a layer of security. Users from Untrusted Zone only have direct access to the DMZ, but not to any other "deeper" NSIC's part.

Trusted Zone with the controlled environment after the internal back-end NGFW is designed for internal-exposed systems (like internal load balancing/performance optimization, device testing, and troubleshooting platforms, application services and management, services with limited access to Cyber Polygon's staff only, etc.).

Restricted Zone contains Cyber Polygon's high risk and/or mission-critical systems (critical services and servers, Data Center with a knowledge base and other DBs with sensitive data, etc.).

Management Zone with Cyber Polygon's assets such as infrastructure services, network devices, traffic telemetry, storage, Data Center with certain computational power and applications supports its centralized functioning, big data processing, virtualization, configuration, changes, patch, backups and IS management, including IS centralized real-time logging, monitoring, reporting, IPTs like NGFW, SIEM and DLP systems, security scanners, analytical tools, regulatory compliance, etc.

We propose to design several subzones within trusted and restricted security zones that enable special cases, listed above under the description of their functional components and additional subzone boundary control.

Sandbox can be regarded as an isolated highly controlled quarantine (a specific example of virtualization), which is used to analyze unknown users' behavior, activity in the CPU of sandboxed hosts and its execution flow at the assembly code level or unverified programs with possible viruses, malicious code or targeted attacks evaded traditional signature-based defense. The main aim is to see if something triggers anything beyond what was normally expected. Hence, in the Cyber Polygon, sandbox appliances support a safe and sheltered environment, where the behavior of processes, programs, memory, files, file descriptors, file system space, unknown users, etc., can be carefully and constantly observed without compromising the entire Cyber Polygon.

Bearing in mind this infrastructure of the Cyber Polygon, we will further describe the site design at MEPhI for its implementation for educational purposes.

5 MEPhI Cyber Polygon Site Project

The objectives of the currently created site of the MEPhI Cyber Polygon are defined as follows:

- Practical training of IS specialists;
- Visual monitoring the Cyber Polygon's information space to identify internal and external threats and assess their level according to certain criteria;
- Analysis and operational exchange of information about new IS threats, timely receipt of information by teams for decision-making;
- Developing scenarios of IS threats and working out approaches to neutralizing and actively countering them;
- Conducting scientific research of software and hardware attacks on the Cyber Polygon's educational facilities and identifying IPTs' vulnerabilities in its network environment;
- Development of proposals for the adoption of effective managerial decisions and the development of interactions with other organizations to ensure IS adaptive to the situation.

On this basis, let us formulate the tasks solved by the MEPhI Cyber Polygon:

- Development of skills to detect computer attacks;
- Development of skills for investigating IS incidents;
- Development of a skill in assessing IS of networks' elements;
- Development of interaction between departments;

- Development of guidelines for the neutralization of computer attacks;
- Development of preventive measures for computer attacks.

Thus, students who have completed practical training at the MEPhI Cyber Polygon will acquire Hard Skills, namely knowledge of attack and defense technologies and knowledge and ability to use the necessary IPTs and other tools, as well as Soft Skills, including teamwork skills, the ability to interact with other members of the response and monitoring teams and time management skills.

The MEPhI Cyber Polygon site is planned to be used within a framework of the "Business Continuity and Information Security Maintenance" Master's degree programme (the "Information Security" training direction) for the following disciplines: "Secure Information Systems", "Objects' Information Security Technologies Maintenance, "Disaster Tolerance of Information Systems", "Fundamentals of Risk Management", "Fundamentals of Incident Management", "Assessment of Information Technology Security" and "Information Security Management". Within these disciplines, time will be allocated for conducting comprehensive group exercises designed for a group of students of up to 16 people.

At present, training scenarios are being prepared, as well as the methodological base. During the exercises, it is supposed to divide the initial group into two groups of 8 people each, in which subgroups of from 2 to 4 people are distinguished. Each training scenario involves the participation of two groups of students representing different user roles (Fig. 3): a group for the preparation and conduct of the assessment of the network security and monitoring attacks (Monitoring Team) and a group of the rapid response to IS incidents and the elimination of their consequences (Response Team).

Fig. 3. Training participants

Assessment of students in the above disciplines at the MEPhI Cyber Polygon should be comprehensive (integral) and objectively reflect the actual level of formation of the relevant competencies. It can consist of several components:

- Assessment of the implementation of practical tasks;
- Assessment of the completeness and significance of identified vulnerabilities in the protection of information at the selected object to be protected;
- Assessment of the completeness and quality of proposals to eliminate IS threats and reduce IS risks;
- Assessment of passing tests and quizzes;
- Estimation of time to complete tasks and work in general;
- Assessment of learning to work in a team;
- Assessment of search skills and analytical work with information, etc.

At present, the hardware and software base of the MEPhI Cyber Polygon site are being developed to create training virtual emulation stands for two objects to be protected:

1. Enterprise's information and telecommunication network (ITCN), including an industrial control system model, an emulator for mobile platforms, and a set of vulnerable applications based on iOS and Android operating systems (Oss);
2. ISP infrastructure.

The basis of the MEPhI Cyber Polygon will be a stand emulating ITCN of a typical organization and including the following set of components:

- Client OSs (Windows, Linux);
- Server OSs (Windows Server, Linux, Unix) with deployed standard secure services (Web, FTP, Database, SSH, etc.);
- Firewall for OS protection (iptables, Checkpoint, etc.);
- Next Generation Firewall (Palo Alto, etc.);
- Security Information and Event Management (SIEM) system Splunk, RUSIEM, etc.);
- Host- and network-based Intrusion Detection and Prevention systems (IDPSs);
- OS and service security analysis tools (OpenVas, Nessus, MaxPatrol, etc.);
- Tools for simulating the network topology of an Internet service provider's (ISP) infrastructure (GNS3, UnetLab, etc.);
- Tools for setting a secure connection between hosts of a protected computer network (SSH, OpenVPN, Vipnet Coordinator, etc.);
- User computers with predefined penetration test tools (Kali Linux).

The elements of the stand will be connected in a single topology corresponding to the general structure of the organization's ITCN, implemented in the MEPhI NSIC: untrusted zone of the external network, DMZ and closed trusted zone.

It is obvious that the stand will include both virtualized IPTs and physical devices (if virtualization is not possible). To prepare our stand, we are going to use virtualization tools, as well as a hardware platform that can support the simultaneous operation of a large number (up to 100) of virtual machines and virtualized switching equipment that make up the model of interaction between the protected information infrastructure and the Internet service provider's infrastructure. Multi-platform application execution requires a hardware hypervisor with the ability to allocate at least 4 GB of RAM to each virtual machine.

The basic currently available hardware of the Cyber Polygon will include the following:

- high-performance server system, including 128 computing cores with the ability to scale up to 512;
- high-performance data storage system (HP 3PAR storage C7200);
- converged switching equipment (HP);
- unidirectional data transmission system (TOK-11 of the Russian Primetech company);
- non-operational thin terminal clients (for example, Dell Wyse P 25, 21 pcs).

ITCN training virtual stand will include:

- Implemented on the basis of NSIC virtual IT infrastructure of typical elements of the enterprise's corporate network: DMZ, protected segment of industrial control system, Data center, IT administration segment, segments of typical departments, its branch, hosted web portal, a set of vulnerable applications based on iOS and Android OSs;
- Typical DMZ services: proxy server, IDPS with sensors and agents, web portal, DNS (public), e-mail, FTP;
- Typical Data center services: domain controller, DNS (private), database management system (DBMS), telephony, backup server, mail server, file exchange server;
- At least 30 typical workstation users of the enterprise's departments (including the branch) with various Oss (Windows XP, Windows 7, Ubuntu, Debian, MacOS) in three NSIC's labs with installed tools of simulating user actions and necessary application software;
- IPTs that take into account typical practices and requirements of regulators: antivirus, IDPS, Web Application Firewall (WAF), a system for collecting and analyzing IPT logs, a VPN gateway.

ISP training virtual stand will include:

- configured network equipment of the ISP infrastructure: configured routing rules, access lists, ssh access, telnet access;
- prepared internal and external ISP infrastructure services: domain name registrar, DNS, billing service, information portal, WHOIS, secure hosting (WAF, IDPS);
- several points of connection of user workstations (trainees) to the ISP network;
- implemented connection of the enterprise virtual infrastructure to the ISP network.

On this basis, the first six basic training scenarios are being developed, which will be substantially expanded in the future:

1. Protection of critical ITCN servers and workstations (for example, against web injection and phishing correspondently);
2. Protection of the ITCN domain controller;
3. Data protection of the Automated Process Control System's segment;

4. Protection of enterprise's databases;
5. Protection of the financial data of the enterprise;
6. Protection of scientific and technical information of the enterprise.

As the site of the MEPhI Cyber Polygon is implemented, these decisions will be refined and adjusted, as well as all its provision will be replaced by more productive hardware and new software versions.

6 Conclusion

After a detailed study of the issue put at the beginning, it is obvious to us that creating a cyber polygon for educational purposes is the urgent need of today for any university, teaching network security. As is clear from the paper, the experience of its wide practical use is not yet available, but soon we hope to get results that we will definitely share with the public.

Based on this study, the following features of cyber polygons can be distinguished:

- The use of cyber polygons is an educational process, not a show;
- They must fill in the gaps in technical and organizational knowledge;
- They should provide an opportunity to look at attacks from different angles: both attacking and defending;
- They should give an experiment with IDPSs and IPTs;
- They should provide reasonable metrics for assessing network security.

The next steps in improving the existing MEPhI Cyber Polygon are to develop the widest possible set of typical scenarios of the studied network attacks, as well as various scenarios of its use for students to learn how to counteract network attacks and respond to IS incidents.

After this, it is advisable to start using samples of not only widely used at present but also promising systems in the educational process.

As the MEPhI Cyber Polygon is being improved, the creation of a project and regulations for connecting it to the FEFU will be required [13–15].

In addition, the development of the short-term professional refresher courses for obtaining additional professional competencies and retraining programs for developing trainees' competencies related to a new type of professional activity is seen as a necessary direction.

And in a more global perspective, the MEPhI Cyber Polygon's infrastructure already tested in practice may be of interest as a promising software and hardware platform for the deployment of competency assessment centers for network security.

Acknowledgement. This work was supported by the MEPhI Academic Excellence Project (agreement with the Ministry of Education and Science of the Russian Federation of August 27, 2013, project no. 02.a03.21.0005).

References

1. Miloslavskaya, N.: Network security intelligence center as a combination of SIC and NOC. Procedia Comput. Sci. **145**, 354–358 (2018). https://doi.org/10.1016/j.procs.2018.11.084. Post proceedings of the 9th Annual International Conference on Biologically Inspired Cognitive Architectures, BICA 2018 (Ninth Annual Meeting of the BICA Society)
2. Miloslavskaya, N.: Network Security Intelligence Centres for Information Security Incident Management. University of Plymouth. Research Theses Main Collection (2019). http://hdl.handle.net/10026.1/14306. Accessed 24 Oct 2019
3. White, G.B., Sward, R.E.: Developing an undergraduate lab for information warfare and computer security. In: Proceeding of the IFIP TC11 WG11.8 First World Conference on Information Security Education, Kista, Sweden, 17–19 June 1999, pp. 163–170 (1999)
4. Armstrong, C.J., Armstrong, H.L.: The virtual campus. In: Proceeding of the IFIP TC11 WG11.8 Second World Conference on Information Security Education, Perth, Australia, 12–14 July 2001, pp. 161–168 (2001)
5. Gritzalis, D., Tryfonas, T.: Action learning in practice: pilot delivery of an INFOSEC university laboratory course. In: Proceeding of the IFIP TC11 WG11.8 Second World Conference on Information Security Education, Perth, Australia, 12–14 July 2001, pp. 169–182 (2001)
6. Hoffman, L.J., Dodge, R., Rosenberg, T., Ragsdale, D.: Information assurance laboratory innovations. In: Proceedings of the 7th Colloquium for Information Systems Security Education, Washington, DC, USA (2003)
7. Miloslavskaya, N., Tolstoy, A.: Network security scientific and research laboratory. In: Proceedings of the 3st World Conference on Information Security Education WISE3, Monterey, USA (2003)
8. Miloslavskaya, N., Tolstoy, A., Migalin, A.: "Network security intelligence" educational and research center. In: Information Security Education for a Global Digital Society. WISE 2017. IFIP AICT, vol. 503, pp. 157–168. Springer (2017)
9. Dodge, R.C., Hay, B., Nance, K.: Using virtualization to create and deploy computer security lab exercises. In: Proceeding of 6th World Conference on Information Security Education (WISE6). IFIP, vol. 278, January 2010
10. Executive Cyber Intelligence Report, 15 July 2014. https://www.tripwire.com/state-of-security/government/executive-cyber-intelligence-report-July-15–2014/. Accessed 08 Dec 2019
11. Cyber Czech 2015 took place on the Cyber polygon in Brno, 7 October 2015. https://www.govcert.cz/en/info/events/2428-cyber-czech-2015-took-place-on-the-cyber-polygon-in-brno/. Accessed 08 Dec 2019
12. Sberbank subsidiary BI.ZONE presents report on Cyber Polygon international training session to World Economic Forum, 13 November 2019. https://www.sberbank.ru/en/press_center/all/article?newsID=d7cdcfa5-c5a9-41ca-b507-62e1fbdb3daa&blockID=1539&regionID=77&lang=en&type=NEWS. Accessed 08 Dec 2019
13. Shmyrov, V.: In Russia, a "Cyber Ppolygon" Will be Created. Who will build it? 25 July 2019. https://safe.cnews.ru/news/top/2019-07-25_vlasti_vydadut_dengi_na_sozdanie_kiberpoligona. Accessed 08 Dec 2019. (in Russian)

14. In Russia there will be a Cyber Polygon, 05 September 2019. https://www.securitylab.ru/news/500837.php. Accessed 08 Dec 2019. (in Russia)
15. “Cyberpolygon” is the Best Source of Trained Professional Personnel in the field of IS, 03 November 2017. https://www.ramax.ru/press-center/articles/112/. Accessed 08 Dec 2019. (in Russain)
16. Miloslavskaya, N.: Security zone infrastructure for network security intelligence centers. In: Postproceedings of the 10th Annual International Conference on Biologically Inspired Cognitive Architectures, BICA 2019 (9th Annual Meeting of the BICA Society). Procedia Computer Science (2019). ISSN 1877-0509

Selection of a Friction Model to Take into Account the Impact on the Dynamics and Positioning Accuracy of Drive Systems

S. Yu. Misyurin[1,2(✉)], G. V. Kreinin[1], N. Yu. Nosova[1,2], and A. P. Nelyubin[1,2]

[1] Blagonravov Mechanical Engineering Research Institute of the Russian Academy of Sciences (MERI of RAN), 4 Mal. Kharitonyevskiy Pereulok, 101990 Moscow, Russia
{ssmmrr,natahys}@mail.ru
[2] National Research Nuclear University MEPhI (Moscow Engineering Physics Institute) (MEPhI), 31 Kashirskoe Shosse, 115409 Moscow, Russia

Abstract. The problem of choosing a friction model for solving the problems of controlling positional systems, primarily with a pneumatic drive, is discussed. Due to their high dynamics, good towing capacity and relatively low price, pneumatic positioning systems are an attractive alternative to electric drives. However, the use of pneumatic systems involves some difficulties caused by the nonlinearities of its individual elements, in particular the flow characteristics of the servo valve, the compressibility of the working fluid, and also the friction acting on the piston. The main goal of this work is to analyze the stability in the interaction of the energy and control units under the influence of friction forces represented by various models. The Karnopp model was considered as one of the models, which has the advantage in describing the interaction with the friction forces in the transition from the state of rest to motion and vice versa.

Keywords: Similarity · Analogousness · Pneumatic drive · Dynamic system · Control system · Dimensionless parameters · Intelligent control · Optimization

1 Introduction

Technical systems with hydraulic and pneumatic piston drives are widely used in various fields of industry. Both differ in high values of specific power. The first ones are used in construction machines, material testing machines, in active suspensions of transport vehicles, mine equipment, simulators, paper making equipment, shipbuilding, robotics, cable manufacturing and many other areas where great work effort and various laws of movement are required. The main field of application of the second is equipment for food, pharmaceutical, electronic, robotic, engineering and some other industries [1, 4]. Comparative characteristics of drive types are given in [5].

One of the most common operations performed by pneumatic facilities is the transportation of materials weighing up to 20 kg over distances of up to 1 m with a power consumption of up to 3 kW. Performing such operations by pneumatic means is

A. V. Samsonovich et al. (Eds.): BICA 2020, AISC 1310, pp. 309–319, 2021.
https://doi.org/10.1007/978-3-030-65596-9_37

economically more advantageous in comparison with electric or hydraulic systems. This is explained not only by the high specific power of pneumatic drives (PD), but also by their high speed, ease of maintenance, the presence of compressed air lines in almost all enterprises, as well as high environmental friendliness. However, the advantages noted above do not solve the problem of PD competition with respect to other types of drives. Nonlinearities such as air compressibility, nonlinearity of the flow rate characteristics of control equipment, the uncertainty of the values of many parameters and, which is especially important, a complex and difficultly predictable model of friction forces affect the dynamics of PD [1].

For these reasons, the use of PD in most cases is limited to operations of moving the object “from lock to lock”, i.e. at a fixed distance within the rigidly established initial and final positions. However, the rapid development of industry, in particular, robotics, requires universal facilities that allow the rapid change in the stroke of the actuator, its boundary positions, as well as the implementation of movement along a given trajectory. As already mentioned above, the solution of these problems, along with other non-linearities, is prevented by the factor of sliding friction in the piston seals. Therefore, further attention will be focused on the study of the influence of friction forces on friction dynamics and control of PD, which is expressed through deviations of the motion processes from given modes, loss of accuracy and stability in working out a given position or a given trajectory, occurrence of limit cycles and other disturbances in normal operation.

2 Models of Friction in the Piston of the Actuator

Piston friction is a non-linear phenomenon that is difficult to analytically describe. The friction characteristics can change over time in an unpredictable way, depending on many factors – medium temperature, quality and quantity of lubricant, and even depending on the direction of movement of the piston. Despite this, the friction model is necessary in the study of the behavior of a piston drive system, as evidenced by many studies (see, for example, [1–4, 6–16]).

1. The most developed model of the process of friction in a piston is the multiphase dynamic model of LuGre. It explains the regularity of friction that occurs at very low piston speeds, including at the preliminary stage of displacement of the bearing surfaces even before the active movement of the piston. This is an important component of the model for drive positioning systems. The physical process of sliding friction in the general case is represented [1, 9] in the form of the following modes: static friction; boundary lubrication movement; partial fluid lubricated movement; friction with full fluid lubrication. Let us consider these modes in detail.

The model of the appearance of **static friction** was considered in various works, for example [1, 13], it is logical and well-founded. Johnson (1987), Dahl (1968, 1976, 1977) from experimental observations of friction concluded that at this stage the contacting rough surfaces protrusions behave like damped springs, i.e. are deformed elastic and plastic. When, under the action of the driving force, the protrusion of the roughness protrusions begins, their elastic deformation admits microscopic displacements preceding the onset of sliding. The displacement values remain approximately proportional to the driving force until the moment when the driving force reaches a critical value, causing the separation of roughness and the beginning of the active sliding of the surfaces. In this mode, the tangential force arising from friction has the form:

$$F_t(x) = -k_t x$$

here F_t is the tangential force, k_t is the tangential contact stiffness, and x is the offset from equilibrium.

The **motion model with boundary lubrication** is characterized by the fact that the system leaves the static friction mode and enters the friction mode with boundary lubrication, which begins to flow on the contact surface.

At a very low sliding speed of contact roughness, the speed is not sufficient to create a liquid film between the surfaces. The boundary layer serves to provide lubrication. It must be hard to withstand contact stress, but have low shear strength to reduce friction. Since there is contact between a solid and a solid, the stick-slip (SS) type effect of sticking solid bodies together arise. With this effect, the shear friction force is usually greater than the subsequent sliding force. Some boundary lubricants do reduce static friction to a level below Coulomb friction and completely eliminate the SS effect.

At high speeds, **a partial fluid lubricated motion model** begins to operate. As the surface slip rate increases, fluid is drawn into the contact zone. The viscosity of the lubricating fluid ensures the formation of a surface film in the contact zone. Its thickness depends on the viscosity of the oil and the sliding speed of the surfaces. The higher the viscosity or speed of movement, the thicker the liquid film will be, and, therefore, the less the force of contact between the bodies, the less the friction force and the greater the acceleration of movement.

When the thickness of the surface film becomes less than the height of the roughness, a certain contact of the roughness of two bodies occurs. With a sufficiently thick film, the two bodies separate, and the load is fully supported by the fluid. All this indicates the complexity of the physical process in this mode, and, consequently, the complexity of its mathematical modeling.

In the model of **friction with full liquid lubrication**, the contact of two solids is completely eliminated, the surfaces interact only through the lubricant, there is a stable friction mode proportional to speed.

2. There are many works devoted to mathematical models of friction at various levels. Depending on one or another statement of the problem, models that differ in level of complexity and non-linearity are used. Figure 1 shows the mathematical models corresponding to different operating modes at different speeds.

The following are descriptions of the friction models in accordance with the numbering shown in Fig. 1.

I. The Coulomb friction model described by the equation:

$$F_t(x) = F_c sign(\dot{x})$$

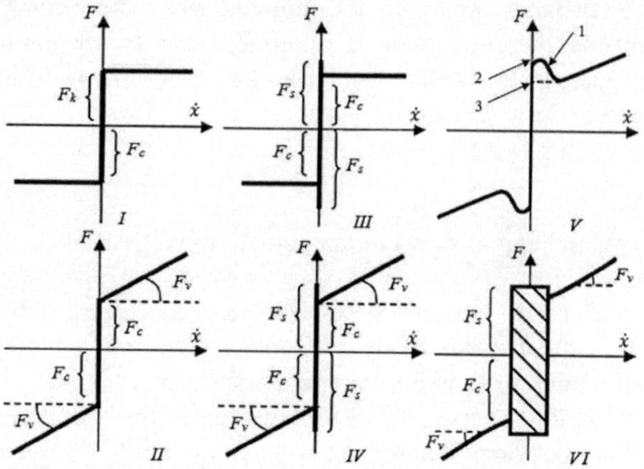

Fig. 1. Friction models.

Regardless of the contact area, Coulomb friction always acts against relative motion and is proportional to the normal contact force. Coulomb friction is a phenomenon of friction that depends only on the direction of speed, and not on its magnitude; F_c is the Coulomb friction coefficient.

II. Viscous friction occurs as a result of the action of a layer of lubricating fluid between two rubbing surfaces. As shown above, viscous friction is presented as a linear function of speed:

$$F = (F_c + F_v \dot{x}) sign(\dot{x})$$

where F_v is the coefficient of viscous friction.

It was experimentally established that the rest friction force can be higher than the driving force or Coulomb friction. In this state, it is necessary to exert an external force equal to or greater than the rest friction force in order to bring the body into motion, that is, into sliding. For Coulomb and viscous friction, mathematical models taking into account the static friction forces are presented in graphs III and IV. These models are rough and unacceptable for describing processes at low speeds.

After numerous experimental studies for such operating modes, a mathematical model of friction was proposed taking into account the Striebeck effect (graph V, 1 – Striebeck effect, 2 – static friction, 3 – Coulomb friction). This is a friction phenomenon that occurs when using liquid lubricant and leads to a decrease in friction with increasing speed at low speed. For low speeds, the friction force decreases with increasing speed.

The Karnopp model (graph VI) has widespread use in controlling drive systems. This model takes into account the SS effect which at low speeds has a significant effect on the positioning of objects. The mathematical model is as follows:

$$F = \begin{cases} F_0 & |\dot{x}| \leq \dot{x}_{min} \; and \; F_0 < F_s \\ F_s sign(\dot{x}) & |\dot{x}| \leq \dot{x}_{min} \; and \; F_0 \geq F_s \\ (F_c + F_v \dot{x}) sign(\dot{x}) & |\dot{x}| > \dot{x}_{min} \end{cases}$$

Here F is the resulting frictional force, F_s is the static friction, F_c is the Coulomb friction, F_0 is the resulting driving force, and F_v is the coefficient of viscous friction.

A more detailed is the mathematical model of LuGre [8, 9], which takes into account tribological effects:

$$\begin{cases} F = \sigma_0 z + \sigma_1 \dot{z} + \sigma_2 \ddot{z} \\ \dot{z} = \dot{x} - \frac{|\dot{x}|}{g(\dot{x})} z \\ g(\dot{x}) = F_c + (F_s - F_c) e^{-\left(\frac{\dot{x}}{v_s}\right)^2} \end{cases}$$

here the internal state variable z, which cannot be measured, describes the deviation of the contact surfaces (preliminary displacement in the contact). The parameter σ_0 is the stiffness coefficient, σ_1 is the damping coefficient, and σ_2 is the viscous friction coefficient. The model parameters are also set by the static friction F_s, Coulomb friction F_c and the Striebeck speed v_s.

In the LuGre model, friction in the contact of two rigid surfaces is visualized in the form of adhesion of elastic links – bristles. It is a popular dynamic model, which is characterized by:

- it covers many features of friction, such as Striebeck effect, variable separation force at the beginning of movement, the phenomenon of delay;
- adequately represents the real friction that occurs in contact with the lubricant of two contacting surfaces (i.e. describes the transition from static to dynamic friction, as a continuous process);
- represents the process of friction at low speeds as an internal process, which is important when choosing an adaptive control law to compensate for the harmful regime (SS).

The Karnopp-type friction model (Fig. 1-IV) is relatively simpler than the LuGre model and, nevertheless, in many cases is sufficient to describe the friction effect [1, 3, 12]. This model is very effective from the point of view of use in numerical calculations and, therefore, is widely used in modeling and control. Three features of the Karnopp model should be noted. The first is that the "stick" phase encompasses a velocity element that is inside a certain, albeit very small, range of speeds on both sides of zero (that is, it goes beyond zero speed, which is usually not accepted). The second feature – the movement in this section can stop if the driving force is less than the friction force. Then it is assumed that in this state the friction force is equal to the sum of all the driving forces of the drive. If the driving forces exceed the limit value, the friction force assumes the previous value, and the system returns to the normal mode of motion under the action of the driving force. The third feature is that the friction force at all phases, including the SS regime, is included in the equation of motion as a full-fledged component of the balance of forces.

The zero speed interval of the piston $|\dot{x}| < \dot{x}_{min}$, included in the model, solves the problem of zero speed and the transition from the sticking stage to the slip stage. The values $\dot{x}, \dot{x}_{min}$ represent here the piston velocities and the zero velocity interval, respectively. Total friction force according to the Karnopp model, including Coulomb and fluid friction.

As follows from Fig. 1-IV, the Striebeck effect is not included in the Karnopp friction model. The Striebeck loop is replaced by a sharp drop in the friction force from the current value of F to the value of F_c at the exit from the zero interval. However, the refusal to fully represent the Striebeck loop in this case is quite justified, since, firstly, this has little effect on the reproduction of SS-type motion and, secondly, the very process of selecting and assigning the parameters of the Striebeck loop presents a particular problem.

3 Formulation of the Problem

The study of the influence of the friction force on the positioning process of the object suggests the presence of fairly common dynamic models of the actuator (Fig. 2), the regulator and the friction forces. The transformed (dimensionless) model developed earlier in [17–20] and formulated in [21–23] is used as a generalized model of the drive itself. The structure and form of this model is determined by the accepted units of measurement of the variables included in the original equations, and the dependencies between the parameters representing its coefficients. The transformed (dimensionless) drive model is obtained as a result of the optimal solution of specially formed transition equations that meet the problem. The nonlinearity caused by high compressibility of the air, the uncertainty of some parameters and characteristics of the dynamic model, caused by the inability to accurately determine them, as well as variability for various reasons, make it difficult to choose the basic models of the drive and controller to control such a model. On the other hand, attempts to take into account all the features of the drive lead to an undesirable complication of the model, which causes additional difficulties in its study. The approach taken in this paper is a compromise solution to this problem. Since in dimensionless models of the drive and controller the parameters are represented by dimensionless quantities, varying them allows one to obtain general estimates of the effect on the dynamics and accuracy of the system of parameters included in the model, taking into account the set limits for their change.

The controller model (1, 2) determines the response of the system to external control actions (Fig. 2). This apparatus is necessary for modeling the behavior of the general drive model in various conditions, analyzing the results of the response to control actions. It is also necessary for choosing rational parameter values first of models, and then real devices. The formation of the initial model of the controller presents significant difficulties, since in most cases when creating positional systems, primarily with pneumatic drives, it is necessary to solve the problem almost every time anew, starting with the formation of a dynamic model and ending with checking whether its properties correspond to the properties of a real object [1, 3, 7].

The significant nonlinearity of the positional **pneumatic** drive does not allow the use of the simplest linear PID controller model as the base controller model [1, 14]. Of

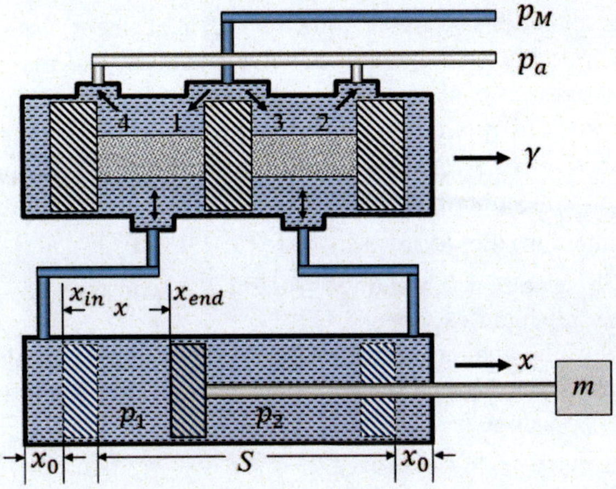

Fig. 2. Pneumatic positional drive model.

the known nonlinear models of regulators in positional systems, the most frequently used models of regulators operating in the so-called sliding mode. Such a controller reproduces a certain basic sliding surface and further keeps the drive in motion as close as possible to this surface. Based on preliminary studies, we selected the original **sinusoidal model of the sliding surface** as a base, supplemented by **feedback** on the differential pressure in the cavities, which is an analog of the feedback on acceleration, but much simpler to implement. The basic equation of the sliding surface in the variables of displacement and speed is:

$$\lambda_B = 0,5 \cdot (1 - \cos(\omega\tau)); \; \dot{\lambda}_B = 0,5 \cdot \omega \cdot \sin(\omega\tau), \tag{1}$$

where $\omega = \pi/\tau_S$, τ_s is dimensionless time, equivalent to the process half-period λ_B. Given the additional term, depending on the pressure in the cavities, the control law is presented in the form:

$$\gamma = \vartheta_1 \cdot (\lambda_B - \lambda) + \vartheta_2 \cdot \left(\dot{\lambda}_B - \dot{\lambda}\right) - \vartheta_3 \cdot (\sigma_1 - \sigma_2), \tag{2}$$

where $\vartheta_1, \vartheta_2, \vartheta_3$ are the feedback gains, $\sigma_{1,2}$ are the pressure values in corresponding drive cavities. A signal γ (called an equivalent element) is a control signal that changes the flow areas of the channels that control air flow. The flow characteristic of the valve (called the robust element), which controls the opening of these channels, is in this case symmetrical, with zero dead zone.

Here is a model of the drive, which is discussed in detail in [17]:

$$\begin{cases} m\ddot{x} = F_1 p_1 - F_2 p_2 + P_f + P_L, \\ \dot{p}_{1,2} = \left(\frac{1}{x_{01,2} \mp x}\right) \cdot \left[\frac{\gamma \cdot f_{1,2}^+}{F} K p_M \varphi\left(\frac{p_{1,2}}{p_M}\right) - \frac{\gamma \cdot f_{1,2}^-}{F} K p_{1,2} \varphi\left(\frac{p_a}{p_{1,2}}\right) \pm \dot{x}\right]. \end{cases} \tag{3}$$

The equations include: t – travel time; x – displacement; p_i – pressure in two cavities (i = 1, 2); γ – degree of opening of valve channels; m – mass of moving parts; F – cylinder cross-sectional area; $F_{1,2}$ – the area of the piston on the side of each cavity; P_f, P_L – forces of friction and external resistance; x_0 – clearance space characteristic; $f^{\pm}$ – effective cross-sections of valve channels at the inlet and outlet; p_M, p_a – supply pressure and atmospheric pressure; $\varphi(\varepsilon)$ – the consumption function (according to Sunville) is taken in the form $\varphi(\varepsilon) = \varphi^*\{1 - [\varepsilon - b/1 - b]^2\}^{1/2}$, if $\varepsilon > b$ and $\varphi(\varepsilon) = \varphi^*$, if $\varepsilon \le b$, here $b = 0.5$ and $\varphi^* \approx 0.26$; K – constant, $K = [2k/(k-1)]^{1/2}$, here $k = 1.4$ is the adiabatic exponent.

By the method of the analogy theory, which was developed for drive systems [17–20], Eqs. (3) are transposed into a dimensionless form by replacing the variables with their dimensionless analogues λ, τ, σ according to the relations: $x = q_1\lambda$; $t = q_2\tau$; $p = q_3\sigma$; $x_{01,2} = q_1\lambda_{01,2}$; $\chi_L = P_L/Fp_M$; $\chi_f = P_f/Fp_M$; $\dot{x}_H = (f^+/F)K$; $\kappa_{1,2} = F_{1,2}/F$.

The general drive model is represented as:

$$\begin{cases} \ddot{\lambda} = \sigma_1 - \sigma_2 + \chi_f + \chi_L, \\ \dot{\sigma}_1 = \left(\frac{\sigma_1}{\lambda_{01} + \lambda}\right)\left[|\gamma| \cdot \left(\frac{\varphi(\sigma_1)}{\sigma_1}\right) - \dot{\lambda}\right], \\ \dot{\sigma}_2 = \left(\frac{\sigma_2}{\lambda_{02} - \lambda}\right)\left[-|\gamma| \cdot \bar{\omega} \cdot \varphi\left(\frac{\sigma_a}{\sigma_2}\right) + \dot{\lambda}\right]. \end{cases} \tag{4}$$

here $\bar{\omega} = f^-/f^+$, and $\sigma_a = p_a/q_3 = p_a/p_M$.

The adopted drive model (4) contains the following variable values: power load $|\chi_L|$, friction force χ_L, initial volumes of cavities λ_1, λ_2 in relative measurement, process duration τ_S. Let us point out the invariance of the initial and final signals of the piston position, which is achieved due to the targeted selection of the appropriate scales of transition to dimensionless variables. Variable values, in addition to the characteristics of the model, also include the characteristics of the controller ϑ_1, ϑ_2, ϑ_3 introduced above.

For researching the drive movement process, a certain initial model was chosen, which was studied initially in the absence of all resistance forces, including friction, in order to obtain a preliminary assessment of the effect of air compressibility on the controllability of the system. The dimensionless form of the model greatly facilitated the visual search for its rational parameters. The operation of the model is characterized by the processes shown in Fig. 3. The initial pressure in both cavities is equal to atmospheric ($\sigma_{10} = \sigma_{20} = \sigma_a = 0.2$).

The process proceeds at a good level, when approximately the same level of pressure in the cavities is provided at the beginning of the process. With a large initial pressure drop, the movement process is accompanied by significant pressure fluctuations, which sharply affects the quality of positioning. As an example Fig. 4 shows the process obtained during the operation of the model with constant power load equal to $\chi_L = -0.1$. The values of the remaining parameters of the model are the same as when modeling dynamics with zero load (Fig. 3).

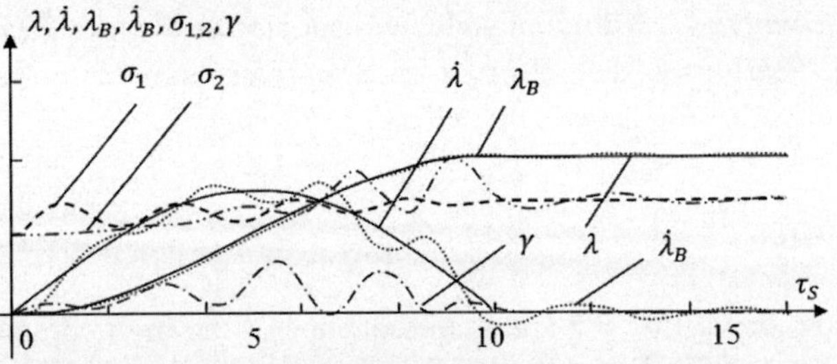

Fig. 3. Assessment of the effect of air compressibility on the controllability of the system in the absence of all resistance forces, including friction.

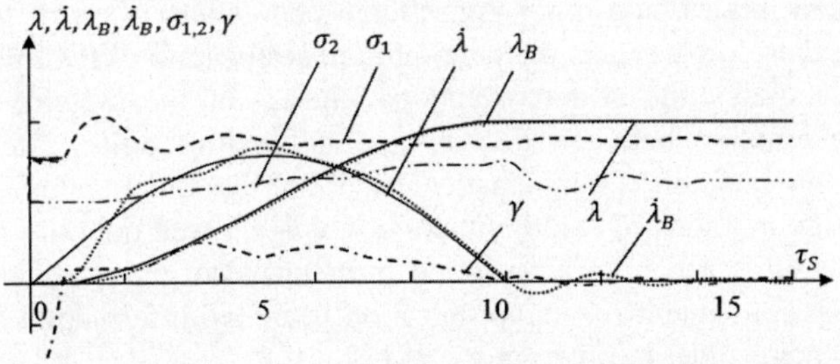

Fig. 4. Operation of the original model with a constant force load χ_L.

However, it can be noted that at initial pressures of 0.2 and 1, the indices of processes with or without power load, as well as the accuracy of working out the final position, are higher than with a combination of 0.2 and 0.2. This should be taken into account in further research.

The constructed drive model was further used for a preliminary study of the effect of friction forces on the process. Due to its physical nature, the positional system is most sensitive to friction in low-speed motion modes, i.e. at the start and finish. We

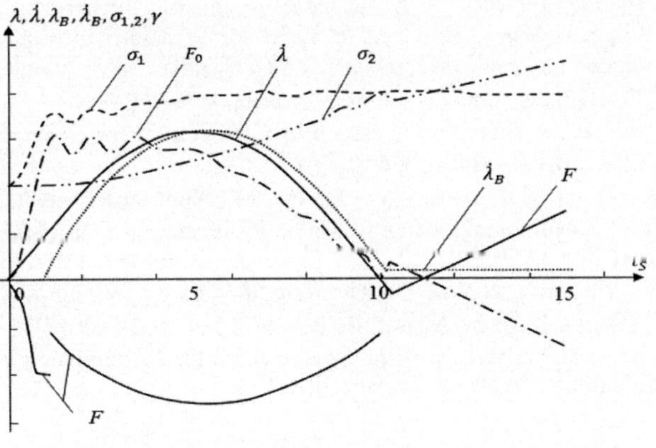

Fig. 5. The influence of friction forces on the positioning process.

give an example where the Karnopp model was adopted as the friction force χ_f (Fig. 5). Here you can clearly see the effect of friction forces on the positioning process, especially the effect of the "stick-slip" effect.

4 Conclusions

Difficulties in the control of PD are a consequence of the high compressibility of the air, the nonlinearity of the flow rate and other dynamic characteristics of the system, the uncertainty of the structure and parameters of the model, as well as the difficulties associated with the influence of the friction process. These factors are expressed in errors in reproducing a given trajectory of motion, in the appearance of limit cycles, "slip-slip" motion modes, and other undesirable phenomena. One of the main directions in solving these problems is the maximum neutralization of the influence of these factors by rationalizing the structure and parameters of the system, increasing the efficiency of interaction between the energy and control units. The basis of this approach is the dimensionless mathematical model of the system adopted in the work, rationalized by the methods of analogy theory, which allowed not only minimizing the number of parameters and criteria included in it, but also giving them a generalized form that is most convenient for its optimization. Detailed information on the features of the energy block model is given in [5, 18].

Acknowledgment. The research was supported by Russian Foundation for Basic Research, project No. 18-29-10072 mk (Optimization of nonlinear dynamic models of robotic drive systems taking into account forces of resistance of various nature, including frictional forces).

References

1. Rahman, R.A.: Dynamical adaptive backstepping – sliding mode control for servo-pneumatic positioning application: controller design and experimental evaluation. A Thesis submitted to the Faculty of Graduate Studies of The University of Manitoba in partial fulfilment of the requirements for the degree of Doctor of Philosophy (2016)
2. Daw, N.A., Wang, J., Wu, Q.H., Chen, J., Zhao, Y.: Parameter identification for nonlinear pneumatic cylinder actuator. In: Zinober, A., Owens, D. (eds.) Nonlinear and Adaptive Control. LNCIS, vol. 281, pp. 77–88. Springer, Heidelberg (2003)
3. Alleyne, A., Liu, R.: A simplified approach to force control for electro-hydraulic systems. Control Eng. Pract. **8**(12), 1347–1356 (2000)
4. Valdiero, A.C., Ritter, C.S., Rios, C.F., Raficov, M.: NonLinear mathematical modeling in pneumatic servo position applications. Math. Problems Eng. **2011** (2011). Article no. 472903, 16 pages
5. Kreinin, G.V., Misyurin, S.Yu.: On choosing the drive type for the power unit of a mechatronics system. J. Mach. Manuf. Reliab. **44**(4), 305–311 (2015)
6. Márton, L., Fodor, S., Sepehri, N.: A practical method for friction identification in hydraulic actuators. Mechatronics **21**(1), 350–356 (2011)

7. Guenther, R., Perondi, E.C., De Pieri, E.R., Valdiero, A.C.: Cascade controlled pneumatic positioning system with LuGre model based friction compensation. J. Braz. Soc. Mech. Sci. Eng. **26**(1), 48–57 (2006)
8. Schindele, D., Aschemann, H.: Adaptive friction compensation based on the LuGre model for a pneumatic rodless cylinder. In: Proceedings of the 35th Annual Conference of the IEEE Industrial Electronics Society, Porto, Portugal, pp. 1432–1437. IEEE (2009)
9. Khayati, K., Bigras, P., Dessaint, L.-A.: LuGre model-based friction compensation and positioning control for a pneumatic actuator using multi-objective output-feedback control via LMI optimization. Mechatronics **19**(4), 535–547 (2009)
10. Olsson, H., Åström, K.J., Canudas de Wit, C., Gäfvert, M., Lischinsky, P.: Friction models and friction compensation. Eur. J. Control **4**, 176–195 (1998)
11. Åström, K.J., Canudas de Wit, C.: Revisiting the LuGre friction model. IEEE Control Syst. Mag. **28**(6), 101–114 (2008)
12. Karnopp, D.: Computer simulation of stick-slip friction in mechanical dynamic systems. J. Dyn. Syst. Meas. Contr. **107**(1), 100–103 (1985)
13. Armstrong-Hélouvry, B., Dupont, P., Canudas de Wit, C.: A survey of models, analysis tools and compensation methods for the control of machines with friction. Automatica **30**(7), 1083–1138 (1994)
14. Canadas de Wit, C., Olsson, H., Åström, K.J., Lischinsky, P.: A new model for control system with friction. IEEE Trans. Autom. Control **40**(3), 419–425 (1995)
15. Dupont, P., Armstrong, B., Hayward, V.: Elasto-plastic friction model: contact compliance and stiction. In: Proceedings of the 2000 American Control Conference, Chicago, IL, USA, pp. 1072–1077. IEEE (2000)
16. Nouri, B.M.Y., Al-Bender, F., Swevers, J., Vanherek, P., Van Brussel, H.: Modeling a pneumatic servo positioning system with friction. In: Proceedings of the 2000 American Control Conference, Chicago, Illinois, USA, pp. 1067–1071. IEEE (2000)
17. Kreinin, G.V., Misyurin, S.Yu., Nosova, N.Yu., Prozhega, M.V.: Parametric and structural optimization of pneumatic positioning actuator. In: Advanced Technologies in Robotics and Intelligent Systems. Mechanisms and Machine Science, vol. 80, pp. 395–403 (2020)
18. Misyurin, S.Yu., Kreinin, G.V., Nosova, N.Yu.: Similarity and analogousness in dynamical systems and their characteristic features. Nonlinear Phys. Mech. **15**(3), 213–220 (2019)
19. Misyurin, S., Kreinin, G., Nelubin, A., Nosova, N.: The synchronous movement of mechanisms taking into account forces of the different nature. IOP J. Phys. Conf. Ser. **1439** (1) (2020). Article no. 012016, 4 pages
20. Kreinin, G.V., Misyurin, S.Yu., Lunev, A.V.: Coordinated interaction of two hydraulic cylinders when moving large-sized objects. IOP J. Phys. Conf. Ser. **937**(1) (2017). Article no. 012023, 3 pages
21. Misyurin, S.Yu., Ivlev, V.I., Bozrov, V.M., Nelyubin, A.P.: Parameterization of an air motor based on multiobjective optimization and decision support. J. Mach. Manuf. Reliab. **42**(5), 353–358 (2013)
22. Kreinin, G.V., Misyurin, S.Yu.: Selection of the scheme for incorporating a drive into the structure of a mechanism in solving problems of kinematic synthesis. J. Mach. Manuf. Reliab. **37**(1), 1–5 (2008)
23. Ivlev, V.I., Misyurin, S.Yu.: Calculated and experimental characteristics of a scroll machine operating in the air motor mode. Dokl. Phys. **62**(1), 42–45 (2017)

Kinematics and Dynamics of the Spider-Robot Mechanism, Motion Optimization

Sergey Yu. Misyurin[1,2(✉)], German V. Kreinin[1], Natalia Yu. Nosova[1,2], and Andrey P. Nelyubin[1,2]

[1] Blagonravov Mechanical Engineering Research Institute of the Russian Academy of Sciences (MERI of RAN), 4 Mal. Kharitonyevskiy Pereulok, 101990 Moscow, Russia
{ssmmrr,natahys}@mail.ru

[2] National Research Nuclear University MEPhI (Moscow Engineering Physics Institute) (MEPhI), 31 Kashirskoe Shosse, 115409 Moscow, Russia

Abstract. In this paper, we consider the kinematics and dynamics of a spider robot mechanism with 18 degrees of freedom (six legs). The equations of kinematics and dynamics are written out; and the issue of optimizing the robot's movement is considered. The robot's gait is analyzed, in which part of the legs is on the ground and supports the robot, and part of the legs moves in the air. At the first stage for solving this problem, one leg is considered separately, as a kinematic system with open kinematics and with three degrees of freedom. The kinematics equations were presented in matrix form using the principle of rotation of the coordinate system. The dynamics equations are based on Lagrange equations of the second kind. The mass of the legs, reduced to the center of gravity, moments of inertia, moments developed by engines were taken into account, and etc. The conclusions were made about the optimal movement of the leg based on the obtained equation of kinetic energy of the robot's leg based on the obtained equation of the kinetic energy of the robot leg.

Keywords: Kinematics · Dynamics · Spider robot · Optimization · General Lagrange equation

1 Introduction

The use of mechanisms with open kinematics is found everywhere. Currently, a huge variety of industrial robots with open kinematics similar to the "Kuka" robots have found application in the industry [1]. The advantage of these robots is their relatively simple kinematic structure and good positioning accuracy. As a driving force, as a rule, electric motors are used, less often rotary pneumatic motors, for example [2]. Similar schemes of open kinematics are used in biorobots that copy the mode of movement of both humans [3] and insects [4, 5], in particular, spiders robotic (Fig. 1). In these mechanisms, each leg has three degrees of freedom. Such insects are quite popular in the search for alternative movement mechanisms on difficult surfaces. Indeed, leg systems theoretically offer the potential for better rugged terrain than traditional wheeled or tracked designs, due to the low ground contact and lack of slippage. The

A. V. Samsonovich et al. (Eds.): BICA 2020, AISC 1310, pp. 320–326, 2021.
https://doi.org/10.1007/978-3-030-65596-9_38

use of robots with a large number of legs implies a better adaptation to uneven terrain, but at the same time a large number of legs inevitably lead to a more complex control system. Of all the legged systems, six-legged robots seem to be the logical trade-off between stability in motion and control complexity, and are therefore of great interest to research engineers. There are many developed gaits (kinematics of motion) spider-robot, but not always sufficiently investigated the dynamics of these "walks", insufficient attention has been paid to the issue of optimization of parameters of the legs, and optimizing movements [6–8].

2 Description of the Mechanism

Let us consider in more detail the structure of the robot's leg (Fig. 2). Let us introduce a coordinate system in the center of the hinge for attaching the leg to the body O_1. This mount is a rotational pair in the plane of the "spider" body. The O_1z axis is directed vertically up along the rotation axis of the actuator. The O_1x axis is directed perpendicular to the O_1z axis, and is directed to the center of mass of the spider's body; the O_1y axis is perpendicular to the O_1zx plane and forms the right coordinate system.

Fig. 1. General view of the spider robot

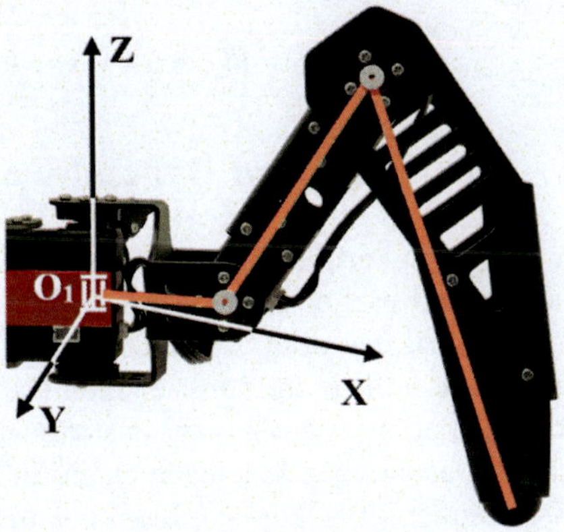

Fig. 2. Spider robot leg.

Let's go to the kinematic description of the robot's leg (Fig. 3). O_1, O_2, O_3 are rotary kinematic pairs with angles of rotation. The O_1O_2 link rotates in the O_1x plane, α is the angle of the link deviation from the O_1x axis. The axis of rotation of the kinematic pairs O_2, O_3 are parallel to the O_1x plane. We introduce the following notation: $|O_1O_2| = l_1, |O_2O_3| = l_2, |O_3O_4| = l_3, |O_1A_1| = \rho_1, |O_2A_2| = \rho_2, |O_3A_3| = \rho_3$; J_{A1}, J_{A2}, J_{A3} are the moments of inertia of links 1, 2 and 3 relative to their centers of mass; $m_{1,2,3}$ are masses of these links; $\rho_{1,2}$ are the coordinates of their centers of mass relative to the kinematic pairs; J_{d1}, J_{d2}, J_{d3} are the moments of inertia of the engine rotors, installed in the joints of the links (if the movement from the actuator to the link is transmitted through the reducer, then the moment of inertia of the is substituted into the equation given to the output shaft of the gearbox); α, β and γ are generalized coordinates – the angles of rotation of the links, calculated according to the scheme in Fig. 3. M_{d1}, M_{d2}, M_{d3} are the moments of engines and M_{c1}, M_{c2}, M_{c3} are the moments of resistances. We introduce the notation for the lengths $|O_1A_2| = L_1$, $|O_1A_3| = L_2$, $|O_2A_3| = L_3$.

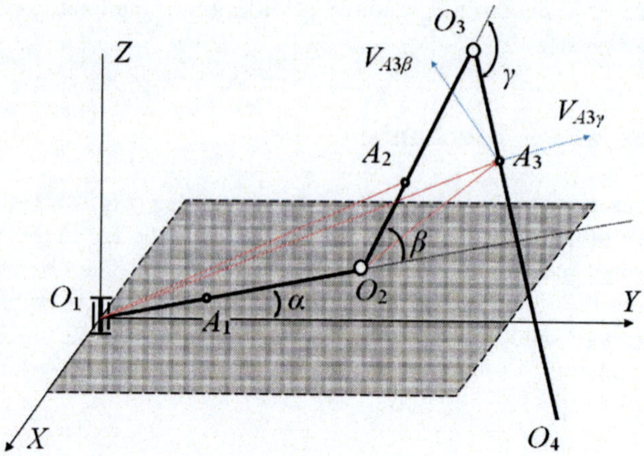

Fig. 3. A spider robot leg kinematic scheme

3 Kinematics and Dynamics of the Mechanism

When solving problems of kinematics and dynamics of a robot with six legs, we rely on the fact that part of the legs moves in the air, while part of the legs are on the ground and supports the robot. It is necessary to find out how each leg moves separately, which movement will be the most economical or fast. In other words, at the first stage, we assume that the robot's body is stationary. According to this assumption, we consider each leg as an open kinematic chain. In accordance with this assumption, the equation of the direct kinematics problem has the form:

O_1xyz:

$$\overrightarrow{O_1O_4} = M_z(\alpha)\left[M_y(\beta)\left[M_y(-\gamma)\begin{pmatrix}0\\ l_3\\ 0\end{pmatrix} + \begin{pmatrix}0\\ l_2\\ 0\end{pmatrix}\right] + \begin{pmatrix}0\\ l_1\\ 0\end{pmatrix}\right]$$

where M_x, M_y, M_z are rotation matrices.

$$M_x(\alpha) = \begin{pmatrix}1 & 0 & 0\\ 0 & \cos(\alpha) & -\sin(\alpha)\\ 0 & \sin(\alpha) & \cos(\alpha)\end{pmatrix},\ M_y(\alpha) = \begin{pmatrix}\cos(\alpha) & 0 & \sin(\alpha)\\ 0 & 1 & 0\\ -\sin(\alpha) & 0 & \cos(\alpha)\end{pmatrix}$$

$$M_z(\alpha) = \begin{pmatrix} \cos(\alpha) & -\sin(\alpha) & 0 \\ \sin(\alpha) & \cos(\alpha) & 0 \\ 0 & 0 & 1 \end{pmatrix}$$

The resulting expression for the vector O_1O_4 determines the position of the output link of the robot's leg (i.e. O_4) in the O_1xyz coordinate system, given the generalized input coordinates α, β and γ.

Let us write out the equation of the mechanism dynamics based on the General Lagrange equation of the second kind:

$$\frac{d}{dt}\left(\frac{\partial L}{\partial \dot{q}_i}\right) - \frac{\partial L}{\partial q_i} = R_i \tag{1}$$

Where $L = T - U$ is the Lagrange function; T and U are the kinetic and potential energy of the system, respectively; q_i is the i-th generalized coordinate $(i = 1, 2, 3)$. Each of the R_i terms on the right side represents the sum of non-potential generalized forces acting on the i-th coordinate, which also include the driving forces or moments of the drives. In our case, we can take $R_i = M_{di} - M_{ci}$, where $i = 1, 2, 3$.

Let's write out the kinetic energy.

$$\boldsymbol{T} = \boldsymbol{T}^{(1)} + \boldsymbol{T}^{(2)} + \boldsymbol{T}^{(3)}; \tag{2}$$

$$\begin{aligned} \boldsymbol{T}^{(1)} &= \tfrac{m_1 V_{A1}^2}{2} + \tfrac{J_{A1}\dot{\alpha}^2}{2} + \tfrac{J_{g1}\dot{\alpha}^2}{2} = (\rho_1\dot{\alpha})^2\tfrac{m_1}{2} + \tfrac{J_{A1}}{2}\dot{\alpha}^2 + \tfrac{J_{g1}}{2}\dot{\alpha}^2 \\ &= \tfrac{1}{2}\left(\boldsymbol{\rho}_1^2 \cdot \boldsymbol{m}_1 + \boldsymbol{J}_{A1} + \boldsymbol{J}_{g1}\right)\dot{\boldsymbol{\alpha}}^2; \end{aligned} \tag{3}$$

$$\boldsymbol{T}^{(2)} = \frac{V_{A2}^2 \cdot m_2}{2} + \frac{J_{A2}}{2}\left(\dot{\alpha}^2 + \dot{\beta}^2\right) + \frac{J_{g2}}{2}\dot{\beta}^2$$

The speed of the point A_2 is a function of the generalized coordinates α and β, thus $V_{A2} = V_{A2}(\alpha, \beta)$, therefore:

$$\begin{aligned} V_{A2}^2 &= \left(V_{A2\alpha} + V_{A2\beta}\right)^2 = \left(L_1\dot{\alpha} + \rho_2\dot{\beta}\right)^2 \\ &= L_1^2\dot{\alpha}^2 + \rho_2^2\dot{\beta}^2 + L_1\dot{\alpha}\cdot\rho_2\dot{\beta}\cdot cos(90) = L_1^2\dot{\alpha}^2 + \rho_2^2\dot{\beta}^2 \end{aligned}$$

$$\boldsymbol{T}^{(2)} = \frac{(\boldsymbol{L}_1^2\dot{\boldsymbol{\alpha}}^2 + \boldsymbol{\rho}_2^2\dot{\boldsymbol{\beta}}^2)\boldsymbol{m}_2}{2} + \frac{\boldsymbol{J}_{A2}}{2}(\dot{\boldsymbol{\alpha}}^2 + \dot{\boldsymbol{\beta}}^2) + \frac{\boldsymbol{J}_{g2}}{2}\dot{\boldsymbol{\beta}}^2; \tag{4}$$

$$\boldsymbol{T}^{(3)} = \frac{V_{A3}^2 \cdot m_3}{2} + \frac{J_{A3}}{2}\left(\alpha^2 + \left(\dot{\beta} + \dot{\gamma}\right)^2\right) + \frac{J_{g3}}{2}\dot{\gamma}^2; \tag{5}$$

The velocity of point A_3 is a function of the generalized coordinates α, β and γ, thus $V_{A3} = V_{A3}(\alpha, \beta, \gamma)$, therefore:

$$V_{A3}^2 = \left(V_{A3\alpha} + V_{A3\beta} + V_{A3\gamma}\right)^2$$
$$= V_{A3\alpha}^2 + V_{A3\beta}^2 + V_{A3\gamma}^2 + 2V_{A3\alpha}V_{A3\beta}cos(90) + 2V_{A3\alpha}V_{A3\gamma}cos(90) + 2V_{A3\beta}V_{A3\gamma}cos(S)$$
$$= L_2^2 \cdot \dot{\alpha}^2 + L_3^2 \cdot \dot{\beta}^2 + \rho_3^2 \cdot \dot{\gamma}^2 + 2L_3\rho_3 \cos(S) \cdot \dot{\beta};$$

Angle S is the angle between the velocity vectors $V_{A3\beta}$ and $V_{A3\gamma}$. In turn, these vectors are perpendicular to the O_2A_3 and O_2O_3 segments, respectively. Consequently, the angle S is equal to the angle $O_2A_3O_3$. From the triangle $O_2O_3A_3$ we have:

$$l_2^2 = L_3^2 + \rho_3^2 - 2L_3\rho_3 \cdot \cos(S) = > 2L_3\rho_3 \cos(S) = L_3^2 + \rho_3^2 - l_2^2;$$

Thus, we get:

$$V_{A3}^2 = L_2^2 \cdot \dot{\alpha}^2 + L_3^2 \cdot \dot{\beta}^2 + \rho_3^2 \cdot \dot{\gamma}^2 + (L_3^2 + \rho_3^2 - l_2^2) \cdot \dot{\beta}\dot{\gamma},$$

and therefore:

$$\boldsymbol{T}^{(3)} = \frac{(\boldsymbol{L}_2^2 \cdot \dot{\boldsymbol{\alpha}}^2 + \boldsymbol{L}_3^2 \cdot \dot{\boldsymbol{\beta}}^2 + \boldsymbol{\rho}_3^2 \cdot \dot{\boldsymbol{\gamma}}^2 + (\boldsymbol{L}_3^2 + \boldsymbol{\rho}_3^2 - \boldsymbol{l}_2^2) \cdot \dot{\boldsymbol{\beta}}\dot{\boldsymbol{\gamma}}) \cdot \boldsymbol{m}_3}{2} + \frac{\boldsymbol{J}_{A3}}{2}\left(\dot{\boldsymbol{\alpha}}^2 + (\dot{\boldsymbol{\beta}} + \dot{\boldsymbol{\gamma}})^2\right) + \frac{\boldsymbol{J}_{g3}}{2}\dot{\boldsymbol{\gamma}}^2,$$

where

$$L_1^2 = l_1^2 + \rho_2^2 + 2l_1\rho_2 cos(\beta);$$

$$L_2^2 = (l_2 sin(\beta) - \rho_3 sin(\gamma - \beta))^2 + (l_2 cos(\beta) + \rho_3 cos(\gamma - \beta) + l_1)^2$$
$$= l_2^2 + \rho_3^2 + 2l_2\rho_3 \cos(\gamma) + l_1^2 + 2l_1l_2 \cos(\beta) + 2l_1\rho_3 \cos(\gamma - \beta);$$

$$L_3^2 = l_2^2 + \rho_3^2 + 2l_2\rho_3 \cdot \cos(\gamma).$$

The potential energy of the system is:

$$U_2 = m_2g \cdot \rho_2 \cdot cos(\beta) + m_2g \cdot l_2 \cdot \cos(\beta);$$
$$U_3 = m_3g \cdot \rho_3 \cdot cos(\beta - \gamma). \tag{6}$$

Let us write out the kinetic energy by substituting expressions 3–5 in 2 and grouping the terms by the derivatives of the generalized coordinates:

$$\boldsymbol{T} = \dot{\boldsymbol{\alpha}}^2 \cdot \boldsymbol{J}^{(1)} + \dot{\boldsymbol{\beta}}^2 \cdot \boldsymbol{J}^{(2)} + \dot{\boldsymbol{\gamma}}^2 \cdot \boldsymbol{J}^{(3)} + \dot{\boldsymbol{\beta}}\dot{\boldsymbol{\gamma}} \cdot \boldsymbol{J}^{(4)}, \tag{7}$$

where

$$J^{(1)} = \frac{1}{2}\left(\rho_1^2 \cdot m_1 + J_{A1} + J_{g1} + J_{A2} + L_1^2 \cdot m_2 + J_{A3} + L_2^2 \cdot m_3\right);$$

$$J^{(2)} = \frac{1}{2}\left(\rho_2^2 \cdot m_2 + J_{A2} + J_{g2} + J_{A3} + L_3^2 \cdot m_3\right);$$

$$J^{(3)} = \frac{1}{2}\left(\rho_3^2 \cdot m_3 + J_{A3} + J_{g3}\right);$$

$$J^{(4)} = \frac{1}{2}\left(L_3^2 + \rho_3^2 - l_2^2 + J_{A3}\right).$$

Substituting expressions 6, 7 into 1, we get the equations for the dynamics of the robot's leg.

4 Conclusions

The obtained dynamic equations of the robot leg are quite difficult to analyze analytically. In general, they do not have an analytical solution. In general, they have no an analytical solution. For numerical solution and analysis, the system must be supplemented by adding the operation of three electric motors to the obtained equations.

The multiplier $J^{(1)}$, facing $\dot{\alpha}^2$, is the moment of inertia of links 1 and the system of the 2nd and 3rd links (considered at each given moment as a rigid structure) relative to the center O_1 of rotation of the first hinge. It is seen that the multiplier $J^{(1)}$ takes on its maximum value when L_1 and L_2 become maximum. This occurs when $\cos(\gamma) = \cos(\beta) = \cos(\gamma - \beta) = 1$. Therefore, the most difficult case for the engine (motor) of the first link will be at $\beta = \gamma = 0$, when three links are stretched into one line. It would seem that for fast movement, when it is easier for the first motor to turn with its foot (movement along the angle α) at angles $\gamma = 180$, $\beta = 90$, but the width of the robot's stroke will be minimal and the robot will make a narrow step. For a more detailed analysis of the choice of optimal movement, for example, for fast movement, it is necessary to conduct a numerical experiment based on a mathematical model of the dynamic process of movement. Based on a numerical experiment, it is necessary to perform multicriteria optimization using methods of visualizing solutions to the problem [6–9].

Acknowledgment The research was supported by Russian Foundation for Basic Research, project No. 18-29-10072 mk (Optimization of nonlinear dynamic models of robotic drive systems taking into account forces of resistance of various nature, including frictional forces).

References

1. https://www.kuka.com/. Accessed 21 June 2020
2. Ivlev, V.I., Misyurin, S.Y.: Calculated and experimental characteristics of a scroll machine operating in the air motor mode. Dokl. Phys. **62**(1), 42–45 (2017)
3. Azar, A.T., Ammar, H.H., Beb, M.Y., Garces, S.R., Boubakari, A.: Optimal design of PID controller for 2-DOF drawing robot using bat-inspired algorithm. In: Proceedings of the International Conference on Advanced Intelligent Systems and Informatics 2019, pp. 175–186. Springer, Cham (2019)
4. Chen, J., Liu, Y., Zhao, J., Zhang, H., Jin, H.: Biomimetic design and optimal swing of a hexapod robot leg. J. Bionic Eng. **11**(1), 26–35 (2014)
5. Hu, Y., Mombaur, K.: Bio-inspired optimal control framework to generate walking motions for the humanoid robot iCub using whole body models. Appl. Sci. **8**(2), 278, 1–22 (2018). 22 pages
6. Misyurin, S.Y., Ivlev, V.I., Bozrov, V.M., Nelyubin, A.P.: Parameterization of an air motor based on multiobjective optimization and decision support. J. Mach. Manuf. Reliab. **42**(5), 353–358 (2013)
7. Misyurin, S.Y., Nelyubin, A.P., Potapov, M.A.: Multicriteria approach to control a population of robots to find the best solutions. In: Samsonovich, A.V. (ed.) BICA 2019. Advances in Intelligent Systems and Computing, vol. 948, pp. 358–363. Springer, Cham (2020)
8. Misyurin, S.Y., Nelyubin, A.P., Galkin, T.P., Galaev, A.A., Popov, D.D., Pilyugin, V.V.: Usage of visualization in the solution of multicriteria choice problems. Sci. Vis. **9**(5), 59–70 (2017)
9. Kreinin, G.V., Misyurin, S.Yu., Nelyubin, A.P., Nosova, N.Yu.: Visualization of the interconnection between dynamics of the system and its basic characteristics. Sci. Vis. **12**(2), 9–20 (2020)

Multiagent Model of Perceptual Space Formation in the Process of Mastering Linguistic Competence

Zalimkhan Nagoev and Irina Gurtueva(✉)

The Federal State Institution of Science Federal Scientific Center, Kabardino-Balkarian Scientific Center of Russian Academy of Sciences, I. Armand Street, 37-a, 360000 Nalchik, Russia
gurtueva-i@yandex.ru

Abstract. We propose a simulation model of early development for language competence. It is a model of phonemic imprinting that describes the process of mapping audio stimuli into classes of elementary units of a language, taking into account the influence of social factors. The supervised machine-learning algorithm used in the model was developed using the features of speech addressed to children. This model will allow us to explore the features of phonetic perception, the cognitive mechanisms that underlie language development, highlight the main factors affecting the duration of the period of plasticity. On that basis we can build perceptual maps, create diagnostic tools to describe the sensitive period, which will facilitate the study of the stages of its opening and closing. The model can also be used to create speech recognition systems that are resistant to a variety of accents and effective when used in noisy environment.

Keywords: Speech recognition · Multi-agent systems · Artificial intelligence · Early development · Speech acquisition

1 Introduction

Automatic speech recognition systems can be generally described as software systems that convert the sound wave of a speech message from an acoustic form into a symbolic one [1, 2]. The range of practical application of speech recognition technologies is extremely wide – from voice typewriters to natural-language control systems of multi-agent robots [1].

Although in recent years, some researchers have stated that the problem of speech recognition has been mainly solved and human parity has been achieved. However, there are still problems with robustness and overlapping speech, the so-called "cocktail party" situation [1, 3]. More than that, existing systems cannot be considered universal [4]. In our hypothesis, a successful solution to this problem is possible if the strategies that children use when mastering language skills are identified. It is necessary to study the critical period for phonetic training in order to create a system based on the cognitive functions that a person uses when decoding audio messages [5].

A. V. Samsonovich et al. (Eds.): BICA 2020, AISC 1310, pp. 327–334, 2021.
https://doi.org/10.1007/978-3-030-65596-9_39

This concept is based on the various results of studies of experimental psycholinguistics and neuro-linguistics, fundamental concepts of developmental psychology, cognitive sciences, and the studies of maternal speech [5–11].

2 Methods

The concepts on the acquisition of phonemic awareness that proposed in this work are based on the most important discovery of P. Eimas, proving that infants have a unique ability to distinguish phonetic contrasts of any language [12]. This result has been confirmed in numerous experiments [5]

The multi-agent [13–17] model of phonemic imprinting in the process of acquiring language skills, as Fig. 1 shows, consists of three stages: registration, evaluation, and placement.

At the pre-processing stage, the audio signal is converted into a set of signatures, on which a matrix is created. The matrix characterizes the acoustic characteristics of the signal. Since the sound wave with sufficient completeness can be characterized by four physical parameters, intramodal differentiation is reduced to the allocation of four layers in the structure of the sound stream: amplitude, spectral composition, duration of sound, and location of the signal source [18]. This information is fed to the input of the multi-agent system [15], where actor-agents are created. Their functionality corresponds to human auditory sensors [19, 20].

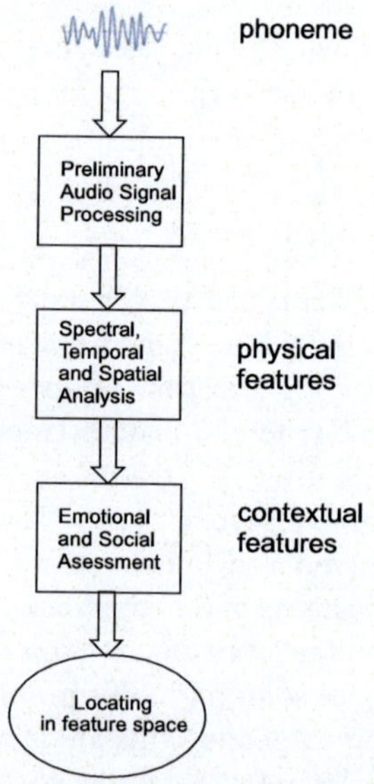

Fig. 1. Stages of processing an audio message by a multi-agent phonemic imprinting model

The first stage – assessment – imitates the activation of neurons in the speech zones of the brain with sound stimuli. In forming ideas about the principles of the functioning of this unit, we relied on studies of maternal speech [7, 21].

The use of maternal speech with very few exceptions is observed in all language communities. Maternal speech is traditionally studied at four levels – phonological, lexical, syntactic and communicative. Researchers set aside four, at least, and more features that distinguish speech addressed to children from speech used in peer communication [6].

For researches in present work, the features of great interest are phonological and prosodic ones. In addition, of course, the principles of expediency that underlie communication with children are in the focus of our attention.

First, there is an exaggeration of the acoustic ranges in motherese, presumably aimed at forming ideas about the permissible limits of speech variability. Primarily, the modulation of the tone level can be attributed to this principle; moreover, the extension of the upper boundary of the frequency range of the fundamental tone is typical [22]. The intonation contours of statements [22, 23] are also exaggerated.

Second, we should note simplifications and repetitions, contributing, at the initial stage of speech development, to facilitate the segmentation of continuous speech, and then accelerate understanding and learning. Namely, labialization, the exclusion of consonants difficult in articulation, onomatopoeia, a decrease in the rate of speech, observed not only in the caregivers' communication with infants, but also in the speech of experienced teachers. In addition, there is a tendency to increase the duration of pauses. The frequency feature of speech addressed to children is also the prolongation of vowel phonemes. According to the results of studies [21], the average duration of pronouncing the syllable core is much longer during communication with a child. The research data [5] document the excess of the prolongation of the sound of vowels realized in the speech of mothers compared with their sound in communication with equal age recipients three times. Repeats of prosodic patterns are also used.

The third principle, which is persistently preserved in maternal speech, can be described as the stimulation of social involvement, aimed at improving the effectiveness of learning and the duration of memorization. Speech addressed to children attracts the attention of the child, gives clear emotional signals, and engages in communication. However, maternal speech does not necessarily sound louder than normal. Interestingly, whispers are often used in the communication of an adult with a child. The most common hypothesis that explains this is that mothers instinctively use whispers to attract attention. In most cases, the final syllables in the last syntagma sound like this, sometimes the whole utterance is pronounced in a whisper. In conversation with two-year-olds, every sixth phrase sounds in a whisper. It was shown in [24] that when interacting with children, repeated rhythmic structures, rhymes, doubling of sounds are used; they provoke infants to synchronous motor activity, which also contributes to a more effective mastery of language skills. Undoubtedly, the greatest influence is exerted by the preference of certain melodic contours, which give a positive coloring to the communicative situation.

A significant influence of social factors in the acquisition of language skills has traditionally been noted in the theory of social learning [25]. However, new studies show that social interaction contributes not only to lexical development, but also to the mastery of the elementary units of the language. Research [26] showed a strong correlation between differences in the social behavior of children during class and the degree of mastery of phonetic contrasts.

Thus, the ability to language acquisition at an early stage of development is a set of general computational skills and extraordinary learning abilities. Therefore, at the assessment stage, the system supplements the acoustic (physical) signs of the signal recorded by the human peripheral auditory system with parameters corresponding to mental representations of them. Namely, it assigns a binary feature (prototype/non-prototype) and emotional coloring based on a comparative assessment of the duration of the maternal vowels and the average duration of the vowels extracted from the speech of adults, as well as using information that low-volume signals attract attention and are evaluated positively.

Finally, at the third stage, the system places the agent in the feature space. Placement is carried out taking into account the conceptual representations of the magnetic theory of native language [5]. Phonemes labeled as prototypical converge in the feature space; stimuli corresponding to the studied categories are processed faster. This will help to reflect the fact that early learning can limit subsequent learning [27]. And also there is the well-known evidence that, having completed teaching the native language, adults learning a new language are not able to implicitly absorb the statistical properties of the phonetic units of the new language. Attention and cognitive efforts are still determined by the structure of the mastered category. Increased attention and mental effort are needed to process stimuli that are sharply inconsistent with existing phonetic categories. θ-rhythms of the brain indicate a steady change in the perception of speech, due to the influence of the linguistic environment [27]. The neural network of the child's brain concentrates on recording high-frequency speech events represented by phonetic categories used in this environment.

3 Result

The speech stream is recorded by a system of microphones. Then, its spectral composition is revealed using the short-time Fourier Transform [1] using the Cooley and Tukey algorithm [1]. Then, the YIN method [28], similar to the autocorrelation function [1], is used to estimate the fundamental frequency, since this method is most effective for extracting the fundamental frequency of monophonic musical instruments and speech. The composition of harmonics is determined by the two-way mismatch method [29]. Thus, at the pre-processing stage, the audio signal is converted to the following set of signatures:

$$\langle F_0, F_1, F_2, \Delta t \rangle, \tag{1}$$

where F_0 - pitch frequency,

F_1, F_2 - the first and second formants,

Δt - sound duration of the investigated phoneme.

The resulting feature vector is fed to the input of a multi-agent recursive cognitive architecture [15], in which the developers pre-formed a set of so-called neural factories

- agents of a special type that dynamically create agents upon request, determine their type and place them in the corresponding space area of the multi-agent system. So, the neurofactory creates an agent responsible for some phoneme. To create such an agent, a special program is used that reads the agent's genome - the starting set of production rules in the agent's knowledge base [15].

The machine-learning process involves the supply to the input of the system a complete set of training material, which is selected taking into account the features of speech used when referring to children [5]. As described in detail above, one of the universal features of motherese is the duration of the vowel phonemes, exceeding the duration of the vowels in speech addressed to adults, more than three times [5]. The multi-agent system, based on a comparative assessment of the duration of the analyzed phoneme with existing statistical estimates of the longitude of the sound of the "mother's" phoneme [5], assigns an additional attribute to the agent - emotional coloring, and also identifies it as a prototype/non-prototype. Since the duration of the sound is not informative beyond this context, this parameter is a binary feature. Emotional assessment is determined in the range from 0 to 1, taking into account experimental data on the positive perception of prolonged vowels and concentration on sounds with a low volume level.

Thus, at this stage, the knowledge base of the agent characterizing the phoneme is the following set of signs:

$$\Phi_{vowel} = (\langle F_0,\ F_1,\ F_2,\ \Delta t\rangle,\ \mathrm{K}_{letter},\ \varepsilon,\ P), \quad (2)$$

where F_0 - pitch frequency,

F_1, F_2 - the first and second formants,

Δt - sound duration of the investigated phoneme.

K_{letter} - classifying contractual relationship with an agent-letter,

$\varepsilon \in [0, 1]$ - emotional assessment,

$P(0, 1)$ - prototype/nonprototype.

Then, the agent, realizing the behavior determined by the rules written in his own genome, in order to find the class to which he belongs, questions the expert. Based on the expert's response, a contract is concluded between the first-level agent and the letter-agent characterizing this class, that is, supervised machine learning is implemented [30].

It is important to note that the prototype/nonprototype attribute and the emotional assessment determine not only the spatial position of the agent in the multi-agent perceptual space, but also the initial life time of the agent in the system, which is then controlled by a stepwise memory function. In the case of a long period of inactivity of the agent, it is expelled from the system. The settings of the agent's life expectancy parameter will allow one to study the problems of the effectiveness of children's mastering new knowledge, as well as the problems of memorization [6].

As Fig. 2 shows, the result of the functioning of the first level of the architecture under development is the creation of agent agents that record the acoustic characteristics of the signal, like human auditory receptors and the formation of sets of agents corresponding to each minimal speech unit of the language and agents corresponding to the phoneme. The proposed algorithm allows to track the mechanism of formation of human auditory patterns.

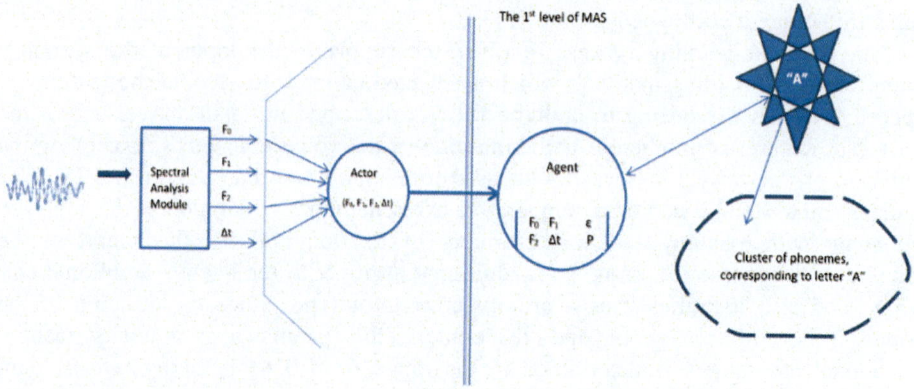

Fig. 2. Multiagent model of perceptual space of vowels for native language

At subsequent levels, to speed up the recognition process, it is planned to apply phonological, grammatical restrictions [31]. It is also planned to introduce feedbacks for correction and refinement of decoding results.

Thus, based on an analysis of experimental data from behavioral studies and model ideas about speech recognition mechanisms from the point of view of psycholinguistic knowledge, a machine learning method with imitation of the formation of a person's phonemic hearing was developed.

4 Conclusion and Discussion

In this paper, we propose the multi-agent model for the early development of phonemic competence and the supervised machine-learning algorithm with an imitation of the mechanism of formation of human neural systems. The proposed model will allow to study the factors affecting the duration of the sensitivity period, its causes and mechanisms, create diagnostic tools to describe the sensitivity period, study the content of the stages of its opening and closing, and create speech systems that are resistant to a variety of accents.

Acknowledgements. The research was supported by the Russian Foundation of Basic Research, grants No. 18-01-00658, 19-01-00648.

References

1. Jurafsky, D., Martin, J.: Speech and Language Processing: An Introduction to Natural Language Processing, Computational Linguistics, and Speech Recognition, 2nd edn. Prentice Hall, New Jersey (2008)
2. Gupta, V.: A survey of natural language processing techniques. Int. J. Comput. Sci. Eng. Technol. (IJCSET) **5**(1), 14–16 (2014)
3. Zion Golumbic, E.M., Ding, N., Bickel, S., et al.: Mechanisms underlying selective neuronal tracking of attended speech at a "cocktail party". Neuron **77**(5), 980–991 (2013). https://doi.org/10.1016/j.neuron.2012.12.037
4. Waibel, A., Lee, K.-F.: Readings in Speech Recognition. Morgan Kaufman, Burlington (1990)
5. Strange, W.: Speech Perception and Linguistic Experience: Issues in Cross-Language Research. York Press, Baltimore (1995)
6. Tseitlin, S. N.: A Child and a Language: Child Speech Linguistics. Humanitarian Publishing Center VLADOS, Moscow (2000). [Tseitlin, S. N.: Yazyk i rebyonok: lingvistika detskoi retchi. Gumanitarny Izdatel'sky Tsentr VLADOS, Moskva (2000)]
7. Chomsky, N.A.: A Review of Skinner's Verbal Behavior. In: Jakobovits, L.A., Miron, M.S. (eds.) Readings in the Psychology of Language. Prentice-Hall, New Jersey (1967)
8. Morozov, V.P., Vartanyan, I.A., Galunov, V.I.: Speech Perception: Problems of Functional Brain Asymmetry. Science, St. Petersburgh (1988). [Morozov, V.P., Vartanyan, I.A., Galunov, V.I.: Vospriyatie Rechi: Voprosy Funkcionalnoi Asimmetrii Mozga. Nauka, Leningrad (1988)]
9. Newell, A.: Unified Theories of Cognition. Harvard University Press, Cambridge (1990)
10. Haikonen, P.: The Cognitive Approach to Conscious Machines. Imprint Academic, Exeter (2003)
11. Schunk, D.H.: Learning Theories: An Educational Perspective. Pearson Merrill Prentice Hall, New York (2011)
12. Pinker, S.: The Language Instinct: How the Mind Creates Language. Harper Perennial, New-York (2007)
13. Kotseruba, I., Tsotsos, J.K.: A review of 40 Years of Cognitive Architecture Research: Core Cognitive Abilities and Practical Applications. https://arxiv.org/abs/1610.08602
14. Wooldridge, M.: An Introduction to Multi-Agent Systems. Wiley, Hoboken (2009)
15. Nagoev, Z.V.: Intellectics, or thinking in living and artificial systems. Publishing House KBSC RAS, Nalchik (2013). [Nagoev, Z. V.: Intellektika ili myshleniye v zhyvych i iskusstvennych sistemach. Izdatel'stvo KBNC, Nal'chik (2013)]
16. De Mulder, W., Bethard, S., Moens, M.-F.: A survey on the application of recurrent neural networks to statistical language modeling. Comput. Speech Lang. **30**(1), 61–98 (2015)
17. Deng, L., Li, X.: Machine learning paradigms for speech recognition: an overview. IEEE Trans. Audio Speech Lang. Process. **21**(5), 1060–1089 (2013)
18. Nagoev, Z., Lyutikova, L., Gurtueva, I.: Model for Automatic speech recognition using multi-agent recursive cognitive architecture. In: Annual International Conference on Biologically Inspired Cognitive Architectures BICA, Prague, Czech Republic. http://doi.org/10.1016/j.procs.2018.11.089
19. Nagoev, Z., Gurtueva, I., Malyshev, D., Sundukov, Z.: Multi-agent algorithm imitating formation of phonemic awareness. In: Samsonovich, A.V. (ed.) BICA 2019. AISC, vol. 948, pp. 364–369. Springer, Cham (2020). https://doi.org/10.1007/978-3-030-25719-4_47

20. Nagoev, Z.V., Gurtueva, I.A., Нагоев, З.В., Гуртуева, И.А.: Fundamental Elements for Cognitive Model of Speech Perception Mechanism Based on Multiagent Recursive Intellect, vol. 3, no. 89, pp. 3–14. News of Kabardino-Balkarian Scientific Center of RAS (2019) [Nagoev, Z. V., Gurtueva, I. A.: Bazovye element kognitivnoi modeli mehanizma vospriyatiya rechi na osnove multiagentnogo rekursivnogo intellekta. Izvestiya Kabardino-Balkarskogo nauchnogo tsentra RAN (89), 3-14 (2019).]
21. Garnica, O.: Some prosodic and paralinguistic features of speech to young children. In: Snow, C., Ferguson, Ch. (eds.) Talking to Children. Cambridge University Press, Cambridge (1977)
22. Fernald, A.: Four-month-old infants prefer to listen to motheress. Infant Behav. Develop. **8**, 181–195 (1985)
23. Fernald, A., Kuhl, P.: Acoustic determinants of infant preference for motherese speech. Infant Behav. Develop. **10**, 279–293 (1987)
24. Moerk, E.L.: Principles of interaction in language learning. Merril-Palmer Q. **18**, 229–257 (1972)
25. Vygotsky, L.S.: Thinking and Speech. Piter, St-Petersburg (2019). [Vygotsky, L. S.: Myshlenie I Rech'. Piter, Sankt-Peterburg (2019)]
26. Conboy, B.T., Kuhl, P.K.: Impact of second-language experience in infancy: brain measures of first- and second-language speech perception. Develop. Sci. **14**, 242–248 (2011). https://doi.org/10.1111/j.1467-7687.2010.00973.x
27. Doupe, A.J., Kuhl, P.K.: Birdsong and human speech: common themes and mechanisms. In: Zeigler, H.P., Marler, P. (eds.) Neuroscience of Birdsong. Cambridge University Press, pp. 5–31 (2008)
28. De Cheveigne, A., Kawahara, H.: YIN, a fundamental frequency estimator for speech and music. J. Acoust. Soc. Am. **111**(4), 1917–1930 (2002). https://doi.org/10.1121/1.1458024
29. Maher, R.C., Beauchamp, J.W.: Fundamental frequency estimation of musical signals using a two-way mismatch procedure. J. Acoust. Soc. Am. **95**, 2254 (1994). https://doi.org/10.1121/1.408685
30. Coates, A., Ng, A.Y.: Learning feature representations with K-Means. In: Montavon, G., Orr, G.B., Müller, K.R. (eds.) Neural Networks: Tricks of the Trade. Lecture Notes in Computer Science, vol 7700. Springer, Berlin, Heidelberg (2012)
31. Pye, C.: Quiché Mayan speech to children. J. Child Lang. **13**(1), 85–100 (1986). https://doi.org/10.1017/S0305000900000313

The Role of Gender in the Prosocial Behavior Mechanisms

Yulia M. Neroznikova and Alexander V. Vartanov(✉)

Lomonosov Moscow State University, Moscow, Russia
Julia-sheyn@mail.ru, a_v_vartanov@mail.ru

Abstract. Prosocial behavior is progressively being studied because of its enormous role in the changing conditions of the modern world. We study the prosocial behavior by analyzing the brain features mechanisms during the custom setting - empathy game "Stone-Paper-Scissors". Observed results for the cohort of 55 participants (28 women and 27 men) were obtained from the test questionnaires and EEG. Our approach relates to the field of medicine and neuroscience, in particular, to a method for studying the activity of individual brain structures predefined by their spatial position according to the scalp multichannel EEG. The results of the questionnaires were tabulated and processed by the factor analysis. Six factors were identified that accounted for 41.45% of the total variance of the data. The factors received the following interpretation: "Altruistic", "Publicity", "Emergency prosocial behavior", "Conformist prosocial behavior", "Anonymous prosocial behavior", "Emotional prosocial behavior". Comparison by Student's T-test showed a significant ($p < 0.05$) difference between men and women in the "Publicity" factor and in the "Emotional prosocial behavior" factor (females are more emotional than males).

Keywords: EEG · Prosocial behavior · Empathy

1 Introduction

The development and correlation of prosocial behavior (i.e., behavior that is beneficial to others) have been an active area of research for the past three decades. Given the importance of understanding behaviors that benefit society, efforts are now just beginning to study prosocial behavior at the level of brain processes and transference to the capabilities of artificial intelligence. In order for Artificial Intelligence (AI) to empathize with human emotions, AI needs to be able to learn about the range of emotions that we experience. With the help of empathy in AI Systems, artificial emotional intelligence can allow us to take the first step toward functional artificial general intelligence that will empower human-kind.

While some measures do exist, they tend to conceptualize prosocial behavior as a global construct. However, researchers have shown that there are different types of prosocial behavior and that they are differently related to theoretically conditioned constructs [1]. Moreover, the existing prosocial measures can be divided into at least two categories: measures that assess global prosocial behavior and measures that assess

A. V. Samsonovich et al. (Eds.): BICA 2020, AISC 1310, pp. 335–341, 2021.
https://doi.org/10.1007/978-3-030-65596-9_40

prosocial behavior in a specific situation. The most common propositional measures are those designed to assess global propositive behavior. Global measures of prosocial behavior are defined as measures that assess personal trends in the manifestation of a range of prosocial behavior in different contexts and motivations [2].

There is an antisocial side of behavior and it can include such serious things as double-crossing play, harming, stealing. Decisions about future actions and behavior are made depending on the expected profit, and inclusion in a particular social context. It is important to recognize that reflection on the origin of prosocial behavior has generated yet another scientific debate about human nature. In reality, we are by origin egoists or prosocial [3].

A lot of social situations could be modeled like non-zero-sum games, in order that all participants can benefit from the interaction, sometimes they are stuck in social dilemma. Social dilemma investigations in psychology have typically based on experimental games where players are interdependent and the choices offered in the game benefits they provide to self and others.

Electroencephalography (EEG) is a technique that has been around much longer and has proven very successful in assessing differences in broad patterns, such as differences between left and right hemisphere functions. It is known that functions related to prosocial behaviors attributed to brain regions. To consider empathic and altruistic behavior we can highlight responsible brain regions, such as 1. The insular cortex (anterior insula (AI) is responsible for homeostatic altering to emotional stimuli, perceiving one's own and other's pain, empathy, fairness considerations, and betrayal aversion 2. The frontal lobe, Anterior cingulate cortex (ACC) – conflict detection and monitoring; recruits lateral PFC to resolve conflicts; link with cognitive control during the empathic response. 3. Subcortical, Ventral striatum (VS) nucleus accumbens - valuation; comparing actual with expected rewards; computing experienced decision utility.

Based on the social role theory, it was demonstrated that the relations between gender and helping differed as a function of the type of helping examined. Specifically, helping that was more heroic or more chivalrous was exhibited more often by young men than young women, whereas helping embedded in a relational context was exhibited by young women more than young men.

The previous analysis did not reveal any differences between the motives for prosocial behavior, while we used the experimental design with two series including the identified empathy motives and the prosocial factors. The present study has a considerable interdisciplinary background, linked with researching the psychophysiological mechanisms of prosocial behavior and comprehension in psychological concepts of altruism and empathy. To summarize, we present the factors of the prosocial personality by examining the intercorrelations among the measures. We created the model that highlighted the main brain structures involved in prosocial reactions and behaviors.

2 Method

2.1 Data Processing

The analysis of the data was performed using the new (author-developed by A.V. Vartanov) method of brain activity localization that can be briefly explained as the "virtually implanted electrode". The method reconstructs the electrical activity from the scalp EEG data, by modeling the electrical activity distribution in the brain with a local field potential ansatz. Such representation is similar to the spatial filtering method (presented in US Patent No. 5263488, authors Van Veen, Joseph, Hecox, 11.23.1993) or more generally to a group of source localization methods, such as "radiation pattern formation" ("Beamforming"). However, the used method converges to an unambiguous and reliable solution, that is not always guaranteed for source localization methods. The method can also be used for filtering the background noise in the EEG data by subtracting the signal of the surrounding points, in particular within a radius of 1 cm.

To study the prosocial behavior we used the two methodic: 1) Test-questionnaires of prosocial behavior: "Measurement of prosocial trends" (G. Carlo and B. A. Randall), the Altruism scale (F. Rushton), Method for the empathy study of I.M. Yusupov 2) An electrophysiological experiment with the registration of a Nineteen-channel Encephalograph (according to the 10–20% system on a Neuro-KM electroencephalograph) consisted of two series. The results of the recording were cleared from the artifact data with the help of the expert analysis. We filtered desired fragments and make it averaged to obtain the evoked potentials (EP). Participants were acquainted with the full information about the current study and gave their consent to the processing.

3 Participants and Procedure

We used the data from 55 participants (28 women and 27 men) in the questionnaire study and the data from 16 participants (6 women and 10 men) in the EEG experiments.

To stimulate brain activity, participants were asked to watch the game "Stone-Paper-Scissors". "Stone-Paper-Scissors" is a hand game usually played between two people, in which each player simultaneously forms one of three shapes with an outstretched hand. These shapes are "rock" (a closed fist), "paper" (a flat hand), and "scissors" (a fist with the index finger and middle finger extended, forming a V). Each of the three basic hand signs (from left to right: rock, paper, and scissors) beats one of the other two and loses to the other.

We showed participants images with different outcomes. The images contain 6 different options of hand signs and the facial expressions of the players. The participant reaction (EP) was tracked over a period of time from 200 ms to 500 ms after the stimulus.

In the first series, the instruction was required to bet on one of the two playing persons. We expected that the participant would be interested in the success and experienced the situation as the "first-person". We can therefore assume that the "First-person" experiences show a person's involvement not only in the empathy process but also in conjunction with altruistic thoughts.

In the second series, the instruction hast told participants to determine the emotionally "close" player and empathize with him throughout the series. We expected that the participant will presumably experience empathy for the player. We can explain such behavior by assuming that a person could remember someone who had always supported him and never betrayed. Incentives were presented 50 times in random order. We used the same images for the first and second series.

Current experiment is approved by the Commission on Ethics and Regulations of the Academic Council Psychology Faculty of Lomonosov Moscow State University.

4 Results

We compared the data of the Questionnaire on Prosocial behavior and registration of EEG potentials for the same participants.

The results of the questionnaires were tabulated and processed by factor analysis. The results of the analysis suggested the 6 factors structure, which accounted for 41.45% of the total variance of the data. We suggest the following interpretation of the factors: F1 - "Altruistic", F2 - "Publicity", F3 - "Emergency prosocial behavior", F4 - "Conformist prosocial behavior", F5 - "Anonymous prosocial behavior", F6 - "Emotional prosocial behavior". Comparison by Student's T-test showed a significant ($p < 0.05$) difference between the men and the women in the "Publicity" factor and in the "Emotional prosocial behavior" factor (females are more emotional than males). These differences are manifested in the entire questionnaire data sample and in the data samples of the EEG experiment (Table 1).

Table 1. Interrelations Among the Prosocial Tendencies Measure Subscales T-tests for gender group 1 – male, 2 – female by Factors (F1–F6).

Variable	T-test; Grouping; Sex group 1-male group 2-female group									
	Male	Female	t-value	df	p	valid N male	valid N female	Std. Dev male	Std. Dev female	F-ratio Variances	p-Variances
"Altruistic"	0	0	0,1	53	0,9	27	28	0,8	1,2	1,9	0,1
"Publicity"	0,3	−0,3	2	53	0	27	28	0,8	1,1	1,7	0,7
"Emergency prosocial behavior"	0,1	−0,1	0,6	53	0,6	27	28	0,9	1,1	1,5	0,3
"Conformist prosocial behavior"	0,2	−0,1	1,1	53	0,3	27	28	0,9	1.1	1.6	0,2
"Anonymous prosocial behavior"	0	0	−0,3	53	0,8	27	28	0.8	1,1	1,8	0,1
"Emotional prosocial behavior"	−0,5	0,5	−3,8	53	0	27	28	0,7	1	1,8	0,1

We use the "virtually implanted electrode" to analyze activity at 29 points selected from the MNI152 atlas in the center of the following structures: Brainstem, Mesencephalon, Hypothalamus, Caput n.Caudati L, Caput n.Caudati R, Medula Oblongata, G. Cingulate Medialis, Globus Pallidus Medialis L, Globus Pallidus Medialis R, Corpus Amygdaloideum L, Corpus Amygdaloideum R, Anterior Cingulate BA32, Dorsomedial prefrontal cortex BA9 L, Dorsomedial prefrontal cortex BA9 R, Hippocampus L, Hippocampus R, Insula L BA13, Insula R BAex Baal, BAex Pariet R, Putamen L, Putamen R, Supramarginal gyrus BA40 L, Supramarginal gyrus BA40 R, Thalamus L, Thalamus R, V1 BA17 L, V1 BA17 R, Ventral Striatum BA25.

5 Discussion

The present study has a considerable interdisciplinary background, linked with researching the psychophysiological mechanisms of prosocial behavior and comprehension in psychological concepts of altruism and empathy. To summarize, we present the factors of the prosocial personality by examining the inter-correlations among the measures. We created the model that highlighted the main brain structures involved in prosocial reactions and behaviors (Fig. 3).

We compared the EP separately for the males and females' group in the experimental series. The statistically significant stimulus reaction (EP) was found in the hypothalamus (Fig. 1), which is more pronounced in the male group in the series requiring personal inclusion compared to the empathy series in time components N100, N250, and N400 (differences are significant, $p < 0.05$). We also observed statistically significant differences ($p < 0.05$) in the female group. These components are less pronounced, but even here in the series for empathy, their amplitude is significantly less ($p < 0.05$ in late latency) than in the series requiring personal participation. Potentials were also found in the right Insula (Fig. 4) and the Anterior Cingulate (Fig. 2). At the same time, they are more pronounced and differ more strongly between the male's series. A certain activity was also found associated with the studied factors of gender and empathy in such structures as the Mesencephalon, Medula Oblongata, left Thalamus, Putamen, left Caudati nucleus, Ventral striatum.

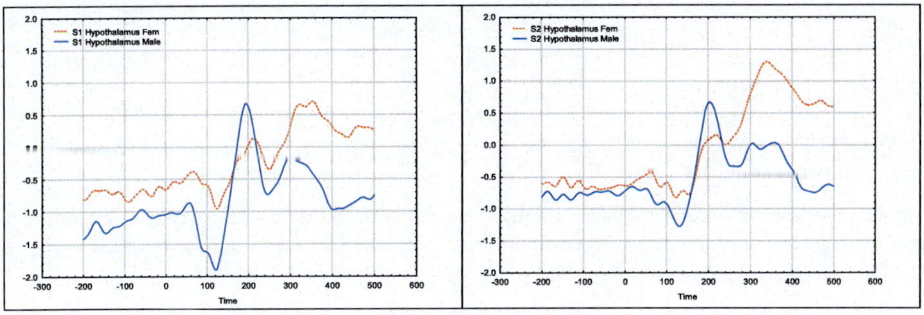

Fig. 1. Averaged evoked potentials, virtually withdrawal from the Hypothalamus, in the group of men (blue solid line) and women (red dotted line) in series 1 (left) and 2 (right).

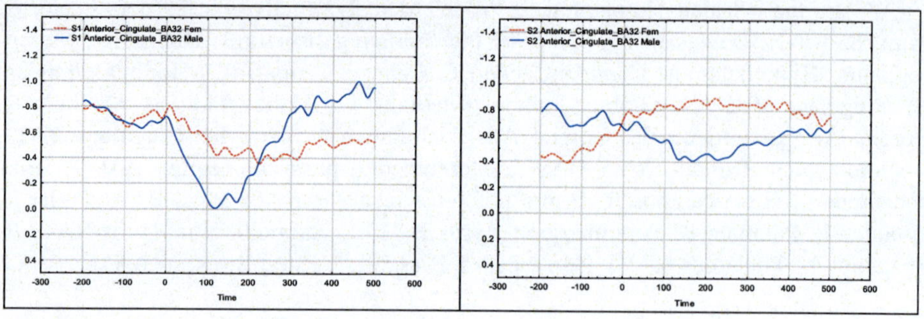

Fig. 2. Averaged evoked potentials that are virtually withdrawal from the Anterior Cingulate cortex (32 BA) in the group of men (blue solid line) and women (red dotted line) in series 1 (left) and 2 (right).

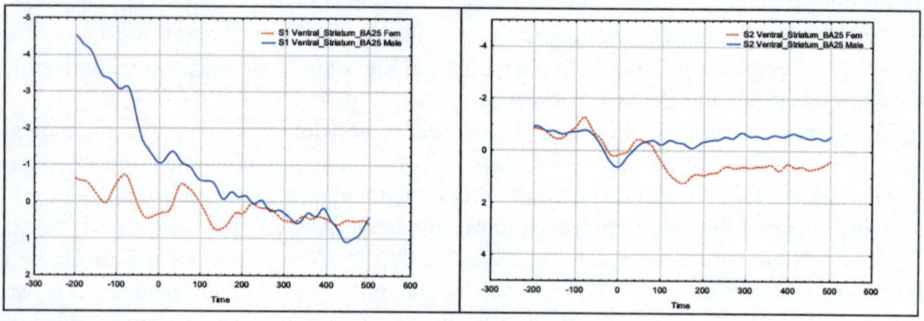

Fig. 3. The averaged evoked potentials virtually withdrawal from the Ventral Striatum (25 BA) in the group of men (blue solid line) and women (red dotted line) in series 1 (left) and 2 (right).

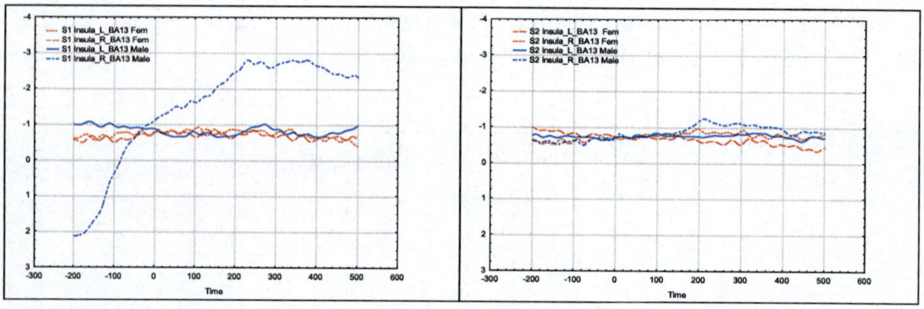

Fig. 4. Averaged evoked potentials virtually withdrawal from the Insula (13 BA) of the left and right hemispheres in the group of men (blue solid line) and women (red dotted line) in series 1 (left) and 2 (right).

6 Conclusion

Our research aimed to study the prosocial behavior by analyzing the brain features mechanisms during the custom setting - empathy game "Stone-Paper-Scissors" and it adds a new view on prosocial behavior from neuroscience and provides a complementary knowledge about the brain structures (using method «virtually implanted electrode") responsible for the processes of prosocial behavior by gender.

According to the questionnaires, we found the six prosocial behavior factors: "Altruistic", "Publicity", "Emergency prosocial behavior", "Conformist prosocial behavior", "Anonymous prosocial behavior", "Emotional prosocial behavior". According to the EEG experiment, we found that the Cingulate Cortex is actively involved in the empathic response in women at latencies of 100–400 ms. We know that the Anterior Cingulate (BA32) is an important part of the limbic system, responsible for the formation and processing of emotions. It combines behavioral outcomes with motivation (for example, if an action triggered a positive emotional response, it promotes learning). We found that the Ventral Striatum participates in the empathic response in males at latencies of 300–400 ms and in females at 200–250 ms. The Ventral Striatum primarily mediates reward, cognition, reinforcement. The Striatum is activated by reward-related stimuli as well as repulsive, new, unexpected, or intense stimuli and cues associated with such events. Striatal dysfunction can lead to a variety of disorders, especially depression and obsessive-compulsive disorder.

The phenomenon of empathy can be considered as one of the keys for artificial intelligence systems. AI is the property of intelligent systems to perform creative functions that are traditionally considered the prerogative of humans and today it is a science and technology for creating intelligent machines, especially intelligent computer programs. Due to the fact that artificial empathy is different from human empathy, this is an interesting perspective in human-machine interaction for future research.

7 Limitations

In future research, we suggested expanding to a large sample.

Acknowledgments. Funding: The research is financially supported by the Russian Science Foundation for Basic Research, Project № 20-013-00834.

References

1. Davis, M.: Measuring individual differences in empathy: evidence for a multidimensional approach. J. Person. Soc. Psychol. **44**(1), 113–126 (1983)
2. Eagly, A.H., Crowley, M.: Gender and helping behavior: a meta-analytic review of the social psychological literature. Psychol. Bull. **100**, 283–308 (1986)
3. Eisenberg, N., Fabes, R.: Prosocial development. In Damon W., Eisenberg, N. (eds.) Handbook of Child Psychology, vol. 3: Social, Emotional, and Personality Development, 5th edn., pp. 701–778. Wiley, New York (1998)

Reflection Mechanisms of Empathy Processes in Evoked Potentials

Yulia M. Neroznikova and Alexander V. Vartanov(✉)

Lomonosov Moscow State University, Moscow, Russia
Julia-sheyn@mail.ru, a_v_vartanov@mail.ru

Abstract. Empathy is the ability to recognize, understand, and share the feelings of another person, including the perception of another person. Empathy promotes prosocial or helping behavior that comes from within, rather than under duress.

Our paper revealed brain activity including empathy and altruistic processes in evoked potentials (EP). To review empathy, we analyze the brain activity characteristics with electrodes during watching the empathic game "Stone-Paper-Scissors". Results based on the nineteen-channel electroencephalography (EEG), according to the 10–20% system on a Neuro-KM electroencephalograph, recordings experiment on a sample of 16 participants (6 women and 10 men). Using the EP method, electrical activity was measured for the situation when the empathy behavior was activated. The evoked potential consists of a sequence of negative and positive deviations from the mainline and lasts 500 ms after the end of the stimulus. In EP, evaluate the amplitude and latent period of occurrence. To register the EP, the same electrodes are used as for the EEG recording, and with the unified observational conditions. We focused on brain structures associated with prosocial behavior, including the cortex, amygdala, and thalamus. A comparison between the empathy situations and the 'first-person' experience has been performed separately for reactions of men and women. We found significant differences in the following leads: T6, P4, O2, Pz in EP obtained in 2 series separately by gender.

Keywords: Evoked potentials · Empathy · EEG

1 Introduction

Human social interactions are greatly influenced by understanding the emotional states of others. Empathy, the affective response that stems from the apprehension or comprehension of another's emotional state or condition, allows for the understanding of what another person is feeling or would be expected to feel. The experience of empathy is a powerful interpersonal phenomenon necessary in everyday social interaction. It facilitates parental care and enables us to live in groups, cooperate, and socialize. The empathy experience also drives the way for the development of moral reasoning and motivates prosocial behavior [1].

However, on this path, there is a significant problem in the study of empathy - the separation of the empathic state from the first person, and empathizing with another.

A. V. Samsonovich et al. (Eds.): BICA 2020, AISC 1310, pp. 342–349, 2021.
https://doi.org/10.1007/978-3-030-65596-9_41

The term "empathy" was introduced by Edward Titchener, who traced the German word "Einfühlung", used in 1885 by Theodor Lipps in the impact context of the art theory. There are distinguished the following types of empathy: emotional empathy, based on the mechanisms of projection and imitation of the motor and affective reactions of another person, cognitive empathy is based on intellectual processes - comparison, analogy, etc., predictive empathy is manifested as the ability of a person to predict the affective reactions of another person in specific situations. Empathy occurs when humans vicariously feel the emotions of others (emotional resonance), explicitly understand the target's states and their sources, and evoke affective communications that motivate these individuals to remove the sources of the target's distress and/or provide comforts, such as empathic concern, sympathy, or compassion. Prosocial behavior includes helping, sharing, collaboration, and voluntary work. These actions motivated by empathy, altruism, or concern for the well-being and rights of others, as well as selfish or practical considerations [2].

There are three known theories of altruism, for example, according to one of them, the theory of social exchange, help, like any other social behavior, is motivated by the desire to minimize costs and optimize rewards. The theory of social norms proceeds from the fact that assistance is associated with the existence of certain rules in society. The reciprocity norm encourages us to respond with good, and not evil, to those who came to our aid. The norm of social responsibility makes us care for those who need it, as long as necessary, even when they are not able to thank us. Evolutionary psychology proceeds from the existence of two types of altruism: al-truism based on the defense of one's own kind, and altruism based on mutual exchange. However, most evolutionary psychologists believe that genes of selfish individuals are more likely to survive than genes of individuals who are prone to self-sacrifice and that therefore society should teach altruism. The third approach to the interpretation of altruism is based on evolutionary theory. It is known that according to the ideas of evolutionary psychologists, the quintessence of life is the preservation of the gene pool. Our genes make us behave in such a way as to create conditions that are most favorable for their survival in the future [3]. The genes that make a person ready to sacrifice himself for the welfare of a stranger had no chance of surviving in the competition of species for existence. However, thanks to genetic egoism, we are prone to special, unselfish altruism, which can even be called sacrificial and manifests itself as a protection of the genus and mutual exchange.

While the development of empathy has been explored for decades in behavioral studies, due to methodological constraints, many functional neuroimaging experiments have been conducted by fMRI. Focusing on brain structures associated with prosocial behavior, including the cortex, amygdala, and thalamus, provides comparing and integrating data. We aimed to study empathy differences in which the given context plays a leading role by EP.

2 Method

2.1 Data Processing

The collection of the data was performed using the EEG. In the EEG measurement, we used electrodes that were superimposed on the skin surface. The electrodes were connected with conductors in the electroencephalograph amplifier panel.

The electrodes we used were coated with chlorinated silver that provided a low transition resistance (less than 3–5 kΩ), small polarization degree, and high corrosion resistance. To fix the electrodes on the scalp using electrodes mesh.

We assumed two methods of EEG registration - monopolar and bipolar. In the bipolar setup potential, the difference was measured between two electrically active areas of the brain (both electrodes were attached to the scalp). Thus, the EEG is captured at the head surface, G. Berger proved that part of the electrical activity was defined by the activity of the brain, and not by the tissues covering its surface. The monopolar setup, the potential difference recorded between electrically active and neutral areas (such as earlobe or nose bridge). Lead electrodes can be attached to various areas of the head that correspond to the projected brain regions. In our experiment, we used the international system for the location of electrodes: 10–20%. "10" and "20" refer to the fact that the actual distances between adjacent electrodes are either 10% or 20% of the total front or back right left distance of the skull. In our experiment measurements are taken at the top of the head, from Nasion to the occipital protuberance. In the measurement setup, each electrode placement area had a letter to identify the brain area: Frontal (F), Temporal (T), Parietal (P), Occipital (O), and Central (C). There are also (Z) sites: "Z" (zero) refers to the electrode located on the midline sagittal plane of the skull (Fz, Cz, Oz, Pz). Also, the numbers (2, 4, 6, 8) refer to the right hemisphere and odd numbers (1, 3, 5, 7) refer to the left hemisphere.

To register the EP, the same electrodes are used as for the EEG recording. The evoked potential consists of a sequence of negative and positive deviations from the mainline and lasts 500 ms after the end of the stimulus. In the EP measurement, we focus on the amplitude and the latent occurrence period. These include the unification of methodological techniques of all series of experiments, conducting them on the same subject, in the same state, using the same stimulation parameters. Data processing includes methods of mathematical and statistical analysis.

To study the prosocial behavior, we used a nineteen-channel encephalography (according to the 10–20% system on a Neuro-KM electroencephalograph) consisted of two series. The results of the recording were cleared from the artifact data with the expert analysis. We filtered desired fragments and averaged them to obtain the EP. Participants were acquainted with the full information about the current study and gave their consent to the data processing.

2.2 Participants and Procedure

We used the data from 16 participants (6 women and 10 men, mean age = 19.6) in the EEG experiments.

To stimulate brain activity, participants were asked to watch the game "Stone-Paper-Scissors". "Stone-Paper-Scissors" is a hand game usually played between two people, in which each player simultaneously forms one of three shapes with an out-stretched hand. These shapes are "rock" (a closed fist), "paper" (a flat hand), and "scissors" (a fist with the index finger and middle finger extended, forming a V). Each of the three basic hand signs (from left to right: rock, paper, and scissors) beats one of the other two and loses to the other.

During the experiment, we showed participants images with different outcomes. The images contained 6 different options of hand signs and the facial expressions of the players. The participant reaction in the form of EP was tracked over a period of time from 200 ms to 500 ms after the stimulus.

In the first series, the instruction was required to bet on one of the two playing persons. We expected that the participant would be interested in the success and experienced the situation as the "first-person". We can therefore assume that the "First-person" experiences show a person's involvement not only in the empathy process but also in conjunction with altruistic thoughts.

In the second series, the instruction hast told participants to determine the emotionally "close" player and empathize with him throughout the series. We expected that the participant will presumably experience empathy for the player. We can explain such behavior by assuming that a person could re-member someone who had always supported him and never betrayed. Incentives were presented 50 times in random order. We used the same image sequences for the first and second series.

Current experiment is approved by the Commission on Ethics and Regulations of the Academic Council Psychology Faculty of Lomonosov Moscow State University.

3 Results

According to the results we obtained, as can be demonstrated from Fig. 1, the largest differences between the selfish first-person response and empathic response are observed at latencies of early components 150–250 ms in the T6 lead. This difference is more pronounced for the women group, while for the men group, the difference between the series was found at a later latency of 400 ms. Thus, the differences in the amplitude of the EP peaks at this time interval (after the presentation of the stimulus) reflect the change level of empathy, which may indicate that men show a higher level of empathy.

In addition, we suggest that it could be explained because men show an empathic reaction more consciously, perhaps semantic it.

At the same time, in Fig. 3 we present the measured signals in the occipital lead O2. We confirm the gender difference found previously in the signal. For the men group, the greatest differences were found in late latency with-in the interval of 300–500 ms, while in the women group differences are found in the earlier components' latency of 150 ms. Also, such latency difference also indicates that female empathic reaction with visual perception manifests itself at an unconscious level and does not differ much from the non-empathic response, i.e. the manifestation of empathy by women is less dependent on the conscious effort in connection with the instruction in the experiment.

The graph in Fig. 2 shows us that the differences between the selfish first-person response and empathic response are observed in the early components of latency 100–150 ms in lead P4. This difference is also more pronounced in the group of women, while in the group of men the difference between the series is also found in later latency - 400 ms.

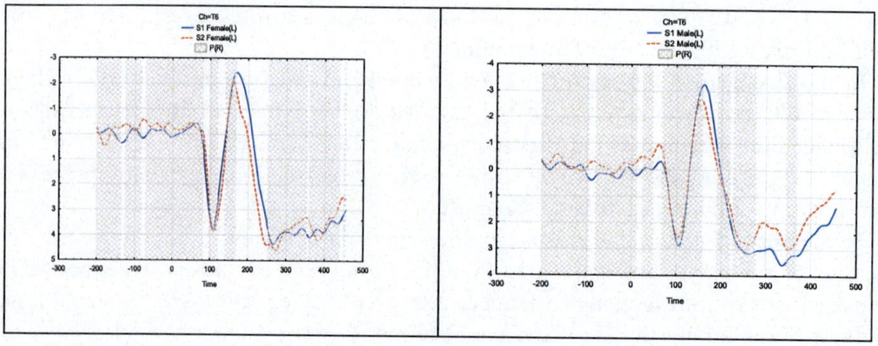

Fig. 1. EP obtained in 2 series (S1 - blue solid line; S2 - red dotted line) separately by gender according to the lead of Channel = T6. Time is recorded in milliseconds (ms)

Moreover, the greatest deviations affecting both the early and late components of the EP in the women group were found in the Pz lead that is shown in Fig. 4. For the men group in this lead, there are no significant differences between the series.

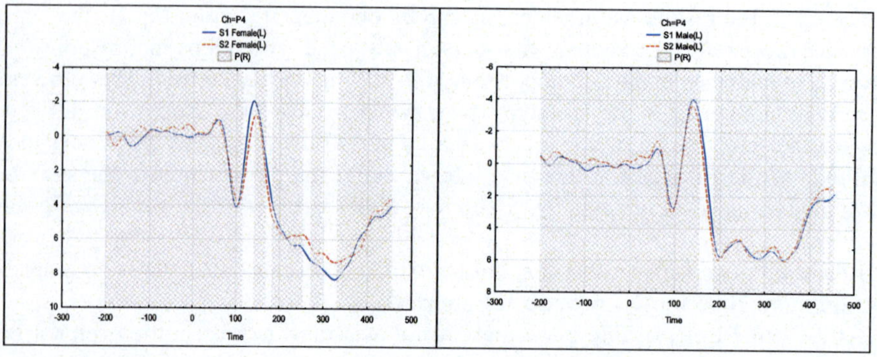

Fig. 2. EP obtained in 2 series (S1 - blue solid line; S2 - red dotted line) separately by gender according to the lead of Channel = P4. Time is recorded in milliseconds (ms)

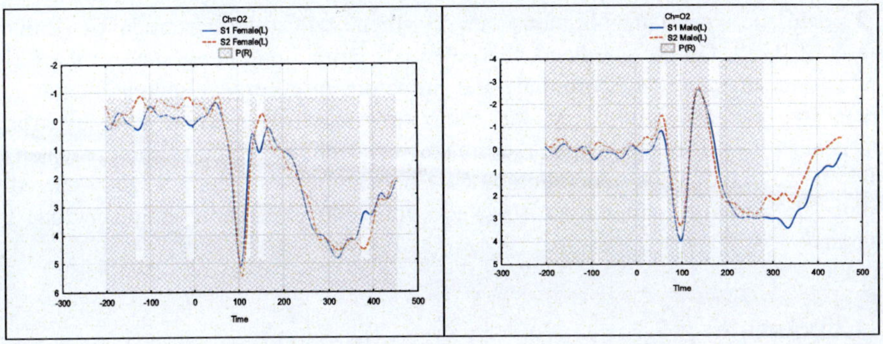

Fig. 3. EP obtained in 2 series (S1 - blue solid line; S2 - red dotted line) separately by gender according to the lead of Channel = O2. Time is recorded in milliseconds (ms)

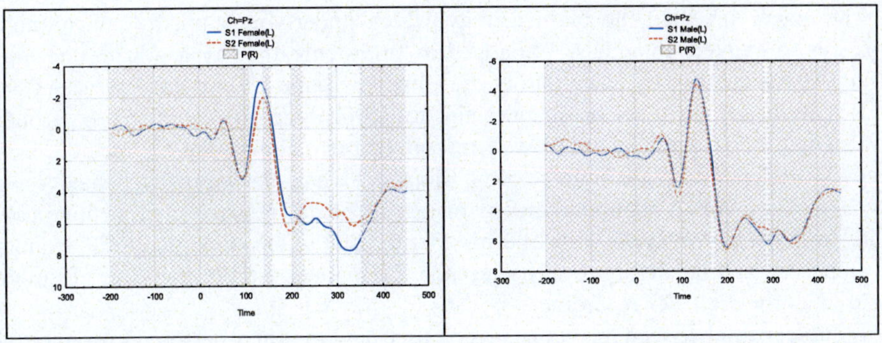

Fig. 4. EP obtained in 2 series (S1 - blue solid line; S2 - red dotted line) separately by gender according to the lead of Channel = Pz. Time is recorded in milliseconds (ms)

4 Discussion

As shown in the figures (Fig. 1, 2, 3 and 4), the sequence of differences in empathy in 2 series with emotional stimulus conditions can be modeled with the wave signal. The signal is then associated with endogenous events occurring in the brain upon recognition of "significant" incentives, retention of events in memory (memorization), events count, and decision making, i.e. with brain functions related to cognitive events.

Using the "odd-ball" technique in 1965 C. Sutton (Sutton et al., 1965) first observed and described in human's cognitive potential of P300. It should be noted that in reality the latency period of the P300 wave can be less than 300 ms, or reach up to 400 ms or more, depending on the characteristics of a particular method.

In the few decades that have passed since the discovery of P300, several explanations of its nature have been put forward. The generally accepted hypothesis by E. Donchin, according to which the P300 reflects the process of the context updating, i.e. reforming the forecast, environ-mental models. Also, existing theories linking the P300 to expectation, memory, and other phenomena.

To date, there is no unambiguous data on which brain structures are involved in the genesis of P300. On the one hand, the potentials large amplitudes, such as the P300, should have extensive, synchronous, and predominantly cortical sources (apparently frontal and parietal cortex). At the same time, recording using microelectrodes immersed in the cortex and application magnetoencephalographic method revealed an additional source of P300 generation in the hippocampus - structure, closely related to memory. In addition, structures associated with P300 generation should include the thalamus.

5 Conclusion

Empathy is the ability to recognize, understand, and share the feelings of another person, including the perception of another person, not just oneself, and promotes prosocial or helping behavior that comes from within, rather than under duress.

This paper revealed the brain activity which underlying empathy and altruistic processes in evoked potentials. We aimed to study empathy, by analyzing the brain activity characteristics by leads during the empathic game "Stone-Paper-Scissors".

Results based on nineteen-channel electroencephalograph recordings experiment on a sample of 16 participants (6 women and 10 men). Using the EP method, brain electrical activity was measured for the situation when the empathy behavior was activated. The evoked potential consists of a sequence of negative and positive deviations from the mainline and lasts 500 ms after the end of the stimulus. In EP, evaluate the amplitude and latent period of occurrence. To register the EP, the same electrodes are used as for the EEG recording.

A comparison between the empathy situations and the 'first-person' experience has been performed separately for gender reactions of men and women. We found significant differences in the following leads: T6, P4, O2, Pz in EP obtained in 2 series separately by gender. 1. This difference is more pronounced in the group of women, while in the group of men, the difference between the series is found at a later latency – 400 ms. 2. This may indicate that men show more empathy. The manifestation of empathy by women is less dependent on the conscious effort in connection with the instruction in the experiment but is manifested constantly. 3. The greatest differences affecting both the early and late components of the EP in the group of women are found in the Pz lead, while in the group of men in this lead there are no significant differences between the series. To sum up, the obtained results bring additional knowledge contribution to the study of empathic processes in neuroscience.

Acknowledgements. The research is financially supported by the Russian Science Foundation for Basic Research, Project № 20-013-00834.

References

1. Decety, J.: Promises and challenges of the neurobiological approach to empathy. Emot. Rev. **3**, 115–116 (2011)

2. Eisenberg, N., Eggum, N.D.: Empathic responding: sympathy and personal distress. In: The Social Neuroscience of Empathy. MIT Press, Cambridge, pp. 71–83 (2009)
3. Waal, F., Preston, S.: Mammalian empathy: behavioural manifestations and neural basis. Nat. Rev. Neurosci. **18**, 498–509 (2017)

Lateralization in Neurosemantics: Are Some Lexical Clusters More Equal Than Others?

Zakhar Nosovets[1,2(✉)], Boris M. Velichkovsky[1,2], Liudmila Zaidelman[1,2(✉)], Vyacheslav Orlov[2], Sergey Kartashov[1,2], Artemiy Kotov[1,2], Vadim Ushakov[2,3,4], and Vera Zabotkina[1]

[1] Russian State University for the Humanities, Moscow, Russia
nosovets@phystech.edu, luda.zaidelman@yandex.ru
[2] National Research Center "Kurchatov Institute", Moscow, Russia
[3] National Research Nuclear University MEPhI, Moscow, Russia
[4] Institute for Advanced Brain Studies, Lomonosov Moscow State University, Moscow, Russia

Abstract. In this study, we have implemented neurosemantic analysis to identify brain's voxel-wise representations of words in Russian spoken narratives and their asymmetries in the brain. 25 subjects listen to five stories, first person narratives of dramatic events, while their brain activation was registered by 3T functional magnetic resonance imaging (fMRI). Seven best subjects in terms of their engagement and objective control of brain reaction were selected for further analysis. Twelve lexical clusters were found, with different semantics – from time-and-space concepts to human actions and mental states. Clusters "experience" and "threat" were ones that on average demonstrated a symmetrical localization. For other clusters, brain localization has a left-sided bias. Our results support the view of non-modular and widely distributed nature of semantic representations, not limited to the activity of structures in the temporal and frontal lobes. These results also demonstrate that the right hemisphere can be involved in representation of mental lexicon. Our findings were broadly consistent with those reported for the English language, which points to the universality of the factors governing brain's lexical representations.

Keywords: Neurosemantics · Narratives · Russian language

1 Introduction

The question of asymmetry of computing processes has an intrinsic value for the build-up of cognitive systems, both natural and artificial. Although hemispheric lateralization is not unique to *Homo sapiens* (Halpern et al. 2005), most research on this issue is dedicated to the anatomy, physiology and pathology of the human brain (Herbert et al. 2005; Pujol et al. 2002). For example, major neurological disorders and psychiatric diseases−schizophrenia, clinical depression, autism, and dyslexia−are known to be accompanied by disturbances in brain symmetry (Carper et al. 2016; Renteria 2012; Sun et al. 2006). Stable differences in lateralization are found in asymmetry of genomic

A. V. Samsonovich et al. (Eds.): BICA 2020, AISC 1310, pp. 350–358, 2021.
https://doi.org/10.1007/978-3-030-65596-9_42

expression for mRNA (Dolina et al. 2017) and regulatory microRNA (Velichkovsky et al. 2020); at least within the frontopolar cortex, which has become famous due to Yakovlevian Torque phenomenon. In this phenomenon, frontal structures anterior to the right Sylvian fissure are 'torqued forward' relative to their counterparts on the left. First described by P.I Yakovlev after WW2, it seems to be supported by modern ontogenetic (Hrvoj-Mihic et al. 2013) and paleoneurological findings (Toga and Thompson 2003).

With respect to functionalities, the established view follows early observations on aphasic disturbances in language production (Luria 1976) and split-brain experiments by Roger Sperry and his school (Sperry 1974; Gazzaniga and LeDoux 1978). A.R. Luria's work on aphasias used the distinction of syntagmatic and paradigmatic relations, which he attributed to anterior and posterior parts of the left hemisphere. According to Sperry, the isolated left hemisphere is engaged in abstract thinking, symbolic relationships, and logical analysis. It can speak, write and make calculations. It is also the more dominant, executive, leading hemisphere to control the behavior and the nervous system. The right hemisphere is mute and generally lacks the ability to communicate with the outside world. Due to its muteness, the right hemisphere is difficult to study experimentally, and thus it is considered as being subordinate to the left hemisphere. Sperry and his colleagues have demonstrated that the right hemisphere is superior to the left hemisphere in its capacity for emotional processing, concrete thinking, spatial representations, and comprehension of complex relationships. However, they still insisted that the right hemisphere is clearly inferior to the left because it lacks the ability to subtract, multiply and divide, to process symbolic relationships and execute the logical analysis of details, particularly of temporal relationships.

These assertions have never been checked in a systematic way, since experimental studies of sematic lateralization are relatively underdeveloped in cognitive neuroscience. Recently we demonstrated that brain representations of meaningful texts cannot be interpreted within the classical framework of syntagmatic vs. paradigmatic relations but rather should involve the ideas of situational semantics (Zaidelman et al. 2020 submitted). With minor differences, this study used the methodology proposed by Alexander Huth and colleagues (Huth et al. 2016). They studied brain representation of the English speech with the help of functional magnetic resonance imaging (fMRI) in seven subjects in response to orally presented narratives. The authors report two results. Firstly, brain mapping of natural language categories on an idealized two-dimensional surface of cortex showed a similarity to the outlines of the Default Mode Network (DMN), a conglomerate of brain structures, which are active in resting state. Secondly, these representations demonstrated broad distribution across the brain with no obvious signs of the initially expected left-sided asymmetry.

Earlier, we applied neurosemantic mapping to Russian-language narratives (Velichkovsky et al. 2020). We also analyzed the linguistic nature of representation within brain's semantic clusters (Zaidelman et al. 2020 submitted). In the current work, the issue of laterality of neurosemantic representations will be considered.

2 Method

2.1 Participants and Stimuli

25 subjects participated in the experiment (mostly students of philology department at the Russian State University for the Humanities, 23 years as an average age, 17 females among them, 23 subjects reported to be right-handed). Informed consent was obtained from each subject prior to the experiment. Participation in the study was unpaid. Ethical approval for this study was provided by the Ethics Committee of the National Research Center "Kurchatov Institute". Subjects were asked to maintain wakefulness with closed eyes during the study.

Stimulus material consisted of five original texts in Russian (1500 wordforms). Each text was a personal story addressing a significant social problem. Participants' involvement in perception and comprehension of stimuli narratives was controlled by a questionnaire and by objective brain imaging data. A professional broadcaster audio recorded the texts, which were presented in a 3T Siemens MRI scanner while whole-brain BOLD (blood-oxygen-level dependent) signal was continuously registered. The scanning process had two stages: capturing high-resolution anatomical data and recording functional data by a parallel scanning protocol with ultrafast EPI-sequence (TR = 1100 ms, TE = 33 ms, 56 slices, slice thickness −2 mm, spatial resolution in each slice -2×2 mm).

2.2 Advanced Pre-processing

Each word was represented as a 997-dimensional vector. These dimensions (features) were numeric estimations: word2vec distances[1] of a text lexeme to the 997 most frequent Russian nouns and verbs. We evaluated the influence of these features on voxels of individual brains' grey matter over the time of texts presentation ($p < 0.05$, uncorrected). Thus, 10000 best voxels in terms of predicting their activation were selected; the time of presentation was divided into 490 epochs (see (Velichkovsky et al. 2020)). This allowed us to build up a transformation matrix of word features vectors into voxels' activity vectors over the presentation time for every subject.

In the next stage of data preparation, we have selected seven participants (all right-handed, five females among them) who were maximally engaged in listening and processing the texts according to the questionnaire and imaging data such as activation of auditory areas, hippocampi and amygdalae related to memory and emotion. Thereafter, a principal component analysis (PCA) on the matrix [Voxels X Features] was performed. The first four components of the PCA scores matrix were used as respective dimensions. The stimulus words were projected into this principal component space by multiplying each word vector by the feature vector in the PCA space. We chose words in this 4D space implying that their connotation was measured by the distance from the center of the word cloud. Within a single step of selection, we have randomly taken 80% of stimulus words and found their convex hull in this space. By

[1] For Word2vec general approach, see: https://code.google.com/archive/p/word2vec/. We used a particular model: *ruwikiruscorpora-nobigrams_upos_skipgram_300_5_2018 model.*

repeating this procedure 1000 times, we constructed a set of words that appeared on the hull at least once. In this way, 163 words were selected for further processing. We proceeded with hierarchical clustering of words within the space. After examining the cluster tree diagram, we have selected a cutoff threshold (the maximum distance for words located within the same cluster) equal to 1.

3 Results

As a result of the final cluster analysis, 12 lexical clusters were obtained with a cardinality varied from 5 to 23 words This selection was further confirmed by a high *post hoc* correlation of individual brain activation, which varied with precision of only one voxel in pairwise comparisons (Pearson correlation) of the seven best subjects with correlations from 62.4 to 78.3%, and average value of 74.4%.

Another validation of the 12 neurosemantic clusters is their relatively straightforward interpretation (Zaidelman et al. 2020 submitted). Indeed, all these clusters were thematically consistent. For example, clusters 2, 4 and 10 were semantically connected to space-and-time of described actions, while clusters 3, 6 and 7 – to topics of threats, overt conflicts, and deprivation. Their names ascribed by a panel of five experts were "city", "event", "space" and "threat", "war", "deprivation", respectively. Notions related to the subjective experience such as "conscious", "intelligence", "feeling", were mapped by the cluster analysis, according to the combined word semantics and brain activity, within the cluster 8, named "experience". Modal lexical frames such as "have to", various action options and goals were concentrated within cluster 9, "goal". The remaining clusters 1, 5, 11 and 12 were related to topics of "rebuilding", "collective", "order", and "try". Note that though the majority of clusters deals with objective circumstances, clusters 3, 7, 9 and 8 also refer to mental states and processes. However, they do it in different ways, which is merely passive suffering, goal-directed planning of action, or reflective contemplation (consciousness) as a part of mental life, correspondingly.

While the linguistic implications of our findings are relatively evident (see (Zaidelman et al. 2020 submitted)), the neurophysiological and neuropsychological significance of our data is more difficult to reveal, even in the case of brain hemispheric lateralization. In Table 1, data for each of seven subjects and for every cluster out of 12 is presented. Number of active voxels in the left (l) and right (r) hemispheres are indicated.

As one can see from the table, brain voxel-wise activation seems to be strongly depended both on the neurosemantic clustering and on the individual differences among subjects. For example, the number of active voxels is maximal for cluster "experience" followed by that for clusters "goal" and "war". To evaluate such influences statistically, we made a three-way analysis of variance, considering factors "subject", "hemisphere", and "cluster" as independent variables for the purposes of this procedure. The analysis of variance demonstrated that all main effects are significant

($p < 0.05$ and better) whereas in the case of hemispheric lateralization left-side localization of active voxels was predominant for the majority of subjects and clusters. Two-way interactions were nearly significant ($p \leq 0.06$). This supports the initial hypothesis that the pattern of brain lateralization may vary across subjects and clusters. In particular, brain structures for clusters "threat" and "experience" on average showed symmetrical or slight right-sided activation. Three-way interaction was insignificant.

Table 1. Number of active voxels for clusters related to subjects and hemisphere.

Subj	Hemi	Rebuilding	City	Threat	Event	Collective	War	Deprivation	Experience	Goal	Space	Order	Try
1	Left	290	458	345	359	341	451	310	469	456	346	383	386
	Right	326	485	396	387	381	478	328	512	520	345	425	428
2	Left	313	452	396	374	364	496	321	479	478	365	416	401
	Right	275	399	336	303	296	392	266	444	425	319	322	367
3	Left	350	492	432	445	434	539	361	589	557	377	454	491
	Right	283	449	375	380	361	491	310	546	491	325	400	424
4	Left	298	446	362	368	365	462	305	472	474	349	382	412
	Right	271	388	325	320	325	398	269	419	407	307	344	346
5	Left	238	360	283	280	290	369	266	359	362	276	316	306
	Right	291	434	335	346	360	449	315	424	451	343	354	374
6	Left	322	502	374	393	369	483	329	495	501	393	404	429
	Right	329	493	464	404	409	572	303	551	516	374	449	430
7	Left	332	487	433	442	398	563	361	560	555	381	458	481
	Right	309	478	383	380	384	491	306	558	522	358	443	402
Mean	Left	306	457	375	380	366	480	322	489	483	355	402	415
Mean	Right	298	447	373	360	359	467	300	493	476	339	391	396

4 Discussion

The neurosemantic paradigm is a new approach in the field of cognitive science, which permits investigation of meaningful texts processing and comprehension by healthy subjects in experimental settings. Are the results of our study in agreement with the contemporary neurocognitive theories of semantic processing and representation in the left and right hemispheres?

Most theories of semantic hemispheric differences stem from clinical observations and studies. The 'hub-and-spoke' model is the authoritative model of conceptual knowledge, which is based on data gathered in patients with semantic dementia. In its classical version (Patterson 2007), this model stipulated that conceptual representations are stored in a unitary "amodal" format in the left and right anterior temporal lobes (ATL), because in semantic dementia the disorder cuts across modalities and categories. Partial specialization occurs across the ATL hub as a consequence of graded differential connectivity across the region. The role of the left hemisphere was repeatedly emphasized due to connectivity with the left-lateralized speech output module in prefrontal and motor cortices (Gainotti 2013; Hoffman 2018). In the same time, the idea placing conceptual representations in such a focal area can easily be wrong since this region is not even part of heteromodal association cortex. Recent discussions emphasize modality-specific components of the semantic representation system (‘spokes’), their interaction with attention, capacity for abstraction, and also considered other clinical models such as Alzheimer’s and Parkinson’s (Chiou and Lambon 2019; Gainotti 2017; Xi et al. 2019; Rice et al. 2018).

Some of our results are in good agreement with the established views. This is, first of all, a generally left-sided lateralization of the activated voxels in the majority of clusters and subjects. Additional analysis of brain structures, where these voxels were located, showed that they are distributed much wider than only in the “core language network” of left frontal and temporal regions, particularly ATL. Prefrontal brain, both left and right, was involved, as well as the regions around temporoparietal junction, precuneus and other structures, even the cerebellum and deep subcortical structures. Similar broad but nearly symmetrical distribution earlier was reported by Huth and colleagues (Huth et al. 2016). No specific relationship to the DMN has been found in our study for the obvious reason that subjects were in the state of active listening.

A novel result of our study is the merely symmetrical localization of activity in case of clusters “experience” and “threat”. This cannot be explained simply by the fact that these clusters refer to mental states because clusters “deprivation” and “goal” can be considered “mental” as well. However, both of these last clusters demonstrated a clear left-sided bias. The pattern reminds us of the results of levels-of-processing studies where semantic encoding was compared with self-and-other referential effects in memory performance (Challis 1996) and where a follow-up positron emission tomography (Craik et al. 1999) showed a prefrontal localization in left hemisphere for semantic categorization and a symmetrical localization or a non-significant trend towards the frontopolar right hemisphere in self-reference. Perhaps, the crucial difference here seems to be the reflexivity or consciousness in perceiving and understanding subject-matter of stimuli narratives. Of interest is the activity on the right pole of prefrontal cortex, which was mentioned earlier as the anatomical site of Yakovlevian Torque. In fact, there was a strong BOLD activation here for “experience” cluster in some of our subjects.

The broad agreement of our findings with those of Huth et al. (2016) for the English language points to the universality of the neural organization of lexical mapping and its relative independence of the peculiarities of specific languages. This finding is important by emphasizing the primacy of the shared world we inhabit and experiences we have, over the linguistic idiosyncrasies of individual languages in determining how lexicon is represented in the brain. The discovery of the mechanisms of conceptual processing requires more than simply mapping to cortical and sub-cortical activity. Models from cognitive science are likely to play a more central role (Barsalou 2017; Bauer 2019). We have argued in a related article (Zaidelman et al. 2020 submitted) that lexical representations of narratives have situational nature. Thus, it seems that the regularities of language and the world that they reflect shape the processing correlations in the brain (Hamilton 2018). However, the real world is reflected in multiple imaginary worlds that are overtly subjective as are stories in most narratives. It is this subjectivity that makes neurosemantic approach so exiting.

5 Conclusion

In this article, we used the neurosemantic approach to find voxel-wise representations of words in Russian spoken narratives and their asymmetries in the brain. All in all, 12 lexical clusters were found, with topics expanding from time-and-space of described actions to merely mental states and actions. Clusters 'threat" and "experience" were ones that on average demonstrated a symmetrical localization of activated voxels. In other clusters, brain localization had a left-sided bias. On the whole, our results support the view of non-modular and widely distributed nature of semantic representations, which is not limited to structures of left temporal and frontal lobes. Our findings were broadly similar to those reported for the English language, which points to the cross-linguistic universality of the principles of lexical mapping in the brain.

Acknowledgments. The study has been in part supported by the Russian Science Foundation, grant 17-78-30029. We thank John Gallant, Elkhonon Goldberg, and Alexander Huth for useful discussions.

References

Barsalou, L.W.: What does semantic tiling of the cortex tell us about semantics? Neuropsychologia **105**, 18–38 (2017). https://doi.org/10.1016/j.neuropsychologia.2017.04.011

Bauer, A.J., Just, M.A.: Brain reading and behavioral methods provide complementary perspectives on the representation of concepts. NeuroImage **186**, 794–805 (2019)

Carper, R.A., Treiber, J.M., DeJesus, S.Y., Muller, R.A.: Reduced hemispheric asymmetry of white matter microstructure in autism spectrum disorder. J. Am. Acad. Child Adolesc. Psychiatry **55**, 1073–1080 (2016). https://doi.org/10.1016/j.jaac.2016.09.491

Challis, B.H., Velichkovsky, B.M., Craik, F.I.M.: Levels-of-processing effects on a variety of memory tasks: new findings and theoretical implications. Conscious. Cogn. **5**(1/2), 142–164 (1996)

Chiou, R., Lambon, R.: Unveiling the dynamic interplay between the hub- and spoke-components of the brain's semantic system and its impact on human behaviour. NeuroImage **199**, 114–126 (2019). https://doi.org/10.1016/j.neuroimage.2019.05.059

Craik, F.I.M., Moroz, T.M., Moscovitch, M., Stuss, D.T., Winocur, G., Tulving, E., Kapur, S.: In search of the self: a positron emission tomography study. Psychol. Sci. **10**(1), 26–34 (1999). https://doi.org/10.1111/1467-9280.00102

Dolina, I.A., Nedoluzhko, A.V., Efimova, O.I., Kildyushov, E.M., Sokolov, A.S., Ushakov, V. L., Khaitovich, P.E., Sharko, F.S., Velichkovsky, B.M.: Exploring terra incognita of cognitive science: lateralization of gene expression at the frontal pole of the human brain. Psychology in Russia: State of the Art **10**(3), 231–247 (2017). https://doi.org/10.11621/pir.2017.0316

Gainotti, G.: The contribution of language to the right-hemisphere conceptual representations: a selective survey. J. Clin. Exp. Neuropsychol. **35**(6), 563–572 (2013). https://doi.org/10.1080/13803395.2013.798399

Gainotti, G.: The differential contributions of conceptual representation format and language structure to levels of semantic abstraction capacity. Neuropsychol. Rev. **27**(2), 134–145 (2017). https://doi.org/10.1007/s11065-016-9339-8

Gazzaniga, M., LeDoux, J.: The Integrated Mind. Plenum Press, New York (1978)

Halpern, M.E., Gunturkun, O., Hopkins, W.D., Rogers, L.J.: Lateralization of the vertebrate brain: taking the side of model systems. J. Neurosci. **25**, 10351–10357 (2005). https://doi.org/10.1523/JNEUROSCI.3439-05.2005

Hamilton, L.S., Huth, A.G: The revolution will not be controlled: natural stimuli in speech neuroscience. Lang. Cogn. Neurosci. **35**(5), 573–582 (2018). https://doi.org/10.1080/23273798.2018

Herbert, M.R., Ziegler, D.A., Deutsch, C.K., O'Brien, L.M., Kennedy, D.N., Filipek, P.A., Bakardjiev, A.I., Hodgson, J., Takeoka, M., Makris, N., Caviness, V.S.J.: Brain asymmetries in autism and developmental language disorder: a nested whole-brain analysis. Brain **128**, 213–226 (2005). https://doi.org/10.1093/brain/awh330

Xi, Y., Li, Q., Gao, N., He, S., Tang, X.: Cortical network underlying audiovisual semantic integration and modulation of attention: an fMRI and graph-based study. PLoS One **14**(8) (2019). https://doi.org/10.1371/journal.pone.0221185

Hoffman, P., Lambon, R.: From percept to concept in the ventral temporal lobes: graded hemispheric specialization based on stimulus and task. Cortex **101**, 107–118 (2018). https://doi.org/10.1016/j.cortex.2018.01.015

Hrvoj-Mihic, B., Bienvenu, T., Stefanacci, L., Muotri, A.R., Semendeferi, K.: Evolution, development, and plasticity of the human brain: from molecules to bones. Front. Hum. Neurosci. **7**, 707 (2013). https://doi.org/10.3389/fnhum.2013.00707

Huth, A.G., De Heer, W.A., Griffiths, T.L., Theunissen, F.E., Gallant, J.L.: Natural speech reveals the semantic maps that tile human cerebral cortex. Nature **532**(7600), 453–458 (2016)

Luria, A.R.: Basic Problems in Neurolinguistics. Mouton, The Hague (1976)

Patterson, K., Nestor, P.J., Rogers, T.T.: Where do you know what you know? The representation of semantic knowledge in the human brain. Nat. Rev. Neurosci. **8**(12), 976–987 (2007). https://doi.org/10.1038/nrn2277

Pujol, J., Lopez-Sala, A., Deus, J., Cardoner, N., Sebastian-Galles, N., Conesa, G., Capdevila, A.: The lateral asymmetry of the human brain studied by volumetric magnetic resonance imaging. NeuroImage **17**, 670–679 (2002). https://doi.org/10.1006/nimg.2002.1203

Rice, G.E., Hoffman, P.I., Binney, R.J., Lambon Ralph, M.A.: Concrete versus abstract forms of social concept: an fMRI comparison of knowledge about people versus social terms. Philos. Trans. Royal Soc. B Biol. Sci. **373**(1752), 20170136 (2018). https://doi.org/10.1098/rstb.2017.0136

Renteria, M.E.: Cerebral asymmetry: a quantitative, multifactorial, and plastic brain phenotype. Twin Res. Hum. Genet. **15**, 401–413 (2012). https://doi.org/10.1017/thg.2012.13

Sperry, R.W.: Lateral specialization in the surgically separated hemispheres. In: Schmitt, F., Worden, F. (eds.) Neurosciences: Third Study Program, vol. 3, pp. 5–19. MIT Press, Cambridge (1974)

Sun, T., Collura, R.V., Ruvolo, M., Walsh, C.A.: Genomic and evolutionary analyses of asymmetrically expressed genes in human fetal left and right cerebral cortex. Cereb. Cortex **16**(Suppl 1), i18–i25 (2006). https://doi.org/10.1093/cercor/bhk026

Toga, A.W., Thompson, P.M.: Mapping brain asymmetry. Nat. Rev. Neurosci. **4**, 37–48 (2003). https://doi.org/10.1038/nrn1009

Vatansever, D., Bzdok, D., Wang, H.-T., Mollo, G., Sormaz, M., Murphy, C., Karapanagiotidis, T., Smallwood, J., Jefferieset, E.: Varieties of semantic cognition revealed through simultaneous decomposition of intrinsic brain connectivity and behaviour. NeuroImage **158**, 1–11 (2017). https://doi.org/10.1016/j.neuroimage.2017.06.067

Velichkovsky, B.M., Nedoluzhko, A.V., Goldberg, E., Efimova, O.I., Sharko, F.S., Rastorguev, S.M., Krasivskaya, A.A., Sharaev, M.G., Korosteleva, A.N., Ushakov, V.L.: New insights into the human brain's cognitive organization: views from the top, from the bottom, from the left and particularly, from the right. Procedia Comput. Sci. **169**, 547–557 (2020)

Velichkovsky, B.M., Zabotkina, V.I., Nosovets, Z.A., Kotov, A.A., Zaidelman L.Y., Kartashov S.I., Korosteleva, A.N., Malakhov, D.G., Orlov, V.A., Zinina, A.A., Goldberg, E., Ushakov, V.L.: Towards semantic brain mapping methodology based on a multidimensional markup of continuous Russian-language texts. STM **12**(2), 14–25 (2020)

Zaidelman, L.Y., Nosovets, Z.A., Kotov, A.A., Ushakov, V.L., Zabotkina, V.I., Velichkovsky, B.M.: Russian-language neurosemantics: clustering of word meaning and sense from the oral narratives. Cogn. Syst. Res. (2020, in press)

Brain Inspiration Is Not Panacea

Pietro Perconti(✉) and Alessio Plebe

Department of Cognitive Science, University of Messina, Messina, Italy
{perconti,aplebe}@unime.it

Abstract. The idea of taking inspiration from how the brain works for designing algorithms has been a fruitful endeavor across domains like cybernetic, artificial intelligence, and cognitive science. However, recent achievements in deep learning has provided some surprising counterevidence, where adopting strategies that are different from those adopted by the brain is successful. We review here the cases of learning rules and vision processing. We suggest two possible justifications of these evidences. It might be that our knowledge of how a problem is solved by the brain is incomplete or lackluster. Therefore, we are not able to translate the genuine brain solution to this problems into the proper algorithm. Or, it might be that the algorithmic solution applied by the brain to a problem is not the most effective for digital computers. Note that the two possibilities are not necessarily mutually exclusive.

Keywords: Computational theory of mind · Deep learning · Artificial neural networks · Convolutional neural networks

1 The Landscape Changed Again

In principle, the design of a given cognitive architecture should be measured only on its performance. If it performs the task it has been designed for, then it is a good cognitive architecture; otherwise not. In building an artificial cognitive architecture, therefore, being inspired by the way human and animal brains face a certain cognitive task could be considered simply in an opportunistic way. If the similarity with the biological working is helpful for the efficacy of the purpose, then it should be considered as a good thing; otherwise it should be done otherwise by creating a different way of dealing with the problem.

This way of reasoning in the last two or three decades has become not very popular. In fact, at least since classical computational psychology has started to show its limits and since connessionism has instead shown its advantages, the idea of designing cognitive architectures that were inspired by the human brain seemed a value in itself. The theoretical perspective to which BICA is inspired is a evidence of this trend. More recently, however, things seem to have changed again and the above mentioned opportunistic attitude seems to be appropriate again. What seems to have significantly changed the landscape in artificial intelligence and the design of particularly effective cognitive architectures has been

A. V. Samsonovich et al. (Eds.): BICA 2020, AISC 1310, pp. 359–364, 2021.
https://doi.org/10.1007/978-3-030-65596-9_43

the success of deep neural networks. The resonance of the successes of deep learning has stirred up intense reflections and discussions within cognitive science and philosophy [11,12,14,19,20,25].

Deep learning is not at all a contrasting perspective with the idea of designing a certain cognitive architecture in a biologically inspired way. On the contrary, it ultimately derives from the central role that connessionism had given to learning and this, in turn, is an idea that is based on the way learning of brains works and that comes from the biological environment. Deep learning is an ecological way of dealing with the phenomenon of knowledge. However, the surprising results that deep neuronal networks have achieved in recent years do not depend on having followed the ecological path in computational design, but – more simply – on taking advantage of a more sophisticated mathematical apparatus than that traditionally used.

2 Learning Rules

Probably one of the most successful borrows from the mind and the brain to the world of digital computation has been the paradigm of learning. Alan Turing [29] first, with extraordinary foresight, cradled the idea of computers that, instead of being rigidly programmed in their tasks, could learn them by themselves through a series of experiences. The inspiration comes directly from the empirical philosophy tradition, according to which the mind is largely the product of experience, and reason gets all its materials from perception. But Turing also took inspiration from neuroscience: he envisioned a machine based on distributed interconnected elements, called *B-type unorganized machine*, where a learning method can "organize" the machine by reinforcing or cutting connections, depending on the experience.

The idea of learning become reality in the '80s with the artificial neural networks, and today their "deep" version is responsible for the fast resurgence of artificial intelligence. So far, so good: brain inspiration has paid off. Actually, things have not turned out exactly like that. As soon as one tried to copy biological learning in more detail, the advantages disappeared.

One of the earliest interpretation of synaptic plasticity was the "perceptron", an electronic device designed by Frank Rosenblatt [22]. The limitations of his learning algorithm offer the opportunity to the anti-empiricist side if artificial intelligence to disparage the entire project of brain imitation, learning included [17].

The later success of artificial neural network [24] was mostly due to a very efficient learning algorithm that departs drastically from biological learning. It is known as *backpropagation*, a term already used by Rosenblatt but for a different method. Being $\mathbf{w}$ the vector of all learnable parameters in a network, and $\mathcal{L}(\mathbf{x}, \mathbf{w})$ a measure of the error of the network with parameters $\mathbf{w}$ when applied to the sample $\mathbf{x}$, backpropagation updates the parameters iteratively, according to the following formula:

$$\mathbf{w}_{t+1} = \mathbf{w}_t - \eta \nabla_w \mathcal{L}(\mathbf{x}_t, \mathbf{w}_t) \tag{1}$$

where t spans over all available samples $\mathbf{x}_t$, and η is the *learning rate.*

This method had an earlier formulation by Paul Werbos [34] in the context of social science statistics, but developed independently for neural networks by Geoffrey Hinton [23] the father of deep learning. The roots of the method have nothing to do with neuroscience, derive from the gradient methods, developed in mathematics for the minimization of continuously differentiable function [3], and found considerable interest in the first half of last century in several engineering applications [13].

Today deep neural networks can still learn with algorithms quite close to the good old backpropagation, even if this term has fallen into disuse, replaced by *Stochastic Gradient Descent* (SGD) [2]. Its basic formulation is the following:

$$\mathbf{w}_{t+1} = \mathbf{w}_t - \eta \nabla_w \frac{1}{M} \sum_i^M \mathcal{L}\left(\mathbf{x}_i, \mathbf{w}_t\right) \tag{2}$$

and it is immediate to see the similarity with the standard backpropagation Eq. (1). Instead of computing the gradients over a single sample t, in Eq. (2) a stochastic estimation is made over a random subset of size M of the entire dataset, and at each iteration step t a different subset, with the same size, is sampled.

Hinton, despite his fundamental contribution in the invention of backpropagation, did not like it that much, because so far from the brain. He worked on an alternative neural model, called *Boltzmann Machine* that used a more plausible learning method [8], but the performances were much behind that of networks learning with backpropagation. Also today, Hinton continues to experiment whether there are variations in SGD in the direction of some sort of biological plausibility, but the results on large scale benchmarks are well behind those of SGD [1].

A well known different domain where learning is used without supervision is self-organization. This domain too is illuminating in showing how brain inspiration can be great, but only at an appropriate distance. The first artificial models of self-organization were developed by Christoph von der Malsburg and David Willshaw, with the purpose of explaining emergent phenomena in the visual system like the formation of ocular dominance strips [33] or the arrangement of neurons responding to oriented edges [32]. The learning rule implemented in these models was the Hebbian law combined with synaptic homeostasis, therefore very close to the brain mechanisms. No artificial application was never derived from these models. Years later, a different learning algorithm was proposed by the Finnish electronic engineer Teuvo Kohonen [9], offering self-organizing properties at a much cheaper price. The algorithm, known as SOM (*Self-Organizing Features Map*), has no relation with brain mechanisms, no resemblance with Hebb's law or synaptic homeostasis, but has been quite successful in artificial intelligence.

3 Vision Processing

Brain imitation has also played a role in the field of vision, but going further does not pay off. The case of vision is intriguing, because it is the domain where deep neural models had their initial success [10], where they soon defeated all competition, and where human capacity is reached by artificial intelligence [30]. All deep models for vision are variations over an architecture known as DCNN (*Deep Convolutional Neural Networks*), derived from the earlier *Neocognitron*, introduced by Kunihiko Fukushima [6]. The Neocognitron was directly inspired by neuroscientific discoveries of the primary visual cortex. It uses units called *S-cell* and *C-cell* in analogy with the classification in simple and complex cells in the primary visual cortex. The S-units act as convolution kernels, while the C-units downsample the images resulting from the convolution by spatial averaging.

Since 1980 there has been a tremendous progress in the understanding of the biological visual system, and several neurocomputational models of vision more advanced than Neocognitron have been proposed, let us mention a couple. Topographica implements lateral inhibitory and excitatory cortical connections together with several other biological details [16]. VisNet is organized into five layers resembling cortical areas, and learning includes a specific mechanism called *trace memory*, aimed at accounting for the natural dynamics of vision, where the invariant recognition of objects is learned by seeing them when moving under various different perspectives [21,28]. Neither Topographica, nor VisNet, nor any other brain plausible neurocomputational model of vision found any artificial applications.

Even more striking is the evolution in the overall conception of natural vision since 1980, where it was dominated by the computational approach outlined by David Marr [15]. One of the earliest papers challenging the classical account of vision is *A critique of pure vision* [4]. The wrong assumptions in the "pure" vision are the idea that goal of vision is to create a detailed model of the world in front of the eyes in the brain, and that vision process consists in the hierarchical extraction of increasingly specific features. Moreover, vision is basically independent from other brain tasks. Since this seminal paper, criticism of "pure" vision has become one of the dominant themes in cognitive science [18], and several artificial vision systems inspired by active and enactive vision were designed [5,31]. None of these attempt reached marketable applications. On the contrary, DCNN are the perfect manifestation of the deprecated "pure" vision: they take static images as their only input, and extract features in a hierarchical way [19]. Yet, DCNN are by large the most efficient and successful algorithms in vision.

4 Discussion

The fact that deep neural networks have been able to achieve some remarkable results in cognitive tasks, even superior to those that natural calculators are capable of, could have two possible justifications. Perhaps it is our knowledge of the human brain that is incomplete. Or, it might be that the algorithmic solution applied by the brain to a problem is not the most effective for digital computers. The fact that the two possibilities are not necessarily mutually exclusive suggests to adopt the opportunistic perspective we talked about at the beginning of this paper. The idea of adopting an opportunistic attitude in designing of cognitive architectures has attracted some significant attention [7,26,27]. We still do not know on what depends the achievement of remarkable results in cognitive tasks by deep neural networks. It could be matter of more sophisticated mathematical apparatuses and brute computational power. Or it could mean that looking for similarities with the way human brains work is the wrong way. After all, there is no assurance that the path followed by natural evolution was the best one. Natural evolution, in fact, does not pursue goals related to the elegance of the solution which is found, unless in the perspective of deep time. The turbulent path of evolution might have selected brains that achieve their goals by implementing styles of cognitive architectures that are not the best from a computational point of view. Feeling free from biological inspiration could lead to greater creativity, and this could result in more elegant effects than those found in biological contexts. But we don't know that yet. That is why, until we have all the elements that are necessary for a definitive judgment, the opportunistic attitude could be the best one to adopt, as it is the most modest from the epistemological point of view.

References

1. Bartunov, S., Santoro, A., Richards, B.A., Marris, L., Hinton, G.E., Lillicrap, T.: Assessing the scalability of biologically-motivated deep learning algorithms and architectures. In: Advances in Neural Information Processing Systems (2018)
2. Bottou, L., LeCun, Y.: Large scale online learning. In: Advances in Neural Information Processing Systems, pp. 217–224 (2004)
3. Cauchy, A.L.: Méthode générale pour la résolution des systèmes d'équations simultanées. Comptes rendus des séances de l'Académie des sciences de Paris **25**, 536–538 (1847)
4. Churchland, P.S., Ramachandran, V., Sejnowski, T.: A critique of pure vision. In: Koch, C., Davis, J. (eds.) Large-Scale Neuronal Theories of the Brain. MIT Press, Cambridge (1994)
5. De Croon, G.C., Sprinkhuizen-Kuyper, I.G., Postma, E.: Comparing active vision models. Image Vis. Comput. **27**, 374–384 (2009)
6. Fukushima, K.: Neocognitron: a self-organizing neural network model for a mechanism of pattern recognition unaffected by shift in position. Biol. Cybern. **36**, 193–202 (1980)
7. Guindon, R.: Designing the design process: exploiting opportunistic thoughts. Hum.-Comput. Interact. **5**, 305–344 (1990)
8. Hinton, G.E., Sejnowski, T.J., Ackley, D.H.: Boltzmann machines: constraint networks that learn. Technical report 84-119. Computer Science Department, Carnegie-Mellon University (1984)
9. Kohonen, T.: Self-Organizing Maps. Springer, Berlin (1995)
10. Krizhevsky, A., Sutskever, I., Hinton, G.E.: ImageNet classification with deep convolutional neural networks. In: Advances in Neural Information Processing Systems, pp. 1090–1098 (2012)

11. Lake, B.M., Ullman, T.D., Tenenbaum, J.B., Gershman, S.J.: Building machines that learn and think like people. Behav. Brain Sci. **40**, 1–72 (2017)
12. Landgrebe, J., Smith, B.: Making AI meaningful again. Synthese, 1–21 (2019). https://doi.org/10.1007/s11229-019-02192-y
13. Levenberg, K.: A method for solution of certain non-linear problems in least squares. Q. Appl. Math. **2**, 164–168 (1944)
14. Marcus, G.: Deep learning: a critical appraisal. CoRR abs/1801.00631 (2018)
15. Marr, D.: Vision: A Computational Investigation into the Human Representation and Processing of Visual Information. W. H. Freeman, San Francisco (1982)
16. Miikkulainen, R., Bednar, J., Choe, Y., Sirosh, J.: Computational Maps in the Visual Cortex. Springer-Science, New York (2005)
17. Minsky, M., Papert, S.: Perceptrons: An Introduction to Computational Geometry. MIT Press, Cambridge (1969)
18. O'Regan, J.K., Noë, A.: A sensorimotor account of vision and visual consciousness. Behav. Brain Sci. **24**, 939–1031 (2001)
19. Perconti, P., Plebe, A.: Deep learning and cognitive science. Cognition **203**, Article 104,365 (2020)
20. Plebe, A., Grasso, G.: The unbearable shallow understanding of deep learning. Mind. Mach. **29**, 515–553 (2019)
21. Rolls, E.T., Stringer, S.M.: Invariant visual object recognition: a model, with lighting invariance. J. Physiol. Paris **100**, 43–62 (2006)
22. Rosenblatt, F.: Principles of Neurodynamics: Perceptron and the Theory of Brain Mechanisms. Spartan, Washington (1962)
23. Rumelhart, D.E., Hinton, G.E., Williams, R.J.: Learning representations by back-propagating errors. Nature **323**, 533–536 (1986)
24. Rumelhart, D.E., McClelland, J.L. (eds.): Parallel Distributed Processing: Explorations in the Microstructure of Cognition. MIT Press, Cambridge (1986)
25. Schubbach, A.: Judging machines: philosophical aspects of deep learning. Synthese, 1–21 (2019). https://doi.org/10.1007/s11229-019-02167-z
26. Siminia, M., Kolodner, J.L.: Opportunistic reasoning: a design perspective. In: Annual Conference of the Cognitive Science Society, vol. 17, p. 78 (1995)
27. Sperber, D., Mercier, H.: Cognitive opportunism. In: Sperber, D., Mercier, H. (eds.) The Enigma of Reason, pp. 76–89. Harvard University Press, Cambridge (2017)
28. Stringer, S.M., Rolls, E.T.: Invariant object recognition in the visual system with novel views of 3D objects. Neural Comput. **14**, 2585–2596 (2002)
29. Turing, A.: Intelligent machinery. Technical report, National Physical Laboratory, London (1948). Raccolto in Ince, D.C. (ed.) Collected Works of A. M. Turing: Mechanical Intelligence, Edinburgh University Press (1969)
30. VanRullen, R.: Perception science in the age of deep neural networks. Front. Psychol. **8**, 142 (2017)
31. Viéville, T.: A Few Steps Towards 3D Active Vision. Springer, Berlin (1997)
32. von der Malsburg, C.: Self-organization of orientation sensitive cells in the striate cortex. Kybernetic **14**, 85–100 (1973)
33. von der Malsburg, C., Willshaw, D.J.: A mechanism for producing continuous neural mappings: ocularity dominance stripes and ordered retino-tectal projections. Exp. Brain Res. **1**, 463–469 (1976)
34. Werbos, P.: The Roots of Backpropagation: From Ordered Derivatives to Neural Networks. John Wiley, New York (1994)

Linear Systems Theoretic Approach to Interpretation of Spatial and Temporal Weights in Compact CNNs: Monte-Carlo Study

Artur Petrosyan(✉), Mikhail Lebedev, and Alexey Ossadtchi

National Research University Higher School of Economics, Moscow, Russia
petrosuanartur@gmail.com, aossadtchi@hse.ru
https://bioelectric.hse.ru/en/

Abstract. Interpretation of the neural networks architectures for decoding the signals of the brain usually reduced to the analysis of spatial and temporal weights. We propose a theoretically justified method of their interpretation within the simple architecture based on a priori knowledge of the subject area. This architecture is comparable in decoding quality to the winner of the BCI IV competition and allows for automatic engineering of physiologically meaningful features. To demonstrate the operation of the algorithm, we performed Monte Carlo simulations and received a significant improvement in the restoration of patterns for different noise levels and also investigated the relation between the decoding quality and patterns reconstruction fidelity.

Keywords: ECoG · Weights interpretation · Limb kinematics decoding · Deep learning · Machine learning · Monte Carlo

1 Introduction

A step towards improving the performance of neurointerfaces is the use of advanced machine learning methods - Deep Neural Networks (DNN). DNNs learn a complete signal processing pipeline and do not require hand-crafted features. Interpretation of DNN solution plays a crucial role to 1) identify optimal spectral and temproral patterns that appear pivotal in providing the decoding quality (knowledge discovery) 2) ensure that the decoding relies on the neural activity and not on the unrelated physiological or external artefacts.

Recently, a range of compact neural architectures has been suggested for EEG, ECoG and MEG data analysis: EEGNet [4], DeepConvNet [10], VAR-CNN and LF-CNN [11]. The weights of these architectures are readily interpretable using the standard linear estimation theoretic approaches [3]. However, a special attention is needed to make the correct weights interpretation in the architectures with simultaneously adaptable temporal and spatial weights.

A. V. Samsonovich et al. (Eds.): BICA 2020, AISC 1310, pp. 365–370, 2021.
https://doi.org/10.1007/978-3-030-65596-9_44

2 Generating Data Model

The data generative model is illustrated in Fig. 1. Neural populations $G_1 - G_I$, which are responsible for movement, generate activity $\mathbf{e}(t) = [e_1(t), \ldots, e_I(t)]^T$ that is further translated into a trajectory of movements with some non-linear transform H, i.e. $z(t) = H(\mathbf{e}(t))$. We assume that there are populations $A_1 - A_J$ with unrelated movement activity. Its activity is mixed into the sensors as well. At each time step t, we observe K-dimensional $\mathbf{x}(t)$ vector of sensor signals instead of the intensity of firing $\mathbf{e}(t)$ of individual populations. Vector $\mathbf{x}(t)$ is traditionally modelled as a linear mixture with $\mathbf{A}$ and $\mathbf{G}$ matrices, reflecting local field potentials $\mathbf{f}(t)$ and $\mathbf{s}(t)$ formed around both populations:

$$\mathbf{x}(t) = \mathbf{A}\mathbf{f}(t) + \mathbf{G}\mathbf{s}(t) = \sum_{j=1}^{J} \mathbf{a}_j f_j(t) + \sum_{i=1}^{I} \mathbf{g}_i s_i(t) \quad (1)$$

The local field potentials (LFPs) result from the nearby populations' activity and their characteristic frequency is typically related to the size [1] of each population. The firing intensity of the proximal neuronal population is approximated by the envelope of LPF. To counter the volume conduction effect, we will seek to obtain the estimates of LFPs as $\hat{\mathbf{s}}(t) = \mathbf{W}^T\mathbf{X}(t)$ and columns of $\mathbf{W} = [\mathbf{w}_1, \ldots, \mathbf{w}_M]$ are referred to as spatial filters.

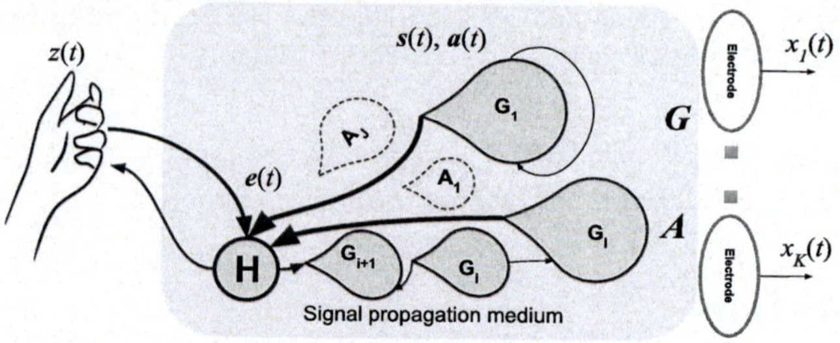

Fig. 1. Phenomenological model

Our regression task is to decode the kinematics $z(t)$ from simultaneously recorded neural populations activity $\mathbf{x}(t)$. Generally, we do not have any true knowledge on $\mathbf{G}$ and other parameters of the forward mapping and transform H, therefore we need to parameterize and learn the entire mapping $z(t) = \mathcal{F}(\mathbf{x}(t))$.

3 Network Architecture

Figure 2 demonstrates our adaptable and compact CNN architecture based on the idea of (1). It consists of spatial filtering, adaptive envelope extractor, and a fully-connected layer. Spatial filtering is done via a pointwise convolution layer

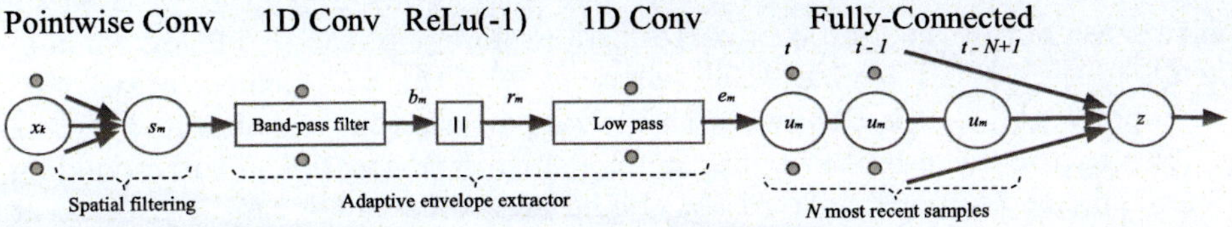

Fig. 2. Proposed compact DNN

used to unmix the sources. Adaptive envelope extractor is in a form of two depth-wise convolutions for bandpass and lowpass filtering, used with non-trainable batch normalization (for explicit power extraction) and absolute value nonlinearity in-between. The fully-connected layer is used in order to model envelope to kinematics $z(t)$ transformation with H as a function of lagged envelopes of layers signals extracted previously. This architecture is implemented using the standard DNN layers. In principle, the temporal filtering layers can be replaced by a sinc-layer [8].

4 Spatial and Temporal Weights Interpretation

Assume that the data are processed in chunks of size N equal to the length of the temporal convolutional layer weights $\mathbf{h}_m$ $\mathbf{X}(t) = [\mathbf{x}(t), \mathbf{x}(t-1), \dots \mathbf{x}(t-N+1)]$. Since the set of envelopes maps isomorphically onto a set of analytic signals [2], perhaps with the accuracy to a sign, the task of tuning the weights of the first three layers of our architecture to predict envelopes $e_m(t)$ can be replaced with a regression problem of learning and correcting spatial and temporal weights to get the analytic signal $b_m(t)$ giving rise to the envelope. Assume that the temporal weights are fixed to their optimum value $\mathbf{h}^*_m$, then the optimal spatial filter weights can be obtained as:

$$\mathbf{w}^*_m = argmin_{\mathbf{w}_m}\{\| b_m(n) - \mathbf{w}^T_m \mathbf{X}(t)\mathbf{h}^*_m \|^2_2\} \tag{2}$$

and therefore assuming statistical independence of the rhythmic LFPs $\{s_m(t)\}$, $m = 1, \dots, M$ the spatial pattern of the underlying neuronal population is [3]

$$\mathbf{g}_m = E\{\mathbf{Y}(t)\mathbf{Y}^T(t)\}\mathbf{w}^*_m = \mathbf{R}^Y_m \mathbf{w}^*_m, \tag{3}$$

where $\mathbf{Y}(t) = \mathbf{X}(t)\mathbf{h}_m$ is a temporally filtered chunk of multichannel data and $\mathbf{R}^Y_m - E\{\mathbf{Y}(t)\mathbf{Y}^T(t)\}$ is a branch-specific $K \times K$ covariance matrix of temporally filtered data, assuming that $x_k(t)$, $k = 1, ..., K$ are zero mean processes.

Symmetrically we can write an expression for the temporal weights interpretation as

$$\mathbf{q}_m = E\{\mathbf{V}(t)\mathbf{V}^T(t)\}\mathbf{h}^*_m = \mathbf{R}^V_m \mathbf{h}^*_m, \tag{4}$$

where $\mathbf{V}(t) = \mathbf{X}^T(t)\mathbf{w}^*_m$ is a piece of spatially filtered data and $\mathbf{R}^V_m = E\{\mathbf{V}(t)\mathbf{V}^T(t)\}$ is a branch specific $N \times N$ covariance matrix of spatially filtered data, assuming that $x_k(t)$, $k = 1, ..., K$ are all zero-mean processes. To make sense out of the temporal pattern we explore it in the frequency domain, i.e. $Q_m[f] = \sum_{k=0}^{k=N-1} q_m[k]e^{-j2\pi fk}$, where $q_m[k]$ if the k-th element of $\mathbf{q}_m$.

Importantly, as it is the case with spatial pattern, that the obtained vectors $\mathbf{g}_m$ can be usually used to fit dipolar models [6] and locate the corresponding source [3], the temporal patterns $\mathbf{h}_m$ found according to (4) can be used to fit dynamical models such as those, for example, implemented in [7].

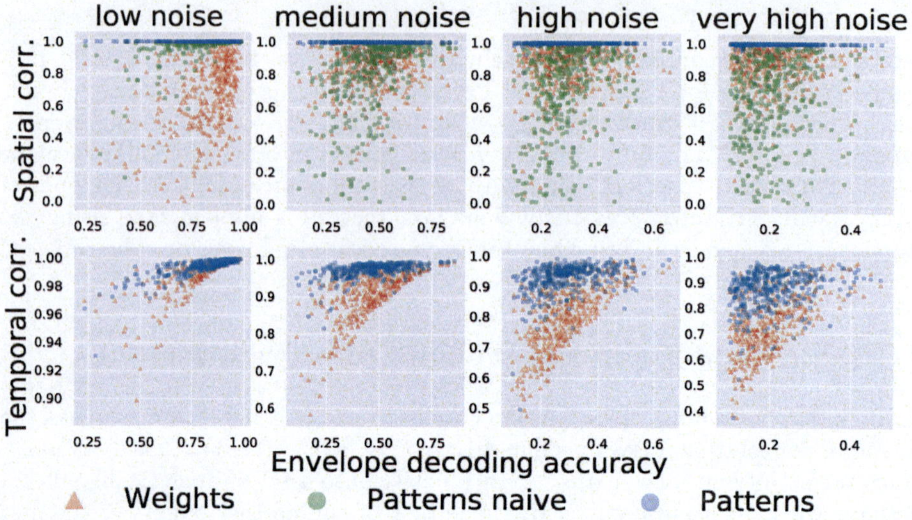

Fig. 3. Monte Carlo simulations. Point coordinates reflect the achieved at each Monte Carlo trial envelope decoding accuracy (x-axis) and correlation coefficient with the true pattern (y-axis). Each point of a specific color corresponds to a single Monte Carlo trial and codes for a method used to compute patterns. *Weights* direct weights interpretation. *Patterns naive* Spatial patterns interpretation without taking branch specific temporal filters into account, *Patterns* - the proposed method

Table 1. Correlation between true and predicted kinematics of the winning solution for BCI competition IV dataset (Winner) and proposed architecture (NET)

Subject 1—2—3	Thumb	Index	Middle	Ring	Little
Winner	.58—.51—.69	.71—.37—.46	.14—.24—.58	.53—.47—.58	.29—.35—.63
NET	.54—.50—.71	.70—.36—.48	.20—.22—.50	.58—.40—.52	.25—.23—.61

5 Comparative Decoding Accuracy

In the context of the electrophysiological data processing, the main benefit of deep learning solutions is their end-to-end learning method which does not require task-specific features preparation [9]. To make sure that our implementation of a simple CNN is capable of learning the needed mapping, we applied it to the collected by Kubanek et al. publicly available data from the BCI Competition IV and compared its performance to that of the winning solution [5].

The results of both algorithms are listed in Table 1. Our simple neural network has comparable decoding quality as the linear model [5] but does not require upfront feature engineering but rather learns the features itself.

6 Monte-Carlo Simulations

We followed the setting described in Fig. 1 to generate the data. We simulated $I = 4$ sources related to the task with rhythmic LFPs $s_i(t)$, occupying the different ranges: 170–220 Hz, 120–170 Hz, 80–120 Hz and 30–80 Hz bands. The target kinematics $z(t)$ was simulated as a linear combination of 4 envelopes of rhythmic LFPs with a vector of random coefficients. We used $J = 40$ unrelated to the task rhythmic LFP sources in the bands of 180–210 Hz, 130–160 Hz, 90–110 Hz and 40–70 Hz. Each band contained ten sources. Matrices $\mathbf{G}$ and $\mathbf{A}$ which model the volume conduction effects at each Monte Carlo trial were randomly generated according to $\mathcal{N}(0, 1)$ distribution. We created 20 min worth of data sampled 1000 Hz.

For neural network training we use Adam optimiser. We made about 15k steps. At 5k and 10k step we halved the learning rate to get more accurate patterns. In total, we have performed more then 3k simulations.

We performed Monte-Carlo study with different spatial configuration of sources at each trial. For each realisation of the generated data we have trained the DNN to predict the kinematic variable $z(t)$ and then computed the patterns of sources the individual branches of our architecture got "connected" to as a result of training.

Figure 3 shows that only the spatial *Patterns* interpreted using branch-specific temporal filters match well the simulated topographies of the true underlying sources. Moreover *Patterns naive* and *Weights* correlation decreasing with noise raises, while *Patterns* is almost perfectly recover true patterns for all noise level settings.

The spectral patterns recovered using the proposed approach also appear to match well with the true spectral profiles of the underlying sources, while directly considering the Fourier coefficients of the temporal convolution layer weights results into erroneous spectral profiles. Using the proper spectral patterns of the underlying neuronal population it is now possible to fit biologically plausible models, e.g. [7], and recover true neurophysiological mechanisms underlying the decoded process.

7 Conclusion

We proposed a theoretically justified method for the interpretation of spatial and temporal weights of the CNN architecture composed of simple envelope extractors. This result extends already existing approaches [3] to weights interpretation. With Monte-Carlo simulations we were able to demonstrate that the proposed approach accurately recovers both spatial and temporal patterns of the underlying phenomenological model for a broad range of signal to noise ratio values.

Acknowledgments. This work is supported by the Center for Bioelectric Interfaces NRU HSE, RF Government grant, ag. No. 14.641.31.0003.

References

1. Buzsaki, G.: Rhythms of the Brain. Oxford University Press, New York (2006)
2. Hahn, S.L.: On the uniqueness of the definition of the amplitude and phase of the analytic signal. Sig. Process. **83**(8), 1815–1820 (2003)
3. Haufe, S., Meinecke, F., Görgen, K., Dähne, S., Haynes, J.D., Blankertz, B., Bießmann, F.: On the interpretation of weight vectors of linear models in multivariate neuroimaging. Neuroimage **87**, 96–110 (2014)
4. Lawhern, V.J., Solon, A.J., Waytowich, N.R., Gordon, S.M., Hung, C.P., Lance, B.J.: EEGNet: a compact convolutional network for EEG-based brain-computer interfaces. arXiv preprint arXiv:161108024 (2016)
5. Liang, N., Bougrain, L.: Decoding finger flexion from band-specific ECoG signals in humans. Front. Neurosci. **6**, 91 (2012). https://doi.org/10.3389/fnins.2012.00091
6. Mosher, J., Leahy, R., Lewis, P.: EEG and MEG: forward solutions for inverse methods. NeuroImage **46**, 245–259 (1999). https://doi.org/10.1109/10.748978
7. Neymotin, S.A., Daniels, D.S., Caldwell, B., McDougal, R.A., Carnevale, N.T., Jas, M., Moore, C.I., Hines, M.L., Hämäläinen, M., Jones, S.R.: Human Neocortical Neurosolver (HNN), a new software tool for interpreting the cellular and network origin of human MEG/EEG data. eLife **9**, e51214 (2020). https://doi.org/10.7554/eLife.51214
8. Ravanelli, M., Bengio, Y.: Speaker recognition from raw waveform with SincNet. In: 2018 IEEE Spoken Language Technology Workshop (SLT), IEEE, pp. 1021–1028 (2018)
9. Roy, Y., Banville, H., Albuquerque, I., Gramfort, A., Falk, T.H., Faubert, J.: Deep learning-based electroencephalography analysis: a systematic review. J. Neural Eng. **16**(5), 051001 (2019)
10. Schirrmeister, R.T., Springenberg, J.T., Fiederer, L.D.J., Glasstetter, M., Eggensperger, K., Tangermann, M., Hutter, F., Burgard, W., Ball, T.: Deep learning with convolutional neural networks for brain mapping and decoding of movement-related information from the human EEG. arXiv preprint arXiv:170305051 (2017)
11. Zubarev, I., Zetter, R., Halme, H.L., Parkkonen, L.: Adaptive neural network classifier for decoding MEG signals. NeuroImage **197**, 425–434 (2019)

The Economic Cross of the Digital Post-coronavirus Economy (on the Example of Rare Earth Metals Industry)

O. Victoria Pimenova[1], Olga B. Repkina[1], and Dmitriy V. Timokhin[1,2(✉)]

[1] National Research Nuclear University MEPhI, Moscow, Russia
dtprepod@yandex.ru
[2] Moscow State University of Humanities and Economics, Moscow, Russia

Abstract. The article contains proposals for the development of modeling of the national industry using IT solutions based on the "economic cross" methodology. The reserves of the digitalization of the planning procedure are considered, taking into account the modern achievements of IT technologies and the development of the global information and communication infrastructure. The conclusions and suggestions contained in the article are formulated taking into account the extrapolation of current trends of nuclear energy development.

Keywords: Rare earth metals industry · Economic cross methodology · Industry 4.0. · Information and communication technologies · Global digital space · Economic forecasting · Modernization

1 Introduction

The phenomenon of the "digital economy" is cross-sectoral in nature. In other words, it is formed at the intersection of the life cycles of several traditional industries. The article presents the concept of the digital economy as a model formed at the intersection of the technological (Kondratieff) and infrastructural (Simon Kuznets) cycles. The aspects concerning the interaction of the participants of these cycles in the digital space are considered, including the aspects of unification of technological standards and terminology. Considerable attention is paid to the practical aspects of the implementation of digital technological solutions in the innovation and production process at the macroeconomic, microeconomic and mesoeconomic levels. The reasons for the insufficient adaptation of transnational business to the digitalization opportunities offered by transnational technological digital platforms are investigated. Both pre-coronavirus and post-coronavirus economies in the context of the formation of Industry 4.0 are considered as the initial material confirming the conclusions of the authors regarding the state of the "economic cross".

A. V. Samsonovich et al. (Eds.): BICA 2020, AISC 1310, pp. 371–379, 2021.
https://doi.org/10.1007/978-3-030-65596-9_45

2 Assessment of the Prerequisites for Closing the Economic Cross of the Digital Economy

A distinctive feature of the digital economy is its trans-sectoral nature. Despite the fact that the IT industry occupies one of the central places in the structure of the digital economy, the development of this type of economy cannot be carried out solely through the development of this industry. An example that proves the validity of this thesis is the situation in the modern economy in 2020 as part of the spread of a new coronavirus infection. Indeed, the economic response to the Covid-19 pandemic was predominantly digital. These measures can be grouped as follows:

- transfer of a part of employees, whose activities are not associated with direct material impact on objects of labor, to the online format (remote form of work);
- introduction of digital control tools for the movement of potential carriers of infection in order to more effectively predict and plan anti-epidemiological measures;
- the use of digital tools for remote interaction with clients in order to avoid both the risks of infection and the additional costs of the organization associated with ensuring epidemiologically safe forms of face-to-face interaction.

From an economic point of view, all three forms of digital interaction did not contain technical obstacles and were not associated with an increase in systemic costs for their implementation.

At the same time, between February and September 2020, there was a significant increase in the costs of enterprises for the introduction of digital technologies.

The analysis of the cost structure for the introduction of digital technologies in the post-coronavirus period, carried out by the authors for 500 organizations from the United States, Great Britain, France and Russia, proves that most of these costs fell on the need for an emergency technological reorientation of business. The results of the analysis carried out by the authors are presented in Fig. 1.

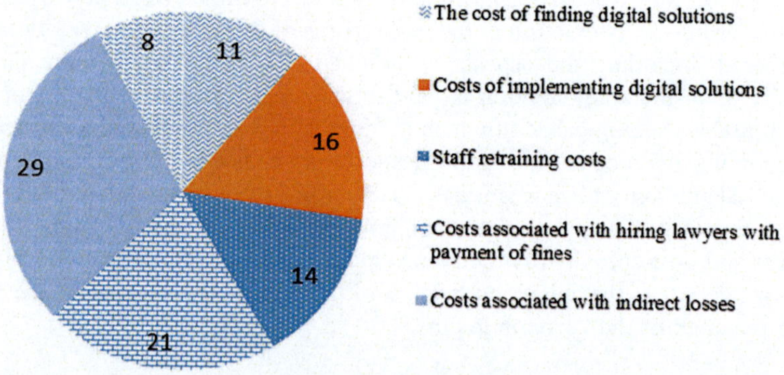

Fig. 1. - Assessment of the ratio of the costs of organizations in the USA, Great Britain, France and Russia on the introduction of digital technologies in the post-coronavirus period. Compiled on the basis of [1–3].

Analysis of the information presented in Fig. 1 suggests that the overwhelming majority of both domestic and foreign organizations for the introduction of digital technologies is of an indirect nature. Most of the relevant categories of costs associated with digitalization are not inevitable. The costs under consideration are primarily due to the unsystematic introduction of digital technologies into production. As a reason for such a non-systematic flow of economic processes of digitalization of business, it should be noted that most companies have no plans for digitalization of business, despite the fact that technological solutions that were relevant for them existed and, moreover, created an opportunity to optimize economic processes. Consequently, the economic losses incurred by a business in the context of the Covid-19 pandemic lie in the plane of ineffective strategic planning.

The validity of the chosen thesis is proved by the statistics of the use of Internet traffic during the pre-coronavirus and post-coronavirus periods by leading innovative companies. The indicators of Internet traffic use by countries considered to be leaders in organizing information and communication interactions between residents are shown in Fig. 2.

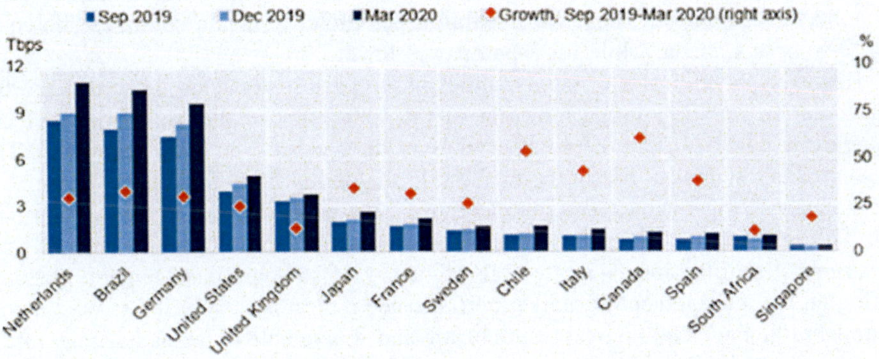

Fig. 2. - Estimation of the median internet exchange point (IEP) peak traffic aggregated for the pre- and post Covid - 19 period for global economy, 2019–2020 [6].

Of particular importance is the fact that, in general, international companies have not abandoned the digital technologies they use in the post-coronavirus period, which confirms the noneconomic the nature of the reasons for ignoring them in 2010–2019.

The company, which is a participant in the market of innovative products, is currently focused on improving the quality of its goods while maintaining the level of prices for products not lower than in the previous period. At the same time, it is important to ensure financial opportunities for the innovative development of the organization while maintaining its existing market share or, ideally, increasing it. All other conditions unchanged, the effectiveness of the innovation process depends on the absolute value of investment, and not on the relative. In other words, in the long term, the winner of the competition is the company that has formed the largest budget for innovative development with a structure that meets the requirements of institutional lenders and investors. A feature of the period 2010–2019. was the emergence of a contradiction between these two conditions.

It should be noted that the heterogeneity of the development of the global digital technology market is largely determined by the heterogeneity of demand for relevant products in different countries. Orientation of a digital product manufacturer to a conditional single global market would create an opportunity to reduce the costs of creating a digital product and improve its quality through scaling. However, the differentiation of national markets in terms of digital product requirements reduces economic scalability and limits the innovative growth of digital product leaders because they need to adapt to the needs of buyers from developing countries with lower requirements for digital product quality.

The compliance of the technological development of the organization with the requirements of scaling the digital product reduced its attractiveness in the eyes of institutional lenders and investors due to the fall in liquidity indicators.

A comparative analysis of the information presented in Fig. 2 gives grounds to assert that there is a certain average level of load on Internet channels, which is relevant for the modern economy. Attention should also be paid to the fact that the differences in the peak load on Internet channels for countries differ not only in terms of their level of economic development. Thus, the maximum peak load from the countries shown in Fig. 2 was for Canada, while for Japan it was lower.

However, the indicator of the dynamics of the peak load cannot be considered the only criterion for the optimal indicator of the sufficiency of the use of digital technologies by business from the point of view of the needs of the global economy. The reason for this statement lies in the fact that in the period from February to September 2020, business was unable to master the entire innovative potential of the digital space [4]. The inability of a business to reorient to digital technologies in such a tight timeframe lies in the insufficiently high efficiency of strategic management. Concentrating on the management of market performance indicators, the business was unable to undergo global technological restructuring and was unable to take advantage of the window of opportunity, the formation of which began in 2010 along with the development of the 6-staged economy.

Let us consider the reasons that determined the economic inefficiency of the strategic planning processes for the innovation of production processes in the context of the formation of Industry 4.0:

- the inability of business to combine technologically effective strategic planning of the life cycle of production systems and the requirements of institutional lenders and investors;
- the lag of the infrastructure component, primarily the labor market and Internet communications coverage of consumers in developed countries, from the needs of the digital economy;
- the lack of a single information space with built-in criteria for selecting digital solutions for manufacturing and logistics businesses;
- the inability or unwillingness of business to reorient itself to functioning models that are relevant in the context of the formation of Industry 4.0.

Optimizing the development of digitalization projects in 2021-2025 is proposed on the basis of the "economic cross" model, which is the closure of technological and production cycles within the framework of a project that creates consumer value. The use of this model ensures the correct distribution of the specified consumer value between all participants in the digitalization process, regardless of their chronological position in the production chain.

3 The Concept of Digitalization Based on the Closure of the Technological and Production Cycles, Implemented by Unleaded Partners

The marketing concept of promoting goods at the beginning of the 21st century involves the formation of a parametric model of a product based on customer requests. This setting can be considered as a constraint on the development of the digital economy, since the consumer is guided mainly by the formalized offers of the manufacturer, which, in turn, are created based on his preferences. Exit from this logical circle by the efforts of one company is impossible, since all other market participants with such behavior of the manufacturer implement the most aggressive competitive strategies in relation to him. The threat of opposition of the competitive coalition to the entrepreneur turns out to be the higher, the higher the entrepreneur's need for interaction with the external market. Examples of such interaction are the interaction of an entrepreneur with the sphere of finance, the sphere of suppliers of rare components, the sphere of advertising and the logistics system for promoting an innovative product in general.

A possible option for the entrepreneur to counter the current collision is the formation of a technology consortium. Such a consortium may be successful when its members are able to meet the needs of an innovative company in the resources that it previously received in the external market. An example of such a consortium entering the market is a consortium that supplies digital and robotic systems.

The expected dynamics of the needs of the global market for such robots is shown in Fig. 3.

Studying the dynamics of the needs of the global economy in digital and robotic systems allows us to draw the following conclusions.

1. The exponential growth of the needs of the global economy in digital and robotic systems makes it possible to speak about the attractiveness of the corresponding market for all participants. At the same time, to date, neither the supposed leader of the corresponding market has been formed, nor the maximum capacity of this market has been estimated. This circumstance speaks of the difficulties faced by the participants in the global digital economy. The manufacturer adapts to the needs of the consumer, while the consumer's requests are poorly formalized from the point of view of strategic management and do not provide systematic requests for the modernization of the manufacturer's technologies. Under these conditions, the manufacturer has to adapt the level and structure of the digital technologies used to

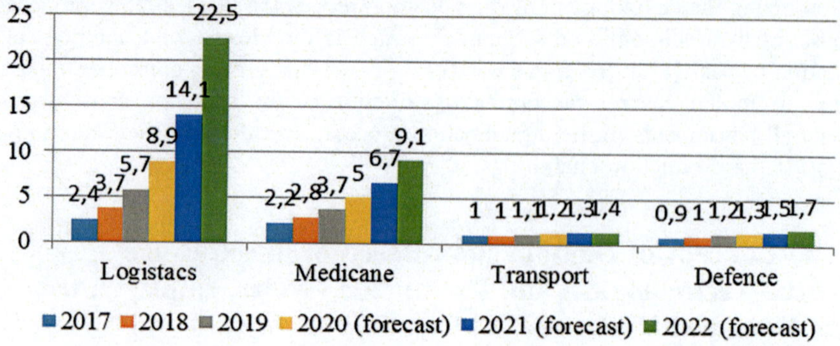

Fig. 3. - Assessment of the needs of various areas of the global economy in digital and robotic systems, 2017–2019 and forecast for 2020–2022, in billions of $ [5]

the needs of a particular institutional buyer, at the expense of the logic of the development of the technological platform in the long term.

2. The digital solutions market in Russia has not fully exhausted the possibilities of forward economic mechanisms. Trade in digital technologies and products is carried out in the spot market format or within the minimum period of time from the point of view of the modern forward market. Expanding the horizon of technological planning for the development of companies offering digital products can contribute to a more rational use of economies of scale both at the level of an individual innovator and at the level of a national innovation system involved in global competition.

Forecasting the development of the digital economy can be carried out on the basis of the economic cross model. Figure 4 shows the closure of the technological and production cycles of the post-coronavirus digital economy.

The main forecast options for the development of the digital economy in the post-coronavirus period, built on the basis of the "economic cross" model, are as follows.

1. Development of digital technologies in the format of the formation of innovative start-ups focused on the point involvement of production in the digital economy; the most widespread variety of this option can be the creation of free economic zones for the needs of the IT industry [12]. In this case, the most probable development is the development of the digital space as a set of independent clusters and start-up projects that implement independent digital solutions in accordance with the diffusion model of innovations by J. Schumpeter [13].
2. Development of the digital component of the national economies of developing countries based on the model of "reverse innovation" with maximum import substitution of digital products of their industrialized countries; this concept is relevant for those areas of the economic space. Which are currently most affected by the geo-economic confrontation between global technology centers. For the period 2021–2025 such a prediction seems to be justified in relation to the centers of development of communication 5G, as well as the field of social networks. If in the period 2014–2019. the confrontation concerned mainly the national dimension of the

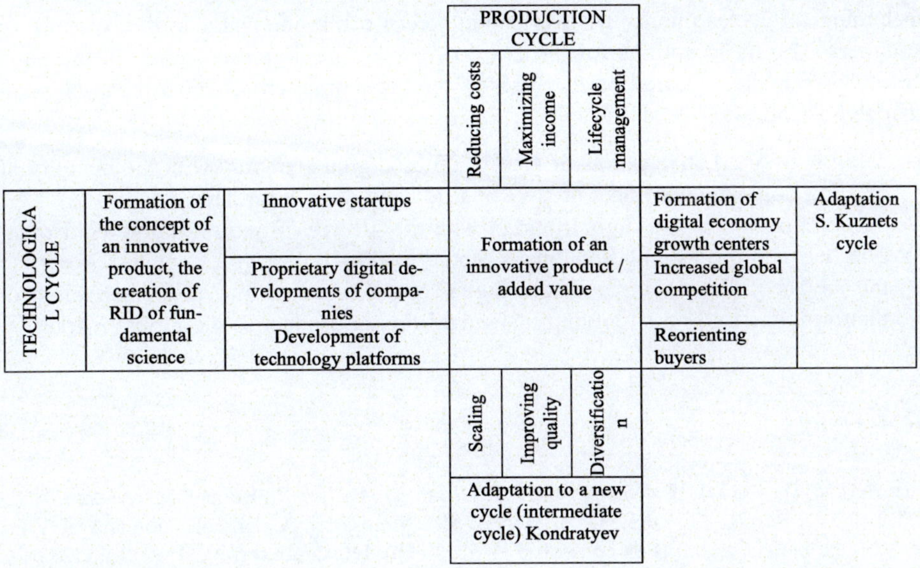

Fig. 4. The economic cross model, built for the purpose of forecasting the development of the digital economy as a closure of the technological and production cycles. Built on the basis of [7–9]

global technology market (confrontation between China and the United States as potential leaders of the technological space of the 6-staged economy), then by 2025. in the spheres affected by the confrontation, one should expect the emergence of international technological alliances, heterogeneous in structure of the countries entering them and possessing their own technological standards.

3. Transnationalization of the digital space on the basis of universal technological platforms that take into account the needs of all market participants and reduce the risks of the innovator's functioning in the digital environment or take them upon themselves for an appropriate fee.

4 Conclusions

Thus, modeling the digital economy based on the "economic cross" model allows one to more accurately predict the options for closing the Kondratieff cycle and the infrastructure (construction) cycle of the blacksmith. Planning the development of innovative companies at the intersection of these two cycles reduces the risks of entering the market for digital technologies and products that have high risks of being in demand. Planning the structure of a digital product within the framework of the proposed "economic cross" model can be carried out in the form of a combination of technological and production solutions known from the state of the art with the subsequent creation of a consortium that ensures the closure of production and technological cycles due to the overlapping needs of the innovator in the structure of suppliers. The basic options for the development of the digital economy in the post-coronavirus reality, formed on the basis of the "economic cross" model, can be summarized as follows:

- creation of alternative points of growth of the digital economy with the inability of existing companies to meet the existing demand at an affordable price;
- decentralization of the digital market with the growth of geo-economic confrontation between the leading economies of the world;
- the advanced development of the digital space based on universal technological platforms focused on supplying a standard digital product to the global market.

References

1. Horgan, D., Hackett, J., Westphalen, C.B., Kalra, D., Richer, E., Romao, M., Andreu, A.L., Lal, J.A., Bernini, C., Tumiene, B., Boccia, S.k., Montserrat, A.: Digitalisation and COVID-19: the perfect storm. Biomed. Hub **5**, 1–23 (2020). https://doi.org/10.1159/000511232
2. Skulmowski, A., Günter, D.: COVID-19 as an accelerator for digitalization at a German university: establishing hybrid campuses in times of crisis. Hum. Behav. Emerg. Technol. **2** (3), 212–216 (2020). https://doi.org/10.1002/hbe2.201
3. Barnes, S.: Information management research and practice in the post-COVID-19 world. Int. J. Inf. Manage. **19**, 102175 (2020) https://doi.org/10.1016/j.ijinfomgt.2020.102175
4. Ting, D., Carin, L., Dzau, V., Wong T.: Digital technology and COVID-19. Nat. Med. **27** (2020) https://doi.org/10.1038/s41591-020-0824-5
5. Khan, Z.H., Siddique, A., Lee C.W.: Robotics utilization for healthcare digitization in global COVID-19 management. Environ. Res. Publ. Health **17**(11), 3819 (2020). https://doi.org/10.3390/ijerph17113819
6. Lopez-Gonzalez, J., Andrenelli, L., Ferencz, J., Sorescu, S.: Leveraging digital trade to fight the consequences of COVID-19 in OECD report. https://read.oecd-ilibrary.org/view/?ref=135_135517-02bikxyksj&title=Leveraging-Digital-Trade-to-Fight-the-Consequences-of-COVID-19. Accessed 30 Sep 2029
7. Timokhin, D., Vorobyev, A., Bugaenko, M., Popova, G.: Formirovanie mekhanizmov ustojchivogo innovacionnogo razvitiya atomnoj otrasli. Tsvennie metallic **3**, 23–35 (2016)
8. Putilov, A.V., Timokhin, D.V.: Innovacionnye vozmozhnosti ispol'zovaniya metodologii ekonomicheskogo kresta v prognozirovanii perspektiv razvitiya dvuhkomponentnoj atomnoj energetiki. Innovacii **1**(255), 12–20 (2020). https://doi.org/10.26310/2071-3010.2020.255.1.002
9. Timokhin, D.V., Bugaenko, M.V., Putilov, A.V.: The use of IT technologies in the implementation of the economic cross methodology in the Breakthrough project of Rosatom. Procedia Comput. Sci. **169**, 445–451 (2020). https://doi.org/10.1016/j.procs.2020.02.227
10. Fraile, F., Sanchis, R., Poler, R., Ortiz A.: Models for digital manufacturing platforms. Appl. Sci. **9**(4433), 102–124 (2019) https://doi.org/10.3390/app9204433
11. Ignatiev, M.B., Karlik, A.E., Iakovleva, E.A., Karlik, E.: Challenges for strategic management of the development of the digital economy and advanced training. In: XVII Russian Scientific and Practical Conference on Planning and Teaching Engineering Staff for the Industrial and Economic Complex of the Region (PTES), St. Petersburg, pp. 197–200 (2018). https://doi.org/10.1109/ptes.2018.86042

12. Morten, M., Zoran, J.: Digital transformation, governance and coordination models: a comparative study of Australia, Denmark and the Republic of Korea. In: The 21st Annual International Conference on Digital Government Researchю, pp. 285–293, June (2020). https://doi.org/10.1145/3396956.3396987
13. Marc, K., Corin, K., Lindeque, J.: Strategic action fields of digital transformation: an exploraon of the strategic action fields of Swiss SMEs and large enterprises. J. Strategy Manage. **1**(13), 81–93 (2019). https://doi.org/10.1108/JSMA-05-2019-0070

Comparative Analysis of Methods for Calculating the Interactions Between the Human Brain Regions Based on Resting-State FMRI Data to Build Long-Term Cognitive Architectures

Alexey Poyda[1(✉)], Maksim Sharaev[2], Vyacheslav Orlov[1], Stanislav Kozlov[1], Irina Enyagina[1], and Vadim Ushakov[1,3,4]

[1] NRC Kurchatov Institute, Akademika Kurchatova pl. 1, 123182 Moscow, Russia
Poyda_AA@nrcki.ru

[2] Skolkovo Institute of Science and Technology, Bolshoy Boulevard 30, Moscow, Russia

[3] NRNU MEPhI, Kashirskoe hwy 31, 115409 Moscow, Russia

[4] Institute for Advanced Brain Studies, Lomonosov Moscow State University, GSP-1 Leninskie Gory, Moscow, Russia

Abstract. In this work, we compared many different methods proposed for calculating the functional interaction of brain regions based on resting-state fMRI data. We compared them according to the criterion of the stability of the results to small changes in the parameters of both the methods themselves and the input data including different levels of noise. By stability, here we mean that small changes in the parameters and level of noise will lead to small changes in the obtained estimates of the interaction. Since fMRI has a temporal resolution of about 2 s, here we focused on long-term architectures (400 s or more). Our study revealed that measures of Correlation and Coherence families show slightly better values for sustainability. This is in a good agreement with the result obtained earlier on synthetic fMRI data when evaluating medium-strength connections, that may be characteristic of resting-state.

Keywords: fMRI · Resting state · Functional connectomes · Effective connectomes

1 Introduction

Research in the field of constructing functional connectomes of the human brain at rest has been carried out over the past several decades. To build a functional connectome, it is necessary to evaluate the level of interaction between the regions that determine its functional elements. To determine this level, it is necessary to determine both the joint work of different brain areas, as well as establish causal relationships between them. Currently, many different methods have been proposed for calculating the interaction of brain regions [1], also during resting-state [2], based on correlation [3], Granger

A. V. Samsonovich et al. (Eds.): BICA 2020, AISC 1310, pp. 380–390, 2021.
https://doi.org/10.1007/978-3-030-65596-9_46

causality [4], transfer entropy [5], coherence [6], mutual information [7], and etc. Moreover, each of the methods depends on a number of parameters, which could affect the obtained result (for example, the size of the sliding window, frequency bands, model order, etc.). Facing such a variety of methods, the question arises - which methods are better suited for the analysis of functional connectivity. It is impossible to compare methods directly by the accuracy of evaluating connections since ground truth about functional connections in the brain is not available to us. For the resting-state, this situation is exacerbated by the lack of external modulating (experimental) effects.

There are studies in which methods are compared on the basis of synthetic data. So, for example, in [8], the authors use 5 different mathematical models to generate synthetic data with a given number of regions, communication strength and noise level, and then evaluate how precisely each method with different parameters reveals predefined ground truth. In this particular study, interesting results were obtained, which we focused on when conducting our research. However, the results obtained on synthetic data may not correspond to the case of real data. This may be due to the fact that the mathematical models used to generate synthetic data do not correspond to the real (unknown) model of activity in the brain. For example, the authors in [8] suggest that the connections between brain regions do not change over time, which is argued in modern theories [9, 10].

In our work, we compared various methods on real fMRI data. Since ground truth about the connections of brain regions is not available to us, we focused on comparing methods by indirect criteria. One is the stability of the results to small changes in the parameters of both the methods (for example, window size) and the input data (for example, different noise levels). By stability, we mean that small changes in the parameters will lead to small changes in the obtained estimates of the interaction. We believe that this is a very important criterion for assessing connections between brain regions on real data at resting-state because, in the absence of a priori knowledge about the true connectivity, the question of the ideal selection of parameters for a particular method could not be practically solved. Thus, if a small change in the parameters leads to significant changes in obtained connection strength, it will be difficult to obtain a robust estimate and prove its validity. At the same time, the severity of this problem can be reduced if small changes in the parameters do not lead to large changes in the calculated estimates.

Currently, there is evidence in favor of the so-called dynamic connectivity of the brain regions, namely, microstates (from milliseconds to tens of seconds) that vary in time. However, since fMRI has a temporal resolution of about 2 s, here we focused on long-term architectures (400 s or more). This can also be explained by a number of reasons:

1) many methods provide good estimates only with a sufficiently large time-series length (from 200 values or more) [8];
2) we can expect the stability of long-term connections characteristics in space and in time because the estimate is calculated by averaging many functional interactions between brain areas that can be observed during the analyzed long-term period;

3) if the method does not show a stable result in the long-term analysis on averaged functional connections, this may indirectly indicate its unsuitability for the analysis of short-term conditions.

We'd like to emphasize that the global aim of our research is the comparison and evaluation of methods for calculating the strength of functional connections between brain regions based on resting-state fMRI data, not only for the long-term but also for short-term connections. But specifically in this work, we focus on the characteristics of the methods for long-term connections calculation. We imply that the best-performing methods should be considered as potential methods for assessing short-term connectivity, which should be the subject of further research.

2 Data Acquisition and Preprocessing

In this work we used fMRI data from 3 subjects acquired according to the protocol described in [11]. Permission to conduct the experiment was granted by the Ethical Commission of the Kurchatov Institute.

The data was obtained using a SIEMENS Magnetom Verio 3 T scanner. To obtain a high-resolution structural image of the brain (T1-weighted image), the following parameters of the rapid gradient echo sequence were used: 176 slices, TR (repetition time) = 1900 ms, TE (echo time) = 2.19 ms, voxel size $-1 \times 1 \times 1$ mm^3, flip angle = 90, inversion time = 900 ms, observation field = 250×218 mm^2.

To obtain fMRI data, the following parameters were used: TR = 2000 ms, TE = 20 ms, number of slices −42, voxel size $-2 \times 2 \times 2$ mm^3. Additionally, field mapping data were obtained to reduce spatial distortions of echo-planar images.

One fMRI data set was selected for further analysis. The selection was performed manually based on visual analysis, based on the criterion of physiological noise minimum.

To select the regions of interest, we used the method for selecting functionally homogeneous regions, which we developed earlier and described in detail in the article [11]. The method is based on distinguishing large spatial regions with a high correlation between the dynamics of the voxels of these regions. There are several sequential steps in the method. At the first step, C-region is determined for each voxel - a spatially connected region around the voxel in analysis, consisting of voxels with dynamics correlating with the dynamics of the initial voxel in analysis, not weaker than with the required coefficient (in our case the cutoff is 0.85). In the second step, C-regions of maximum size are selected. If the C-region of one voxel is smaller than the C-region of another voxel, and this voxel with smaller C-region is covered by a larger C-region, then smaller C-region is excluded. In the third step, all voxels are distributed between maximum C-regions: if the voxel falls into several C-regions, then a C-region is selected by the level of correlation with its "center". At last, small C-regions are excluded - those regions, where the number of voxels is less than the specified value (we used a value of 10). The remaining C-regions are taken as the resultant functionally homogeneous regions.

After applying our method to fMRI data, 280 regions ranging in size from 10 to 33 voxels were identified.

Next, the dynamics of all voxels in each region were averaged, resulting in 280 initial time courses - one per region.

To speed up calculations (which is relevant for some methods, such as Granger causality), the number of regions was reduced to 20 (see Fig. 1). Regions were selected based on the following principles: first, by their occurrence or partial intersection with anatomical regions of interest, such as resting-state networks: DMN (LIPC, RIPC regions and big information hubs PCC, MPFC [12], as well as distinct regions as Frontal Pole (Brodmann area 10), VLPFC, Hippocampus, Parahippocampal Gyrus; and second, by correlation level: we selected the regions in a way that the correlation coefficients between their dynamics covered the largest possible range of values from −0.4 to 1.

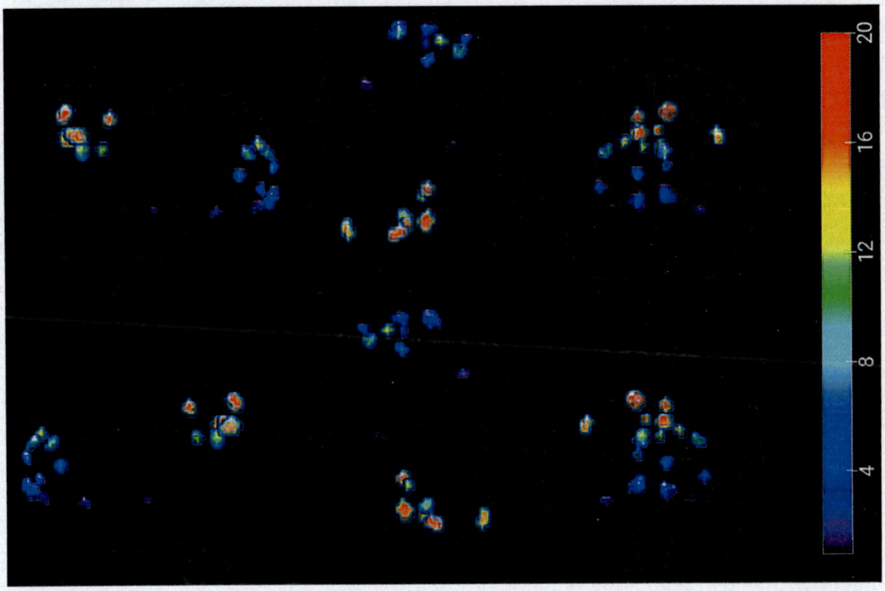

Fig. 1. Selected 20 regions encoded with colors: 1 – Parahippocampal Gyrus L, 2 – Hippocampus R, 3 – VLPFC L, 4 – MPFC R, 5 – MPFC R, 6 – MPFC L, 7 – FP (BA10 R), 8 – FP (BA10 R), 9 – FP (BA10 L), 10 - PCC L, 11 – FP (BA10 R), 12 – FP (BA10 L), 13 – PCC L, 14 – PCC R, 15 – FP (BA10 R), 16 – LIPS, 17 – PCC L, 18 – PCC R, 19 – PCC R, 20 – PCC L.

3 Methods

Our study consists of two parts. First, we aimed at assessing the stability of the measures in relation to the initial choice of the time interval. This stability suggests that a small shift of a time window (by a few percent in relation to its total length) should not lead to a significant change in the values of the estimated connection strength, even if we assume dynamic connectivity.

Second, we assessed the stability in relation to the level of noise. This stability assumes that smooth changes of the noise level in the signal lead to smooth changes of estimated connections strength. This is important for real data since the real fMRI signal contains both physiological and hardware noise the level of which is difficult to estimate.

A more detailed description of the methods for evaluating both time interval shift stability and noise stability is given in Sects. 3.2 and 3.3 while analyzed measures are described in Sect. 3.1.

3.1 Functional Connectivity Measures and Program Tools

To select and evaluate the functional connectivity measures, we used the results obtained in [8], where various functional connectivity measures were evaluated using synthetic fMRI and EEG data. We chose methods that showed good results on synthetic fMRI data. Thus, we chose 14 methods from 4 families (here and next we use the same notation as in [8]):

- Transfer entropy family includes measures based on Transfer entropy [5]: BTED, PTED, BTEU and PTEU. The prefix B- corresponds to the binary measure between each pair of regions without taking into account other connections while the prefix P- corresponds to a partial connection, in which the influence of other connections is removed. The suffixes U and D correspond to directional and undirectional cases.
- The Correlation family includes measures based on Pearson's correlation [3]: BCorrD, and BCorrU, where prefix B- and suffixes D, U have the same meanings as for the transfer entropy.
- The Granger causality family includes measures based on Granger causality [4]: GC (classic Granger causality), CondGC (conditional Granger causality [13]) and PGC (partial Granger causality).
- Coherence family includes measures calculated in the frequency domain [14, 15]: BCohF, PCohF, BCohW, and PCohW, where suffix W corresponds to wavelet transforms, while the suffix F corresponds to the Fourier transform. We also included the Instantaneous coherence measure [16] which was not used by Weng et al. but previously showed good performance.

For our experiments, we used a modified open-source MULAN framework developed by the authors of [8] in the MATLAB programming language. Almost all the chosen measures are already implemented in MULAN. We added Instantaneous coherence measure, new metrics for results estimation, a noise-adding module as well as program modules for batch calculations to automate numerical experiments and get more statistics.

3.2 Time Interval Shift Stability

To assess the stability of the measures in relation to the initial choice of the time interval, we: 1) split the entire time interval into overlapping windows of equal length;

2) calculated the functional connections in each time window with all methods and received for each connection a series of estimates on sequential time windows; 3) calculated the parameters characterizing the variability of the obtained series for each measure. Let us consider each step in more detail.

The entire time interval, which consists of 1000 samples, was split into windows of 250 samples (500 s) in length with a 95% overlap (the windows are shifted by 5%), thus resulting in 61 windows for each connection between all pairs of selected regions.

In each window, the strength of the functional connection was calculated by each of the investigated measures, thus 15 series of 61 points were obtained for each connection, where each series is composed of the values obtained by the given measure in sequential time windows. It should be noted that the obtained series have different limits of variation of the values. So, for example, the correlation varies from −1 to +1, and transfer entropy, for example, is strictly non-negative. Therefore, in order to be able to compare the obtained series with each other, we normalized them so that the average over all values (over all time windows and all connections) was equal to zero, and the standard deviation was equal to 1. This normalization allows us to align the average value and the level of variation of values for all measures.

To assess the variability of the obtained normalized series, we used the following measures:

- The standard deviation of each series makes it possible to estimate the variability of functional connectivity values over time.
- The standard deviation of a derivative of series calculated on adjacent values. This measure allows us to compare the differences in the calculated connectivity values between adjacent time windows. The smaller the value of this parameter, the less the connectivity values change when the windows are shifted.
- Two measures listed above are finally averaged over all connections making it possible to describe all connections with one value for each of the aforementioned measures. In order to minimize the influence of outliers, we used the median value instead of average.

3.3 Noise Stability

To assess the stability of the connection estimates in relation to noise level, we added the noise to the original signal (obtained by averaging the fMRI data in each region) and then calculated the connection estimates for the noisy data. Finally, we compared the "noisy" results with the values of the same estimates calculated for the original signal that we consider "true". A more formal description of the calculations is as follows:

1. Calculate the connectivity estimates on original data for all considered measures. Set the SNR value (the ratio of the source signal power to the noise component power [17]) equal to 1.
2. For each region, create a (white) noise component with a given SNR. Add this noise to the original data.

3. On a set of noisy signals, calculate the connectivity estimates on the whole time-series length (1000 samples) resulting in a single value for each pair of regions in case of unidirectional measure (two values if the connection is directional).
4. Repeat steps 2–4 for 99 times with different noise realizations, resulting in 100 realizations of noisy data with a given SNR and the same number of connectivity estimates on them.
5. Calculate the average of 100 values of connectivity estimates for each connection and subtract the "true" value of connectivity estimates from this average value.
6. Normalize the obtained difference to the standard deviation calculated for all connections of the original signal. This normalization makes it possible to compare the sets of differences obtained for different measures.
7. Average the obtained normalized differences for all connections.
8. Repeat steps 2–7 for SNRs from 1 to 40 with a step of 1 resulting in a numerical series NS (SNR) consisting of 40 values for SNR for each considered connectivity measure.

Obviously, when the SNR tends to infinity, the influence of noise tends to 0 and NS should tend to 0. Therefore, at SNR = 40, we expect to get NS values of about 0 for all measures. The subject of the study is the behavior of NS at the SNR near 1. If measure is independent of noise (the ideal case), then the NS should remain about 0 for any SNR value. In real cases, adding the noise leads to a change in the NS values, and the sharper and greater the change, the less considered measure is robust to noise. Therefore, one of the characteristics that we evaluate is integral (discrete sum) of the function NS. If the integral is small the change of NS is also small as SNR tends to 1.

4 Result

Table 1 shows the results of the time interval shift stability analysis. The measures are ordered by increasing values of the STD of derivatives. We assume this parameter to be more important: it shows the amplitudes of estimate fluctuations, while STD of values show the range of changes over the full-time interval.

We can see that according to the STD of derivatives the measures could be divided into three groups: measures which showed the best result about 0.2 (BCorrD, B CohW, instcoherence, BCorrU); measures which showed a result in the range of 0.3–0.4, and measures which showed a result about 0.5 and worse. Considering that we normalized the data in such a way that standard deviation of data was equal to 1, the average of STD of derivatives equal to 0.5 indicates fluctuations with an amplitude of 50% or more of the normalization interval.

The lowest measures of the STD of values correspond to the group of measures that showed the smallest values of the STD of derivatives. The large values indicate that connections strength changes over time while small values indicate that the connection strength remains stable over time.

Figure 2 shows a graph of the NS function for all considered connectivity measures.

Table 1. The results of the time interval shift stability analysis.

Method	STD of values	STD of derivatives
BCorrD	0,27	0,12
BCohW	0,33	0,17
Instcoherence	0,33	0,19
BCorrU	0,35	0,23
PCohW	0,49	0,29
BTEU	0,63	0,35
BTED	0,67	0,35
BCohF	0,44	0,36
PTED	0,74	0,40
PTEU	0,74	0,42
GC	0,81	0,53
PGC	0,85	0,54
CondGC	0,87	0,57
PCohF	0,64	0,59

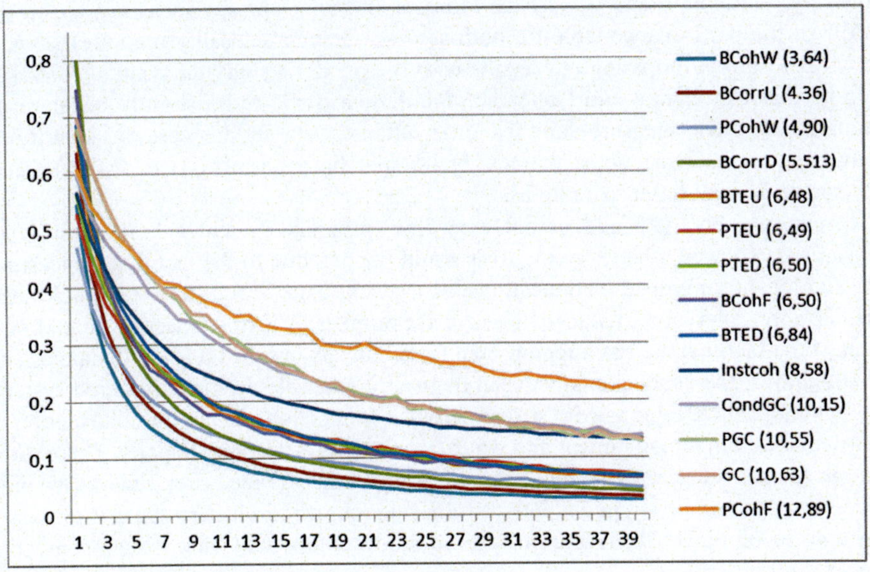

Fig. 2. Noise stability of measures at various SNR levels. Along the X-axis are SNR values, along the Y-axis are values of NS function. The value of the integral under the NS curve is in parentheses.

The graph shows that NS of the measures are similar: for small SNR the NS values are high reaching 0.8. It means that for a given noise level the average deviation of "noise measure" from the "true measure" is 80% of the standard deviation of all connections. As the SNR increases the NS values gradually decrease. In the legend, the measures are ordered by the increasing area under the graph the value of which (the discrete sum) is indicated in brackets.

Analyzing the graph we can note that:

1. For SNR of 1, all measures have deviations from true value 0.45 to 0.8.
2. Starting from SNR = 3 measures can be divided into two groups: the PCohF, GC, PGC, CondGC measures converge more slowly than others, and the instcoherence measure joins them from SNR = 10.
3. Other measures behave similarly and converge to the "true value" with increasing SNR.

5 Discussion

Among the measures that showed the best results in both studies, we should highlight BCohW, BCorrD, BCorrU, PCohW. Measure instcoh showed robustness to choosing a time interval, but was not the best (compared to other methods) in noise stability test.

Comparing the results for measure families, it can be noted that the best result was shown by Correlation and Coherence families. Transfer entropy measures are in the middle of the performance table for both studies. Granger causality measures showed the worst results. Comparing our results with the previous study on synthetic data [8], we note that Correlation and Coherence families also showed a slightly better result compared to other measure families on synthetic data in the case of identifying medium-strength connections, that might be specific for resting-state fMRI data in comparison to task-based paradigms.

Comparing bivariate and partial measures (prefixes B- and P-), in most cases measures B- showed a better result. This could happen due to the fact that P-measures give significantly more zero connections, as a result, their overall STD (and hence the normalization interval) is reduced. The greater number of zero connections could be a result of the fact that the brain regions can form clusters in terms of their dynamics and the strength of the connections between them. For example, in the considered system, we could have three regions with strong pairwise connections due to the contribution of a particular common component and weak connections with other regions. These three regions form a cluster with high internal connections. B-measures will show high values of connections strengths between the regions of the cluster, while P-measures will tend to eliminate the impact of the common component thus reducing its connection strength.

The features of the applied methods and the ways of their improvement should also be taken into account. First, to normalize the data, we used the standard deviation calculated from a measure values on all connections. We assume that this approach characterizes this measure with respect to the whole brain. But if we have many weak connections ("zero links"), their standard deviation is also small, so the normalization coefficient. This could lead to bad accuracy estimates of a selected measure. By choosing 20 regions with connections from the entire range, we tried to mitigate this problem, but selected regions were chosen only by the correlation measure. Therefore, the task of investigating "zero links" for different methods is still relevant. We also believe that this task is closely related to the task of identifying a binarization threshold for each measure, which determines whether there is a connection between regions or not.

We tested all methods on a single subject data. In the future, the obtained results should be validated on a larger sample.

6 Conclusion

Our studies showed that measures of Correlation and Coherence families demonstrate slightly better values of the stability of the results for the initial choice of the time interval and noise level when estimating functional connectivity between the human brain regions based on resting-state fMRI data to build long-term cognitive architectures. This result is in good agreement with the previous result obtained in [8] on synthetic fMRI data evaluating medium-strength connections, which might be the case of resting-state brain activity.

Nevertheless, we consider it necessary to conduct further studies, in particular, studies aimed at analyzing "zero-links", as well as studies of these measures to assess short-term relationships in a framework of dynamic connectivity.

Acknowledgements. This work was supported by the Russian Foundation of Basic Research: grant RFBR 18-29-23020 mk "Investigation of functional architecture of the human brain's resting state networks as the basic model of energy-efficient information processes of consciousness". Data acquisition was performed within the grant RFBR 17-29-02518 ofi-m supported by the Russian Foundation of Basic Research.

References

1. Pereda, E., Quiroga, R.Q., Bhattacharya, J.: Nonlinear multivariate analysis of neurophysiological signals. Prog. Neurobiol. **77**, 1–37 (2005). https://doi.org/10.1016/j.pneurobio.2005.10.003
2. Sharaev, M., Ushakov, V., Velichkovsky, B.: Causal interactions within the default mode network as revealed by low-frequency brain fluctuations and information transfer entropy. In: Biologically Inspired Cognitive Architectures (BICA) for Young Scientists. Advances in Intelligent Systems and Computing, Proceedings of the First International Early Research Career Enhancement School (FIERCES 2016), pp. 213–218 (2016). https://doi.org/10.1007/978-3-319-32554-5_27
3. Rodgers, J.L., Nicewander, W.A.: Thirteen ways to look at the correlation coefficient. Am. Stat. **42**, 59–66 (1988). https://doi.org/10.2307/2685263
4. Granger, C.W.J.: Investigating causal relations by econometric models and cross-spectral methods. Econom. J. Econom. Soc. **37**, 424–438 (1969). https://doi.org/10.2307/1912791
5. Schreiber, T.: Measuring information transfer. Phys. Rev. Lett. **85**, 461–464 (2000). https://doi.org/10.1103/PhysRevLett.85.461
6. Lopes da Silva, F., Pijn, J.P., Boeijinga, P.: Interdependence of EEG signals: linear vs. nonlinear associations and the significance of time delays and phase shifts. Brain Topogr. **2**, 9–18 (1989). https://doi.org/10.1007/BF01128839

7. Grassberger, P., Schreiber, T., Schaffrath, C.: Nonlinear time sequence analysis. Int. J. Bifurc. Chaos **1**, 521–547 (1991). https://doi.org/10.1142/S0218127491000403
8. Wang, H.E., Bénar, C.G., Quilichini, P.P., Friston, K.J., Jirsa, V.K., Bernard, C.: A systematic framework for functional connectivity measures. Front. Neurosci. **8**, 405 (2014). https://doi.org/10.3389/fnins.2014.00405
9. Friston, K.J., Harrison, L., Penny, W.: Dynamic causal modelling. Neuroimage **19**, 1273–1302. https://doi.org/10.1016/s1053-8119(03)00202-7
10. Fox, M.D., Snyder, A.Z., Vincent, J.L., Corbetta, M., Van Essen, D.C., Raichle, M.E.: The human brain is intrinsically organized into dynamic, anticor- related functional networks. Proc. Natl. Acad. Sci. U.S.A. **102**, 9673–9678 (2005). https://doi.org/10.1073/pnas.0504136102
11. Kozlov, S., Poyda, A., Orlov, V., Malakhov, D., Ushakov, V., Sharaev, M.: Selection of functionally homogeneous brain regions based on correlation-clustering analysis. Procedia Comput. Sci. **169**, 519–526 (2020). https://doi.org/10.1016/j.procs.2020.02.215
12. Sharaev, M., Orlov, V., Ushakov, V.: Information transfer between rich - club structures in the human brain. Procedia Comput. Sci. **123**, 440–445 (2018). https://doi.org/10.1016/j.procs.2018.01.067
13. Seth, A.K.: A MATLAB toolbox for granger causal connectivity analysis. J. Neurosci. Methods **186**, 262–273 (2010). https://doi.org/10.1016/j.jneumeth.2009.11.020
14. Hinich, M.J., Clay, C.S.: The application of the discrete Fourier transform in the estimation of power spectra, coherence, and bispectra of geophysical data. Rev. Geophys. **6**, 347–363 (1968). https://doi.org/10.1029/RG006i003p00347
15. Grinsted, A., Moore, J.C., Jevrejeva, S.: Application of the cross wavelet transform and wavelet coherence to geophysical time series. Nonlinear Process. Geophys. **11**, 561–566 (2004). https://doi.org/10.5194/npg-11-561-2004
16. Laird, A., Carew, J., Meyerand, M.: Analysis of the instantaneous phase signal of a fMRI time series via the Hilbert transform, vol. 2, pp. 1677–1681 (2001). https://doi.org/10.1109/acssc.2001.987770
17. Widrow, B., Kollár, I.: Quantization Noise: Roundoff Error in Digital Computation, Signal Processing, Control, and Communications, p. 778. Cambridge University Press, Cambridge (2008). ISBN-13 9780521886710, ISBN 0521886716

The Use of the Economic Cross Method in IT Modeling of Industrial Development (Using the Example of Two-Component Nuclear Energy)

Aleksandr V. Putilov[1], Dmitriy V. Timokhin[1,2(✉)], and Marina V. Bugaenko[1]

[1] National Research Nuclear University MEPhI, Moscow, Russia
dtprepod@yandex.ru
[2] Moscow State University of Humanities and Economics, Moscow, Russia

Abstract. The article contains proposals for the development of modeling of the national industry using IT solutions based on the "economic cross" methodology. The reserves of the digitalization of the planning procedure are considered, taking into account the modern achievements of IT technologies and the development of the global information and communication infrastructure. The conclusions and proposals contained in the article are formulated taking into account the extrapolation of the successful experience of digitalization in leading global companies. The reserves of using the experience and best practices in the implementation of information and communication technologies that emerged from companies during the Covid-19 pandemic, as well as the proposals of economists - supporters economic national development. The research results contain an analysis of the possibilities for expanding the current practice of using the "economic cross" based on expanding the range of information technologies used in planning and realizing the potential of the "bid data" methodology by expanding the coverage and saturation of information flows. The theoretical proposals contained in the article are adapted to the current tasks facing the state corporation Rosatom in the framework of the implementation of the "Breakthrough" concept.

Keywords: Method of "economic cross" · Digitalization · Two-component nuclear energy · Economic modeling · Forecasting information processing · Industry 4.0

1 Introduction

A feature of the innovation process in the 21st century is the significant dependence of the innovative solutions implemented by individual organizations on the quality of technological platforms, within which the corresponding innovative solutions are implemented. The parameters of technological platforms determine the following indicators of the innovation process:

A. V. Samsonovich et al. (Eds.): BICA 2020, AISC 1310, pp. 391–399, 2021.
https://doi.org/10.1007/978-3-030-65596-9_47

- the structure of the participants in the innovation process and the depth of the division of labor, which, in turn, form indicators of the economic efficiency of the innovation process;
- adaptability of the innovation process to changes in the global market and the results of the scientific and technical process, which affects the value of systemic non-hedged risks of the innovation process and determines the competitiveness of the issuer - innovator in the investment and credit markets;
- limiting the speed of business processes associated with the exchange of information between participants in the innovation process and the implementation of legally significant interactions between them;
- the intensity of the use of the basic technologies of Industry 4.0, including digital technologies;
- the attractiveness of the innovation process for external market participants of innovative technologies, including the innovator's partners in the outsourcing network; this indicator is especially important when planning the organization of the innovation process in the form of a startup;
- compliance of the innovation process with the trends of long-term economic, technological development of the global market, that is, the relevance of innovative technology from the point of view of the theory of the life cycle of technology [5].

The experience of business development during the Covid-19 pandemic is of significant importance from the point of view of assessing the effectiveness of modern planning practices for the innovation process and its compliance with the capabilities of technology platforms relevant for 2020 in the context of the formation of Industry 4.0. Despite the relatively small, compared to the experience of earlier epidemics, direct losses from the spread of the Covid - 19 virus, losses, the global economy suffered significant damage. Thief many reasons for this phenomenon lie in the planning of technological and economic processes. A review of foreign publications on the relevant topic reveals the following systemic shortcomings of strategic economic and technological planning for business development at the beginning of the XXI century, which led to its vulnerability in 2020:

- the lack of strategically acceptable planned alternatives for interacting with employees and contractors in the online environment of more than 85% of all organizations (calculated for the USA, Canada, Great Britain and Italy), while national and global technology platforms provided appropriate alternatives [1];
- the difficulty of finding alternative partners according to the criteria of the terms of the contract (price, terms of delivery, urgency of the contract, volumes, responsibility of the parties, etc.) in a short time for a significant number of national businesses in the presence of a global information space [2];
- vulnerability of organizations to incomplete, inaccurate or deliberately distorted inaccurate information and the limited ability to work in the "big date" format [3];
- problems of a secondary nature associated with the introduction of innovative solutions, such as the unpreparedness of personnel to use remote forms of work, the inability of the management level to function in conditions of high economic and informational volatility of the markets [4];

The analysis clearly demonstrates the limited possibilities of modern forecasting methods to ensure the optimal choice of technological solutions for the long-term strategic development of industries. To solve this problem, the authors have proposed a model of the "economic cross", devoid of the above drawbacks of other analytical tools and allowing to take into account, when planning the development of production, the possibilities of modern IT platforms and take into account their probable vectors of technological development.

2 The Model of the "Economic Cross" and the Concept of Its Application in IT Modeling of Industrial Development

The authers considered the main distinctive features of the concept of strategic planning at the beginning of the XXI century from the corresponding practice in the XX century. These features, considered in [7, 8] in relation to various types of industry activities, can be classified into the following consolidated groups:

- the planning process is undergoing more and more automation, based on the more and more intensive use of information and communication technologies, the obsolescence cycle of which is decreasing throughout the period 2000–2020;
- with intra-sectoral planning, the role of inter-sectoral factors significantly increases, primarily the role of technological innovative solutions presented in Fig. 1.

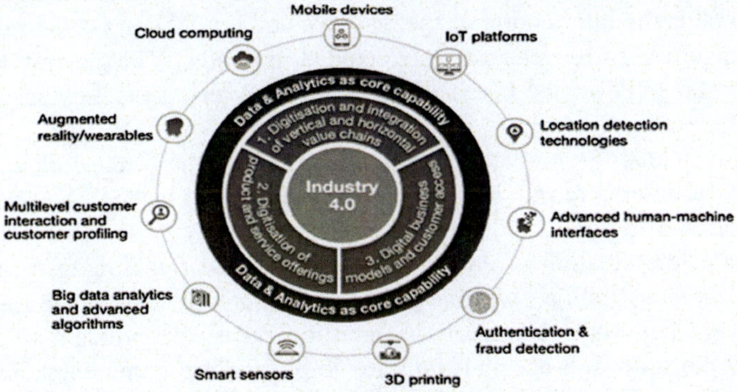

Fig. 1. - Aggregated technological groups of information and communication innovations requiring integration into the production process in the strategic planning of sectoral development [6]

Innovative solutions included in the groups proposed in Fig. 1, as a rule, contain built-in universal feasibility studies for their effectiveness when used within the format proposed by the developer. The complex nature of the innovative solutions offered by the modern market and their strict binding to the technological platforms chosen by the developer and the set of used complementary goods, planning facilitates the planning process, and the formalization of the key parameters of an innovative solution in digital format increases the possibility of its automation through the use of computer technologies. Along with this, the process of strategic planning is complicated by the problems of choosing the structure and completing the proposed groups of innovative solutions and is accompanied by a choice between competing technological platforms. A striking example of realized risks associated with insufficiently effective planning are the problems of technology companies that arose in 2019–2020. In connection with US restrictions on the use of resources of Chinese technology platforms.

In connection with the above features, the concept of strategic planning for innovative sectoral development can be based on the modular principle, or the "constructor principle". Distinctive features of this principle are, on the one hand, higher planning efficiency and virtually an infinite set of possible combinations. On the other hand, the new concept increases the information load at the initial stage of the planning process - the stage of selecting basic innovative solutions. The information space for the selection of initial options is the global space of the world market for innovative technologies. It should be noted the heterogeneity of this information space and the absence of uniform standards of functionality and quality for many technological solutions.

The economic cross methodology is the closure of the economic processes of production cycles at the stages of production of added value that are common for these cycles.

The simplest model of the "economic cross" is the closure of the resource and production cycle, which are implemented in parallel by economic entities not affiliated with each other. The added value formed by direct contact between the seller and the buyer is formed at the intersection of the resource and production cycle in the center of the "economic cross", while costs are collected at its "ends". The general model of the "economic cross" built for the intersection of the resource and fuel cycles is shown in Fig. 2.

The task of planning innovative development within the framework of building a model of the "economic cross" is to determine the optimal technologies for using the resources indicated in Fig. 1.

In this case, the creation of added value requires the functioning of at least two independent suppliers, while only one of the participants is engaged in direct contact with the buyer. This formulation of the question raises the problem of choosing a technological platform and optimal tools for implementing communications between organizations. At the same time, it is important to ensure the reliability of the data that falls into the cycle of information interaction, since any, even minor errors and distortions multiply over time in proportion to the number of connections and participants.

The solution to the stated problem of strategic planning of sectoral development is possible for the design of information and communication support for the business processes declared within the framework of the concept of the "economic cross", as shown in Fig. 3.

The vertical component of the "economic cross" corresponds to the levels of organization of business processes, and the horizontal one corresponds to the stages of processing the resource used within the projected production technology. The most appropriate standard recommended for use by businesses when designing a digital model of the "economic cross" is IEF 62890.

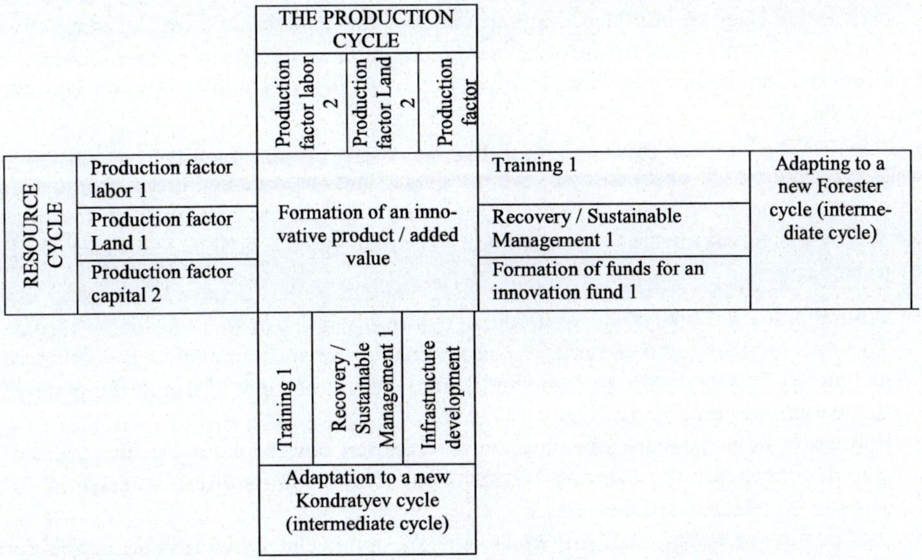

Fig. 2. The basic model of the economic cross, proposed for a comprehensive assessment of commodity and cash flows in the implementation of different-level innovative technologies in the framework of an innovative project. Compiled by the authors.

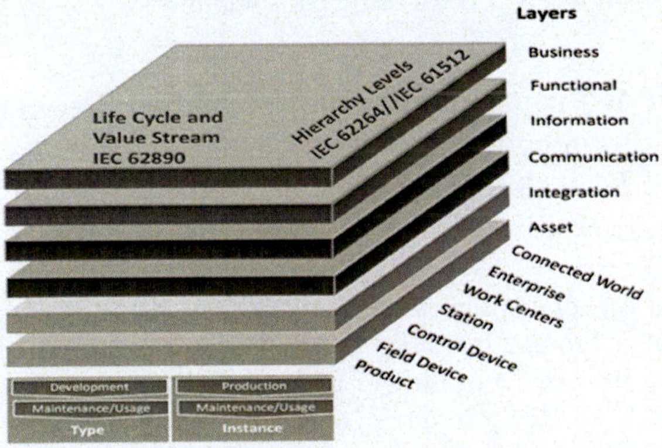

Fig. 3. The optimal format for digital design of the economic cross, recommended when planning long-term industrial development [10]

It is proposed to consider products with real consumer value as stages of intersection of the resource and technological cycles. Within the digital (virtual) model of the "economic cross", they can be described as the result of the interaction of production and processing technologies. All solutions that have a consumer value can be divided into levels, at each of which technologies that are conventionally considered basic are used.

The initial stage of building a digital model of the "economic cross" of innovative sectoral development should be considered the finding of one or more innovative solutions that can be presented as a result of the interaction of solutions known from the state of the art.

The role of digital support for the formation of the "economic cross" at the initial stage is to index each of the known technologies of the resource and production cycles according to predetermined parameters. Given the significant volume of solutions offered by the global innovative market, the most demanded digital technologies at this stage are:

1. Programs for the analysis of heterogeneous information in the "big data" format. The task of using the appropriate programs is the most complete selection of technical solutions that are promising from the point of view of use in the planned innovation process.
2. Formation of a database of solutions in a format corresponding to the tasks of automated generation of innovative technical solutions through the intersection of previously selected technologies.
3. Presentation of an automatically generated solution package in a format suitable for subsequent comparison and processing by both artificial intelligence and humans (the coincidence of formats for the specified tasks is not necessary.

Further formation of the "economic cross" is a sequence of actions for the automated formation of secondary possible intersections of technologies (economic crosses) based on previously formed technical solutions.

3 Adaptation of Proposals for the Use of the "Economic Cross" Methodology in IT - Design of Sectoral Development for the Formation of Two-Component Nuclear Energy

Strategic design of economic processes of sectoral reorganization required by the state corporation "Rosatom" for the formation of an efficient two-component nuclear energy in accordance with the Decree of the President of the Russian Federation V.V. Putin No. 270 dated April 16, 2020, is one of the priorities of national innovative development. The conditions for effective strategic design from the point of view of the conjuncture of the global market are:

The economic cross model published by the authors in the author's earlier works for the development of two-component energy [7, 9] and the modernization of related backbone areas, such as education [8], requires significant IT support at the level of both infrastructural and functional support.

The model of integration of individual IT technologies within the framework of a unified production system of nuclear power with a closed fuel cycle is shown in Fig. 4.

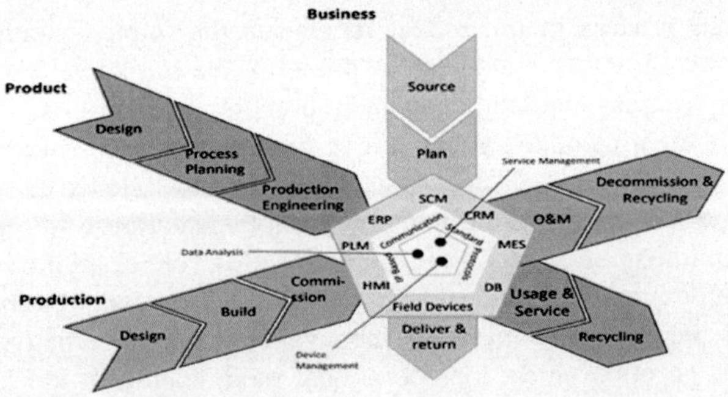

Fig. 4. - Model of interaction of production and IT - technologies when closing the nuclear fuel cycle

Information and communication problems facing the state corporation Rostat at the present stage are:

– ensuring the readiness of all structural divisions to use planned for the introduction of innovative technologies;
– ensuring the scaling of the economic effect from the use of technologies created or purchased by the state corporation Rosatom;
– elimination of duplication of the use of technologies to solve similar solutions;
– ensuring the compatibility of the use of technologies and means of their implementation in different areas of the "economic cross" of two-component nuclear energy.

Parametric indexing of all used technologies within the framework of a unified model of the "economic cross" allows solving the problem of time lags in the implementation of technologies in various structural divisions of Rosatom State Corporation for different stages of implementation of the economic cross. Consider the "economic cross" model recommended for the state corporation Rosatom in Fig. 5.

		PROCESS B. Phase 2.1 Establish a smart control room for Rosatom				
		b1	b2	b3		
PROCESS A. Step 1.1 Development of online retraining measures for Rosatom staff	al a2 a3	Stage 2 - optimization of the use of the fund of working time			a4 a5 a6	PROCESS A. Stage 1, 3 Organizational losses due to failures of the timing construction model
		b4	b5	b6		
		PROCESS B. Stage 2, 3 Losses of the organization due to ineffective performance (non-performance) of uncovered ICT - control functions				

Fig. 5. Model of the economic cross of digital support of the process of monitoring the timing indicators of working time (a1–a6, b1–b6 - intensive, moderate and extensive implementation of an innovative technological solution)

The vertacular process in the model represents the implementation of a "smart control room" (infrastructure management process), the horizontal process - online - employee training (human capital management process). Both processes themselves are not yet capable of creating added value, but at their intersection the company is able to make a profit. The use of the "economic cross" model allows us to compare the possible costs and savings from the integrated implementation of the proposed information and communication solution for the entire life cycle of an innovative solution. In addition, the indexing of each innovative technology (innovative solution) according to the specified parameters, subject to the availability of alternative technological solutions, creates an opportunity for operational modification of the information and communication interaction system with minimal losses in economic efficiency.

4 Conclusions

Thus, the use of the "economic cross" model in planning sectoral development helps to expand the potential coverage of information. Early identification of the needs for external technologies (technologies from suppliers not affiliated with the manufacturer) allows timely indexing of the collected information and ensuring the readiness of structural units for modernization and adaptation in advance. The presentation of IT - technologies offered for use within the framework of industry development, as compatible according to the parameters declared in the project "economic cross" creates an opportunity to ensure the uninterrupted technological modernization of the state corporation. The conclusions and proposals contained in the article are of interest from the point of view of optimizing the transaction costs of forming a system of digital support for planning the development of two-component nuclear power.

References

1. Barua, S.: Understanding coronanomics: the economic implications of the coronavirus (COVID-19) pandemic. SSRN (4) (2020). http://dx.doi.org/10.2139/ssrn.3566477
2. Goodell, J.W.: COVID-19 and finance: agendas for future research. Finan. Res. Lett. (2020). https://doi.org/10.1016/j.frl.2020.101512
3. Nicola, M., Alsafi, Z., Sohrabi, C., et al.: The socio-economic implications of the coronavirus pandemic (COVID-19). Rev. Int. J. Surg. **78**, 185–193 (2020). https://doi.org/10.1016/j.ijsu.2020.04.018
4. Hsiang, S., Allen, D., Annan-Phan, S., et al.: The effect of large-scale anti-contagion policies on the COVID-19 pandemicю. Nature **584**, 262–267 (2020). https://doi.org/10.1038/s41586-020-2404-8
5. Bartik, A., Bertrand, M., Cullen, Z., Glaeser, E., Luca, M., Stanton, C.: The impact of COVID-19 on small business outcomes and expectations. Proc. Natl. Acad. Sci. **117**(30), 17656–17666 (2020). https://doi.org/10.1073/pnas.2006991117
6. Oztemel, E.: Literature review of industry 4.0 and related technologies. J. Intell. Manuf. **3** (2018). https://doi.org/10.1007/s10845-018-1433-8

7. Timokhin, D.V., Bugaenko, M.V., Putilov, A.V.: The use of IT technologies in the implementation of the "economic cross" methodology in the Breakthrough project of Rosatom in Procedia Computer Science, 169, pp. 445–451. (2020) https://doi.org/10.1016/j.procs.2020.02.227
8. Putilov, A.V., Bugaenko, M.V., Timokhin, D.V.: Revisiting the modernization of the educational process IT component in Russia on the basis of the model of "economic cross". In: AIP Conference Proceedings 1797, 020014 (2017). https://doi.org/10.1063/1.4972434
9. Putilov, A.V., Timokhin, D.V., Pimenova, V.O.: Adaptation of the educational process to the requirements of the global nuclear market according the concept of «economic cross» through its digitalization. Procedia Comput. Sci. **169**, 452–457 (2020). https://doi.org/10.1016/j.procs.2020.02.226
10. Fraile, F., Sanchis, R., Poler, R., Ortiz, A.: Models for digital manufacturing platforms. Appl. Sci. **9**(20), 4433 (2019). https://doi.org/10.3390/app9204433
11. Gökalp, E., Şener, U., Eren, P.: Development of an assessment model for industry 4.0: industry 4.0-MM. In: International Conference on Software Process Improvement and Capability Determination (2017). https://doi.org/10.1007/978-3-319-67383-7_10
12. Matt, D., Modrák, V., Zsifkovits, H.: Industry 4.0 for SMEs: Challenges, Opportunities and Requirement. Palgrave Macmillan, London (2020) https://doi.org/10.1007/978-3-030-25425-4
13. Ignatiev, M.B., Karlik, A.E., Iakovleva, E.A., Karlik, E.M.: Challenges for strategic management of the development of the digital economy and advanced training In: XVII Russian Scientific and Practical Conference on Planning and Teaching Engineering Staff for the Industrial and Economic Complex of the Region (PTES), pp. 197–200 (2018). https://doi.org/10.1109/ptes.2018.86042
14. Timokhin, D, Vorobyev, A., Bugaenko, M., Popova, G.: Formirovanie mekhanizmov ustojchivogo innovacionnogo razvitiya atomnoj otrasli in Tsvennie metalliю **3**, 23–35 (2016)
15. Putilov, A.V., Timokhin, D.V.: Innovacionnye vozmozhnosti ispol'zovaniya metodologii ekonomicheskogo kresta v prognozirovanii perspektiv razvitiya dvuhkomponentnoj atomnoj energetiki. Innovacii. **1**(255), 12–20 (2020). https://doi.org/10.26310/2071-3010.2020.255.1.002

Check for updates

Intelligence - Consider This and Respond!

Saty Raghavachary(✉)

University of Southern California, Los Angeles, CA 90089, USA
saty@usc.edu

Abstract. Regarding intelligence as a 'considered response' phenomenon is the key notion that is presented in this paper. Applied to human-level intelligence, it seems to be a useful definition that can lend clarity to the following related aspects as well: mind, self/I, awareness, self-awareness, consciousness, sentience, thoughts and feelings, free will, perception, attention, cognition, expectation, prediction, learning. Also, embodiment is argued to be an essential component of an AGI's agent architecture, in order for it to attain grounded cognition, a sense of self and social learning - via direct physical experience and mental processes, all based on considered response.

Keywords: AGI · Artificial general intelligence · Artificial intelligence · Evolution · Adaptation · Intelligence · Response · Embodiment

1 Introduction

AI is a venerable field, quite broad in scope, ambitious, and, over 60 years old, [1–3]. Even at the 1956 Dartmouth Conference that 'birthed' AI [4], there was not an agreed upon definition of intelligence; instead, the program description included this: 'The study is to proceed on the basis of the conjecture that every aspect of learning or any other feature of intelligence can in principle be so precisely described that a machine can be made to simulate it.'

A formal lack of definition has meant that the field diverged in its goal, scope, techniques - today we are the benefactors of rich set of these, and lately, a handful of techniques (based on Deep Learning, a subset of Machine Learning, a subset of AI) have begun transforming societies worldwide. Nevertheless, the achieving of human-level intelligence and beyond, generally termed AGI, has elusively and tantalizingly remained beyond reach thus far.

In this paper, we make a modest attempt at reframing the notion of 'intelligence', to make it be applicable to the diversity of it in the natural world - this is in contrast with the more common descriptions that pertain mostly to human-level intelligence (e.g. involving logical reasoning, planning, problem-solving, natural language, etc.) on which AI has been focused.

In turn, this lets us provide working definitions of associated concepts such as consciousness, perception etc., which have also not been mainstream AI topics.

Embodiment, which has been largely ignored by AI for the most part, is considered in light of the proposed intelligence reframing, as an essential aspect that could lead to AGI.

A. V. Samsonovich et al. (Eds.): BICA 2020, AISC 1310, pp. 400–409, 2021.
https://doi.org/10.1007/978-3-030-65596-9_48

2 Postulates on Intelligence and Related Phenomena

The following 15 items (axioms) are presented, without proof or even strong evidence/argument; nevertheless, taken together, they do provide a unifying view of a variety of terms that are used in AI, cognitive science and philosophy. The idea is to be able to use them for constructing biologically-inspired AGIs; the descriptions for the items are rooted in biological (higher animal) existence - obtained from observing others (including animals), wide-ranging study of existing literature (involving neurological studies), and old-fashioned self-examination. In addition, they do seem to be useful launch points for BICA-AGI-related discussions. Here are the items:

- **Intelligence:** all form of **intelligence is considered response** (often referred to as $C \rightarrow R$, or simply, 'CR', in what follows). For instance, intelligence is a situated agent's considered response to its environment. The considerations are viewed as **processes**, whose **responses** constitute intelligence. The rest of the items that follow (except for the last one) are described in terms of consideration processes and/or responses, and apply primarily to higher-level (animal-like) intelligence.
- **Mind:** the mind constitutes a collection of independent, interacting **processes** that carry out consideration tasks.
- **Self/I:** the 'self' is a privileged, meta ('higher') level **process** which (learns over time that it) 'owns' the rest of the processes as well as owning the body + brain in which it resides - the owner is the 'self', the 'self' is the owner.
- **Awareness:** the 'self' process just **responding** that is 'ready'/'on' [but not 'for' anything in particular], is 'awareness'.
- **Self-awareness:** the 'self' process **responding** to (acknowledging) itself is 'self awareness' - 'I exist'.
- **Consciousness:** the 'self' process **responding** to (being aware of, i.e. considering and acknowledging) sensation is said to be 'conscious'.
- **Thoughts and feelings:** thoughts and feelings are **responses** - of consideration processes that usually involve access to memory (in which is stored the agent's ongoing life experiences, skills, facts and more). In the brain, both thoughts and feelings are manifested as electrical activity (occurring in neural populations), or chemical activity (via neurotransmitters); feelings can also trigger, or be triggered by, physical responses which we term 'emotions'.
- **Sentience:** the 'self' process considering and acknowledging its feelings is said to be 'sentient' - it is a subjective **response**
- **Free will:** the 'self' process considering a situation and creating response choices, being aware of those choices (i.e. considering, acknowledging them) and picking one of them as its **response**, is said to exercise 'free will'. Free will is somewhat of an illusion - it seems to exist, given the lack of predictability on an agent's part, from the points of view of other agents (since they do not have access to the free-will-exhibiting agent's consideration processes and memory contents).
- **Perception:** perception is the **process** of considering sensory data (which involves more than merely labeling it), using prior knowledge and experience stored in memory, in order to subjectively interpret it.

- **Attention:** attention involves selective consideration/**processing**, by the self, of whatever is salient at the moment, within the agent and/or its environment; attention (focus/concentration) is how the self exerts control over ongoing CR processes - attention strength can vary, and attention can be split.
- **Cognition:** these are the **processes** underlying various forms of consideration.
- **Expectation:** expectation is an internal **response**, to a consideration process that involves perception of the environment (which can include an agent's own body), and subsequent matching against an internal, subjective model of the world. The agent's considerations and/or responses are adjusted to minimize mismatches - this is often the goal of intelligent behavior.
- **Prediction:** given its personal 'version' of the world and the current state of it to consider, the brain is constantly predicting the next state - it is also a **response**. Prediction and expectation are often conflated, but it useful to look at prediction as a lower-level consideration process, which feeds into the expectation process (which can involve cognition in addition).
- **Learning:** learning is an ongoing accumulation in memory, for the most part (with corrections and deletions as well), of what to consider, and how (and where pertinent, why), for the purposes of responding. Learning is augmented by 'experience' - it consists of not only objective knowledge about the world, but also, personal (subjective) aspects such as skills and capabilities, beliefs, habits, etc.

In summary: in order that higher level, (e.g. human-like) intelligence can be exhibited, evolutionary adaptation has resulted in an electrochemical processor (brain), which, coupled with the body it is housed in, can consider and respond to externally and internally generated inputs, via a set of ongoing, interacting 'CR' processes that make use of, and contribute to, accumulated experience - mediated by the self when necessary.

3 Discussion

In the following subsections, we are going to consider what can follow from the 'considered response' notion, aspects of human intelligence, and a case for embodiment.

3.1 Intelligence - A Continuum of Responses

A wide variety of definitions for intelligence exists, but almost all are focused on higher level brain processes such as reasoning and problem-solving. For example, after collecting more than 70 definitions, Shane and Hutter summarize them like so: 'intelligence measures an agent's ability to achieve goals in a wide range of environments' [5].

The diversity of intelligence in the natural world (including simple reactive creatures, swarm intelligence that emerges in collections of bees, termites [6], fish, and birds [7], alien-like octopus brain 'wiring', animals and birds with a wide range of sensing, capabilities and behaviors, and of course, humans) suggests that there is a common underlying principle, one that is simple, elegant, and extremely widely applicable, and is likely tied to evolution! It is such reasoning that leads us to the following realization: intelligence is a biological phenomenon tied to evolutionary adaptation, meant to aid an agent survive and reproduce in its environment by interacting with it appropriately - it is one of considered response.

To make an analogy with computation, consideration is equivalent to executing a function or process, whose algorithm handles inputs, possibly accesses additional stored data in memory, and returns (responds with) a result - consideration is the processing, and response is the result.

The notion of intelligence being a response, helps us place it in the following response continuum, based on structure-property consideration.

Simply put, structures, from the microscopic to human level to cosmic level, organic and inorganic, exhibit ('respond with') phenomena on account of their spatial and temporal arrangements, under conditions external to the structures. This can sometimes result in new structures that display phenomena appropriate to them, and this process of structure+phenomena creation could continue.

For example, molecules in ice vibrate when heated, eventually turning into liquid water - a different structure which undergoes its own phenomenon (flowing); sand dunes form wave patterns in response to wind, which then result in sound production on account of the periodicity - structures exhibiting phenomena could result, as a response, in further structures that exhibit their own phenomena. Not only can the formation of the Milky Way, the solar system and the Earth be considered responses (to universe-scale matter, long and short range forces, space, time and energy), so can the origin of life itself, on our planet: given ideal conditions that exist here, primordial carbonaceous structures must have undergone phenomena that resulted in life-bearing forms that exhibit the phenomena of survival and reproduction!

The continued evolution of simple life into multi-cellular organisms, sea creatures, and land dwellers, and their underlying adaptations are of course, responses to environmental factors, guided by an unseen hand as it were. In this scheme, intelligence becomes another type of response, exhibited by a variety of biological structures in a variety of ways, for purposes of survival and reproduction. More generally, life can be regarded as being comprised of evolved structures and processes that involve information, energy and matter based interactions with their environment, for purposes of survival and reproduction; the information-processing ('considered response') aspect is what can be termed intelligence. In this view, intelligence is seen as a requisite component of all forms of life.

In plants, intelligence is manifested in the design of their immobile structure - strong roots, thick leaves, thorns, bright flowers, etc. In lower animals, the structures that exhibit reactive intelligence are their simple nervous systems. In collective animal colonies (ants, termites, bees, birds, fish... as noted above), the evolved design response seems to be a two-stage one - evolution of simple actions to be performed using simple brains, which are able to result in global ('emergent') behavior, which could be considered to be the response of a colony-level, self-organized structure. In addition, mimicry, camouflage, aposematism, symbiosis, stygmergy... can all be seen as intelligence-related design responses in form or the other, meant to aid the predator, prey, individual or the colony.

Moving on to higher animals including humans, the intelligent response by the individual stems from a more developed brain (and better co-evolved bodies) that can form memories, develop complex skills, develop communication, etc.

In all cases, it is structures that exhibit specific phenomena, which we regard as their response, intelligent or otherwise - note that we only attribute intelligence to specific biological structures designed (evolved) chiefly for survival and reproduction.

3.2 Human Intelligence

Now we look at how the notion of considered response can shed light on the complex phenomenon that is human (and other higher animal's) intelligence.

Figure 1 shows what consideration and response might look like, in a human brain.

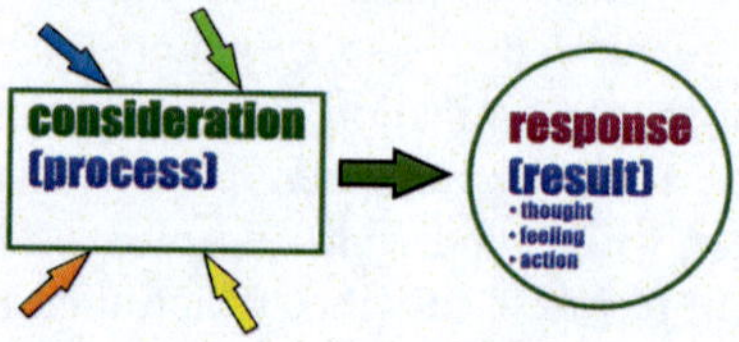

Fig. 1. Consideration and response.

Consideration produces as response the following: thoughts, feelings (physically manifested as emotions), and actions (including vocalizations (which encompass verbalizations)).

The consideration process gets as its inputs, sensations from the environment (which includes the agent's own body), contents retrieved from long term memory, as well as responses from other ongoing and prior considerations. These operate in a continuous, parallel, interacting, shared-memory fashion (where they retrieve as well as write to long term memory, and working memory) - this is precisely what gives rise to the complexity of human thought, feeling and action (including language). This is illustrated in Fig. 2 below.

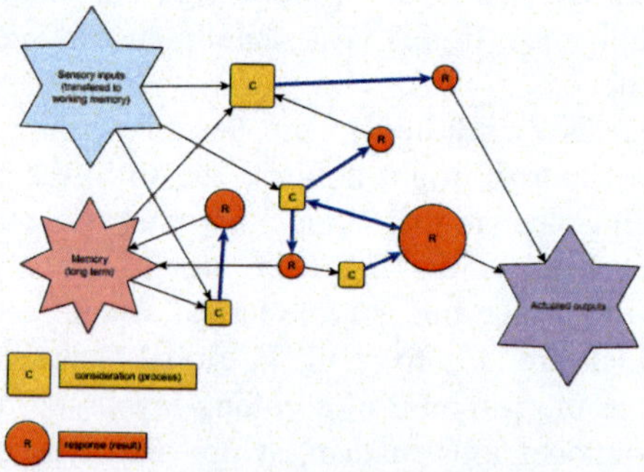

Fig. 2. Interacting CR processes.

3.3 Embodiment - A Necessity for Human-Level AI

An AGI agent without a body (virtual or physical) is, for all practical purposes, a 'brain in a jar'. We note that in nature, given all the biological diversity, there is not a single example of such an entity - brains are always housed in bodies, in exchange for which they help nurture and protect the body in numerous ways (depending on the complexity of the organism). The only disembodied being is one worshiped by billions, an omnipotent, omnipresent and omniscient God!

An AGI brain does not always need to be housed in a AGI body - but the point is that it would need a body, in order to attain certain characteristics that are otherwise unattainable. Let us confine our discussion to a human-like AGI, although the ideas and lessons should transfer to lesser forms (simpler AGIs) as well.

A body provides a brain, the following advantages:

- separation from the environment - when the torso moves, so does the brain, creating a sense of ownership and situatedness
- identity - the body parts are controllable by the brain (thanks to the integrated, matched 'wiring' that enables this; in humans, this starts at birth, in fact in the womb during late stages of pregnancy) - this, is conjunction with social conditioning (e.g. a mother using a baby's name repeatedly), helps develop a sense of self
- 'grounding', via direct physical experience - from being able to pick up light, small things (but not big, heavy ones) and rolling down a slope, to learning the passage of time to stereopsis to being able to tilt the head and view the world sideways, etc., the experience and knowledge gained is subjective, first-hand and incremental
- agency - this is a big advantage; the agent does not need to rely on another, to carry out possible actions, instead, the agent can just do them on its own (also, this facility of direct first-person-oriented action cannot be compensated by a disembodied agent having available to it, rich multi-sensory 'data' from the world - this is about active engagement with the environment, only possible via a physical form)
- agency also leads to the development of 'free will' - "I can control what I can or will not do"
- theory of mind - "looking at your actions, I can tell you are like (or not like) me"
- empathy - "I see you howling in pain after you fell, I know how you feel, I fell in the exact same spot earlier and it hurt"
- social/imitation learning - thanks to mirror neurons, a kid would be able to imitate (gracelessly at first, till reinforcement learning kicks in) her pre-school teacher's hokey-pokey dance routine; same goes for language - understanding words by associating them with things (nouns), actions (verbs) and qualities (adjectives) become trivial when these are shown, told, read to, overheard etc.
- model-free learning - experiencing the world first, then being able to fill in the reasoning (e.g. for why a rainbow forms, or why metal gets hot, etc.)
- number sense, grouping, abstraction etc. - by using tangible objects and having features subtracted (or singled out) verbally and by demonstration, the brain is able to grasp underlying concepts that are abstract and non-immediate, such as opposites, colors, groups, hierarchies, numbers, etc. (pre-school board books are a

testament to this) - this offers the brain a way to learn to create and use symbols, from within.

As is evident from the above list, a human AGI without a body is bound to be, for all practical purposes, a disembodied 'zombie' of sorts, lacking genuine understanding of the world (with its myriad forms, natural phenomena, beauty, etc.) including its human inhabitants, their motivations, habits, customs, behavior, etc., the agent would need to fake all these. To put it differently, a disembodied agent is forever left ungrounded, reduced to deriving meaning from symbols and/or data that the agent did not develop or gather first-hand.

As discussed earlier, an embodied AGI architecture needs to be 'matched' in body and brain - the brain architecture needs to be designed in such a way that it is integrated into the body so as to provide optimal control of the body, monitor and regulate it well, and obtain appropriate sensory inputs from it. Further, the body+brain would need to be designed so as to function optimally in the environment which they will inhabit.

As for a human-level AGI brain architecture, we need a design that involves continuously running, interacting, consideration-response processes that draw from two shared memory stores - a short-term working memory one (for processing sensory inputs, i.e. perception, and details related to current processing), and a long-term memory that houses accumulated experiences, memories of events, semantic information, thoughts, feelings, etc. (Fig. 3):

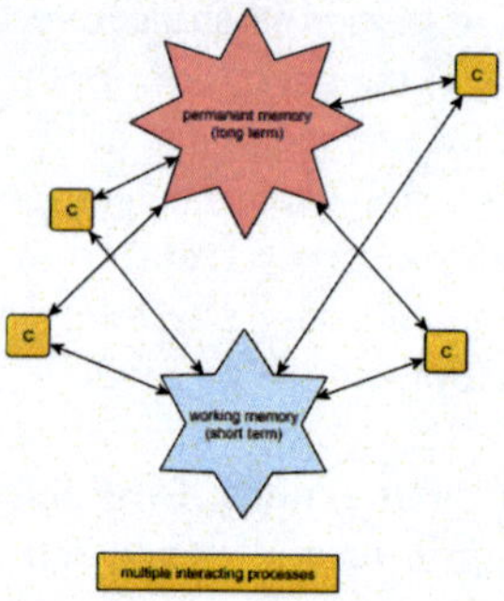

Fig. 3. Use of long and short term memories, by CR processes.

As shown earlier, the responses that are generated can serve as inputs for further considering and responding, and results of such interactions can be stored away as memories for future consideration. All this makes for a rich web of memory store that would comprise of an ever-changing mix of content and procedural knowledge, thoughts, feelings and experiences. Further, these items are hypothesized to be stored as key:value associations, connected together in the form of deeply and richly linked, flexibly indexed hypergraphs for efficient retrieval by ongoing CR processes.

We can use '**4E/6EXP**' to mnemonically describe human-level brain function. As for '**4E**', the brain is '**e**'mbodied (situated in a suitably co-evolved body), '**e**'mbedded (in the environment, via the body), '**e**'xtended (where it makes uses of the environment for some of its consideration processes), and '**e**'nacted (its primary purpose is to support body functioning). The '**6EXP**' aspects are as follows. Figure 4 shows how we can consider human intelligence as comprising three components, as it relates to a suitably embodied agent negotiating its world: a memory store that contains an agent's ongoing, accumulated '**exp**'erience for future use; 'regular' CR processes that interact among themselves and with short-term memory as they access and possibly modify the memory store, during routine, day-to-day '**exp**'eriencing; a supervising/'self'/"meta-CR"/owner/'**exp**'eriencer process that directs the regular CR processes when necessary - for conscious attention, imagining, simulating, planning, problem-solving, reasoning - thereby exerting explicit control over the agent's responses. The stored experience and knowledge, coupled with ongoing experiencing, lets the brain '**exp**'lain away its reality, '**exp**'ect (anticipate/predict) what is about to happen, and let the agent '**exp**'ress itself (respond) via thoughts, feelings, words and deeds.

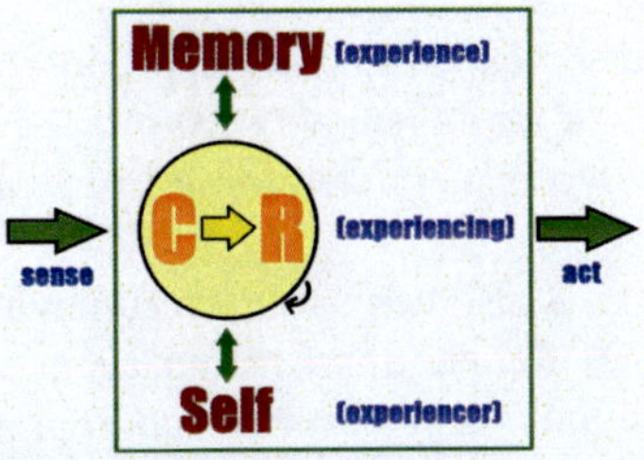

Fig. 4. Experience-oriented aspects of human intelligence.

Any form of AI will be limited in its CR, unless it is embodied in a way that permits subjective and active first-hand exploration that provides for direct physical experience, interaction with other similar agents, remembering and recalling experiences, continuous learning and growing (mentally and even physically). In the absence of directly gained '3D' knowledge of the world (which includes forms, patterns, a variety of phenomena, events, experiences of different places, etc.), abduction (i.e. the 'frame' problem of generating hypotheses or justifying existing ones) will remain a difficult issue, because there is simply no way for the system to know what is 'beyond' (even if combined with semantic knowledge graphs, which will help avoid the problem somewhat, but will not entirely eliminate it). **A suitably paired (brain, body) embodiment that is also matched to its environment, is key to 'fluid' intelligence (C → R) that can continue to grow and adapt** - in short, the experiences, experiencing, and experiencer depicted in Fig. 4, all need to be body-oriented.

In sharp contrast to standard AI, AGI architectures tend to be more biologically-based, with the presumption that a biological underpinning helps them have advantages similar to biological beings including humans. A variety of AGI architectures are documented in Samsonovich [8] and kept up-to-date at the associated website [9]; see [10] for recent work on humanoid robots. Research is underway to imbue AGI agents created using architectures we just noted, with characteristics such as consciousness, self knowledge, ethics, empathy, episodic memory, emotions [11], creativity [12], etc.

This is a promising area of research, going forward - if the goal of AI is to imitate (and possibly surpass) human intelligence, a good functional basis for it would be the brain, considered alongside the body in which it is housed. Depending on the architecture, the C in these systems would include episodic and semantic memory, stored experiences, and inputs from sensors and receptors; R would include thoughts, feelings, vocalizations and other actuations (even possibly including creative pursuits such as art and music). Again, embodiment is likely to be a key requirement in such implementations.

We briefly note the following: in general, to create embodied agents, we have two choices - we could use physical robots, as was done by Goertzel [13], or we could use virtual agents, as advocated in [14], and possibly create embodiments of the virtual versions later.

4 Conclusions

The 'intelligence as considered response' notion proposed in this paper, seems uniformly applicable to a variety of intelligence(s), from reactive to distributed to enactive human-level - differences between them stemming from what is considered and how, and what the response is.

Applied to human-level AI, it also helps reinforce viewing biological intelligence as an evolutionary adaptation phenomenon, where an agent's brain that has co-evolved along with the body, helps the agent survive and reproduce (in a broader sense, negotiate its environment), by carrying out ongoing considering and responding, with or without active involvement of the self. The 'consideration as a process' idea helps provide working descriptions of several other terms (such as awareness, thought…) which have been rather difficult to be conceptualized in a simple and implementable manner.

To achieve human-level AI, a strong case is made for embodiment - this would provide agency, which in turn would help create a sense of self (and prove invaluable in multiple other crucial aspects related to learning, mentioned earlier). The embodied agent's direct physical experience would provide it grounded, ongoing, actively and subjectively acquired meaning of the world ('experience'); this experience is used in considering and responding (actively guided and directed by the self, when necessary) - in short, to exhibit intelligence.

References

1. Moravec, H.: Mind Children: the Future of Robot and Human Intelligence. Harvard University Press, Cambridge (1988)
2. Nilsson, N.: The Quest for Artificial Intelligence: A History of Ideas and Achievements. Cambridge University Press, Cambridge (2010)
3. Lungarella, M., Iida, F., Bongard, J., Pfeifer, R.: 50 Years of Artificial Intelligence: Essays Dedicated to the 50th Anniversary of Artificial Intelligence. Springer, Heidelberg (2007). https://doi.org/10.1007/978-3-540-77296-5

4. Moor, J.: The Dartmouth college artificial intelligence conference: the next fifty years. AI Mag. **27**(4), 87 (2006)
5. Legg, S., Hutter, M.: A collection of definitions of intelligence. Front. Artif. Intell. Appl. **157**, 17–24 (2007)
6. Resnick, M.: Turtles, Termites, and Traffic Jams: Explorations in Massively Parallel Microworlds. MIT Press, Cambridge (1994)
7. Reynolds, C.: Flocks, herds and schools: a distributed behavioral model. SIGGRAPH Comput. Graph. **21**(4), 25–34 (1987). https://doi.org/10.1145/37402.37406
8. Samsonovich, A.V.: Toward a unified catalog of implemented cognitive architectures. BICA **221**, 195–244 (2010)
9. https://bicasociety.org/mapped/. Accessed July 2020
10. Chella, A., Cangelosi, A., Metta, G., Bringsjord, S. (eds.).: Consciousness in Humanoid Robots. Frontiers Media, Lausanne (2019). https://doi.org/10.3389/978-2-88945-866-0
11. Samsonovich, A.V.: Emotional biologically inspired cognitive architecture. BICA **6**, 109–125 (2013)
12. Turner, J., DiPaola, S.: Transforming kantian aesthetic principles into qualitative hermeneutics for contemplative AGI agents. In: Iklé, M., Franz, A., Rzepka, R., Goertzel, B. (eds.) Artificial General Intelligence. AGI 2018. Lecture Notes in Computer Science, vol. 10999. Springer, Heidelberg (2018)
13. Goertzel, B., et al.: OpenCogBot: achieving generally intelligent virtual agent control and humanoid robotics via cognitive synergy (2010)
14. Raghavachary, S., Lei, L.: A VR-based system and architecture for computational modeling of minds (2020). https://doi.org/10.1007/978-3-030-25719-4_55

Check for updates

Simple Model of Origin of Feeling of Causality

Vladimir G. Red'ko[1,2(✉)]

[1] Scientific Research Institute for System Analysis, Russian Academy of Sciences, Moscow, Russia
vgredko@gmail.com
[2] National Research Nuclear University MEPhI (Moscow Engineering Physics Institute), Moscow, Russia

Abstract. A simple computer model of a feeling of causality of autonomous agents has been created and investigated in the current article. The model of the evolution of a population of agents is considered. The population includes agents of two kinds: 1) agents with a feeling of causality and 2) agents without such a feeling. Each agent has its internal resource. Agents with a feeling of causality remember causal relationships between situations in the external environment. It is shown that agents with a feeling of causality in the process of evolution can displace agents without a feeling of causality from the population. So, the current model can be considered as the model of origin of a feeling of causality.

Keywords: Feeling of causality · Autonomous agents · Evolution · Agent's internal resource

1 Introduction

In 1748 David Hume wrote "Philosophical essays concerning human understanding" [1], where he called into question the notion of causality. He asked: why do we conclude that one event *A* is the cause of another event *B* when we observe that the event *B* many times appears after the event *A*? For example, if we observe many times that the sun illuminates the stone (the event *A*) and after this, the stone becomes warm (the event *B*), we conclude that the sunlight is the cause of the heating of the stone.

Actually, Hume asked: what are reasons of our conclusions about events in nature? What are origins of such conclusions? He tried to understand what is the reason of such conclusions: "the event *A* is the cause of the event *B*". Hume wrote that if we analyze thoroughly the issue of our conclusions about causal relations between the cause and the effect, we can establish that the only reason for such conclusions is custom or habit. According to Hume, the notion of causality is due to some our internal feeling or habit to form the idea of causal relation after the observation of multiple event pairs.

In this paper, we propose a model of the internal feeling of causality of autonomous agents. The model is built in the most general form in order to characterize the most general properties of the concept of causality. The model is investigated by means of computer simulation.

A. V. Samsonovich et al. (Eds.): BICA 2020, AISC 1310, pp. 410–416, 2021.
https://doi.org/10.1007/978-3-030-65596-9_49

The model analyzes the properties of the feeling of causality in the most general form, in which cause-and-effect relationships of the form $A \rightarrow B$ are remembered. The model directly corresponds to Hume's approach.

2 Description of the Model

We assume that there is an evolving population of agents. Each agent has its own resource R. Consider the world (the external environment of agents) in which there are different situations $S_i(t)$. Time t is discrete. We assume that the situations are mostly random and alternate in time randomly. Nevertheless, there are causal relationships between certain situations. We believe that successively alternating situations $S_{1i}(t)$, $S_{2j}(t + 1)$ appear episodically:

$$\{S_{1i}(t) \rightarrow S_{2j}(t+1)\}. \tag{1}$$

These pairs represent causal relationships of the external world. Some of the consequence situations S_{2j} (the second elements of the pairs) are essential for agents. In these situations, agents can acquire or lose a significant part of their resource $R_k(t)$. We will call these situations S_{2j} favorable if the agent acquires a resource in them, and unfavorable if the agent loses a resource in such a situation.

Consider the evolution of a population of agents. If during the life of an agent its resource becomes less than zero ($R_k(t) < 0$), then such an agent dies. If the agent's *resource has exceeded a certain threshold, then this agent can "give birth" to a child, in this case the parent agent gives half of its resource to the child agent.*

We assume that there are two types of agents: agents of the first type and agents of the second type. Agents of the first type have no feeling of causality; agents of the second type have a feeling of causality.

We assume that when new agents are born, the child agent has the same form as the parent agent, i.e. descendants of agents with a feeling of causality also have a feeling of causality. The descendants of agents without a feeling of causality do not have this feeling.

Let us describe the properties of agents of the first type (without a feeling of causality). These agents accidentally find themselves in significant situations and, if the situation is favorable, then the agent acquires the resource Drp_1 ($R_k(t + 1) = R_k(t) + Drp_1$) in this situation, if the situation is unfavorable, then the agent loses the resource Drm_1 ($R_k(t + 1) = R_k(t) - Drm_1$).

Let us describe the properties of agents of the second type (with a feeling of causality). We believe that these agents can be untrained and trained for certain causal relationships $\{S_{1i}(t) \rightarrow S_{2j}(t + 1)\}$. Resource changes for untrained agents of the second type occur in the same way as for agents of the first type. We also assume that the trained agent of the second type, having memorized the causal relationship for the sequential pair $\{S_{1i}(t) \rightarrow S_{2j}(t + 1)\}$, has the property to prepare for the essential situation $S_{2j}(t + 1)$.

The training process for agents of the second type is as follows. Suppose that the agent observes a pair of situations $\{S_{1i}(t) \rightarrow S_{2j}(t + 1)\}$, and the second element of the pair S_{2j} is essential for the agent. That is the situation S_{2j} is a favorable or unfavorable situation. After this, the agent enters this pair into its knowledge base, in which it remembers the pair $\{S_{1i} \rightarrow S_{2j}\}$ and memorizes the number of observations of this pair $N(S_{1i} \rightarrow S_{2j})$. Naturally, that after the first observation of a given pair, $N(S_{1i} \rightarrow S_{2j}) = 1$. For each repeated observation of this pair, the number $N(S_{1i} \rightarrow S_{2j})$ increases by 1. When the number of observations of a certain pair exceeds a certain threshold: $N(S_{1i} \rightarrow S_{2j}) > N_{th}$, then the agent is trained: it has solid memory about this causal relationship.

After such training, when a situation $S_{1i}(t)$ appears, an agent that has memorized a certain cause-and-effect relationship makes a prediction $S_{1i}(t) \rightarrow S_{2j}(t + 1)$ and prepares for the emergence of an essential situation $S_{2j}(t + 1)$. If the situation S_{2j} is favorable, then due to the preparation the agent acquires the resource Drp_2 that is greater than the increase in the resource Drp_1 for the agent of the first type, which does not have the ability to predict this situation and the ability to prepare for this situation. As an analogue of prediction and preparation for a favorable situation, one can recall the experiments by Ivan Pavlov, in which, after the development of a conditioned reflex in the dog, before the appearance of food, preparation for food (salivation) took place. If the situation S_{2j} is unfavorable, then the trained agent of the second type also prepares for this situation (for example, the agent can run away from danger). Due to prediction and preparation, the decrease in the resource of the agent Drm_2 is less than for the agent of the first type Drm_1, which does not remember and does not prepare in advance for an unfavorable situation.

Thus, the agent with the feeling of causality memorizes in its knowledge base the cause-and-effect relationships between repeated pairs of situations $S_{1i} \rightarrow S_{2j}$ and, after training, prepares in advance for the appearance of favorable and unfavorable situations S_{2j}. Due to this, the resource of such agents becomes larger than that of agents without a feeling of causality.

It is clear that agents with the feeling of causality will recruit the resource faster in the population of agents. Agents of this type will more often give birth to offspring and can displace agents without a feeling of causality from the population.

Let us emphasize that agents with the feeling of causality in this model directly cognize the laws of nature, namely, laws that do not depend on the actions of the agents themselves. The knowledge of these agents directly corresponds to the ideas of Hume [1].

Agents with the feeling of causality are clearly using their knowledge about the laws of the external world. Due to the property of prediction, agents of the second type can adapt to future events and, due to this, have advantages over agents without a feeling of causality. Due to this, the resource of agents of the second type should grow faster than that of agents of the first type. Although results of our model are not so simple. Due to the birth of descendants (to whom half of the resource is given), the resource of agents of the second type can decrease and they have no obvious advantages as compared with agents of the first type. The details of competition between agents of the first and second types in populations were analyzed by computer modeling.

3 Computer Simulation Results

The model was analyzed by means of computer simulation. The main parameters of the simulation were determined as follows.

The maximum total population of two types of agents is 100. The total number of situations in the external world is 100. The number of significant pairs of situations, i.e. the number of causal relationships of the form (1) in the external world is 10. Changes in the resource of agents are as follows. The increase in the resource of agents of the first type and untrained agents of the second type when falling into a favorable situation is equal to $Drp_1 = 0.1$, the decrease in the resource of agents of the first type and untrained agents of the second type when falling into an unfavorable situation is equal to $Drm_1 = 0.4$. For trained agents of the second type, the increase and decrease in the resource when entering a favorable or unfavorable situation is equal to $Drp_2 = 2.0$ or $Drm_2 = 0.02$, respectively. We assume that an agent of the second type becomes trained for a certain pair, i.e. remembers a certain causal relationship if it observed this relationship more than $N_{th} = 10$ times. The trained agent can then prepare in advance for the occurrence of the corresponding favorable or unfavorable situation S_{2j}. An agent can give birth to a child if the agent's resource is greater than a certain threshold R_{th} . The value of this threshold in computer simulations was varied. The initial numbers of agents of the first and second types were approximately the same. The initial resource of any agent was random, uniformly distributed in the interval [0, 10].

Typical dependences of the number of agents of the first and second types, as well as the total population size, on time are shown in Figs. 1, 2.

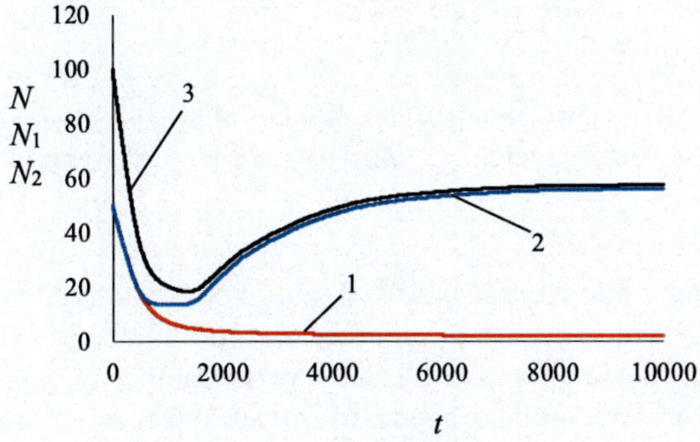

Fig. 1. Dependences of the number of agents of the first type N_1 (curve 1) and the second type N_2 (curve 2), as well as the total population size N (curve 3) on time t. $R_{th} = 100$. Results are averaged over 1000 different calculations.

The threshold resource required for reproduction was $R_{th} = 100$ (Fig. 1) and $R_{th} = 10$ (Fig. 2).

An example of the dynamics of agents' resources is shown in Fig. 3. The presented dependences were averaged over 1000 independent calculations; due to averaging, the calculation error did not exceed 3%.

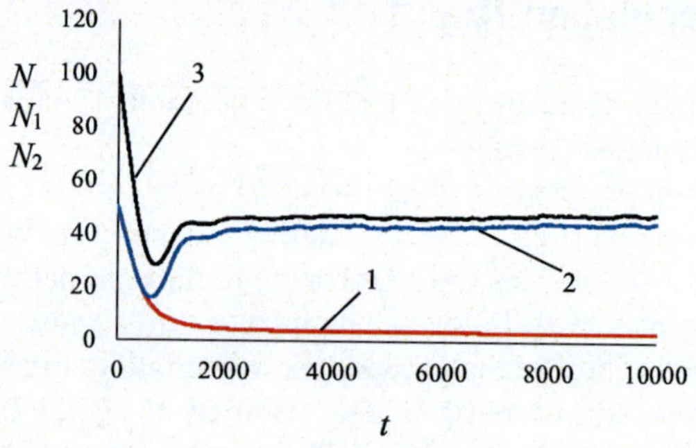

Fig. 2. Dependences of the number of agents of the first type N_1 (curve 1) and the second type N_2 (curve 2), as well as the total population size N (curve 3) on time t. R_{th} = 10. Results are averaged over different 1000 calculations.

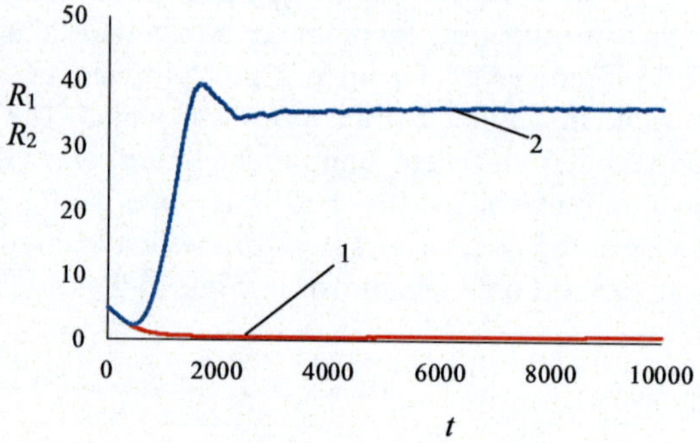

Fig. 3. Dependences of the population-average resource of agents of the first type R_1 (curve 1) and the second type R_2 (curve 2) on time t. R_{th} = 100. Results are averaged over 1000 different calculations.

These figures show that there is an initial period of population development (about 1000 time steps), in which both the resources and the number of sub-populations of agents of the first and second types will change in the same way, regardless of the type of agents. This means that initially agents of the second type are not yet trained, but then, after training (for $t > 2000$), agents of the second type have a clear selective advantage over agents of the first type: the resource of agents of the second type grows and they actively give birth to offspring. Moreover, at $t \approx 2000$ (when agents of the second type are just beginning to give birth to descendants, see Fig. 1) their resource is maximum, but then, due to the fact that these agents give half of their resource to descendants, their resource decreases slightly.

The result of population evolution is following: agents of the second type (with the feeling of causality) generally displace agents of the first type from the population. Although, a small number of agents of the first type remains in the population.

If the threshold resource for reproduction R_{th} increases, then the advantage of agents of the second type becomes more pronounced, since agents of the first type do not have time to acquire the resource required for the birth of offspring, and practically do not reproduce (the number of agents of the first type decreases significantly, see Fig. 1). If the threshold R_{th} decreases, then the agents begin to actively give birth to children, give them their resource, and their resource can significantly decrease. For example, the calculation with $R_{th} = 1$ shows that at this threshold all agents quickly give birth to offspring, give them half of their resource. As a result, the resource of agents drops sharply to zero and the population dies out at about 1000 time cycles. That is, in the present model, the value of the threshold resource required for reproduction R_{th} plays an important role.

4 Conclusion

Thus, the model of the feeling of causality of autonomous agents has been constructed and analyzed. The evolution of a population of agents of two types is considered: 1) agents without a feeling of causality and 2) agents with the feeling of causality. Agents with the feeling of causality memorize causal relationships between significant situations in the outside world. Due to this memorization, agents with the feeling of causation after training are able to anticipate favorable and unfavorable situations and prepare in advance for these situations. After such training, agents with the feeling of causality increase or maintain their resource. Agents with a sufficiently large resource give birth to offspring.

Agents with the feeling causality are explicitly exploiting this feeling. They adapt to changes in the external environment. Their resource grows faster than for agents without a feeling of causality. As a result of population evolution, agents with the feeling of causality can displace agents without a feeling of causation from the population.

It should be emphasized that the use of the property of causality, the use of predictions is one of the key properties of cognition of the laws of the external world. The study of this property is important at modeling of cognitive evolution, the evolution that resulted in our thinking, which we use during the scientific cognition of nature [2, 3].

Additionally, we note that it is reasonable to use the property of causality at the investigation of cognitive autonomous agents and cognitive architectures [4–9].

Funding. This work was supported by the Russian Science Foundation, grant numbers 18-11-00336.

References

1. Hume, D.: Philosophical Essays Concerning Human Understanding. A. Millar, London (1748). See also: Hume, D.: Philosophical Essays Concerning Human Understanding. General Books LLC (2010)

2. Turchin, V.F.: The Phenomenon of Science: A Cybernetic Approach to Human Evolution. Columbia University Press, New York (1977)
3. Red'ko, V.G.: Modeling of Cognitive Evolution: Toward the Theory of Evolutionary Origin of Human Thinking. KRASAND/URSS, Moscow (2019)
4. Xue, J., Georgeon, O.L., Guillermin, M.: Causality reconstruction by an autonomous agent. In: Samsonovich, A.V. (ed.) Proceedings of International Conference on Biologically Inspired Cognitive Architectures, pp. 347–354. Springer (2018)
5. Vernon, D., von Hofsten, C., Fadiga, L.: Desiderata for developmental cognitive architectures. Biol. Inspired Cogn. Archit. **18**, 116–127 (2016)
6. Samsonovich, A.V.: On a roadmap for the BICA challenge. Biol. Inspired Cogn. Archit. **1**, 100–107 (2012)
7. Samsonovich, A.V.: Schema formalism for the common model of cognition. Biol. Inspired Cogn. Archit. **26**, 1–19 (2018)
8. Chella, A., Pipitone, A.: A cognitive architecture for inner speech. Cogn. Syst. Res. **59**, 287–292 (2020)
9. Vityaev, E.E.: Consciousness as a logically consistent and prognostic model of reality. Cogn. Syst. Res. **59**, 231–246 (2020)

Extending the Intelligence of the Pioneer 2AT Mobile Robot

Michael A. Rudy(✉), Eugene V. Chepin, and Alexander A. Gridnev

Institute of Cyber Intelligence Systems, National Research Nuclear University MEPhI (Moscow Engineering Physics Institute), Moscow, Russia
michaelrudy95@gmail.com

Abstract. This paper describes the process of expanding the intellectual capabilities of a Pioneer 2AT skid-steer mobile robot by developing a module that allows using the ROS (Robotic Operating System) and its capabilities. This software and hardware system allow to quickly expand the capabilities of an outdated mobile robot, giving it access to modern algorithms e.g. SLAM. It also granted the support for various attachments - lidars, stereo cameras, etc., autonomous navigation algorithms, data collection for training CNNs, as well as simulating the operation of these algorithms in the Gazebo. The module consists of two parts, the first is launched on the Arduino board, to which the wheelbase with the associated sensors is connected, the second part is the ROS node, which translates all the data coming from the Arduino into a format corresponding to the ROS interface.

Keywords: Mobile robotics · SLAM · ROS · Skid-steer · Differential drive

1 Introduction

The main goal of the project was to upgrade the hardware of the Pioneer 2AT mobile robot. The upgrade of this robot in many years of operation was carried out twice. The purpose of those modifications was, in fact, the installation of a up-to-date on-board computer. The objectives of this upgrade are wider:

- Replacing an outdated on-board computer with a modern PC;
- Replacement of the burnt control board as a result of natural wear and its unreliability, by developing its Arduino-analogue while maintaining the speed and motors specifications of the robot;
- Orientation to support the ROS interface.

In fact, it was necessary to develop a new version of the mobile robot, which could provide a significant improvement in the development and application of intelligent algorithms and software:

- Support a modern common in the industry framework - ROS;

A. V. Samsonovich et al. (Eds.): BICA 2020, AISC 1310, pp. 417–424, 2021.
https://doi.org/10.1007/978-3-030-65596-9_50

- Support SLAM algorithms, in particular: Gmapping, Rtabmap, as well as autonomous navigation algorithms;
- Widespread implementation of data collection for training for, in particular, the use of convolutional neural networks (CNNs);
- Support simulations in the Gazebo;
- Provide with a ROS-compatible interface all installed sensors, as well as additional equipment that was not previously on board the robot.

The four-wheeled skid-steer Pioneer 2AT, had initially the following specifications: wheelbase - 4 wheels (diameter 220 mm, wheel width 75 mm), with maximum speed of 2.5 km/h, 17,000 ticks of the encoder per rotation (counts/rev), robot base 40 cm; batteries – 3 × 12 V 7.2Ah; lidar - LMS200 (range 80 m, FOV 180, number of scans 181, frequency 75 Hz); 2005 on-board energy-saving PC. After the upgrade, the on-board PC had the following specs: 8 GB ddr4 RAM, intel CPU g4620, SSD 120 GB, added stereo camera support: FOV 80, each camera resolution is 640 × 480, the distance between cameras is 6 cm, color depth is 8 bytes. Motor specifications, despite the replacement of internal electronics, remained the same.

2 Extending the Intelligence of Mobile Robot

The hardware upgrade made it possible to fundamentally change the situation using modern software technologies. First of all, this concerns the possibility of using the ROS framework, and therefore the entire arsenal that it possesses namely SLAM algorithms that were originally built into the system, such as: gmapping, hectormap, rtabmap and in-built autonomous navigation algorithms. Secondly, due to the unified interface between ROS nodes, now there is no need to develop protocols for various tasks - this is standardized at the system level by message types.

Next, we will take a look at the tasks that were solved by upgrading and go directly to the developed software and hardware system. That module allows to quickly connect and configure the wheel base of robot to the on-board computer with ROS, thereby expanding its intellectual functionality.

One of the main goals of the project was to support modern SLAM algorithms. SLAM is one of the most important areas of modern robotics, which, however, is considered resolved at the theoretical and conceptual level [1]. The undoubted advantage of SLAM is the provision of a built and globally harmonized environmental map using both robot motion measurements and loop closure, without loop closure, SLAM is reduced to odometry [2].

Each year, new implementations and papers appear in this area, so you can not do any research without access to the already proven algorithms, new developments, as well as independent research in this area. In modern SLAM, two different approaches are most often used - methods based on filtering and based on key frames. [3]. Since the modeling of the environment by the robot depends both on its complexity and on the limitations of the sensors [4]. Based on the above, as well as the available equipment on the robot - lidar and stereo cameras, the following algorithms were selected as testing the final functionality after the upgrade:

- Gmapping is an algorithm based on Rao-Blackwellized Particle Filters. The particle filter converges to a solution that well represents the environment, especially when there is loop closure [5].
- Hectormap, which, however, cannot be called a full SLAM, since it does not detect loop closure, which does not allow the map to be corrected when visiting the area again, however, it generates localization with a very low drift in real-world autonomous navigation scenarios, and in the compartment with the lack of need for odometry and low computational costs, this algorithm is very promising to use [6].
- Rtabmap is an algorithm whose main advantage is not depending on the odometry approach used, that is, it can work as data coming from cameras, from lidars, or even just by wheel odometry [7].

All of them can work with lidar, and rtabmap is initially ready with a stereo camera. For the first two algorithms, the output of the stereo camera data, that is, the point cloud can be converted into messages such as ROS laser scanning and thus use gmapping and hectormap with a stereo pair.

An important and relevant methodology of today is a large set of techniques and technologies - convolutional neural network (CNNs). In particular, the use of CNNs [8], to solve the problems of detecting objects in images, which is one of the popular approaches for increasing the intelligence of control systems for mobile robots. For example, the YOLO system (You Only Look One) is an intelligent technology for computer vision technology that provides pattern recognition, and is also used to create a visual intelligence system for robots or used for a neural network approach of vision useful for social robots [9] for testing needs large amounts of visual data. In this paper, the latest methods for detecting moving objects in video sequences captured by a moving camera are discussed [10].

For training, as well as testing CNN in real conditions, large amounts of data are needed - ensuring the possibility of obtaining such data was also one of the goals of the project. In addition to the data obtained directly from real sensors, with the installation and configuration of ROS software, it is possible to simulate the work of mobile robot in the Gazebo. Simulation allows to create video sequences that are subsequently used for testing and training CNNs, as well as research and debug various development of robot‘s motion algorithms developed in the laboratory. Also, simulation allows research on the accuracy and performance of various "assemblies" of attachments (lidar together with a stereo camera, single lidar, single camera, etc.), and all this without the need to be directly in the laboratory with a robot. Subsequently, all tests are transferred to a real robot due to compatibility with the specs of the model.

3 Theoretical Assumptions

Before description of the developed software and hardware system, backstage theory need to be discussed. ROS uses only two velocity components to describe motion - linear u_s and angular u_ω. Such a model for describing the motion of mobile robot is called the Unicycle model [11], which literally translates as unicycle motion.

$$\dot{x} = u_s cos\theta$$

$$\dot{y} = u_s sin\theta$$
$$\dot{\theta} = u_\omega \quad (1)$$

However, the Pioneer 2 AT has four wheels, and robot is a skid-steer robot, so the wheels on each side rotate at the same speed. We use the motion model for a differential two-wheeled robot [11], where r is the radius of the wheel, L is the distance between the centers of the wheels, ω_l, ω_r are the speeds of rotation of the left and right wheels.

$$\dot{x} = \frac{r}{L}(\omega_r + \omega_l)cos\theta$$
$$\dot{y} = \frac{r}{L}(\omega_r + \omega_l)sin\theta$$
$$\dot{\theta} = \frac{r}{L}(\omega_r - \omega_l) \quad (2)$$

Combining (1) and (2), we obtain ω_l, ω_r - the angular rotation speeds for each side.

$$\omega_l = \frac{2u_s - u_\omega L}{2r}, \omega_r = \frac{2u_s + u_\omega L}{2r} \quad (3)$$

Conclusion (3) is often used [11, 12], this allows us to simplify the control logic of various types of robot: differential drive, 4 wheeled skid-steer, but, with any simplification, limitations appear that affect the accuracy of the model. When turning wheels of skid-steer robots inevitably slip [13–16]. There are many ways to deal with this: in [14], a method for introducing the slip coefficient into the kinematic model of robot is described, but such a coefficient will have to be calculated for each space in which robot will be operated, however, this is the easiest way to improve the quality of robot odometry without complications of the mathematics. To achieve better positioning accuracy without using external or internal devices, the Instantaneous Center of Curvature (ICC) [17] can be used to improve the accuracy of positioning when turning, it needs to be known where ICC_r for the right side and ICC_l for the left, which will be used in Eq. (2) instead of L, now it will be the distance between ICC_r and ICC_l [18], to calculate the ICC the lateral and longitudinal speeds are needed $\dot{x}, \dot{y}$ of the robot from (2), or speeds wheels ω_l, ω_r from (3) [17, 19]. A model using ICC will subsequently be added as an alternative (3). Also, to improve the accuracy of positioning a robot with an onboard turn, the following internal systems can be used: traction control system, special tires [20]. Using external and internal devices, the most useful in this case will be the use of an accelerometer and gyroscope, as well as GPS [20]. A complete analysis of the mathematical apparatus, which includes the kinematic and dynamic models of a robot with an onboard turn, can be found in these scientific papers.

To complete the model, we need to convert the data received from the wheelbase to odometry, according to the specifics of ROS. After the transformations from the real ω_l, ω_r and the interval Δt for which these speeds are calculated, we obtain the distances that the right and left wheel of the MPC have passed over Δt, namely D_r and D_l.

Then from [25] we get:

$$x(t_i) = x(t_{i-1}) + \frac{D_l + D_r}{2} cos(\theta(t_i)),$$

$$y(t_i) = y(t_{i-1}) + \frac{D_l + D_r}{2} sin(\theta(t_i)),$$

$$\theta(t_i) = \theta(t_{i-1}) + \frac{D_r - D_l}{L}$$

where $\Delta t = t_i - t_{i-1}$. With the coordinates of the robot in the interval t_i and the previous speed t_{i-1}, speeds are calculated as the coordinate difference divided by Δt.

4 Software System: Results

The developed intelligent control software system is divided into two independently executable modules: the first works on the Arduino board, the second is the ROS node. Two modes of functioning of the system are possible. In this work, for the Pioneer 2AT, an implementation is used in which the modules use the speeds in ticks of the encoder per second (tick/s) when communicating with each other. It allows to reduce the delay and reduce the number of computational operations on the Arduino board, without losing accuracy by using integer calculations. In an alternative implementation, the speed between the modules is transmitted in m/s, in this case the modules can be used separately and for various tasks, for example, only for processing data from the wheelbase or only for converting odometry coming from the signal processing board of the wheelbase to a format, corresponding ROS.

To implement the developed software solution into a specific mobile robot model, you need to know the following robot specifications: the radius of the robot wheels, the distance between the wheel centers, the number of ticks of the encoder with a full wheel turn, the minimum and maximum speed of the robot, the maximum current for motors. Additional equipment is connected using the appropriate ROS nodes.

4.1 Algorithm

1. ROS node sets the necessary movement speeds of the robot in the Unicycle model form;
2. Linear and angular speeds are converted to rotational speeds of the right and left wheels of the robot in tick/s;
3. Speeds transfer to the Arduino board via the serial port;
4. Translation of the obtained speeds into PWM and their application to the wheelbase motors;
5. In specified period of time (standard 100 ms), values are read from the encoders - these are real speeds in tick/s;

    a. The speeds are adjusted by the PID to those specified by ROS node in (1);
    b. Speeds are sent via the serial port to the ROS node;

6. At the ROS node, the speeds are converted to the speeds of the right and left wheels in m/s;
7. Odometry is calculated: the position of the robot relative to the last known position, speed and slippage coefficients are applied to improve the positioning of the robot
8. The received information is published in the ROS topic.

5 Practice Results

To test the intelligence of mobile robot, a fairly well-known, widely used in practice problem was chosen: SLAM. Testing was carried out in two stages. At the first stage, the gmapping, hectormap, rtabmap algorithms were launched in the Gazebo simulation environment on a model that corresponds to the specifications of the Pioneer 2 AT mobile robot. In the course of robot testing, it was necessary to create a room map, having passed the path from point A to B, for each experiment this path was the same.

At first, the algorithms were tested for working with lidar (see Fig. 1). In this case, all the algorithms provided fairly high-quality maps, according to which further autonomous navigation is possible. Next, tests were carried out with a stereo camera (see Fig. 2), and if rtabmap was originally designed to work with a stereo camera, then for two other algorithms it was necessary to convert a cloud of 3D points from a disparity map into a two-dimensional contour emulating the behavior of a lidar. This transformation imposed some significant limitations: a low scanning range, a small viewing angle, low frequency, point drift, in general, all those properties that, in general, a real lidar does not have.

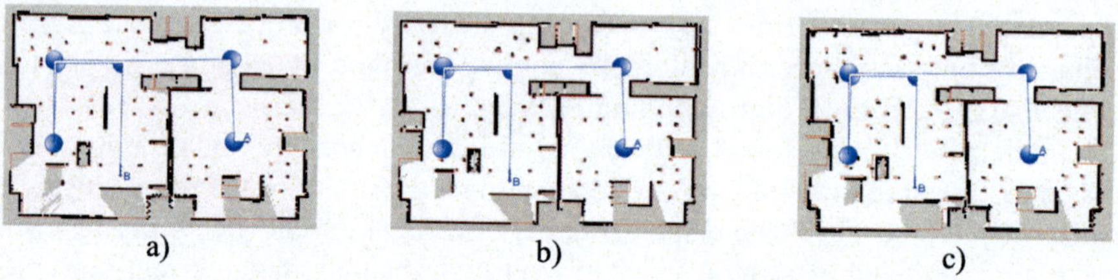

Fig. 1. SLAM results with lidar a) gmapping b) hectormap c) rtabmap

Fig. 2. SLAM results with stereo camera: a) gmapping b) rtabmap

When constructing a map using gmapping, satisfactory results were achieved only with a halving of the speed of movement of the robot model (linear speed 0.09 m/s, angular 0.19 rad/s) and a long selection of internal parameters. Hectormap failed to achieve adequate results. At the same time, rtabmap, which was initially launched with a stereo camera, produced a map of comparable quality, giving an undeniable bonus as a three-dimensional display of the surrounding space, despite the fact that the speed of model did not change from tests conducted with lidar (linear speed 0.19 m/s, angular 0.39 glad/s).

The second stage of testing was planned to be carried out on a real Pioneer 2 AT robot, and it would consist in repeating the experiments conducted on the model, in the room according to which a room model was created from the experiments of the first stage. Unfortunately, the circumstances surrounding the pandemic did not allow a full-fledged final experiment on the real mobile robot, so only the results obtained in the simulation are given below.

6 Conclusion

The results presented in this article allow us to state that if the mechatronic base of the wheeled mobile robot is well preserved, it is possible, due to the relatively simple hardware and software upgrade of the on-board electronics and computer equipment, in some cases only software upgrade, to bring the robot to the modern level its intellectual properties. The article describes in detail the process of such an upgrade for mobile robot Pioneer 2AT. A similar technique can be applied to a wide range of robots of this type.

References

1. Durrant-Whyte, H., Bailey, T.: Simultaneous localization and mapping: part I. IEEE Robot. Autom. Mag. **13**(2) (2006). https://doi.org/10.1109/mra.2006.1638022
2. Cadena, C., Carlone, L., Carrillo, H. (eds): Past, present, and future of simultaneous localization and mapping: toward the robust-perception age. IEEE Trans. Robot. **32**(6) (2016). https://doi.org/10.1109/tro.2016.2624754
3. Perera, S., Barnes, N., Zelinsky, A.: Exploration simultaneous localization and mapping (SLAM). In: Computer Vision, pp. 268–275 (2014). https://doi.org/10.1007/978-0-387-31439-6_280
4. Bailey, T., Durrant-Whyte, H.: Simultaneous localization and mapping (SLAM): Part II. IEEE Robot. Autom. Mag. **13**(3) (2006). https://doi.org/10.1109/mra.2006.1678144
5. Grisetti, G., Stachniss, C., Burgard, W.: Improved techniques for grid mapping with rao-blackwellized particle filters. IEEE Trans. Robot. **23**(1) (2007). https://doi.org/10.1109/tro.2006.889486
6. Kohlbrecher, S., von Stryk, O., Meyer, J., Klingauf, U.: A flexible and scalable SLAM system with full 3D motion estimation. In: IEEE International Symposium on Safety, Security, and Rescue Robotics (2011). https://doi.org/10.1109/ssrr.2011.6106777
7. Labbe, M., Michaud, F.: RTAB-Map as an open-source lidar and visual simultaneous localization and mapping library for large-scale and long-term online operation. Journal Field Robot. **36**(3) (2018). https://doi.org/10.1002/rob.21831
8. Verbitsky, N.S., Chepin, E.V., Gridnev, A.A.: Study on the possibility of detecting objects in real time on a mobile robot. In: Misyurin, S., Arakelian, V., Avetisyan, A. (eds) Advanced Technologies in Robotics and Intelligent Systems. Mechanisms and Machine Science, vol 80. Springer, Cham (2020)
9. Kulik, S.D., Shtanko, A.N.: Experiments with neural net object detection system YOLO on small training datasets for intelligent robotics. In: Misyurin, S.Y., Arakelian, V., Avetisyan, A.I. (eds.) Advanced Technologies in Robotics and Intelligent Systems. MMS, vol. 80, pp. 157–162. Springer, Cham (2020). https://doi.org/10.1007/978-3-030-33491-8_19
10. Yazdi, M., Bouwmans, T.: New trends on moving object detection in video images captured by a moving camera: a survey. Comput. Sci. Rev. **28**, 157–177 (2018)
11. LaValle, S.M.: Planning Algorithms, pp. 727–730. Cambridge University Press, Cambridge (2006)

12. Abolore Yekinni, L., Dan-Isa, A.: Fuzzy logic control of goal-seeking 2-wheel differential mobile robot using unicycle approach. In: IEEE International Conference on Automatic Control and Intelligent Systems (I2CACIS 2019) (2019). https://doi.org/10.1109/i2cacis.2019.8825082
13. Caracciolo, L., De Luca, A., Iannitti, S.: Trajectory tracking control of a four-wheel differentially driven mobile robot. In: IEEE International Conference on Robotics & Automation (1999). https://doi.org/10.1109/ROBOT.1999.773994
14. Domski, W., Mazur, A.: Slippage influence in skid-steering platform trajectory tracking quality with unicycle approximation. In: 24th International Conference on Methods and Models in Automation and Robotics (MMAR), (2019).https://doi.org/10.1109/mmar.2019.8864720
15. Jian, Z., Shuang-shuang, W., Hua, L., Bin, L.: The Sliding mode control based on extended state observer for skid steering of 4-wheel-drive electric vehicle. In: 2nd International Conference on Consumer Electronics, Communications and Networks (CECNet) (2012). https://doi.org/10.1109/cecnet.2012.6201445)
16. Shuang, G., Cheung, N.C., Eric Cheng, K.W., Lei, D., Xiaozhong, L.: 7th International Conference on Power Electronics and Drive Systems (2007). https://doi.org/10.1109/peds.2007.4487913
17. Pentzer, J., Brennan, S., Reichard, K.: The use of unicycle robot control strategies for skid-steer robots through the ICR kinematic mapping. In: IEEE/RSJ International Conference on Intelligent Robots and Systems (IROS 2014) (2014). https://doi.org/10.1109/iros.2014.6943006
18. Wang, T., Wu, Y., Liang, J., Han, C., Chen, J., Zhao, Q.: Analysis and experimental kinematics of a skid-steering wheeled robot based on a laser scanner sensor sensors **2015**(15), 9681–9702 (2015). https://doi.org/10.3390/s150509681
19. Hellström, T.: Kinematics Equations for Differential Drive and Articulated Steering, 28 August 2011. Umeå University (2011)
20. Barrero, O., Tilaguy, S., Nova, Y.M.: Outdoors trajectory tracking control for a four wheel skid-steering vehicle. In: IEEE 2nd Colombian Conference on Robotics and Automation (CCRA) (2018). https://doi.org/10.1109/ccra.2018.8588153
21. Kozłowski, K., Pazderski, D.: Modeling and control of a 4-wheel skid-steering mobile robot 2014. Int. J. Appl. Math. Comput. Sci. **14**(4), 477–496 (2004)
22. Pazderski, D., Kozłowski, K., Gawron, T.: A unified motion control and low level planning algorithm for a wheeled skid-steering robot. In: IEEE 20th Conference on Emerging Technologies & Factory Automation (ETFA) (2015). https://doi.org/10.1109/etfa.2015.7301490
23. Barrero, O., Murcia, H.F.: Modeling and parameter estimation of a 4-wheel Mobile Robot (2016) https://www.researchgate.net/publication/314091229_Modeling_and_parameter_estimation_of_a_4-wheel_Mobile_Robot. Accessed 15 Apr 2020
24. Ailon, A., Cosic, A., Zohar, I., Rodic, A.: Control for teams of kinematic unicycle-like and skid-steering mobile robots with restricted inputs: analysis and applications. In: 15th International Conference on Advanced Robotics (ICAR) (2011). https://doi.org/10.1109/icar.2011.6088541
25. LabVIEW Robotics Programming Study Guide - Kansas State University – Polytechnic Campus. http://faculty.salina.k-state.edu/tim/robotics_sg/index.html. Accessed Apr 2020

On the Regularity of the Bias of Throughput Estimates on Traffic Averaging

Victor A. Rusakov(✉)

National Research Nuclear University MEPhI, (Moscow Engineering Physics Institute), 31 Kashirskoe Shosse, 115409 Moscow, Russia
VARusakov@mephi.ru

Abstract. The interaction of intelligent agents implies the existence of an environment to support it. The usual representations of this environment are graphs with certain properties. Like reliability, throughput is one of the most important characteristics of such graphs. When evaluating throughput in the analysis and synthesis of graphs, a reasonable combination of heuristic and strict approaches is used. In practice, this leads to the use of graph metrics. Usual shortest paths are widely used as part of the various multi-colour flow distribution procedures. The analytical capabilities of the Euclidian metric can achieve much more than just obtaining such a distribution. Such a metric allows us to introduce an abstract measure of the (quadratic) proximity of an arbitrary graph to a complete graph. This measure can be used as a single indicator of the reliability and throughput of the graph. Other conditions for tasks on graphs can be attributed to restrictions. A traffic matrix is one of these conditions. Non-stationarity of traffic when averaging over time can significantly reduce the accuracy of the estimates of throughput. The described dependencies of the throughput on the traffic's non-stationarity can be used in the analysis and synthesis of the communication environment when organizing the structure of the interaction of intelligent agents in the conditions of limited resources. These dependencies are verified by the results of numerical experiments.

Keywords: Graph metrics · Throughput analysis and synthesis of graphs · The Euclidian graph metric · The Moore-Penrose pseudo inverse of the incidence matrix · Traffic non-stationarity

1 Introduction

Graphs are widely used as a model of real objects. A packet-switched computer network is one of the types of such objects. There is a clear correspondence between the communication nodes and graph vertices and also between the trunk communication channels and the graph edges in the model.

From now on a graph will be understood as a finite undirected graph without loops and multiple edges having k vertices and h edges.

Information flows enter the network nodes from the outside and load the channels during their movement through intermediate nodes to the destination nodes. These nodes transmit flows outside the network.

A. V. Samsonovich et al. (Eds.): BICA 2020, AISC 1310, pp. 425–434, 2021.
https://doi.org/10.1007/978-3-030-65596-9_51

There are flows of different k colours in the graph each of which is determined by the destination vertex. So each vertex t is a receiver of the flow of colour t and also it's a source of the flows having colours 1, …, $t - 1$, $t + 1$, …, k. A flow is a sequence of packets in the packet-switched computer network. In the real world it takes random amounts of time for any packet to go through the network from its origin node to the destination one.

The simplest probabilistic models are often used to get the average amount of such time estimated for the whole network, not for the single packet or for some pair or subset of nodes. A number of supplementary assumptions and terms are incorporated into these models. Some of these additions are quite unrealistic, but in practice the simplest models give not bad results due to the helter-skelter nature of the flows in the network. Using this heuristic approach, the average packet delay τ during transmission over the network was obtained, $\tau = \gamma^{-1} \sum_i \lambda_i \tau_i$, where γ is the total intensity of the packet arrival in the network, λ_i is the same value for the i-th trunk communication channel, and τ_i is the average delay of the packet having length μ^{-1} on this channel with a capacity σ_i, $\tau_i = (\mu\sigma_i - \lambda_i)^{-1}$ [1].

In practice, the selection of channel capacities must be made from a small finite set [2]. In many procedures either these capacities are assigned before the routing is performed, or a routing is first performed and then the capacities assigned [3]. Following the second approach, we will further assume that channel capacities are the same.

Large amounts of accurate data are very rare [4, 5]. The situation is the same for the network's draft stage [6]. At this stage throughput estimates are usually obtained under the condition that equal opportunities are provided to all pairs of nodes [2, 7]. That's why the matrix of the so-called uniform traffic is often used. Such a matrix has all equal entries except the zero ones on the main diagonal. These entries are time-average intensities of the input flows. Such entries are equal to 1 in a normalized traffic matrix.

There is a huge number of graphs even for modest k and h. So the problem of finding the distribution of a multi-coloured flow has to be solved many times when synthesizing networks that have the best throughput index. Using an averaged traffic matrix significantly reduces the complexity of obtaining estimates of graph throughput. However, this approach leaves open the question of the accuracy of such estimates. The importance is emphasized in [8, 9] of the robustness of the design solutions with respect to the low certainty and non-stationarity of traffic. An ARPANet insensitivity property *was established* that justifies the use of traffic averages for network design. A *supposition* was made that this property is typical for distributed networks [2, 10].

Numerical simulation of non-stationary traffic is described in [9] using ARPANet examples. A slight *decrease* (10–13%) in the estimate of the throughput for non-stationary traffic is noted compared to the case of an averaged traffic matrix. There were no justifications or generalizations of such behavior of these estimates.

2 Linear Metric Models. The Best Quadratic Approximations

The τ expression can be used as an objective function of the problem of finding the distribution of a multi-coloured flow for a given traffic matrix and a set of other constraints. The distribution found should minimize the value of the average delay τ. A dual problem statement is also used. The distribution found should maximize the value of the $f = \gamma\mu^{-1}$ total flow through the network for a given set of constraints with $\tau \leq \tau_{\max}$ among them. The traffic matrix may vary proportionately when searching for the distribution.

In practice, reasonable simplifying approaches are used to find distributions under the conditions described in the Introduction. They are based on the linearization of the objective function and the splitting/recomposition of the multi-colour flow. Linearization of the τ expression and splitting the flow into interacting pairs of vertices means a transition to *metric* tasks on the graph. The *recomposition* of the obtained edge flows, taking into account the flow's colour, gives the final distribution [11]. This happens in the following usual way. When simultaneously viewing two or more flows of *one and the same* colour, for some edges the flows can go in opposite directions. In this case the magnitudes of the flows having opposite signs are subtracted from one another. But there is no subtraction for the flows having *different* colours in such a situation. These flows are considered to each simply belong to their direction of the edge [12].

The use of the usual shortest paths here is traditional and, at the same time, paradoxical. To bypass overloaded edges in their composition, you have to modify the graph and over and over again reuse other shortest paths in procedures such as "cut saturation" [3, 13]. The Euclidian metric does not exclude such an application. But due to its quadratic nature, it makes the use of such procedures practically unnecessary [11]. For notes on the octahedral metric (shortest paths) and cubic metric (minimal cuts) see [11, 14, 15].

Let X and Y be vector spaces with dimensions h and k correspondingly over the field of real numbers. The arbitrarily introduced *orientations* of the edges in the absence of limits to their capacity turns the incidence matrix A of an undirected graph having ± 1 as non-zero elements in each column into a useful instrument for the *linear* transformation $Ax = y$ of the edge flows $x \in$ X into the vertex flows $y \in$ Y. From now on the incidence matrix A will be understood to be such a matrix.

Let $\{x_t \in \mathrm{X} \mid Ax_t = y_t\}$ where y_t has an entry t equal to $1 - k$, and the remaining entries are equal to +1. Any of the vectors x_t represent the transfer of a flow with a magnitude of 1 and of the colour t from s_r, $r = 1, \ldots, (k - 1)$, to t along the edges of the graph. Then, without additional designations one should search for the extreme norm along all such x_t.

The subscript indices are used not only to indicate the colour of the flow. Therefore, to visually distinguish the meaning of the indices, the vertex flow of the colour t can be denoted below as $y(t) = y_t$ (edge flow as $x(t) = x_t$). To explicitly denote time, we use the notation τ and never t.

Further «′» and «$^+$» symbolize a transposition and the Moore-Penrose pseudo inverse correspondingly. Let C designate the incidence matrix of a complete graph. The subscript indices «c» and «a» together with $x(t)$ are used to indicate the following: $x_c(t) = C^+y(t)$ and $x_a(t) = A^+y(t)$.

Here and below let N(H) and R(H') mean the H kernel and the H' image of an arbitrary matrix H, respectively. Also let (u, w) mean the scalar (dot) product of vectors u and w.

Insert zero columns into the matrix A at the places of the missing edges in an arbitrary graph. Matrix A^+ will have zero rows with the same numbers. Let $\forall\, y_0 \in$ R(A). Clearly, $y_0 \in$ R(C).

Lemma 1 [11, 14, 16]. $(C^+y_0, A^+y_0 - C^+y_0) = 0$. Proof. [11, 14, 16]. □

Figure 1 illustrates Lemma 1. The vectors $d = A^+y_0 - C^+y_0$ and C^+y_0 for $y_0 = y(t)$ are the legs of a right-angled triangle with the hypotenuse A^+y_0 for the same y_0.

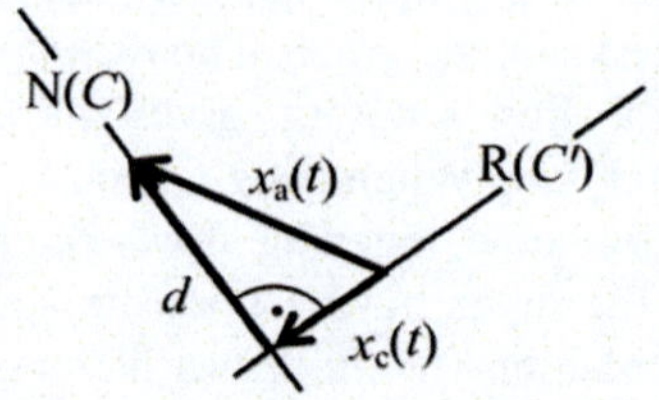

Fig. 1. The spaces and vectors of Lemma 1 for $y_0 = y(t)$

Obviously, for a given k and $y_0 = y(t)$, $t = 1, \ldots, k$, the vector $x_c(t) = C^+y(t)$ is constant. Then, according to the Pythagorean theorem, the Euclidian lengths of vectors d and $x_a(t) = A^+y(t)$ are minimal *simultaneously* for an arbitrary graph with the incidence matrix A. According to a well-known [17, 18] property of the Moore-Penrose pseudo inverse, the Euclidian norm of the vector $x_a(t) = A^+y(t)$ will be minimal among $\{x(t) \in X \mid Ax(t) = y(t)\}$. This means that the distribution along the edges of the flow $x_a(t)$ of any colour t in an arbitrary graph is the best quadratic approximation to the similar flow $x_c(t)$ in the complete graph [11, 14].

Remark. Column number t of the matrix of uniform normalized traffic was previously described as a vector $y_t \in$ R(A). Of course, the above is true not only for y_t, but also for any given $y_0 \in$ R(A). We will use this further when considering the edge flow of the colour t, corresponding to the t-th column of a *random* traffic matrix.

Here we recall that the idea of *approximation* is not so simple in nature. On one hand, the closer an approximating object is to the object of approximation, the more the first possesses the *qualities* of the second. On the other hand, if in the next moment the second object *has changed*, then we will have to make more significant *changes* in the first one. After all, we continue to maintain the specified *power* of approximation.

The situation is well known to those involved in the processing of measurement results. Under conditions of significant error there is no point in achieving a neat passage of the curve almost through all the points obtained, in fact going over to *interpolation*. For another set of measurement results, the curve will have to *change* significantly. But if the *power* of the approximant is relatively small, then its description and properties will practically not change. And this will continue, despite significant changes occurring again and again in other sets of measurement results.

Let's go back to the flow distribution. Figure 2(a) gives an example ($k = 8$) of the edge flow $C^{+}y_0$ of the colour t in the complete graph for $y_0 = y_t$. Figure 2(b) shows a similar flow for the vector $y_0 \in R(A)$, corresponding to the t-th column of a random traffic matrix. The flow is represented by thick, thin and broken lines depending on its magnitudes.

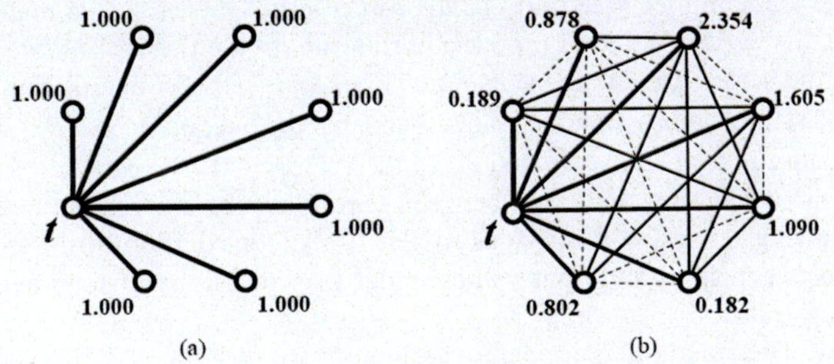

Fig. 2. The flow of the colour t in the complete graph for (a) uniform averaged and (b) random traffic

By [11, 14, 16] $x_c = C^{+}y = k^{-1}C'y$, $\forall\ y \in R(A)$. By Lemma 1, this simply means that the edge flow of an arbitrary graph $x_a = A^{+}y$ is the best quadratic approximation to the elementary transformed column of the traffic matrix. The column number is the colour of the flow. In the terms described above, Fig. 2(a) corresponds to measurements without errors, and Fig. 2(b) corresponds to one of the possible sets of random deviations.

The aforesaid allows us to formulate the following *rule* 1 [6].

Let us take a graph with *greater throughput* (with a *better* flow distribution) for the time-averaged traffic matrix, Fig. 2(a). Then we get *greater changes* in this index (in this distribution) with random traffic, Fig. 2(b).

What these changes lead to will be seen in the next section.

3 Normed Linear Spaces. Positive Definiteness of the Matrix Z

We will clarify the systematic changes in throughput using the formal basis of normed spaces. A linear space with a scalar product is a normed space when equating the vector norm with its length [19–21].

In linear models, we use nonzero vectors $y \in R(A)$, for which all entries except the t-th, are non-negative, and the t-th entry is equal to the sum of the others with a minus sign, $t = 1, \ldots, k$. An example of such a vector for time-averaged uniform normalized traffic was described above, this is the vector y_t. We further assume that any random traffic matrix is formed by the vectors $y \in \mathrm{R}(A)$ with the properties just described.

Let each such random traffic matrix be a member of a finite temporal series. We total up this matrix series and divide the sum by the number of members in the series. It is clear that by using the appropriate distribution of random values (entries of traffic matrices) and the length of the series, it is always possible to achieve the desired proximity of the resulting matrix to the matrix of *uniform normalized traffic*. Its t-th column is the vector $y_t = y(t)$. Denote the norm of this vector by $\|y(t)\|$.

Now we calculate the same norm of the t-th column of each of the matrices of the described matrix series. We total up these norms and divide the sum by the number of members in the series. Denote the obtained value as $m_{\|\cdot\|}(t)$. By axiomatics of normed spaces [19–21] the values obtained are bound by the inequality $m_{\|\cdot\|}(t) \geq \|y(t)\|$, $\forall$ t. This inequality is formed by the *lengths* of the vectors $\in \mathrm{R}(A)$. Next we will turn our attention to similar correspondences between the *lengths* of the vectors $\in \mathrm{R}(A')$.

The following lemma will allow us to clarify the contribution of the graph to the correlation between its throughput values under the conditions of time-averaged and non-stationary traffic.

Lemma 2. The matrix Z is a positive definite matrix.

Proof. The explicit form of A^+ is described in [11, 14–16, 22]. So it's easy to see, that $A^{+\prime}A^+ = pZAA^+ = pZ(I-F) = pZ(I-F)^2 = pZ'(I-F)'(I-F) = p[(I-F)Z]'(I-F) = p(I-F)Z(I-F)$. Matrices Z and F, as well as the normalizing factor p, are described in the same references, and matrix I is the identity matrix. For an arbitrary matrix H, the matrix HH^+ is a projector on $\mathrm{R}(H)$ [17], thus, the matrix $AA^+ = (I-F)$ is a projector on $\mathrm{R}(A)$. We have taken advantage of this above in the chain of equalities. Thus $\forall\, u, v \in \mathrm{R}(A)$ $(A^+u, A^+v) = (u, pZv)$.

Also we note that $\mathrm{R}(H^+) = \mathrm{R}(H')$ [17], and we get $\mathrm{R}(A^+) = \mathrm{R}(A')$. Our interest in the aforementioned vectors $\in \mathrm{R}(A')$ comes from this.

The matrix Z is symmetric. Therefore, it has a complete system of orthonormal eigenvectors.

$Z\xi = \xi$, where ξ is a vector having all entries equal to 1 [23]. Therefore, one of the eigenvectors of Z (with an eigenvalue $\lambda_0 = 1$) is proportional to ξ. The dimension of $\mathrm{R}(A)$ is $k-1$. This subspace of the k-dimensional space Y contains all *nonzero* vectors (*nonzero* vertex flows). For each of them the sum of the entries is 0. The meaning of this equality is obvious – the sum of flows going into the vertices of the graph from outside is equal to the sum of flows going outside from the vertices of the graph. In terms of the scalar product, such an equality for any vector $\in \mathrm{R}(A)$ means its orthogonality to the vector ξ. Hence we conclude that the remaining $k-1$ *nonzero* eigenvectors of Z make the *orthonormal* basis of the subspace $\mathrm{R}(A)$.

Let y_i be any of such eigenvectors with an eigenvalue λ_i, $i = 1, \ldots, k-1$. Since $AA^+y_i = y_i \neq \theta_Y$, and $A\theta_X = \theta_Y$, then $A^+y_i \neq \theta_X$. Here θ_X and θ_Y are the zero vectors of the spaces X и Y. Therefore $0 < (A^+y_i, A^+y_i) = (y_i, pZy_i) = (y_i, p\lambda_i y_i) = p\lambda_i$. Since $0 < p$, then from here we immediately get $0 < \lambda_i\ \forall\, i$. □

The hypotenuse lengths of interest to us, Fig. 1, are the lengths of the same t-th columns of traffic matrices considered above. Although not in the "usual" I-metric, but in the pZ-metric with the *positive* definite *symmetric* matrix Z [24, 25]. For such a metric, by the same axiomatics of normed spaces, the same inequality remains valid as the one considered above for the usual metric [26].

Due to the aforementioned interpretation of hypotenuses *lengths*, the fulfillment of this inequality for *lengths* is of great practical significance. This interpretation is described in the following *rule* 2 [24, 25].

For any graph throughput is *no less* for a time-averaged traffic matrix than the averaged throughput for random traffic.

The entire variety of graphs fits into the spectra of their Z matrices. It is appropriate to recall here that for a complete graph (with a normalizing factor $p = k^{-1}$) its matrix $Z = I$ [11, 14, 16].

4 Numerical Experiments

Numerical experiments were carried out to verify theoretical correspondences. Three graphs having equal resources were taken. Identical conditions of uniform time-averaged and random traffic were used for each of them. Flow distribution was performed using the Euclidian metric. For a detailed description of it, see [11, 14, 27]. The values of throughput were calculated, and their simple statistical processing was carried out [6].

For each of the three graphs $k = 23$, $h = 28$, and all edges have the same capacity. The first graph is a star with 6 edges added in a random way. The second graph is the result of a synthesis having throughput as the object function and taking into account traffic non-stationarity, see, for example, [11, 28]. And the third one is a given. It's a well-known version of one of the stages of the ARPA network [3, 8, 9]. The graphs are drawn in Fig. 3 and 4.

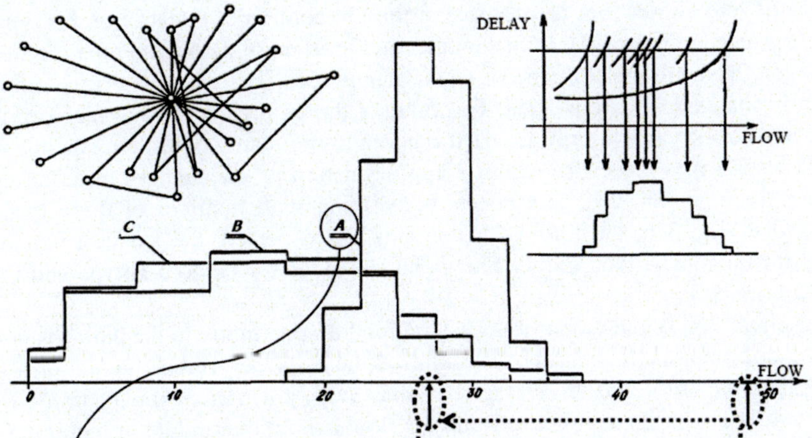

A	*S/M*, %	*Non-stationary traffic, M/k*	*Averaged traffic, f/k*
	11.9	27.0	48.3

Fig. 3. The scheme of experiments, graph and results of its tests

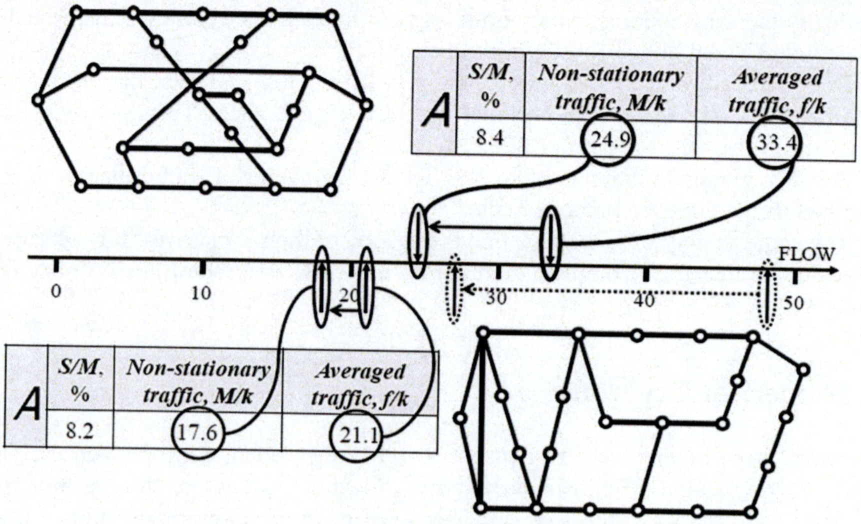

Fig. 4. Graphs and their test results

The outline of the experiments is shown in Fig. 3 in the upper right corner. Without excessive detail, their implementation was as follows. A multi-coloured flow was distributed on each graph and traffic matrix. The traffic matrix was proportionally changed each time to achieve the equality $\tau = \tau_{\max}$. The obtained flow value was used in statistical processing.

Random traffic was generated based on uniformly distributed pseudorandom numbers. Its three types, A, B and C, differ in the ways in which the entries of traffic matrices are formed from these numbers. A-type is characterized by the practical absence of zero entries. For an example of the *t*-th column of a matrix of this type, see Fig. 2(b). A random drawing of the number and location of zero entries was carried out for B-type. Nonzero entries were of equal magnitude. These values were additionally randomly changed for C-type [29]. The value of the throughput flow is included in the statistics with the same weight for each random traffic matrix. This corresponds to the equal intervals in a piecewise-constant approximation of the changes in traffic matrix entries over time [24, 25]. The results of working with matrices of these types are presented in Fig. 3 in the form of histograms. As expected, the smallest variance is typical for random A-type traffic. The variance values for random B-type and C-type traffic are much larger.

For graph Fig. 3 and random A-type traffic, the same figure in the table gives more detailed information. It presents the value of the flow per vertex for uniform time-averaged traffic (*f*/*k*), and an averaged similar value for non-stationary traffic (*M*/*k*). These values are indicated by arrows on the flow axis in the middle of the figure. The table also gives the ratio of the standard deviation *S* to *M*.

The results for the two remaining graphs are given in Fig. 4. These results are collected in two tables next to the graphs and are also indicated by arrows on the flow axis. The arrows of the first graph from Fig. 3 are moved to the same axis for a clear comparison. They are ringed with the same dashed lines as in Fig. 3.

The figures clearly demonstrate the fulfillment of both rules 1 and 2. Note there is an almost twofold decrease in the throughput for random traffic (A-type) compared to the use of time-averaged traffic for the graph in Fig. 3. Random traffic of types B and C gives an even greater decrease.

5 Conclusion

Like reliability, throughput is one of the most important characteristics of graphs. For many reasons we end up using heuristics and simplifications when estimating throughput. Time-averaged traffic is one of these simplifications. Thanks to the analytical capabilities of the Euclidian metric, we have found out that this is not always justified. The reckless desire to achieve greater throughput when using averaged traffic leads to a significant overrating of the graph throughput since the real traffic is non-stationary. Theoretical correspondences have been verified by the results of numerical experiments.

References

1. Kleinrock, L.: Analytic and simulation methods in computer network design. In: Proceedings of the AFIPS Conference, SJCC, vol. 36, pp. 569–579 (1970)
2. Frank, H., Kahn, R.E., Kleinrock, L.: Computer communication network design – experience with theory and practice. In: AFIPS Conference Proceedings, SJCC, pp. 255–270 (1972)
3. Frank, H., Chou, W.: Topological optimization of computer networks. Proc. IEEE **60**(11), 1385–1397 (1972)
4. Hamming, R.W.: Numerical Methods for Scientists and Engineers. McGraw-Hill, N.Y. (1962)
5. Voevodin, V.V.: Computational Foundations of Linear Algebra. Nauka, Moscow (1977)
6. Rusakov, V.A.: Synthesis of computer network structures and the problem of small certainty of initial values. USSR AS's Scientific Council on Cybernetics. In: Proceedings 5th All-Union School-Seminar on Computing Networks, 1, VINITI, Moscow-Vladivostok, pp. 112–116 (1980)
7. Frank, H., Frisch, I.T., Chou, W.: Topological considerations in the design of the ARPA computer network. In: AFIPS Conference Proceedings, SJCC, pp. 581–587 (1970)
8. Van Slyke, R., Frank, H., Chou, W.: Avoiding simulation in simulating computer communication networks. In: Proceedings of the AFIPS Conference, 4–8 June NCCE, pp. 165–169 (1973)
9. Frank, H., Chou, W.: Network properties of the ARPA computer network. Networks **4**, 213–239 (1974)
10. Gerla, M., Kleinrock, L.: On the topological design of distributed computer networks. IEEE Trans. Commun. **COM-25**(1), 48–60 (1977)
11. Rusakov, V.A.: Using metrics in the throughput analysis and synthesis of undirected graphs. In: Antipova, T. (ed.) ICIS 2020. LNNS, vol. 136, pp. 277–287. Springer, Cham (2021). https://doi.org/10.1007/978-3-030-49264-9_25

12. Hu, T.C.: Integer Programming and Network Flows. Addison-Wesley, Menlo Park-London (1970)
13. Chou, W., Frank, H.: Routing strategies for computer network design. In: Proceedings of the Symposium on Computer Communications Networks and Teletraffic, Polytechnic Institute of Brooklyn, 4–6 April, pp. 301–309 (1972)
14. Rusakov, V.A.: Analysis and Synthesis of Computer Network Structures. Part 1. Analysis. Moscow Engineering Phys. Inst. Report: All-Union Sci. Tech. Inform. Center No Б796153 (1979)
15. Rusakov, V.A.: On Markov chains and some matrices and metrics for undirected graphs. In: Antipova, T. (ed.) ICIS 2019. LNNS, vol. 78, pp. 340–348. Springer, Cham (2020). https://doi.org/10.1007/978-3-030-22493-6_30
16. Rusakov, V.A.: Using metrics in the analysis and synthesis of reliable graphs. In: Samsonovich, A.V. (ed.) BICA 2019. AISC, vol. 948, pp. 438–448. Springer, Cham (2020). https://doi.org/10.1007/978-3-030-25719-4_58
17. Albert, A.E.: Regression and the Moore-Penrose Pseudoinverse. Academic Press, New York (1972)
18. Beklemishev, D.V.: Additional Chapters of Linear Algebra. Nauka, Moscow (1983)
19. Horn, R., Johnson, C.: Matrix Analysis, 2nd edn. Cambridge University Press, New York (2013)
20. Voevodin, V.V.: Linear Algebra. Nauka, Moscow (1980)
21. Gantmakher, F.R.: Matrix Theory, 3rd edn. Nauka, Moscow (1967)
22. Rusakov, V.A.: Matrices, shortest paths, minimal cuts and Euclidian metric for undirected graphs. Procedia Comput. Sci. **145**, 444–447 (2018). https://doi.org/10.1016/j.procs.2018.11.104
23. Kemeny, J., Snell, J.: Finite Markov Chains. University series in undergraduate mathematics. Van Nostrand, Princeton, NJ (1960)
24. Rusakov, V.A.: The impact of traffic changes in the tasks of graphs optimization of the backbone data communication network. USSR AS's Scientific Council on Cybernetics. In: Proceedings of the 9th All-Union School-Seminar on Computing Networks, 3.1, VINITI, Moscow-Pushchino, pp. 80–85 (1984)
25. Rusakov, V.A.: On the regularity of the displacement of the mean estimate for the throughput with non-stationary traffic. USSR AS's Scientific Council on Cybernetics. In: Proceedings of the 9th All-Union School-Seminar on Computing Networks, 1.2, VINITI, Moscow-Pushchino, pp. 48–52 (1984)
26. Gelfand, I.M.: Lectures on Linear Algebra, 4th edn. Nauka, Moscow (1971)
27. Rusakov, V.A.: On the On the Moore-Penrose inverse of the incidence matrix for weighted undirected graph. Procedia Comput. Sci. **169**, 147–151 (2020). https://doi.org/10.1016/j.procs.2020.02.126
28. Rusakov, V.A.: A technique for analyzing and synthesizing the structures of computer networks using Markov chains. Computer networks and data transmission systems, Znaniye, Moscow, pp. 62–68 (1977)
29. Rusakov, V.A.: Non-stationary flows in computer networks: a linear basis for their study. USSR AS's Scientific Council on Cybernetics. In: Proceedings of the 7th All-Union School-Seminar on Computing Networks, 1, VINITI, Moscow-Erevan, pp. 118–124 (1983)

Virtual Convention Center: A Socially Emotional Online/VR Conference Platform

Alexei V. Samsonovich(✉) and Arthur A. Chubarov

National Research Nuclear University "MEPhI", Kashirskoe Hwy 31, Moscow 115409, Russia
Avsamsonovich@Mephi.Ru, Osgilatl7@Gmail.Com

Abstract. A cognitive model, producing believable socially emotional behavior, is used to control the Virtual Actor behavior in an online scientific conference paradigm. For this purpose, a videogame-like platform is developed, in which Virtual Actors are embedded as poster presenters, receptionists, etc. The expectation is that the combination of somatic factors, moral appraisals and rational values in one model has the potential to make behavior of a virtual actor more believable, humanlike and socially acceptable. Implications concern future intelligent cobots and virtual assistants, particularly, in online conferencing and distance learning platforms and in intelligent tutoring systems.

Keywords: Virtual environment · Online conference · Intelligent virtual agents · Gaming platform · Socially acceptable AI

1 Introduction

Today more and more scientific conferences and meetings use online formats. The same is true about virtually all kinds of events: educational, business, cultural, political, etc. In 2020 the majority of them became completely virtual due to the COVID19 pandemic. Independently of the pandemic, however, there is a general trend associated with the rapidly increasing role of computer-based communication technologies in the society. As a consequence, there is a high demand for platforms that support online conferencing like Zoom, to give one example. At the same time, the huge potential of the Internet and modern computers capable of supporting intelligent communication platforms remains largely unexplored, presenting a challenge for developers of Artificial Intelligence (AI).

The present work addresses this challenge. Here we introduce our Virtual Convention Center (VCC): an intelligent platform for online conferencing, implemented as a high-end videogame-like 3D virtual environment. The platform enables professional social multiuser interactions using both, traditional user-computer interface and VR/MR gadgets. It enables multimodal interactions among participants, allowing for such modalities as body language and gaze control, and makes it easy to involve intelligent virtual actors (bots) in communications at a professional participant level. VCC also supports rich behavioral data collection that can be used for off-line analysis.

The VCC platform is intended to deliver to participants nearly the same experience as they get participating in a real scientific conference on site, without the need to leave

A. V. Samsonovich et al. (Eds.): BICA 2020, AISC 1310, pp. 435–445, 2021.
https://doi.org/10.1007/978-3-030-65596-9_52

their home or office. The list of event formats supported by VCC includes forums, seminars, workshops, conferences, public lectures, concurrent sessions, roundtables and discussion panels, as well as educational courses, seminars and individual tutoring.

The following list of general requirements for the platform is targeted by VCC.

- Scalability to large event sizes and large volumes of presented media.
- Reliability, protection from unwanted intrusions, extensibility and the ease of use.
- Compatibility with most popular powerful hardware, from Windows PC to VR sets.
- Ability to stream live sessions to virtually any communication device or platform.
- Delivery of high-quality, near-real-life experience, excluding unpleasant sensations.
- Support for multimodal social interactions (body language, gaze and facial expression control, voice intonation, etc.), using an alternate ego represented by the avatar.
- The ability to involve custom-built intelligent virtual actors in interactions at the social level on equal with human participants, using the same multimodal interface.
- The ability to collect and process large volumes of multimodal behavioral data.

The urgent need for a platform characterized by these features is evidenced by the large number of conferences that use analogous, yet insufficient platforms, such as Mozilla Hubs or Gather. Town and the based on it Virtual Chair (e.g., CogSci 2020, Neutrino-2020 [1], AGI-20 [2], IVA'20). Despite the rapid growth in development of such platforms, we did not find an implemented close analog of VCC, as argued in the next Section.

The rest of the paper is organized as follows. First, we overview the state of the art in the field, looking at existing platforms from the point of view of the above list of requirements. Then we describe technical details of VCC implementation and the outcome of its recent practical usage for hosting two mid-size scientific conferences.

2 Related Work: Analogs and State of the Art

To date, there are no platforms that meet all of the specified requirements 1–7 in the public domain. The closest analogs based on 3D virtual environments are tools like Mozilla Hubs [3], AltspaceVR [4], VRChat [5] and the like (overviewed below), however, they have significant limitations. Most importantly, these platforms are not built specifically for hosting scientific conferences. On the other hand, most platforms designed for hosting scientific and educational events are limited to 2D simplistic (if any) virtual environments and do not extend to VR/AR/MR. Well-known popular examples are Zoom, Skype, WebEx, Meet, Discord, and many others.

One important example among these platforms that rapidly becomes popular today is *Gather.Town* with its latest development called *Virtual Chair* (https://www.virtualchair.net). It is designed for hosting scientific conferences and indeed is widely used for this purpose today. It is sufficiently scalable, reliable and safe, supports multimodal communications including live participant video and audio, whiteboards, chat, etc. and runs in a browser, which makes it compatible with almost any platform. However, Virtual Chair does not support features like an alternative participant ego, intelligent virtual actor involvement in social interactions, and behavioral data collection. Given its simplistic 2D design, it does not create a feeling of presence in a virtual environment and therefore does not deliver a near-real-life experience. Also, it is not compatible with VR. Therefore, it does not provide a solution to the challenge. Selected other close analogs of VCC are overviewed below.

AltspaceVR [4] is a social VR platform with a wide range of tools for customizing and adding interactive objects, developed by Microsoft. The platform provides a virtual reality meeting space where users can chat, watch videos, play games, and browse the Internet. Avatars on AltspaceVR can automatically mimic the user's body language using Microsoft Kinect. The platform also supports eye tracking if the required hardware is available. However, the inclusion of custom-developed virtual actors in it is difficult. Also among disadvantages is a rather low level of graphics and a limited scalability.

VRChat [5] is a Unity-based social VR project that also lends itself well to customization. Player models are capable of supporting audio lip syncing, eye tracking and blinking, as well as full range of motion. The platform also has a desktop version, but this version has limitations. Including custom-built bots in VRChat is also difficult. This platform is not suitable for scientific events because of the too game-like appearance and missing necessary functionality. It also has limited scalability.

Mozilla Hubs [3] is a web portal that combines the ease of use (compatible with platforms such as Skype, Discord, etc.) and VR compatibility. Using the basic functionality of the platform is possible from almost any modern computer or mobile device. Customization of the room environment, however, leads to problems like huge delays on less powerful hardware. Therefore, this platform is not yet suitable for solving the challenge, although it has good prospects. Also, like many other platforms, it is not designed for hosting scientific conferences, and needs additional customization. Rooms have a soft size limitation of approximately 24 participants.

Engage (https://engage2vr.com/en/aboutus) is a realistic VR-based platform that works exclusively on helmets connected to a computer. The platform was intended as an educational and scientific VR project, and it proved scalable: in the Spring of 2020, the annual VIVE Ecosystem Conference was held on the basis of Engage. Engage has a user-friendly interface and allows for custom solutions. Among disadvantages is the high cost of usage, which makes holding a large-scale conference very expensive.

Other similar tools using VR or VE include VRFocus, Vive Sync [6], MeetinVR, Glue, Connec2, MeetingRoom, Dream, VSpatial, OMS (Oxford Medical Simulation), Rumii, Acadicus, WondaVR, and many others [7]. The paradigms for using these tools for training are promising and include corporate adaptation, training in critical skills, etc. Many other analogous tools rapidly emerge (e.g., Teooh: https://www.teooh.com/large-fireside-chat) that cannot be all reviewed here. Very few of them allows the use of independently developed intelligent virtual actors, and therefore do not meet the specified above requirements.

The last, but not the least category of platforms to be mentioned here are Intelligent Tutoring Systems (ITS) using affective synthetic characters that do not use VR: Aleks, MetaTutor, Betty's Brain, AutoTutor, DeepTutor, etc. [8]. These software tools, with all due respect to their far advanced functionality, cannot be adapted for other purposes, such as hosting scientific conferences, or allow their modification. In conclusion, none of the well-known existing platforms provide a solution to the challenge defined above.

3 Materials and Methods

The basis of our approach is the creation of a fundamentally new platform, tentatively called the Virtual Convention Center (VCC), with the multiplayer possibility for hosting scientific conferences online. The platform is based on a videogame-like virtual environment with multi-user capabilities, connecting up to 200 users simultaneously. VCC is implemented on top of the SpatialOS platform, which provides a good level of abstraction for developers. In the course of working with VCC, users identify themselves both by the entered personal data and by the personal choice of one of the avatars offered to them upon login. The choice can be made either when entering the environment or in advance, during registration.

VCC has many multimedia capabilities. In the interior of the environment, any graphical and video materials can be placed in the form of posters, stands, screens, or other kinds of objects. This makes it easy to run virtual poster sessions, virtual demo sessions and slide presentations. A special type of stands works interactively. These stands allow the user to switch pre-prepared images, thus enabling the presentation of slides or changing the poster exposition. Another type of a stand is a virtual web browser window. It can be allocated anywhere in the environment, allowing users to display any content from the Internet with interactive capabilities, including images, YouTube video stream and sound, directly to the environment.

Each of the created avatars supports a range of capabilities required for a conference participant. The participant has the ability to navigate and select actions in the environment using the keyboard and mouse. Examples include the ability to point a laser pointer at any location in the 3D space of the environment, the ability to initiate avatar animations such as raising an arm, sitting down on a virtual chair, applause, nodding, etc.

These features allow participants to have experience approaching that of participation in a real conference. They also allow users to implement virtual actors connected to participants via a multimodal social-emotional interface.

VCC is developed using the game engine Unreal Engine 4. Accordingly, VCC potentially has all the capabilities that the engine supports (Fig. 1). In particular, it makes it possible to integrate into the multiplayer environment, along with user-controlled avatars, also non-player characters (NPC): such as intelligent virtual actors, or bots, implemented as avatars controlled by artificial intelligence controllers.

Bots can be implemented using additional software tools or by means of their own structures written using Blueprints and C++ code: these languages are supported by the Unreal Engine 4 editor. In the second case (C++), hierarchical state machines are usually used that group preprogrammed behaviors by substates. In this case, elements of the bot's behavior can be executed either one at a time, or in sequence. The current behavior is reviewed at a certain frequency, e.g., 10 Hz. In this case, the entire behavior tree is evaluated in each cycle. If another behavior different from the current one should be selected, then the current behavioral state changes. In the Unreal Engine, behavior trees are easily eimplemented, providing the basis for implementation of various artificial intelligence models, much more advanced and transparent compared to finite automata. In practice, a neatly constructed behavior tree makes visual debugging easier.

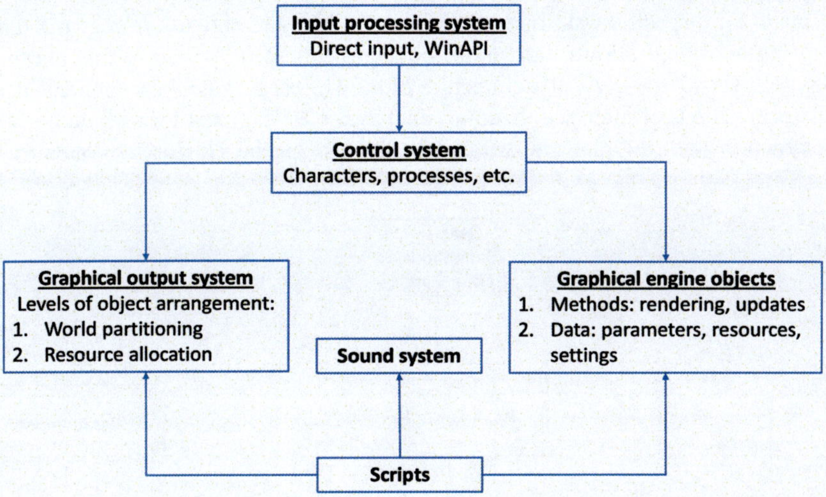

Fig. 1. Logical scheme of the platform: functional connections among the modules.

Any objects available for replication at all participant terminals require synchronization over the network. Traditionally, a server-client architecture (Fig. 2) is used for this purpose. In this case, each of a number of clients communicates over the Internet with one server. As a result, the system has limitations in the number of entities and the size of the virtual environment, which are determined by the weak link (Fig. 2a). The alternative approach used in SpatialOS is based on changing this classical scheme as shown in Fig. 2b.

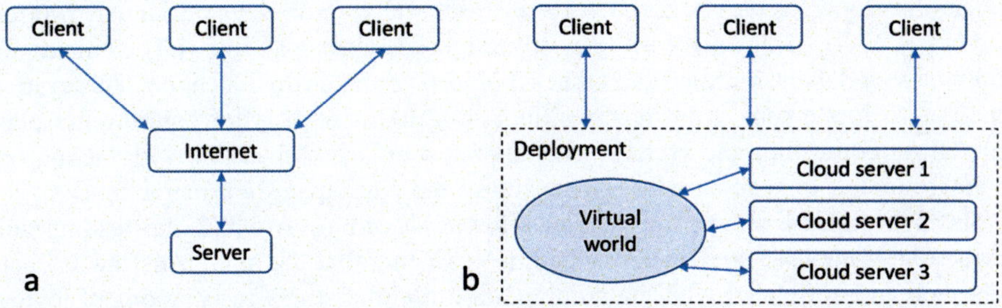

Fig. 2. Schemes for implementing synchronization over the network: (a) the classical scheme for organizing multi-user interaction; (b) the architecture of the SpatialOS services.

It is sufficient for the developer to make the client and server assemblies of the application, and then upload them to a special service (Fig. 2b: Deployment). After the launch, this service performs balancing and selection of the optimal servers for the existing load and size of the virtual world. Thus, the user experience is not determined by one single server or the complex architecture of many servers interacting with each other, made for a specific task. Instead, Spatial OS uses the network framework used in Unreal Engine 4, and allows various network projects to be ported to this platform.

Figure 3 shows the general architecture of the Unreal Engine 4 components used in the platform. The user interface (head-up display, or HUD) used for user authorization and interaction with the platform was made using a special Unreal subsystem.

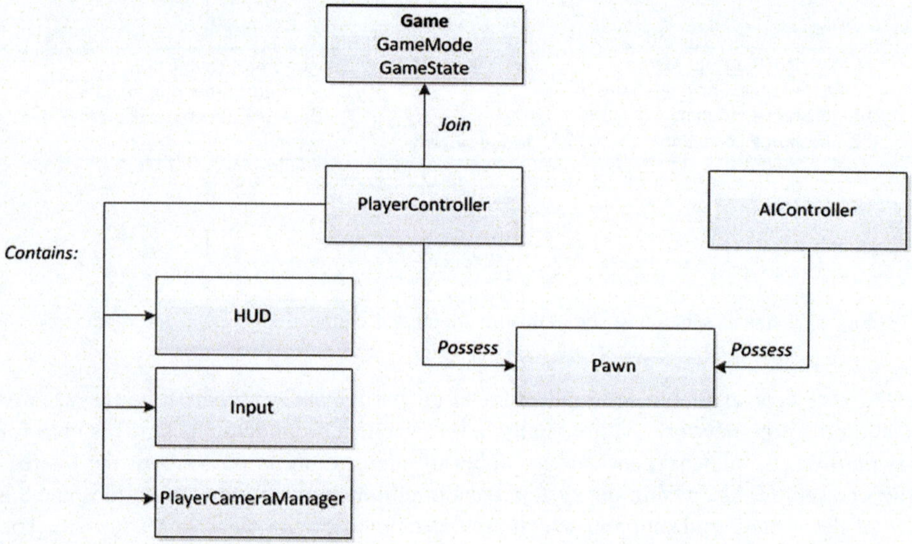

Fig. 3. General architecture of the Unreal Engine 4 components used in the VCC platform.

The developer creates a user interface in Unreal Engine 4 using Unreal Motion Graphics (UMG). UMG makes it easy to design any user interface (UI) elements by simply moving labels, buttons, and other UI objects to the desired location. This system helped us to implement many features, including login screens for entering the platform, UI for controlling the avatar, switching between layers, working with the map of the environment, as well as interfaces enabling the participant-bot interactions.

The figure above shows the general scheme according to which the reading and distribution of the user input is performed in VCC. Initially, the input from the controls is detected and correlated with the configuration file that stores associations of logical operations within the platform and buttons as well as axes of virtual objects used as interface devices. Then the input is transferred to the game log in accordance with the given scheme. In this case, specific classes can be characterized as follows. PlayerCameraManager is used to initialize the initial position of the camera relative to the avatar and also to switch the perspective from a first-person view to a third-person view and back. Pawn and PlayerController are used as special classes for receiving and processing of player commands and their translation into the corresponding actions in accordance with predefined features. GameMode and GameState classes control all avatars and the rules for their interaction in the environment (Fig. 4).

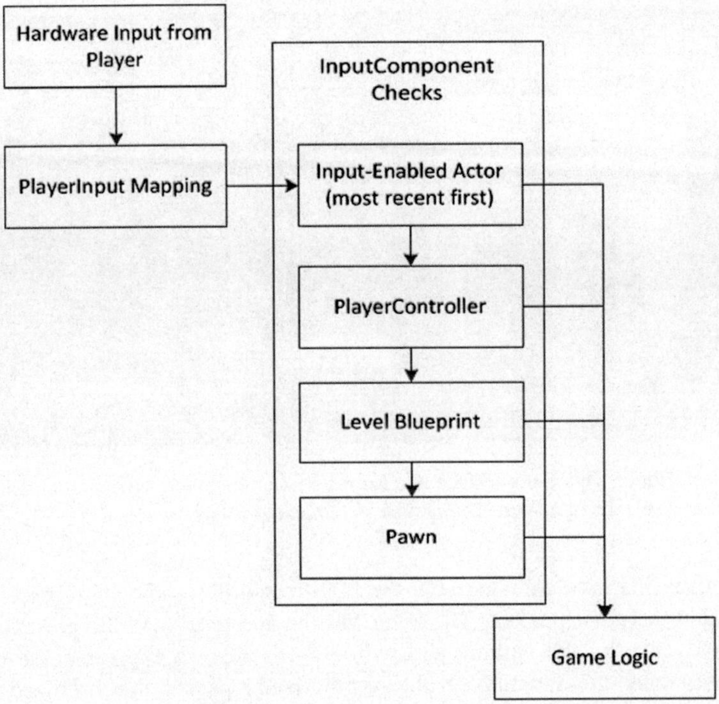

Fig. 4. Unreal Engine 4 control system architecture.

Virtual poster sessions in a mid-size conference require allocation of a large amount of graphical content in the virtual environment in such a way that it is searchable, easily accessible, and each poster can be examined and discussed privately by a small group of participants in convenient settings, in parallel with other posters. This task poses a challenge for developers. To satisfy these requirements, a poster room may have at most 5 to 10 posters in it. Accordingly, if there are thousands of posters presented at the conference, then one needs hundreds of virtual rooms to allocate them. In this case, both the size of the environment and its navigation become problems. The solution that we found is based on the usage of multiple layers (or "levels", in the terminology of UE4) of the environment, as described below.

In order to be able to use a large amount of graphical content, such as posters and associated with them presenter bots, an interface for loading posters together with bots into one and the same room was implemented as switching between the room layers. When a new layer is selected, the filling of the room changes. This includes not only posters and presenter bots standing next to them, but also other participant avatars that may move from layer to layer at their own discretion (Fig. 5).

Thus, dozens of rooms with different content were organized as one multi-layer room, which made it possible not to create large in-game spaces that would take a significant time to load as well as to find the desired poster.

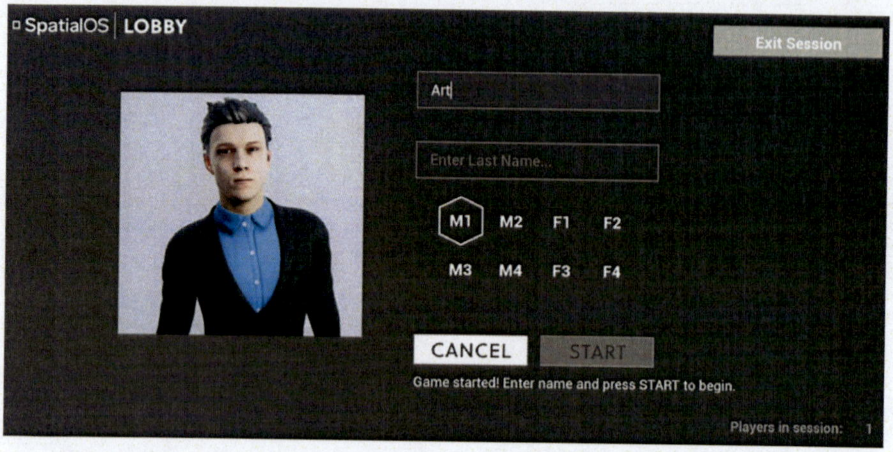

Fig. 5. Screenshot of the login interface: the Lobby. Using this interface, a participant enters her or his name and selects an avatar. The choice is replicated to all participants.

Each poster instance contains both the high-resolution graphical representation of the poster and the presenter bot. The latter has the ability to read the presentation text written in advance by the authors and, while doing this, to express emotions, using facial expressions, the tonality of the synthesized speech, synchronized with lip movements, the gaze control, and also the body language expressed via animations of body movements of the avatar. The model and the algorithm controlling these expressive modalities will be described elsewhere. The general idea is that socially-emotional behavior of the bot should be believable and socially acceptable even more than it needs to be a copy of a typical human behavior in this paradigm. Therefore, the expressed emotion is a function of the content of the text, immediate sensory input, and all the history of interaction with this and other poster attendees, plus, it can be a function of the individual personality of the bot.

The bot is implemented as an interactive social agent. When an attendee participant is approaching the poster, the bot offers to give a presentation, using different wording every time. During the presentation, the attendee may interrupt the bot by hand raising, then there is a possibility to ask questions, ask the bot to repeat what was said, or continue. Similarly, a short dialogue occurs at the end of the presentation. In principle, the bot can take questions and communicate them to the authors. At the end the bot asks to evaluate her job on a five-point scale. The participant input is logged, as well as all events in the environment, for off-line analysis (Fig 6).

This technical description of the VCC platform and methods of its development is far from complete. Missing details can be found in our other publications [9–12].

The platform was tested many times during its development. Participants in test sessions were NRNU MEPhI college students, colleagues from other scientific institutions in Moscow and in other countries, and participants of the conference BICA*AI 2020. In total, approximately 100 volunteers participated in the testing of VCC.

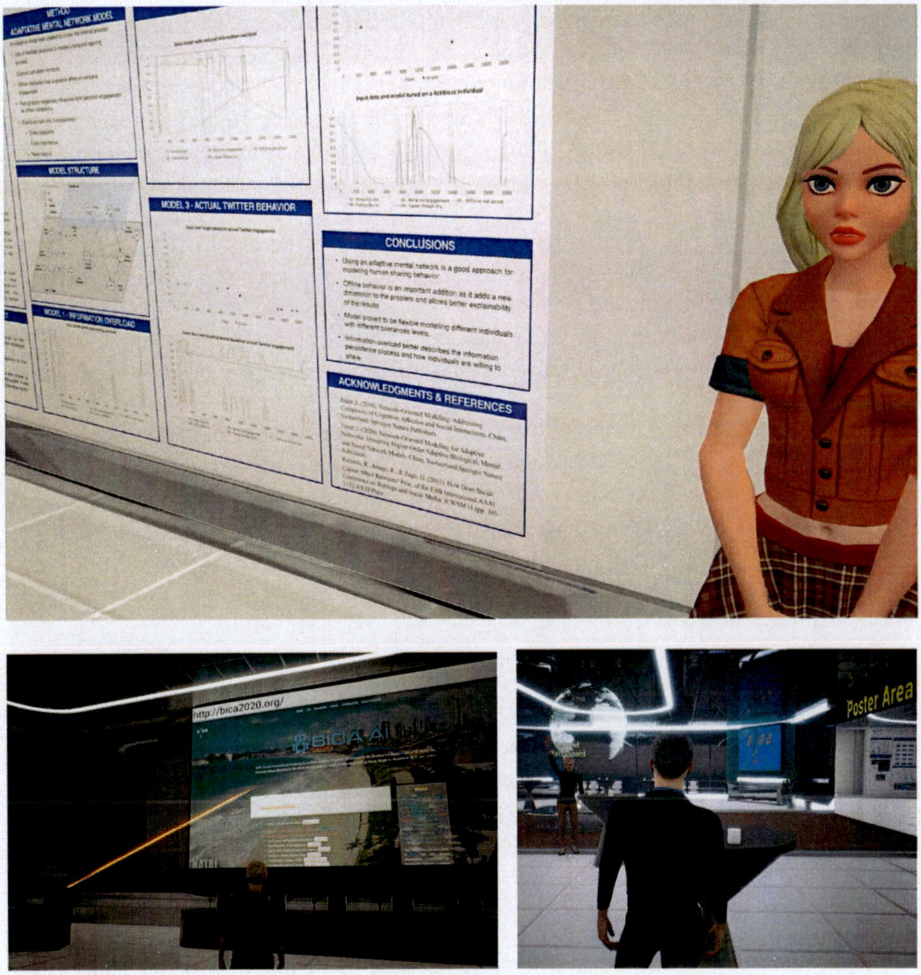

Fig. 6. Top: The presenter bot standing next to her poster after she delivered the presentation. Bottom left: a screenshot from the Grand Ballroom, a place to run plenary sessions. Bottom right: a coffee break screenshot.

4 Practical Usage Outcome - Discussion and Conclusions

The VCC platform was developed, implemented and tested as described above. It was improved incrementally during multiple test sessions, until it became suitable for practical usage. Eventually VCC was used for hosting two mid-size scientific events. One of them is the 2020 Annual International Conference on Brain-Inspired Cognitive Architectures for Artificial Intelligence (BICA*AI 2020: https://bica2020.bicasociety.org), also known as the Eleventh Annual Meeting of the BICA Society, the conference to which this volume of Proceedings belongs. Another is a Russian multiconference: First National Congress on Cognitive Research, Artificial Intelligence and Neuroinformatics (CAICS 2020, https://caics.ru/en), unifying four conferences: RCAI-2020, Ontercognsci-2020, Neuroinformatics-2020 and Physio-2020, also including a set of workshops, symposia and special sessions. Numbers of actual online participants were approximately 100 in BICA*AI 2020 and 1000 in CAICS 2020. The working language and, accordingly, the language of the application was English in the first case and Russian in the second. BICA*AI 2020 was hosted entirely using VCC, while in the case of CAICS 2020 the VCC platform was used only for poster sessions. In both cases it was a success.

The model used for hosting plenary sessions, coffee breaks and other social events was based on a combination of VCC and Zoom, the latter was used for audio connection and for streaming the video to those participants who do not have the minimal hardware to run VCC.

The model used for hosting a poster session was based on a combination of VCC and Mozilla Hubs, the latter was linked to VCC (clicking on a special banner in the environment caused a transition from VCC to Mozilla Hubs). The main purpose of including Mozilla Hubs was to enable participation for those participants who do not have the minimal hardware to run VCC. Another reason was the better quality of local voice chat available in Mozilla Hubs, which is important for mass social events like parties or poster sessions.

Multiple formats of virtual conference sessions were supported by VCC, including plenary sessions, discussion panels, poster sessions, coffee breaks, and informal discussions. Another mode in which VCC was used was a standalone application without multiplayer, that allowed a participant to explore the media on display and listen to bot presentations at any convenient time before or after live sessions. Many more formats are potentially available, including concurrent sessions, breakout groups, and welcome receptions. In addition, after necessary adaptation, VCC can be used in education.

4.1 Concluding Remarks

One key advantage of hosting a scientific conference in Virtual or Mixed Reality is the possibility to use Virtual Actors controlled by Artificial Intelligence as conference participants, in such roles as a Virtual Poster Presenter, a Discussion Panel Moderator, a Lightning Session Chair, and a Virtual Party Servant. All these roles require human-level socially emotional functionality and can be implemented using one approach, which is based on the emotional Biologically Inspired Cognitive Architecture (eBICA [12]). Details will be presented elsewhere.

In conclusion, here we presented the new concept, its implementation and practical usage outcome of a virtual-environment-based platform called Virtual Convention Center (VCC) that is intended to solve one challenge of our time outlined in the Introduction. From another perspective, this platform can be very useful for scientific research by allowing one to test models like the eBICA framework [12] and see its advantages in comparison with alternatives [13].

Acknowledgments. The authors are grateful to Andrey Yu. Shedko, Vladislav Shadrin and many other NRNU MEPhI students for their help with reviewing existing VR/VE conferencing technologies and for their participation in the testing of VCC. This work was supported by the Russian Science Foundation Grant # 18-11-00336.

References

1. Neutrino2020, Fermi National Accelerator Laboratory, 5 July 2020. https://conferences.fnal.gov/nu2020/
2. Artificial General Intelligence, 7 July. http://agi-conf.org/2020/ (2020)
3. Hubs by Mozilla, 10 May. https://hubs.mozilla.com/ (2020)
4. AltspaceVR. In Wikipedia, 10 May, https://en.wikipedia.org/wiki/AltspaceVR/ (2020)
5. VRChat. In Wikipedia, 10 May. https://en.wikipedia.org/wiki/VRChat/ (2020)
6. Sync, 10 May. https://sync.vive.com/login (2020)
7. Lang, B.: 34 VR Apps for Remote Work, Education, Training, Design Review, and More. Road to VR, 3 November. https://www.roadtovr.com/vr-apps-work-from-home-remote-office-design-review-training-education-cad-telepresence-wfh/ (2020)
8. Van Lehn, K.: The relative effectiveness of human tutoring, intelligent tutoring systems, and other tutoring systems. Educ. Psychol. **46**, 197–221 (2011). https://doi.org/10.1080/00461520.2011.611369
9. Chubarov, A, Azarnov, D.: Modeling behavior of virtual actors: a limited Turing test for social-emotional intelligence. In: First International Early Research Career Enhancement School on Biologically Inspired Cognitive Architectures, pp. 34–40. Springer, Cham (2017)
10. Chubarov, A.A., Tikhomirova, D.V., Shirshova, A.V., Veselov, N.O., Samsonovich, A.V.: Virtual listener: a turing-like test for behavioral believability. Procedia Comput. Sci. **169**, 892–899 (2020)
11. Eidlin, A.A., Chubarov, A.A., Samsonovich, A.V.: Virtual listener: emotionally-intelligent assistant based on a cognitive architecture. Advances in Intelligent Systems and Computing, vol. 948, pp. 73–82. Springer, Cham (2020). https://doi.org/10.1007/978-3-030-25719-4_10
12. Samsonovich, A.V.: Socially emotional brain-inspired cognitive architecture framework. Cogn. Syst. Res. **60**, 57–76 (2020)
13. Marsella, S.C., Gratch, J.: EMA: a process model of appraisal dynamics. Cogn. Syst. Res. **10** (1), 70–90 (2009)

Check for updates

Ensembling SNNs with STDP Learning on Base of Rate Stabilization for Image Classification

Alexander Sboev[1,2(✉)], Alexey Serenko[1], Roman Rybka[1,2], and Danila Vlasov[1,2]

[1] National Research Centre "Kurchatov Institute", Moscow, Russia
sag111@mail.ru

[2] MEPhI National Research Nuclear University, Moscow, Russia

Abstract. In spite of a number of existing spiking neural network models for image classification, it still remains relevant from both methodological and practical points of view to develop a model as simple as possible, while at the same time applicable to classification tasks with various types of data, be it real vectors or images. Our previous work proposed a simple spiking network with Spike-Timing-Dependent-Plasticity (STDP) learning for solving real-vector classification tasks. In this paper, that method is extended to image recognition tasks and enhanced by aggregating neurons into ensembles. The network comprises one layer of neurons with STDP-plastic inputs receiving pixels of input images encoded with spiking rates. This work considers two approaches for aggregating neurons' output activities within an ensemble: by averaging their output spiking rates (i.e. averaging outputs before decoding spiking rates into class labels) and by voting with decoded class labels. Ensembles aggregated by output frequencies are shown to achieve a significant accuracy increase up to 95% (by F1-score) for the Optdigits handwritten digit dataset, and is comparable with conventional machine learning approaches.

Keywords: Spiking neural networks · Spike-Timing-Dependent Plasticity · Ensembles · Image recognition

1 Introduction

Of the methods for applying spiking neural networks (SNNs) to machine learning tasks [1], the majority mimic the structures of formal networks. Recently, however, especial relevance is shifting towards the problem of creating methods that would be based on Spike-Timing-Dependent Plasticity (STDP) [2] as a mechanism underlying learning, since STDP, being local, has some prospective potential of being hardware-implemented in energy-efficient memristive devices [3,4]. A general goal of ours is the creation of network topologies simple and as generic as possible, utilizing characteristic phenomena of STDP.

A. V. Samsonovich et al. (Eds.): BICA 2020, AISC 1310, pp. 446–452, 2021.
https://doi.org/10.1007/978-3-030-65596-9_53

The existing models with local learning mechanisms are usually developed for the image recognition task, and employ a large number of neurons, each of which extracts specific high-level features such as directions if strokes or some image parts [5]. Another approach [6] proposes a topology of one layer of excitatory neurons, which under STDP learn in an unsupervised manner to be selectively sensitive to certain shapes in input images. In order for different neurons to memorize different shapes, that approach requires competition among neurons via inhibitory connections, thus necessitating the adjustment of a large number of parameters.

Our previous work [7] proposed a learning algorithm based on the effect of stabilising the mean output spiking rate of a neuron under STDP [8]. In this algorithm, each neuron is trained on its own class and then is to distinguish its own class from the other classes. An advantage of the algorithm is its robustness in a wide range of neuron and STDP parameters, owing to the robustness of the underlying effect of spiking rate stabilisation. In spite of using extremely few neurons, just one neuron per class, the algorithm has proved its efficiency on the benchmark classification tasks of Fisher's Iris and Wisconsin breast cancer.

The simplicity of the model, along with the low number of neurons used in the algorithm, leaves the potential for improving its accuracy by employing more neurons, while still keeping the numbers of adjustable parameters and neurons relatively low compared to other approaches. The aim of this work is to study on a more complex dataset, what accuracy increase an ensemble of several networks with rate-stabilisation-based STDP learning achieve in comparison with the accuracy of a single network.

2 Materials and Methods

2.1 Dataset

As a benchmark to test the principal possibility of image classification using simple SNN topologies, we use the Optdigits [9] dataset of 8×8 images of handwritten digits. We use the 1,797-digit subset of this dataset that is available as part of the scikit-learn package under the name of Digits.

2.2 Models

The SNN scheme proposed is based on Leaky Integrate-and-Fire neurons and STDP plasticity synapses. The optimal values of the neuron constants and synapse models have been adjusted by the MultiNEAT [10] genetic algorithm, the adjustment procedure is described in the preliminary work [11].

The dynamics of the neuron membrane potential V is

$$\frac{dV}{dt} = -\frac{V(t) - V_{\mathrm{rest}}}{\tau_{\mathrm{m}}} + \frac{1}{C_{\mathrm{m}}} \sum_{i} \sum_{t_{\mathrm{sp}}^{i}} w_i \left(t_{\mathrm{sp}}^{i}\right) \frac{q_{\mathrm{syn}}}{\tau_{\mathrm{syn}}} e^{-\frac{t - t_{\mathrm{sp}}^{i}}{\tau_{\mathrm{syn}}}} \Theta\left(t - t_{\mathrm{sp}}^{i}\right),$$

where $V_{rest} = -70$ mV, $\tau_m = 10$ ms, C_m of different networks in the ensembles varied around the genetically-found optimal value of 1.2 fC. In the postsynaptic current, the first sum is over the incoming neuron synapses, and the second sum is over incoming spikes arriving at the i-th synapse. $\tau_{syn} = 5$ ms, $q_{syn} = 5$ fC, Θ is the Heaviside step function; $w_i(t)$ is the weight of the i-th synapse. When the potential exceeds $V_{th} = -54$ mV, the neuron emits a spike, and the potential is clamped to V_{rest} for the refractory period $\tau_{ref} = 2$ ms.

The synaptic plasticity is additive STDP, in which synaptic weight change is

$$\Delta w = \begin{cases} -\alpha\,\lambda \cdot \exp\left(-(t_{pre} - t_{post})/\tau^-\right) & \text{if} \quad t_{pre} - t_{post} > 0; \\ \lambda \cdot \exp\left(-(t_{post} - t_{pre})/\tau^+\right) & \text{if} \quad t_{pre} - t_{post} < 0. \end{cases} \tag{1}$$

Here $\lambda = 0.001$ is effectively the learning rate. The constants $\alpha = 1.5$, $\tau^+ = 80$ ms and $\tau^- = 4.3$ ms have been found with the genetic algorithm. t_{pre} is the moment of arrival of an incoming spike at the synapse, t_{post} is the moment when a neuron emits a spike. The restricted symmetric spike pairing scheme is used, for which the rate stabilization effect has been shown previously [12]: the rule (1) takes into account only pairs of consecutive spikes between which there are no other input spikes (from this input) or output spikes.

To prevent the weight from exceeding its maximum value of 1 or falling below 0, an additional constraint is applied to each weight change after calculating Δw: if $w + \Delta w > w_{max}$, then $\Delta w = w_{max} - w$; if $w + \Delta w < 0$, then $\Delta w = -w$.

2.3 Network Scheme and Learning Algorithm

Each network consists of 10 neurons (one for each input class) not connected to each other. The input synapses of each neuron receive pre-processed and rate-encoded vectors of input images. The preprocessing stage consists of normalizing each vector so that its Euclidean norm equals 1, and then of processing it with receptive fields [13]; thus, the dimension of the vector is increased from 64 to $64 \cdot K$, where K is the number of receptive fields. Presenting the pre-processed vectors to the inputs of the network, they are encoded by spiking rates: a component x_i of a pre-processed vector corresponds to a Poisson spike sequence with the mean frequency $\nu_{low} + x_i \cdot \nu_{high}$ presented to the i input synapse of a neuron during 1 s. The encoding parameters $K = 16$, $\nu_{low} = 4$ Hz, $\nu_{high} = 427$ Hz were also selected with the genetic algorithm.

At the training stage, each neuron in the network receives training samples of only one class chosen for it in advance. Initial synaptic weights are drawn randomly from a uniform distribution from 0 to 1. The learning stage for the neuron is finished when its output frequency is stabilized as the result of STDP. After training, the weights are fixed (plasticity is disabled), and the neurons' spiking rates in response to train samples of their own classes are recorded.

At the testing stage, class labels are decoded by the output frequencies of the neurons in response to an image, using the algorithm developed earlier [7]: the image is assigned to the class whose neuron spikes in response to it with the frequency the closest to the mean frequency of this neuron on its own class' training set vectors.

3 Experiments

An ensemble is formed of several networks (trained independently but on the same data), the topology and learning procedure of which is described in Sect. 2.3. Before that, search for optimal network parameters – input encoding parameters, neuron and synapse constants – has been conducted for a single network model with the help of the genetic algorithm.

Two ways of ensembling are tried: either 1) the networks within an ensemble have all parameters identical (the networks thus differ only by different random initial weights), or 2) each network has its own value of neuron membrane capacity C_m. In the first case, C_m has the value of 1.2 fC as found by the genetic algorithm. In the second case, C_m is varied around this value; the range of this variation is chosen so that the accuracy of the network decreases just slightly. In total, 15 values of C_m have been used, ranging from 0.6 fC to 3.6 fC with the step of 0.2 fC. Networks with different C_m values are sorted in the order of decreasing their performance on the training set (see Fig. 1, lower plot), and incorporated into the ensemble best-first.

Aggregating outputs of trained networks within an ensemble is performed in two ways.

Averaging output rates: output spiking rates of neurons corresponding to the same classes are summed, and the resulting total rates are passed to the decoding procedure described in Sect. 2.3 as if it were outputs of a single network.

Averaging class labels: the output rates of each network are decoded into class labels under the procedure described in Sect. 2.3, and the resulting label of an image is the label outputted by the majority of the networks in the ensemble.

Training and testing of ensembles is conducted with stratified 5-fold cross-validation. The learning performance is measured by the F1-macro score, and the results below present the mean and deviation range of F1-macro over cross-validation folds.

4 Results

A single SNN with the optimal parameters adjusted by the genetic algorithm achieves the F1-score of 85% to 88%, the deviation over cross-validation folds and over 15 independent runs being roughly the same.

Ensembles of up to 15 identical networks do not improve accuracy compared to a single network, as shown by the "Identical networks" curves in Fig. 1. However, ensembling networks with different C_m values increases the accuracy to 95 to 96%. Notably, this increase is achieved by aggregating outputs of several networks, but not because of incorporating some very high-performing network, because the accuracy of each individual network remains roughly the same (see Fig. 1, lower plot). Note, however, that the accuracy increase is only observed

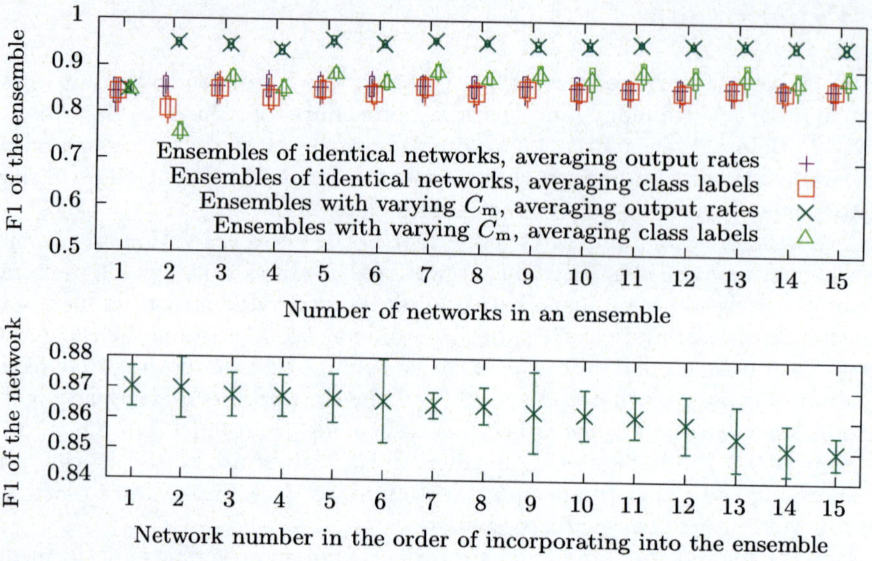

Fig. 1. Upper plot: the F1-macro score of an ensemble in dependence of the number of networks in it, for the two ways of aggregating the outputs of networks within an ensemble (described in Sect. 3). Lower plot: the F1-macro score of individual networks in the order in which they were incorporated into the ensemble. The deviation ranges in both plots show the minimum-to-maximum range of the score over cross-validation folds.

when aggregating network outputs is performed *by averaging output rates*, i.e. if network outputs are averaged before rather than after their decoding.

The F1-macro score achieved is on par with the 98% ± 1% result of the conventional Gradient Boosting classifier from the scikit-learn library, as well as with 98% reported by the dataset maintainers, achieved by the k-nearest-neighbour algorithm on the full version of the dataset [14].

5 Conclusion

This work presents the results of studying spiking neural networks of quite a simple topology, aggregated into an ensemble and applied to an image recognition task. Aggregating networks with different values of neuron membrane capacity allows to improve the classification performance by about 9% by the F1-macro score. The results obtained open a prospective possibility for simple spiking networks to solve not only vector classification tasks as shown earlier, but also image recognition tasks, and pave the way for solving more complex tasks. Thus, this study is a step in the research direction of developing spiking networks based on local STDP plasticity that would be implementable in memristive devices.

Acknowledgments. This work has been carried out using computing resources of the federal collective usage center Complex for Simulation and Data Processing for Mega-science Facilities at NRC "Kurchatov Institute", http://ckp.nrcki.ru/.

References

1. Paugam-Moisy, H., Bohte, S.M.: Computing with spiking neuron networks. In: Rozenberg, G., Back, T., Kok, J. (eds.) Handbook of Natural Computing, pp. 335–376. Springer, Heidelberg (2012). https://doi.org/10.1007/978-3-540-92910-9_10, http://homepages.cwi.nl/~sbohte/publication/paugam_moisy_bohte_SNNChapter.pdf
2. Feldman, D.E.: The spike-timing dependence of plasticity. Neuron **75**(4), 556–571 (2012). https://doi.org/10.1016/j.neuron.2012.08.001, http://www.sciencedirect.com/science/article/pii/S0896627312007039
3. Saïghi, S., Mayr, C.G., Serrano-Gotarredona, T., Schmidt, H., Lecerf, G., Tomas, J., Grollier, J., Boyn, S., Vincent, A.F., Querlioz, D., La Barbera, S., Alibart, F., Vuillaume, D., Bichler, O., Gamrat, C., Linares-Barranco, B.: Plasticity in memristive devices for spiking neural networks. Front. Neurosci. **9**, 51 (2015). https://doi.org/10.3389/fnins.2015.00051
4. Serrano-Gotarredona, T., Masquelier, T., Prodromakis, T., Indiveri, G., Linares-Barranco, B.: STDP and STDP variations with memristors for spiking neuromorphic learning systems. Front. Neurosci. **7**, 2 (2013)
5. Kheradpisheh, S.R., Ganjtabesh, M., Thorpe, S.J., Masquelier, T.: STDP-based spiking deep convolutional neural networks for object recognition. Neural Netw. **99**, 56–67 (2018). https://doi.org/10.1016/j.neunet.2017.12.005, http://www.sciencedirect.com/science/article/pii/S0893608017302903
6. Diehl, P.U., Cook, M.: Unsupervised learning of digit recognition using spike-timing-dependent plasticity. Front. Comput. Neurosci. **9**, 99 (2015)
7. Sboev, A., Serenko, A., Rybka, R., Vlasov, D.: Solving a classification task by spiking neural network with STDP based on rate and temporal input encoding. Mathematical Methods in the Applied Sciences (2020). https://doi.org/10.1002/mma.6241, https://onlinelibrary.wiley.com/doi/abs/10.1002/mma.6241
8. Kempter, R., Gerstner, W., Hemmen, J.L.v.: Intrinsic stabilization of output rates by spike-based Hebbian learning. Neural Comput. **13**(12), 2709–2741 (2001). http://infoscience.epfl.ch/record/97800/files/Kempter01.pdf
9. Alpaydin, E., Kaynak, C.: Optical recognition of handwritten digits data set (1995). http://archive.ics.uci.edu/ml/datasets/Optical+Recognition+of+Handwritten+Digits
10. Stanley, K., Miikkulainen, R.: MultiNEAT – a portable software library for performing neuroevolution. http://multineat.com/index.html
11. Sboev, A., Serenko, A., Rybka, R., Vlasov, D., Filchenkov, A.: Estimation of the influence of spiking neural network parameters on classification accuracy using a genetic algorithm. In: Postproceedings of the 9th Annual International Conference on Biologically Inspired Cognitive Architectures (BICA), vol. 145, pp. 488–494 (2018). https://doi.org/10.1016/j.procs.2018.11.111, http://www.sciencedirect.com/science/article/pii/S1877050918323998

12. Sboev, A., Rybka, R., Serenko, A., Vlasov, D., Kudryashov, N., Demin, V.: To the role of the choice of the neuron model in spiking network learning on base of spike-timing-dependent plasticity. In: Klimov, V.V., Samsonovich, A.V., (eds.) 8th Annual International Conference on Biologically Inspired Cognitive Architectures (BICA), vol. 123, pp. 432–439. Elsevier BV (2018). https://doi.org/10.1016/j.procs.2018.01.066, https://www.sciencedirect.com/science/article/pii/S187705091830067X
13. Wang, J., Belatreche, A., Maguire, L., McGinnity, T.M.: An online supervised learning method for spiking neural networks with adaptive structure. Neurocomputing **144**, 526–536 (2014). https://doi.org/10.1016/j.neucom.2014.04.017, http://www.sciencedirect.com/science/article/pii/S0925231214005785
14. Kaynak, C.: Methods of combining multiple classifiers and their applications to handwritten digit recognition. MA thesis, Institute of Graduate Studies in Science and Engineering, Bogazici University (1995)

Mathematical Methods for Solving Cognitive Problems in Medical Diagnosis

Yuri Kotov[1] and Tatiana Semenova[2(✉)]

[1] Keldysh Institute of Applied Mathematics, Moscow, Russia
[2] National Research Nuclear University MEPhI (Moscow Engineering Physics Institute), Moscow, Russia
contreraskosha@yandex.ru

Abstract. Complex system exploring at initial stage includes fixation of basic data blocks, problem categories, and other. At this stage, an adequate description uses gestalt-like patterns for the future system, its components and their functioning. In the structure of such system, functional and logical relationships can be found between the components or processes. Such a complex system is a living organism (for example, a patient). A skilled patient-treating physician can discover functional and logical associations between the components or processes in the studied system. In some difficult cases, doctors cannot explain their decisions and actions. They might use it practically, but cannot formulate verbally. For these cases, mathematician I.M. Gelfand has proposed a method of diagnostic games. The game is a cognitive research to reveal the doctor's intuitive action plan for specific case of the patient's treatment. Working together, the mathematician and physician can formulate a verbal case description. The objective of this study is to extract strict formalized elements from the doctor's gestalt perception: the rules of diagnostic decision. In order to analyze the intuitive actions of the specialists, the authors propose a mathematical language and technologies based on non-numerical statistics and three-valued logic. The language helps us detect and solve such cognitive problems. The collaboration with doctors allows us to create clear diagnostic rules based on the latent knowledge of an experienced specialist. The article provides a brief description of the method used for solving the problems of practical medicine.

Keywords: Gestalt · Diagnostic game · Formalization of the Doctor's knowledge · Non-numerical statistics · Three-valued logic

1 Introduction

This paper presents the mathematical methods for solving the cognitive problems of information processing helping to increase the fixation accuracy of the doctor's decision.

Consider the simplest diagnostic classification problem. Let us have a patient's group with known diagnosis. We intend to build the rules for predicting the treatment results (choice of classes) for new patients using available information array about the early stages of process. This is important for the selection and correction of the treatment methods.

A. V. Samsonovich et al. (Eds.): BICA 2020, AISC 1310, pp. 453–465, 2021.
https://doi.org/10.1007/978-3-030-65596-9_54

At the beginning of the joint work with the doctor, we have a classification of some patients (standard classes) and formulate the prototypes of classes – diagnostic rules. Next we check the obtained rules on the same array of patients, comparing the properties of each patient with standard classes.

The medical data have their own particularities.

First, the doctor knows more than he can express in words. Therefore, these latent knowledge (skills) should be identified and translated into verbal form.

Second, many medical and biological numerical parameters do not obey the Gaussian distribution (the so-called "normal"). "However, in reality, the normal distribution of data in biomedical research occurs slightly more often than never, which is why non-parametric methods form the basis of the mathematical apparatus of medical statistics" [1]. Therefore, the study of numerical data should be carried out by non-parametric statistics.

Third, a significant part of the doctor's information is non-numerical (doctor's opinions, verbal assessments, patient's organism characteristics). That part of the data must be processed by special statistical methods – methods of non-numerical statistics that do not use arithmetic operations (e.g., addition) [2].

Fourth, medical information may contain unremovable gaps due to inaccessibility or loss of a part of data. Therefore, data analysis methods should, as far as possible, be insensitive to some data absence.

An organism is a very complex system and consists of many closely related subsystems. Therefore, no instantaneous changes in the its state indicators can be considered neither independent nor random. Changes cannot be independent, since all subsystems interact with each other. Events in the body cannot be accidental, since the organism is trying all the time to restore its balance with the environment. This leads to the conclusion that it is impossible to use classical statistics, which is applicable only to independent random variables.

On the other hand, in medicine, tables of standard values of the parameters of healthy organisms are often used. These tables are usually based on information from large groups of people, processed by the classical statistical methods. The real organism, even with a mild illness, may have deviated dependencies between the parameters compared to the parameters of a healthy one. Using standard normative tables to assess the current state of the patient and predict his behavior might lead to the erroneous result.

This task seems completely insoluble.

But we know that the doctor treats the patient. He knows how to influence this organism so that his state of health does not worsen. A physician's own experience consists of at least two parts. One part is what he was taught at the university. The second part is his personal experience with patients that he is currently treating, considering the peculiarities of the living conditions of the indigenous population. We only know some part of his experience, that can be verbally expressed either in textbooks or in constant conversations with colleagues. Therefore, in order to understand the diagnosis he set to the patient, it is necessary to examine the actual current work of the doctor as accurately as possible.

This work can be done with a diagnostic game [3]. For a diagnostic game session, we use a real medical history of a patient previously cured by this doctor. The doctor is offered to assess the patient's condition by asking the guide questions which can be answered "YES" or "NO" only.

Often the guide role is played by a mathematician who works with this doctor on this problem. The mathematician finds the relevant information in the medical history and answers the doctor's question about the patient. It is forbidden to provide patient-identifying information (personal name, nosology, etc.). As soon as the doctor has formed his concept of the patient, he reports the result. At this point, the diagnostic game session is terminated. If the doctor recognized the patient's name, the session is considered unsuccessful and the information is not taken for processing. The entire dialogue of the game is recorded on a tape. The doctor is not informed whether the diagnosis made during the game matches the old one. These sessions are repeated using the medical records of other patients with the same disease. If the doctor comes to a firm conclusion, the game is considered successful. In several game sessions with an experienced doctor, we get a set of dialogues that capture doctor's point of view on the course of the disease and treatment results.

The game result is presented in the form of logical statements. For further analysis, it is possible to use questions and statements that are often repeated in different patients. Identified essential statements can be used as elements of a dictionary for describing patient's condition or primary gestalt pattern (the first archetype [4, 5]). The next stage consists in building a structure for patient's description using the dictionary. The physicians usually solve this problem using variants of structures from his past experience. This task can be facilitated for him and the solution can be made more accurate using the structure-construction tools available in mathematics.

Mathematician's purpose is to help physician in structuring and compressing the information. Mathematician can to build more detailed model (another archetype) using the logical structure of the doctor's knowledge and non-numeric statistical methods [2]. For each patients class a prototype of the class is developed on basis of the training material – the primary diagnostic rule that helps formally classify some patients into this class, and then they are checked for coincidence with the a priori class ("training exam"). If the predictive rules work well in this test and the prior classification is validated, then they are suitable for the new patient population.

2 Mathematical Methods

2.1 Logical Symptom Method

The main data element in formal processing is the logical symptom [6]. A logical symptom (three-valued variable) contains a statement about some property of a particular patient's organism or about a specific feature of the treatment process. It can display a verbal statement, the numerical measurement result, formulas, inequalities, the timing of various treatment stages, treatment methods, prescribed medications, a combination of numbers and verbal statements or other formal texts that have well-defined meanings.

The logical symptom is a variable of the extended version of Lukasiewicz's three-valued logic [7]. Symptom definition includes a header ("@symptom") that identifies the numeric symptom code and formal condition. Here are some examples of possible logical symptoms for medical problems.

@symptom 026 premature maturation of placenta villi
@symptom 027 arterial hypertension was registrated
@symptom 028 heparin therapy was performed
@symptom 029 alt >= 66 && alt< 72 && gest <= 10 && gest <= 12
@symptom 030 CREA >= 53 && CREA <= 68 && gest >= 35 && gest <= 37.

Symptoms 026 – 028 are verbal statements. Symptom with code 026 contains the conclusion of a morphologist who examined the mother's placenta after childbirth. Symptom with code 027 contains the underlying disease diagnosis of pregnant patient – arterial hypertension. Symptom with code 028 means that the patient was treated with heparin. Symptoms with codes 029, 030 contain complex statements describing observations of the patient's organism condition at certain periods of the treatment process (gestational intervals are set). They reflect the numerical values of the enzyme ALT (alanine aminotransferase) and creatinine (CREA) concentrations in the blood at the corresponding values of gestational age.

The complete set of symptoms for particular patient forms a vector (row) of logical symptoms corresponding to one observation. Each research object (patient) is described by a logical vector:

$$\mathbf{F}_j = \{x_{j1}, x_{j2}, x_{j3}, x_{j4}, \ldots, x_{jn}\}, \tag{1}$$

where coordinates are logical symptoms, *jk* is patient index, *k* – order position in vector. The vector dimensions ("*n*") are the same for all patients and equal to the number of symptoms in vector. A vector can contain tens or hundreds of symptoms, depending on the problem being solved. If the patient has been examined several times, he can be described by several logical vectors (one logical vector for each inspection).

The general set of logical symptoms for the entire patient study group forms a matrix of observations.

2.2 Three-Valued Logic

Boolean logic algebra allows only two values of a variable ("true" and "false"). This is not enough to display the medical data. The dataset for a patient is usually not complete. Some studies were not performed, the patient could not remember some information from his anamnesis or could not show up in time for next study, etc. Therefore, there may be gaps in the logical vectors structure. The gap presence hampers a rigorous decision-making procedure and can lead to dangerous errors in the development of diagnostic rule.

For mathematical processing of heterogeneous data with gaps, the three-valued Lukasiewicz's logic is used [7]. In three-valued logic, logical symptoms can take on three meanings: "True (Yes)" (encoded with an "X"), "False" (No) (encoded with a "="), and "Unknown" (encoded with a "." (Dot)). Fig. 1 presents a small fragment of the logical symptom matrix. The top three lines indicate the symptom codes. They are read downwards from top to bottom. The letters in the left column of Fig. 1 indicate the classes to which the patients belong, the next column contains the patient codes.

```
       *                              *
       0000000000000000000000000000000000000000000000000
       0000000000111111111122222222223333333333444444444
       0123456789012345678901234567890123456789012345678
A 4002 .........................============..........==
B 4014 =================........======X=====..........==
A 1021 ==X=X========...........................===....X=
D 4079 ==X=========X.........=X==X=====X====.XX.......X=
D 4099 ==X======X==X....=X=X==X==X==========.==.......X=
C 4059 ==X=X=======X=XX==X==X=X====X==X=====XXX...===X=X
```

Fig. 1. Fragment of logical symptom matrix

For example, the symptom 000 (marked with an asterisk) in five patients has the "false" value (symbol "="), but the patient A 4002 has unknown value of symptom 000 (symbol "."). Patients A 4002, D 4079 and D 4099 have value of symptom 031 (also marked) the meaning "false" (symbol "="), but patient A 1021 his the value unknown.

The three-valued logic allows a uniform formal analysis of medical observations arrays.

2.3 Classes Prototypes Building

We now have a matrix describing the entire patient population and the tools to process it. The next important part of the cognitive problem is to create a formal prototype (mask) for each patient class. The prototype may be considered as a first approximation to the prognostic rule. We derive this rule by using a sample of classes from patients who participated in the diagnostic game (patients previously cured by the doctor). The prototype will build in the format of symptom vector (1) and will be composed of symptoms that more often show a YES (X) value for patients of one class.

For each symptom position in vector we count "yes" values for objects associated with class A (pa) and for objects associated with all other classes ("against class A") (oa). The procedure is also repeated for "no" values in both class groups ("pb" and "ob"). Next the four numbers (pa, oa, pb, ob) are compared using Fisher's exact method. If resulting score is less then selected significance level we put letter "X" to the mask position if $pa*ob > oa*pb$ or "=" in opposite case.

Then we compare the number of "yes" responses in favor of, for example, class A and the number of responses against it. Using the criterion of Fisher's exact method, we obtain the division of the symptom set into a subset "for class A" and "against class A". The procedure is repeated for each symptom.

Selecting new significance level and repeating the general procedure, we can achieve better class separation. Symptoms chosen in this way form a new logical vector (class mask). The next mandatory procedure is the preliminary classification control. It is necessary to check that each patient belongs to assigned class.

2.4 Vector Indistinguishability

At this process stage all information about patients is reduced to a single format (matrix of logical symptoms). The logical symptoms are non-numerical objects, therefore methods of non-numerical statistics must be used in processing [2]. In non-numerical statistics there are no arithmetic operations on data, only metrics (distances between objects) can be used. In the paper we use the vector indistinguishability as metrics [6]. To compare two symptom vectors we count the number $L(x)$ of positions in which symbols for each order position in the first and second vectors are not simultaneously unknown, meaning that symptom values at these positions may be "*X*" or "=", but not "." (Dot). This number $L(x)$, called "the number of known (measured) coordinates" in vector pair, always lies within $0 <= L <= N$, where N is the vector dimension. Let us also calculate the position number $J(x)$ in which the symbols coincide ($x_{1k} = x_{2k}$) but not "." (Dot). This number lies in the range $0 <= J <= N$ and $J <= L$ also. The ratio $J(x)/L(x)$ will be called indistinguishability "u" of compared vectors. Obviously,

$$0 < = u < = 1.$$

Let's give a simple example of calculating indistinguishability (Fig. 2).

object A	= X = . . . = . X X = X . . . . X X	
object B	X X = ... = X = . . . XX = =	
matches	V V V V	J = 4
non-empty symptoms in both vectors	V V V V V V V V	L= 8

Fig. 2. An example of indistinguishability calculating for two objects A and B.

Figure 2 shows the segments of two logical symptoms vectors A and B (two upper lines). In the third line, the matching symptoms are marked with the letter "V" (4 matches, J = 4). In the fourth line symbols indicate all known symptoms in both segments (8 coordinates, L = 8). The indistinguishability **u** between objects A and B is equal to the ratio: **u** = 4/8 = 0.50.

2.5 Indistinguishability of Patient's Vector from the Mask

An example of comparison of logical vectors symptoms to a class mask is shown in Fig. 3 for obstetrics area. The matrix rows contain logical symptoms obtained during the ultrasound examination of the fetus. The complete mask contains 26 ultrasound symptoms. The patients were divided into 4 a priori classes. Class C consists of patients with a threat of preeclampsia (pregnancy complication). Healthy pregnant women were assigned to class Z, and classes A and B included a moderate risk of complications [8].

Figure 3 shows the logical symptom matrix and the mask, constructed to compare class C with the combined classes ZAB (C: ZAB). Mask contains the most severe patients of class C. Mask is framed at the top of Fig. 3. The last column of the matrix (also in the frame) indicates the indistinguishability of the logical vector of the patient

```
mask ─ X.X...=XX=X.X.=.X...XXX.XX
                                        u
C 181      X=...==XX=.=...====X=XX==X   0.73
C 182      X=...==XX=.=...=XXX==X..=X   0.80
C 183      X======XX=.=X===X==X=XX==X   0.79   The indistinguishability
C 340      XXX..==XX===X======X=X..==   0.64   of vector fragment from
C 359      X=X....X..X.=======X=XX=X=   0.67   a mask fragment
C 360      ===..=X==X===X==XXX==X..X=   0.29
C 389      .X.....=............==..X=   0.20
C 390      ===....=..X.====X=X====XX=   0.33
C 10       .=.....=............X=..X=   0.40
C 11       .=.....=............X=..X=   0.40
C 178      ==...==X==X=...=X====X..=X   0.64
C 179      XX.....XX=X=...X===X=XX==X   0.73
C 180      XXX....=..=.X==X===X=X..=X   0.55
           00000000000000000000000000
 C : ZAB   00000000001111111111222222
           01234567890123456789012345
```

Fig. 3. Matrix fragment and mask with values indistinguishability **u** between vectors and mask

from the mask. In this example, the greatest indistinguishability is observed in class C patients (codes 181, 182, 183 and 179).

In the same contingent of pregnant women (as in [8]) with the same division of classes, but with a different symptom set, another problem associated with the risk of preeclampsia was solved [9]. Each patient's vector consisted of 124 logical symptoms, containing information about the anamnesis, blood parameters, concomitant diseases, etc. Particularly, it includes symptoms that reflected the doctor's findings about complications during childbirth, treatment measures in the antenatal period, and complications from diabetes. The result of comparing of combined classes B and C (patients with moderate and high severity) with class Z (healthy) was considered. For this, a new Z:BC mask was built and the indistinguishability of the symptom vectors of each patient from the mask was calculated. Fig. 4 shows the distribution of patient indistinguishability from the new mask for all classes. The two-samples Smirnov test [10] indicates the boundaries that separate the distributions from each other, which allows to check the quality of the preliminary classification at the training stage.

The upper part on Fig. 4 shows the distribution diagrams of indistinguishability **u** for patient classes Z, A, B, C. The general scale for indistinguishability values is placed above the diagram. Below, 4 lines of diagram are placed. For each class a line contains: median indistinguishability values (asterisk), interquartile range (dashes). Under the diagram there is a table of statistical parameters of the distributions for each class: number of patients N, minimum and maximum indistinguishability values, median (med) and quartiles (q1, q2). Interquartile interval of Z-class does not intersect with C-class interval.

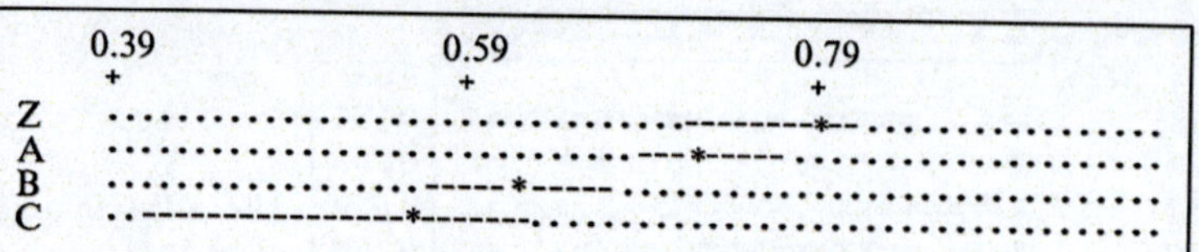

	N	min	q1	med	q2	max
(Z)	60	0.57	0.72	0.79	0.82	0.89
(A)	15	0.56	0.69	0.72	0.78	0.87
(B)	15	0.46	0.57	0.62	0.67	0.70
(C)	5	0.39	0.41	0.57	0.62	0.64

Fig.4. The indistinguishability distribution diagram across all patients classes separation.

These results are shown in Fig. 5, where the patient distributions by classes are presented and the thresholds for separation by class (calculated using the Smirnov criterion [10]) are given. Both Fig. 4 and 5 illustrate good of the heavier classes B and C and the lighter A and healthy Z. It is more important for the doctor to notice the heavier patients. Additional attention to such patients will not harm them.

Thresholds u	<= 0,64	<= 0,72	<= 0.75	>0.75
Z	2	11	4	43
A	2	4	3	6
B	8	7	0	0
C	4	1	0	0

Fig. 5. Patients classification and thresholds separating classes.

We can present one more example of a joint work between mathematicians and a morphologist. A morphologist who works with obstetricians studies the structure of the mother's placenta. The morphologist's conclusion is very important for the treating physician. The disorders in the placenta structure indicate to certain diseases in the patient.

In the articles [11–13] we studied the placentae of 124 women with chronic kidney disease. All pregnant women were divided into 5 classes: class Z – healthy (13 women), A – with a mild chronic kidney disease (70 women), B, C – with severe forms (14 and 22 women), D – with the most severe degree of disease (5 women) (Fig. 6). The study identified 4 important logical symptoms characterizing class D: arterial hypertension, severe preeclampsia, premature maturation of the villi and a low total cross-sectional area of the chorionic villi capillaries (1.7×10^3 μ^2). The patients with severe disease have the area about in 5 times smaller than healthy ones.

Results obtained by the morphologist help the physician adjust the treatment of future patients. Especially for cases when possible fetal hypoxia suspected.

Symp. code	Symptom name	Class	Z	A	B	C	D
		Object number	13	70	14	22	5
028	Obliteration angiopathy		8%	-	-	-	-
034	Intervillous hematoma		31%	-	-	-	-
030	Numerous syncytial nodes		-	37%	-	-	-
041	Heparin therapy		-	41%	-	-	-
033	The basal lamina pathology		-	-	14%	9%	-
035	Placenta infarction		-	-	22%	5%	-
032	Common fibrinoid. Pseudoinfarctions		-	-	28%	27%	-
020	Circumvallate placenta / extrachorial		-	-	-	32%	-
038	Arterial hypertension		-	17%	64%	70%	100%
025	Premature villi maturation		-	3%	-	-	100%
040	Severe preeclampsia		-	-	-	-	40%
S	The total cross-sectional area of capillaries in one villus, 10^3 μ^2		9	3	1.5	2	1.7

Fig.6. The placental symptom distribution in women with chronic kidney disease according to the disease severity

2.6 Some Dynamic Models

The above methods for cognitive problem solving related mainly to static symptoms. Modelling of dynamic processes in living organisms is a more complicated task. Data processing requires large volumes of databases and more complex methods. But in some cases, the dynamic model gives a successful solution for medical problem.

Often and often use of sliding standards helps solve a problem. Below we give some examples.

Obstetricians use the "clinically contracted pelvis" diagnosis when preparing women for childbirth. If the diagnosis was set, the pregnant woman cannot give birth spontaneously. It was necessary to clarify the definition of the term "clinically contracted pelvis". Then a sliding standard of growth of the biparietal size of the fetal head during pregnancy was built (see Fig. 7 [6]). Biparietal (BPR) size is the distance between the two temporal bones of the fetus's head. During the fetus's growth the size of the head increases. And if at the time of birth, according to the sliding standard, the biparietal size exceeds the sizes of the mother's pelvis, spontaneous childbirth is impossible. In Fig. 7 BPR medians, based on two-week measurements, are depicted by black belts. The light rectangles near the belts indicate quartile boundaries. The solid central line is the median dependence on gestational age. The upper and lower solid lines are the quartile lines. Using this sliding standard, we can count BPR at the moment of delivery and compare it to pelvis sizes. If BPR is larger than some sizes of the pelvis the woman cannot give birth spontaneously. Thus, an exact mathematical criterion of the "clinically contracted pelvis" concept was obtained. Described standard used only numeric variables.

The concept of indistinguishability in cognitive tasks has another important property. It can be used to construct informative sliding standards in the space of heterogeneous logical variables.

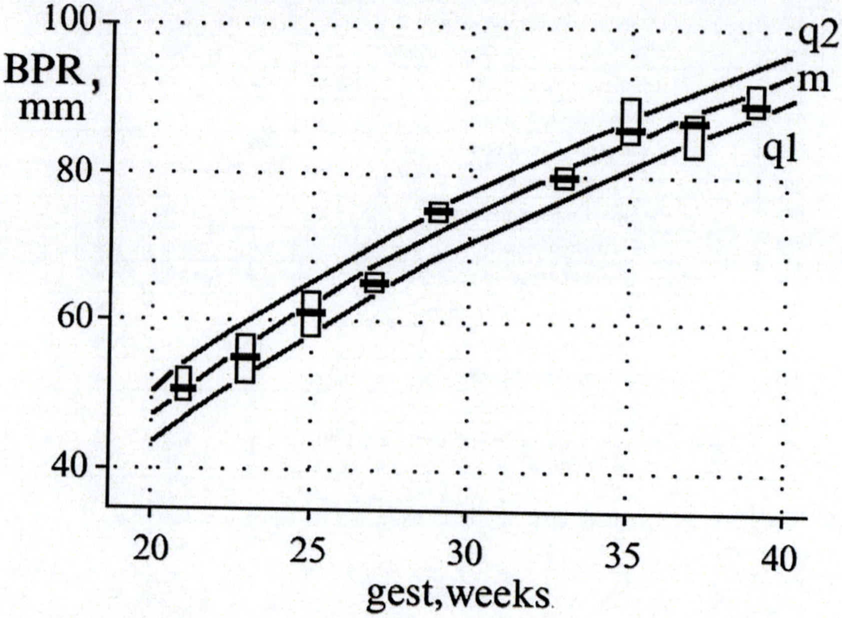

Fig. 7. Sliding standard for BPR fetal head depending on gestational age

We analyzed some dynamic processes during pregnancy using the sliding standards in the space of logical symptoms. Thus, in [8] ultrasound measurements of the fetus were carried out for pregnant women with preeclampcia threat. The matrix of logical symptoms was built for 89 patients divided in 4 classes and mask C: ZAB was created: class C of severe patients versus combined classes Z (healthy), A and B (with milder degrees of disease). Article [8] gives a calculated sliding standard for indistinguishability between patients and mask. A prognostic rule was obtained: if the patient's indistinguishability percentile is greater than 93 units during pregnancy, the patients belong to heaviest class.

In [14] was formulated the prediction rule of delivery result for mother with type 2 diabetes. The symptom matrix of 86 patients contained 374 vectors with 262 logical symptoms each. Women were a priori divided into 4 classes. Class A (60 persons, 249 vectors) was formed by mothers who gave birth to healthy children. Class B (11 persons, 64 vectors) consisted from the mothers with children having minor deviations in health. Class C (7 persons, 25 vectors) was formed by mothers with more serious deviations. Class D (8 persons, 36 vectors) – mothers who gave birth to seriously ill children with severe disorders of the central nervous system.

In this problem, for the 374 × 262 matrix, a CD:AB mask was constructed for unify class pair C and D with the most severe complications in children against united classes A and B for almost healthy ones. In Fig. 8, a sliding standard of indistinguishability for all patients of classes C and D from this mask is constructed. Medians and quartiles of distributions are plotted at two-week intervals. It is curious that the maximum indistinguishability of women in classes C and D from the mask corrtsponds to 28–30 weeks of pregnancy.

Thus, we found that this interval is the critical. Early in pregnancy there are usually few measurements (patients rarely come to the clinic). But in the period of 26–33 weeks, the threat to the condition of the fetus is growing. Doctors begin to treat women intensively, and the indistinguishability gradually decreases. Fig. 8 clearly shows the success of the treatment. The predictive results of the study are given in [14].

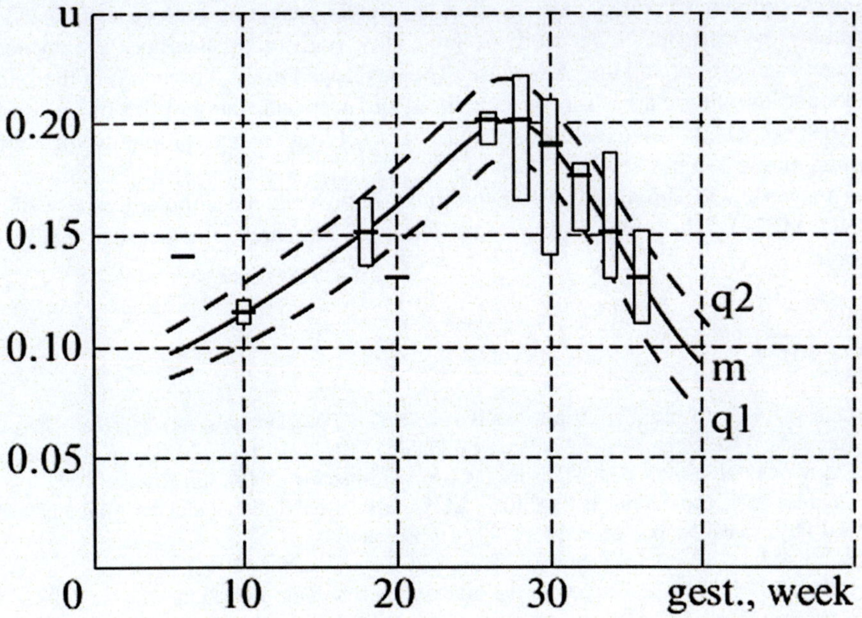

Fig. 8. Sliding standard for indistinguishability of patient's classes C and D from mask CD:AB

And one more remark: disease is a dynamic process. As the disease develops, many properties of the organism can change. Recently, we mainly worked with obstetricians. Pregnancy is not a disease, but it is a vivid example of the high variability of the organism. As the fetus develops, the mother's organism is rebuilt radically. Hormones are produced, there are reactions to immunological and pathological disturbances. These changes lead to both the successful result of the entire organism restructuring, and to possible complications. If these changes are taken into account and the patient is examined in different phases of the disease, we can get a picture of the dynamics of pregnancy in different cases. An example of such a study arose when we worked with physicians who were leading the pregnant women with type 1 diabetes. It turned out that some of these pregnancies end in the death of the fetus or newborn. A deeper study of the dynamics of pregnancy using our method showed that these mothers had a small short-term disturbance of the blood coagulation system during pregnancy. This disturbance could be overlooked and then disappear. But for majority of patients who had it, the pregnancy ended in the death of the child [6].

3 Conclusion

The proposed data analysis technology has a clear structure designed to solve the cognitive problems of medicine – to extract the implicit knowledge of an experienced doctor and formulate diagnostic rules. The new medical research creates a dictionary containing terms and unambiguous interpretation of their contents. The patient's condition is presented in the language of logical symptoms, processing of which is carried out by the methods of non-numerical statistics and three-valued logic. The cognitive problem of patient classification is solved by creating prototypes (masks) for each class of patients. A measure of proximity (indistinguishability) is introduced between each patient and the class mask. The developed tools allow varying the sets of masks and dividing classes into groups in accordance with the purpose of the study.

The proposed language is suitable for solving many problems of object classification.

The work was supported by the program improving the competitiveness of the NRNU MEPhI. The authors thank the National Research Nuclear University for support.

References

1. Giltsov, I.V., Semenov, V.M., Zenkova, S.K.: Fundamentals of Medical Statistics Biomedical Research Design: A Practical Guide. VSMU, Vitebsk (2014). (in Russian)
2. Orlov, A.I.: Non-numerical Statistics. MZ-Press, Moscow (2004). (in Russian)
3. Gelfand, I.M., Rosenfeld, B.I., Shifrin, M.A.: Essays on Collaboration of Mathematicians and Physicians. Nauka, Moscow (1989). (in Russian)
4. Solso, R.L.: Cognitive Psychology. Allyn and Bacon, Boston (2011)
5. Gelfand, I.M.: Two archetypes in the mankind psychology. Ecol. Life **2**, 8–15 (1981). (in Russian)
6. Kotov, Yu.B.: New Mathematical Approach to Medical Diagnostic Problems. Editorial URSS, Moscow (2004). (in Russian)
7. Karpenko, A.S.: Lukasiewicz's Logic and Prime Numbers. Nauka Publisher, Moscow (2000). (in Russian)
8. Guryeva, V.M., Kotov, Yu.B., Morohotova, L.S., Semenova, T.A.: Classification of patients with gestosis threat by ultrasound measurements. Vestnik NIYaU MIFI **6**, 548–554 (2017). https://doi.org/10.1134/S2304487X17060025. (in Russian)
9. Guryeva, V.M., Kotov, Yu.B., Morohotova, L.S., Semenova, T.A.: Technology of mask formation in the method of logical symptoms. Vestnik NIYaU MIFI. **6**, 368–374 (2017). https://doi.org/10.1134/S2304487X17040058. (in Russian)
10. Orlov, A.I.: Nonparametric criteria for the agreement of Kolmogorov, Smirnov. Omega-square and errors in their application. Res. J. KubSAU. **97**(03), 1–29 (2014). (in Russian)
11. Barinova, I.V., Guryeva, V.M., Kotov, Yu.B., Semenova, T.A.: Mathematical methods for structuring and formalization of medical experience. Procedia Comput. Sci. **145**, 84–88 (2018). https://doi.org/10.1016/j.procs.2018.11.014. (in Russian)
12. Barinova, I.V., Nikolskaja, I.G., Kotov, Yu.B., Prokopenko, E.I., Kondrikov, N.I.: Placenta in pregnant women with chronic kidney disease. Obstet. Ginecol. **1**, 169–177 (2020). https://doi.org/10.18565/aig.2020.1.169-177. (in Russian)

13. Kotov, Yu.B., Barinova, I.V., Nikolskaja, I.G., Semenova, T.A.: Statistical analysis of the relationship between placenta condition at chronic kidney disease. Vestnik NIYaU MIFI **9**, 262–269 (2020). https://doi.org/10.1134/S2304487X20020078. (in Russian)
14. Barinova, I.V., Guryeva, V.M., Kotov, Y.B., Morohotova, L.S., Semenova, T.A.: Mathematical forecasting of the newborn's poor state for mothers with type 2 diabetes. Vestnik NIYaU MIFI **8**, 179–187 (2019). https://doi.org/10.1134/S2304487X19020032. (in Russian)

Brain Cognitive Architectures Mapping for Neurosurgery: Resting-State fMRI and Intraoperative Validation

M. Sharaev[1(✉)], T. Melnikova-Pitskhelauri[2], A. Smirnov[2], A. Bozhenko[1], V. Yarkin[1], A. Bernshtein[1], E. Burnaev[1], P. Petrov[1], D. Pitskhelauri[2], V. Orlov[3], and I. Pronin[2]

[1] Skolkovo Institute of Science and Technology, Moscow, Russia
m.sharaev@skoltech.ru

[2] Federal State Autonomous Institution "N.N. Burdenko National Medical Research Center of Neurosurgery" of the Ministry of Health of the Russian Federation, Moscow, Russia

[3] National Research Center "Kurchatov Institute", Moscow, Russia

Abstract. Despite the importance of experimental confirmation, the ability of wide range of brain mapping methods to discover brain cognitive architectures in most studies can't be evaluated directly. Only in rare cases, when due to medical need, it is possible to conduct experiments during neurosurgical operations, is it possible to assess the accuracy of certain approaches. In this paper we evaluate, how well we can reveal brain cognitive architectures structure by established and novel approaches based on fMRI data, with special focus on resting-state fMRI, and how well these findings match Direct Cortical Stimulation mapping data, which is a gold standard in neurosurgery. We illustrate our approach on three representative examples with different motor and cognitive architectures mapping. We found a significant correspondence between predicted maps and intraoperative data for both brain networks. This indicates that resting-state fMRI could be used as an additional source of information for neurosurgical planning, though its applicability to exploring and describing the whole variety of brain architectures for research purposes should be investigated in future.

Keywords: Cognitive architectures · Brain mapping · Constrained ICA · Resting-state fMRI · Neurosurgery

1 Introduction

Nowadays, there is much evidence that large-scale neuronal networks are responsible for brain functions, instead of separate local regions as was thought before. This includes low-level sensory functions, as well as motor, speech, memory and cognition. Thus, the ability to discover, explore and describe different properties of brain neuronal networks is essential for contemporary neuroscience and fields of knowledge that draw information from it. One example is the field of creating cognitive architectures based on biological prototypes.

A. V. Samsonovich et al. (Eds.): BICA 2020, AISC 1310, pp. 466–471, 2021.
https://doi.org/10.1007/978-3-030-65596-9_55

There are different methods of "brain mapping", i.e. localizing distinct brain areas, connections between them [1], activation patterns [2] and information flow [3] in different states of consciousness or cognitive tasks. These methods are usually based on experimental data like electroencephalography (EEG), functional MRI (fMRI) and others. However, the ability of brain mapping methods to discover functional architectures in the brain usually can't be evaluated directly. For example, if a new method of brain mapping based on (indirect) resting-state fMRI (rs-fMRI) data is developed, in most cases it can only be compared with existing brain mapping methods based on (also indirect) resting-state or task-based fMRI (tb-fMRI), see for example [2]. Only in a few cases, for example, in neurosurgical practice, we can assess the precision of developed mapping method directly. This could be done by Direct Cortical Stimulation mapping (DCS mapping) technique – an invasive procedure of direct stimulation of brain cortex and measuring the response in different tasks, like motor, visual, language and others. Here brain motor and cognitive architectures are mapped directly, sometimes with white matter tracts connecting distinct cortical areas. Moreover, the knowledge of ground truth, i.e. the real functional architecture structure, might help to improve methods or predictive models for brain mapping based on non-invasive imaging techniques. Thus, working with neurosurgical cases is beneficial both for clinics (we develop and improve new methods of pre-surgical planning) as well as for theoretical and applied neuroscience (we can make direct inferences on cognitive architectures based on invasive data).

In this paper we evaluate, how well we can predict tb-fMRI activation pattern (i.e. conventional method for brain eloquent areas mapping in neurosurgery) based on resting-state fMRI data and how well both predictions match DCS mapping data, i.e. could these techniques be used in neurosurgical practice. We illustrate our approach on three representative examples with different brain architectures mapping: motor network and language network, namely Broca and Wernicke areas.

2 Methods

2.1 Subjects and Data

We examined 3 patients with intracerebral tumors (gliomas) located near brain eloquent areas: brain motor area (patient A), Broca (patient B) and Wernicke (patient C) areas. Patients underwent brain scanning at MRI 3.0T Signa HDxt (GE). Tb-fMRI was conducted using a standard block paradigm for the identification of language and motor areas, during rs-fMRI scanning patients were asked to keep eyes closed and try not to think of anything in particular.

All patients underwent axial FSPGR BRAVO T1-weighted (thickness 1 mm for structural MRI). For the tb-fMRI protocol T2*-gradient echo-planar sequence was used with the following parameters: TR/TE/FA-3000 ms/35 ms/90o, FOV-240, matrix size 64×64, slice thickness 3, gap 0, in-plane pixel size 2,9 mm $\times$ 2,9 mm and axial slices 50. We used well-known tapping and language tests for tb-fMRI. The block tb-fMRI was repeated 5 times (5-with task, 5-rest, $t = 30$ s) for each patient. For rs-fMRI we used the T2*-gradient echo-planar sequence with the following parameters

TR/TE/FA-2500 ms/30 ms/90°. Matrix size 64 × 64, FOV-240 mm, slice thickness 3, no gap, in-plane pixel size 2,9 mm × 2,9 mm and axial slice number - 50, slice order - interleaved mode and time acquisition for rs-fMRI - 12,5 min.

2.2 Data Preprocessing and Analysis

In tb-fMRI study, activation zones were obtained on a General Electric AW VolumeShare 5 workstation in the BrainWavePA 2.0.00 program (GE Healthcare, USA). For each fMRI sample, a threshold for significance level was manually selected by an experienced neuroradiologist. Then, the obtained images of the activation zones were saved in DICOM format and exported to a PC with MATLAB v.2017b (MathWorks, USA). Activation zones and their corresponding structural brain images were exported in NIFTI format for further analysis.

In rs-fMRI study, functional images were preprocessed with a common pipeline, including slice-timing correction, bad slices and all spike noise suppression was used Artrepair toolbox for spm (http://cibsr.stanford.edu/tools/humanbrain-project/artrepair-software.html), motion artifacts correction using a least square approach and a 6 parameter (rigid body) spatial transformation. Finally, fMRI data were normalized to bring structural and functional images to the coordinates of the MNI. Pre-processed fMRI data were passed to GIFT software [4] to create both blind and semi-blind spatially-constrained ICA [5] for rs-fMRI. The details of constrained ICA analysis could be found in [6]. Brain functional architectures were derived from independent components, and their spatial maps were analyzed. To measure similarity between tb-fMRI and rs-fMRI results we used Dice coefficient, defined by formula:

$$D = \frac{2|X \cap Y|}{|X| \cup |Y|} \tag{1}$$

where X refers to active voxels of the first map, Y active voxels of the second map. Thus, Dice coefficient shows the degree of overlap between active voxels in both maps. It is worth noting, that the overlap between the rs-fMRI and tb-fMRI activation maps was evaluated within the cerebral cortex binary mask without taking into account subcortical structures and white matter. In order to calculate Dice coefficient, spatial maps should be binarized, i.e. each voxel is either active 1 or not active 0. For tb-fMRI the binarization threshold was selected by experienced neuroradiologist based on standard clinical pipeline, but for rs-fMRI this threshold is not known a priori. In order to show the dependence of Dice coefficient on the threshold and find its optimal value, we plot Dice values as a function of threshold for a wide range of values. Finally, as intraoperative DCS-positive mapping points reveal true activation areas, we evaluate distances between them and both fMRI-derived maps.

3 Results and Discussion

The results for three patients are shown on Fig. 1. In all cases, the Dice graph has a clear maximum value, and individual patient maps are shown with corresponding optimal threshold. Optimal thresholds are very close for motor and Broca areas maps, (2.13 and 2.40 correspondingly) for Wernicke area there are several close to optimal thresholds ranging from 0.7 to 2.0 with maximum Dice in 1.0 point.

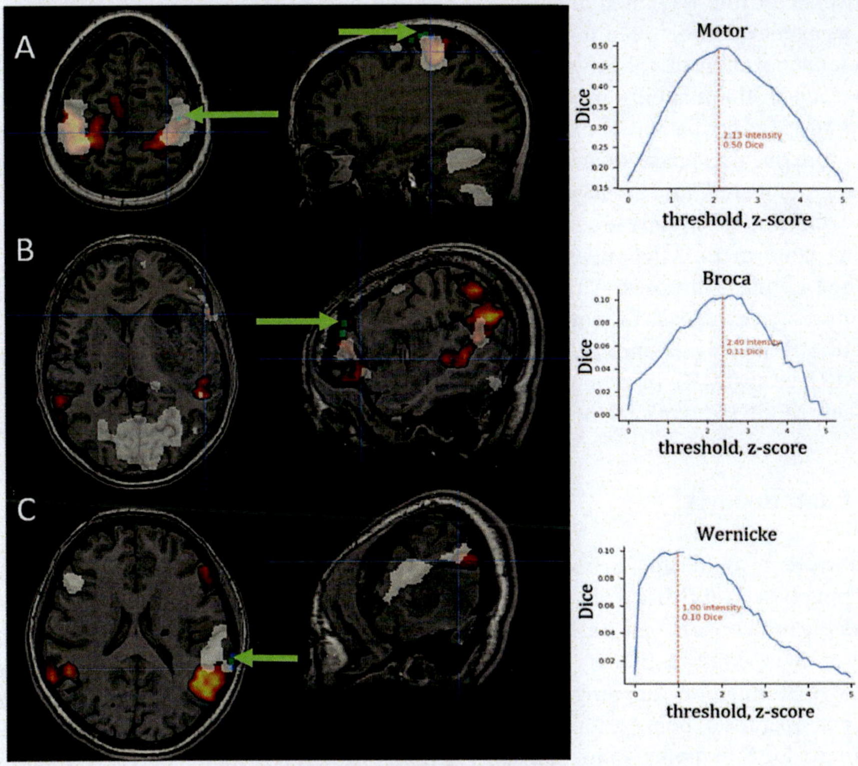

Fig. 1. Comparison between tb-fMRI, rs-fMRI maps and DCS-positive mapping points. *To the left*: A) Patient A, brain motor network mapping; B) Patient B, brain language network mapping (Broca); C) Patient C, brain language network mapping (Wernicke). Tb-fMRI maps are in grey color; rs-fMRI maps are shown by red heatmaps; green arrows indicate green DCS-positive mapping points. *To the right*: corresponding Dice plots, optimal thresholds are marked.

It can be seen both from spatial maps and Dice graph, that for patient A the overlap between tb-fMRI and rs-fMRI maps is pretty high, however, rs-fMRI map reveals additional structures of brain motor network, for instance, part of supplementary motor area. From the other side, tb-fMRI shows activity in cerebellum, which might be relevant in terms of brain motor network architecture, but cannot be evaluated in patient A with DCS mapping due to cortical tumor localization. Two out of three DCS-positive mapping points fall into both tb-fMRI and rs-fMRI maps, the third point (see sagittal view on Fig. 1) is less than 10 mm from predicted activation. This result confirms good predictive ability of rs-fMRI with semi-blind ICA for brain motor network localization.

Language network maps obtained from tb-fMRI and rs-fMRI do not show high level of concordance, which is in line with many previous studies [7, 8]. However, in patient B, rs-fMRI activation is perfectly confirmed by DCS mapping of Broca area, while tb-fMRI activity is somewhat far from the upper DCS-positive mapping point (green arrow on saggital view, see Fig. 1). In patient C the DCS-positive mapping point falls directly into the small area

of intersection between tb-fMRI and rs-fMRI maps adjacent to Wernicke area. This result cannot indicate the advantage of one or another method, but shows their general adequacy in the mapping problem. A small intersection of maps also can be seen on saggital view for patient C exactly in the Wernicke area.

From our three case studies it can be seen that maps for motor brain network being denser compared to language networks, are characterized with higher Dice value. Nevertheless, true activation represented by intraoperative DCS-positive mapping points confirm rs-fMRI maps suggesting the potential applicability of rs-fMRI for applied clinical purposes. This also means that to some extent brain network architectures, at least in case of motor and language functions, could be mapped based on rs-fMRI. Still it remains unclear, whether it could be possible to intraoperatively validate rs-fMRI or tb-fMRI mappings of other brain cognitive architectures, primarily, the Default Mode Network (DMN) which is of special interest in cognition studies.

4 Conclusions

This work is dedicated to the analysis of brain motor and cognitive architectures mapping based on fMRI with specific tasks and on resting-state fMRI. Here we showed the degree of concordance between spatial maps obtained from tb-fMRI and rs-fMRI as well as their confirmation by the gold standard for intraoperative mapping – Direct Cortical Stimulation mapping. DCS mapping can be performed only during neurosurgical operation, being an invaluable source of information about the functioning of the brain. DCS mapping validation showed that obtained rs-fMRI maps correctly reveal spatial information about brain networks of interest and in some practical cases the quality of rs-fMRI maps is enough to perform initial preoperative planning.

5 Funding Information

The reported study was supported by RFBR according to the research project 18-29-01032

References

1. Friston, K.J.: Functional and effective connectivity: a review. Brain Connect. **1**, 13–36 (2011). https://doi.org/10.1089/brain.2011.0008
2. Tavor, I., Jones, O.P., Mars, R.B., Smith, S.M., Behrens, T.E., Jbabdi, S.: Task-free MRI predicts individual differences in brain activity during task performance. Science (80-.)**352**, 216–220 (2016). https://doi.org/10.1126/science.aad8127
3. Sharaev, M., Orlov, V., Ushakov, V.: Information transfer between rich - club structures in the human brain. Procedia Comput. Sci. **123**, 440–445 (2018). https://doi.org/10.1016/j.procs.2018.01.067
4. Calhoun, V.D., Adali, T., Pearlson, G.D., Pekar, J.J.: A method for making group inferences from functional MRI data using independent component analysis. Hum. Brain Mapp. **14**, 140–151 (2001). https://doi.org/10.1002/hbm.1048
5. Lin, Q.H., Liu, J., Zheng, Y.R., Liang, H., Calhoun, V.D.: Semiblind spatial ICA of fMRI using spatial constraints. Hum. Brain Mapp. **31**, 1076–1088 (2010). https://doi.org/10.1002/hbm.20919
6. Sharaev, M., Smirnov, A., Melnikova-Pitskhelauri, T., Orlov, V., Burnaev, E., Pronin, I., Pitskhelauri, D., Bernstein, A.: Functional Brain Areas Mapping in Patients with Glioma Based on Resting-State fMRI Data Decomposition. In: 2018 IEEE International Conference on Data Mining Workshops (ICDMW), pp. 292–298. IEEE (2018). https://doi.org/10.1109/ICDMW.2018.00049
7. Lu, J., Zhang, H., Hameed, N.U.F., Zhang, J., Yuan, S., Qiu, T., Shen, D., Wu, J.: An automated method for identifying an independent component analysis-based language-related resting-state network in brain tumor subjects for surgical planning. Sci. Rep. **7**, 1–16 (2017). https://doi.org/10.1038/s41598-017-14248-5
8. Smirnov, A.S., Sharaev, M.G., Melnikova-Pitskhelauri, T. V., Zhukov, V.Y., Bikanov, A.E., Sharova, E.V., Pogosbekyan, E.L., Turkin, A.M., Fadeeva, L.M., Pitskhelauri, D.V., Kornienko, V.N., Pronin, I.N.: Resting state fMRI in pre-surgical brain mapping. Literature review. Med. Vis. 6–13 (2018). https://doi.org/10.24835/1607-0763-2018-5-6-13

Machine Learning Based on the Principle of Minimizing Robust Mean Estimates

Z. M. Shibzukhov[1,2](✉)

[1] Institute of Mathematics and Informatics of Moscow Pedagogical State University, Moscow, Russia
intellimath@mail.ru
[2] Institute of Physics and Technology, Moscow, Russia

Abstract. The classical approach in machine learning, based on the principle of minimizing the arithmetic mean of parameterized functions, is vulnerable if training is carried out on the basis of data containing outliers. The article considers the approach to the construction of robust methods and machine learning algorithms, which are based on the principle of minimizing estimates of average values that are insensitive to outliers. Proposed machine learning algorithms are based on the principle of iterative reweighting. Illustrative examples show the ability of the proposed approach and algorithms to overcome the effect of outliers.

Keywords: Machine learning · Robust estimate · Iteratively reweigting · Pattern recognition · Data analysis

1 Introduction

Machine learning tasks are often formulated as tasks of minimizing the weighted sums of parameterized functions:

$$\mathcal{Q}(\mathbf{w}) = \sum_{k=1}^{N} v_k \ell_k(\mathbf{w}), \tag{1}$$

where $\mathcal{Q}$ is the minimized target function, $\mathbf{w}$ is unknown parameter vector of the model to be optimized, $v_1, \ldots, v_N$ are sample weights, $\ell_1, \ldots, \ell_N$ are parameterized basic functions that evaluate the contribution of the objects of the training sample. If the sample weights are not specified, then $v_1 = \cdots = v_N = 1/N$ is usually selected. In this case

$$\mathcal{Q}(\mathbf{w}) = \frac{1}{N} \sum_{k=1}^{N} \ell_k(\mathbf{w}). \tag{2}$$

The optimal parameter vector $\mathbf{w}^*$ should minimize the objective function $\mathcal{Q}$:

$$\mathcal{Q}(\mathbf{w}^*) = \min_{\mathbf{w}} \mathcal{Q}(\mathbf{w}). \tag{3}$$

A. V. Samsonovich et al. (Eds.): BICA 2020, AISC 1310, pp. 472–477, 2021.
https://doi.org/10.1007/978-3-030-65596-9_56

1.1 Outliers Problem

In the standard setting with minimizing the *arithmetic mean* (AM) (2), a problem may arise if the empirical distribution of the values of the basis functions $\{z_1, \ldots, z_N\}$, where $z_k = \ell_k(\mathbf{w})$ $(1 \leqslant k \leqslant N)$, contains outliers. This can usually happen for the following reasons:

1) the training sample contains highly distorted data, which makes them inadequate in relation to the problem being solved;

2) the training sample contains data that are substantially inadequate for the selected model.

Under these conditions, minimization of the objective function $\mathcal{Q}(\mathbf{w})$, constructed over the entire training sample, usually leads to a distortion of the desired model parameters. This is due to the fact that the AM is not a robust estimate of the mean with respect to outliers. There are some classic approaches to overcome influence of the outliers on the base of robust statistics [1,2]. Here we propose a different approach.

2 The Principle of Minimizing Robust Average Estimates

To overcome this issue, it is advisable to use estimates of the average value, which are insensitive to outliers. Moreover, it is highly desirable that such estimates be differentiable. Otherwise, it becomes impossible to use gradient minimization procedures for the function $\mathcal{Q}$.

Let $\mathsf{M}\{z_1, \ldots, z_N\}$ be one of such estimates of the average value, insensitive to outliers. The function $\mathcal{Q}$ is defined as follows [3,4]:

$$\mathcal{Q}(\mathbf{w}) = \mathsf{M}\{\ell_1(\mathbf{w}), \ldots, \ell_N(\mathbf{w})\}. \tag{4}$$

The gradient $\mathcal{Q}$ will look like:

$$\nabla \mathcal{Q}(\mathbf{w}) = \sum_{k=1}^{N} \frac{\partial \mathsf{M}}{\partial z_k} \nabla \ell_k(\mathbf{w}),$$

where $z_k = \ell_k(\mathbf{w})$. Computing $\partial \mathsf{M}/\partial z_k$ is usually not cheap. Therefore, it is advisable to use the method of *iterative reweighting*.

It builds the sequence $\mathbf{w}_t$ $(t = 0, 1, 2, \ldots)$. At the beginning, $\mathbf{w}_0$ is selected. The rest $\mathbf{w}_t$ $(t > 0)$ are calculated according to the following rule:

$$\mathbf{w}^{t+1} = \arg\min_{\mathbf{w}} \sum_{k=1}^{N} v_k^t \ell_k(\mathbf{w}),$$

where $v_k^t = \partial \mathsf{M}\{z_1^t, \ldots, z_N^t\}/\partial z_k$, $z_k^t = \ell_k(\mathbf{w}^t)$.

2.1 Outliers Insensitive Average Estimates

Almost all known functions for estimating the average value can be defined as the solution to the problem of minimizing the penalty function:

$$\bar{z} = \mathsf{M}\{z_1, \ldots, z\} = \arg\min_u \mathcal{S}(z_1 - u, \ldots, z_N - u), \tag{5}$$

where $\mathcal{S}(x_1, \ldots, x_N)$ is a convex penalty function (penalty for deviation from the average value). It should be noted that in the general case, the set of values of u at which the minimum is reached in (5) can be a connected set with the boundaries $a < b$. In such cases, by convention, the value $(a + b)/2$ is taken as the value $\mathsf{M}\{z_1, \ldots, z_N\}$.

You can use the iterative procedure to find the value $\bar{z}$:

$$u_{t+1} = \frac{\sum_{k=1}^N \varphi_k(z_1 - u_t, \ldots, z_N - u_t) z_k}{\sum_{k=1}^N \varphi_k(z_1 - u_t, \ldots, z_N - u_t)},$$

where $\varphi_k(x_1, \ldots, x_N) = \frac{1}{x_k} \frac{\partial \mathcal{S}(x_1, \ldots, x_N)}{\partial x_k}$. Usually $\varphi_k(x_1, \ldots, x_N) \geqslant 0$.

As a result, the average value $\bar{z}$ can be represented as a weighted arithmetic mean:

$$\bar{z} = \sum_{k=1}^N v_k(z_1, \ldots, z_N) z_k,$$

where $v_k(z_1, \ldots, z_N) = \dfrac{\varphi_k(z_1 - \bar{z}, \ldots, z_N - \bar{z})}{\sum_{l=1}^N \varphi_l(z_1 - \bar{z}, \ldots, z_N - \bar{z})}$.

Based on this view, we can construct another version of the iterative reweighting method to minimize $\mathcal{Q}$:

$$\mathbf{w}^{t+1} = \arg\min_{\mathbf{w}} \sum_{k=1}^N v_k^t \ell_k(\mathbf{w}),$$

where $v_k^t = v_k(z_1^t, \ldots, z_N^t)$, $z_k^t = \ell_k(\mathbf{w}^t)$.

Next, consider various estimations of mean insensitive to outliers.

M-averages. Let $z_1, \ldots, z_N$ be a finite set of real numbers. The center point $\bar{z}_\rho = \mathsf{M}_\rho\{z_1, \ldots, z_N\}$ is defined as follows:

$$\mathsf{M}_\rho\{z_1, \ldots, z_N\} = \arg\min_u \sum_{k=1}^N \rho(z_k - u), \tag{6}$$

where $\varrho(r)$ is a convex function such that $\min \rho(r) = \rho(0) = 0$.

For example,

1) if $\rho(r) = r^2$, then $\bar{z}_\rho$ is arithmetic mean value;

2) if $\rho(r) = (\alpha - [\![r < 0]\!])r$, then $\bar{z}_\rho$ is value of α-quantile[1] (median for $\alpha = 0.5$).

[1] If the logical expression S is true, then $[\![S]\!] = 1$, otherwise $[\![S]\!] = 0$.

If $\varrho(r)$ is twice differentiable, then $\mathsf{M}_\rho\{z_1,\ldots,z_N\}$ has partial derivatives:

$$\frac{\partial \mathsf{M}_\rho}{\partial z_k} = \frac{\rho''(z_k - \bar{z})}{\rho''(z_1 - \bar{z}) + \cdots + \rho''(z_N - \bar{z})}.$$

Smoothly Median Analogues. Using a median instead of AM helps to cope with the problems of fitting linear regression with a large number of outliers (up to 50%) in the data. Gradient-type algorithms are not applicable in this case, since the partial derivatives of the median have properties unsuitable for gradient algorithms. Smoothed versions of the median with continuous partial derivatives can be defined using the parameterized functions $\rho_\varepsilon(z)$, such that:

1) $\lim\limits_{\varepsilon\to 0} \rho_\varepsilon(z) = |z|$; 2) $\lim\limits_{\varepsilon\to 0} \rho'_\varepsilon(z) = \operatorname{sign} z$; 3) $\lim\limits_{\varepsilon\to 0} \rho''_\varepsilon(z) = \delta(z)$.[2]

For example, $\rho_\varepsilon(r) = \sqrt{\varepsilon^2 + r^2} - \varepsilon$.

In order to define smoothed variants of α-quantiles let's define

$$\rho_{\varepsilon,\alpha}(r) = \alpha\rho_\varepsilon(r)[\![r \geqslant 0]\!] + (1-\alpha)\rho_\varepsilon(r)[\![r < 0]\!].$$

Smoothly Winzorized Averages. Winzorized AM is a variant of the AM, which is less sensitive to outliers. Since the values of the basic functions are non-negative and large residual values are associated with outliers, in this case it can be considered as the AM, when calculating which values exceeding $\bar{z}_\alpha$ $(\alpha > 0.5)$ are replaced by $\bar{z}_\alpha$. We will replace the value of the α-quantile$\bar{z}_\alpha$ with a smoothed analogue:

$$\mathsf{WM}_{\rho_\alpha}\{z_1,\ldots,z_N\} = \frac{1}{N}\sum_{k=1}^{N} \max\{z_k, \bar{z}_{\rho_\alpha}\},$$

where ρ_α is the dissimilarity function for constructing a smoothed version of the α-quantile. Partial derivatives are of the form:

$$\frac{\partial \mathsf{WM}_{\rho_\alpha}}{\partial z_k} = \frac{1}{N}[\![z_k \leqslant \bar{z}_{\rho_\alpha}]\!] + \frac{m}{N}\frac{\partial \mathsf{M}_{\rho_\alpha}}{\partial z_k},$$

where m is the number value $z_k > \bar{z}_{\rho_\alpha}$.

Smoothly Truncated Averages. Let's define

$$\tilde{z}_\alpha = \arg\min_u \Big\{ \sum_{|z_k - u| \leqslant c_\alpha} (z_k - u)^2 + \sum_{|z_k - u| > c_\alpha} c_\alpha^2 \Big\},$$

where $c_\alpha^2 = \mathsf{M}_{\rho_\alpha}\{(z_1 - u)^2, \ldots, (z_N - u)^2\}$.

From the definition we get a recurrence relation for calculating $\tilde{z}_\alpha$:

$$u_{t+1} = \frac{1}{N}\sum_{k=1}^{N}\Big([\![|z_k - u_t| \leqslant c_{\alpha,t}]\!] + m\frac{\partial \mathsf{M}_{\rho_\alpha}}{\partial v_{k,t}}\Big) z_k,$$

[2] Dirac's δ-function.

where $c_{\alpha,t}^2 = \mathsf{M}_{\rho_\alpha}\{(z_1-u_t)^2,\ldots,(z_N-u_t)^2\}$, m is the number of values $z_k : |z_k - u_t| > c_{\alpha,t}$.

In the limit, we obtain the weighted AM

$$\tilde{z}_\alpha = \frac{1}{N}\sum_{k=1}^{N}\Big([\![|z_k - \tilde{z}_\alpha| \leqslant \tilde{c}_\alpha]\!] + m\frac{\partial \mathsf{M}_{\rho_\alpha}}{\partial v_k}\Big) z_k,$$

where $\tilde{c}_\alpha^2 = \mathsf{M}_{\rho_\alpha}\{(z_1-\tilde{z}_\alpha)^2,\ldots,(z_N-\tilde{z}_\alpha)^2\}$.

In this situation, to minimize the objective function (4), it is advisable to use the following version of the iterative reweighting method:

$$\mathbf{w}_{t+1} = \arg\min_{\mathbf{w}} \sum_{k=1}^{N}\Big(\frac{1}{N}[\![|z_{k,t} - \tilde{z}_{\alpha,t}| \leqslant \tilde{c}_{\alpha,t}]\!] + \frac{m}{N}\frac{\partial \mathsf{M}_{\rho_\alpha}}{\partial v_k}\Big)\ell_k(\mathbf{w}),$$

where $\tilde{c}_{\alpha,t}^2 = \mathsf{M}_{\rho_\alpha}\{(z_{1,t}-\tilde{z}_{\alpha,t})^2,\ldots,(z_{N,t}-\tilde{z}_{\alpha,t})^2\}$, $z_{k,t} = \ell_k(\mathbf{w}_t)$.

3 Illustrative Examples

Linear Regression

The data sets correspond to almost the dependence $y = 4x + 1$ with little noise. In addition, outliers are added, so that they are 29%, 38%, 49% of the total number of points, respectively. Outliers are added so that the classic robust linear regression methods will obviously produce distorted results.

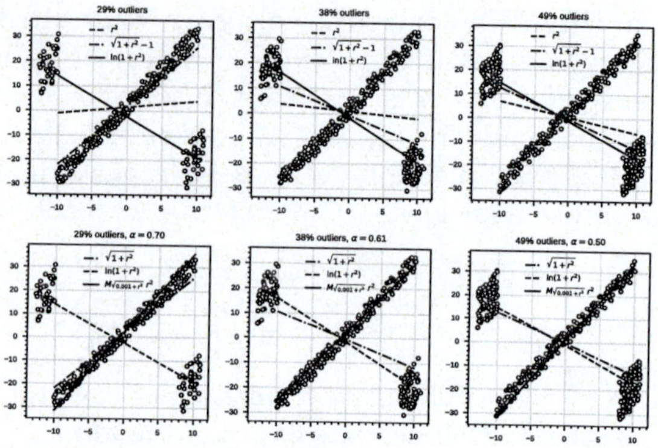

Clustering Example

Consider datasets S3–S4 for clustering from [5,6]. They contain 5000 points, 15 clusters. On each figure, on the left side there is the result of the robust algorithm, and on the right side there is the classical one. During the training, a robust mean estimate was used with $\tilde{z}_\alpha$, $\varepsilon = 0.001$, $\alpha = 0.96-0.97$. It is easy to see that the robust algorithm allows one to find more adequate positions of the centers of clusters and the shape of the variance matrices.

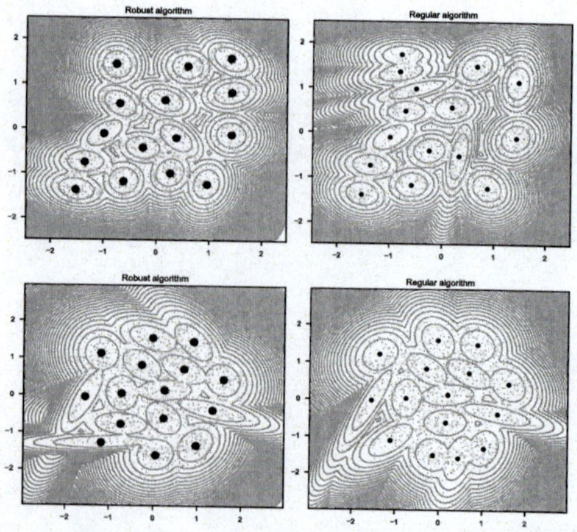

Classification

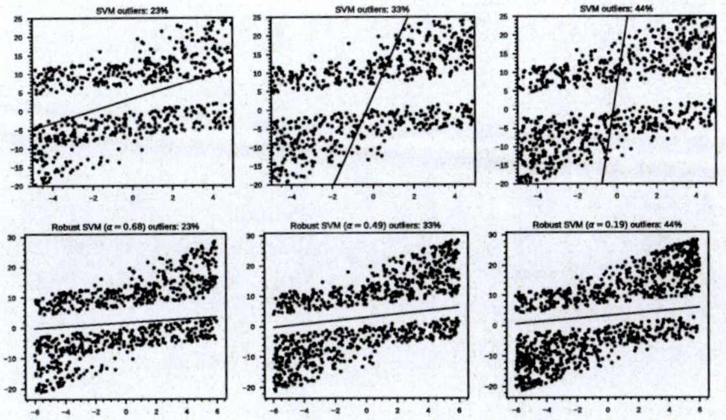

Two classes are generated in such a way that the standard SVC method obviously leads to the rotation of the line separating the classes. A robust version of the SVC using a smoothed quantile, instead of the arithmetic mean allows you to get a line that confidently separates the classes.

Conclusion

Thus, a new approach to the construction of robust machine learning algorithms is proposed, which is based on replacing the arithmetic mean in the objective function with a differentiable estimate of the average value that is insensitive to outliers. The proposed mathematical formulation of the problem initially lays the foundation for the robustness of solving the problem of finding the optimal parameters of the model. The proposed scheme of iterative reweighing differs in that the sample weights are calculated as partial derivatives of a robust estimate of the mean value. As a result, the samples are assigned weights based on the distance from the mean. This helps to increase the robustness of this iterative reweighting scheme in cases where the values of the minimum errors are significantly separated from zero. Illustrative examples of the problems of regression, classification, and clustering convincingly show the stability of this approach to large volumes of outliers.

This work was supported by grant of RFBR No. 18-01-00050.

References

1. Maronna, R., Martin, R., Yohai, V.: Robust Statistics: Theory and Methods. Wiley, New York (2006)
2. Huber, P.J.: Robust Statistics. Wiley, New York (1981)
3. Shibzukhov, Z.M.: On the principle of empirical risk minimization based on averaging aggregation functions. Dokl. Math. **96**(3), 494–497 (2017). https://doi.org/10.1134/S106456241705026X
4. Shibzukhov, Z.M., Kazakov, M.A.: Clustering based on the principle of finding centers and robust averaging functions of aggregation. In: Proceedings of V International Conference Information Technology and Nanotechnology – ITNT 2019. Journal of Physics: Conference Series (2019). https://iopscience.iop.org/article/10.1088/1742-6596/1368/5/052010/pdf
5. Fränti, P., Sieranoja, S.: K-means properties on six clustering benchmark datasets. Appl. Intell. **48**(12), 4743–4759 (2018)
6. Clustering basic benchmark. http://cs.joensuu.fi/sipu/datasets/

Parameterized Families of Correctly Functioning Sigma-Pi Neurons

Z.M. Shibzukhov[1,2(✉)]

[1] Institute of Mathematics and Informatics of Moscow Pedagogical State University, Moscow, Russia
intellimath@mail.ru

[2] Moscow Institute of Physics and Technology, Moscow, Russia

Abstract. The ΣΠ-neuron is a biologically inspired formal model for logical information processing. The ΣΠ-neuron model adequately reflects information processing processes in the cerebral cortex and in the dendritic trees of neurons. The advantage of the ΣΠ-neuron model is the ability to accurately represent any Boolean function and the possibility of constructive learning (direct construction) in a single pass of the training sample. Another possibility is the direct construction of an ensemble of ΣΠ-neurons that function correctly on the training sample. This article discusses a new algorithm for constructing an ensemble of ΣΠ-neurons in parameterized form. This form can also be easily represented as a single ΣΠ-neuron with a hidden layer of linear and threshold linear units. In some cases, this makes it easier to retrain on new inputs by setting the appropriate control parameter values.

1 Introduction

The ΣΠ-neuron is highly biologically inspired formal model for logical information processing. The ΣΠ-neuron model adequately reflects the processes of information processing in the cerebral cortex [1] and in dendritic trees of neurons [2,3]. ΣΠ-neuron is able to represent arbitrary partial boolean function. The advantage of the ΣΠ-neuron model is the possibility of its constructive training (direct construction) in one pass of the training sample [4]. Using the constructive training procedure, it is also possible to build ensembles of ΣΠ-neurons. At the same time, each of them will function correctly on the training sample [5]. The ΣΠ-neuron model is very effective in correcting Boolean algorithms. [6,7]. Effective electrical circuit design for ΣΠ-neurons was discussed in [8]. Using ΣΠ-neurons can reduce the number of layers in deep networks architecture [9].

The ΣΠ-neuron model define a boolean function in the following form:

$$y = \mathrm{spn}(\mathbf{x}) = \mathrm{sgn}\big(\mathrm{sp}(\mathbf{x})\big),$$

where $y \in \{0,1\}$, $\mathbf{x} \in \{0,1\}^n$, $\mathrm{sgn}(s) = [\![s \geqslant 0]\!]$[1],

$$\mathrm{sp}(\mathbf{x}) = \theta + \sum_{k=1}^{M} w_k \prod_{i \in \mathbf{i}_k} x_i, \tag{1}$$

[1] $[\![S]\!] = 1$ if S is true and $[S] = 0$ if S is false.

A. V. Samsonovich et al. (Eds.): BICA 2020, AISC 1310, pp. 478–483, 2021.
https://doi.org/10.1007/978-3-030-65596-9_57

$\theta \in \mathbb{R}$ is a free constant, $w_k \in \mathbb{R}$ is a weight of the k-th synaptic cluster formed by inputs with indices from the *multi-index* $\mathbf{i}_k \subset \{1, \ldots, n\}$.

There is a constructive procedure for constructing (possibly large) ensembles of ΣΠ-neurons that function correctly on their training set [5]. This is very useful when building majority-voting procedures that help increase stability and generalization ability of the ΣΠ-neuron model. The possibility of parametric representation of such ensembles could allow us to solve the problems of searching for ΣΠ-neurons that meet certain criteria that can be formulated in terms of these parameters (for example, find ΣΠ-neurons with a potentially minimal number of multiplicative terms). It can also simplify the retraining process for new inputs.

In this paper, we propose a method for constructive training of parametric families of ΣΠ-neurons that function correctly on a training set. By setting the values of free parameters, you can get all ΣΠ-neurons from this family. In addition, it will be shown that the values of the weights of ΣΠ-neurons can be calculated using linear and threshold-linear neurons with integer weights, the input of which receives the values of free parameters. The example shows the possibility of easy additional training by setting the values of free parameters.

2 The Problem

Let $\mathbf{X} = \{\mathbf{x}_1, \ldots, \mathbf{x}_N\}$ be a finite set of Boolean n-dimensional vectors, $\mathbf{Y} = \{y_1, \ldots, y_N\}$ be a set of Boolean values ($y_k \in \{0, 1\}$) corresponding to the set $\mathbf{X}$. We consider Boolean functions $f(\mathbf{x})$ such that $y_k = f(\mathbf{x}_k)$ for all $k = 1, \ldots, N$. It is required to construct the representation of $f(\mathbf{x})$ in the form of a ΣΠ-neuron. Since for $N \ll 2^n$ there is a very large number of such representations, it is necessary to obtain some parametric description of this family, which would make it possible to extract ΣΠ-neurons from it that satisfy some optimization criteria.

To describe the algorithm for constructing a ΣΠ-neuron, we define the relation $\not\geq$ on the set of Boolean vectors.

Definition. For $\mathbf{x}', \mathbf{x}'' \in \{0, 1\}^n$ and $\mathbf{i} \subseteq \{1, \ldots, n\}$ we have relation $\mathbf{x}' \not\geq_{\mathbf{i}} \mathbf{x}''$ if there is $i \in \mathbf{i}$ such that $x_i' < x_i''$ (i.e., $x_i' = 0$ and $x_i'' = 1$).

We define the relation of *triangular ordering* on the set of Boolean vectors.

Definition. The sequence $\mathbf{X} = \{\mathbf{x}_1, \ldots, \mathbf{x}_N\}$ is triangularly ordered with respect to the sequence of multi-indices $\mathbf{I} = \{\mathbf{i}_1, \ldots, \mathbf{i}_N\}$ if

1) for all $k = 1, \ldots, N$ and all $i \in \mathbf{i}_k$ there is $x_{ki} = 1$;

2) for all pairs $j < k$: $\mathbf{x}_j \not\geq_{\mathbf{i}_k} \mathbf{x}_k$.

The concept of triangular ordering has the following consequences when using the operator of the product of Boolean values.

Fact. If $\mathbf{X}$ is triangularly ordered with respect to $\mathbf{I}$, then

1) for all $k = 1, \ldots, N$: $\prod\limits_{i \in \mathbf{i}_k} x_{ki} = 1$;

2) for all pairs $j < k$: $\prod\limits_{i \in \mathbf{i}_k} x_{ji} = 0$.

This fact serves as the basis for the direct construction of ΣΠ-neurons.

3 Construction of ΣΠ-Neurons

Here is the constructive learning procedure for the ΣΠ-neuron, which is a slightly modified version of the standard learning procedure [4,5]. To describe it, we introduce the notation of the poly-linear function at p-th step:

$$\mathrm{sp}^p(\mathbf{x}) = \theta + \sum_{k=1}^{p} w_k \prod_{i\in\mathbf{i}_k} x_i.$$

The procedure for constructing a ΣΠ-neuron RCL-SPN (Recurrent Constructive Learning of SPN) from the sequence $\mathbf{X}$, triangularly ordered relative to $\mathbf{I}$, and the corresponding sequence of values $\mathbf{Y}$, has the form:

```
def RCL-SPN(I, X, Y):
    sp^0(x) ← θ
    for k in 1...N:
        s ← sp^{k-1}(x_k)
        if sgn(s) = y_k:
            sp^k(x) ← sp^{k-1}(x)
        else:
            w_k ← ỹ_k − 1 − s
            sp^k(x) ← sp^{k-1}(x) + w_k ∏_{i∈i_k} x_i
    spn(x) ← sgn(sp^N(x))
    return spn(x)
```

It is easy to establish that applying the RCL-SPN procedure to $\mathbf{X}, \mathbf{Y}$ we get correctly functioning ΣΠ-neuron $\mathrm{spn}(\mathbf{x})$, i.e. for all $k = 1, \ldots, N$: $y_k = \mathrm{spn}(\mathbf{x}_k)$.

4 Construction of a Parameterized Family of ΣΠ-Neurons

The set of all sequences of multi-indices with respect to which the sequence $\mathbf{X}$ is triangularly ordered, we denote $\mathcal{I}(\mathbf{X})$ and introduce the relation $\prec$.

Definition. $\mathbf{I}' \prec \mathbf{I}''$, if

1) for all $k = 1, \ldots, N$: $\mathbf{i}'_k \subseteq \mathbf{i}''_k$;
2) there is k such that $\mathbf{i}'_k \subset \mathbf{i}''_k$.

Note that for any sequence $\mathbf{I} \in \mathcal{I}(\mathbf{X})$: $\mathbf{I} \prec \mathbf{I}(\mathbf{X})$ or $\mathbf{I} = \mathbf{I}(\mathbf{X})$. Let $\mathcal{I}^*(\mathbf{X})$ denote the set of minimal elements in $\mathcal{I}(\mathbf{X})$ with respect to the relation $\prec$.

Fact. If $\{\mathbf{i}_1, \ldots, \mathbf{i}_N\} \in \mathcal{I}^*(\mathbf{X})$, then for all $k = 1, \ldots, N$ multi-index $\mathbf{i}_k \in \mathcal{I}_k(\mathbf{X})$, where $\mathcal{I}_k(\mathbf{X})$ is the set of minimal multi-indices $\mathbf{i} \subseteq \mathbf{i}(\mathbf{x}_k)$ such that

1) for all $i \in \mathbf{i}$: $x_{ki} = 1$;
2) for all $j < k$: $\mathbf{x}_j \not\succcurlyeq_{\mathbf{i}} \mathbf{x}_k$.

For each $1 \leqslant k \leqslant N$ we define a multi-linear function

$$P_k(\mathbf{x}) = \sum_{\mathbf{i}\in\mathcal{I}_k(\mathbf{X})} \gamma_{\mathbf{i}} \prod_{i\in\mathbf{i}} x_i, \tag{2}$$

where $\gamma_{\mathbf{i}} \geqslant 0$ and

$$\sum_{\mathbf{i} \in \mathcal{I}_k(\mathbf{X})} \gamma_{\mathbf{i}} = 1. \tag{3}$$

Fact. $P_k(\mathbf{x})$ has the following properties:

1) for all $k = 1, \dots, N$: $P_k(\tilde{\mathbf{x}}_k) = 1$;
2) for all $j < k$: $P_k(\tilde{\mathbf{x}}_j) = 0$.

The set $\mathcal{I}_k(\mathbf{X})$ can be constructed using a procedure from [5].

Define a parameterized multi linear functions:

$$\mathrm{sp}^p(\mathbf{x}) = \theta + \sum_{k=1}^{p} w_k P_k(\mathbf{x}), \quad 1 \leqslant p \leqslant N.$$

The following algorithm constructs a parameterized ΣΠ-neuron.

```
def RCL-SPN2(I, X, Y):
|   sp^0(x) ← θ
|   for k in 1...N:
|   |   s ← sp^{k-1}(x_k)
|   |   if sgn(s) = y_k:
|   |   |   sp^k(x) ← sp^{k-1}(x)
|   |   else:
|   |   |   w_k ← y_k − 1 − s
|   |   |   sp^k(x) ← sp^{k-1}(x) + w_k P_k(x)
|   spn(x) ← sgn(sp^N(x))
|   return spn(x)
```

It is easy to establish that applying the RL-SPN2 procedure to $\mathbf{X}, \mathbf{Y}$ we obtain a parameterized ΣΠ-neuron $\mathrm{spn}(\mathbf{x})$ such that for any set of parameter values for all $k = 1, \dots, N$: $y_k = \mathrm{spn}(\tilde{\mathbf{x}}_k)$ under the condition (3) for all $\gamma_{\mathbf{i}}$, where $\mathbf{i} \in \mathcal{I}_k(\mathbf{X})$.

5 Illustrative Example

Consider the training set given by the table. In addition to the columns $x_1, \dots, x_4$ and y, which specify the values at the inputs and outputs, the column $\mathcal{I}_k$, which defines the minimum allowed multi-indices, and the column P_k, defining the corresponding multi-linear functions (2)–(3).

№	x_1	x_2	x_3	x_4	y	$\mathcal{I}_k$	P_k
1	0	0	0	0	0		1
2	1	1	0	0	1	{1}, {2}	$ax_1 + (1-a)x_2$
3	0	0	1	1	1	{3}, {4}	$bx_3 + (1-b)x_4$
4	1	0	1	0	1	{1,3}	x_1x_3

№	x_1	x_2	x_3	x_4	y	$\mathcal{I}_k$	P_k
5	0	1	0	1	1	{2,4}	x_2x_4
6	0	1	1	0	1	{2,3}	x_2x_3
7	1	0	0	1	1	{1,4}	x_1x_4

The sequence of multi-linear functions $\mathrm{sp}_k(\mathbf{x})$ that are obtained during the construction process is as follows:

$\mathrm{sp}_1(\mathbf{x}) = -1$
$\mathrm{sp}_2(\mathbf{x}) = \mathrm{sp}_1(\mathbf{x}) + ax_1 + (1-a)x_2$
$\mathrm{sp}_3(\mathbf{x}) = \mathrm{sp}_2(\mathbf{x}) + bx_3 + (1-b)x_4$
$\mathrm{sp}_4(\mathbf{x}) = \mathrm{sp}_3(\mathbf{x}) + (1-a-b)_+x_1x_3$.[a]

[a] $(s)_+ = \max\{s, 0\}$.

$\mathrm{sp}_5(\mathbf{x}) = \mathrm{sp}_4(\mathbf{x}) + (a+b-1)_+x_2x_4$
$\mathrm{sp}_5(\mathbf{x}) = \mathrm{sp}_4(\mathbf{x}) + (a+b-1)_+x_2x_4$
$\mathrm{sp}_7(\mathbf{x}) = \mathrm{sp}_6(\mathbf{x}) + (b-a)_+x_1x_4$

Here $0 \leqslant a, b \leqslant 1$. As a result, we obtain a family of multi-linear functions:

$$\begin{aligned}\mathrm{sp}(\mathbf{x}; a, b) = -1 &+ ax_1 + (1-a)x_2 + bx_3 + (1-b)x_4 + \\ &+(1-a-b)_+x_1x_3 + (a+b-1)_+x_2x_4 + \\ &+(a-b)_+x_2x_3 + (b-a)_+x_1x_4.\end{aligned}$$

For clarity, we give several variants:

a	b	$\mathrm{sp}(\mathbf{x})$
0	0	$-1 + x_2 + x_4 + x_1x_3$
0	1	$-1 + x_2 + x_3 + x_1x_4$
1	0	$-1 + x_1 + x_4 + x_2x_3$

a	b	$\mathrm{sp}(\mathbf{x})$
1	1	$-1 + x_1 + x_3 + x_2x_4$
$\frac{1}{2}$	$\frac{1}{2}$	$-1 + \frac{1}{2}x_1 + \frac{1}{2}x_2 + \frac{1}{2}x_3 + \frac{1}{2}x_4$

Since the minimum number of terms corresponds to combinations of Boolean values a and b, in the above table we can see such minimal ΣΠ-neurons.

It is easy to see that if the parameters a, b are considered as additional control parameters, then the ΣΠ-neuron takes the following form:

$$\begin{aligned}\mathrm{sp}(\mathbf{x}, \mathbf{u}) = &-1 + u_1x_1 + u_2x_2 + u_3x_3 + u_4x_4 + \\ &+ u_5x_1x_3 + u_6x_2x_4 + u_7x_2x_3 + u_8x_1x_4,\end{aligned}$$

where

$$\begin{aligned}&u_1 = a, \quad u_2 = 1-a, \; u_3 = b, \; u_4 = 1-b, \; u_5 = (1-a-b)_+, \\ &u_6 = (a+b-1)_+, \; u_7 = (a-b)_+, \; u_8 = (b-a)_+.\end{aligned}$$

As a result, we get a ΣΠ-neuron, in which the synaptic parameters are calculated using linear and ReLU control elements. In essence, it can be considered as single ΣΠ-neuron, which, in addition to the inputs $x_1, \ldots, x_4$, also has inputs from an intermediate layer consisting of threshold-linear neurons.

In some cases it will be easy to adapt the parameterized ΣΠ-neuron without retraining on 2 new inputs. For example, let $(1, 0, 0, 0)$ be new input with expected value $y = 1$: $\mathrm{sp}(1, 0, 0, 0; a, b) = -1 + ax_1$. So we may set $a = 1$ and get new parameterized ΣΠ-neuron with

$$\mathrm{sp}(\mathbf{x}, b) = -1 + x_1 + bx_3 + (1-b)x_4 + bx_2x_4 + (1-b)x_2x_3,$$

where $0 \leqslant b \leqslant 1$. On next new input $(0, 0, 1, 0)$ with $y = 1$: $\mathrm{sp}(0, 0, 1, 0; b) = -1 + b$. So we may set $b = 1$ and get

$$\mathrm{sp}(\mathbf{x}) = -1 + x_1 + x_3 + x_2x_4.$$

Thus, parameterized ΣΠ-neurons can adapt to new inputs by setting values to the corresponding free control parameters.

Conclusion

Thus, we have constructed a constructive procedure that allows us to present a parametric representation for a family of ΣΠ-neurons that function correctly on the training set. This family is represented by a single ΣΠ-neuron with an intermediate layer of linear and threshold linear elements that calculate the weights of any ΣΠ-neuron from this family for given values of input control parameters. This is especially useful if the ΣΠ-neurons must participate in some sequential information processing process. The parametric representation of the ΣΠ-neuron family that functions correctly on the training set allows us to search for optimal combinations of control parameters within the framework of various optimization criteria. A simple illustrative example explains that it can also simplify the retraining process for new inputs.

References

1. Mel, B.W.: The sigma-pi column: a model of associative learning in cerebral cortex. Technical Report CNS Memo 6, Computation and Neural Systems Program, California Institute of Technology (1990)
2. Jadi, M.P., Behabadi, B.F., Poleg-Polsky, A., Schiller, J., Mel, B.W.: An augmented two-layer model captures nonlinear analog spatial integration effects in pyramidal neuron dendrites. Proc. IEEE **102**(5), 782–798 (2014)
3. Mel, B.W.: Toward a simplified model of an active dendritic tree. In: Dendrites, 3rd edn. Chap. 16, pp. 405-486. Oxford University Press (2016)
4. Timofeev, A.V., Pshibikhov, V.K.: Learning algorithms and minimizing the complexity of polynomial recognizing systems. Izv. AN SSSR. Tech. Cybern. **5**(7), 214–217 (1974). (in Russian)
5. Shibzukhov, Z.M.: Constructive learning methods of ΣΠ-neural networks. MAIK Nauka (2006). (in Russian)
6. Shibzukhov, Z.M.: Correct aggregation operations with algorithms. Pattern Recogn. Image Anal. **24**(3), 377–382 (2014)
7. Lyutikova, L.A.: Sigma-Pi neural networks: error correction methods. Procedia Comput. Sci. **145**, 312–318 (2018)
8. Potapov, A.A., Zhang, W., Feng, T., Rekhviashvili, S.Sh.: Sigma-pi neural unit: switching circuit simulation. In: Proceedings of the International Conference on Artificial Intelligence, Information Processing and Cloud Computing, AIIPCC 2019. Article No. 21, pp. 1–5 (2019). https://doi.org/10.1145/3371425.3371477
9. Delalleau, O., Bengio, Y.: Shallow vs. deep sum-product networks. In: Advances in Neural Information Processing Systems 24, pp. 666-674 (2011)

The Loop of Nonverbal Communication Between Human and Virtual Actor: Mapping Between Spaces

Vladimir R. Shirokiy[1,2](✉), Daria V. Tikhomirova[2], Roman D. Vladimirov[1], Sergei A. Dolenko[1], and Alexei V. Samsonovich[2]

[1] D.V. Skobeltsyn Institute of Nuclear Physics, M.V. Lomonosov Moscow State University, Moscow 119991, Russian Federation
shiroky@srd.sinp.msu.ru
[2] National Research Nuclear University MEPhI, Kashirskoye Shosse, 31, Moscow 115409, Russian Federation

Abstract. There is a question about the appropriate emotional and expressive language of a virtual actor. In this paper we study facial expressions. We investigate the transformations between the space of Action Units and the standard affective space in the loop of nonverbal communication between a person and a virtual actor using facial expressions [1]. We are mapping both dimensions into each other using various machine learning algorithms. Action Units space was mapped into emotional space directly using artificial neural networks. Emotional space was mapped into Action Units space with help of dimensionality reduction followed by clusterization of the latter. After the final synthesis, the facial expression of virtual actor can be determined.

Keywords: Valence · Arousal · Action units · Machine learning · Dimensionality reduction

1 Introduction

When a person interacts with a virtual actor, there is a question about the appropriate emotional and expressive language of the latter. We are talking about such expressive, communicative ways of expressing emotions as gestures, facial expressions and pantomime. In this paper we study facial expressions. With the help of this approach, such emotions as joy, sadness, anger, disgust, fear, contempt, etc. are clearly manifested.

Let's look at the loop of nonverbal communication between a person and a virtual actor using facial expressions [1]. Facial expression and gaze of the participant can be analyzed, e.g., using an IOS device with a True-Depth Camera or software such as FaceReader by Noldus. The outcome of this analysis is a representation of the entire facial configuration using Action Units, described by the Facial Action Coding System (FACS) [2].

However, it would be difficult to find an elegant model based on first principles, such that it can be successfully applied to 42-dimensional vectors. Therefore, it is

A. V. Samsonovich et al. (Eds.): BICA 2020, AISC 1310, pp. 484–489, 2021.
https://doi.org/10.1007/978-3-030-65596-9_58

necessary to reduce the dimensionality. For this purpose, a standard affective space model like PAD (Pleasure-Arousal-Dominance) can be used [3, 4]. The PAD emotional state model uses three numerical dimensions, pleasure, arousal and dominance to represent all emotions (Fig. 1). The pleasure-arousal part of PAD was developed into a circumplex model of emotion experience, and those two dimensions were termed "core affect". The dominance part of PAD was re-conceptualized as a part of the appraisal process in an emotional episode (a cold cognitive assessment of the situation eliciting the emotion). The pleasure-displeasure scale measures how pleasant or unpleasant one feels about something. The arousal-non-arousal scale measures how energized or soporific one feels. The dominance-submissiveness scale represents the controlling and dominant versus controlled or submissive one feels [3].

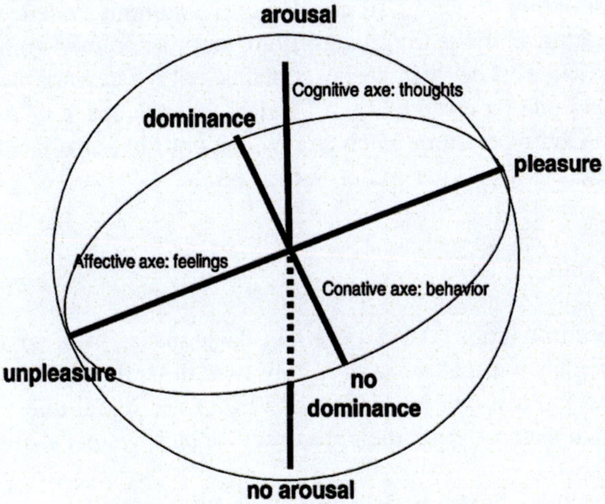

Fig. 1. PAD emotional state model.

The result of compression can consist of 4 numbers: two coordinates in the affective space – pleasure (valence) and arousal, plus the two eye fixation coordinates [4]. This 4-vector (in this study we consider only the 2-vector valence-arousal without eye fixation coordinates) and its past trajectory can now be analyzed by the eBICA model [5], using an appropriate Moral Schema [6]. Based on this information processing using the Moral Schema, the system generates an adequate response, which is also represented by a 4-vector, and then gets expanded again into the set of Action Units.

After the final synthesis, the facial expression of virtual actor is determined and enacted. The participant visually perceives the resultant expression, which closes the loop.

2 Materials and Methods

The study involved 30 volunteers who were students of the National Research Nuclear University MEPhI (Moscow Engineering Physics Institute) at the age from 22 to 24. The task for them was to express consistently emotional states on the face, for 10 s each: neutral expression, joy, sadness, anger, disgust, fear, contempt, satisfaction, excitement.

During the experiment, the participants' faces were videotaped with a webcam using the Bandicam program. At the same time, 52 face parameters were logged in real time using an iPad. The videos were further processed using the emotion recognition tool "FaceReader" by Noldus and OpenFace.

As a result of the experiment, we have created an annotated database of emotions that includes information about 42 Action Units, 7 emotions, valence and arousal.

We used several machine learning algorithms in order to map emotional space into the Action Units space. The first space contains only two variables - valence and arousal. The second one represented by 42 variables is the space of Action Units from The Facial Action Coding System. Each one corresponds to some facial muscle such as lid raiser, lid tightener, lip corner puller and so on.

3 Data Analysis

The dataset we used contains 31595 points in each space. In order to determine the accuracy of our models, our dataset was split in two, the train dataset, which is 80% of the initial one, and the test dataset, which is 20% of the initial one. All models were trained on the train dataset, and their accuracy values were calculated on the test dataset.

As baseline, we used linear models trained on the emotional space to predict each of the Action Units.

The Action Units space is a high-dimensional space, so in order to map it into a two-dimensional one, first we used dimensionality reduction. To achieve this we used principal component analysis (PCA) [7] with a wide range of components, and checked the explained variance on each iteration as a reference to information loss.

After dimensionality reduction we used clusterization of the new space in order to cut it into small pieces and to map each point in the emotional space to one of these clusters in the new space. As clusterization algorithm, we used k-means algorithm [8].

To predict the cluster number we used simple decision tree classifier which is quite capable in dimensions of such order [9].

In order to map Action Units space into emotional space, multilayer perceptron (MLP) [10] and linear regression models were trained, and their results were compared.

4 Results

The dataset used had the initial size of 31595 samples. Each sample had 42 floating point features of Action Units space and 2 floating point features of emotional space. After splitting into train and test datasets, the former had 25276 samples and the latter had 6319 samples.

We trained baseline models to predict the Action Units values on the test dataset and calculated the R^2 score. The resulting histogram is shown in Fig. 2.

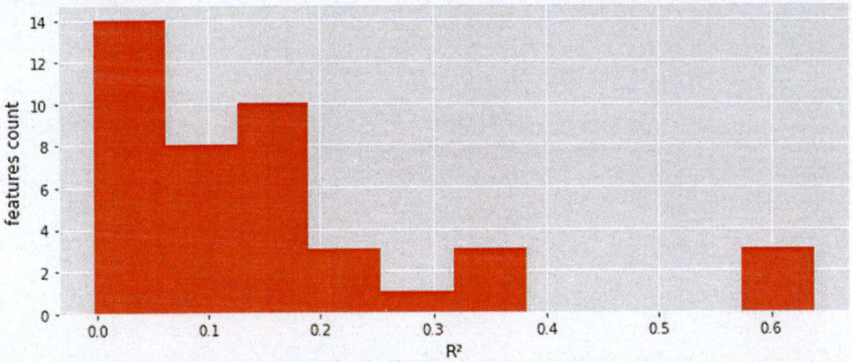

Fig. 2 Histogram of the scores of the linear regressions of Action Units on the test dataset. Scores ranges along the *x* axis and Action Units amount in each range along the *y* axis.

In order to perform dimensionality reduction, we obtained PCA models with the number of components in the range from 2 to 19. In addition to the explained variance, we also checked linear regression of each component of the emotional space trained in the PCA spaces obtained as additional reference of information loss with lower number of components.

The results of the PCA models obtained are shown in Fig. 3.

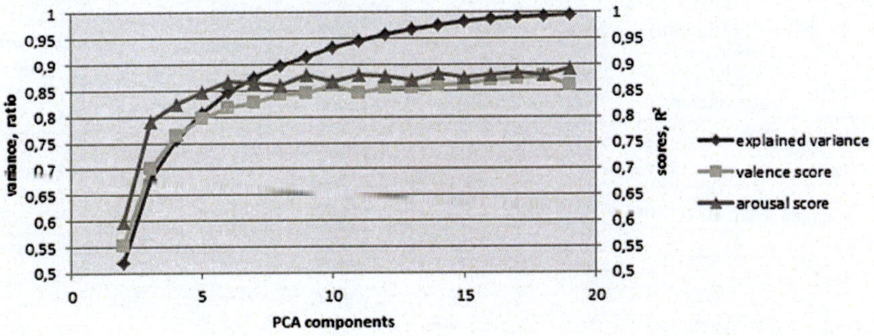

Fig. 3. Explained variance and linear models scores on trained on transformed test datasets.

As we can see, linear models scores achieve plateau simultaneously near 6 or 7 components of PCA space, and the explained variance at this point is greater than 80%. Please refer to Table 1 for more information.

Table 1. Some values of the explained variance and the accuracy of linear models of valence and arousal trained in PCA space

PCA components	Explained variance	Valence test score	Arousal test score	Classification accuracy
2	0.519	0.537	0.560	0.559
3	0.682	0.678	0.781	0.554
4	0.757	0.772	0.825	0.574
5	0.806	0.800	0.839	0.571
6	0.847	0.817	0.867	0.574
7	0.876	0.848	0.876	0.571
10	0.932	0.849	0.872	0.575
15	0.983	0.868	0.882	0.583
19	0.996	0.881	0.889	0.602

After dimensionality reduction we performed clusterization using K-means with 100 clusters. We obtained a cluster model for each of the PCA spaces from previous step, and applied the model to the test set.

Each point in the emotional space was mapped to one of the clusters using the decision tree algorithm. The classification accuracy values of these models on the test set are shown in the last column of Table 1. As one can see, we achieve plateau after 4 PCA components.

To predict the points in the emotional space from Action Units space, we trained multilayer perceptrons (MLP), whose results were compared with linear regression results as a baseline. An MLP model had 20 and 10 neurons in the two hidden layers, tanh activation function and learning rate of 0.001. The results are shown in Table 2.

Table 2. Results of machine learning models mapping Action Units space into emotional space

Model	Valence, R^2 on train dataset	Arousal, R^2 on train dataset	Valence, R^2 on test dataset	Arousal, R^2 on test dataset
Linear regression	0.761	0.849	0.754	0.848
MLP	0.855	0.889	0.851	0.886

5 Conlusion

We proved that we can easily predict emotional space from Action Units space using machine learning models. MLP preforms quite well in comparison with linear models. R^2 on the test dataset exceeds 0.85.

Action Units space can be easily reduced from the dimension of 42 features into a space of 7 features using PCA. In this case, information loss is less than 15%, and it is decreasing as the number of components of PCA increases. Thus, facial expression can be characterized by a space of a given dimension.

It is possible to map the emotional space into a new space of transformed Action Units using machine learning algorithms, K-means and decision trees, with classification accuracy of 0.571 using 7-dimensions space of transformed PCA with 100 clusters, and with accuracy of 0.602 using 19-dimensions space of transformed PCA with 100 clusters. After this the facial expression of virtual actor can be determined.

Acknowledgements. This study has been performed at the expense of the grant of Russian Science Foundation (project no.18-11-00336). The authors are grateful to all NRNU MEPhI students who participated in the study.

References

1. Eidlin, A.A., Chubarov, A.A., Samsonovich, A.V.: Virtual listener: emotionally-intelligent assistant based on a cognitive architecture. In: Biologically Inspired Cognitive Architectures Meeting, pp. 73–82. Springer, Cham (2019)
2. Tian, Y.I., Kanade, T., Cohn, J.F.: Recognizing action units for facial expression analysis. IEEE Trans. Pattern Anal. Mach. Intell. **23**(2), 97–115 (2001)
3. Mehrabian, A.: Basic Dimensions for a General Psychological Theory: Implications for Personality, Social, Environmental, and Developmental Studies, vol. 2. Oelgeschlager, Gunn & Hain, Cambridge (1980)
4. Russell, J.A.: A circumplex model of affect. J. Pers. Soc. Psychol. **39**(6), 1161–1178 (1980)
5. Samsonovich, A.V.: Emotional biologically inspired cognitive architecture. BiolInsp. Cogn. Arch. **6**, 109–125 (2013).https://doi.org/10.1016/j.bica.2013.07.009
6. Samsonovich, A.V.: Schema formalism for the common model of cognition. BiolInsp. Cogn. Arch. **26**, 1–19 (2018).https://doi.org/10.1016/j.bica.2018.10.008
7. Karl Pearson, F.R.S.: LIII. On lines and planes of closest fit to systems of points in space. Lond. Edinburgh Dublin Phil. Mag. J. Sci. **2**(11), 559–572 (1901)
8. Jin, X., Han, J.: K-means clustering. In: Sammut, C., Webb, G.I. (eds.) Encyclopedia of Machine Learning. Springer, Boston (2011). https://doi.org/10.1007/978-0-387-30164-8_425
9. Breiman, L., Friedman, J.H., Olshen, R., Stone, C.: Classification and Regression Trees. Wadsworth & Brooks, Pacific Grove (1984)
10. Haykin, S.: Neural Networks: A Comprehensive Foundation. Macmillan (1994)

Preliminary Experiment on Emotion Detection in Illustrations Using Convolutional Neural Network

Alexander Shtanko[1(✉)] and Sergey Kulik[1,2]

[1] National Research Nuclear University MEPHI, Kashirskoe Shosse 31, Moscow 115409, Russia
shtanko-mephi@yandex.ru, sedmik@mail.ru
[2] Moscow State University of Psychology and Education (MSUPE), Sretenka St. 29, Moscow 127051, Russia

Abstract. The paper describes an experiment on emotion detection in images, specifically illustrations and cartoon images. Usually, detection and classification of emotions are performed on human faces so the algorithm can learn, for example, what a "happy human face" looks like. These algorithms probably can't transfer their understanding of happiness features onto different types of objects, like animals or cartoon illustrations. We, humans, can recognize and detect signs of emotions (although maybe falsely) in new and unusual objects. Developing an algorithm capable of recognizing emotions in objects it wasn't trained on would allow for better human-like robots and systems. This is a preliminary study on how well knowledge gained by a typical neural network detection system on a set of objects transfers to new, unknown objects. The neural network detection system used in this study is YOLO. We collected small training datasets using cartoon illustrations of several animals of two categories: happy and sad. We tested the trained network on a set of illustrations depicting a different animal the network hasn't seen in training. The best performance achieved is 0.69 F_1-score.

Keywords: Machine learning · Object detection · Convolutional neural networks · Emotion

1 Introduction

Computer vision methods have been developing for many years and successfully used for solving various tasks. In recent years machine learning and neural networks rise in popularity and variety of use. Computer vision is one of the fields that adopted convolutional neural networks. Application of neural networks in computer vision extends to such classical tasks as classification, object detection, segmentation and so on.

The field of computer vision has been dominated by convolutional neural networks completely, most of the classical tasks now have more efficient neural network solutions. Various fields of application have visual information that needs to be analyzed. Thus, convolutional neural networks proved to be useful in autonomous driving [1],

A. V. Samsonovich et al. (Eds.): BICA 2020, AISC 1310, pp. 490–494, 2021.
https://doi.org/10.1007/978-3-030-65596-9_59

remote sensing [2], forensic science [3], etc. This shows the vast capabilities of convolutional neural networks. In one of our previous works, we discussed the importance of personnel training in neural networks [4].

The topic of this paper is object detection systems. Usually, target objects for detection are solid, constant in appearance items. However, the visual world is more complicated than that. Some objects carrying important information are abstract. One of such abstract objects is emotions. Emotions are an important target for intelligence systems to understand and adequately react to. One side of the task is to express appropriate emotions depending on circumstances, another side of the task is to detect emotions and recognize them in the environment of the system in question.

Frequently, emotion recognition is performed in two steps. First, the target object (head) position is found using an object detection algorithm, then a classification algorithm determines expressed emotion. The goal of this research is to see how well this task can be performed using an object detection algorithm directly.

Humans can detect the signs of emotions in abstract shapes and inanimate objects (due to the human tendency to seek patterns), even in never seen objects (or in unusual representations of the objects). A system able to find areas expressing emotions without an attachment to specific objects it was trained on would be a step towards more human-like artificial intelligence.

Emotions are an important target for various data processing systems to handle. Research on emotions includes emotion detection [5], emotion recognition [6], agents with emotional behavior [7], cognitive models [8], etc.

In this paper, we're going to train a convolutional neural network on a dataset containing animal illustrations. The training goal is to locate emoting characters and classify expressed emotion into 2 categories: sad and happy. The performance of the network is then assessed on a testing dataset. However, emoting objects (cartoon animals) are different in training and testing dataset so we'll see how good the network understands emotion features rather than memorizing the emotion features of that specific cartoon animal. Training will be done on several datasets of different sizes and containing different animal types. Additionally, in the datasets there're illustrations of different styles which poses an extra challenge to the neural network.

Numbers of training examples in the datasets are relatively small because it's also important to test the learning ability on small datasets, considering in many applications it's hard to collect large amounts of training data.

In the future, we're going to use the developed detection systems in web-crawlers implemented in an agent-based fashion [9] using regular agent technologies [10].

The neural network used in our research is a well-known YOLO detection system first proposed in 2016 [11], later expanded to YOLO9000 [12], with the latest iteration called YOLOv3. It's a fast and accurate one-stage object detection system.

2 Datasets

We used datasets consisting of several base sets of images. We used images of six categories: images of cartoon bears, birds, bunnies, dogs, pigs, and cats. For each category, we collected 60 images with animals emoting happy expression and 60 images with sad animals. Specifically, images associated with queries like “sad dog illustration” or “happy bear illustration”. All images were annotated with each emoting character localized (by bounding box around their head associated with corresponding class “happy” or “sad”). Image distribution is described in Table 1. Thus, all cat images are designated as testing dataset. Other images are split into training and validation datasets. Dataset 1 is the smallest one and contains images of only one type of animal. The next dataset contains the first one and also adds images of the second type of animal. Datasets expand further and further until the largest dataset – dataset 5. This is done in order to see how increasing the variety and size of the dataset affect performance.

Table 1. Details of the datasets.

Dataset	Type	Objects	Training images	Validation images	Testing images
1	Training	Bear	100	20	0
2	Training	Bear, bird	200	40	0
3	Training	Bear, bird, bunny	300	60	0
4	Training	Bear, bird, bunny, dog	400	80	0
5	Training	Bear, bird, bunny, dog, pig	500	100	0
6	Testing	Cat	0	0	120

In all datasets, there’re two classes to detect: happy and sad emotion. But training and testing datasets contain different characters (objects) emoting these expressions. So, the performance on testing datasets shows how good knowledge train on a set of objects performs on new, unseen types of objects.

3 Results

We trained 5 YOLOv3 networks on each dataset and tested their performance on the testing dataset. Tiny version of YOLO was used because of its fast training and good results on small datasets. The results are shown in Table 2.

On dataset 1, when there’s only one type of object to learn from on testing dataset, the neural network performs pretty well when it comes to localization of emoting characters however classification is at a level of random guessing. Sometimes the system registers two detections for each class in the same position. With increasing the size of the dataset and including new types of animals, classification becomes better but still, the network makes a lot of mistakes. Thus, the best performance we managed to achieve on new, unseen types of objects is 0.69 F_1-score.

Table 2. Networks performance on testing dataset.

Metric	Dataset 1	Dataset 2	Dataset 3	Dataset 4	Dataset 5
Class “sad”: true positive	20	30	35	39	35
Class “sad”: false positive	14	27	23	20	18
Class “happy”: true positive	27	37	39	42	43
Class “happy”: false positive	16	20	17	11	17
Precision	0.61	0.59	0.65	0.72	0.69
Recall	0.38	0.54	0.60	0.65	0.63
F_1-score	0.47	0.56	0.62	0.69	0.66

4 Conclusions

Thus, we investigated the ability of object detection system YOLO to detect emotions. Specifically, how well knowledge about emotions’ features learned on one set of objects applies to a different set of objects. We collected necessary images for datasets, trained neural networks and measured their performance. Depending on the training dataset results went from random guessing to 0.69 F_1-score.

Acknowledgements. This work was supported by Competitiveness Growth Program of the Federal Autonomous Educational Institution of Higher Education National Research Nuclear University MEPhI (Moscow Engineering Physics Institute).

References

1. Arnold, E., Al-Jarrah, O.Y., Dianati, M., Fallah, S., Oxtoby, D., Mouzakitis, A.: A survey on 3D object detection methods for autonomous driving applications. IEEE Trans. Intell. Transp. Syst. **20**(10), 3782–3795 (2019)
2. Cheng, G., Han, J.: A survey on object detection in optical remote sensing images. ISPRS J. Photogram. Remote Sens. **117**, 11–28 (2016)
3. Kulik, S.D., Shtanko, A.N.: Recognition algorithm for biological and criminalistics objects. In: Samsonovich, A.V. (ed.) Biologically Inspired Cognitive Architectures 2019. Proceedings of the Tenth Annual Meeting of the BICA Society, AISC, vol. 948, pp. 283–294. Springer, Heidelberg (2020)
4. Shtanko, A.N., Kulik, S.D.: Scientific personnel training in convolutional neural networks for the implementation of research projects of the MegaScience class. In: Personnel Training and Legal Support for the Implementation of Research Projects of the MegaScience Class, Journal of Physics: Conference Series, vol. 1406, no. 1, pp. 012014. IOP Publishing (2019)
5. Abdul-Mageed, M., Ungar, L.: Emonet: fine-grained emotion detection with gated recurrent neural networks. In: Proceedings of the 55th annual meeting of the association for computational linguistics, vol. 1, pp. 718–728. Association for Computational Linguistics (2017)
6. Albanie, S., Nagrani, A., Vedaldi, A., Zisserman, A.: Emotion recognition in speech using cross-modal transfer in the wild. In: Proceedings of the 26th ACM international conference on Multimedia, pp. 292–301. Association for Computing Machinery (2018)

7. Bogatyreva, A.A., Sovkov, A.D., Tikhomirova, S.A., Vinogradova, A.R., Samsonovich, A. V.: Virtual pet powered by a socially-emotional BICA. Procedia Comput. Sci. **145**, 564–571 (2018)
8. Samsonovich, A.V.: On semantic map as a key component in socially-emotional BICA. Biol. Insp. Cogn. Arch. **23**, 1–6 (2018)
9. Ananieva, A., Onykiy, B., Artamonov, A., Ionkina, K., Galin, I., Kshnyakov, D.: Thematic thesauruses in agent technologies for scientific and technical information search. Procedia Comput. Sci. **88**, 493–498 (2016)
10. Artamonov, A., Onykiy, B., Ananieva, A., Ionkina, K., Kshnyakov, D., Danilova, V., Korotkov, M.: Regular agent technologies for the formation of dynamic profile. Procedia Comput. Sci. **88**, 482–486 (2016)
11. Redmon, J., Divvala, S., Girshick, R., Farhadi, A.: You only look once: unified, real-time object detection. In: 2016 IEEE Conference on Computer Vision and Pattern Recognition (CVPR), pp. 779–788. IEEE (2016)
12. Redmon, J., Farhadi, A.: YOLO9000: better, faster, stronger. In: 2017 IEEE Conference on Computer Vision and Pattern Recognition (CVPR), pp. 7263–7271. IEEE (2017)

AGI Components for Enterprise Management Systems

Artem A. Sukhobokov[1,2](✉) and Liubov I. Lavrinova[2]

[1] SAP DBS CIS,
Kosmodamyanskaya nab., 52/7, Moscow 115054, Russian Federation
artem.sukhobokov@yandex.ru
[2] Bauman Moscow State Technical University,
Ul. Baumanskaya 2-ya, 5, Moscow 105005, Russian Federation

Abstract. As a continuation of previous work, a functional architecture of a full-fledged AGI is proposed for working in an enterprise management system. Since the creation of such an AGI will require more than one decade, an approach is proposed that allows combining individual more ready-made blocks from a full-fledged AGI with traditional enterprise management systems. Such systems including one or two AGI blocks and Consciousness are called Lean AGI. For these systems, it is possible to use not full Consciousness, but some its limited industry-specific edition, located in the interval between the minimal and full Consciousness. Such editions are called Lean Consciousness. In the paper, the cases for the possible use of Lean AGI as part of enterprise management systems are shown. For one of these cases, the Lean AGI architecture based on the Goal analytics block is proposed for working on top of traditional enterprise management applications.

Keywords: AGI · Consciousness · Lean AGI · Lean consciousness · Cognitive architecture · Goal analytics · Enterprise management applications

In a review [1], 84 cognitive architectures were considered, among which a small group of architectures was singled out, the development of which was aimed to create a General Artificial Intelligence (AGI): Soar [2], ACT-R [3], NARS [4], LIDA [5], SiMA (formerly ARS) [6], Sigma [7] and CogPrime [8–10]. The recently proposed ICOM [11, 12] and the architecture developed by Michael S. P. Miller [13–15] can be added to this list.

For at least one of the listed architectures, an attempt is made to use the created AGI at the top level of company management over ERP systems [16]. Such a solution corresponds to the highest stage of business intelligence tools' development [17] and promises very good prospects. However, time constraints must be considered. Experts expect the completion of the creation of AGI only in the 2040 s [18]. Therefore, it seems appropriate, without waiting for the emergence of a full-fledged AGI, to consider the possibility of using its individual components in combination with traditional applications that are parts of corporate management systems.

In order to be able to use individual AGI components, it is necessary to have a relatively complete architecture from which you can select the necessary components.

A. V. Samsonovich et al. (Eds.): BICA 2020, AISC 1310, pp. 495–500, 2021.
https://doi.org/10.1007/978-3-030-65596-9_60

The architecture from [19], in which a set of expansions were made, was used as the basis for such an AGI architecture. Characteristic features of the architecture from [19]:

- the existence of two relatively independent but interconnected components – Consciousness and Subconsciousness which has the ability to solve a wide range of identical tasks;
- Consciousness uses metagraph conscious memory [20] and solves all tasks using inference mechanisms (such as, for example, are proposed in [21]) or probabilistic inference mechanisms described in [22];
- Subconsciousness uses metagraph unconscious memory [20] and various models of narrow AI for decision making. The choice of models is made by meta-learning mechanisms.

The extensions made to increase the intelligence of the architecture:

- the composition of the functional blocks Consciousness and Subconsciousness is supplemented by the components of Time Analytics and Spatial Analytics;
- Social Analytics, Ethical Analytics and Worldview Analytics blocks are included in Consciousness without adding them to Subconsciousness.

The resulted AGI architecture is shown in Fig. 1.

It can be hoped that agents with the architecture shown in Fig. 1, in the future after long-term learning, will be comparable to the AGI or will approach to their capabilities. However, even now, without waiting for the appearance of fully functional agents, we propose using separate blocks from the proposed architecture in combination with limited consciousness. For example:

- a bunch of time-spatial analytics blocks could be used in supply chain planning;
- Attention analytics and Abstraction analytics blocks could be used in processing incoming streams of text information;
- the Goal analytics block could be used in conjunction with various enterprise applications (ERP, EAM, CRM, SCM, CPM, etc.) for planning target values of upper-level indicators.

A key solution for all of these cases is to build a limited artificial consciousness. To date, many methods have been proposed for implementing Consciousness in AI in the form of a computable procedure, for example [23–26]. Additionally, about 10 different models of Consciousness can be found in early reviews [27, 28]. When constructing limited artificial consciousness, it is important that consciousness be limited by specific knowledge from the subject area, necessary for solving specific problems and not include unnecessary extraneous knowledge. The work [29], which proposed an architecture that supports the minimal machine consciousness, allows us to hope for the possibility of constructing the necessary simplified versions of Consciousness which we will name Lean Consciousness.

Of the three listed cases for using individual AGI blocks, the most promising is the third, which uses a Goal Analytics. The choice of the goal of the next action is a classic problem of cognitive architectures. In one way or another, the problem of choosing the next step is solved in each architecture. But the operations with goals are most fully developed in specialized architectures, such as MIDCA [30–32], GRIM [33, 34] and

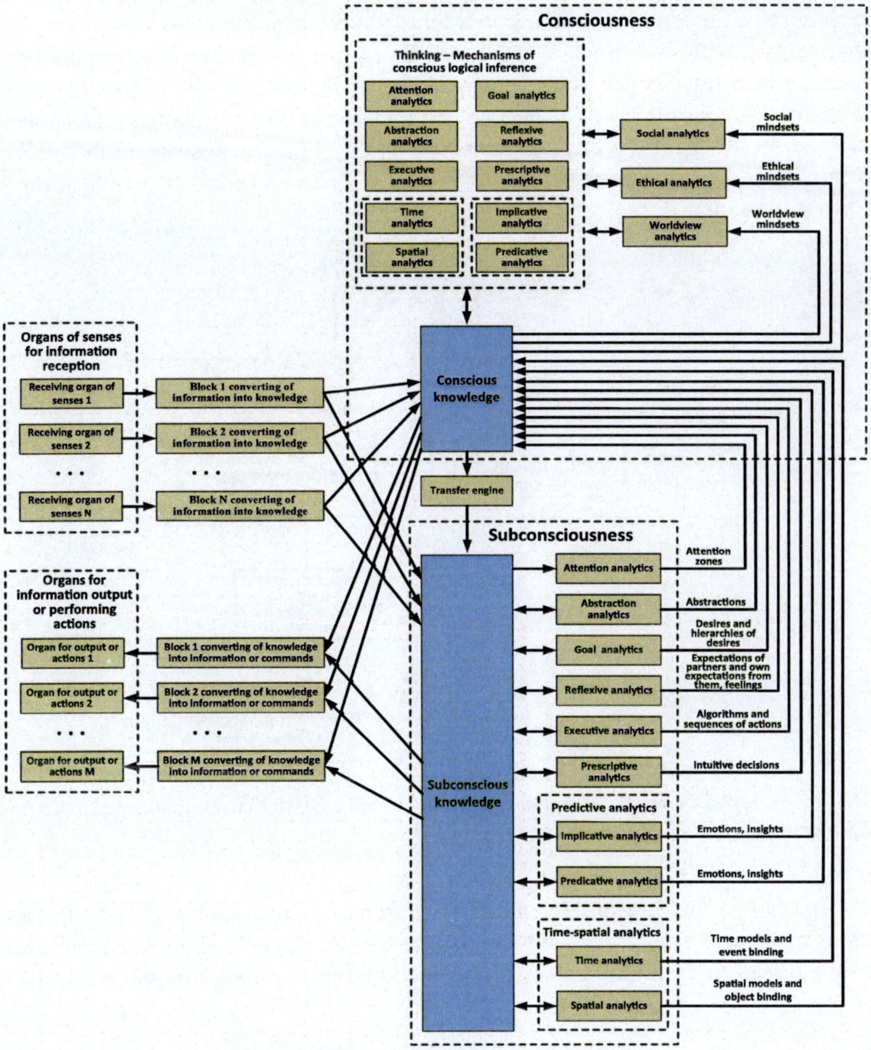

Fig. 1. Proposed full-fledged AGI architecture.

earlier ARTUE [35, 36]. In addition, a number of methods for working with goals were developed outside of cognitive architectures in various control systems for autonomous devices (autonomous robots and autonomous systems moving in various environments - on the ground, on water, in the air, in space), for example, [37, 38].

The architecture that allows combining the AGI components (the Consciousness and the Goal Analytics block) with the applications of the enterprise management system, is shown in Fig. 2. In order for the integrated solution to be self-sufficient, it must include a Data Collection block on the state of the external environment in which the enterprise operates (products, markets, including the personnel market and the banking services market, territorial conditions, government agencies, regulators, customers, competitors etc.). Such a block will be equivalent to the sense organs in the architecture of a full-fledged AGI, shown in Fig. 1. And within the framework of the same analogy, a controller will be needed that transforms the information collected into knowledge. In order for the created Lean AGI can evaluate its position in the external environment and plan actions in it, the system will need a block of analytics on the external environment.

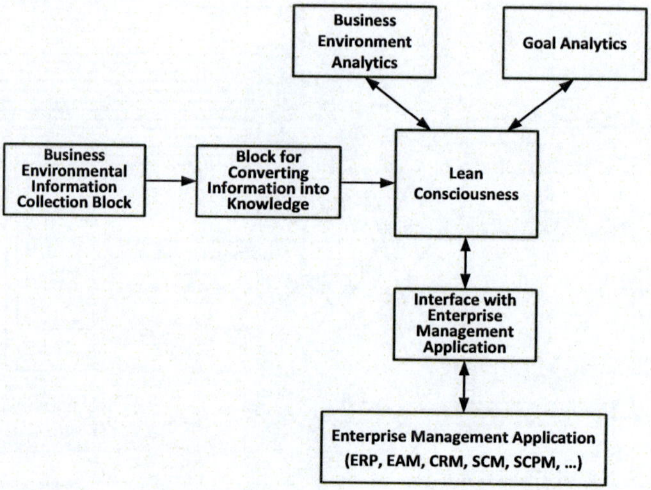

Fig. 2. An example of Lean AGI for integrating goal analytics with traditional enterprise management applications.

The proposed architecture is aimed at improving the quality of management decisions made at the enterprise and automating the process of making such decisions. This should lead to an improvement in the economic performance of the enterprise.

References

1. Kotseruba, I., Tsotsos, J.K.: A review of 40 years in cognitive architecture research core cognitive abilities and practical applications. arXiv:1610.08602v3 [cs.AI], 13 January 2018
2. Laird, J.E.: The Soar Cognitive Architecture. MIT Press, Cambridge (2012)
3. Lebiere, C., Pirolli, P., Thomson, R., Paik, J., Rutledge-Taylor, M., Staszewski, J., Anderson, J.R.: A functional model of sensemaking in a neurocognitive architecture. Comput. Intell. Neurosci. Article ID 921695 (2013). https://doi.org/10.1155/2013/921695
4. Wang, P.: Natural language processing by reasoning and learning. In: Kühnberger, K.-U., Rudolph, S., Wang, P. (eds.) AGI 2013. LNCS, vol. 7999, pp. 160–169. Springer, Heidelberg (2013)

5. Faghihi, U., Franklin, S.: The LIDA model as a foundational architecture for AGI. In: Wang, P., Goertzel, B. (eds.) Theoretical Foundations of Artificial General Intelligence, pp. 103–121. Atlantis Press, Paris, France (2012)
6. Schaat, S., Wendt, A., Kollmann, S., Gelbard, F., Jakubec, M.: Interdisciplinary development and evaluation of cognitive architectures exemplified with the SiMA approach. In: EuroAsianPacific Joint Conference on Cognitive Science. CEUR-WS, vol. 1419 (2015)
7. Pynadath, D.V., Rosenbloom, P.S., Marsella, S.C.: Reinforcement learning for adaptive theory of mind in the sigma cognitive architecture. In: Goertzel, B., Orseau, L., Snaider, J. (eds.) AGI 2014. LNCS, vol. 8598, pp. 143–154. Springer, Cham (2014)
8. Goertzel, B.: From abstract agents models to real-world AGI architectures: bridging the gap. In: Everitt, T., Goertzel, B., Potapov, A. (eds.) AGI 2017. LNCS, vol. 10414, pp. 3–12. Springer, Cham (2017)
9. Goertzel, B., Pennachin, C., Geisweiller, N.: Engineering General Intelligence, Part 1: A Path to Advanced AGI via Embodied Learning and Cognitive Synergy. Atlantis Press, Paris, France (2014)
10. Goertzel, B., Pennachin, C., Geisweiller, N.: Engineering General Intelligence, Part 2: The CogPrime Architecture for Integrative, Embodied AGI. Atlantis Press, Paris, France (2014)
11. Kelley, D.J., Twymon, M.A.: Independent core observer model (ICOM) theory of consciousness as implemented in the ICOM cognitive architecture and the associated consciousness measures. In: 2019 Towards Conscious AI Systems Symposium, vol. 2287. CEUR-WS (2019)
12. Waser, M.R., Kelley, D.J.: Implementing a seed safe/moral motivational system with the independent core observer model (ICOM). Proc. Comput. Sci. **88**, 125–130 (2016)
13. Miller, M.S.P.: Building minds with patterns. In: Video from 10th Annual International Conference on Biologically Inspired Cognitive Architectures BICA 2019, 16–18 August 2019, Redmond, WA, USA (2019). https://www.youtube.com/watch?reload=9&v=kqicbyONxO8
14. Miller, M.S.P.: Building Minds with Patterns (2018)
15. Miller, M.S.P.: Coding Artificial Minds (2020)
16. Kelley, D.J.: Independent core observer model (ICOM) cognitive architecture-based system. In: Video from 10th Annual International Conference on Biologically Inspired Cognitive Architectures BICA 2019, 6–18 August 2019, Redmond, WA, USA (2019). https://vimeo.com/384904907
17. Sukhobokov, A.A.: Business analytics and AGI in corporate management systems. Proc. Comput. Sci. **145**, 533–544 (2018). https://doi.org/10.1016/j.procs.2018.11.118
18. Potapov, A.: Technological singularity: what do we really know? Information **9**(4), 82 (2018). https://doi.org/10.3390/info9040082
19. Sukhobokov, A.A., Gapanyuk, Y.E., Chernenkiy, V.M.: Consciousness and Subconsciousness as a means of AGI's and narrow AI's integration. In: Samsonovich A. (ed.) BICA 2019. AISC, vol. 948, pp. 515–520. Springer, Cham (2020)
20. Chernenkiy, V., Gapanyuk, Y., Revunkov, G., Kaganov, Y., Fedorenko, Y.: Metagraph approach as a data model for cognitive architecture. In: Samsonovich, A.V. (ed.) BICA 2018. AISC, vol. 848, pp. 50–55. Springer, Cham (2019)
21. Varlamov, O.O.: Wi!Mi expert system shell as the novel tool for building knowledge-based systems with linear computational complexity. IREACO **11**(6), 314–325 (2018)
22. Tarassov, V.B.: Development of fuzzy logics: from universal logic tools to natural pragmatics and non-standard scales. Proc. Comput. Sci. **120**, 908–915 (2017). https://doi.org/10.1016/j.procs.2017.11.325

23. Rolls, E.T.: Consciousness, decision-making and neural computation. In: Cutsuridis, V., Hussain, A., Taylor, J.G. (eds.) Perception-Action Cycle: Models, Architectures, and Hardware, pp. 287–333. Springer, New York (2011)
24. Taylor, J.G.: Solving the Mind-Body Problem by the CODAM Neural Model of Consciousness?. Springer, Dordrecht (2013)
25. Bringsjord, S., Bello, P., Govindarajulu, N.S.: Toward axiomatizing consciousness. In: Jacquette, D. (ed.) The Bloomsbury Companion to the Philosophy of Consciousness, pp. 289–324. Bloomsbury Academic, London (2018)
26. Govindarajulu N.S., Bringsjord S.: Towards a computable & harnessable model of consciousness. In: 2019 Towards Conscious AI Systems Symposium, vol. 2287. CEUR-WS (2019)
27. Seth, A.: Models of consciousness. Scholarpedia **2**(1), 1328 (2007)
28. Taylor J.G.: A review of models of consciousness. In: Cutsuridis, V., Hussain. A., Taylor, J. G. (eds.) Perception-Action Cycle: Models, Architectures, and Hardware, pp. 335–357. Springer, New York (2011)
29. Wiedermann, J., van Leeuwen, J.: Finite state machines with feedback: an architecture supporting minimal machine consciousness. In: Manea, F., Martin, B., Paulusma, D., Primiero, G. (eds.) CiE 2019. LNCS, vol. 11558, pp. 286–297. Springer, Heidelberg (2019)
30. Cox, M.T., Alavi, Z., Dannenhauer, D., Eyorokon, V., Munoz-Avila, H., Perlis, D.: MIDCA: a metacognitive, integrated dual cycle architecture for self regulated autonomy. In: Proceedings of the Thirtieth AAAI Conference on Artificial Intelligence, vol. 5, pp. 3712–3718. AAAI Press, Palo Alto (2016)
31. Cox, M.T., Dannenhauer, D., Kontrakunta, S.: Goal operations for cognitive systems. In: Proceedings of the Thirty-first AAAI Conference on Artificial Intelligence, pp. 4385–4391. AAAI Press, Palo Alto (2017)
32. Kondrakunta, S., Gogineni, V.R., Molineaux, M., Munoz-Avila, H., Oxenham, M., Cox, M. T.: Toward problem recognition, explanation and goal formulation. In: Working Notes of the 2018 IJCAI Goal Reasoning Workshop. IJCAI (2018). https://onedrive.live.com/?authkey=%21APF92MPxKVFvhLU&cid=FDD01A401DA67D10&id=FDD01A401DA67D10%214040&parId=FDD01A401DA67D10%21105&o=OneUp. Accessed 10 June 2020
33. Johnson, B., Floyd, M.W., Coman, A., Wilson, M.A., Aha, D.W.: Goal reasoning and trusted autonomy. In: Abbass, H.A., Scholz, J., Reid, D.J. (eds.) Foundations of Trusted Autonomy, pp. 47–66. Springer, Cham (2018)
34. Johnson, B., Roberts, M., Apker, T., Aha, D.W.: Goal reasoning with informative expectations. In Finzi, A., Karpas, E. (eds.) ICAPS 2016. Planning and Robotics: Proceedings of the 4th Workshop on Planning and Robotics (PlanRob), pp. 93–102. AAAI Press, Palo Alto (2016)
35. Wilson, M., Auslander, B., Johnson, B., Apker, T., McMahon, J., Aha, D.W.: Towards applying goal autonomy for vehicle control. Knexus Research Corporation, Springfield, VA (2013). https://apps.dtic.mil/dtic/tr/fulltext/u2/a610457.pdf. Accessed 10 June 2020
36. Powell, J., Molineaux, M., Aha, D.W.: Active and interactive discovery of goal selection knowledge. In: Murray, R.C., McCarthy, P.M. (eds.) Proceedings of the 24th International FLAIRS Conference, pp. 413–418. AAAI Press, Palo Alto (2011)
37. Roberts, M., Vattam, S., Alford, R., Auslander, B., Apker, T., Johnson, B., Aha, D.W.: Goal reasoning to coordinate robotic teams for disaster relief. In: Finzi, A., Ingrand, F., Orlandini, A.A. (eds.) ICAPS 2015. Planning and Robotics: Proceedings of the 3rd Workshop on Planning and Robotics (PlanRob), pp. 127–138. AAAI Press, Palo Alto (2015)
38. Rabideau, G., Chien, S.A., Mclaren, D.: Tractable goal selection for embedded systems with oversubscribed resources. J. Aerosp. Comput. Inf. Commun. **8**(5), 151–169 (2011)

Eligibility of English Hypernymy Resources for Extracting Knowledge from Natural-Language Texts

Oleg Sychev(✉) and Yaroslav Kamennov

Volgograd State Technical University, Volgograd, Russia
oasychev@gmail.com

Abstract. A common subtask of knowledge acquisition from natural-language texts is classifying words and recognizing entities and actions in the text. It is used in the analysis of both scientific and narrative texts. Thesauri and lexical databases containing hypernymy relationship between synsets may be a useful resource for entity and action recognition. In this study, we compared the performance of three major English thesauri containing hypernymy relationship in different forms - WordNet, Roget's Thesaurus, and FrameNet - on 6 word-meaning categories that are used for the analysis of narrative and scientific natural-language texts. The results show that WordNet contains more words than FrameNet, and is more suitable for scientific texts, but FrameNet contains better-defined hypernyms and shows better precision for many narrative natural-language tasks, especially for verbs. Roget's Thesaurus performance is average between WordNet and FrameNet in most word-meaning categories Enhancing FrameNet by adding more lexical units to existing frames would allow creating a powerful resource for entity and action recognition in text analysis. Fixing WordNet problems require revising its system of hypernyms.

Keywords: Computational linguistics · Hypernymy · Natural language processing · Knowledge acquisition

1 Introduction

Hypernymy as a relationship between generic and more specific terms is used in computational linguistics for determining semantic similarity between texts, anaphora resolution, and language translation. It can also be used while analyzing natural-language texts for entity recognition and action classification which is useful both in narratology and knowledge acquisition. This can be seen as a word-classification task. The hypernymy relationship is the basis for many knowledge bases [7]. One of the approaches to named entity recognition is determining the entities using trigger words. These words can be found not only through synonymy but also as hypernyms or hyponyms [8].

The reported study was funded by RFBR, project number 20-07-00764.

A. V. Samsonovich et al. (Eds.): BICA 2020, AISC 1310, pp. 501–507, 2021.
https://doi.org/10.1007/978-3-030-65596-9_61

There are two approaches to hypernymy allocation: automatic allocation or manual database compilation. The key convenience in the case of automatic selection is the volume of processed words [12]. The widely-known lexical databases for English created manually are WordNet, FrameNet, and Wiktionary. Using these databases allows simplifying training neural networks for entity recognition [4]. Knowledge base OntoWordNet [5] was created based on WordNet database. Some projects like [11] combine the information in existing databases, however, these combinations are often mechanical and doesn't add new structural information. Connections built in these networks can be used to simplify the understanding of complex concepts [13].

The research question of this study is: can hypernymy relationships in manually-created lexical databases be efficiently used for word classification in narratology and knowledge acquisition fields?

2 Method

Initially, we choose four sources of information about hypernymy: WordNet, FrameNet, Roget's Thesaurus and Wiktionary. WordNet [9] is an extensive lexical database (or thesaurus) where English words are organized in synonym sets (synsets) which are linked by conceptual relations, one of which is hypernymy. Wiktionary is a free dictionary developed by collaborative efforts. It contains thesaurus with hypernyms and hyponyms for words. FrameNet [1] is a lexical database of word valence that doesn't store hypernymy relationship. However, it stores word senses as Lexical Units and a frame can be considered a hypernym for its lexical units. Frames also have an inheritance relationship which can be considered as further hypernymy. Roget's Thesaurus is an old but still widely-used English thesaurus with the system of classes derived from epistemological works that is regularly updated and can be used in natural language processing [6].

To conduct this study, 6 groups of words with particular meanings were created. Four of them are used in narratology: Personal Relations (nouns), Leadership (nouns), Professions (nouns), and Verbs of speech. The words belonging to the groups Personal Relations, Leadership nouns, and Professions are often used in the narrative text instead of names to denote the speaker or actor (i.e. in phrases like "The father said..." or "The fisherman went..."). Determining the characters' professions, leadership positions and relations is necessary for attributing direct speech fragments [14] and actions to a particular character, and the first step in doing this is identifying such words in the text. Verbs of speech may be used to improve character identification because modern methods of named entity recognition have limited precision discerning persons from the other named entities [3]. Using a hypothesis that anyone who speaks is a person will allow increasing the precision of named entity recognition in narrative texts. The two last groups of words - Social Sciences and Geometry - were common terms chosen from different scientific areas to study the usefulness of hypernymy in lexical databases for knowledge acquisition in natural language texts.

Searching the thesauri for the relevant words we found that Wiktionary contains too few words that have hypernymy relationships defined (see Table 1) so it was excluded from the further study.

Table 1. The number of words from test samples with hypernymy relationships in each lexical database

Thesaurus	Personal Relations	Verbs of speech	Leadership	Professions	Social sciences	Geometry
Test sample	88	67	166	83	93	141
WordNet	63	67	136	66	85	141
FrameNet	50	64	164	37	44	22
Roget	40	63	111	36	66	58
Wiktionary	13	1	8	6	2	1

For each group of word meaning, a set of hypernyms with the best coverage (by F-measure) was determined for each lexical database. As a common task is testing whether a given word in the text belongs to the semantic group or not, a set of independent hypernyms will perform as well as one hypernym, so the lack of one hypernym covering the entire group is not significant. Precision, Recall, and F-measure were calculated for each hypernym set. However, not every word was present in each lexical database. Depending on the software using these hypernym sets, this can lead to false-negative results or to words marked as undetermined. So we calculated two measures of coverage: the first considered missing words as false negatives, the second ignored them, calculating coverage only for the words present in the lexical database. These hypernym sets are presented in Table 2.

3 Results

3.1 Hypernym Sets Performance

The high amount of hypernyms in the set shows the lack of high-level hypernyms for these semantic categories in the database. The table shows that Roget's Thesaurus has the necessary words spread over many categories, WordNet is organized better for nouns, but very poorly for verbs (17 hypernyms were necessary to cover verbs of speech with significant recall), while FrameNet requires fewer hypernyms on average.

Table 3 summarizes the results of the experiment for each word meaning. The numbers before the slash sign show the hypernym-sets performance when the words missing from the lexical database were considered false negatives while the numbers after slash show the situation when the missing words were omitted which led to higher recall with almost the same precision. The table shows that for most categories, hypernym sets have good precision (more than 80%) but recall varies widely. The main source of false positives for WordNet are polysemous words that have rarely used senses (like "cat" applied to a woman).

It can be seen that Roget's Thesaurus usually has average F-measure between the winner and the loser for each word meaning, except verbs of speech where

Table 2. Hypernym sets covering word meanings for thesauri

Word meaning	Thesaurus	Hypernyms
Personal Relations	WordNet	a person related by blood or marriage
	Roget	\\11 Consanguinity; \\943 Marriage; \\174 Posterity; \\173 Paternity; \\945 Divorce
	FrameNet	Kinship; Personal_relationship
Verbs of speech	WordNet	17 hypernyms
	Roget	\\555 Disclosure; \\512 Assent; \\561 Affirmation; \\490 Evidence; \\482 Answer; \\552 Information; \\535 Prediction; \\796 Request; \\799 Promise; \\877 Lamentation
	FrameNet	Statement
Leadership	WordNet	a person who rules or commands
	Roget	\\767 Government; \\722 Direction; \\914 Nobility; \\1040 Clergy; \\776 Master; \\723 Director
	FrameNet	Leadership
Professions	WordNet	a person who works at a specific occupation; a person engaged in one of the learned professions
	Roget	\\719 Agent; \\279 Traveler; \\723 Director; \\585 Artist; \\836 Treasurer; \\1008 Jurisdiction; \\777 Servant; \\789 Consignee; \\280 Mariner; \\1011 Lawyer
	FrameNet	People_by_vocation
Social sciences	WordNet	(psychology) the performance of some composite cognitive activity; (plural) any group of human beings (men or women or children) collectively
	Roget	\\767 Government; \\654 Business; \\965 Dueness; \\79 Class; \\384 Mankind; \\737 Opposition; \\967 Duty; \\813 Property; \\721 Conduct; \\766 Authority
	FrameNet	Aggregate, Organization, Leadership, Locale_by_ownership
Geometry	WordNet	a closed plane figure bounded by straight sides; the property of relative size or extent
	Roget	\\254 Angularity; \\257 Circularity; \\209 Length; \\211 Thickness; \\99 Four
	FrameNet	Dimension; Shapes

it performs on par with FrameNet, giving better recall with worse precision. WordNet covers scientific areas significantly better but it performed poorly for leadership and verbs-of-speech categories. Verbs of speech are particularly poorly defined in WordNet, scattered among 17 different hypernyms. The category "Professions" has no clear better thesaurus: FrameNet gives better precision while WordNet has better recall, containing almost twice as many words of this category as FrameNet.

The manual analysis shows that FrameNet has better-defined hypernyms for narrative word meanings than WordNet which is supported by consistently better precision. However, the smaller dictionary of lexical units sometimes gives FrameNet worse recall (see Personal Relations or Professions). Also, FrameNet performs worse for scientific word categories: mathematics and social sciences.

Table 3. Words coverage by hypernym sets (missing words are treated as errors/missing words are excluded)

Word meaning	Thesaurus	Precision	Recall	F
Personal Relations	WordNet	0.91/0.91	0.88/0.96	0.89/0.93
	Roget	0.80/0.80	0.72/0.94	0.76/0.87
	FrameNet	0.93/0.93	0.67/0.99	0.78/0.96
Verbs of speech	WordNet	0.81/0.81	0.89/0.89	0.85/0.85
	Roget	0.94/0.94	0.97/0.98	0.96/0.96
	FrameNet	0.99/0.99	0.93/0.94	0.96/0.96
Leadership	WordNet	0.84/0.85	0.32/0.38	0.47/0.53
	Roget	0.92/0.92	0.91/0.96	0.91/0.94
	FrameNet	0.90/0.90	0.97/0.98	0.93/0.94
Professions	WordNet	0.90/0.90	0.97/0.96	0.93/0.93
	Roget	0.92/0.92	0.93/0.99	0.92/0.95
	FrameNet	0.98/0.98	0.71/0.94	0.82/0.91
Social sciences	WordNet	0.90/0.90	0.99/0.99	0.94/0.94
	Roget	0,89/0,89	0,91/0,95	0,90/0,92
	FrameNet	0.86/0.86	0.74/0.87	0.80/0.87
Geometry	WordNet	0.96/0.96	0.96/0.97	0.96/0.96
	Roget	0.85/0.85	0.67/0.90	0.75/0.87
	FrameNet	0.90/0.90	0.27/0.81	0.41/0.86

3.2 Problematic Hypernyms

Hypernyms and their relationships in WordNet showed a big amount of inconsistency and lack of systematization. It's common to have two hypernyms with almost the same meaning, for example, the hypernym "a female relative" doesn't include words “wife”, “daughter” and “widow” which are hyponyms of “a person who belongs to the sex that can have babies”. The latter doesn't contain the words “aunt” and “sister” which are linked to “a female relative”. This can be resolved by making “a person who belongs to the sex that can have babies” a hypernym of “a female relative”.

Similarly, “a human being” and “a living organism characterized by voluntary movement” are two different hyponyms of “a living thing that has (or can develop) the ability to act or function independently” while human beings are characterized by voluntary movement.

Professions requiring college degrees like “teacher” or “doctor” have hypernym “a person engaged in one of the learned professions” which is separate from hypernym “a person who works at a specific occupation”: the first common hypernym for them both is “a human being” as if teaching isn't an occupation. It would work better if “a person who works at a specific occupation” would be a hypernym for “a person engaged in one of the learned professions.”

The main problem of Rodget Thesaurus is his historical taxonomic structure which depth is limited, so it usually takes many hypernyms to cover high-level word meaning useful in modern research.

FrameNet turned out to be more systematically structured which resulted in high precision for narrative word groups. But FrameNet's hypernymy hierarchy is much shallower as few frames have an inheritance relationship. The main disadvantage of FrameNet is a significantly lower number of lexical units compared to the words indexed in WordNet, so FrameNet often gives lower recall. As [1] states, avoiding incorporating technical terms and proper nouns was a design limitation of FrameNet. So by itself, it performs poorly in scientific applications like knowledge acquisition. FrameNet extensions for some domains are available to counter this problem [2,10].

4 Conclusion

Based on the data obtained, it can be concluded that WordNet and FrameNet dictionaries have the potential to highlight words of certain categories. WordNet is more useful for scientific knowledge acquisition because of its large vocabulary while FrameNet contains better-developed classes of common words and so is more useful for studying narrative texts. Roget's Thesaurus can also be used, but it mostly performs worse than the winner in each category. The study showed that WordNet's hypernymy structure is broad and deep but not systematic and prone to errors. FrameNet contains fewer words and shallower hypernymy hierarchy, but is more systematic and gives better precision in its area of expertise. Adding more lexical units to FrameNet will turn it into a powerful tool for classifying words in texts.

References

1. Baker, C.: FrameNet: a knowledge base for natural language processing. In: Proceedings of Frame Semantics in NLP: A Workshop in Honor of Chuck Fillmore (1929–2014), Baltimore, MD, USA, pp. 1–5. Association for Computational Linguistics, June 2014. https://doi.org/10.3115/v1/W14-3001. https://www.aclweb.org/anthology/W14-3001
2. Bertoldi, A., de Oliveira Chishman, R.L.: Developing a frame-based lexicon for the Brazilian legal language: the case of the criminal_process frame. In: Palmirani, M., Pagallo, U., Casanovas, P., Sartor, G. (eds.) AI Approaches to the Complexity of Legal Systems. Models and Ethical Challenges for Legal Systems, Legal Language and Legal Ontologies, Argumentation and Software Agents, pp. 256–270. Springer, Heidelberg (2012). https://doi.org/10.1007/978-3-642-35731-2_18
3. Chiu, J.P., Nichols, E.: Named entity recognition with bidirectional LSTM-CNNs. Trans. Assoc. Comput. Linguist. **4**, 357–370 (2016). https://doi.org/10.1162/tacl_a_00104. https://www.aclweb.org/anthology/Q16-1026
4. Deschacht, K., Moens, M.F.: Efficient hierarchical entity classifier using conditional random fields. In: Proceedings of the 2nd Workshop on Ontology Learning and Population: Bridging the Gap between Text and Knowledge, Sydney, Australia, pp. 33–40. Association for Computational Linguistics, July 2006. https://www.aclweb.org/anthology/W06-0505
5. Gangemi, A., Navigli, R., Velardi, P.: The OntoWordNet project: extension and axiomatization of conceptual relations in WordNet. In: Meersman, R., Tari, Z., Schmidt, D.C. (eds.) On The Move to Meaningful Internet Systems 2003: CoopIS, DOA, and ODBASE, pp. 820–838. Springer, Heidelberg (2003)
6. Jarmasz, M.: Roget's Thesaurus as a Lexical Resource for Natural Language Processing. Master's thesis, University of Ottawa, July 2003. https://arxiv.org/abs/1204.0140
7. Liang, J., Zhang, Y., Xiao, Y., Wang, H., Wang, W.Y., Zhu, P.: On the transitivity of hypernym-hyponym relations in data-driven lexical taxonomies. In: AAAI (2017)
8. Magnini, B., Negri, M., Prevete, R., Tanev, H.: A WordNet-based approach to named entities recognition, pp. 1–7, January 2002. https://doi.org/10.3115/1118735.1118744
9. Miller, G.A.: WordNet: a lexical database for English. Commun. ACM **38**(11), 39–41 (1995). https://doi.org/10.1145/219717.219748
10. Schmidt, T.: The Kicktionary: combining corpus linguistics and lexical semantics for a multilingual football dictionary, pp. 11–21. The Linguistics of Football, Narr, Tübingen (2008). http://nbn-resolving.de/urn:nbn:de:bsz:mh39-23491
11. Shi, L., Mihalcea, R.: Putting pieces together: combining FrameNet, VerbNet and WordNet for robust semantic parsing. In: Gelbukh, A. (ed.) Computational Linguistics and Intelligent Text Processing, pp. 100–111. Springer, Heidelberg (2005)
12. Shwartz, V., Goldberg, Y., Dagan, I.: Improving hypernymy detection with an integrated path based and distributional method, pp. 2389–2398, March 2016. https://doi.org/10.18653/v1/P16-1226
13. Swain, D., Tambe, M., Ballal, P., Dolase, V., Agrawal, K., Rajmane, Y.: Lexical Text Simplification Using WordNet, pp. 114–122, July 2019. https://doi.org/10.1007/978-981-13-9942-8_11
14. Sychev, O., Kamennov, Y., Shurlaeva, E.: Approach to automatic determining of speakers of direct speech fragments in natural language texts. In: Samsonovich, A.V. (ed.) Biologically Inspired Cognitive Architectures 2019, pp. 527–531. Springer, Cham (2020)

The Use of Digital Tools in the Formation of Two-Component Nuclear Energy on the Base of Economic Cross Method

D. V. Timokhin(✉)

National Research Nuclear University MEPhI, Moscow 115409, Russia
dtprepod@yandex.ru

Abstract. The article contains proposals for the development of modeling of the national industry using IT solutions based on the "economic cross" methodology. The reserves of the digitalization of the planning procedure are considered, taking into account the modern achievements of IT technologies and the development of the global information and communication infrastructure. The conclusions and suggestions contained in the article are formulated taking into account the extrapolation of current trends of nuclear energy development.

Keywords: Information and communication technologies · Two-component nuclear power · Strategic planning · Economic modeling · Foresight analysis · Sectoral economics

1 Introduction

Strategic design of economic processes of sectoral reorganization required by the state corporation "Rosatom" for the formation of an efficient two-component nuclear energy in accordance with the Decree of the President of the Russian Federation V.V. Putin No. 270 dated April 16, 2020, is one of the priorities of national innovative development. The conditions for effective strategic design from the point of view of the conjuncture of the global market are:

- the maximum possible degree of virtualization of modeling of the designed processes and their implementation;
- the maximum degree of use of intersectoral convergent processes for the most complete involvement of innovative resources for the development of related industries in the technological processes of the formation of two-component nuclear energy;
- outstripping replacement of information technologies traditionally used in the 2000s–2020s by their more modern counterparts to ensure outrunning industry development;
- integration of economic effects from the use of heterogeneous information and communication technologies and engineering solutions at the level of projected technological platforms of two-component nuclear energy.

A. V. Samsonovich et al. (Eds.): BICA 2020, AISC 1310, pp. 508–516, 2021.
https://doi.org/10.1007/978-3-030-65596-9_62

Standing in the context of the decree of the President of the Russian Federation V.V. Putin No. 270 of April 16, 2020, the tasks determine the following features of the strategic economic development of two-component nuclear energy, namely:

- long periods of implementation of chains of innovative projects, comparable in duration to the secular economic cycles of J. Forester;
- the need for preliminary framework agreement on the price parameters of the key resources for all stages of the development of two-component nuclear energy;
- creation of conditions for ensuring financial support of social and economic conditions for supporting the long-term functioning of the nuclear power industry;
- the relevance of creating conditions for the functioning of a transparent, barrier-free, inclusive and competitive environment that ensures optimal management of the process of creating a two-component nuclear energy based on the use of modern information and communication technologies.

The above conditions cause an increase in the need for a design system to expand the use of both software and hardware tools for information and communication support. The modern market of information and communication technologies provides an assortment variety of digital products sufficient to achieve the goals set by the state corporation Rosatom, and the use of foreign experience allows for shorter than the nearest competitors of the state corporation Rosatom (AREVA corp., Westinghouse Electric Corp., Urenco corp., Etc. .). At the same time, the stake placed by the state corporation Rosatom on advanced development limits the use of foreign experience in the selection of the most advanced information and communication solutions for the 2020–2021 technology level. In addition, the crisis phenomena of 2014–2020. raised a question regarding the structural choice of software and hardware digital support for the design and creation of a two-component nuclear power industry in terms of such parameters as: economically safe, both in the long-term and in the short-term, the structure of suppliers, hardware and software information compatibility of the information and communication solutions used, contractual and economic the form of using hardware and software solutions (licensing, leasing, acquisition of property rights to the results of intellectual activity and rights to trade them and sublease) and other parameters, for a comprehensive determination of which the capabilities of the methodology of modern economic science are limited.

2 The Statement of the Problem: Strategic Priorities for the Development of Two-Component Nuclear Energy in Russia

The concept of two-component nuclear energy represents the coexistence of both traditional nuclear energy producers and nuclear energy producers with a closed fuel cycle within the framework of a single national energy system. Traditional manufacturers are represented by NPPs for 2020, the fuel cycle of which ends with the formation of spent nuclear fuel (SNF), which requires disposal and creates significant environmental risks. The presence of such risks is the main disadvantage of the nuclear power industry, in which both the experts themselves - atomic scientists and representatives of the civil society agree. The presence of systemic risks of force majeure circumstances and the sensitivity of modern nuclear power to the possibility of long-term (more than one Forester cycle) storage of spent nuclear fuel (SNF) reduces the attractiveness of this type of energy as a priority in the post-industrial society. In terms of other socio - economic parameters, such as the production cost of 1 kW/h, the ability to solve energy problems in a short time, compliance with the cynological needs of society, nuclear energy is significantly ahead of both traditional (thermal and hydro-power) and alternative (solar energy, wind energy, energy of tides, etc.) types of electricity production.

Multiple reprocessing of spent nuclear fuel due to the closure of the nuclear cycle is and the repeated use of enriched uranium up to the reduction of its radioactivity to a level comparable to the natural level is a solution to the problem of SNF risk management. The fuel cycle closure model proposed by the state corporation Rosatom for the period up to 2035 is shown in Fig. 1.

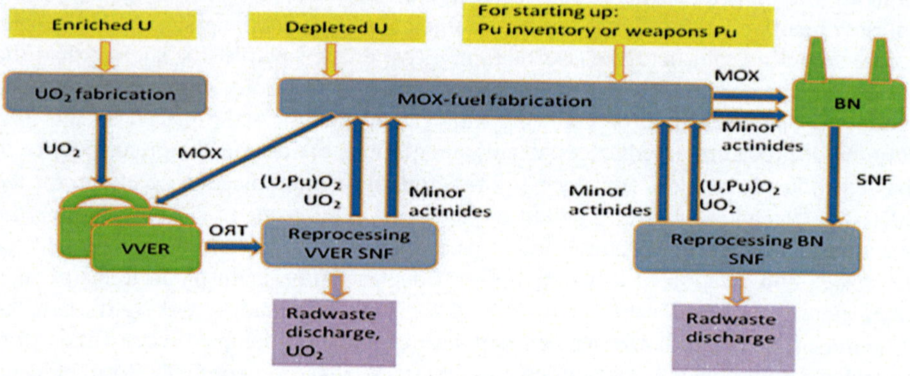

Fig. 1. Algorithm for closing the nuclear fuel cycle in the framework of the concept of two-component nuclear power [1]

At the transitional stage of closing the nuclear fuel cycle, traditional power engineering can be considered as a source of financial resources for the development of closed-cycle nuclear power (the "cow" model in accordance with the classification of the Boston Group's analytical matrix). In addition, in the period 2020–2050. Substantial replacement of traditional nuclear power by reactors with closed cycles (Model VVER-1200 according to the classification of the State Atomic Energy Corporation Rosatom) is impossible due to the extreme resource intensity of creating such plants and infrastructural and cyclical constraints on the widespread use of this technology in the Russian economy.

An increase in the share of BN-1200 reactors in the Russian power industry is planned for the period 2065–20115, as shown in Fig. 2.

Compiled on the basis of data from the official website of the state corporation "Rosatom", presentation by A.V. Putilova Consortium for the development of a

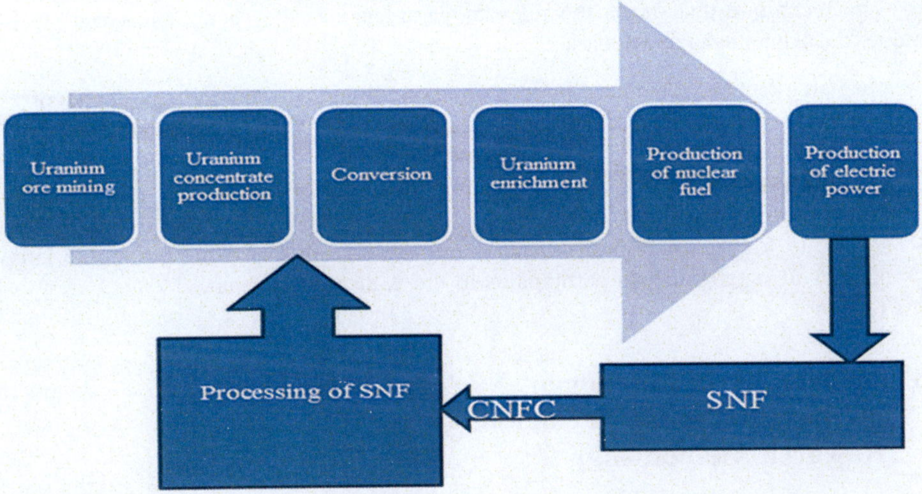

Fig. 2. The structure of the business processes of the nuclear closed nuclear fuel cycle - energy in accordance with the model of the economic cross. Developed by the author based on [2]

scientific and technological forecast for the sustainable development of the real sector of the economy (on the example of two-component nuclear energy).

Part of the needs of the BN-1200 reactors for nuclear fuel will be met not only by new enriched nuclear fuel, but also by using:

- spent nuclear fuel from Russian NPPs; - spent nuclear fuel from countries that do not have their own closed nuclear fuel cycle;
- previously transferred for temporary storage, but not utilized spent nuclear fuel.

Planning within the time frame indicated in Fig. 2 requires the development of an innovative planning methodology that will meet the following conditions:

- the possibility of expanding the number of initial parameters without the problem of multicollinearity;
- linking long-term target parameters of sectoral development with the fundamental economic parameters of the external environment, which will ensure the adaptability and flexibility of strategic development plans;
- the possibility of a multi-level iterative selection of potential participants in the planning process, creation and modernization of a two-component national energy sector;
- maximum throughput, that is, the ability to process the maximum amount of information at the initial (qualifying) stage, combined with the ability to visualize intermediate analytical results.

The expansion of the amount of information presupposes the maximum use of the required analytical method by means of its digital processing and the involvement of big data tools in the analytical process.

The technique that meets the relevant parameters is the "economic cross" technique, the advantages of which are:

- open architecture of the used information processing algorithm, a high degree of its integrability with other analytical tools used in planning in the nuclear power industry;
- the ability to use in the planning period, in terms of comparable duration with Forester cycles;
- suitability for the analysis of financial and resource flows arising between a large number of heterogeneous participants in the value added chain.

3 Model of Two-Component Nuclear Energy, Built on the Basis of the "Economic Cross" Methodology Research Methodology

The concept of closing the nuclear fuel cycle (CLFC) based on the economic cross methodology assumes:

- coordination of existing and projected material, financial and information flows in time on the basis of ensuring the balance of discounted outgoing and outgoing flows;
- the distribution of the added value formed through the sale of the bed product at the intersection of the fuel and resource cycles between all participants in the "economic cross" in time;
- determination of financial and economic reserves and additional financial requirements for various sections of the life cycle of the closed nuclear fuel cycle technology with their subsequent distribution among interested beneficiaries of the economic process in proportion to their expected share in the final economic result (added value).

Let us present the sequence of business operations included in the closed fuel cycle economic cross model in Fig. 2.

By using the "economic cross" model in the strategic planning of two-component nuclear power, it is possible to analyze a variety of alternative options for the interaction of counterparties within the framework of constructing a nuclear fuel cycle closure loop. Net value added producers should be used as the "center" of the economic cross, while participants in inbound and outbound business processes are viewed as investment or financially dependent in the period under review. The general model of the "economic cross" for two-component nuclear power is shown in Fig. 3.

The definition of the specific structure of a set of tools for information and communication support of processes should be carried out taking into account the current trends in the spatial development of information and production space of Industry 4.0.

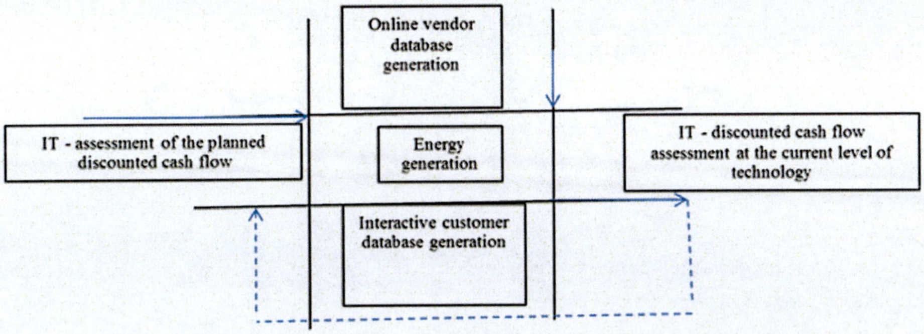

Fig. 3. The structure of the business processes of the nuclear closed nuclear fuel cycle - energy in accordance with the model of the economic cross; developed by the author based on [2]

4 Development of a Set of Measures for Digital Support of Strategic Planning Processes for Two-Component Nuclear Energy Based on the "Economic Cross" Methodology

The key areas of use of information and communication technologies in order to increase the efficiency of strategic planning of the "economic cross" of two-component nuclear energy are:

- automation of the processes of collection, processing, systematization, systematization and interpretation of information;
- digital modeling of technological platforms for the development of two-component nuclear energy;
- information and communication support for the promotion of two-component nuclear energy as an element of a post-industrial society;
- increasing the economic efficiency of standard technological processes forming a closed fuel cycle through automation based on modern information and communication solutions;
- expansion of logistics capabilities due to the maximum automation of distribution systems, control and planning.

The structure of information and communication support for the strategic development of two-component nuclear energy based on the "economic cross" model has a pyramidal shape, as shown in Fig. 4.

At present, the distribution of the use of technologies of the 6th economic order within the design of industry 4.0 as a whole within the framework of the technological pyramid presented in Fig. 4 is tilted towards expanding space towards digital interaction. Thus, IoT, Big Data, Cloud computing technologies account for 57% of all expenses of nuclear companies on the development of internal IT technologies. The distribution of financing by large companies for the development of IT technologies that are relevant for use within the framework of the proposed information and communication support for the "economic cross" of two-component nuclear energy is shown in Fig. 5.

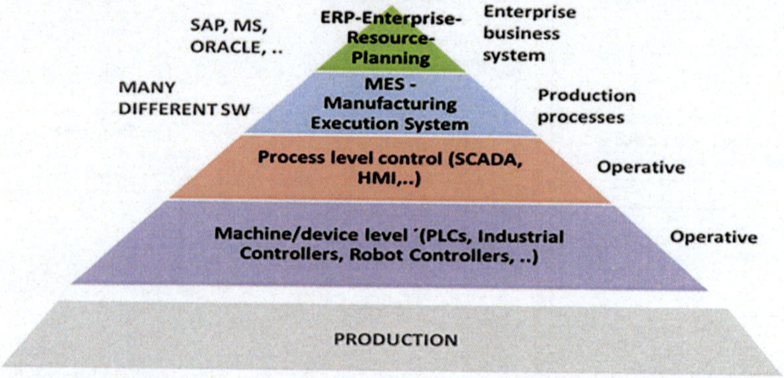

Fig. 4. Model of pyramidal structuring of digital components of the development of two-component nuclear power with a closed fuel cycle, recommended when using the model of the "economic cross" in strategic planning [5]

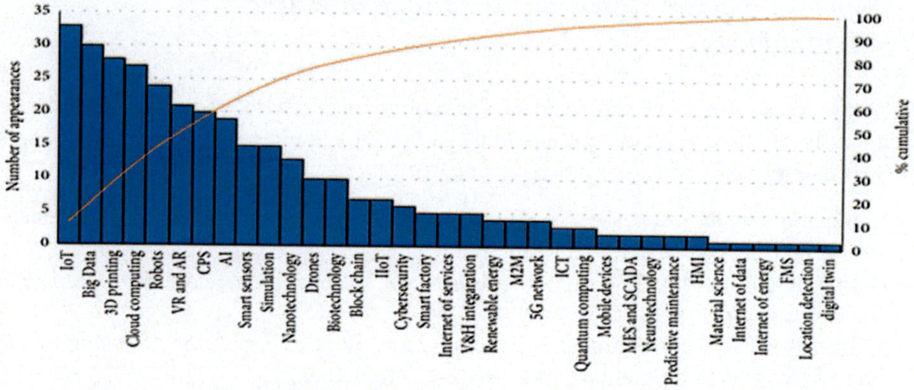

Fig. 5. Model of pyramidal structuring of digital components of the development of two-component nuclear power with a closed fuel cycle, recommended when using the model of the "economic cross" in strategic planning [6]

A promising direction in the development of a digital support system for strategic planning based on the "economic cross" model is to expand the practice of using databases and knowledge bases. For the period 2020–2050 the most relevant of the least demanded for non-economic reasons in 2010–2020. Technologies for digital support for the strategic development of two-component energy based on the "economic cross" model are:

- introduction of quantum computers to form time-distributed models of the "economic cross";

- wider introduction of neural networks adapted for the development of nuclear energy in the framework of the analysis of long-term models of development of two-component nuclear energy
- the formation of internal networks with the subsequent formation of the "Internet of energy".

5 Conclusions

Thus, the tasks of providing digital support for planning processes for the development of a two-component nuclear power industry in Russia require expanding the productivity of the configurations of analytical computer networks existing for 2020. The economic cross model is an effective tool for long-term planning. This model, which is the design of alternative economic long-term solutions as intersections of key economic processes, is recommended for use in the design of two-component nuclear power in the format of intersection of resource and production cycles. To provide hardware and software support for modeling the development of two-component nuclear energy based on the "economic cross" methodology, a more active introduction of quantum computers can be recommended to form distributed over time models of the "economic cross", a wider introduction of neural networks adapted for the development of nuclear energy within the framework of the analysis long-term models for the development of two-component nuclear energy and the formation of internal networks with the subsequent formation of the "Internet of energy". The main limitation on its use in the planned activities of the state corporation "Rosatom" is the limited information flows coming to the state corporation from the external environment and the complexity of its comprehensive interpretation.

References

1. Petrov, A., Shutikov, A., Ponomarev-Stepnoy, N., Bezzubtsev, V., Bakanov, M., Troyanov, V.: Prospects of creation of the two-component nuclear energy system. Nucl. Energy Technol. **5**, 265–271 (2019)
2. Timokhin, D.V., Putilov, A.V.: Innovative possibilities of using the "economic cross" methodology in predicting the development prospects of two-component nuclear energy. Innov. J. Volune **255**(1), 12–20 (2020). https://doi.org/10.26310/2071-3010.2020.255.1.002
3. Bugaenko, M.V., Timokhin, D.V., Putilov, A.V.: The use of IT technologies in the implementation of the economic cross methodology in the Breakthrough project of Rosatom. Proc. Comput. Sci. **169**, 445–451 (2020). https://doi.org/10.1016/j.procs.2020.02.227
4. Anirudh, A.: Incorporating industry 4.0 in firm strategy in AOM specialized conference: big data and managing in a digitalized economy. University of Surrey (2020). https://doi.org/10.4018/978-1-5225-3468-6.ch009
5. Andreja, R.: Industry 4.0 concept: background and overview. iJIM **11**(5), 77–90 (2017). https://doi.org/10.3991/ijim.v11i5.7072
6. Ocident, B., Gilibrays, G., Ocen, C., Oyondi, N., Musinguzi A., Omara T.: Exponential disruptive technologies and the required skills of industry 4.0. J. Eng. **11**(2020). Article ID 4280156, https://doi.org/10.1155/2020/4280156

7. Veile, J.W., Kiel, D., Müller, J.M., Voigt, K.-I.: Lessons learned from industry 4.0 implementation in the German manufacturing industry. J. Manuf. Technol. Manage. **1** (2019). https://doi.org/10.1108/JMTM-08-2018
8. Slipac, G., Zelijko, M., Šljivac, D.: Importance of reliability criterion in power system expansion planning. Energies **12**(9), 1714 (2020). https://doi.org/10.3390/en12091714
9. Tsitsimpelis, J., Taylor, J., Lennox, B., Joyse, M.: A review of ground-based robotic systems for the characterization of nuclear environments. Prog. Nucl. Energy (111), 109–124. https://doi.org/10.1016/j.pnucene.2018.10.023
10. Jalilian–Marian, J.: Particle Production at High Energy: DGLAP, BFKL and Beyond in Selected Papers from the 7th International Conference on New Frontiers in Physics (ICNFP 2018) – Universe, vol. 5, no. 2, p. 62 (2019). https://doi.org/10.3390/universe5020064
11. Cantore, N., Nussbaumet, P., Wei, M., Kammen, D.: Promoting renewable energy and energy efficiency in Africa: a framework to evaluate employment generation and cost effectiveness, vol. 12 IOP Publishing (2017). https://doi.org/10.1088/1748-9326/aa51da
12. Valdes, J., Bastiánpoque, A., Camargo, R., Fenández, V., Macia, M., Dorner, W.: Industry, flexibility, and demand response: applying German energy transition lessons in Chile. Energy Res. Soc. Sci. **54**, 12–25 (2019). https://doi.org/10.1016/j.erss.2019.03.003
13. Timokhin, D., Vorobyev, A., Bugaenko, M., Popova, G.: Formirovanie mekhanizmov ustojchivogo innovacionnogo razvitiya atomnoj otrasli. Tsvennie metalli **3**, 23–35 (2016)
14. Jungjohann, A., Besnard, M., Buser, M., Fairlie, I., MacKerron, G., Macfarlane, A., Matyas, E., Marignac, Y., Sequens, E., Swahn, J., Wealer, B.: World nuclear waste report. Focus Europe (2019). https://doi.org/10.25530/03552.42

Complex Objects Identification and Analysis Mechanisms

Mikhail Ulizko[1(✉)], Larisa Pronicheva[2], Alexey Artamonov[2], Rufina Tukumbetova[1], and Evheniy Tretyakov[3]

[1] Plekhanov Russian University of Economics, Stremyannyy Pereulok, 36, 115093 Moscow, Russia
mulizko@kaf65.ru

[2] National Research Nuclear University MEPhI (Moscow Engineering Physics Institute), Kashira Hwy, 31, 115409 Moscow, Russia

[3] Moscow Center of Fundamental and Applied Mathematics, Leninskie Gory, 1, Moscow 119991, Russian Federation

Abstract. Currently the volume of information on the Internet global network is inclined to increase. It affects various areas of activity. The paper presents the issue of identifying and analyzing complex objects in the context of information sources devoted to project activity of the US National Institutes of Health (NIH). As part of the solution to the problem, information sources relevant to the provided topic were found, and the data were downloaded with the use of agent techniques. The methods of primary analysis and data processing were developed to create a data storage structures in SQL and NoSQL models. The analysis of the presented database models was conducted, their advantages and disadvantages were revealed. As a result software tools have been developed that provide data representation of a complex object and organization of work with it by web interface.

Keywords: Web development · Web scrapping · Software agent · NoSQL · Data processing

1 Introduction

The information on the Internet is characterized with the following properties – heterogeneity, distribution, growing amounts. The heterogeneity is that the Internet consists of the text, tables, images, videocontent, etc. [10]

Distribution is about the fact that there is no unified standard to create web resources and post information on them. Information is posted randomly (it is not possible to differentiate a particular cluster). Therefore, it undermines search for data.

Growing amounts of information is about the traffic increase. According to Cisco report, Internet traffic has been significantly increased and it is expected to continue increasing over the next few years. According to projections, IP traffic will reach 4.8 ZB per year by 2022 [1].

A. V. Samsonovich et al. (Eds.): BICA 2020, AISC 1310, pp. 517–526, 2021.
https://doi.org/10.1007/978-3-030-65596-9_63

The user deals with such tasks as search for relevant sources, automated data processing, express data analysis, keep information up-to-date (information updating) [2, 6].

By complex object of the Internet is meant an entity, which possess a large number of properties, which are also heterogeneous. The paper presents the mechanisms of data collection and data analysis of the National Institutes of Health projects. The project data is presented on two resources – Federal RePORTER searchable database and RePORT NIH website, which provides the tools for data analysis of the National Institutes of Health, including writing search queries in research projects [3, 4].

2 Methodology

2.1 Agent Technology

The articles of the authors [11, 12] present data collection techniques and primary data analysis methods with the use of agent technologies. An agent is a software and/ or hardware item, that can function demandingly to perform tasks on behalf of its user [8].

Web scraping is one of the ways to extract data on the Internet. It is often used with HTTP and the GET method. The agent goes on site pages with the use of unique URLs and collect the content of these HTML pages [5].

There is a document in the web space – "The robots exclusion standard" [13]. "The robots exclusion standard" (robots.txt) is a file used by websites to restrict on agents accessing website content. The robots.txt file must be in the top-level directory of the host (have a path relative to the site name/robots.txt). Incompliance with this standard does not bear any responsibility, but the robots.txt file content govern the predisposition of the resource to agent activity.

Agents can perform tasks periodically on schedule, by this solving the problem of continuous updating of information on the Internet.

A significant number of articles are devoted to solving the tasks software implementation of agents [7, 14, 15]. Currently, one of the languages predisposed for the rapid writing of agents and solving data analysis problems is Python. To interact with Internet resources, a driver is also required, that provides remote behavior by a web browser. Selenium WebDriver is the driver. Using it with the help of Python makes it possible to control browser actions and collect data from web pages on schedule.

2.2 Data Unification

Resources in web space store data in accordance with their own requirements, which vary from resource to resource, so the representation of the same information on two Internet resources may differ. In addition to differences in the data itself, differences can be either in file formats or in the names of the corresponding fields. As a result, there is a need to unify and/ or merge data for subsequent storage.

Usually XML, CSV and JSON file formats are used in data processing. Using tags in XML documents enables store data, pass it and carry out data requests conveniently, but in some cases such use is unnecessary. The CSV file format is more universal and understandable, but does not enable the use of certain characters (‘,’, ‘;’, ‘\ n’), which limits its functionality for working with text data.

The most predisposed file format for solving merge tasks is JSON, a data interchange format that is convenient for reading and writing by both a person and a computer. JSON text is (in encoded form) one of two structures:

- a set of key pairs: value, where the key is a string, value - any form (string, date, integer, etc.);
- an ordered set of values (for example, an array).

A JSON file can store information about objects of any kind: educational organization, user, published work, patent, etc. A sample file of this format are shown below (see Table 1).

Table 1. JSON file example

```
{
"Project": {
  "ProjectId": ...,
  "ProjectNum": "...",
  "CoreProjNumber": "...",
  "ProjectTitle": "...",
  "ContactPi": "...",
  "AdditionalPiNames": null,
  "PiOrgName": null,
  "Ic": null,
  "Agency": "...",
  "Department": "...",
  "Fy": 2010,
  "FyTotalCostAmt": ...,
  "IrdbApplId": null,
  "SubProjectId": null,
  "SmOrgName": "...",
   ...
```

The JSON data format has been chosen for unification, as it has the following advantages:

- it is a light weighted format;
- it is easy for humans to read and write and for machines to understand and generate;
- it is often used as a format for data view in NoSQL database.

Applying agent technology and JSON file format for combining data we get an algorithm diagram (see Fig. 1). Software agents in this case has the same structure, differing among themselves by the behavior of the Selenium driver depending on the information source.

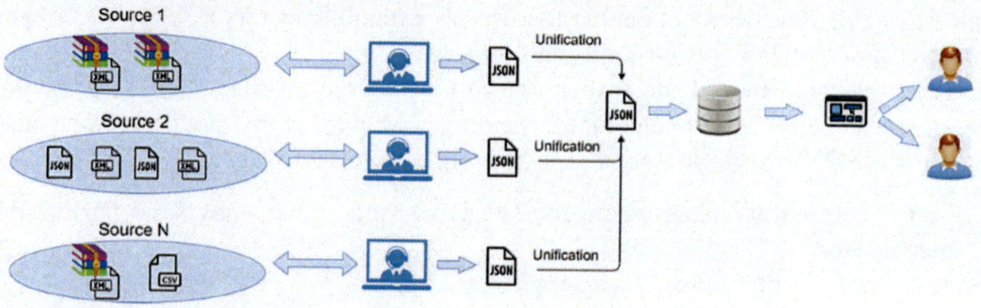

Fig. 1. Algorithm

2.3 Data Storing

An important aspect in the agents development is the maintenance of effective data storing in terms of volumes and processing speed, especially with a large number of entities, as the US National Institutes of Health projects (about 1 million projects).

Information systems that implement automated data collection, processing and manipulation, depending on the structure of this data and the relationships between them, use different types of databases.

A database (storage or database) is an information model that enables to store data about a group of objects that have the same set of properties orderly. A database model is a formal representation, that is, the logical structure of the database, as well as methods for processing and organizing data.

There are three main types: hierarchical, graph, relational. The hierarchical database is based on vertical communication and the ordering of elements in the record, one element is considered the main, the rest are subordinates. A graph database, also called a network database, is a modification of the hierarchical representation, where the elements can be connected regardless of the structure of the subordinate and the main element and do not have a strictly specified sequence, and it is also possible to establish horizontal connections in addition to vertical hierarchical connections.

At the same time, the database models themselves are divided into two large groups: SQL and NoSQL. Relational databases are included in the group of SQL databases, and the graph and hierarchical databases are included in the NoSQL group, therefore, we consider the description and fundamental differences of these groups (see Table 2).

For the SQL model, first of all it is necessary to develop a database schema. To design a database schema, the following steps are due to be taken:

1. identify objects, information of which is displayed in the database;
2. determine the relationship between the objects;
3. determine the objects attributes;
4. normalize relationships.

If a description of objects does not exist in advance, the first point is carried out by the analyst or database developer. To perform the second stage it is necessary to consider the types of relationships. Database tables can be related to each other by the

Table 2. Description of the differences between SQL and NoSQL databases

Criterion	SQL	NoSQL
Data storage view	The database consists of tables that contain columns and rows. All rows of one table have a single structure that defines the data types of each column	The database can be presented in the form of a graph, storage of the "key-value" type, or in the form of a document-oriented system
Changing the structure and data types	The database uses strictly defined database schemas and data types, the change of which leads to a change in the database as a whole	The database does not impose restrictions on the types of stored data, unstructured data is allowed, it allows you to change the structure without editing existing data
Data access	Data is created, updated, deleted, and queried using SQL Structured Query Language	Data is created, updated, deleted, and queried using an API method call, some implementations provide SQL-like syntax
Data integrity	Data that does not satisfy the database schema cannot be stored in it; this restriction promotes data integrity	There are no restrictions on the data structure, therefore, data integrity control lies entirely with the software being developed
Scalability	Vertical scalability: an increase in the load on each individual server is carried out, CPU and RAM capacities are increased	Horizontal scalability: the load is distributed due to the separation of data and the addition of new servers, the use of cloud technology

following types: one-to-one, one-to-many, many-to-many. To determine the attributes of objects, it is necessary to decide which of the table's fields are key – they are designated as PK (Primary Key), which fields are the keys of related objects – they are designated as FK (Foreign Key), the remaining fields are non-key. Also, each field of the database table must have a standardized data type: string, integer, real number, boolean (true or false), date, time, etc.

There are three properties of normal database forms, with each next normal form being better than the previous form. Also, during the transformation to the next normal form, the properties of the previous normal properties are preserved. The first three normal forms are considered the basic ones. The first normal form is that any table field contains indivisible information and a primary key is defined in the table. The rule of the second normal form is that the scheme must satisfy the first normal form and any non-key field must be uniquely identified by key fields. Third normal form is that the circuit must satisfy the second normal form and none of the non-key fields should be uniquely identified by the value of another non-key field.

Now it is necessary to consider the implementation of the proposed methodology based on an analysis of the information sources of the US National Institutes of Health.

Target information (project activity) is contained in the two information sources mentioned. The information in Federal RePORTER is taken from several existing databases - eRA, IRDB and databases of institutions. Federal RePORTER system unifies data between these heterogeneous sources. Federal RePORTER presents data on scientific projects of the Department of Defense, the Department of Education, the Department of Health and Human Services and other institutions. The source is predisposed to agent activity.

RePORTER contains information on in-patient and outpatient research projects funded by the NIH, research papers, and patents and clinical research supported by the NIH grants. The information is taken from several existing databases - eRA, Medline, PubMed Central, and the NIH internal database. When downloading data, the developed agent takes into account that according to the Robots Exclusion Protocol, the source is not predisposed to agent activity (an IP address can be blocked).

A database schema is constructed based on the analysis of the initial data. The scheme consists of 13 tables, the part of the scheme related to Federal RePORTER is presented below (see Fig. 2).

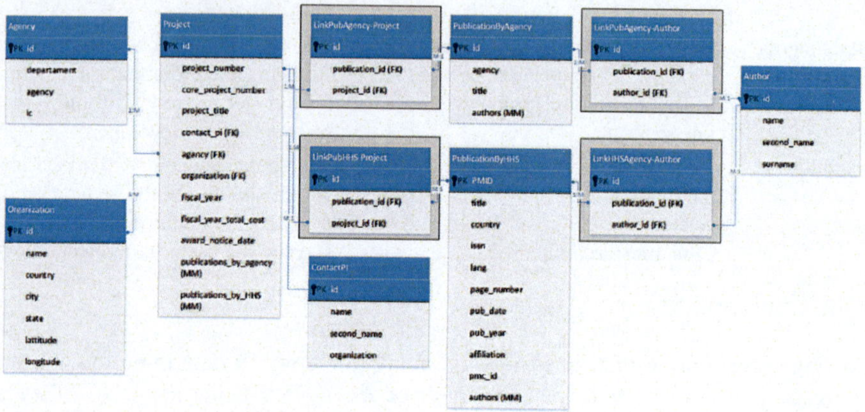

Fig. 2. Database diagram

In the case of using a non-relational database (NoSQL), it is sufficient to define the DBMS. For example, if you use a key-value database, it is enough to have values (JSON files), the rest of the work will be on the DBMS. One such example is ElasticSearch, an open source distributed software search engine that has a JSON REST API for interacting with downloaded data.

Below the main accepted concepts for a distributed software search system are considered [9]:

- index - a logical grouping of document types, index name must be unique and consist of lowercase letters;
- type - a logical separation of various documents types;
- node - an isolated amount of memory (server) that stores part of the data;
- cluster - a group of nodes that process requests;
- shard (segment) - data instance.

Based on the basic concepts, it can be noted that Elasticsearch is a distributed storage system. A graphical representation of nodes, clusters and shards is presented below (see Fig. 3).

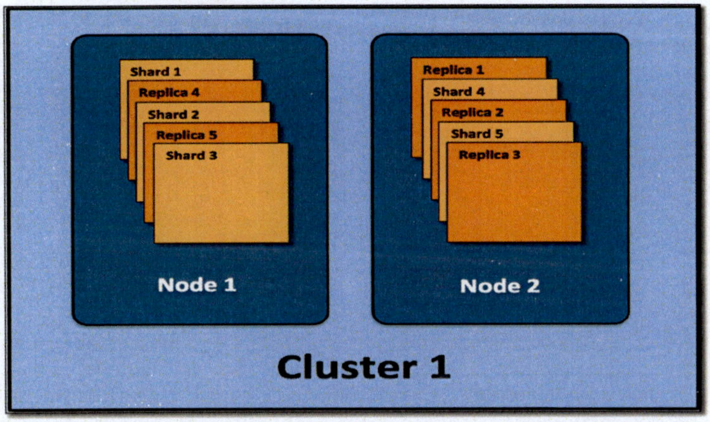

Fig. 3. The graphical representation of memory allocation in Elasticsearch

2.4 Visualization and Analysis

The information obtained is interesting because it can be analyzed. The relational database is accessed using SQL queries. Since the initial data for queries is a collection of tables, the return value will also be a table (or a number). An important role in the analysis is the clarity in data representation, so developed systems with databases often use visualization tools. An example is the library of ChartJS, Plotly and others can be an example of such visualization tools. These libraries was used for constructing bar charts that reflect the average funding per project among the institutes of the National Institutes of Health in $million (see Fig. 4) and the total funding in $ billion (see Fig. 5) for the period of 2010–2018.

To visualize data in ElasticSearch, there is Kibana, an open source data visualization and exploration platform, which is a powerful analysis tool. An example of analysis is given below (see Fig. 6).

The advantage of Kibana is the availability of various visualization forms. In addition to standard diagrams, the platform has tag clouds and geographical maps. All diagrams can be combined in a dashboard, which reflects all the necessary attributes, which will simultaneously change when the search query is refined. Thus, Kibana allows you to quickly evaluate raw data and subsequently perform in-depth data analysis based on the conditions of the task. This initial analysis of the 2013 RePORTER data showed that majority of the projects were research grants on genetics, clinical research and cancer.

In the course of the work, Federal RePORTER and RePORTER information sources, which have data on the project activities of NIH and other divisions of HHS and other departments, are examined. The designed database is filled with data from

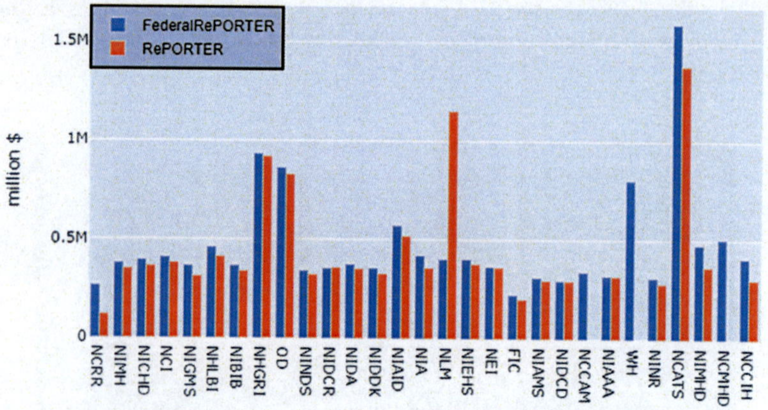

Fig. 4. Distribution of funding between health institutes

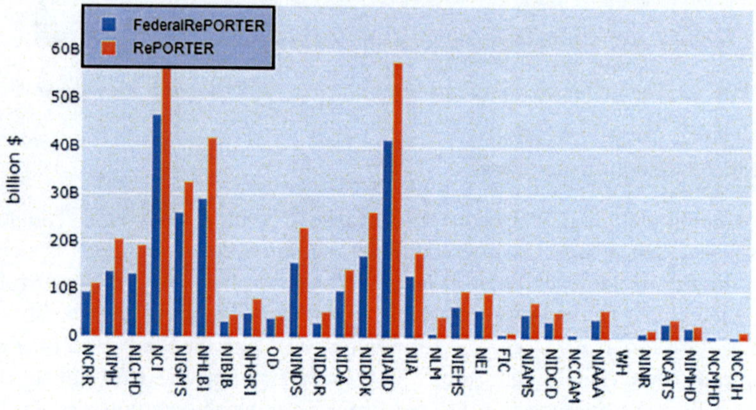

Fig. 5. Average cost of one project

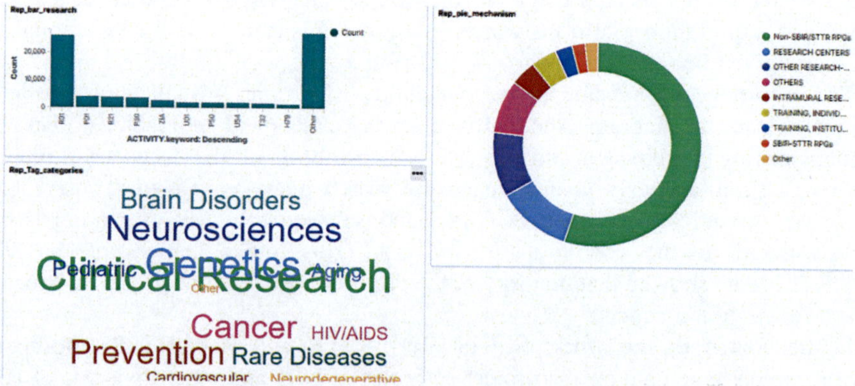

Fig. 6. Dashboard Research for RePORTER in 2013.

these sources. To interact with data, a web application was developed in the PyCharm Django framework in the Python programming language and a database was built using Elasticsearch and Kibana. The analysis of project activities was carried out in four ways: conducting initial analysis using scripts in Python, building interactive Plotly graphs, building a standard information retrieval system (backend SQL query), building a search database using Elasticsearch and Kibana. The relational database consists 13 tables with more than 3 million records. A distributed database is recommended due to the high speed of system development and data processing and visualization.

3 Conclusion

Due to the constant amount of information on the Internet, to carry out the intelligent search is constantly becoming more complicated, objects become more multifaceted, filling up with hidden connections that a person cannot identify by himself. In this regard, there is a need to identify relevant information sources and build intelligent search systems to identify these hidden connections and determine development trends in a particular area.

Within the framework of this problem, on the basis of the proposed methodology, the task of collecting and analyzing data on the National Institutes of Health has been solved. It is concluded that scientific research is mostly devoted to genetics, cancer, clinical research and is implemented in the form of grants.

This approach can also be used to analyze complex objects such as organizations involved in project activities.

Acknowledgments. The study was carried out at the expense of the Russian Science Foundation grant (project No. 19-71-30008).

References

1. Cisco. https://www.cisco.com/c/dam/m/en_us/solutions/service-provider/vni-forecast-highlights/pdf/Global_2022_Forecast_Highlights.pdf. Accessed 29 July 2019
2. Kulik, S.D., Protopopova, J.: Genetic algorithm and software tools for solving optimization problems in intelligent robotics. Mech. Mach. Sci. **80**, 171–178 (2020)
3. Federal RePORTER. https://federalreporter.nih.gov. Accessed 30 Jul 2019
4. RePORTER. https://report.nih.gov/. Accessed 30 July 2019
5. Ionkina, K., Svistunov, A., Galin, I., Onykiy, B., Pronicheva, L.: MIAS database semantic structure. Procedia Comput. Sci. **145**, 254–259 (2018)
6. Kulik, S.D.: Factographic information retrieval for biological objects. Bio. Insp. Cogn. Arch. **2019**, 277–282 (2019)
7. Filatova, T., Verburg, P.H., Parker, D.C., Stannard, C.A.: Spatial agent-based models for socio-ecological systems: challenges and prospects. Environ. Modell. Softw. **45**(7), 1–7 (2013)
8. A Brief Introduction to Software Agent Technology. https://link.springer.com/chapter/10.1007%2F978-3-662-03678-5_2. Accessed 10 Aug 2019

9. Elastic: Terminology. https://www.elastic.co/guide/en/elastic-stack-glossary/current/terms.html. Accessed 01 Aug 2019
10. CMPUT690 Principles of Knowledge Discovery in Databases. https://www.exinfm.com/pdffiles/intro_dm.pdf. Accessed 30 July 2019
11. Ananieva, A.G., Artamonov, A.A., Galin, I.U., Tretyakov, E.S., Kshnyakov, D.O.: Algorithmization of search operations in multiagent information-analytical systems. J. Theor. Appl. Inf. Technol. **81**(1), 11–17 (2015)
12. Onykiy, B., Artamonov, A.A., Tretyakov, E.S., Ionkina, K.V.: Visualization of large samples of unstructured information on the basis of specialized thesauruses. Sci. Visual. **9**(5), 54–58 (2017)
13. A Standard for Robot Exclusion. http://www.robotstxt.org/orig.html. Accessed 20 Jan 2020
14. Knox, S., Meier, P., Yoon, J., Harou, J.J.: A python framework for multi-agent simulation of networked resource systems. Environ. Modell. Softw. **5**(103), 16–28 (2018)
15. Schreinemachers, P., Berger, T.: An agent-based simulation model of human–environment interactions in agricultural systems. Environ. Modell. Softw. **7**(26), 845–859 (2011)

To Help or Not to Help: A Network Modelling Approach to the Bystander Effect

Joey van den Heuvel and Jan Treur(✉)

Social AI Group, Vrije Universiteit Amsterdam, Amsterdam, The Netherlands
joeyhilll@hotmail.com, j.treur@vu.nl

Abstract. This paper focuses on defining and simulating behavioural outcomes of the bystander effect. These insights were modeled by temporal-causal networks. Typical patterns of bystander behaviour were translated into three requirements and seven simulated scenarios of the bystander effect. All scenarios were simulated to showcase the main bystander effect dynamics and its accordance with the literature. Unknown parameters of the effect were further estimated by a Simulated Annealing algorithm. In the end, the created model shows the potential to simulate the bystander effect in different and new scenarios. The created model adhered to the stated three requirements and shows potential to verify the model predictions independently and for new bystander situations.

Keywords: Bystander effect · Network modelling · Simulated annealing

1 Introduction

Helping other people occurs more frequently when alone than in the company of passive bystanders [3]. This phenomenon is called the bystander effect, a prosocial behaviour affected by the altruistic nature of people; e.g., [10], p. 328. Here, help is defined as bystander intervention and occurs when a bystander offers support to the victim(s). This support is either expressed verbally and/or physically, in such that it affects the victim's cognitive or behavioural state over time. Cases with no help offered at all, are called bystander apathy. This was first investigated in the murder of Kitty Genovese [14], which was the main driver in literature to investigate why bystanders show these behavioural patterns [9, 14, 19].

With decades of subsequent research on the bystander effect topic (see Table 1 in Fischer et al. [8]), this study builds upon earlier research by combining causal relationships. Many relationships and patterns were found in the research of [8] but only focus on earlier research. By choosing some moderator variables from this research, the combined impact of moderator variables within (new) bystander situations can be determined. Within this paper, the following variables were chosen: Proportional Dominance Effect (PDE), In-Group Effect (IGE), Effect Evaluation (E), Dangerous emergency (DEm) and Distress (D) perception. These variables were firstly chosen due to its strong and unambiguous character within the bystander effect [1, 8]. Secondly, helping is often in line with the cost-reward model as the decision to act or not comes with benefits and drawbacks [7, 18]. With balancing variables in the cost-reward

A. V. Samsonovich et al. (Eds.): BICA 2020, AISC 1310, pp. 527–540, 2021.
https://doi.org/10.1007/978-3-030-65596-9_64

model, potential bystander intervention and apathy models are more likely to be simulated both.

As these five variables have an impact on the overall bystander effect, it is still necessary to define how the bystander effect works. Therefore, a conceptual model of Latané and Darley [15] is used as starting point for the bystander effect dynamics (see Fig. 1). While this model shows conceptually how bystanders make decisions, it does not clear where and how each moderator variables is incorporated. Also, with numerous other variables available, it is hard to pinpoint what the realistic combined total impact would be for every variable over time. Although, incorporating all found variables will result in a too large model that lacks focus. By using an initial solution that is in line with the cost-reward model, answers can still be found to gaps in the literature (e.g. unclarity what the in-between effects are between variables in [6]). The total combined impact of the five variables can be modeled within the temporal-causal networks by combination functions. With this in mind, simulating the effects between the five variables can be realised. The final goal of this paper is to determine what settings of the network characteristics simulate the bystander effect most realistically.

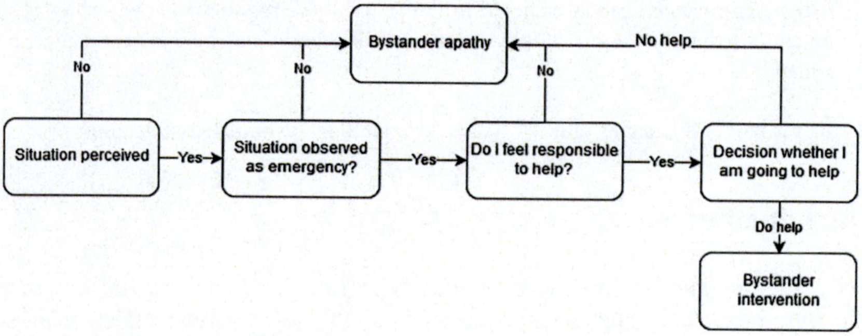

Fig. 1. Procedure for determining whether to help or not to help

2 The Network-Oriented Modelling Approach

First, the Network-Oriented Modelling approach is explained in some detail. The approach was described in [20] and addresses how mental and behavioural states change over time. For each state, characteristics are given to calculate the state value over time, which leads to the opportunity of model complex processes in an easy, structured and intuitive manner. This approach allows to incorporate temporal complexity into causal graphs. In total, three network structure characteristics are used for a model; together they define a temporal-causal network model:

- **Connectivity:** connection weights $\boldsymbol{\omega}_{X,Y}$ from a state X to a state Y
- **Aggregation:** a combination function for each state Y, denoted by $\mathbf{c}_Y(..)$
- **Timing:** a speed factor for each state Y, denoted by $\boldsymbol{\eta}_Y$

This is a model where the connections between states and other network characteristics are defined in a declarative mathematical manner. By the developed dedicated modeling environment (see [21], Chap. 9), the mentioned network characteristics are transformed automatically into a numerical representation that can be used for simulation, which has the definitions shown in Table 1.

Table 1. Numerical representations for a temporal-causal network model

Concept	Representation	Explanation
State values over time t	$Y(t)$	At each time point t each state Y in the model has a real number value in [0, 1]
Single causal impact	$\mathbf{impact}_{X,Y}(t) = \boldsymbol{\omega}_{X,Y}\, X(t)$	At t state X with connection to state Y has an impact on Y, using connection weight $\boldsymbol{\omega}_{X,Y}$
Aggregating multiple impacts	$\mathbf{aggimpact}_Y(t)$ $= \mathbf{c}_Y(\mathbf{impact}_{X1,Y}(t), \ldots, \mathbf{impact}_{Xk,Y}(t))$ $= \mathbf{c}_Y(\boldsymbol{\omega}_{X1,Y}X_1(t), \ldots, \boldsymbol{\omega}_{Xk,Y}X_k(t))$	The aggregated causal impact of multiple states X_i on Y at t, is determined using a combination function $\mathbf{c}_Y(V_1, \ldots, V_k)$ and applying it to the k single causal impacts
Timing of the causal effect	$Y(t+\Delta t) = Y(t) + \boldsymbol{\eta}_Y\, [\mathbf{aggimpact}_Y(t) - Y(t)]\, \Delta t = Y(t) + \boldsymbol{\eta}_Y\, [\mathbf{c}_Y(\boldsymbol{\omega}_{X1,Y}X_1(t), \ldots, \boldsymbol{\omega}_{Xk,Y}X_k(t)) - Y(t)]\, \Delta t$	The causal impact on Y is exerted over time gradually, using speed factor $\boldsymbol{\eta}_Y$; here the X_i are all states from which state Y has incoming connections

3 The Introduced Temporal-Causal Network Model

In Fig. 1 an overview of the model for the bystander effect is shown. In total, four areas were identified. For each area, an in-depth literature analysis was conducted to find which variables belong to which area and/or which states X_k can be defined. Starting with a situation where a potential bystander effect could first occur, is the perception of the event. When an event is not perceived, the individual will not intervene at all. This perception is dependent upon the individual sensory capabilities; e.g., [20], p. 213. In this model, various variables were perceived from the world, processed and interpreted by the sensory system, and given meaning to. The meaning is determined by making a trade-off of all input variables and decide to act upon or not. This process is highly similar to Fig. 1, where conceptually it is shown how the bystander effect works. Next, the position of found states X_k require to be defined. In total, five variables were defined which are all handled as a chain of states within a person. This chain means that each variable has its perception of the world, the collection and interpretation of the sensory input, leading to a certain meaning and belief to it. This chain starts with specified values from the given variables, which are the costs and rewards. The following three variables were identified: Identifiable Victim Effect (IVE), In Group Effect (IGE), and the Propositional Dominance Effect (PDE) [5, 6, 13]. With clear effects in the latter two, Fischer et al. [8] argue that the effect of helping someone familiar is not different from an unknown person. Therefore, IVE remains unambiguous as a variable input variable. With this information, PDE and IGE remain within the scope of the research and more clarity is needed on the IVE aspect. For PDE, Erlandsson et al. [6] state:

> 'First, a within-subject contrast test showed that there was a linear trend on helping motivation, $F(1,38) = 55.67$, $p < 0.001$, $\eta^2 = 0.594$. As predicted, helping motivation increased as the victim reference-group got smaller.' (p. 5).

For IGE, the following is stated in the same paper:

> 'First, a within-subject contrast test showed that there was a linear trend on helping motivation, $F(1,38) = 105.58$, $p < 0.001$, $\eta^2 = 0.735$. As predicted, helping motivation increased as the victim became more part of one's in-group' (p. 6).

This shows that PDE and IGE are both strong determinants with IGE having a larger effect. In both a cost and rewarding perspective, Dangerous Emergencies (DEm) play a role in either one perspective dependent on how dangerous the situation becomes. Fischer et al. [7] show that dangerous situations are perceived better and lead to intervening more often than in non-dangerous situations. Therefore, when the emergency need is low, one will show apathy more frequently (which is depicted in Fig. 1 as well in step two). Also, the need to help will increase over time when the situation becomes more dangerous. In the paper of Fischer et al. [8], it is stated:

> 'In the present meta-analysis, perceived danger is reflected by the following coded variables: (a) emergency danger (high vs. low), (b) perpetrator present versus absent (…) As expected, non-emergencies ($g = -0.47$, $SE = 0.041$, $Z = -11.42$, $p < 0.001$, $N = 22$) yielded a larger bystander effect than dangerous emergencies ($g = -0.30$, $SE = 0.048$, $Z = -6.34$, $p < 0.001$, $N = 65$) and potentially dangerous villain acts ($g = 0.29$, $SE = 0.101$, $Z = 2.82$, $p < 0.005$, $N = 14$)' (p. 7).

Therefore, a key influence to the DEm is the presence of a fierce perpetrator [8]. With no fierce perpetrator present, bystander apathy will more likely occur and intervention when that person is present. From a cost perspective only, Distress (D) is added as it affects the bystander effect [2]. When the distress level is enhanced, an avoidance or freezing dominated bystander apathy effect occurs [12]. Furthermore, in this paper is stated that 'only personal distress predicted the negative effect of bystanders during an emergency.' (p. 252). Determining the costs and rewards is explained by the other states in the model. An individual evaluates (E) the situation all the input and prepares an answer to the situation. For example, Fischer et al. [7] state:

> 'Dangerous emergencies should be associated with a clearer and earlier recognition of the emergency, resulting in an increased degree of empathic arousal, increased attribution of personal responsibility, and thus a greater willingness to accept increased costs for helping.' (p. 270).

This means that the costs and rewards are weighed. Dovidio, Piliavin, Gaertner, Schroeder, and Clark III [4] show that when the costs of helping are low, people will more likely intervene when the net costs are high. The outcomes do not outweigh the potential rewards from it, where the Effect Evaluation (E) plays a decisive role. This intervention has a specific effect on the overall variables and whether it is smart to help in a certain situation. All insights lead to Fig. 2, which show a temporal-causal model for the bystander effect. In this case, only one individual is shown and determined for whether they will show bystander apathy or intervention. All abbreviations mentioned in Fig. 2 are explained in a Table 2. Summarized, this network model contains state variables (called states) starting with a world state ws_x that affects the sensory state ss_x for a stimulus x. This leads to a sensory representation state srs_x and a certain belief b_x of the bystander situation. Based on literature, other states have been added. The belief of all discussed states influence the balance between positive influence p_{inf} and negative influence n_{inf}. Next, a decision is prepared as ps_a and finally the actual action execution is performed, modeled as es_a.

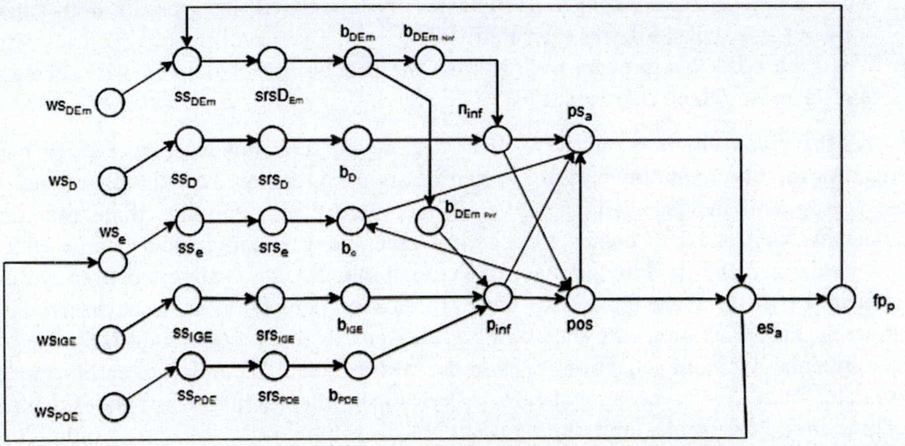

Fig. 2. Connectivity of the temporal-causal network for an individual determining to help a victim. The states and causal connections provide an overview of the factors that play a role in the causation for the decision to help or not to help.

Table 2. Explanation of the states

State Nr	Name	Explanation
X_{1-5}	ws_x	World state for stimulus *x*
X_{6-10}	ss_x	Sensor state for stimulus *x*
X_6	ss_D	Sensor state for stimulus Distress (D)
X_7	ss_{DEm}	Sensor state for stimulus Dangerous Emergency (DEm)
X_8	ss_{IGE}	Sensor state for stimulus In-Group Effect (IGE)
X_9	ss_{PDE}	Sensor state for stimulus Proportional Dominance Effect (PDE)
X_{10}	ss_e	Sensor state for effect e
X_{11-15}	srs_x	Sensory representation state for *x*
X_{16-22}	b_x	Interpretation of sensory representation for *x*
X_{21}	$b_{DEmpinf}$	Belief state for rewarding influence of a dangerous emergency situation
X_{22}	$b_{DEmninf}$	Belief state for cost influence of a dangerous emergency situation
X_{23}	n_{inf}	Interpretation of costs of incoming variables
X_{24}	p_{inf}	Interpretation of rewards of incoming variables
X_{25}	pos	Prior ownership state for action a with incoming belief b_e
X_{26}	ps_a	Preparation state for response action a
X_{27}	es_a	Execution state for action a
X_{28}	fp_p	Presence of a fierce perpetrator

4 Simulation Results

In order to determine the quality of the simulation, several patterns were found in the previous section. These were written in the form of model requirements to show which outcomes the bystander effect:

1. An increasing dangerous emergency situation will positively influence the bystander intervention [8].
2. When a person has a strong IGE/PDE, the bystander will intervene in both situations but IGE will be faster than PDE [6].
3. When the costs overpowers the rewards, the probability of helping will decrease (and increase when vice versa) [4].

All the given requirements were divided in seven scenarios to showcase the patterns. For the first requirement, three scenarios are used to show how the likelihood of intervening will increase with a higher danger perception. For this, three different parameters were used to depict a dangerous emergency situation: low danger (0.2), medium danger (0.5) and high danger (0.8). Each initial value is always chosen within the scale of [0,1], using 0.5 as the person's relative perception as what an average danger is. The simulations are shown in Fig. 3. Here, as in the other graphs, time is on the horizontal axis and activation level on the vertical axis. According to earlier found research, there should be an increase in bystander intervention when the situation becomes more dangerous until tipping point T. This tipping point leads to an individual showing bystander intervention and not apathy anymore. In Fig. 3, the three respective scenarios are depicted with an additional letter A for the intervention.

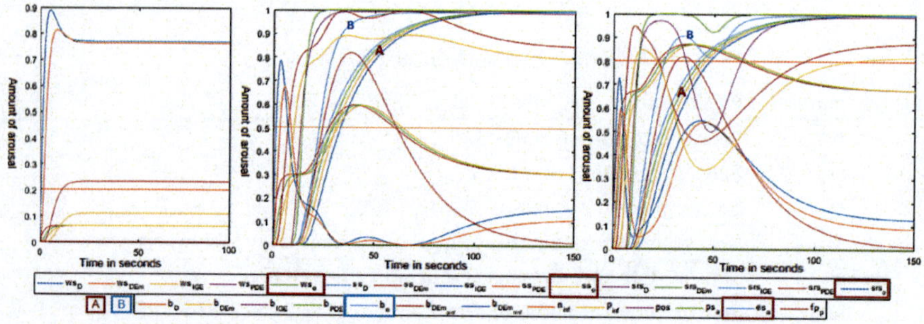

Fig. 3. Three different scenarios with DEm = 0.2 (left), DEm = 0.5 (middle), DEm = 0.8 (right). Here A indicates the intervention.

This letter shows four lines (depicted as A) that show the pattern of bystander intervention as it slowly converges to value 1 in the latter two graphs. Generally, when the belief b_e of the effect e is high, one will act in the situation (line depicted with B). These lines are the most interesting of all depicted variables: they depict a positive effect evaluation and thus a higher possibility of acting represented by es_a. In the dangerous situation of DEm = 0.2, no intervention is depicted as the individual will show bystander apathy. In the middle graph, the faster intervention takes place as the situation is more dangerous than the left one, which is according to the found pattern. Interestingly, DEm = 0.8 does not yield a much faster intervention than DEm = 0.5. This is most likely due to the higher costs of helping. For the second requirement, the descriptive statistics in the research of Berggren et al. [1] were used. Both IGE and PDE variables have an bystander intervention pattern when either one or both of the values are high. The expected pattern would be that PDE leads to a faster intervention due to $\eta^2 = 0.735$ than the score of IGE with $\eta^2 = 0.594$.

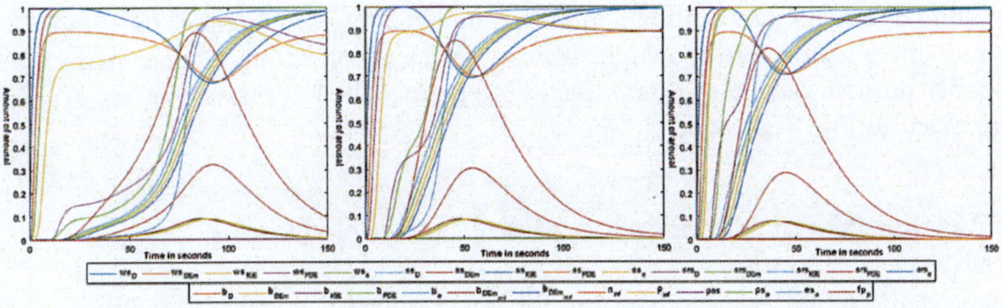

Fig. 4. Simulations of different PDE/IGE values with left PDE = 0 and IGE = 1, in the middle PDE = 1 and IGE = 0, and right both value one. As shown the bystander intervention occurs faster from right to left.

In Fig. 4, three scenarios are given with PDE = 0 and IGE = 1, PDE = 1 and IGE = 0, and both variables with value one. In the last requirement, negative influences should overpower the positive ones. A clear case to depict this, is a very distressed situation (D = 1). This simulation is shown in Fig. 5.

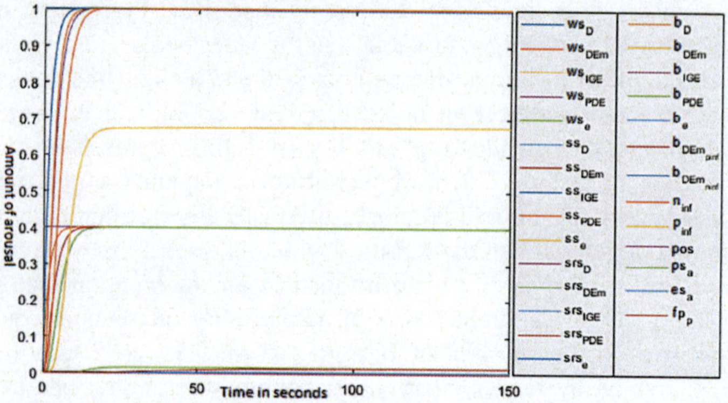

Fig. 5. Simulations of a very distressed, dangerous emergency situation. The state of Es_a reaches at highest a value of 0.02.

In the figure, no intervention and almost no dynamics are found in this graph. The costs of the situation are too high, leading to no action at all. The scenario where primarily positive outcomes do not require to be modelled again, as various examples were given in Fig. 4 already.

5 Mathematical Verification of the Network Model

Based on the differential equation in Table 1, stationary points or equilibria of the created model can be verified by calculating for states X_i in which $dX_i/dt = 0$ the $\mathbf{aggimpact}_{Xi}(t)$ for X_i and comparing the deviation of this to the simulation outcome $X_i(t)$; see Table 3. For this verification, the last graph of Fig. 3 (DEm = 0.8) was used to calculate the mathematical validity of the model. For the chosen graph, the intervention degree is almost in an equilibrium at time step $t = 150$. By doubling this amount to $t = 300$, it is assured that the equilibrium is certainly approximated. The following procedure described in [20], p. 327, is used for the mathematical verification:

Table 3. Mathematical verification of the created bystander effect model, based on the metrics of the last graph of Fig. 3 at time point $t = 300$.

State X_i	State name	$X_i(t)$	$\text{aggimpact}_{Xi}(t)$	Deviation
X_1	ws_D	0	0	0
X_2	ws_{DEm}	0.8000	0.8000	0
X3	ws_{IGE}	0	0	0
X_4	ws_{PDE}	0	0	0
X_5	ws_e	0.9990	0.9990	$5.16 * 10^{-8}$
X_6	ss_D	0	0	0
X_7	ss_{DEm}	0.6734	0.6734	$-1.33 * 10^{-7}$
X_8	ss_{IGE}	0	0	0

(continued)

Table 3. *(continued)*

State X_i	State name	$X_i(t)$	aggimpact$_{Xi}(t)$	Deviation
X_9	ss_{PDE}	0	0	0
X_{10}	ss_e	0.9990	0.9990	$5.72812 * 10^{-8}$
X_{11}	srs_D	0	0	0
X_{12}	srs_{DEm}	0.6734	0.6734	$-1.482 * 10^{-7}$
X_{13}	srs_{IGE}	0	0	0
X_{14}	srs_{PDE}	0	0	0
X_{15}	srs_e	0.9990	0.9990	$6.36 * 10^{-8}$
X_{16}	b_D	0	0	0
X_{17}	b_{DEm}	0.6734	0.6734	$-1.65 * 10^{-7}$
X_{18}	b_{IGE}	0	0	0
X_{19}	b_{PDE}	0	0	0
X_{20}	b_e	0.9929	0.9933	0.0038
X_{21}	$b_{DEmpinf}$	0.8798	0.8798	$2.54 * 10^{-7}$
X_{22}	$b_{DEmninf}$	0.1202	0.1202	$-2.54 * 10^{-7}$
X_{23}	n_{inf}	0.0820	0.0820	$-2.24 * 10^{-7}$
X_{24}	p_{inf}	0.8205	0.8205	$2.101 * 10^{-7}$
X_{25}	pos	0.98999	0.99248	0.0025
X_{26}	ps_a	0.99991	0.99997	$6.35 * 10^{-5}$
X_{27}	es_a	0.99899	0.99904	$4.994 * 10^{-5}$
X_{28}	fp_p	0.004	0.004	$-1.85 * 10^{-6}$

1. The simulation is generated of DEm = 0.8.
2. For a number of states Y identify stationary points with their time points t and state values $Y(t)$.
3. For each of these stationary points for a state Y at time t identify the values $X_1(t)$, .., $X_k(t)$ at that time of the states X_1, ..., X_k connected toward Y.
4. Substitute all these values $Y(t)$ and $X_1(t)$, .., $X_k(t)$ in the criterion $\mathbf{c}_Y(\omega_{X1,Y}X_1(t), \ldots, \omega_{Xk,Y}X_k(t)) = Y(t)$ derived from Table 1.
5. If the equation holds (for example, with an accuracy $<10^{-2}$), then this test succeeds, otherwise it fails.
6. If this test fails, then it has to be explored were the error can be found.

The outcomes are provided in Table 3. As shown, the accuracy of each state is very high with a deviation rate of $<10^{-3}$. Therefore, the model shows evidence that it is mathematically valid.

6 Validation Using Empirical Data and Parameter Tuning

In the simulation results section, three requirements with seven simulations were already depicted. While this output shows interesting insights, many different scenarios and other dependencies remain untested. The requirements were already based on literature and clear effects were found between several moderating variables. However, the requirements on itself could lead to a certain clash, which is still interesting to further investigate. For example, on the one hand the first requirement is stated that an increasing dangerous emergency will lead to more bystander intervention. On the other hand, the third requirement states that costs will overpower the rewards, the probability of helping will decrease. In situations where emergencies become more than highly dangerous (DEm = [0.83,1]), investigating this further will further tune the overall model. Outcomes of such models are depicted in Fig. 6.

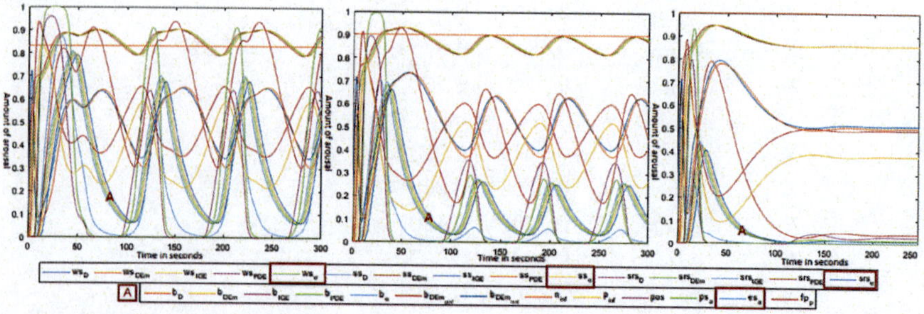

Fig. 6. Three different scenarios with DEm = 0.83 (left), DEm = 0.9 (middle), DEm = 1.0 (right). Here A depicts the intervening.

In these three new scenarios, it is shown that primarily requirement three is still true. The costs outweigh the rewards and thus the amount of rewards and intervening (A) slowly becomes less likely. This brings two possible outcomes, which are either:

1) Requirement one remains true at all times, thus a very dangerous situation of DEm = 1, never outweighs the costs, or
2) Requirement two remains true, thus the rewards are outweighed by the costs.

Both scenarios are tested with an algorithm called Simulated Annealing, which tries to find what the most probable answer would be and which parameters come along. For this, the middle scenario of DEm = 0.9 is chosen due to being a very dangerous situation but remains an average very dangerous one. In total N = 14000 iterations with the algorithm were performed, consisting of four runs of n = 3500. The parameter tuning was executed by using the initial values of the states X_{23}–X_{27}. These values were determined until the first possible execution in state X_{27}, in such that the final decision can be determined by the algorithm (see line B in Fig. 7). The chosen network characteristics were tuned in multiple ways: the threshold τ and speed values η of states were tuned in all runs. Also, two different approaches were chosen, which were the tuning of the connection weights $\omega_{X,Y}$:

1) Only tuning the connection weights to state X_{27} (es_a), which led to 12 characteristics being tuned, and
2) In addition tuning all connection weights that start as incoming connection from X_{23} till X_{27}.

This led to 16 characteristics being tuned. In Table 4 below, the results are shown in terms of optimal values per scenario and what the root mean square error (RMSE) is. This showcases an estimate of how the SA algorithm worked.

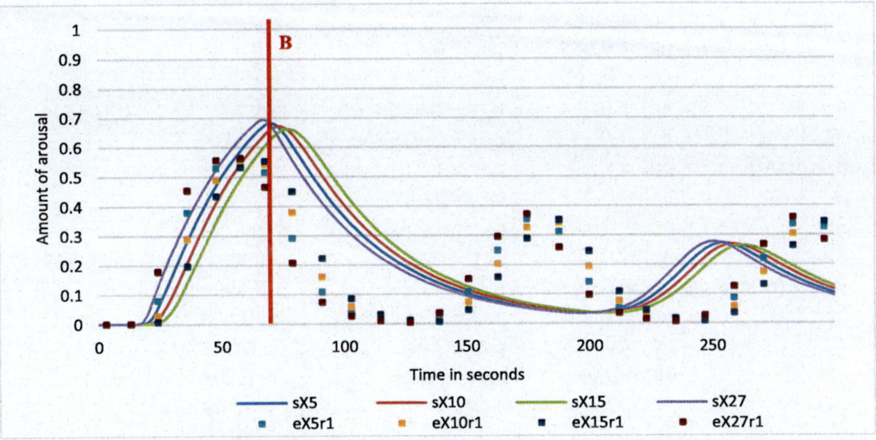

Fig. 7. RMSE errors over time for Run 1 in Table 3 with best scoring RMSE = 0.096

In Fig. 7, it is clear that requirement 3 overpowers the first requirement. The costs are higher than the rewards, thus based on these parameters, the outcome would require the first requirement to be adjusted: 'An increasing dangerous emergency will positively influence the bystander intervention until tipping point T has reached. When this point is reached, there is an increasing possibility of bystander apathy'. Finally, the other runs in Table 4 were taken into account as well. It shows that run 3 (r3) and 4 (r4) contradict the new requirement by stating requirement one remains true (see Fig. 8). Further research is needed to state what happens when the situation is very dangerous. When this is clear, the set of parameters can be used in Table 4.

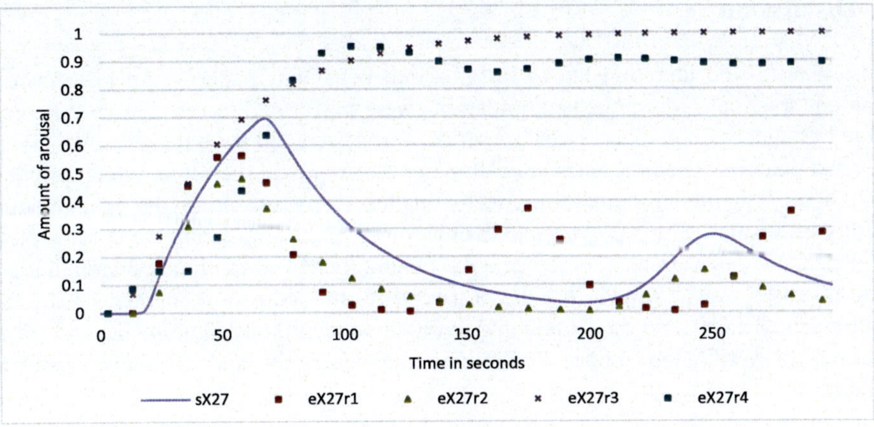

Fig. 8. Deviations over time for Run 1 in Table 3 with best scoring RMSE = 0.096

Table 4. RMSE and outcomes of the SA-algorithm

Run	Speed factors η	Thresholds τ	Connection weights ω	RMSE
Run 1 - 12 characteristics (i = 3500)	$n_{inf} = 0.715$ $p_{inf} = 0.23$ $pos = 0.9659$ $ps_a = 0.689$ $es_a = 0.168$	$n_{inf} = 0.297$ $p_{inf} = 0.10$ $pos = 0.23$ $ps_a = 0.47$ $es_a = 0.24$	$\omega_{pos,esa} = 0.82$ $\omega_{psa,esa} = 0.174$	0.096326
Run 2 - 12 characteristics (i = 3500)	$n_{inf} = 0.134$ $p_{inf} = 0.272$ $pos = 0.55$ $ps_a = 0.511$ $es_a = 0.063$	$n_{inf} = 0.446$ $p_{inf} = 0.916$ $pos = 0.10$ $ps_a = 0.6214$ $es_a = 0.0003$	$\omega_{pos,esa} = 0.79$ $\omega_{psa,esa} = 0.09$	0.108578
Run 3 - 19 characteristics (i = 3500)	$n_{inf} = 0.743$ $p_{inf} = 0.84$ $pos = 0.999$ $ps_a = 0.145$ $es_a = 0.05$	$n_{inf} = 0.407$ $p_{inf} = 0.693$ $pos = 0.554$ $ps_a = 0.076$ $es_a = 0.409$	$\omega_{be,\ pos} = 0.599$ $\omega_{be,\ psa} = 0.599$ $\omega_{ninf,\ pos} = 0.603$ $\omega_{ninf,\ psa} = 0.603$ $\omega_{pinf,\ pos} = 0.904$ $\omega_{ninf,\ psa} = 0.904$ $\omega_{pos,\ psa} = 0.033$ $\omega_{pos,\ esa} = 0.911$ $\omega_{psa,\ esa} = 0.913$	0.132568
Run 4 - 19 characteristics (i = 3500)		$n_{inf} = 0.44$ $p_{inf} = 0.711$ $pos = 0.076$ $ps_a = 0.756$ $es_a = 0.734$	$\omega_{be,\ pos} = 0.134$ $\omega_{be,\ psa} = 0.134$ $\omega_{ninf,\ pos} = 0.71$ $\omega_{ninf,\ psa} = 0.71$ $\omega_{pinf,\ pos} = 0.1624$ $\omega_{ninf,\ psa} = 0.1624$ $\omega_{pos,\ psa} = 0.494$ $\omega_{pos,\ esa} = 0.827$ $\omega_{psa,\ esa} = 0.8317$	0.155321

7 Discussion

The review of the literature led to insights into behaviour patterns. By transforming these patterns into requirements, simulations were performed to test certain outcomes of the bystander effect. In total, ten scenarios were tested and gave insights into several expected patterns. The outcomes are following the empirical literature, which provides a possibility to model more complicated patterns. By checking the mathematical validity of the model, it is proven that the outcomes of the simulation are correct. Next, the dependency of variables was checked by using the SA algorithm and simulating N = 14000 cases. It was shown that the best scoring run, led to a refinement in the first requirement. The created model adhered to the three identified requirements and shows potential to verify the model predictions independently and for new bystander situations.

Certain limitations in the work require more attention. In terms of the temporal-causal model itself, the bystander effect is largely affected by other moderating non-depicted variables as well. This provides a potential to incorporate these, such as group size [3, 8, 16, 17], competency of bystanders [8], harm avoidance [11], and the further investigation of IVE [1, 8]. Another interesting issue is the investigation of weak connection weights between states. In Table 2, only relatively stronger relations are shown, while multiple small impacts could alter the overall outcome. Such weak connection weights could also be expressed in the way different individuals communicate with each other. Now, only one bystander is simulated while in real-life situations, multiple people are present. This leads to a certain shared responsibility, as shown in Fig. 1. By defining the impact of different bystanders, more complex insights can be modelled. Another way is to give more characteristics to the fierce perpetrator or the victim as these are not within the focus of this paper. Lastly, the paper had a focus on real-life situations while the bystander effect could occur online as well. Investigating this in an online setting has an easier empirical data collection possibility than in real life.

References

1. Berggren, N., Blonievsky, T., Derakshan, N.: Enhanced visual detection in trait anxiety. Emotion **15**(4), 477 (2015)
2. Cialdini, R.B., Schaller, M., Houlihan, D., Arps, K., Fultz, J., Beaman, A.L.: Empathy-based helping: is it selflessly or selfishly motivated? J. Pers. Soc. Psychol. **52**(4), 749 (1987)
3. Darley, J.M., Latané, B.: Bystander intervention in emergencies: diffusion of responsibility. J. Pers. Soc. Psychol. **8**(4, Pt.1), 377–383 (1968)
4. Dovidio, J.F., Piliavin, J.A., Gaertner, S.L., Schroeder, D.A., Clark III, R.D.: The Arousal: Cost-Reward Model and the Process of Intervention: A Review of the Evidence. Sage Publications, Inc., Newbury Park (1991)
5. Erlandsson, A., Björklund, F., Bäckström, M.: Perceived utility (not sympathy) mediates the proportion dominance effect in helping decisions. J. Behav. Decis. Mak. **27**(1), 37–47 (2014)
6. Erlandsson, A., Björklund, F., Bäckström, M.: Emotional reactions, perceived impact and perceived responsibility mediate the identifiable victim effect, proportion dominance effect and in-group effect respectively. Organ. Behav. Hum. Decis. Process. **127**, 1–14 (2015)
7. Fischer, P., Greitemeyer, T., Pollozek, F., Frey, D.: The unresponsive bystander: are bystanders more responsive in dangerous emergencies? Eur. J. Soc. Psychol. **36**(2), 267–278 (2006)
8. Fischer, P., Krueger, J.I., Greitemeyer, T., Vogrincic, C., Kastenmüller, A., Frey, D., Kainbacher, M.: The bystander-effect: a meta-analytic review on bystander intervention in dangerous and non-dangerous emergencies. Psychol. Bull. **137**(4), 517–537 (2011)
9. Gansberg, M.: Thirty-eight who saw murder didn't call the police. New York Times, 27 (1964)
10. Glassman, W.E., Hadad, M.: Approaches to Psychology, 6th edn. McGraw-Hill Education (UK), Maidenhead (2013)
11. Hofmann, S.G., Bitran, S.: Sensory-processing sensitivity in social anxiety disorder: relationship to harm avoidance and diagnostic subtypes. J. Anxiety Disord. **21**(7), 944–954 (2007)

12. Hortensius, R., de Gelder, B.: From empathy to apathy: the bystander effect revisited. Curr. Dir. Psychol. Sci. **27**(4), 249–256 (2018)
13. Jenni, K., Loewenstein, G.: Explaining the identifiable victim effect. J. Risk Uncertain. **14** (3), 235–257 (1997)
14. Latané, B., Darley, J.M.: Group inhibition of bystander intervention in emergencies. J. Pers. Soc. Psychol. **10**(3), 215–221 (1968)
15. Latané, B., Darley, J.M.: The Unresponsive Bystander: Why Doesn't He Help?. Appleton-Century-Crofts, New York (1970)
16. Latané, B., Nida, S.: Ten years of research on group size and helping. Psychol. Bull. **89**(2), 308–324 (1981)
17. Liebst, L.S., Philpot, R., Bernasco, W., Dausel, K.L., Ejbye-Ernst, P., Nicolaisen, M.H., Lindegaard, M.R.: Social relations and presence of others predict bystander intervention: evidence from violent incidents captured on CCTV. Aggressive Behav. **45**(6), 598–609 (2019)
18. Schwartz, S.H., Gottlieb, A.: Bystander reactions to a violent theft: crime in Jerusalem. J. Pers. Soc. Psychol. **34**(6), 1188–1199 (1976)
19. Stengs, I.: Ephemeral memorials against 'senseless violence': materialisations of public outcry. Etnofoor **16**(20), 26–40 (2003)
20. Treur, J.: Network-Oriented Modeling: Addressing Complexity of Cognitive, Affective and Social Interactions. Springer, Cham (2016)
21. Treur, J.: Network-Oriented Modeling for Adaptive Networks: Designing Higher-Order Adaptive Biological, Mental and Social Network Models. Springer, Cham (2020)

A Model of Top-Down Attentional Control for Visual Search Based on Neurosciences

Natividad Vargas(✉), Juan Luis del Valle-Padilla, Juan P. Jimenez, and Félix Ramos

Department of Computer Science, Center for Research and Advanced Studies of the National Polytechnic Institute (CINVESTAV IPN) Unidad Guadalajara, Guadalajara, Jalisco, Mexico
Natividad.Vargas@cinvestav.mx

Abstract. Visual attention is an essential and critical mechanism that allows humans to select the most relevant visual information of potential interest to focus on certain aspects of the environment. There are several proposals to model visual attention. However, those models only describe behaviors in simple tasks like free visualization. In more complex tasks such as visual search, it requires attention processes to guide behavior. Attentional control provides these mechanisms. Through which, top-down (goal-directed) information is represented and configured according to the various constraints and dynamics of task processing. In this article, we describe a task-dependent approach to model attentional top-down control based on neuroscience. We present a general conceptual model of visual attention. We describe its three main components and their relationship with other cognitive functions. Also, we show more detailed information about the flow of information that our model follows using a simple guided search case study. Our proposal intends to be the basis to treat top-down attentional information in a broader cognitive architecture. We find that the existence of IPS templates provide a general and biologically inspired representation for relevant objects. Our results show that the proposed model is significantly more consistent and explanatory in information processing compared to other state-of-the-art models.

Keywords: Bioinspired model · Visual attention · Cognitive architectures · Guided search · Top-down control · Goal-driven behavior

1 Introduction

In this article, we show a computational model of attentional control that resembles the way humans do it and contributes to a better understanding of the mechanisms and functions involved in visual search tasks. Petersen and Posner [4] show the existence of a model of three attentional networks. These three networks are the *vigilance* or *alert* network which is responsible for producing and

A. V. Samsonovich et al. (Eds.): BICA 2020, AISC 1310, pp. 541–546, 2021.
https://doi.org/10.1007/978-3-030-65596-9_65

maintaining optimal surveillance and performance during task execution. The *orienting* network focuses on the ability to prioritize stimuli input by selecting a modality or location. The last one network is the *attentional control*, it involves he preparation and application of the selection of stimuli and goal-oriented responses (top-down). Some cognitive architectures have implemented visual attention models. Some **ACT-R** implementations integrate an attention module to preselect information [3]; five basic features characterize visual objects: color, shape, shading, orientation, and size. These features are chosen because of their pop-out nature and importance in guiding visual attention [9]. Each of those features can have a wide range of values, such as red and green for color, and oval, and rectangle for shape. However, an absence of a general cognitive theory makes it hard to investigate top-down influence, for example, in search or discrimination tasks. The **LIDA** CA solves routine problems with novel content. The current percept is written to the workspace, which roughly plays the same role as the preconscious buffers of human working memory. Attention codelets (see below) watch what is written in the workspace to react to it. Upon no suitable situation, an attention codelet will increase its activation as a function of the match between the situation and its preferences [5]. At the end of this review, we could say that despite the advantages observed in the different visual attention models, the problem is that most of them focus on the bottom-up processing of visual attention. Due to, in that approach, there is a lack of essential details about the transformation information because they do not specify high-level operations. In this way, we propose the convenience of feasible computational operations on independent components, and the description of a general cognitive system (in this case of attentional control), which can be refined with the addition of more components and operations (of the same or another involved function) [6].

The structure of this paper as follows: in section "Related models of top-down attention processing" we describe some other bioinspired models for visual attention to review the state of the art of other works that model visual attention. The "Neuroscientific Evidence" section explains the neuroscientific basis of attentional control. The section "Bio-inspired model of top-down attentional control" presents the proposed model and describes in detail the different processes considered. In "Case of Study" we show the experiment to validate and to evaluate the model and explain the implementation and results of the experiment. The "Discussion and Conclusions" section presents a discussion on the contributions and limitations of the model and future work.

2 Bio-Inspired Model of Top-Down Attentional Control

In this section, we present a neuroscientific inspired computational model to solve a visual search task using a guided search paradigm, which shows the interaction between attentional control processes and other cognitive functions, such as planning and decision-making, working memory, sensory and others. Figure 1 shows the brain structures involved in the attentional control process mediated

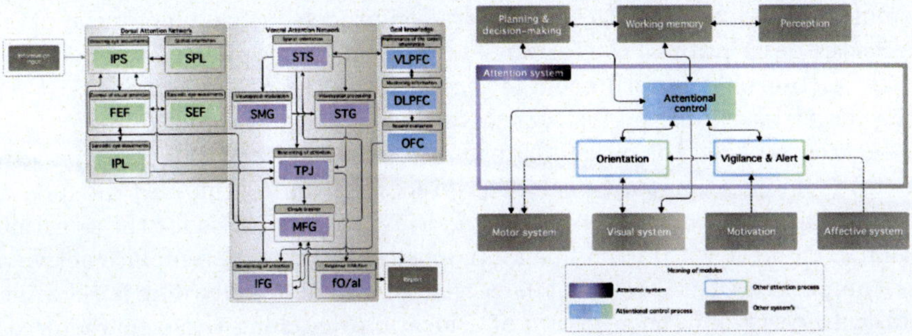

(a) Conceptual architecture (b) Conceptual model

Fig. 1. (a) Conceptual architecture: shows brain structures involved in attention networks. Dorsal Attention Network (DAN) and Ventral Attention Network (VAN). (b) Conceptual model of the attention system.

by top-down information. **Dorsal Attention Network**. The core regions of this network identified include, bilaterally, dorsal parietal cortex along the medial wall of the ventral and posterior intraparietal sulcus (IPS) and in the superior parietal lobule (SPL), and dorsal frontal cortex along the precentral sulcus, near or at the intersection with the superior frontal sulcus. Additional regions include the motion-sensitive area MT in the visual cortex and regions along the medial wall near/at the supplementary eye field (SEF). These regions driven by covert and overt (eye movements) attention shifts [2], consistent with the notion that shifts of spatial attention are functionally related to eye movement plans [2]. **Ventral attention network**. In contrast to the fairly focal pattern of activation observed in dorsal frontoparietal regions during the preparatory interval between an attention cue and a target, a much broader set of brain areas is activated by target detection [2]. The validity of the target modulates a more restricted subset of regions; that is, whether a target appears at an unattended (invalidity cued) or attended (validity cued) location. These regions include, in addition to dorsal regions SPL/precuneus, IPS, and FEF, the posterior sector of the superior temporal sulcus (STS) and gyrus (STG) and the ventral part of the supra marginal gyrus (SMG), and middle frontal gyrus (MFG), inferior frontal gyrus (IFG), frontal operculum, and anterior insula. All these regions respond more strongly to invalid than valid targets and show lateralization to the right hemisphere. [2] proposed that ventral frontoparietal regions TPJ and VFC form a ventral frontoparietal network specialized in stimulus-driven reorienting that co-activates with the dorsal network when a novel behaviorally relevant event occurs in the environment. A region at the junction of the inferior frontal sulcus with precentral sulcus also shows task-evoked properties that are common to both networks [7].

Figure 2 exposes the five primary activities involved in a search task. The first lane shows the *internal state* of the system. In this activity, information useful to solve the task goes into a process determining the task set. The task-set information serves as input to the preparation process (second lane); the task set contains the description of the searched target. The *preparation activity* includes two subfunctions. The first in charge of extracting the characteristics that make up the target based on the task-set, resulting in an attentional-template useful in the search process; the second thread is related to the *response preparation*. That sends information to the selection process so that it can bias the following processes. Later, we turn to the activity *initiation* responsible for the selection of a region of interest within the image. It must keep the care template active and make comparisons between what is currently available and what is relevant to the task, to carry out the selection of a location matching these characteristics. Once it makes the selection, the *generation* activity takes that information, and move the fixation point to this probable location of the target characteristics. The *execution* activity then determines whether the selection is a distractor, or whether it is the target. In the first case, send a message to end the search. Otherwise, make a new selection until it finds the target. It is essential to mention that the performance monitoring process sends feedback information to the generation activity to know if the current selection is a distractor to improve the search in the future.

3 Case of Study

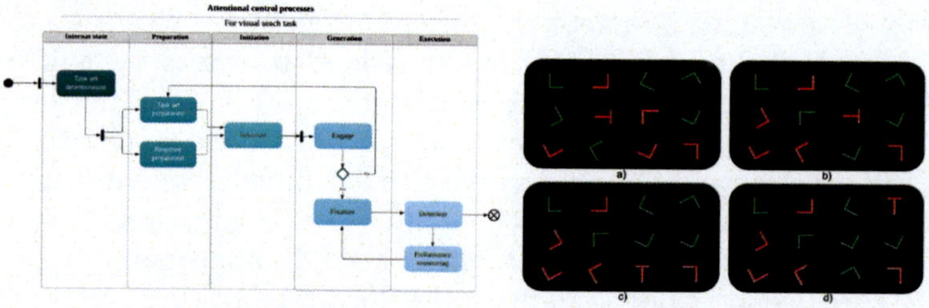

(a) Execution flow (b) Example

Fig. 2. (a) Execution flow: shows five different activities to carry out a visual search task. (b) The basic visual search paradigm. A target (in this case a red T) is presented among a variable number of distractors. Search efficiency can be indexed by the function relating reaction time to the visual set size.

The purpose of this article is to find that subset of information that is relevant to the search and to show how the attention control processes help its performance. The top-down guide is based on the match between a stimulus and the desired properties of the target. Therefore, this top-down approach will be the basis for the development of the case study. We tested the viability of the attentional control model for visual search proposed using a simple Guided search paradigm [8]. The case of study aims to show if the model is capable of detecting an object in the presence of distractors, using top-down information, and with this to show how well the entire process flow works.

The method is as follows: the search objective is present in the working memory. The goal description is in the task-set. The visual object keeps in working memory throughout the search process. Task preparation process receives as input the task-cueing from which an attentional-template will be obtained. A set of attributes (orientation = vertical or horizontal, color = green or red) constitute the attentional template. This attentional template will be guiding for the search by characteristics. The attentional template passes to the initiation process. The initiation process establishes an activation map containing the target information and the environment information. This process assigns higher weights to these attributes (orientation and color), skewing the visual system towards those attributes. The search process begins with the entry of an image with seven visual objects at its center. Because Image processing is not in the scope of this article, we assume that the feature-map derives from that processing. From the conjunction of this feature-map with the information of the attentional-template, the corresponding activation map is obtained. This activation-map generates a fixation point (a region in the image), having the highest activation, and that possibly correspond with the target. Once the fixation point is generated, the execution process determines if it is more similar to the target or is a distraction. If it is the target, it performs a second confirmation by reviewing the attentional template, to finish the search; otherwise, it labels the fixation point as a distractor, and the search process continues from the beginning.

Despite the promising results, some flaws were evident, such as the effects of color on static image search. Since the target's color is previously known, when the image is presented for the first time, the color preview of the distractors results in an anticipatory set that facilitates search performance when the target is known. However, the color preview results in an inhibitory set when the target is unknown and is the same color as the distractors seen above. These effects of an anticipatory or inhibitory set provide evidence of dual, top-down attentional mechanisms that use information from color previews.

4 Discussion and Conclusions

In this section, we summarize the most relevant aspects of this article. First, say that our system addresses the biological guided search of visual objects by modeling brain structures involved in this process according to neuroscience evidence. This research follows the methodology proposed in our research group, resulting in a modular approach. We believe that this modular and general approach, together with the results of this simple experiment, contributes to a better understanding of the mechanisms and functions of attentional control involved in the guided search process for visual objects, specifically, a top-down approach. Due to this modular modeling approach, and the use of a distributed framework for implementation [1], future integration of more cognitive functions is possible.

We are aware of the current limitations; some of them remarked in the previous section. In addition to the discoveries made in the experiment, some modules are missing from our proposal. First, implementation of the activation map generation process is straightforward and could be improved while retaining the benefits of using target-related information. Furthermore, due to the case study used, not all functions in the prefrontal cortex are present, especially some operations of the working memory [7]; we are sure its inclusion will impact memory usage and system performance. Finally, we conclude that future research might benefit from the interactions and operations outlined in this work.

Acknowledgments. This work's research was possible thanks to the grants and support from Consejo Nacional de Ciencia y Tecnología to the authors and SEP-CINVESTAV FIDSC2018/93.

References

1. Hernandez, A.C.: Development environment for cognitive architectures. Master's thesis, Centro de Investigación y de Estudios Avanzados del I.P.N. Unidad Guadalajara (2015)
2. Corbetta, M., Shulman, G.L.: Control of goal-directed and stimulus-driven attention in the brain. Nat. Rev. Neurosci. **3**, 201–215 (2002)
3. Langley, P., Laird, J., Rogers, S.: Cognitive architectures: research issues and challenges. Cogn. Syst. Res. **10**, 141–160 (2009). https://doi.org/10.1016/j.cogsys.2006.07.004
4. Petersen, S., Posner, M.: The attention system of the human brain: 20 years after. Annu. Rev. Neurosci. **35**, 73–89 (2012). https://doi.org/10.1146/annurev-neuro-062111-150525
5. Ramamurthy, U., Baars, B., D'Mello, K., Franklin, S.: LIDA: a working model of cognition. In: Proceedings of the 7th International Conference on Cognitive Modeling, January 2006
6. Torres, G.A.: Metodología para el desarrollo de arquitecturas cognitivas basada en neurociencias. Ph.D. thesis, Cinvestav unidad Guadalajara (2013)
7. Wang, M., Yang, P., Wan, C., Jin, Z., Zhang, J., Li, L.: Evaluating the role of the dorsolateral prefrontal cortex and posterior parietal cortex in memory-guided attention with repetitive transcranial magnetic stimulation. Front. Hum. Neurosci. **12**, 236 (2018). https://doi.org/10.3389/fnhum.2018.00236
8. Wolfe, J.: Guided search 4.0: current progress with a model of visual search. In: Gray, W. (ed.) Integrated Models of Cognitive Systems. Oxford University Press, New York (2013)
9. Wolfe, J., Horowitz, T.: What attributes guide the deployment of visual attention and how do they do it? Nat. Rev. Neurosci. **5**, 495–501 (2004). https://doi.org/10.1038/nrn1411

Comparison of Brain Induced Potentials in Internal Speech in Studied and Unknown Languages

Alexander V. Vartanov and Alisa R. Suyuncheva(✉)

Lomonosov Moscow State University, Moscow, Russia
a_v_vartanov@mail.ru, suyuncheva.a.r@gmail.com

Abstract. Significant differences were found in the bioelectric responses of the brain to words with the same meaning in Russian and Japanese in different series in the electrophysiological experiment on a sample of 18 people. Two groups were used: those who knew Japanese and Russian, and those who knew only Russian as their native language. Differences in mutual articulation of these words were found due to differences in evoked potentials of the late latency (300–500 MS) in the frontal, central and parietal leads, and also in the earlier - 200 MS - latencies in the central, temporal and frontal leads). When comparing between groups (knows Japanese or does not know), differences in the components of EP 200 and P300 were obtained by speaking in Russian, however, there is no difference in primary and semantic processing (P100, N400). The main significant differences over the channels C3, F3, P3 given in this chapter are the amplitude and latency of latency in N200, as well as the decrease in amplitude N400. These data are correlated with earlier studies.

Keywords: EEG · EP · Non-existent words · Covert speech

1 Introduction

Speech research from the very beginning was aimed on solving practical problems that were often associated with the improper development of the speech apparatus, or even its complete (innate or acquired) dysfunction. With the development of technology and the advent of methods for decoding and representing muscle and brain signals, the range of potentially solved problems was growing dramatically [1].

Due to the relevance of the problem of reconstruction of internal speech, in the last ten years, such a direction as the development and improvement of BCI (Brain Computer Interface) is actively developing, which allows you to interact with the human brain directly with a computer, set-top box, or phone (excluding intermediary in the form of muscle activity), interact with the outside world, move, talk.

Recent studies [2–5], using the example of creating invasive and non-invasive interfaces, provide evidence of the existence of such a phenomenon as "inner speech", which, with correctly selected processing methods, can allow a sufficiently high level to cope with such a task as, for example, controlling a drone with six direction commands, or compiling words from letters presented on the screen, counting component P300.

A. V. Samsonovich et al. (Eds.): BICA 2020, AISC 1310, pp. 547–552, 2021.
https://doi.org/10.1007/978-3-030-65596-9_66

Unfortunately, the scope of these developments was currently rather narrow. As a rule, these are laboratory prototypes created in order to test new hypotheses. Very rarely do they get in for testing in clinics or entering the marke due to a number of problems that need to be solved.

One such problem is the choice of a potential diversion technology that achieves optimal spatiotemporal resolution and eliminates invasiveness. The subject's learning capacity also plays a role (according to studies, only 19% of people managed to achieve recognition accuracy of 80–100%). Another problem is the choice of phonologic categories, the recognition of which will be carried out in the experiment. In addition, the possibility of reconstructing words in various languages, as well as reproducing non-existent words (neologisms), remains little understood. For example, based on studies, it is known that word recognition is entirely less effective than phoneme recognition.

In addition, there is information on the influence of the intonation and properties of the native language of the speaker on the word recognition process, which can also serve as an important factor for the speech characteristics of the word to itself.

In this work, it is planned to consider the features of brain responses to words in Japanese (stimuli are presented with transcription) and Russian language in two groups: learning this language, and perceiving it as words that do not make sense.

2 Method

In this electrophysiological study, we defined that internal speech is the silent production of words (without articulation) associated with an efferent copy. The study involved 18 people (M = 10, F = 8) Aged from 18 to 25 years. All participants of the experiment had no brain injuries and were right-handed.

Brain electrical activity was recorded in a monopolar manner using the "Neuro-KM" 19-channel electroencephalograph. The electrodes were positioned in a "10–20%" system with two mostoids. The Presentation v.18.0 was used to present stimuli. Brainsys was used to process the raw EEG data.

To present the sound stimulus, the subjects were wearing headphones in which they listened to stimuli pre-recorded on the recorder. As a visual incentive, a cross was presented in the center of the screen, on which it was necessary to fix the look. Total time of experiment from 40 to 55 min due to different time for presentation of instruction.

The experimental scheme corresponds to the standard schemes described in similar studies: within 700 ms, the test subject is shown a visual or sound stimulus; then a fixing cross appears on the screen, on which the subject should focus his/her gaze and which is the starting command of internal speech (1500 ms); the fixing cross ceases to be displayed on the screen and a pause occurs during which the test subject does not perform internal speech and prepare for the next stimulus.

3 Results

Student's T-test for independent groups was selected to find the difference between two experimental groups. All obtained EPs were averaged by test group.

The differences in internal articulation are best seen on central fronal and, in some cases, parietal leads. As an example here, I will give the data from the Cz electrode for both visual series. In the Fig. 1 with the native language of respondents, differences in P100, N100, P200, N200, as well as in P300 (latency shift) and N400 are significant according to the criterion.

Similar results can be found on temporal and parietal leads, but they are more noisy. The results obtained on the left hemisphere (C3, F3, Fz, Pz, P3, F7) were more significantly different.

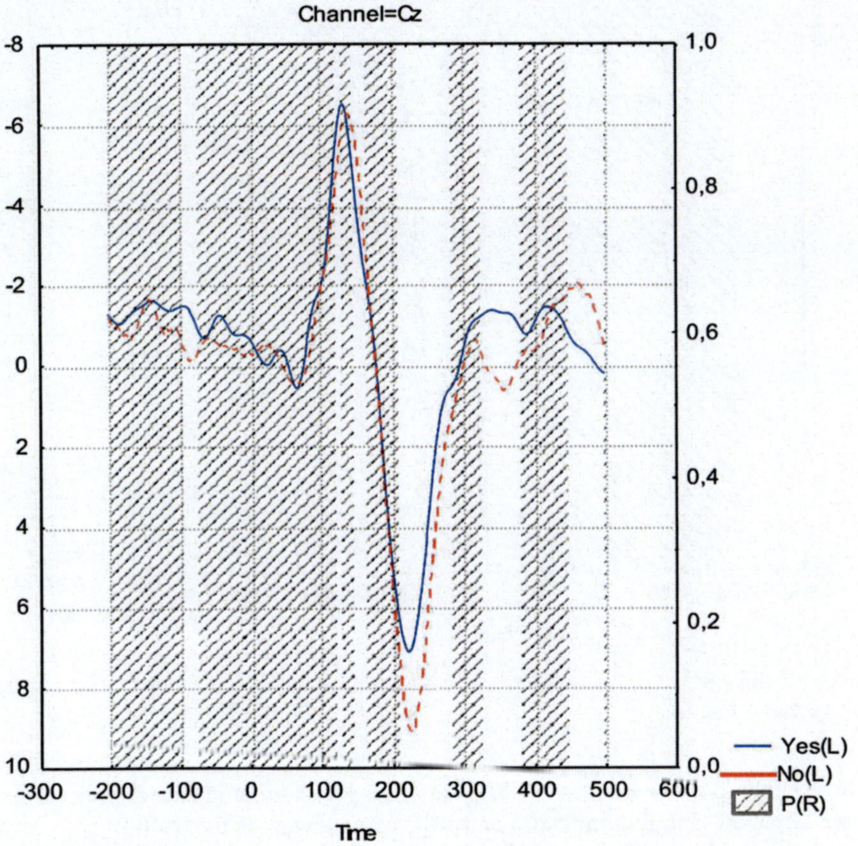

Fig. 1. EP obtained in the Russian language series for two groups (Yes – know Japanese, No – does not know Japanese)

In this study we compared EP in respondents who had previously studied Japanese versus those who had studied English as a second language, i.e. perceived the stimuli presented as neologisms. The main significant differences over the channels C3, F3, P3 given in this chapter are the amplitude and latency of latency in N200, as well as the decrease in amplitude N400. The amplitude of the P100 varies greatly depending on the location of the diversion on the scalp, which is associated with the activation of the auditory cortex (Fig. 2).

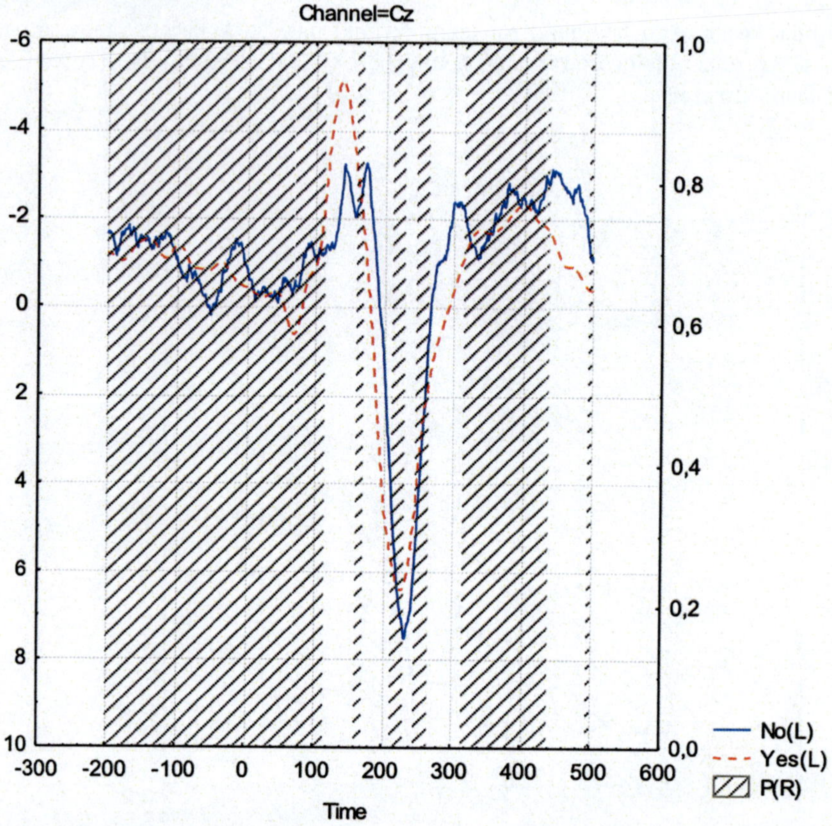

Fig. 2. EP obtained in the Japanese language series for two groups (Yes – know Japanese, No – does not know Japanese)

According to the central leads C3, C4 and Cz, very similar in latency and amplitude EP are obtained, differing on N200 and later components of EP (N400).

The results show significant differences in the components P1, N1 and N2, P2, as well as in the area of the component P300, compared to the control series and the Russian language speech. A strong decrease in amplitude and a shift in latency along the P1, N1 and N2, P2 indicate a reduced activation of the auditory cortex and a weaker process of implementing articulatory sub-preparation.

When comparing between groups (knows Japanese or does not know), differences in the components of EP 200 and P300 were obtained by speaking in Russian, however, there is no difference in primary and semantic processing (P100, N400). These data are correlated with earlier studies [1–3]. The results of experiments on the group of children with dyslexic disorders and the control group showed a higher delay for control children compared to dyslexic and weak reading groups, as well as large N1 amplitudes for control infants, while N2 amplitudes were larger in dyslexic and poor read- ing groups. The poorly read group also generated a higher amplitude characteristic P2.

When comparing between groups (knows Japanese or does not know) in Japanese, differences are obtained in the components N1 and P2 and N2, as well as P3 and later components. Significant differences in the P300 component here relate to semantic processing processes. A group who did not know Japanese showed results similar to early Molfes studies. And differences across later components in the present study may also be associated with reflecting a decline in cognitive processing of stimuli. Previous studies have shown that the attenuation of the amplitude of the P300 during the presentation of states outside the task contrasts with states when performing the task, these results are interpreted as cognitive deviation of the processing of external stimuli.

4 Conclusion

It has been experimentally proved [2, 4] that the late components of in internal speech are similar, regardless of the conditions of initiation, the knowledge of the language, that is, the semantic component, plays a role. Significant differences between EP during perception and speech were found. Significant differences in EP were found when speaking meaningful and non-meaningful words (in groups that know and do not know Japanese).

This processing is initially reflected in the region of the first large solid component of ERP (N100), which interacts with the auditory cortex in both hemispheres to distinguish voiced from vocal consonants.

Subsequently, after about 300 ms, there was a time-varying response, which appeared to arise as a result of brain processes in the right temporal lobe to carry out similar discrimination along the boundaries of the phonological category.

Acknowledgments. The research is financially supported by the Russian Science Foundation, Project № 20-18-00067

References

1. Andrews, E.: Neuroscience and Multilingualism. Cambridge University Press, Cambridge (2014)
2. Electrophysiological dynamics of Chinese phonology during visual word recognition in Chinese-English bilinguals (2007)

3. Thierry, G., Wu, Y.J.: Brain potentials reveal unconscious translation during foreign-language comprehension (2017)
4. Kiroy, V.N., Bakhtin, O.M., Minyaeva, N.R., Lazurenko, D.M., Aslanyan, E.V., Kiroy, R.I.: Electrographic Correlations of Inner Speech (2015)
5. Molfese, D.L.: Predicting Dyslexia at 8 years of age using neonatal brain responses. Brain Lang. **72**(3), 238–245 (2000)

Analysis of Using of Neural Networks for Real-Time Process Control

V. S. Volodin(✉) and A. O. Tolokonskij

National Research Nuclear University MEPhI,
31 Kashirskoe Shosse, 115409 Moscow, Russia
VSVolodin@mephi.ru

Abstract. Machine learning is one of the key technologies of the current scientific and technological revolution. Despite the fact that research in the field of "intelligent" control systems began in the last century, real-time control systems based on machine learning, specifically neural networks, began to be actively implemented only in the past decade. In this paper, the authors analyze the current state of the problem of using real-time control systems based on neural networks and using machine learning in real-time control systems.

Keywords: Neural network · Control system · Real-time control system

1 Introduction

Artificial Intelligence (AI) is one of the key technologies in Industry 4.0, along with robotics, the Industrial Internet of Things (IIoT) and Big Data. The rise in popularity of machine learning came after the publication of an article by Jeffrey Hinton et al. in 2006, in which they showed how to train a deep neural network [1]. This method is used to solve a large number of tasks - from image classification tasks to tasks of developing a car autopilot. This technology finds its application in complex technical objects with real-time control systems. In combination with other technologies of Industry 4.0, the following trends of application of AI in industry can be distinguished [2, 3]: digital twins of technological objects, data mining coupled with IIoT, and computer vision.

2 Digital Twins of Technological Objects

A digital twin (DT) is a software analogue of a physical device that simulates internal processes, technical characteristics and behavior of a real object under the influence of interference and the environment. An important feature of the digital twin is that information from sensors of a real device operating in parallel is used to set input influences on it. DT improves the process of managing a complex system at all stages of its life cycle: from design to decommissioning.

With the help of information collected by the elements of the IIoT of a complex system (for example, as Tesla does with their cars), it is possible to predict failures and

A. V. Samsonovich et al. (Eds.): BICA 2020, AISC 1310, pp. 553–557, 2021.
https://doi.org/10.1007/978-3-030-65596-9_67

estimate the residual resource of equipment, optimize the operation process (for example, by adjusting automatic regulators). Full-scale mathematical models describing the object, or mathematical models based on machine learning are used in solutions of these problems. Application of AI allows avoiding the time-consuming process of building an accurate model without loss of accuracy, but large data sets are required to train it (Big Data). One of the ways to get around this difficulty is to train the neural network on accurate mathematical models using powerful graphics accelerators.

Digital twins are designed and used by such large companies as General Electric [4], Tesla [5], Siemens [6], Rosatom [7]. Figure 1 shows a block diagram of the use of the DT of battery of an electric vehicle in order to optimal control its charge.

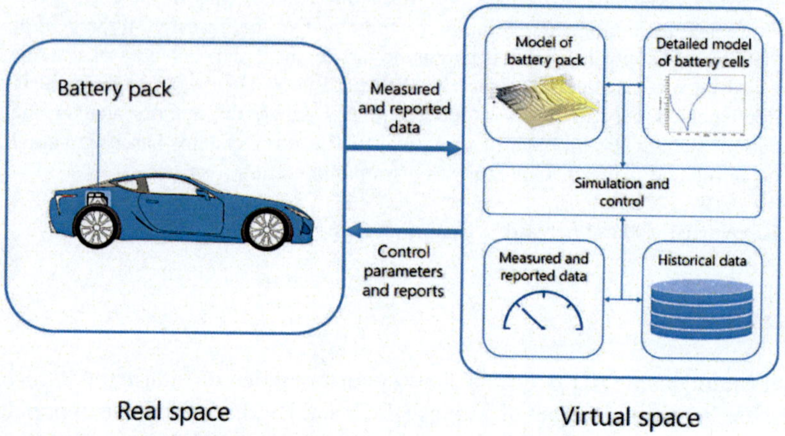

Fig. 1. Digital twin of Tesla battery pack

3 Intelligent Data Analysis Together with IIoT

Industrial Internet of Things, IIoT - Internet of Things for industry application - a system of interconnected computer networks and connected industrial facilities with built-in sensors and software for collecting and exchanging data, with the possibility of remote monitoring and control in an automated mode, without human participation. There is IIoT architecture on Fig. 2.

The principle of operation of the technology is as follows: initially, sensors, actuators, controllers and human-machine interfaces are installed on key parts of the equipment, after that information is collected, which subsequently allows the company to acquire objective and accurate data on the state of the enterprise. The processed data is delivered to all departments of the enterprise, which helps to establish interaction between employees of different departments and make informed decisions. Since a large number of sensors are used, data processing is carried out using machine learning.

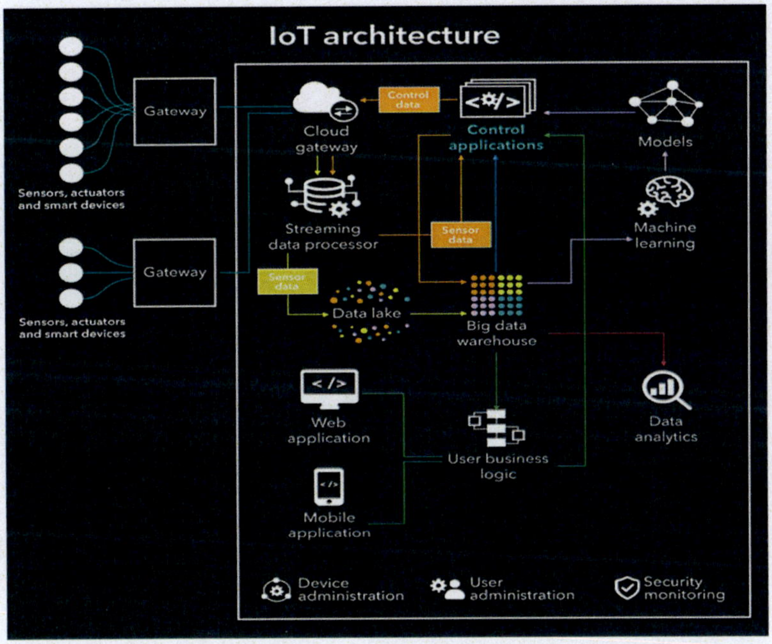

Fig. 2. IIoT architecture

This information is commonly used to prevent unplanned downtime, equipment breakdowns, reduce unscheduled maintenance and supply chain management failures, thereby enabling the enterprise to function more efficiently.

As an example of IIoT, we can cite "smart" or "smart" networks (smart grids) and meters (smart meters).

4 Computer Vision

Computer vision is the theory and technology of creating machines that can detect, track and classify objects.

Machine vision is a scientific direction in the field of artificial intelligence, in particular, robotics, and related technologies for obtaining images of real-world objects, processing them and using the obtained data to solve various kinds of applied problems without the participation of (full or partial) humans.

In industry, computer vision is usually used to classify raw materials on production lines [8]. However, control systems with computer vision are used not only in industry [9] - for example, the existing developments on autopilot are based on the recognition of AI objects on the road. Figure 3 shows a perspective view of the NVIDIA autopilot.

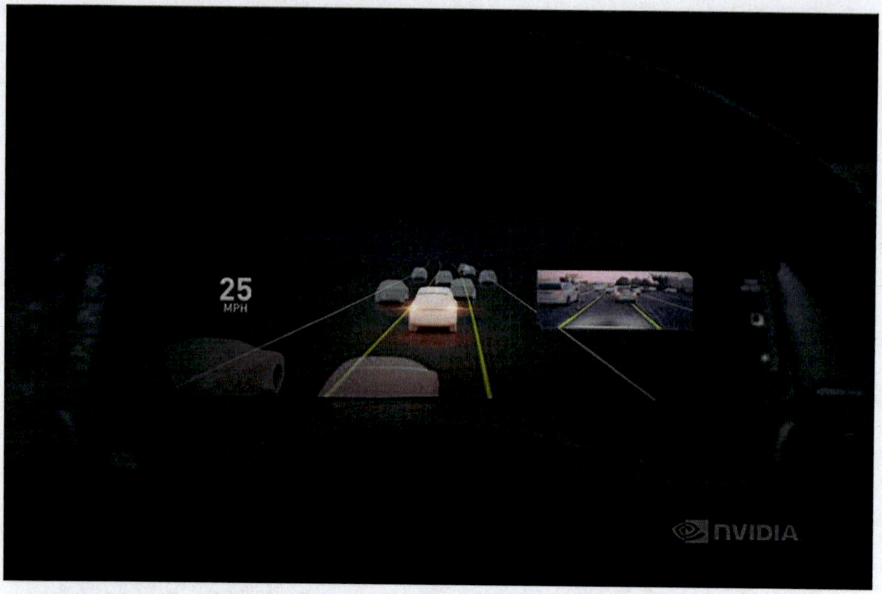

Fig. 3. Computer vision in NVIDIA autopilot

5 Conclusion

Artificial Intelligence is one of the key technologies in Industry 4.0. This technology finds its application in complex technical objects with real-time control systems. In combination with other technologies of Industry 4.0, the following trends of application of AI in industry: digital twins of technological objects, data mining coupled with IIoT, and computer vision.

The use of neural networks as control systems for real equipment is hindered by the following fundamental difficulty, which is relevant at the moment - no one knows how a neural network with a large number of hidden layers makes decisions. It is the interaction of computations within a deep neural network that is crucial for making complex decisions, but these computations are a web of mathematical functions and variables [10]. We cannot use the network as a control system for a real complex system (for example, an NPP power unit) due to safety reasons. At the moment, this problem remains unresolved.

References

1. Geron, A.: Hands-On Machine Learning with Scikit-Learn and TensorFlow. O'Reilly Media Inc., Boston (2017)
2. Patel, A.R., Ramaiya, K.K., Bhatia, C.V., Shah, H.N., Bhavsar, S.N.: Artificial intelligence: prospect in mechanical engineering field—a review. In: Kotecha, K., Piuri, V., Shah, H., Patel, R. (eds.) Data Science and Intelligent Applications. Lecture Notes on Data Engineering and Communications Technologies, vol. 52, pp. 267–282. Springer, Singapore (2021)
3. Swami, M., Verma, D., Vishwakarma, V.P.: Blockchain and industrial internet of things: applications for industry 4.0. Bansal, P., Tushir, M., Balas, V., Srivastava, R. (eds.) Proceedings of International Conference on Artificial Intelligence and Applications. Advances in Intelligent Systems and Computing, vol. 1164, pp. 279–290. Springer, Singapore (2021)
4. GE Digital. Digital twin. https://www.ge.com/digital/applications/digital-twin
5. Tesla. About Tesla. https://www.tesla.com/about
6. Cambashi, Industry Knowledge for Business Advantage. https://cambashi.com/siemens-goes-all-in-on-digital-twin/amp/
7. Atomic Expert and Atomic Energy journal, Digital twins in nuclear industry. http://atomicexpert.com/virtual_npp_rosatom
8. Nasiri, A., Omid, M., Taheri-Garavand, A.: An automatic sorting system for unwashed eggs using deep learning. J. Food Eng. **283**, 110036 (2020)
9. NVIDIA DRIVE Autopilot. https://nvidianews.nvidia.com/news/nvidia-introduces-drive-autopilot-worlds-first-commercially-available-level-2+-automated-driving-system
10. Knight, W.: The Dark Secret at the Heart of AI MIT Technology Review May/June 2017. https://www.technologyreview.com/2017/04/11/5113/the-dark-secret-at-the-heart-of-ai/. Accessed 11 Apr 2017

Comparison Between Coordinated Control and Interpretation Methods for Multi-channel Control of a Mobile Robotic Device

Timofei I. Voznenko(✉), Alexander A. Gridnev, Eugene V. Chepin, and Konstantin Y. Kudryavtsev

Institute of Cyber Intelligence Systems, National Research Nuclear University MEPhI, Moscow, Russia
snaipervti@gmail.com

Abstract. Control methods of the mobile robotic device can be divided into single-channel and multi-channel. A single-channel control method is a control method that uses a single data channel. Each control channel has its own operational features that can affect the quality of control. In order to take advantage of multiple control channels, multi-channel control is used. In multi-channel control information comes from different, heterogeneous, independent control channels. The task of the control system is to make a decision and choose valid (the most probable) command. There are two approaches to this problem: coordinated control and decomposition of multi-channel control into single-channel control. The decomposition method implies ignoring of command information from not selected channels and commands. The interpretation method allows using such command information to improve the quality of control of a mobile robotic device. In this paper, we consider coordinated control and interpretation methods and also compare implementations of these control methods for multi-channel control of a mobile robotic device.

Keywords: Robotics · Multi-channel control · Decision-making · Decomposition method · Interpretation method

1 Introduction

Control methods of the mobile robotic device can be divided into single-channel and multi-channel [1]. A single-channel control method is a control method that uses a single data channel. Each control channel has its own operational features, and in practice distracting factors may arise which may affect the quality of control [2,3]. In order to prevent negative influence of distracting factors on the quality of control, it is necessary either to develop algorithms that can work taking into account distracting factors [4] or to develop methods that detect distracting factors that negatively affect the accuracy of control and switch to another control method that is not subject to specific external factor [5].

A. V. Samsonovich et al. (Eds.): BICA 2020, AISC 1310, pp. 558–564, 2021.
https://doi.org/10.1007/978-3-030-65596-9_68

In order to take advantage of multiple control channels, multi-channel control is used. In this case, the command information comes from different, heterogeneous, independent from each other data channels from the user to a certain system. The purpose of this system is to provide valid (the most probable) command information at the output, i.e. it is necessary to solve the problem of decision making. The problem of choosing a valid output command is that since there are several independent data channels, the commands competing for the same resources necessary for their execution can come from them. To solve the problem of choosing a valid command in this case, there are two main approaches for the system implementation: implementation of coordinated control and decomposition of multi-channel control method into single-channel control.

2 Related Works

In the case of multi-channel control, to solve the coordinated control problem [1], it is necessary to implement a decision-making system (DMS), which will make decisions about which of the commands now needs to be invoked and which is not. One of the approaches of DMS implementation is development of own algorithm based on solving the task of linear programming. For example, if a linear input-output model can be constructed for single-channel control, then this model can be expanded for a multi-channel control system [6,7]. For this, for each data channel and command, certain parameters are introduced that characterize the reliability of the execution, and at certain points in time, the decision-making system solves the problem of choosing the most probable command to invoke it.

The advantage of coordinated control is the ability to control using multiple control channels. In other words, if there is a situation (external factor) when one of the control channels will work worse or fail, the operator still has the possibility to use other control channels to control the robotic device. The disadvantage of coordinated control is possible negative influence on the control quality caused by type I and type II errors from each control channel.

Another approach to the implementation of multi-channel control is to solve the decomposition problem. The decomposition problem is to eliminate the mutual influence of the system channels [1]. This is achieved using appropriate control algorithms, i.e. selection of control actions.

The main difference between the decomposition and interpretation algorithms is that the interpretation algorithm [8] uses the command information that the decomposition algorithm ignores. That is, if the decomposition algorithm implies that we find and use the best control method, and ignore other methods, then the interpretation algorithm uses this ignored information to improve the system quality.

In the case of multi-channel control, decomposition and interpretation reduce the multi-channel control to a single-channel control method. Thus, it is necessary to solve the problem of selecting the most efficient method of robotic control based on the available control channels. For this, a preliminary test is carried out, during which the test program asks the operator to call various commands for execution, in the process records the number of correct recognition (when the command was recognized, when the operator called it for execution) and type I and type II errors. There are many parameters that can be used to determine the criterion of control quality and evaluation of the system as a whole.

The advantage of decomposition and interpretation is the transformation into a single-channel control method and, therefore, minimizing the potential negative impact of other channels. The disadvantage is the lack of redundancy, i.e. if an external factor arises in which the quality of the control channel decreases, then it will not be possible to use another control channel. To solve this problem, it is possible to develop methods that detect distracting factors that have a negative impact on the accuracy of control, and switch to another control method that is not subject to this external factor [5]. However, in this case, the problems of choosing a list of distracting factors and implementing a detection method for each of them arise. This paper proposes to study the possibility of using the interpretation method for various external factors.

3 Comparison of Control Methods

Let's compare different methods to control a robotic device: single-channel control, coordinated multi-channel control, multi-channel control decomposition, and multi-channel control interpretation.

3.1 Resistance to External Factors

The single-channel control may be subject to external factors that may negatively affect the quality of the control system of a robotic device. In case of multi-channel control decomposition or interpretation, the situation may be the same, as it all comes down to single-channel control. In the case of coordinated multi-channel control, there are redundant control channels that are not affected by external factors.

3.2 Presence of System Errors

The presence of system operation errors means the presence of possible type I and type II errors when different control channels are used. Again, it depends on the implementation of control channels, but in the case of coordinated multi-channel control of a robotic device, type I and type II errors may occur from all control channels, i.e. the number of these errors is summed up. And the more control channels there are, the more potential cases will be when a robotic device is exposed to errors caused by external factors or control channel imperfections.

3.3 The Presence of Ignored Information

This characteristic means the presence of command information, which the DMS must ignore. In the case of coordinated multi-channel control there is no such information, i.e. the operation of each of the control channels is taken into account. In the case of multi-channel control decomposition, there is such information because the best control combination of commands and channels is defined before the operation, and then other command information is ignored. In the case of single-channel control system operation, there is no such information because all control commands are transmitted to a robotic device. The interpretation algorithm, unlike the decomposition algorithm, uses this information to improve the quality of the robot control system.

4 Experiment

For testing multi-channel control, 2 control channels (gesture and voice) were selected with the ability to invoke 5 commands (stop, forward, backward, left, right).

The experiment took place in two stages. In the first stage, the operator had to invoke all commands from all control channels 200 times. This number was chosen because in 200 tests the percentage change of the obtained value from the reference one will be within 5% [9]. These commands were invoked in 3 situations:

1. Without external factors;
2. With external factor F_1: loud noise, which reduces the quality of the voice control channel;
3. With external factor F_2: camera shake (due to uneven surface), which reduces the quality of the gesture control channel.

Based on the results obtained, an error matrix was compiled for each of the situations. The decomposition algorithm [10] was applied to this matrix, which allowed obtaining the best combination of channel commands in 4 cases:

1. Without taking into account the external factors (CMB_0);
2. Only taking into account the external factor F_1 (CMB_1);
3. Only taking into account the external factor F_2 (CMB_2);
4. Taking into account external factors F_1 and F_2 ($F_1 \& F_2$) (CMB_3).

In the second stage, the operator needs to invoke 100 random commands (each of 5 commands 20 times randomly). This testing was carried out in 3 situations: when factor F_1 is present, when factor F_2 is present and when factors $F_1 \& F_2$ are present. For each of these situations, the operator must invoke commands using only two sets of channel commands defined in first stage:

- Without taking into account the external factors (CMB_0);
- Taking into account external factors (CMB_i, $i = 1, 2, 3$).

For the combination without taking into account external factors (CMB_0), the interpretation algorithm [8] was used, which allows one to understand how to interpret unused channel commands (not included in the best channel commands). For combination i (CMB_i), both coordinated control and interpretation were applied. Thus, the results were obtained for 3 cases INT_0, DMS_i, and INT_i. For each of these cases, the value of $MTnP$ parameter [10] was calculated for all commands

$$MTnP \triangleq TPR \times PPV,$$

where TPR – true positive rate of command, PPV – positive predictive value of command.

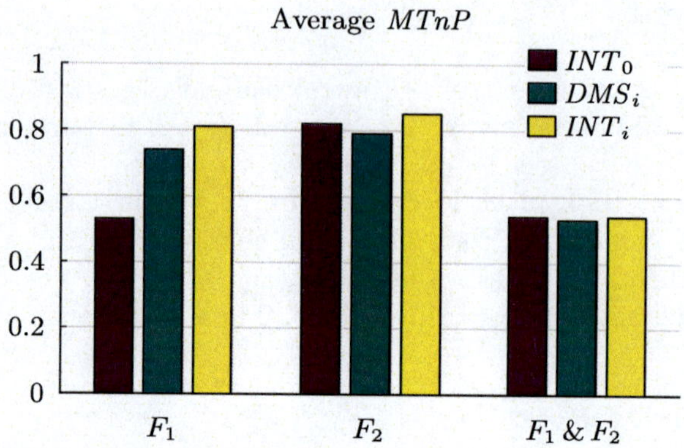

Fig. 1. The average value of $MTnP$ for each case

Table 1. F and p values for different external factors

	F-value	p-value
F_1	**6.46769**	**0.002518**
F_2	0.57274	0.566334
F_1 & F_2	0.07449	0.92828

5 Conclusion

The test results are shown in Fig. 1. It shows the average value of the $MTnP$ parameter for various external factors (F_1, F_2, F_1 & F_2). For each of the situations with external factors (F_1, F_2, F_1 & F_2) the One-Way Repeated Measures ANOVA was applied for the cases INT_0, DMS_i, and INT_i (Table 1). As a result, it was obtained that statistically significant differences are present only for factor F_1 ($p < 0.01$). Fig. 1 shows an increase in the average $MTnP$ value at INT_1 (0.81) relative to INT_0 (0.53). The statistically insignificant result for F_2 is explained by the fact that mainly the voice channel was used to control, so CMB_0 and CMB_2 were similar. And the statistically insignificant result for F_1 & F_2 is explained by the fact that both control channels were negatively affected because of which their average $MTnP$ parameter for all commands was about 50%.

One of the disadvantages of multi-channel control interpretation is low resistance to external factors affecting the control channels. If interpretation is carried out taking into account various factors and the control methods (combination) are switched taking into account external factors, it is possible to achieve results comparable to the coordinated control under these factors. Figure 1 shows that the obtained average $MTnP$ values for INT_i ($i = 1, 2, 3$) cases are comparable with the results of the coordinated control DMS_i. In the future it is planned to increase the number of control channels and external factors and to continue research in this area.

References

1. Miroshnik, I.V.: Teorija avtomaticheskogo upravlenija. Linejnye sistemy (in Russian). Piter (2005)
2. Bhattacharjee, A., Khan, A.I., Haider, M.Z., Fattah, S.A., Chowdhury, D., Sarkar, M., Shahnaz, C.: Bangla voice controlled robot for rescue operation in noisy environment. In: 2016 IEEE Region 10 Conference (TENCON), pp. 3284–3288. IEEE (2016). https://doi.org/10.1109/TENCON.2016.7848659
3. Singh, C., Venter, M., Muthu, R.K., Brown, D.: A real-time DSP-based system for voice activity detection and background noise reduction. In: Intelligent Speech Signal Processing, pp. 39–54. Elsevier (2019). https://doi.org/10.1016/B978-0-12-818130-0.00003-9
4. Po-Yi Shih, Po-Chuan Lin, Wang, J.F., You-Zen Chen: Improving real-time voice activity detection for perceptual robotic control in noisy environment. In: TENCON 2011 - 2011 IEEE Region 10 Conference, pp. 1040–1044. IEEE (2011). https://doi.org/10.1109/TENCON.2011.6129269
5. Plotnikova, N.V.: Expert robot control system. Komp'juternye tehnologii, upravlenie, radiojelektronika, pp. 195–197 (2014). (in Russian)
6. Mitrishkin, J.: Linejnye modeli upravljaemyh dinamicheskih sistem. Ch. 1. Uravnenija "vhod-vyhod" i "vhod-sostojanie-vyhod". M.: Izd-vo MGTU im. NJe Baumana (2008). (in Russian)
7. Gridnev, A.A., Voznenko, T.I., Chepin, E.V.: The decision-making system for a multi-channel robotic device control. Procedia Comput. Sci. **123**, 149–154 (2018). https://doi.org/10.1016/j.procs.2018.01.024
8. Voznenko, T.I., Gridnev, A.A., Chepin, E.V., Kudryavtsev, K.Y.: The command interpretation in decomposition method of multi-channel control for a robotic device. Procedia Comput. Scie. **169**, 152–157 (2020). https://doi.org/10.1016/j.procs.2020.02.127

9. Kudryavtsev, K.Y., Cherepanov, A.V., Voznenko, T.I., Dyumin, A.A., Gridnev, A.A., Chepin, E.V.: The method of statistical estimation of the minimum number of tests for reliable evaluation of the robotic multi-channel control systems quality. In: Biologically Inspired Cognitive Architectures 2018, pp. 203–208. Springer (2019). https://doi.org/10.1007/978-3-319-99316-4_27
10. Voznenko, T.I., Gridnev, A.A., Kudryavtsev, K.Y., Chepin, E.V.: The decomposition method of multi-channel control system based on extended bci for a robotic wheelchair. In: Biologically Inspired Cognitive Architectures 2019, pp. 562–567. Springer (2020). https://doi.org/10.1007/978-3-030-25719-4_73

Combinator-as-a-Process for Representing the Information Structure of Deep Computing

Viacheslav Wolfengagen[1(✉)], Larisa Ismailova[1], and Sergey Kosikov[2]

[1] National Research Nuclear University "Moscow Engineering Physics Institute", Moscow 115409, Russian Federation
jir.vew@gmail.com, lyu.ismailova@gmail.com
[2] NAO "JurInfoR", Moscow 119435, Russian Federation
kosikov.s.v@gmail.com

Abstract. In the present work, the main attention is paid to elucidating the technological features of computations with individual generators, or objects. Their interaction is considered in an applicative environment, which allows us to elucidate the internal structure of ordinary operations, the knowledge of which allows us to understand their properties. The choice of initial constant generators, considered initial and expressed by *combinators*, is discussed. These initial generators are used as the main "building blocks" that enter into the applicative medium in interaction with each other. As a result of the interaction, constructions arise that give representative sets of ordinary operators and immersed computing systems.

Keywords: Combinatory logic combinator-as-a-process · Information process · Applicative structure · Artificial computing · Natural computing

Introduction

The abundance of information services provided makes us take a fresh look at what functions they actually perform. The idea of the internal structure of services is usually absent, and their computational behavior remains not completely clear. Attempts to narrow the scope of phenomena and to specialize the tools used to special sections of programming translate the discussion into the plane of implementation, piling up minor details and moving away from the important mechanisms that underlie any computation. It seems that the difficulties of creating computational models [3,14] remained largely unsolved, and in return there appeared other areas of research, which for a while put aside the old questions and offered to tackle models that were not yet disappointing, the boundaries of their application for a while remain unclear.

Computing issues are left to the will of development companies that focus on the needs of the sales market, and not on the search for really promising

A. V. Samsonovich et al. (Eds.): BICA 2020, AISC 1310, pp. 565–571, 2021.
https://doi.org/10.1007/978-3-030-65596-9_69

concepts [4]. The connections between technological and natural computing are only being outlined [12], and there is no need to speak about the existing conceptual basis, although there are strong arguments that applicative computing and combinatorial logic apply for their role [18]. In the best case, a completely pragmatic – and technological – point of view prevails, which takes the Web as the main challenge to the development of computing, which is seen as a homogeneous and highly connected set of computing resources, which can be considered as some kind of metacomputer [1]. On the way to characterizing the computational environment associated with this rather hypothetical than even a virtual computer, a number of fundamental difficulties arise that could not be overcome [2]. It is not clear what kind of programming languages this metacomputer needs to be equipped with, since it is not possible to obtain a unified computational model that includes the observed effects of the Web, and the sentences are limited to very special cases [13]. Researchers have not decided on the choice between von-Neumann and non von-Neumann architecture, strong structuring of databases and weak structuring of hypertext representations of information, semantic networks and ontologies, semantics of computations with or without states, establishing or blurring the boundaries between programs and data. If all this is additionally superimposed by constant shifts of boundaries between separate disciplines of computing, then fundamental advances and progress may seem more like an exception than a rule, which in the field of information technology creates a favorable basis for an avalanche-like increase in the flow of innovation [13]. These alternatives are far from exhausted, but they help the problematic person navigate the state of research.

Only with the use of applicative computing technologies, as known, is it possible to construct a general theory of computations that operates with an acceptable idea of the individual object in practice. This method was applied to the solution of semantic modeling problems [8,10,14,16] with the formation of a dynamic computational model of the applicative type [15]. An important advance in the construction of a computational model that universally characterizes the interaction of information processes was made in [7,9], where a layered discussion of the interaction is proposed. This approach in this paper was expressed in the equational construction of the theory of computation [6]. The applicative computational models and technologies obtained on its basis are important for the development of modern programming [11], especially in the context of growing interest in the practical use of the functional approach to programming [2,5,17].

In this paper, an attempt is made to elucidate the technological foundations of "pure" computing, especially those of them that relate to calculations with objects. Their interaction is considered in an applicative environment, which allows us to elucidate the internal structure of ordinary operations, the knowledge of which allows us to understand their properties. The choice of initial constant entities, considered initial and called *combinators*, is discussed. These initial entities are used as the main "building blocks" that enter into the applicative environment in interaction with each other. As a result of the interaction, constructions arise that give representative sets of ordinary operators and immersed computing systems.

1 Applicative Environment: Interaction of Objects

We are all used to using operations and operators in everyday computing. From an intuitive point of view, they are used as *elementary*, but occasionally the question still arises, where did they come from. And here the point of view does not always save that operations are the simplest constructions. Is it possible by means of computing to see what their internal structure is?

To compare some objects with others, we need to define some (binary) relation. Usually this relationship is called *conversion* and is defined by *postulates* (CL):

$$(\mathsf{I})\quad \mathsf{I}a = a,\quad (\mathsf{K})\quad \mathsf{K}ab = a,\quad (\mathsf{S})\quad \mathsf{S}abc = ac(bc),$$

$$(\sigma)\quad a = a,\quad (\rho)\quad \frac{a = b}{b = a},\quad (\tau)\quad \frac{a = b,\ b = c}{a = c},$$

$$(\mu)\quad \frac{a = b}{ca = cb},\qquad (\nu)\quad \frac{a = b}{ac = bc},$$

where $a,\ b,\ c$ denotes arbitrary objects, and '=' is the relation of conversion.

Let us try to answer this question using the explanatory system of such computing, the only primitive operation of which is the application – applying – of one object to another. Objects enter into *interaction* with each other, and what happens during the implementation of their *applications* is shown from an intuitive point of view in the Fig. 1. The only thing that can be seen from this figure is the fact of applying, or, in another terminology, the em applying of the object a to the object b. In those applications that are currently reasonably understood, it is believed that a plays the role of *function*, and b plays *argument*, and the interaction of objects is represented by computation. Considering such an application as a *operation*, and even better – as a *meta-operation*, one can raise the question of the *result* of this operation. Following all the well-known traditions of operating with functions, we assume that $a(b)$ is considered the value of applying the object a to the object b. For reasons of symmetry of objects a and b, in practice, $a(b)$, (ab) and ab are used interchangeably by records of the result of their interaction.

Only with applicative forms it is not possible for many useful constructions be formed. Some objects interact with others, they must be exposed by their neighbors in applicative forms, and also act on other objects themselves. As a result of interaction, transformations of objects should occur, and these expected intuitive representations of the object are shown in the Fig. 1. The object is supplied with a unique name identifying it in the subject area. In fact, the object is considered as a pair

$$\langle \text{ redex, contract } \rangle.$$

A *redex* is considered to be a source object with a sequence of *actual* parameters used, and a *contract* is the result of an object interacting with its *possible* parameters. Redex represents the object *before* of interaction with the environment, while the contract represents the object *after* of interaction with the environment. Redex is the place where the original object is born, starting its apparent existence in the environment. In the course of inter-object interaction, a contract is formed, and the original object in the general case is not contained in it, ceasing its explicit existence.

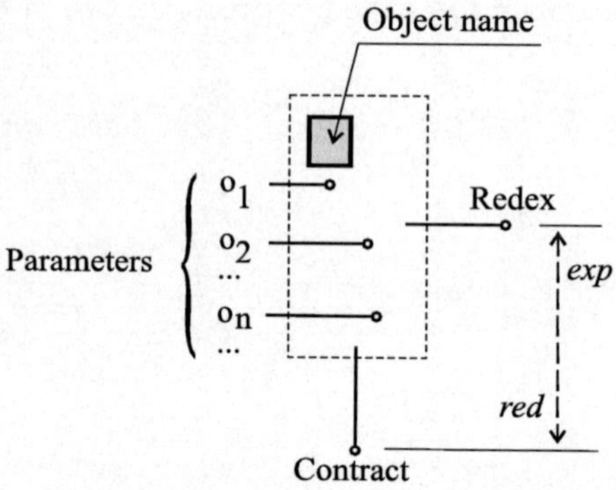

Fig. 1. Diagram of an object.

Interaction with the environment is carried out through parameters, which, in turn, are objects. The number of parameters is not limited, and their effect on the object appears as they are used in calculations. The process leading to the construction of the contract is called *reduction* and is symbolically denoted by *red*, and the process leading to the construction of the redex is *expansion* and is denoted by *exp*. Redex has the greatest computing potential, and a contract has the smallest potential. The contract contains a *virtual* representation of the original object, while the redex has its *actual* representation.

The most important point is that the object is in the process of *conversion*, that is, in the bidirectional reduction-expansion process - the transition from redex state to contract state and vice versa. The reduction of the object is associated with the construction of the computational result, and the expansion is associated with the formation of an object that can lead to the existing computational result. For the feasibility of these transformations, which actually occur, we have to assume for simplicity that there is a *fixed* supply of primitive objects that can cause changes in the applicative environment. Ideally, in the pure theory of objects – *combinatory logic* – usually two such combinatory objects are taken – S and K (see [2]).

Conclusion

Behind the variety of forms in which computing is now manifesting itself, it is not easy to see the unity of information entities and the few general mechanisms that drive the flows of digital information. There is a clear tendency toward the restructuring of computing, the previous forms of which have generated more questions than they have answered. Starting with simple and directly perceived ideas about object diagrams, an extensible computing environment has been built. Structurally, it reproduces itself by cascading component diagrams, and computer control is carried out by a finite set of initial combinatory objects. The "resolution" of the computing system also turns out to be adjustable, allowing not only to analyze the internal structure of the initial operations, which is also a lot, but also to synthesize objects with pre-planned computational behavior.

Acknowledgments. This research is supported in part by the Russian Foundation for Basic Research, RFBR grants 20-07-00149-a, 19-07-00326-a, 19-07-00420-a, 18-07-01082-a.

References

1. Bergman, M.K.: White paper: the deep web: Surfacing hidden value. J. Electron. Publishing **7**(1) (2001). https://doi.org/10.3998/3336451.0007.104
2. Bimbo, K.: Combinatory Logic: Pure, Applied and Typed, 1st edn. Discrete Mathematics and Its Applications, CRC (2012)
3. Curry, H.B.: Foundations of Mathematical Logic, 2, revised edn. Dover Books on Mathematics, Dover Publications (2010)
4. Herbert, W., Franke, H.Z., et al.: Irreducibility and Computational Equivalence: 10 Years after Wolfram's a new Kind of Science, 1 edn. Emergence, Complexity and Computation 2. Springer, Heidelberg (2013)
5. Ismailova, L., Kosikov, S., Kucherov, I., Zhuravleva, O.: Tools of algebraic type for manipulating methodologically oriented cognitive information. Procedia Comput. Sci. **169**, 23–30 (2020). https://doi.org/10.1016/j.procs.2020.02.111. http://www.sciencedirect.com/science/article/pii/S1877050920302349. Postproceedings of the 10th Annual International Conference on Biologically Inspired Cognitive Architectures, BICA 2019 (Tenth Annual Meeting of the BICA Society), Seattle, Washington, USA, 15–19 August 2019
6. Ismailova, L., Kosikov, S., Zinchenko, K., Mikhailov, A., Bourmistrova, L., Berezovskaya, A.: Equationally expressed evaluation. In: 9th International Workshop on Functional and Logic Programming, WFLP 2000, Benicassim, Spain, 28–30 September 2000, pp. 135–143 (2000)
7. Ismailova, L., Wolfengagen, V., Kosikov, S.: Hereditary information processes with semantic modeling structures. Procedia Comput. Sci. **169**, 291–296 (2020). https://doi.org/10.1016/j.procs.2020.02.181. http://www.sciencedirect.com/science/article/pii/S1877050920303045. Postproceedings of the 10th Annual International Conference on Biologically Inspired Cognitive Architectures, BICA 2019 (Tenth Annual Meeting of the BICA Society), Seattle, Washington, USA, 15–19 August 2019

8. Ismailova, L., Wolfengagen, V., Kosikov, S., Dohrn, M.: Capturing push-processing using enriched semantic mesh equipped with functionals-and-hops model. Procedia Comput. Sci. **169**, 590–596 (2020). https://doi.org/10.1016/j.procs.2020.02.205. http://www.sciencedirect.com/science/article/pii/S1877050920303288. Postproceedings of the 10th Annual International Conference on Biologically Inspired Cognitive Architectures, BICA 2019 (Tenth Annual Meeting of the BICA Society), Seattle, Washington, USA, 15–19 August 2019
9. Ismailova, L., Wolfengagen, V., Kosikov, S., Maslov, M., Dohrn, J.: Semantic models to indicate post-truth with fake news channels. Procedia Comput. Sci. **169**, 297–303 (2020). https://doi.org/10.1016/j.procs.2020.02.182. http://www.sciencedirect.com/science/article/pii/S1877050920303057. Postproceedings of the 10th Annual International Conference on Biologically Inspired Cognitive Architectures, BICA 2019 (Tenth Annual Meeting of the BICA Society), Seattle, Washington, USA, 15–19 August 2019
10. Ismailova, L.Y., Kosikov, S.V., Wolfengagen, V.E.: Computational model for the construction of cognitive maps. In: Silvera-Roig, M., Azcárate, A.L.V. (eds.) Cognitive and Intermedial Semiotics, chap. 9. IntechOpen, Rijeka (2020). https://doi.org/10.5772/intechopen.90173
11. Kosikov, S., et al..: Indexical structures to enable knowledge mining tasks. Procedia Comput. Sci. **169**, 284–290 (2020). https://doi.org/10.1016/j.procs.2020.02.180. http://www.sciencedirect.com/science/article/pii/S1877050920303033. Postproceedings of the 10th Annual International Conference on Biologically Inspired Cognitive Architectures, BICA 2019 (Tenth Annual Meeting of the BICA Society), Seattle, Washington, USA, 15–19 August 2019
12. MacLennan, B.J.: Universally programmable intelligent matter summary. In: Proceedings of the 2nd IEEE Conference on Nanotechnology, pp. 405–408 (2002)
13. Scott, D.S.: Lambda calculus then and now. In: ACM Turing Centenary Celebration, ACM-TURING 2012. ACM, New York (2012). https://doi.org/10.1145/2322176.2322185
14. Shaumyan, S.K.: Applicational Grammar as a Semantic Theory of Natural Language. University of Chicago Press, Chicago (1977)
15. Wolfengagen, V., Ismailova, L., Kosikov, S.: Capturing information processes with variable domains. Procedia Comput. Sci. **169**, 276–283 (2020). https://doi.org/10.1016/j.procs.2020.02.177. http://www.sciencedirect.com/science/article/pii/S1877050920303008. Postproceedings of the 10th Annual International Conference on Biologically Inspired Cognitive Architectures, BICA 2019 (Tenth Annual Meeting of the BICA Society), Seattle, Washington, USA, 15–19 August 2019
16. Wolfengagen, V., Ismailova, L., Kosikov, S., Dohrn, J.: Superimposing semantic mesh to prevent information processes entanglement. Procedia Comput. Sci. **169**, 645–651 (2020). https://doi.org/10.1016/j.procs.2020.02.197. http://www.sciencedirect.com/science/article/pii/S1877050920303203. Postproceedings of the 10th Annual International Conference on Biologically Inspired Cognitive Architectures, BICA 2019 (Tenth Annual Meeting of the BICA Society), Seattle, Washington, USA, 15–19 August 2019

17. Wolfengagen, V., Kosikov, S., Ismailova, L., Zaytsev, V.: Semantic framework for data flow control in the network of information graphs. Procedia Comput. Sci. **169**, 16–22 (2020). https://doi.org/10.1016/j.procs.2020.02.110. http://www.sciencedirect.com/science/article/pii/S1877050920302337. Postproceedings of the 10th Annual International Conference on Biologically Inspired Cognitive Architectures, BICA 2019 (Tenth Annual Meeting of the BICA Society), Seattle, Washington, USA, 15–19 August 2019
18. Wolfram, S.: A New Kind of Science, 1 edn. Wolfram Media (2002)

Applicative Model to Bring-in Conceptual Envelope for Computational Thinking with Information Processes

Viacheslav Wolfengagen[1](✉), Larisa Ismailova[1], and Sergey Kosikov[2]

[1] National Research Nuclear University "Moscow Engineering Physics Institute", Moscow 115409, Russian Federation
jir.vew@gmail.com, lyu.ismailova@gmail.com
[2] NAO "JurInfoR", Moscow 119435, Russian Federation
kosikov.s.v@gmail.com

Abstract. Computational activity is now recognized as a natural science, and computational and informational processes have been discovered in the deep structures of many areas. Computations in the natural world were present long before the invention of computers, but a remarkable shift in understanding its fundamental nature occurs, in fact, before our eyes. The present moment, in fact, is a transition from the concept of computer science as an artificial science to the understanding that information processes are abundant in nature. Computing is recognized as a natural science that studies natural and artificial information processes.

All object constructions are considered as virtual objects, which are possibly combinators. They are connected by "combinatory glue" and form an applicative prestructure.

Keywords: Information process · Cognitive interference · Quantum shift learning · Natural computing

1 Introduction

In computer science, the computation of value (evaluation) is one of the central concepts being developed, which is discussed in the theory of computing and leads to the construction of numerous computational models. However, to date, there has not been a single idea about which logical-mathematical apparatus will be most suitable for taking into account the features of the arising computational effects. This question is especially acute in the case of symbolic computations, which more and more have to be dealt with in modern software. The development of the theory of computing has led to the establishment of semantics of programming languages, which discusses the techniques and methods of evaluating expressions. Such efforts lead to success, but the emerging computational models turn out to be cumbersome, abundant in numerous details, and, moreover, poorly amenable to mathematical study. In the present work, an attempt

A. V. Samsonovich et al. (Eds.): BICA 2020, AISC 1310, pp. 572–577, 2021.
https://doi.org/10.1007/978-3-030-65596-9_70

is made to single out the main fundamental features underlying the variety of computational models and form a unified and self-contained explanatory system. This is important for the establishment and development of the fundamental basics of information technology, which is recognized as *relevant*.

Computational models are usually considered within the framework of classical logic-mathematic means, operating with free and bound variables. This most often leads to the use of predicate logic and, in practice, requires special techniques to indicate the ranges over which the variables run. In particular, the resulting layout of computational behavior is complicated if the ranges of one way or another overlap, change over time, arise and cease to exist during the evolution of a compound computational model. In practice, this is a typical mode of operation of an information system and/or application of information technology. A good example of this kind is provided by the technologies of semantic Web [16]. Advancement in understanding the dynamics of computations can be achieved by highlighting computational *invariants*, which requires the development of special logical and mathematical tools of a non-classical type. A relatively new approach is proposed in the paper, connected with the establishment of systems of invariants, their complexing and combination, and also the introduction of characteristic equalities for them. The emerging theory of computation has *new*, previously unexplored properties, allowing us to generate objectively interesting and meaningful principles of computation and build one or another target computational model on their ground.

One of the goals of this work is to revise the well-known ideas that how to get the value of an expression using an applicative computing system. Computations and/or symbolic transformations have undoubtedly become one of the dominant trends in modern computer research. In particular, this refers to the implementation of the processes of the derivation of an object with given properties, which is studied with the formation of sets of equalities, which together lead to a computational model. At the same time, the comprehensive theory of computations receives a targeted equational description, making the process of studying its properties and capabilities rather constructive.

Similar questions in computer science arose at the initial stages of the formation of common formal systems [2], but usually the study was limited to the study of substitutional systems, which encountered difficulties in implementing restrictions on the types of variables and expressions used [3]. Only with the use of applicative computing technologies, as was shown in it is possible to construct a general theory of computations that operates with an acceptable idea of the individual object in practice. This method was applied to the solution of semantic modeling problems [7,12,14], [9] with the formation of a dynamic computational model of the applicative type [10,13]. An important advance in the construction of a computational model that universally characterizes the interaction of information processes was made in [6,8], where a layered discussion of the interaction is proposed. This approach in this paper was expressed in the equational construction of the theory of computation [5]. The applicative computational models and technologies obtained on its basis are important for the development of modern programming [11], especially in the context of growing interest in the practical use of the functional approach to programming [1,4,15].

2 Applicative System and Model

Let I, K, S be (mathematical) objects. In addition, a set of indeterminants is added that gives sufficient stock of *variables.* All variables are included in the feature class. These objects define a set of generic objects, or *atoms.* The definition of derivatives is given by induction on their construction as follows.

Definition (Objects). *(i)* I, K, S *and variables are objects. (ii) If* a, b *are objects, then* $(a\ b)$ *is also an object.*

The induction step needs an intuitive justification. Thus, the binary operator of *applications* $(\cdot\ \cdot)$ is introduced:

$$(\cdot\ \cdot) : \underbrace{\text{object}}_{\text{function}} \times \underbrace{\text{object}}_{\text{argument}} \to \underbrace{\text{object}}_{\text{value}}.$$

This is an object-generating operator that ranges over objects. The first object is considered as a ‘function’, and the second as an ‘argument’. Therefore, the application operator gives a function that, as a result of applying to the argument generates some new object, or the result of the evaluation (in another terminology: application) functions to their argument, and without performing such an application, we cannot get the resulting *value.*

To compare some objects with others, we need to define some (binary) relation. Usually this relationship is called *conversion* and is defined by *postulates* (CL):

$$(\mathsf{I})\quad \mathsf{I}a = a, \quad (\mathsf{K})\quad \mathsf{K}ab = a, \quad (\mathsf{S})\quad \mathsf{S}abc = ac(bc),$$

$$(\sigma)\quad a = a, \quad (\rho)\quad \frac{a = b}{b = a}, \quad (\tau)\quad \frac{a = b,\ b = c}{a = c},$$

$$(\mu)\quad \frac{a = b}{ca = cb}, \qquad (\nu)\quad \frac{a = b}{ac = bc},$$

where a, b, c denotes arbitrary objects, and ‘=’ is the relation of conversion.

2.1 Combinatory Glue

The formulation of the system $(CL\eta\xi)$ is given equationally, i.e. the binary relation ‘=’ of conversion can be used as a form of equivalence. It turns out that additional equalities make it possible to take into account a larger number of features of practically useful objects. So far, the review has been conducted purely at the syntax level. Now let’s try to build ‘embedded applications’ that are essentially semantic.

For purely mathematical reasons, additional (and not original) combinators that simplify computations will be introduced. Here are some of such combinators that are axiomatized by the following equalities:

$$\mathsf{I}x = x,\ \mathsf{C}xyz = xzy,\ \mathsf{B}xyz = x(yz),$$
$$\mathsf{K}xy = x,\ \mathsf{S}xyz = xz(yz),$$
$$\mathsf{D}xy \equiv [x, y] \equiv \lambda r.rxy,\ < f, g > \equiv \lambda t.[ft, gt],$$
$$\Phi xyzw = x(yw)(zw),\ \Psi xyzw = x(yz)(yw),$$
$$\mathsf{B}^2 \equiv \mathsf{BBB},\ \mathsf{C}^2 xyzw = xwyz,$$
$$Curry \equiv \lambda h.\lambda xy.h[x, y],\ p[x, y] = x,\ q[x, y] = y.$$

Further they will be used to improve the properties of the source shell.

2.2 Extracting the Valuation Envelope

Consider a set of $(CL\eta\xi)$ postulates with additional equalities used as the restrictions: $\mathsf{B} = \Psi(\Phi\ \mathsf{I})$, $\mathsf{CB}^2\mathsf{D} = \Psi(\Phi\ \mathsf{D})$, $\mathsf{V}\mathcal{K}\rho = \mathcal{K} = \mathsf{K}\mathcal{K}$, $Curry = \Psi(\Phi\mathsf{D})$ which are not obviously mutually independent.

We argue that they generate the Eval-system for evaluation the expressions in applicative computational system principles in accordance with Table 1.

Table 1. Computational envelope for applicative Eval-system

Equation	Notation	Principle
$\mathsf{B} = \Psi(\Phi\ \mathsf{I})$	$(\cdot(\cdot) =$	Evaluation of an application
$\mathsf{CB}^2\mathsf{D} = \Psi(\Phi\ \mathsf{D})$	$([\cdot, \cdot] =)$	Evaluation of an ordered pair
$\mathsf{V}\mathcal{K}\rho = \mathcal{K} = \mathsf{K}\mathcal{K}\rho$	′ encapsulation	Evaluation of a constant
$Curry = \Psi(\Phi\mathsf{D})$	$([\cdot, \cdot] =)$	Evaluation of an ordered pair

The verification is slightly troublesome but straightforward and is left for reader.

Conclusion

The main results can be briefly systematized as follows.

1. Varying from research to research, the nootion of the essence of an 'object' can be expressed from a computational point of view, presenting and embedding it in some kind of primitive frame. The scheme operates within the framework of function theory related to *combinatory logic* and generates the original conceptual envelope.
2. Combinators provide a semantically consistent *basis* to build a model of data objects. Used objects inherit both syntactic and semantic features and features of the original representation of the object. This leads to computations of the form *object-as-functor* and generates an improvement process that allows one to take into account *methods* for specific objects.

3. As can be shown, concepts are embedded in the envelope and inherit the logical properties of objects. The theory of higher orders is used (with *definite descriptions*).
4. The improvement process satisfies equationally expressed conditions. Separate methods should be studied in an equational framework.

The work is an outlook of the results that are associated with the construction of a generalized computational model and obtained at different times during the implementation of projects partially supported by RFBR grants.

Acknowledgments. This research is supported in part by the Russian Foundation for Basic Research, RFBR grants 20-07-00149-a, 19-07-00326-a, 19-07-00420-a, 18-07-01082-a.

References

1. Bimbo, K.: Combinatory Logic: Pure, Applied and Typed, 1st edn. Discrete Mathematics and Its Applications, CRC (2012)
2. Curry, H.B.: Foundations of Mathematical Logic, 2, revised edn. Dover Books on Mathematics, Dover Publications (2010)
3. Hindley, J.R., Lercher, B., Seldin, J.P.: Introduction to Combinatory Logic, London Mathematical Society Lecture Note Series, vol. 7. Cambridge University Press, Cambridge (1972)
4. Ismailova, L., Kosikov, S., Kucherov, I., Zhuravleva, O.: Tools of algebraic type for manipulating methodologically oriented cognitive information. Procedia Computer Science **169**, 23–30 (2020). https://doi.org/10.1016/j.procs.2020.02.111. http://www.sciencedirect.com/science/article/pii/S1877050920302349. Postproceedings of the 10th Annual International Conference on Biologically Inspired Cognitive Architectures, BICA 2019 (Tenth Annual Meeting of the BICA Society), held 15–19 August 2019 in Seattle, Washington, USA
5. Ismailova, L., Kosikov, S., Zinchenko, K., Mikhailov, A., Bourmistrova, L., Berezovskaya, A.: Equationally expressed evaluation. In: 9th International Workshop on Functional and Logic Programming, WFLP 2000, Benicassim, Spain, 28–30 September, 2000, pp. 135–143 (2000)
6. Ismailova, L., Wolfengagen, V., Kosikov, S.: Hereditary information processes with semantic modeling structures. Procedia Comput. Sci. **169**, 291–296 (2020). https://doi.org/10.1016/j.procs.2020.02.181. http://www.sciencedirect.com/science/article/pii/S1877050920303045. Postproceedings of the 10th Annual International Conference on Biologically Inspired Cognitive Architectures, BICA 2019 (Tenth Annual Meeting of the BICA Society), held 15–19 August 2019 in Seattle, Washington, USA
7. Ismailova, L., Wolfengagen, V., Kosikov, S., Dohrn, M.: Capturing push-processing using enriched semantic mesh equipped with functionals-and-hops model. Procedia Comput. Sci. **169**, 590–596 (2020). https://doi.org/10.1016/j.procs.2020.02.205. http://www.sciencedirect.com/science/article/pii/S1877050920303288. Postproceedings of the 10th Annual International Conference on Biologically Inspired Cognitive Architectures, BICA 2019 (Tenth Annual Meeting of the BICA Society), held 15–19 August 2019 in Seattle, Washington, USA

8. Ismailova, L., Wolfengagen, V., Kosikov, S., Maslov, M., Dohrn, J.: Semantic models to indicate post-truth with fake news channels. Procedia Comput. Sci. **169**, 297–303 (2020). https://doi.org/10.1016/j.procs.2020.02.182. URL http://www.sciencedirect.com/science/article/pii/S1877050920303057. Postproceedings of the 10th Annual International Conference on Biologically Inspired Cognitive Architectures, BICA 2019 (Tenth Annual Meeting of the BICA Society), held 15–19 August 2019 in Seattle, Washington, USA
9. Ismailova, L.Y., Kosikov, S.V., Wolfengagen, V.E.: Computational model for the construction of cognitive maps. In: Silvera-Roig, M., Azcárate, A.L.V. (eds.) Cognitive and Intermedial Semiotics, chap. 9. IntechOpen, Rijeka (2020). https://doi.org/10.5772/intechopen.90173
10. Descles, J.-P., Zlatka Guentchva, S.S.: Theoretical Aspects of Passivization in the Framework of Applicative Grammar. Pragmatics & Beyond. John Benjamins Publishing Company (1985)
11. Kosikov, S., Ismailova, L., Wolfengagen, V., Marenkov, A., Ermak, M., Slieptsov, I., Zaytsev, V., Shedko, A.: Indexical structures to enable knowledge mining tasks. Procedia Comput. Sci. **169**, 284–290 (2020). https://doi.org/10.1016/j.procs.2020.02.180. http://www.sciencedirect.com/science/article/pii/S1877050920303033. Postproceedings of the 10th Annual International Conference on Biologically Inspired Cognitive Architectures, BICA 2019 (Tenth Annual Meeting of the BICA Society), held 15–19 August 2019 in Seattle, Washington, USA
12. Shaumyan, S.K.: Applicational Grammar as a Semantic Theory of Natural Language. University of Chicago Press, Chicago (1977)
13. Wolfengagen, V., Ismailova, L., Kosikov, S.: Capturing information processes with variable domains. Procedia Comput. Sci. **169**, 276–283 (2020). https://doi.org/10.1016/j.procs.2020.02.177. http://www.sciencedirect.com/science/article/pii/S1877050920303008. Postproceedings of the 10th Annual International Conference on Biologically Inspired Cognitive Architectures, BICA 2019 (Tenth Annual Meeting of the BICA Society), held 15–19 August 2019 in Seattle, Washington, USA
14. Wolfengagen, V., Ismailova, L., Kosikov, S., Dohrn, J.: Superimposing semantic mesh to prevent information processes entanglement. Procedia Comput. Sci. **169**, 645–651 (2020). https://doi.org/10.1016/j.procs.2020.02.197. http://www.sciencedirect.com/science/article/pii/S1877050920303203. Postproceedings of the 10th Annual International Conference on Biologically Inspired Cognitive Architectures, BICA 2019 (Tenth Annual Meeting of the BICA Society), held 15–19 August 2019 in Seattle, Washington, USA
15. Wolfengagen, V., Kosikov, S., Ismailova, L., Zaytsev, V.: Semantic framework for data flow control in the network of information graphs. Procedia Comput. Sci. **169**, 16–22 (2020). https://doi.org/10.1016/j.procs.2020.02.110. http://www.sciencedirect.com/science/article/pii/S1877050920302337. Postproceedings of the 10th Annual International Conference on Biologically Inspired Cognitive Architectures, BICA 2019 (Tenth Annual Meeting of the BICA Society), held 15–19 August 2019 in Seattle, Washington, USA
16. Wu, W.: The deep web: Woven to catch the middle ground. In: Proceedings of the 4th International Workshop on Web-Scale Knowledge Representation Retrieval and Reasoning, Web-KR 2013, pp. 5–8. Association for Computing Machinery, New York (2013). https://doi.org/10.1145/2512405.2512408

Computational Model for Capturing the Interdependencies on Information Processes

Viacheslav Wolfengagen[1(✉)], Larisa Ismailova[1], and Sergey Kosikov[2]

[1] National Research Nuclear University "Moscow Engineering Physics Institute", Moscow 115409, Russian Federation
jir.vew@gmail.com, lyu.ismailova@gmail.com
[2] NAO "JurInfoR", Moscow 119435, Russian Federation
kosikov.s.v@gmail.com

Abstract. Current resource integrity techniques such as sub-resource integrity (SRI) do not apply to unknown content, and mitigating attacks along this vector remains an open issue. Complex interdependencies on the Internet have implications that go beyond security, leading to the creation of unwanted manipulation of information processes. Cognitive interference, based on the deliberate entanglement of information processes, is increasingly spreading on the Web. Today, there is still no generally accepted understanding of the mechanism of entanglement of information processes. In this paper, we propose a cognitive architecture and a computational model that allow representing an information process with a variable property, opening up opportunities for studying the forms of its behavior. Entangled processes are considered to occur simultaneously, as coexisting processes that have a common source of origin, and the state of their system is represented by a superposition of components. As it turns out, the integrity of resources is violated when the conditions for the emergence of architecture with a variable property arise in a tangle of process dependencies. This creates entangled information processes. This occurs in the emergence of vulnerabilities in the Web environment such as cross-site scripting XSS. The models and mechanisms of complex interdependencies on the Internet still remain poorly understood.

Keywords: Information process · Cognitive interference · Tangled processes · Computational model

1 Introduction

Since its inception in 1989, the Internet community has decided to allow HTTP pages to upload third-party content to the [9] world wide web.

Multiple information processes arise, for which it is problematic to establish the dependency graph that underlies websites. Processes affect each other, but

A. V. Samsonovich et al. (Eds.): BICA 2020, AISC 1310, pp. 578–584, 2021.
https://doi.org/10.1007/978-3-030-65596-9_71

in order to secure websites and ensure an acceptable level of their performance, it is necessary to install a dependency graph architecture that restricts access to malicious sites using access policies [14]. The resulting tangle of process dependencies is accompanied by the explosive development of advertising and tracking technologies and an increase in dependence on common platforms [15].

As the analysis of [12] shows, [1], popular sites download an average of 73 resources, 23 of which are downloaded from external domains – each double the corresponding figure five years earlier. However, a small number of third parties serve resources for a significant portion of the sites that provide these remote resources. Shared platforms deliver content to many sites – Content Distribution Networks (CDNs) serve 60% of the Top Million, and Cloud Service Providers serve 37% of the Top Million.

The increase in the number of external downloads and their distribution creates an attractive attack vector for compromising large numbers of clients through common dependencies and platforms used on popular sites. Approximately 33% of a million sites download unknown content indirectly through at least one third-party organization, exposing users to resources that the site operators have no relationship with. Combined with the observation that 87% of sites are executing active content (such as JavaScript) from an external domain, a confusing picture of attacks emerges.

Scripting vulnerabilities are notable, and are destructive to web application development: deceptively simple to reason and fix, yet surprisingly difficult to prevent in large-scale web development. This is a well-known deliberate entanglement effect known as XSS (Cross Site Scripting), which occurs when insufficient validation, sanitization, or escaping of data in a web application allows an attacker to cause malicious JavaScript to execute on the browser side in the application context. This injected code can then do whatever the attacker wants by using the victim's [7] privileges, [8]. Exploiting XSS errors results in a complete (though not necessarily permanent) compromise of the victim's session with the vulnerable application.

Current resource integrity techniques such as Sub-Resource Integrity (SRI) are not applicable to unknown content, and mitigation of such attacks remains open [11]. Complex Internet interdependencies have implications that go beyond vulnerabilities [13].

Today, there is still no common understanding of the mechanism of entanglement of information processes. In this paper, we propose a cognitive architecture and a computational model that allows to support an information process with a variable property. Entangled processes are considered to be occurring simultaneously, coexisting processes that have a common source of occurrence, but, possibly, different properties. As it turns out, the integrity of resources is violated when the conditions for the emergence of an architecture with a variable property arise in a tangle of process dependencies. This creates tangled information processes. This occurs in the arising of vulnerabilities in the Web environment, similar to XSS cross-site scripting. The complex interdependencies on the Internet have implications that go beyond security that remain poorly understood. In Sect. 2 we construe a generalized example of a process dependency architecture where an information process exhibits a variable property. As shown, the emerging presentation complexities are fundamental. In this work, the method of indexed expressions [10] is used, which allows the semantic model to take into account the stages of knowledge and their development. The resulting information graphs [6], by analogy with [19], can be interpreted as a data flow propagation environment, which in this sense can control their interference. On their basis, a toolkit for the management of cognitive information [2] can be developed. In addition, in this work, the ideas of modeling information processes by variable domains [17] are largely inherited. The features of process inheritance proposed in [3] have been taken into account. The model solutions used are also based on the idea of channelized propagation of [5] processes. Finally, the possibilities of jump transitions of processes [4], as well as the effects of entanglement of information processes [18], potentially causing cognitive interference and its propagation along the network channels, are taken into account.

2 Computational Model of the Information Process with Variable Property

Let us give a generalized example of conditions when an information process with a variable property arises. A similar example was discussed earlier in structures like the semantic net. The construction of a mathematical model is known as the "Problem of quick (agile) Smith" [16] and illustrated by the situation of the marriage g of Smith s, changing his name r to Jones j, and it is known that A already had Jones j, who was already married g. It turns out that the model presented below also characterizes the type of quasi-entanglement, similar to the known Schrödinger's cat model.

Figure 1 shows entanglement. In the model, A is fixed, and f, B are variables.

The mappings $g : T \to T^+$ and *not-g* $: T \to T^-$ give transitions between the "properties" of T on the one hand and the "properties" of T^+ and T^- on the other hand, respectively, and each such transition "restricts" the elements corresponding to T to the elements corresponding to T^+ and T^- "along" the mappings g and *not-g* respectively.

The mappings $r : T^+ \to T^-$ give "jump-like" transitions between T^+ and T^- "along" the mappings r.

The maps $f : B \to A$ give us transitions between "stages" A and "later" stages B; and each such transition "restricts" elements corresponding to A to elements, corresponding to B "along" the map f. Consider the maps

$$\begin{aligned} h &: A \to T, \\ s &: A \to T^-, \\ j &: A \to T^+. \end{aligned}$$

Furthermore,

$$\begin{aligned} s_f &: B \to C_f^- \subseteq T^-, \\ j_f &: B \to C_f^+ \subseteq T^+, \\ r &: T^- \to T^+, \end{aligned}$$

and in the model turns out that

Tangle $$r \circ s_f = j_j.$$

This means that the f-images of the processes s and j arise simultaneously and are treated as *entangled*, which is characterized by the equation **Tangle**.

Let h be a process in A, the property of which can be verified in the future, but remains undefined. Let s be a process in A, whose property can be verified in A as *not-g*. Let j be a process under fixed conditions A that possesses the verified property g.

Mathematically, this is expressed by

$$\begin{aligned} s &= h \circ \textit{not-g} \\ j &= h \circ g \end{aligned}$$

Thus, it is only when interacting with the environment that conditions arise for checking the *observable* of h properties, one or another.

As it turns out, an interpretation arises according to which the system of processes remains in superposition until it interacts with the external world or is discovered by it. When this happens, the superposition collapses into one or another of the possible defined states.

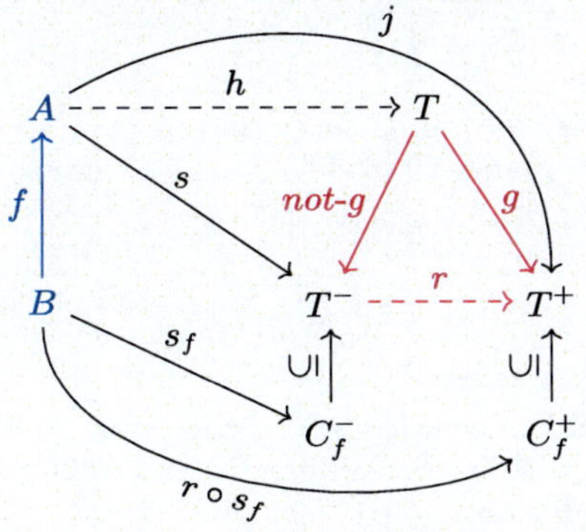

Fig. 1. Entanglement of processes

In the folk model of the language, when constructing a linguistic interpretation, we obtain the following mathematical characteristic:

$$T^- = T \setminus T^+,$$

$$r \circ s_f = j_f = \text{“began to behave like } j_f\text{”},$$

$$not\text{-}g \circ h = s = \text{“behave like } s\text{”},$$

$$g \circ h = j = \text{“behave like } j\text{”},$$

which gives a variant of the phenotypic characterization of entanglement. The folk model of language is usually understood as a way of distinguishing between form and meaning in a language. In this case, it is possible to establish a mathematical characterization that expresses such a distinction.

3 Conclusions

The main results are as follows.

1. The observed increase in the amount of information on the Web gives rise to multiple dependencies of information processes, which gives rise to a tangle of dependencies.
2. As it turns out, the integrity of resources is violated when in a ball process dependencies, conditions are created for the occurrence of variable property architectures. This generates entangled information processes and architecture of the Schrödinger's cat model.
3. In practice, in particular, in the Web environment, vulnerabilities such as XSS cross-site scripting are generated.
4. Models and mechanisms of complex interdependencies of information processes on the Internet remain insufficiently studied, and the complexities of their representation are not only mathematically complex, but also have fundamental significance.

Acknowledgments. This research is supported in part by the Russian Foundation for Basic Research, RFBR grants 20-07-00149-a, 19-07-00326-a, 19-07-00420-a, 18-07-01082-a.

References

1. Byrne, M.D., John, B.E., Wehrle, N.S., Crow, D.C.: The tangled web we wove: a taskonomy of www use. In: Proceedings of the SIGCHI Conference on Human Factors in Computing Systems, CHI 1999, pp. 544–551. Association for Computing Machinery, New York (1999). https://doi.org/10.1145/302979.303154
2. Ismailova, L., Kosikov, S., Kucherov, I., Zhuravleva, O.: Tools of algebraic type for manipulating methodologically oriented cognitive information. Procedia Comput. Sc. **169**, 23–30 (2020). https://doi.org/10.1016/j.procs.2020.02.111. http://www.sciencedirect.com/science/article/pii/S1877050920302349. Postproceedings of the 10th Annual International Conference on Biologically Inspired Cognitive Architectures, BICA 2019 (Tenth Annual Meeting of the BICA Society), held August 15–19, 2019 in Seattle, Washington, USA

3. Ismailova, L., Wolfengagen, V., Kosikov, S.: Hereditary information processes with semantic modeling structures. Procedia Comput. Sci. **169**, 291–296 (2020). https://doi.org/10.1016/j.procs.2020.02.181. http://www.sciencedirect.com/science/article/pii/S1877050920303045. Postproceedings of the 10th Annual International Conference on Biologically Inspired Cognitive Architectures, BICA 2019 (Tenth Annual Meeting of the BICA Society), held August 15–19, 2019 in Seattle, Washington, USA
4. Ismailova, L., Wolfengagen, V., Kosikov, S., Dohrn, M.: Capturing push-processing using enriched semantic mesh equipped with functionals-and-hops model. Procedia Comput. Sci. **169**, 590–596 (2020). https://doi.org/10.1016/j.procs.2020.02.205. http://www.sciencedirect.com/science/article/pii/S1877050920303288. Postproceedings of the 10th Annual International Conference on Biologically Inspired Cognitive Architectures, BICA 2019 (Tenth Annual Meeting of the BICA Society), held August 15–19, 2019 in Seattle, Washington, USA
5. Ismailova, L., Wolfengagen, V., Kosikov, S., Maslov, M., Dohrn, J.: Semantic models to indicate post-truth with fake news channels. Procedia Comput. Sci. **169**, 297–303 (2020). https://doi.org/10.1016/j.procs.2020.02.182. http://www.sciencedirect.com/science/article/pii/S1877050920303057. Postproceedings of the 10th Annual International Conference on Biologically Inspired Cognitive Architectures, BICA 2019 (Tenth Annual Meeting of the BICA Society), held August 15–19, 2019 in Seattle, Washington, USA
6. Ismailova, L.Y., Kosikov, S.V., Wolfengagen, V.E.: Computational model for the construction of cognitive maps. In: Silvera-Roig, M., Azcárate, A.L.V. (eds.) Cognitive and Intermedial Semiotics. IntechOpen, Rijeka (2020). https://doi.org/10.5772/intechopen.90173
7. Kern, C.: Securing the tangled web. Queue **12**(7), 40–55 (2014). https://doi.org/10.1145/2639988.2663760
8. Kern, C.: Securing the tangled web. Commun. ACM **57**(9), 38–47 (2014). https://doi.org/10.1145/2643134
9. Korman, K.: A tangled web. NetWorker **6**(4), 3 (2002). https://doi.org/10.1145/1008573.1008577
10. Kosikov, S., Ismailova, L., Wolfengagen, V., Marenkov, A., Ermak, M., Slieptsov, I., Zaytsev, V., Shedko, A.: Indexical structures to enable knowledge mining tasks. Procedia Comput. Sci. **169**, 284–290 (2020). https://doi.org/10.1016/j.procs.2020.02.180. http://www.sciencedirect.com/science/article/pii/S1877050920303033. Postproceedings of the 10th Annual International Conference on Biologically Inspired Cognitive Architectures, BICA 2019 (Tenth Annual Meeting of the BICA Society), held August 15–19, 2019 in Seattle, Washington, USA
11. Kumar, D., Ma, Z., Durumeric, Z., Mirian, A., Mason, J., Halderman, J.A., Bailey, M.: Security challenges in an increasingly tangled web. In: Proceedings of the 26th International Conference on World Wide Web, WWW 2017, pp. 677–684. International World Wide Web Conferences Steering Committee, Republic and Canton of Geneva, CHE (2017). https://doi.org/10.1145/3038912.3052686
12. Moglen, E.: The tangled web we have woven. Commun. ACM **56**(2), 20–22 (2013). https://doi.org/10.1145/2408776.2408784
13. Page, X., Knijnenburg, B.P., Kobsa, A.: What a tangled web we weave: lying backfires in location-sharing social media. In: Proceedings of the 2013 Conference on Computer Supported Cooperative Work, CSCW 2013, pp. 273–284. Association for Computing Machinery, New York (2013). https://doi.org/10.1145/2441776.2441808

14. Paige, M.: The tangled web: a guide to securing modern web applications by Michal Zalewski. SIGSOFT Softw. Eng. Notes **38**(4), 39–40 (2013). https://doi.org/10.1145/2492248.2492253
15. Vartapetiance, A., Gillam, L.: Deception detection for the tangled web. SIGCAS Comput. Soc. **42**(1), 34–47 (2012). https://doi.org/10.1145/2422512.2422517
16. Wolfengagen, V.: Semantic modeling: computational models of the concepts. In: 2010 International Conference on Computational Intelligence and Security, CIS 2010, Nanning, Guangxi Zhuang Autonomous Region, China, 11–14 December 2010, pp. 42–46 (2010). https://doi.org/10.1109/CIS.2010.16
17. Wolfengagen, V., Ismailova, L., Kosikov, S.: Capturing information processes with variable domains. Procedia Comput. Sci. **169**, 276–283 (2020). https://doi.org/10.1016/j.procs.2020.02.177. http://www.sciencedirect.com/science/article/pii/S1877050920303008. Postproceedings of the 10th Annual International Conference on Biologically Inspired Cognitive Architectures, BICA 2019 (Tenth Annual Meeting of the BICA Society), held August 15–19, 2019 in Seattle, Washington, USA
18. Wolfengagen, V., Ismailova, L., Kosikov, S., Dohrn, J.: Superimposing semantic mesh to prevent information processes entanglement. Procedia Comput. Sci. **169**, 645–651 (2020). https://doi.org/10.1016/j.procs.2020.02.197. http://www.sciencedirect.com/science/article/pii/S1877050920303203. Postproceedings of the 10th Annual International Conference on Biologically Inspired Cognitive Architectures, BICA 2019 (Tenth Annual Meeting of the BICA Society), held August 15–19, 2019 in Seattle, Washington, USA
19. Wolfengagen, V., Kosikov, S., Ismailova, L., Zaytsev, V.: Semantic framework for data flow control in the network of information graphs. Procedia Comput. Sci. **169**, 16–22 (2020). https://doi.org/10.1016/j.procs.2020.02.110. http://www.sciencedirect.com/science/article/pii/S1877050920302337. Postproceedings of the 10th Annual International Conference on Biologically Inspired Cognitive Architectures, BICA 2019 (Tenth Annual Meeting of the BICA Society), held August 15–19, 2019 in Seattle, Washington, USA

Imposing and Superposing the Information Processes over Variable Concepts

Viacheslav Wolfengagen[1](✉), Larisa Ismailova[1], and Sergey Kosikov[2]

[1] National Research Nuclear University "Moscow Engineering Physics Institute", Moscow 115409, Russian Federation
jir.vew@gmail.com, lyu.ismailova@gmail.com
[2] NAO "JurInfoR", Moscow 119435, Russian Federation
kosikov.s.v@gmail.com

Abstract. A conceptual basis for modeling the information process is given in the study of the features of its propagation in case of varying the determining property. The system of conceptual dependencies over variable abstract sets is used as a relativized basis for processing semantic information. Variable domains and mapping are used to control the progress of the process, as well as recognition mappings and reversal mapping of the process property. It is shown that in the representing functorial category they can be interpreted as cognitive architectures that are significant for solving semantic modeling problems. In this case, the superposition of process states can be taken into account.

Keywords: Information process · Variable sets · Category theory · Superposition of states · Cognitive architecture · Schrodinger's cat

1 Introduction

Cognitive interference refers to the loss of the information process, which occurs when ideas/processes generated by other participants interfere with the activity of generating their own ideas belonging to a given individual. It is usually assumed that knowledge is representable by language structures, and reasoning is a computational process that rearranges these structures in accordance with formal rules. Similar representations dominate cognitive science when *computational notions* are applied to understand human cognition.

Modern computational notions are based on drawing ideas from *category theory*. S. MacLane gives convincing examples that illustrate the main thesis: logic is a branch of mathematics, and not vice versa [9]. His reasoning relies on functors. In turn, F. Lawvere proposes a functorial construction *variable set* [8]. Computational models based on systems related to the lambda calculus have been studied by D. Scott [11], especially the ability of parameterization. Similar ideas were developed by G. Cousineau, P.-L. Curien and M. Mauny and brought

A. V. Samsonovich et al. (Eds.): BICA 2020, AISC 1310, pp. 585–590, 2021.
https://doi.org/10.1007/978-3-030-65596-9_72

it to implementation – a computational agent in the form of a categorical abstract machine [1] and the CAML programming language based on it.

All this testifies to a certain proportion of incompleteness of the semantic modeling of information processes, and the available proposals are fragmented. At the same time, the key point is the development of a computational model that really represents the interaction of information processes. To overcome the difficulties and contradictions that arise, we will use the conceptual framework of variable domains, which allows us to develop a rather productive parameterization.

In this work, the method *indexical expressions* [7] is used, which allows the semantic model to take into account the stages of knowledge and their unfolding. The resulting information graphs [6], by analogy with [15], can be interpreted as an environment for the propagation of data streams, which in this sense can control their interference. On their basis, the cognitive information management toolkit [2] can be developed. In addition, in this work, the ideas of modeling information processes by variable domains [13] are largely inherited. The features of process inheritance proposed in [3] are taken into account. The model solutions used are also based on the idea of channelized propagation of [5] processes. Finally, the possibilities of jump transitions of the processes [4], as well as the effects of confusion of information processes [14], potentially causing cognitive interference and its propagation along the network channels, are taken into account.

In order to avoid an excessive level of abstraction, we will build a computational model using a representative end-to-end example. A similar example has been used in the semantic modeling literature, but remained at the declarative level of semantic networks.

2 Modeling a Process with Variable Property

Agile Smite Problem. Let us give a generalized example of conditions when an information process with a variable property arises. A similar example was discussed earlier in structures such as the Semantic Nets. The construction of a mathematical model is known as the "Problem of quick (agile) Smith" [12] and illustrated by the situation marriage of g Smith s, changing r his name to Jones j, and it is known that A already had Jones j, who was already married g.

Problem of Schrödinger's Cat. It turns out that the model presented below also characterizes the type of quasi-entanglement, similar to the Schrödinger's cat model.

E. Schrödinger's proposed an imaginable experiment that explains the quantum superposition [10]. The essence of the experiment is as follows: in a box with a conventional cat there is an atom of a radioactive element and a flask with acid, which will break after the radioactive element decays. If the flask breaks, the cat will die, but we do not know for sure whether the atom of the radioactive element will decay, and no one, including the cat, can influence this.

Consequently, the cat is both alive and dead at the same time, which is called quantum superposition.

The prevailing theory, called the Copenhagen Interpretation, states that a quantum system remains in superposition until it interacts with or is discovered by the outside world. According to this interpretation, the cat remains both alive and dead until the condition is measured by *an external observer.* We will be able to determine the exact state of the cat only after we open the box and, thus, the state of the Schrödinger's cat in the quantum world is determined precisely by the fact of our observation, a kind of cognitive interference.

2.1 Modeling of Change

Let's use a case example to keep it intuitive. Interpretation is reduced to the description of the process of superposition of states, in which there may be variable properties of the individual.

2.2 Variable Domains

Metatheorem (partially interpreted in terms of the Schrödinger's cat problem).

Let us show that the indexical expression method works in the case of a variable property (the cat survived/died).

Let be $h : A \to T$ – individual (cat); $f : B \to A$ – evolvent (experimentation); mappings $g^- : T \to T^-$, $g^+ : T \to T^+$ – recognition functions; $r : T^- \to T^+$ – reversing the property of; $g^- \circ h : A \to T^-$ – recognition of property (cat alive), $g^+ \circ h : A \to T^+$ – recognition of property (cat dead). Let also be $H_T(A) = \{h | h : A \to T\}$ (before experiment), $H_T(f) : h \mapsto h \circ f$ – variable domains (experimentation); $C_f^{g^-}$ (survival), $C_f^{g^+}$ (death) – variable individual concepts, that are determined by $C_f^{g^-} = \{g^- \circ h_f | h : A \to T\}$ (cat's survival), $C_f^{g^+} = \{g^+ \circ h_f | h : A \to T\}$ (cat's death); $H_r(B) : C_f^{g^-} \to C_f^{g^+}$ (process-interference, death in experimentation); $H_{g^-}(f) : h \mapsto g^- \circ h_f$, $H_{g^+}(f) : h \mapsto g^+ \circ h_f$.

- --→ From $H_T(A)$:

$$H_{g^+}(f) = H_r(f) \circ H_{g^-}(A) \text{ – dies in experiment}$$
$$H_{g^+}(f) = H_r(f) \circ H_{g^-}(A) \text{ – dies in experiment}$$
$$H_{g^-}(f) = H_{g^-}(B) \circ H_T(f) \text{ – survives in experiment}$$

- --→ From $H_T(B)$:

$$H_{g^+}(B) = H_r(B) \circ H_{g^-}(B) \text{ – dies in experiment}$$

- --→ From $C_A^{g^-}$:

$$H_r(f) = H_r(B) \circ H_{g^-}(B) \text{ – dies in experiment}$$

- ⇠ Into $C_f^{g^-}$:

$$H_{g^-}(f) = H_{g^-}(B) \circ H_T(f) \text{ – survival}$$
$$H_{g^-}(f) = H_{T^-}(f) \circ H_{g^-}(A) \text{ – survival}$$

- ⇠ Into $C_f^{g^+}$:

$$H_{g^+}(f) = H_{g^+}(B) \circ H_T(f) \text{ – death}$$
$$H_{g^+}(f) = H_r(f) \circ H_{g^-}(A) \text{ – path to death}$$
$$H_r(f) = H_r(B) \circ H_{T^-}(f) \text{ – death}$$

Then the given conceptual dependencies characterize the uncertainty of the state of the individual.

Proof. Derived from the commutative diagram shown in Fig. 1, which enforces the specified conceptual dependencies. □

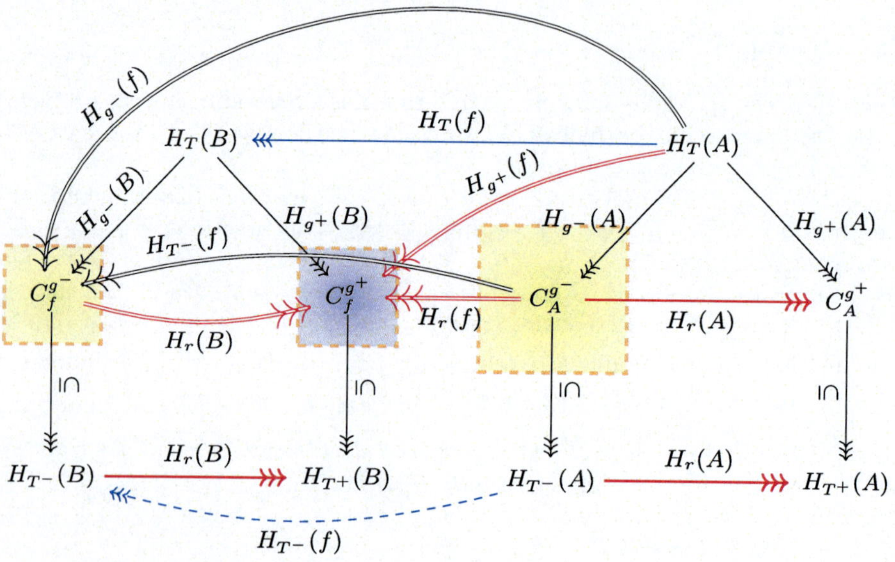

Fig. 1. Variable conceptual dependencies over the variable domains.

In a similar way, the diagram can be interpreted in terms of the "quick Smith" problem.

3 Conclusion

1. A new semantic model for the propagation of information processes based on variable domains is presented. Variable concepts are generated that can be used to model cognitive interference – the propagation of an intervention process.

2. The method of indexical expressions allows to operate with the state/property uncertainty.
3. A computational frame is proposed, written in the form of a system of conceptual dependencies, which establishes a connection between variable domains and variable concepts.
4. The computational frame is implemented as a functor model.

Acknowledgments. This research is supported in part by the Russian Foundation for Basic Research, RFBR grants 20-07-00149-a, 19-07-00326-a, 19-07-00420-a, 18-07-01082-a.

References

1. Cousineau, G., Curien, P.L., Mauny, M.: The categorical abstract machine. In: Proceedings of a Conference on Functional Programming Languages and Computer Architecture, pp. 50–64. Springer, Heidelberg (1985)
2. Ismailova, L., Kosikov, S., Kucherov, I., Zhuravleva, O.: Tools of algebraic type for manipulating methodologically oriented cognitive information. Procedia Comput. Sci. **169**, 23–30 (2020). https://doi.org/10.1016/j.procs.2020.02.111. http://www.sciencedirect.com/science/article/pii/S1877050920302349. Postproceedings of the 10th Annual International Conference on Biologically Inspired Cognitive Architectures, BICA 2019 (Tenth Annual Meeting of the BICA Society), held August 15–19, 2019 in Seattle, Washington, USA
3. Ismailova, L., Wolfengagen, V., Kosikov, S.: Hereditary information processes with semantic modeling structures. Procedia Comput. Sci. **169**, 291–296 (2020). https://doi.org/10.1016/j.procs.2020.02.181. http://www.sciencedirect.com/science/article/pii/S1877050920303045. Postproceedings of the 10th Annual International Conference on Biologically Inspired Cognitive Architectures, BICA 2019 (Tenth Annual Meeting of the BICA Society), held August 15–19, 2019 in Seattle, Washington, USA
4. Ismailova, L., Wolfengagen, V., Kosikov, S., Dohrn, M.: Capturing push-processing using enriched semantic mesh equipped with functionals-and-hops model. Procedia Comput. Sc. **169**, 590–596 (2020). https://doi.org/10.1016/j.procs.2020.02.205. http://www.sciencedirect.com/science/article/pii/S1877050920303288. Postproceedings of the 10th Annual International Conference on Biologically Inspired Cognitive Architectures, BICA 2019 (Tenth Annual Meeting of the BICA Society), held August 15–19, 2019 in Seattle, Washington, USA
5. Ismailova, L., Wolfengagen, V., Kosikov, S., Maslov, M., Dohrn, J.: Semantic models to indicate post-truth with fake news channels. Procedia Comput. Sci. **169**, 297–303 (2020). https://doi.org/10.1016/j.procs.2020.02.182. http://www.sciencedirect.com/science/article/pii/S1877050920303057. Postproceedings of the 10th Annual International Conference on Biologically Inspired Cognitive Architectures, BICA 2019 (Tenth Annual Meeting of the BICA Society), held August 15–19, 2019 in Seattle, Washington, USA
6. Ismailova, L.Y., Kosikov, S.V., Wolfengagen, V.E.: Computational model for the construction of cognitive maps. In: Silvera-Roig, M., Azcárate, A.L.V. (eds.) Cognitive and Intermedial Semiotics. IntechOpen, Rijeka (2020). https://doi.org/10.5772/intechopen.90173

7. Kosikov, S., Ismailova, L., Wolfengagen, V., Marenkov, A., Ermak, M., Slieptsov, I., Zaytsev, V., Shedko, A.: Indexical structures to enable knowledge mining tasks. Procedia Comput. Sci. **169**, 284–290 (2020). https://doi.org/10.1016/j.procs.2020.02.180. http://www.sciencedirect.com/science/article/pii/S1877050920303033. Postproceedings of the 10th Annual International Conference on Biologically Inspired Cognitive Architectures, BICA 2019 (Tenth Annual Meeting of the BICA Society), held August 15–19, 2019 in Seattle, Washington, USA
8. Lavwere, F.W.: Variable quantities and variable structures in topoi. In: Heller, A., Tierney, M. (eds.) Algebra, Topology, and Category Theory, pp. 101 – 131. Academic Press (1976). https://doi.org/10.1016/B978-0-12-339050-9.50014-6. http://www.sciencedirect.com/science/article/pii/B9780123390509500146
9. Mac Lane, S.: The lambda calculus and adjoint functors. In: Logic, Meaning and Computation: Essays in Memory of Alonzo Church, pp. 181–184. Springer, Netherlands, Dordrecht (2001)
10. Schrödinger, E.: Die gegenwärtige situation in der quantenmechanik. Naturwissenschaften **23**(48), 807–812 (1935). https://doi.org/10.1007/BF01491891
11. Scott, D.: Relating theories of the λ-calculus. In: Hindley, J., Seldin, J. (eds.) To H.B. Curry: Essays on Combinatory Logic, Lambda-Calculus and Formalism, pp. 403–450. Academic Press, Berlin (1980)
12. Wolfengagen, V.: Semantic modeling: computational models of the concepts. In: 2010 International Conference on Computational Intelligence and Security, CIS 2010, Nanning, Guangxi Zhuang Autonomous Region, China, 11–14 December 2010, pp. 42–46 (2010). https://doi.org/10.1109/CIS.2010.16
13. Wolfengagen, V., Ismailova, L., Kosikov, S.: Capturing information processes with variable domains. Procedia Comput. Sci. **169**, 276–283 (2020). https://doi.org/10.1016/j.procs.2020.02.177. http://www.sciencedirect.com/science/article/pii/S1877050920303008. Postproceedings of the 10th Annual International Conference on Biologically Inspired Cognitive Architectures, BICA 2019 (Tenth Annual Meeting of the BICA Society), held August 15–19, 2019 in Seattle, Washington, USA
14. Wolfengagen, V., Ismailova, L., Kosikov, S., Dohrn, J.: Superimposing semantic mesh to prevent information processes entanglement. Procedia Comput. Sci. **169**, 645–651 (2020). https://doi.org/10.1016/j.procs.2020.02.197. http://www.sciencedirect.com/science/article/pii/S1877050920303203. Postproceedings of the 10th Annual International Conference on Biologically Inspired Cognitive Architectures, BICA 2019 (Tenth Annual Meeting of the BICA Society), held August 15–19, 2019 in Seattle, Washington, USA
15. Wolfengagen, V., Kosikov, S., Ismailova, L., Zaytsev, V.: Semantic framework for data flow control in the network of information graphs. Procedia Comput. Sci. **169**, 16–22 (2020). https://doi.org/10.1016/j.procs.2020.02.110. http://www.sciencedirect.com/science/article/pii/S1877050920302337. Postproceedings of the 10th Annual International Conference on Biologically Inspired Cognitive Architectures, BICA 2019 (Tenth Annual Meeting of the BICA Society), held August 15–19, 2019 in Seattle, Washington, USA

Improvement of the Technology of fMRI Experiments in the Concealed Information Paradigm

Yuri I. Kholodny(✉), Sergey I. Kartashov, Denis G. Malakhov, and Vyacheslav A. Orlov

National Research Center “Kurchatov Institute”, Moscow, Russia
kholodny@yandex.ru

Abstract. This work is a continuation of research on the creation of a forensic method for MRI diagnostics of hidden information in a person. The article presents some results of the first (technological) stage of these studies, during which the method of fMRI experiments was improved by using an MRI compatible polygraph (MRIcP) within the framework of the information concealment paradigm. Numerous experiments show: the efficiency of MRIcP during fMRI procedure; the usefulness of the methodological technique created through the use of MRIcP that have improved the performance of fMRI experiments; prospects of application of certain tests (in compliance with the strict requirements of forensic science) in the context of the fMRI study.

Keywords: MRI compatible polygraph · fMRI · Forensic diagnostics · Studies using a polygraph

1 Introduction

Psychophysiological studies using a polygraph (SUP) in order to diagnose person’s hidden information has existed since the 1930s and are currently used in about a hundred countries. Despite the long history of SUP, the brain processes that make it possible for a person to reveal information about hidden events in the past have not been studied enough, although by today about two dozens of so called “polygraph theories” has been proposed, trying to explain the work of the brain during the SUP [1]. Therefore, revealing of neurocognitive mechanisms underlying the forensic diagnosis of a person’s hidden information during the SUP is important.

At the beginning of the XXI century, the first studies were conducted on the possibility of using functional magnetic resonance imaging (fMRI) to diagnose information about hidden events of the past in a person [2, 3, etc.]. It was believed that fMRI would allow to visualize the diagnostic process and bring closer the creation of a scientifically based theory of SUP. The results of such fMRI studies were promising, and within five years in the United States two companies were established that provided services on a commercial basis to detect a person’s hidden information using fMRI (in American terminology – “fMRI based lie detection”).

A. V. Samsonovich et al. (Eds.): BICA 2020, AISC 1310, pp. 591–597, 2021.
https://doi.org/10.1007/978-3-030-65596-9_73

However, hopes for rapid success from the use of fMRI were not justified. Scientists have criticized the work carried out by these American companies, and meta-analyses of the results of scientific research in recent years have shown that to create a practically useful method of fMRI diagnostics of hidden information (fMRI diagnostics), a special technology for conducting fMRI research must be developed. And now "most publications report only group analyses, which makes them poorly suited to answering this question" [4, p. 125] about the forensic use of neuroimaging.

Although fMRI based lie detection has been commercialized and is used by some organizations for real-world applications, research on this topic began as a form of basic science with the goal of identifying the neural systems involved in deception.

When conducting research on the "fMRI based lie detection", foreign experts usually use many tens of stimuli during tests and, in an effort to do the test quickly, present stimuli to the subjects at intervals of less than 10 s [5]. Foreign experts perform experiments on small groups of subjects (sometimes – on 3 people [6]), and the subjects give answers by pressing buttons [5, 7].

In our opinion, these and many other experiments were conducted without taking into account the strict methodological requirements for the "fMRI based lie detection" that criminalistics has established for its methods, and applied research on fMRI diagnostics is not given enough attention.

Research on the creation of a complex of fMRI diagnostics was started at the National Research Center "Kurchatov Institute" (NRC "Kurchatov Institute") in 2018 [8]. When conducting forensic research using fMRI, it was hypothesized about the possibility of improving the accuracy of fMRI data by monitoring behavior of subjects and their response to various influences: the control was proposed to be carried out by using simultaneous registration of fMRI and vegetative reactions (VR) of a person.

To test the hypothesis that it is possible to improve the quality of fMRI data, a prototype of a computer MRI compatible polygraph (MRIcP) was created at the NRC "Kurchatov Institute". This device made it possible to observe in real time the fast-flowing VR of a person that arose in response to the action of external or internal factors, including the stimuli presented to the person. For the purposes of fMRI quality control, we used VR parameters that are traditionally used in criminology for the diagnosis of hidden information in humans, namely, the amplitude and/or frequency characteristics of the respiratory system, the cardiovascular system, and the human skin galvanic response (GSR) [9]. MRIcP helped to bring the new technology of fMRI diagnosis to the existing technologies of SUP.

Research on creating a complex of MRI diagnostics includes two components:

- a "technological" stage aimed at developing methodological and technical means to control the response of a person to various stimuli during fMRI research, and
- a "research" stage aimed at detecting areas and neural networks of the brain ("markers") that indicate the concealment of information by a person.

This article describes the results of the "technological" stage of the research.

Taking into account the requirements of forensic diagnostics, the concealed information paradigm was chosen as the methodological basis for experiments.

To test the usefulness of using MRIcP during fMRI and to develop methodological tools for conducting fMRI-experiments, tests were selected that are widely used in forensic practice for SUP. In particular, the test with a concealed name (TCN), the silent response test (SRT) and the "guilty knowledge" test (GKT) were selected as tests of the concealed information paradigm.

During experiments, it was found that MRIcP allowed to see the response of each person during fMRI and thus to divide people into groups with a high and low degree of physiological response to the stimuli presented to them [10].

The use of these tests and the monitoring (with the help of MRTcP) of the dynamics of a person's response to perceived stimuli together opened up new opportunities for studying neurocognitive processes, including for the purpose of diagnosing a person's hidden information. And as will be shown below, the "technological" stage of the improvement process has created conditions for improving the quality of fMRI data registration.

2 Materials and Methods

During the "technological" stage on the creation of a complex of fMRI-diagnostics, the above mentioned hypothesis was studied in experiments on healthy volunteers, 21–35 y. o. in the fMRI procedure using MRIcP and a set of tests borrowed from criminalistics – TCN, SRT and GKT, respectively, on 73, 12 and 49 subjects.

Permission to conduct the fMRI experiments was obtained from the ethics committee of NRC "Kurchatov Institute".

Experiments at the NRC "Kurchatov Institute" have shown:

1. The selected tests (TCN, SRT and GKT) can be successfully applied during fMRI procedure in accordance with the strict requirements of forensic science. Each test lasted 7–8 min, contained 25–30 stimuli, of which 5–6 were significant, and the subjects gave in the tests not motor, but speech responses. At the same time, the tests provided the necessary amount of fMRI data to solve the research tasks.

2. The MRI data was acquired using a 3 T SIEMENS Magnetom Verio MR scanner. The T1-weighted sagittal three-dimensional magnetization-prepared rapid gradient-echo sequence was acquired with the following imaging parameters: 176 slices, TR = 1900 ms, TE = 2.19 ms, slice thickness = 1 mm, flip angle = 9°, inversion time = 900 ms, FOV = 250 mm × 218 mm^2.

At the initial stage of experiments, fMRI data were acquired with the following parameters: 30 slices, TR = 2000 ms, TE = 25 ms, slice thickness = 3 mm, flip angle = 90°, FOV = 192 × 192 mm^2.

Functional and structural MRI data were processed using the SPM8 software package (available at http://www.fil.ion.ucl.ac.uk/spm/software/spm8/). After converting DICOM files to NIFTI format, all images were manually centered to anterior commissure. The EPI images were corrected for magnetic field inhomogeneity using the FieldMap toolbox for SPM8. Then, the fMRI data signals were temporarily corrected. The need to apply this correction is explained by the use of an event-related scheme of the experiment. Anatomical and functional data were normalized in the ICBM stereotactic reference system. T1 images were divided into 3 tissue maps (gray, white matter, and cerebrospinal fluid). Functional data was smoothed using a 6 × 6 × 6 mm^3

FWHM Gaussian filter. Statistical analysis was performed using the Student t-statistics ($p < 0.05$ with correction for multiple comparisons (FWE)).

To increase the volume of MRI data and the sensitivity of the technique, it was considered appropriate to switch from classical scanning sequences (TR = 2000 MS) to ultra-fast ones (TR = 1110 MS). Figure 1 shows the group statistical maps obtained when analyzing the TCN for (a) the classical sequence (TR = 2000 ms), and (b) the ultrafast one (TR = 1110 ms).

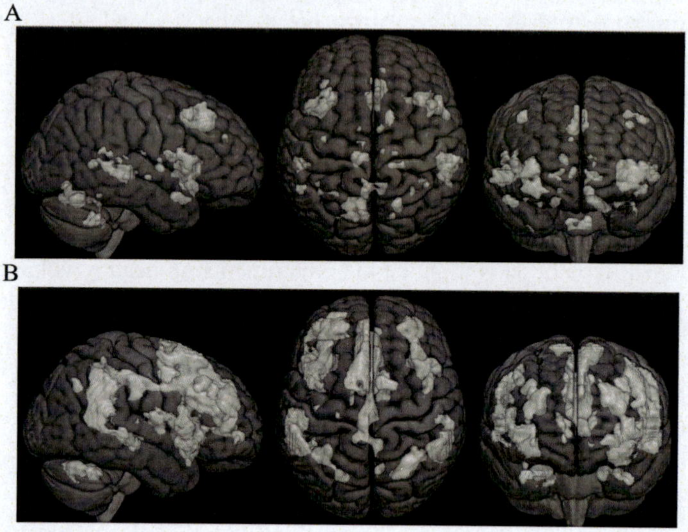

Fig. 1. Group statistical maps drawn on a three-dimensional translucent model of the human brain: (a) TR = 2000 ms; (b) TR = 1110 ms.

The ultra-fast sequences allow the signal to be excited and picked up from multiple slices simultaneously, significantly increasing the signal-to-noise ratio.

3. The use of MRIcP allowed us to monitor the real-time dynamics of physiological reactions of subjects to different influences, and record artifacts, the neglect of which would worsen the correctness of the recorded fMRI data, during fMRI procedure. For example, one of the artifacts is "physiological noise", during which the correct registration of fMRI data is impossible (Fig. 2).

Often, during experiments, there were spontaneous reactions to stimuli that were not marked as significant for a particular person under the conditions of the test (Fig. 3).

The use of MRIcP made it possible to create a method for rejecting artifacts: fMRI data that were registered when spontaneous reactions or "physiological noises" appeared were excluded from the entire collection of fMRI-data obtained during a specific test. When using the rejection technique, a rule was introduced: a fifth part of the total fMRI data was always excluded ("noised" by artifacts data).

3 Results

Summing up the results of the "technological" stage, we can state that the selected tests (TCN, SRT and GKT) are a reliable methodological basis for further research on fMRI-diagnostics.

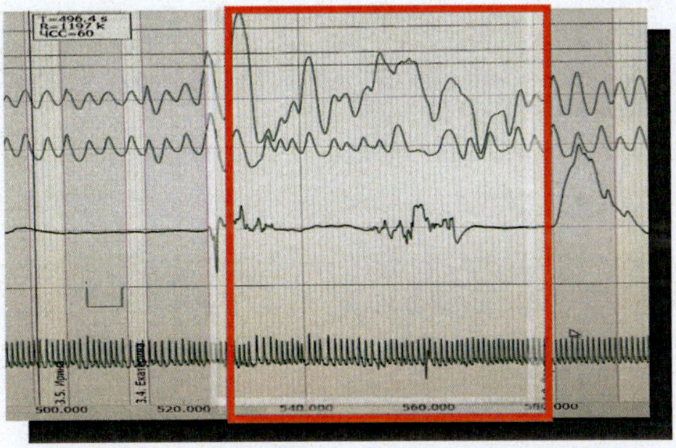

Fig. 2. "Physiological noise" during TCN.

The figure shows a fragment of the experiment (marked with a red border), during which the subject suddenly began to take deep breaths (on the polygram – the first and second graphs from above) and move his hand, which was reflected in the graph of GSR (the third graph from above).

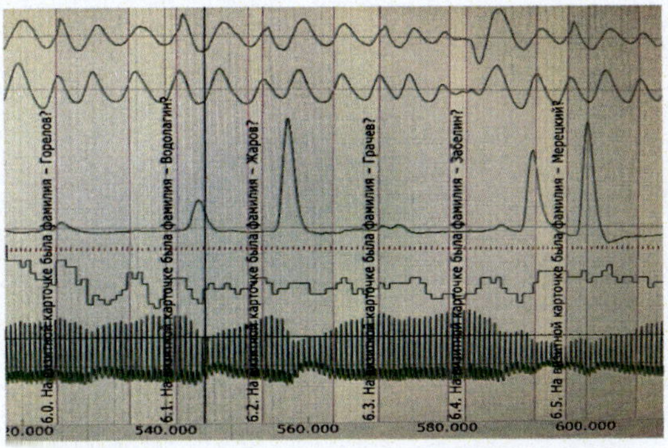

Fig. 3. Spontaneous reaction to an insignificant stimulus in the GKT.

During the sixth presentation of the GKT, the subject had spontaneous reactions in the GSR and photoplethysmogram (on the polygram, respectively, the third and fifth graph from above) to the 2nd (insignificant) stimulus, which are almost equal in magnitude to similar reactions to the 5th (significant) stimulus.

Experiments based on TCN, SRT and GKT have clearly shown the effectiveness of the transition from classical to ultra-fast scanning sequences. Representative samples of the same size showed a tenfold increase in detected statistically significant voxels when using an ultrafast sequence compared to the classical one.

The use of MRIcP and the developed method of rejecting artifacts during MRI, combined with the transition to ultra-fast scanning sequences, allowed, on the one hand, to improve the quality of recorded fMRI data transmitted to subsequent processing, and, on the other, to increase the number of voxels allocated as a result of such processing. So, experiments performed on 104 healthy men at TR = 2000 ms and TR = 1110 ms confirmed the effectiveness of using the method of rejecting artifacts. This technique helped to increase the number of statistically significant voxels (Student statistics $p < 0.001$) by 52% and 17%, respectively, while the volume of fMRI data was reduced by 20%.

At present, almost all of our neurocognitive experiments are conducted with the mandatory use of MRIcP to control the quality of fMRI data registration.

Taking into account the fact that the main physiological process in the rejection of artifacts was the dynamics of GSR, and this improved the quality of selection of active brain zones, it seems correct to study the detected dependence in depth.

4 Conclusion

The results presented in the article show the prospects for continuing experimental work on creating fMRI diagnostics. In the future, the developed set of methodological and technical tools will be supplemented by methods of objective quantitative assessment of physiological data recorded by the MRIcP.

Acknowledgements. The research presented above is an initiative internal research conducted by NRC «Kurchatov Institute» (order no. 1363 of June 25, 2019 «Biomedical technologies», paragraph 4.15).

References

1. Kholodny, Yu.I.: Some aspects of the practice and technology of forensic studies using a polygraph. Invest. Crimes Probl. Ways Solutions **4**(18), 185–192 (2017)
2. Spence, S.A., et al.: Behavioural and functional anatomical correlates of deception in humans. Neuroreport **12**(13), 2849–2853 (2001)
3. Ganis, G., et al.: Neural correlates of different types of deception: an fMRI investigation. Cereb. Cortex **13**(8), 830–836 (2003)
4. Farah, M.J., et al.: Functional MRI-based lie detection: scientific and societal challenges. Nat. Rev. Neurosci. **15**(2), 123–131 (2014)
5. Ofen, N., et al.: Neural correlates of deception: lying about past events and personal beliefs. Soc. Cogn. Affect. Neurosci. **12**(1), 116–127 (2017)
6. Hakun, J.G., et al.: fMRI investigation of the cognitive structure of the Concealed Information Test. Neurocase **14**, 59–67 (2008)

7. Peth, J.: Memory detection using fMRI – does the encoding context matter? Neuroimage **113**, 164–174 (2015)
8. Kovalchuk, M.V., Kholodny, Y.I.: Functional magnetic resonance imaging augmented with polygraph: new capabilities. Advances in Intelligent Systems and Computing (eBook), pp. 260–265 (2019). https://doi.org/10.1007/978-3-030-25719-4
9. Ushakov, V.L., Malakhov, D.G., Orlov, V.A., Kartashov, S.I., Kholodny, Y.I., Kovalchuk, M.V.: Research of neurocognitive mechanisms of revealing of the information concealing by the person. In: Biologically Inspired Cognitive Architectures 2018. Proceedings of the Ninth Annual Meeting of the BICA Society. Advances in Intelligent Systems and Computing, vol. 848, pp. 310–315. Springer (2018)
10. Orlov, V.A., Kholodny, Y.I., Kartashov, S.I., Malakhov, D.G., Kovalchuk, M.V., Ushakov, V.L.: Application of registration of human vegetative reactions in the process of functional magnetic resonance imaging. Advances in Intelligent Systems and Computing(eBook), pp. 393–399 (2019). https://doi.org/10.1007/978-3-030-25719-4_51

Modelling Metaplasticity and Memory Reconsolidation During an Eye-Movement Desensitization and Reprocessing Treatment

Lennart Zegerius and Jan Treur(✉)

Social AI Group, Vrije Universiteit Amsterdam, Amsterdam, Netherlands
{l.zegerius,j.treur}@vu.nl

Abstract. In this paper, an adaptive computational network model is presented to simulate the effect of Eye Movement Desensitization and Reprocessing (EMDR) therapy on persons affected by a Post-Traumatic Stress Disorder (PTSD). During EMDR therapy, within the brain resource competition helps to improve stress regulation. This revitalizes extinction learning and memory reconsolidation. The introduced adaptive network model and its simulation confirms the functionality of the neural processes and effective treatment results of EMDR.

Keywords: Computational model · Hebbian learning · Metaplasticity · EMDR · PTSD · Memory reconsolidation · Resource competition

1 Introduction

A lack of neurological evidence has consistently haunted the craft of psychotherapists; the dogma states 'true is what works'.

> 'In several writings, Freud has noted - sometimes with undisguised regret - that neuroscience was not yet able to provide explanations of psychological functioning.' [46], p. 676

Nowadays, psychological theories and treatments are continually researched in an attempt to validate them in the neurological domain. Francine Shapiro [37, 38] observed that saccadic eye movements reduce the disturbance caused by stressful thoughts and memories. This experience encouraged Shapiro to methodically investigate this phenomenon. In the ensuing years standardized procedures were developed and an effective treatment for Post-Traumatic Stress Disorder (PTSD) was born, termed Eye Movement Desensitization and Reprocessing (EMDR) [37]. EMDR is a treatment for persons who, after a traumatic event, experience regular cognitive distress associated with traumatic memories. During EMDR therapy, the traumatized person actively recollects these traumatic memories while being accompanied by a therapist performing the eye-movement intervention. Due to the neurological mechanisms triggered by eye-movement intervention combined with the therapeutic concepts that encompass EMDR therapy, affective distress is relieved, psychological arousal is reduced and the traumatic memory is reconsolidated in a less toxic form [39].

A. V. Samsonovich et al. (Eds.): BICA 2020, AISC 1310, pp. 598–610, 2021.
https://doi.org/10.1007/978-3-030-65596-9_74

However, the contributive factor of EMDR as a treatment for psychological traumas led to controversy, as some studies showed that the therapeutic concepts that are applied during EMDR therapy would already contribute to the resulting trauma relief and eye movement interventions do not add to treatment outcome [6, 9]. Successive research showed that EMDR provides changes in memory traces, allowing consolidation into the 'normal' memory system without associated extreme emotionality [18], and eye-movements arouse brain activity related to working memory and emotionality [44]. EMDR might relearn how to deploy attention and re-evaluate sensory evidence from the memorized narrative, also by suppressing the bodily reaction to memorization, contributing to improved regulation [7].

In this study, a network-oriented modelling approach [41, 42] is used to introduce a (second-order) adaptive temporal-causal network model in order to simulate EMDR treatment and its effects. The traumatic event, its resulting emotionality, stress regulation, the relevant plasticity and metaplasticity [1], and the effects of EMDR therapy, such as resource competition, are all covered by the introduced network model.

2 Neurological Findings

The neurological turmoil consequences of traumatic events that symptomize PTSD have been studied extensively; e.g., [19, 36]. According to the U.S. National Center for PTSD [43], between 7–8% of the U.S. population develop PTSD. There are many factors increasing the risk of PTSD, such as gender, age and education [34]. Fortunately, the plasticity of the brain and the characteristics of stressful memories have allowed discoveries implying neuronal changes that can be addressed by treatment [24, 25, 46]. Extremely stressful events are so disturbing that they disrupt the information processing system in the brain, in particular the memory traces. Such events may cause overpotentiation of amygdalar synapses that record the events, possibly filling all 84 AMPA-receptor-binding sites [11]. Overpotentiation of an emotional memory trace causes problems when attempting to merge with a cognitive memory trace. As a result, the emotional memory trace remains in a dishevelled state, which could imply various PTSD symptoms, such as a tumult of negative emotion. This is confirmed by studies demonstrating that the amygdala has a weakness in pathological processing of highly stressful memories, which might be due to its undeveloped filtering and modulating systems [21]. A disturbance in assigning emotional charge to memories could explain why a treatment focused on reassigning emotionality, such as EMDR, works.

The discovery and innovative nature of EMDR provoked new interest in the cohesion of eye movement intervention, attention, (working) memory and emotional processing. To dissect the neural process of EMDR, it has been studied extensively. A study displayed that eye-movement intervention enhances cognitive control by deactivating the amygdala [44]. For a person receiving treatment, trauma recall during eye movements is beneficial due to memory recollection suppressing fear exertion by the amygdala and the attenuation of fear responses during working-memory tasks [10, 45]. Cognitive control is impaired in patients with PTSD [5], linked to postmortem brain analyses of depressed patients showing size reductions in the prefrontal cortex [31].

During EMDR therapy, traumatic memory recall is initiated which raises arousal levels. Initial eye movements rapidly decreases these levels [13, 35]. In contrast, during subsequent eye movements arousal levels still rise [33]. This seems like an unforeseen finding, because of the deactivated amygdala, but could be functional in the reconsolidation process. These findings show that EMDR may provide extra support for the inadequate brain networks responsible for cognitive control, which could be an explanation of its effectiveness. In addition, the downregulated amygdala enables the trauma recall to go more in-depth, which possibly aids effective therapy.

An explanation for amygdala deactivation can by explained as being be a result of *resource competition* between reciprocal brain networks [15, 22]. The system occupied by sensory, high concentration tasks such as eye-movement intervention mainly uses the Dorsal Fronto-Parietal Network (DFPN) [30]. During eye-movement intervention, competition between some specific brain-networks occurs. Stress engages the amygdala but impairs the DFPN functioning and DFPN activation deactivates the amygdala [22, 44]. Coupled pathways were found that imply mediation between the ventromedial prefrontal cortex (vmPFC), which is involved in memory recording and extinction learning, and dorsolateral PFC-amygdala interactions. The study by de Voogd et al. found that eye-movements are sufficient to engage the pathways responsible for the regulation of emotion. Thus, stress is regulated due to resource competition caused by stress, autobiographical (traumatic) memory recollection and the initiation of eye-movement intervention.

In the hippocampus, the brain-derived neurotrophic factor (BDNF) is known to support neuronal survival and function [27, 28]. Stress or depression can decrease hippocampal expression of BDNF, thereby blocking synaptic plasticity [17]. Also, Garcia shows that in the hippocampus, long-lasting changes in synaptic strength are prevented by long-term depression, and vice versa. This implies a form of plasticity of synaptic plasticity, which has been termed *metaplasticity* [1], and can lead to either 'negative' or 'positive' metaplasticity, named with respect to their influence on cognitive functioning. Negative metaplasticity represents the shift in synaptic plasticity that occurs due to decreasing levels of BDNF in the hippocampus, and is symptomized by impaired cognitive functions such as learning. Positive metaplasticity is achieved by increasing BDNF levels, which is restorative for impaired cognitive functions.

For the current context of EMDR, the resource competition makes that EMDR activation suppresses the stressfull amygdala activation. Due to this, according to Garcia [17], the negative metaplasticity caused by the stress ultimately turns into positive metaplasticity. Therefore, cognitive functioning is improved, in particular the plasticity of the stress regulation system, so the person can learn again to adapt to a stressful situation by strengthening stress regulation. The ability to learn is a key factor in relieving psychotraumatic symptoms; this is sometimes called *fear extinction learning*. Fear extinction learning is applied in the treatment of fear memories, which is a common symptom for PTSD. Note that fear 'extinction' takes place in an indirect manner by compensation (through regulatory connections) of the existing fear provoking connections (which actually themselves do not extinguish). The consolidation of fear extinction can be strongly enhanced by exposure therapy [16]. EMDR therapy is an effective tool for exposure therapy and is shown to improve fear extinction learning in PTSD patients [32].

3 The Computational Method Used

The model proposed in this study was designed using the adaptive temporal-causal network modelling approach as described in (Treur, 2020a); in this section this approach is briefly summarised. In the network modelling approach adopted for this study, temporal-causal networks are used of the relevant connections and their causality and adaptivity. The temporal aspect is characterized by states that vary over time, and the causal aspect is characterized by how states affect each other over time via their connections. Adaptation is addressed by connections that change over time. The basic relations in temporal-causal networks are represented mathematically with directed relations between states $X \rightarrow Y$, representing causal impacts. The main elements of a temporal-causal network structure are the network structure characteristics for *connectivity*, *aggregation* and *timing*; see also [41, 42].

Connectivity Characteristics. Connectivity of the network is specified by connection weights $\boldsymbol{\omega}_{X,Y} \in [-1, 1]$ for each connection between states $X \rightarrow Y$. The connections are directed and either negative or positive, implying its effect.

Aggregation Characteristics. Each network model is assigned a sequence of basic combination functions $\mathbf{c}_j(..)$, j = 1, .., m. They are used for aggregation in the network. These basic combination functions are from an available library presented in [42].

For each state Y within the network, there are combination function weights $\boldsymbol{\gamma}_{j,Y}$ that specify by a weighted average of the combination functions $\mathbf{c}_j(..)$, j = 1, .., m for Y the aggregation of incoming single causal impacts $\boldsymbol{\omega}_{Xi,Y}X_i(t)$ of the states X_1, .., X_k from which Y receives incoming connections.

For each state Y and combination function $\mathbf{c}_j(..)$, j = 1, .., m there are combination function parameters $\boldsymbol{\pi}_{i,j,Y}$.

Timing Characteristics. Each state Y is assigned a speed factor $\boldsymbol{\eta}_Y \geq 0$.

A numerical representation of the network model is obtained in a canonical manner from the above characteristics as folllows in difference or differential format; this can be used for simulation and mathematical analysis:

$$Y(t+\Delta t) = Y(t) + \boldsymbol{\eta}_Y[\mathbf{c}_Y(\boldsymbol{\omega}_{X1,Y}X_1(t), \ldots, \boldsymbol{\omega}_{Xk,Y}X_k(t)) - Y(t)]\Delta t \tag{1}$$

$$\mathbf{d}Y(t)/\mathbf{d}t = \boldsymbol{\eta}_Y[\mathbf{c}_Y(\boldsymbol{\omega}_{X1,Y}X_1(t), \ldots, \boldsymbol{\omega}_{Xk,Y}X_k(t)) - Y(t)]$$

where X_1, …, X_k are the states from which state Y gets incoming base connections. Here $\mathbf{c}_Y(..)$ is the weighted average of the basic combination functions $\mathbf{c}_j(..)$, j = 1, .., m by weights $\boldsymbol{\gamma}_{j,Y}$ and with parameters $\boldsymbol{\pi}_{i,j,Y}$.

Examples of combination functions are the Euclidean combination function and the advanced logistic combination function:

$$\mathbf{eucl}_{n,\lambda}(V_1, \ldots, V_k) = \sqrt[n]{\frac{V_1^n + \ldots + V_k^n}{\lambda}} \tag{2}$$

$$\mathbf{alogistic}_{\mathbf{s,t}}(V_1,\ldots,V_k) = \left[\frac{1}{1+\mathrm{e}-\boldsymbol{\sigma}(V_1+\ldots+V_k-\boldsymbol{\tau})} - \frac{1}{1+\mathrm{e}^{\boldsymbol{\sigma\tau}}}\right](1+\mathrm{e}^{-\boldsymbol{\sigma\tau}}) \quad (3)$$

Here V_1, ..., V_k stand for the incoming single impacts from other states, parameter $\boldsymbol{n}$ is the order, parameter $\boldsymbol{\lambda}$ the scaling factor, parameter $\boldsymbol{\sigma}$ steepness and $\boldsymbol{\tau}$ excitability threshold. A combination function used for independent input from the external environment is the function $\mathbf{stepmod}_{\boldsymbol{\rho},\boldsymbol{\delta}}$(..), where $\boldsymbol{\rho}$ is a repetition parameter indicating (for repetitive input) after how long the input should be repeated and $\boldsymbol{\delta}$ the duration parameter indicating after how long the input should occur.

To model an adaptive network, the (base) network model must interact with some adaptation model. The latter model characterizes the adaptiveness of the base network. Usually such an adaptation model is specified in an algorithmic or procedural manner and for each type of adaptation requires specific, separate programming to get a working simulation. As an alternative approach, an adaptive network can be modelled by adding socalled *reification states* to the base network (Treur, 2020a), resulting in a reified network. By this approach to adaptive networks, adaptation can be specified by network modelling, without any specific procedural or algorithmic specification or programming. Another advantage is that network reification can easily be iterated to obtain higher-order adaptation: the reified network itself is again a network and not some procedural specification. The adaptive network model introduced below uses this network reification approach. For base network states, additional first-order adaptive network states $\mathbf{W}_{X,Y}$ are introduced and for these $\mathbf{W}_{X,Y}$ states second-order adaptive network states $\mathbf{H}_{\mathbf{W}X,Y}$ are added:

Connectivity Characteristics Reification: W-states Reification states $\mathbf{W}_{Xi,Y}$ are added for each of the adaptive connections $X_i \rightarrow Y$, representing an adaptive connection weight $\boldsymbol{\omega}_{Xi,Y}$.

Timing Characteristics Reification: H-states Reification states $\mathbf{H}_{\mathbf{W}X,Y}$ are added representing an adaptive speed factor $\boldsymbol{\eta}_{\mathbf{W}Xi,Y}$ of state $\mathbf{W}_{Xi,Y}$.

4 The Designed Second-Order Adaptive Network Model

The nomenclature of the model is described in Table 1; for a picture of the model's connectivity, see Fig. 1. The states termed ws are called 'world states' because they represent external input from the world. The world states are the traumatic event $\mathrm{ws_{te}}$, trigger stimulus $\mathrm{ws_{tr}}$ and EMDR therapy $\mathrm{ws_{emdr}}$. There are also sensor states termed ss which sense input provided by the world states. The sensor states are $\mathrm{ss_{te}}$, $\mathrm{ss_{tr}}$ and $\mathrm{ss_{emdr}}$. Following are the sensory representation states termed srs which define the intensity at which a person processes a stimulus, which is variable per person. The sensory representation states are $\mathrm{srs_{te}}$, $\mathrm{srs_{tr}}$ and $\mathrm{srs_{emdr}}$. Upon activation, the sensory representation states $\mathrm{srs_{te}}$ and $\mathrm{srs_{emdr}}$ trigger preparation states that, by prediction links, activate other sensory representation states. The preparation state $\mathrm{ps_b}$ is authoritative in predicting situations before actually expressing emotion b, which represents negative emotion and stress. Depending on its activation level, the preparation state may activate execution state es_b, which is the expression state for emotion b, where high levels indicate high levels of distress. Increasing levels of b calls forth the control state cs_b, which models emotion or stress regulation. By its incoming connections, the control state has a monitoring function for the preparation and feeling for b. The feeling state fs_b represents amygdala activity. Disturbing levels of fs_b are detected (via the incoming connections of cs_b) and downregulated by cs_b (via its outgoing connections). The internal body loops are modelled in a context where the srs_{te} and srs_{emdr} states trigger ps_b and its effect partly journeys through fs_b to cs_b and partly directly to cs_b, where also fs_b in turn affects ps_b (Damasio, 2003).

Table 1. Nomenclature of the model

State	Description
ss_{te}	Sensor state for traumatic event te in the world
ss_{tr}	Sensor state for trigger stimulus in the world
ss_{emdr}	Sensor state for EMDR therapy
srs_{te}	Sensory representation state for traumatic event te
srs_{tr}	Sensory representation state for trauma recall trigger tr
srs_{emdr}	Sensory representation state for EMDR therapy emdr
ps_b	Preparation state for emotional response b
$bs_{te,b}$	Belief state for feeling b due to traumatic event te
cs_b	Cognitive control state for b
fs_b	Feeling state for b
as_{te}	Awareness (flashback) state for traumatic event representation for te
es_b	Expression or execution state for emotion b
$es_{te,b}$	Expression or execution state for belief of feeling b due to traumatic event te
$\mathbf{W}_{psb,csb}$	Representation state for adaptive weight of connection from ps_b to cs_b
$\mathbf{W}_{fsb,csb}$	Representation state for adaptive weight of connection from fs_b to cs_b
$\mathbf{H}_{\mathbf{W}psb,csb}$	Representation state for adaptive speed factor for $\mathbf{W}_{psb,csb}$
$\mathbf{H}_{\mathbf{W}fsb,csb}$	Representation state for adaptive speed factor for $\mathbf{W}_{fsb,csb}$

The awareness state as_{te} acts as a flashback experience that may be activated by higher activation levels of srs_{tr} and srs_{te}. This flashback raises the awareness of the person of his or her trauma, shouts for reinforcement of cs_b, and increases the traumatic experience in itself.

The mutual increase of srs_{te} and as_{te} indicates the intensity of a flashback. A person with PTSD does not necessarily know their symptoms are related to their trauma, since the symptoms of the affected cognitive malfunction may seem unrelated to the traumatic event. Finally, the belief state $bs_{te,b}$ leads to the eventual expression $es_{te,b}$ of the person's knowledge of his or her emotion, which is not a direct representation of amygdala levels from fs_b but rather a representation of the person's expression of knowledge.

A graphical representation of the base model is shown in Fig. 1 in the base plane. The other planes depict the connectivity of the reification levels. The 3D representation

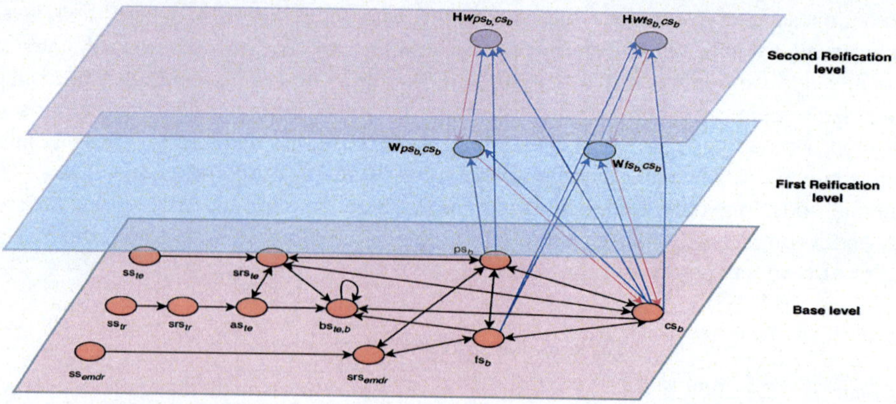

Fig. 1. The three-dimensional graphical representation of the network connectivity of the second-order adaptive network model

shows how network reification is used in this model, representing the structure of a second-order adaptive network [41, 42]. The first reification level of the adaptive network is used for adaptive links and modelled using Hebbian Learning [20] for the connections $X \rightarrow Y$. More specifically, the designed reified network structure incorporates the adaptability of the base network for the connections states $\mathrm{ps_b} \rightarrow \mathrm{cs_b}$ and $\mathrm{fs_b} \rightarrow \mathrm{cs_b}$ by the connection weight reification states $\mathbf{W}_{\mathrm{psb,csb}}, \mathbf{W}_{\mathrm{fsb,csb}}$, respectively, in the first reification level plane. These states follow Hebbian learning principles for adaptivity, defined by the following combination function they use:

$$\mathbf{hebb}_{\boldsymbol{\mu}}(V_1, V_2, W) = V_1 V_2 (1 - W) + \boldsymbol{\mu} W \tag{4}$$

where W represents the weight of the considered connection, $\boldsymbol{\mu}$ the persistence parameter and V_1 and V_2 the activation values for the connected states. The resulting adaptability allows the person to learn to improve his or her stress regulation.

The speed factor reification states $\mathbf{H}_{\mathbf{W}_{\mathrm{psb,csb}}}$ and $\mathbf{H}_{\mathbf{W}_{\mathrm{fsb,csb}}}$ are responsible for changing synaptic plasticity, thus modelling metaplasticity. The values of these **H**-states are low for high values of the feeling state $\mathrm{fs_b}$ and preparation state $\mathrm{ps_b}$ and increase when the values of these feeling state $\mathrm{fs_b}$ and preparation state $\mathrm{ps_b}$ decrease. This polarity change means the reification states are negatively impacted by $\mathrm{fs_b}$ and $\mathrm{ps_b}$, which describes stress-induced negative metaplasticity [17]. As by resource competition an EMDR therapy leads to lower values of $\mathrm{fs_b}$ and $\mathrm{ps_b}$, through this mechanism the **H**-states get higher values an therefore positively affect the plasticity of the connections $\mathrm{ps_b} \rightarrow \mathrm{cs_b}$ and $\mathrm{fs_b} \rightarrow \mathrm{cs_b}$ by which the emotion regulation improves. In this way, the levels of the second-order reification states are responsible for (un)learning the trauma and restoring adequate emotional and cognitive functioning. During EMDR, the trauma is re-experienced using the aforementioned protective measures that eye-movement intervention and therapy provide.

In the adaptive network-oriented modelling approach used, *role matrices* specify the network characteristics according to their roles. In the Appendix at https://www.researchgate.net/publication/341709093, the role matrices are presented; they provide a full specification of the design of the model.

Connectivity characteristics are represented by role matrices **mb** and **mcw**. Role matrix **mb** specifies the base connectivity of the model and **mcw** specifies the weight of the connections as indicated in **mb**. Note that in the darker red cells in **mcw** it is indicated that states X_{12} (= $\mathbf{W}_{psb,csb}$) and X_{13} (= $\mathbf{W}_{fsb,csb}$) play the role of the adaptive incoming connection weights for state cs_b.

Aggregation is attained by using combination functions as specified in **mcfw** and **mcfp**. Role matrix **mcfw** specifies the used combination function weights per state, each value 1 defines its use. It can be seen that the **W**-states use the function $\mathbf{hebb}_{\mu}(..)$ (see (4)) and all other states use $\mathbf{alogistic}_{\sigma,\tau}(..)$ (see (3)), except the sensor states ss_{te} and ss_{tr} which use $\mathbf{eucl}_{n,\lambda}(..)$ (see (2)) and ss_{emdr} which uses $\mathbf{stepmod}_{\rho,\delta}(..)$. Moreover, **mcfp** specifies the parameter values used for each state's combination function, for example: $\mu = 1$, $n = 1$, $\lambda = 1$, $\rho = 140$, $\delta = 60$, and some values for σ and τ. Role matrix **ms** specifies the speed factors for all states. Note that here in the darker red cells for X_{12} (= $\mathbf{W}_{psb,csb}$) and X_{13} (= $\mathbf{W}_{fsb,csb}$) it is indicated that states X_{14} (= $\mathbf{H}_{\mathbf{W}psb,csb}$) and X_{15} (= $\mathbf{H}_{\mathbf{W}fsb,csb}$) play the role of adaptive speed factor. Finally, the initial values are found in matrix **iv**, specifying the levels of certain values at the beginning of the simulation. In the role matrices **mcw** and **ms** the cells marked red indicate the reification states used for adaptive connection weights and speed factors.

5 Simulation Results

In this section, one of the simulations performed is discussed and the results are analyzed. In Fig. 2, a graph of the example simulation is shown displaying the temporal behaviour of the adaptive network model's states. Initially ss_{te} and ss_{tr} already have a level 1. All the sensory representation states srs_{te} and srs_{tr} start rising gradually. The rising sensory representation states activate internal body and emotion responses, represented by the rise of ps_b and fs_b. In their mutual cycle, the preparation state ps_b and the feeling state fs_b affect each other by a predictive as-if body loop [8], before performing an expression action. Once EMDR starts at 60, the simulated person has an amygdala level represented by fs_b above 0.8. In the simulation, the EMDR therapy ends at time 150, which is an average for the individual duration of EMDR therapy; usually between a few weeks and a few months [38]. As a result of the EMDR treatment, a downregulated amygdala makes the **H**-states rise after time 110, initiating a shift from negative to positive metaplasticity by increasing hippocampal expression of BDNF. As a result, by the restored plasticity, cognitive control enhances in an adaptive manner. The improvement of cognitive control authorizes the **W**-states to rise, meaning connection weights become adaptive, eventually allowing synaptic plasticity (Fig. 2).

This leads to increase of the **W**-states after time 140. This in turn causes the feeling state fs_b to remain low. At this point, fear extinction learning and memory reconsolidation is achieved. Trauma-related states srs_{te} and as_{te} also show a significant reduction, also after the EMDR therapy finished, meaning that the intensity and awareness of the

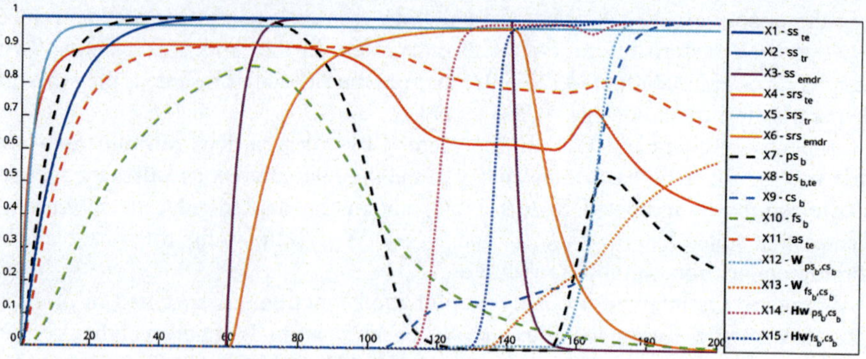

Fig. 2. Simulation results of the second-order adaptive network model.

trauma is not so intrusive anymore and memory reconsolidation is achieved. The trauma can still be remembered, but will stop causing psychotraumatic symptoms.

6 Discussion

In this paper, a computational model has been presented for the process of an occurring traumatic event, its resulting PTSD, the treatment done by EMDR and the effect of the treatment. This model is based on a neurological foundation as presented in its corresponding section. Knowledge of the neural basis of EMDR is still in its infancy but is showing great prospect. The model introduced here has some elements in common with the model described in [14]. However, the focus on EMDR is new, and also addressing both plasticity and metaplasticity. By the new approach to adaptive networks by network reification and it dedicated generic software environment [41, 42], in contrast to [14] it became possible to design the introduced adaptive network model at the modelling level without doing any programming. Moreover, in the same way this enabled to easily add (again without any programming) second-order adaptation to model metaplasticity that plays a crucial role in stress-related disorders, as shown by Garcia [17]. Also this second-order adaptation was not addressed in [14].

The obtained computational model provides insights in EMDR as a treatment for PTSD and explains how traumatic memories are resolved by applying EMDR. A number of the most relevant phenomena in the brain were modelled, providing a model for treating adaptive changes in synaptic plasticity, resource competition and memory reconsolidation. EMDR therapy was used as a treatment that mends malfunctioning memory processes that occur during the recording of traumatic events.

The doors are open for a plethora of further research. Studies by Bergmann [2–4] displayed interesting brain activity, such as the resetting of pacemaker cells in the setup, known for their use in the greater hippocampal system [2]. In another study, Bergmann evaluates evidence that EMDR triggers processing in the cerebellum [3]. Furthermore, he postulates that EMDR increases thalamic activity, which is decreased in PTSD [4]. The study done in this paper can be extended by incorporating the mentioned responsible brain areas. Adding these to the model will provide more detailed insights into the influence of the cerebellum and/or thalamus on resource competition and shifting synaptic plasticity.

Furthermore, saccadic eye-movements play a role in non-visual cognitive tasks [12]. The study mentions the 'Looking at nothing' phenomenon. This phenomenon occurs when individuals periodically look away during conversations. It is postulated that they actively search for visual fields with no useful information, which brought forth more research on visual eye-movements and memory. To support this, an explanation surfaced that studies found the rates of non-visual eye movements increase in long-term memory and episodic recall. Also, it is shown that restriction of eye movements impairs memory performance [23, 26].

Finally, eye-movements are deliberately used to search and retrieve memories, which is an important characterization of REM sleep [7]. The model presented in this study can also be extended for a comparison of the traumatic memory state before and after a night's sleep. This can be done by adding sleeping states representing REM and slow-wave sleep, with neurological evidence provided in [29] and the hypothetical basis provided in [40].

References

1. Abraham, W.C., Bear, M.F.: Metaplasticity: the plasticity of synaptic plasticity. Trends Neurosci. **19**(4), 126–130 (1996). https://doi.org/10.1016/S0166-2236(96)80018-X
2. Bergmann, U.: Speculations on the neurobiology of EMDR. Traumatology **4**(1), 4–16 (1998). https://doi.org/10.1177/153476569800400102
3. Bergmann, U.: Further thoughts on the neurobiology of EMDR: the role of the cerebellum in accelerated information processing. Traumatology **6**(3), 175–200 (2000). https://doi.org/10.1177/153476560000600303
4. Bergmann, U.: The neurobiology of EMDR: exploring the thalamus and neural integration. J. EMDR Pract. Res. **2**(4), 300–314 (2008). https://doi.org/10.1891/1933-3196.2.4.300
5. Bomyea, J., Amir, N., Lang, A.J.: The relationship between cognitive control and posttraumatic stress symptoms. J. Behav. Ther. Exp. Psychiatry **43**(2), 844–848 (2012). https://doi.org/10.1016/j.jbtep.2011.12.001
6. Cahill, S.P., Carrigan, M.H., Frueh, B.C.: Does EMDR work? And if so, why?: a critical review of controlled outcome and dismantling research. J. Anxiety Disord. **13**(1-2), 5–33 (1999). https://doi.org/10.1016/S0887-6185(98)00039-5
7. Chamberlin, D.E.: The predictive processing model of EMDR. Front. Psychol. **10**, 1–14 (2019). https://doi.org/10.3389/fpsyg.2019.02267
8. Damasio, A.R.: Looking for Spinoza: Joy, Sorrow, and the Feeling Brain. Vintage Books, London (2003)
9. Davidson, P.R., Parker, K.C.H.: Eye movement desensitization and reprocessing (EMDR): a meta-analysis. J. Consult. Clin. Psychol. **69**(2), 305–316 (2001). https://doi.org/10.1037/0022-006X.69.2.305

10. Denkova, E., Dolcos, S., Dolcos, F.: Neural correlates of 'distracting' from emotion during autobiographical recollection. Soc. Cogn. Affect. Neuro-Sci. **10**(2), 219–230 (2015). https://doi.org/10.1093/scan/nsu039
11. Earnshaw, B.A., Bressloff, P.C.: Biophysical model of AMPA receptor trafficking and its regulation during long-term potentiation/long-term depression. J. Neurosci. **26**(47), 12362–12373 (2006). https://doi.org/10.1523/JNEUROSCI.3601-06.2006
12. Ehrlichman, H., Micic, D., Sousa, A., Zhu, J.: Looking for answers: eye movements in non-visual cognitive tasks. Brain Cogn. **64**(1), 7–20 (2007). https://doi.org/10.1016/j.bandc.2006.10.001
13. Elofsson, U.O.E., von Schèele, B., Theorell, T., Söndergaard, H.P.: Physiological correlates of eye movement desensitization and reprocessing. J. Anxiety Disord. **22**(4), 622–634 (2008). https://doi.org/10.1016/j.janxdis.2007.05.012
14. Formolo, D., van Ments, L., Treur, J.: A computational model to simulate development and recovery of traumatised patients. Biol. Inspired Cogn. Archit. **21**, 26–36 (2017). https://doi.org/10.1016/j.bica.2017.07.002
15. Fox, M.D., Snyder, A.Z., Vincent, J.L., Corbetta, M., Van Essen, D.C., Raichle, M.E.: From the cover: the human brain is intrinsically organized into dynamic, anticorrelated functional networks. Proc. Natl. Acad. Sci. **102**(27), 9673–9678 (2005). https://doi.org/10.1073/pnas.0504136102
16. Furini, C., Myskiw, J., Izquierdo, I.: The learning of fear extinction. Neurosci. Biobehav. Rev. **47**, 670–683 (2014). https://doi.org/10.1016/j.neubiorev.2014.10.016
17. Garcia, R.: Stress, metaplasticity, and antidepressants. Curr. Mol. Med. **2**(7), 629–638 (2002). https://doi.org/10.2174/1566524023362023
18. Harper, M.L., Rasolkhani-Kalhorn, T., Drozd, J.F.: On the neural basis of EMDR therapy: insights from qEEG studies. Traumatology **15**(2), 81–95 (2009). https://doi.org/10.1177/1534765609338498
19. Hayes, J.P., Hayes, S.M., Mikedis, A.M.: Quantitative meta-analysis of neural activity in posttraumatic stress disorder. Biol. Mood Anxiety Disord. **2**(1), 1–13 (2012). https://doi.org/10.1186/2045-5380-2-9
20. Hebb, D.O.: The Organization of Behavior: A Neuropsychological Theory. Wiley, Hoboken (1949)
21. Hendler, T., Rotshtein, P., Yeshurun, Y., Weizmann, T., Kahn, I., Ben-Bashat, D., Malach, R., Bleich, A.: Sensing the invisible: differential sensitivity of visual cortex and amygdala to traumatic context. NeuroImage **19**(3), 587–600 (2003). https://doi.org/10.1016/S1053-8119(03)00141-1
22. Hermans, E.J., Henckens, M.J.A.G., Joëls, M., Fernández, G.: Dynamic adaptation of large-scale brain networks in response to acute stressors. Trends Neurosci. **37**(6), 304–314 (2014). https://doi.org/10.1016/j.tins.2014.03.006
23. Johansson, R., Holsanova, J., Johansson, M., Dewhurst, R., Holmqvist, K.: Eye movements play an active role when visuospatial information is recalled from memory. J. Vis. **12**(9), 1256 (2012). https://doi.org/10.1167/12.9.1256
24. Kandel, E.R.: A new intellectual framework for psychiatry. Evol. Psychiatr. **67**(1), 12–39 (2002). https://doi.org/10.1016/S0014-3855(02)00104-4
25. Kolb, B., Whishaw, I.Q.: Brain plasticity and behavior. Annu. Rev. Psychol. **49**(1), 43–64 (1998). https://doi.org/10.1146/annurev.psych.49.1.43

26. Laeng, B., Bloem, I.M., D'Ascenzo, S., Tommasi, L.: Scrutinizing visual images: the role of gaze in mental imagery and memory. Cognition **131**(2), 263–283 (2014). https://doi.org/10.1016/j.cognition.2014.01.003
27. Lindholm, D., Carroll, P., Tzimagiorgis, G., Thoenen, H.: Autocrine-paracrine regulation of hippocampal neuron survival by IGF-1 and the neurotrophins BDNF, NT-3 and NT-4. Eur. J. Neurosci. **8**(7), 1452–1460 (1996). https://doi.org/10.1111/j.1460-9568.1996.tb01607.x
28. Mocchetti, I., Spiga, G., Hayes, V., Isackson, P., Colangelo, A.: Glucocorticoids differentially increase nerve growth factor and basic broblast growth factor expression in the rat brain. J. Neurosci. **16**(6), 2141–2148 (1996). https://doi.org/10.1523/jneurosci.16-06-02141.1996
29. Pagani, M., Amann, B.L., Landin-Romero, R., Carletto, S.: Eye movement desensitization and reprocessing and slow wave sleep: a putative mechanism of action. Front. Psychol. **8**, 1–7 (2017). https://doi.org/10.3389/fpsyg.2017.01935
30. Ptak, R.: The frontoparietal attention network of the human brain: action, saliency, and a priority map of the environment. Neuroscientist **18**(5), 502–515 (2012). https://doi.org/10.1177/1073858411409051
31. Rajkowska, G., Halaris, A., Selemon, L.D.: Reductions in neuronal and glial density characterize the dorsolateral prefrontal cortex in bipolar disorder. Biol. Psychiatry **49**(9), 741–752 (2001). https://doi.org/10.1016/s0006-3223(01)01080-0
32. Rousseau, P.-F., El Khoury-Malhame, M., Reynaud, E., Boukezzi, S., Cancel, A., Zendjidjian, X., Guyon, V., Samuelian, J.-C., Guedj, E., Chaminade, T., Khalfa, S.: Fear extinction learning improvement in PTSD after EMDR therapy: an fMRI study. Eur. J. Psychotraumatol. (2019). https://doi.org/10.1080/20008198.2019.1568132
33. Sack, M., Lempa, W., Steinmetz, A., Lamprecht, F., Hofmann, A.: Alterations in autonomic tone during trauma exposure using eye movement desensitization and reprocessing (EMDR)-results of a preliminary investigation. J. Anxiety Disord. **22**(7), 1264–1271 (2008). https://doi.org/10.1016/j.janxdis.2008.01.007
34. Sayed, S., Iacoviello, B.M., Charney, D.S.: Risk factors for the development of psychopathology following trauma. Curr. Psychiatry Rep. **17**(8), 1–7 (2015). https://doi.org/10.1007/s11920-015-0612-y
35. Schubert, S.J., Lee, C.W., Drummond, P.D.: The efficacy and psychophysiological correlates of dual-attention tasks in eye movement desensitization and re-processing (EMDR). J. Anxiety Disord. **25**(1), 1–11 (2011)
36. Scott, J.C., Matt, G.E., Wrocklage, K.M., Crnich, C., Jordan, J., Southwick, S.M., Krystal, J. H., Schweinsburg, B.C.: A quantitative meta-analysis of neurocognitive functioning in posttraumatic stress disorder. Psychol. Bull. **141**(1), 105–140 (2015)
37. Shapiro, F.: Efficacy of the eye movement desensitization procedure in the treatment of traumatic memories. J. Trauma. Stress **2**(2), 199–223 (1989)
38. Shapiro, F.: Eye Movement Desensitization and Reprocessing (EMDR) Therapy: Basic Principles, Protocols, and Procedures, 3rd edn. Guilford Publications, New York (2017)
39. Solomon, R.M., Shapiro, F.: EMDR and the adaptive information processing model potential mechanisms of change. J. EMDR Pract. Res. **2**(4), 315–325 (2008). https://doi.org/10.1891/1933-3196.2.4.315
40. Stickgold, R.: EMDR: a putative neurobiological mechanism of action. J. Clin. Psychol. **58**(1), 61–75 (2002). https://doi.org/10.1002/jclp.1129

41. Treur, J.: Network-Oriented Modeling for Adaptive Networks: Designing Higher-Order Adaptive Biological, Mental and Social Network Models. Springer, Cham (2020)
42. Treur, J.: Adaptive networks at the crossroad of ai and formal, biological, medical and social sciences. In: Rezaei, N. (ed.) Integrated Science. Springer, Cham (2020, in press)
43. U.S. National Center for PTSD: How Common is PTSD in Adults? https://www.ptsd.va.gov/understand/common/common_adults.asp
44. de Voogd, L.D., Kanen, J.W., Neville, D.A., Roelofs, K., Fernández, G., Hermans, E.J.: Eye-movement intervention enhances extinction via amygdala deactivation. J. Neurosci. **38**(40), 8694–8706 (2018). https://doi.org/10.1523/jneurosci.0703-18.2018
45. Vytal, K., Cornwell, B., Arkin, N., Grillon, C.: Describing the interplay between anxiety and cognition: from impaired performance under low cognitive load to reduced anxiety under high load. Psychophysiology **49**(6), 842–852 (2012). https://doi.org/10.1111/j.1469-8986.2012.01358.x
46. Zegerius, L., Waldinger, M.D.: The neurological basis of psychoanalysis. Tijdschrift voor Psychiatrie **42**(9), 675–683 (2000). (in Dutch)

Author Index

A
Abdulov, R. E., 265
Alyushin, Alexander M., 1
Alyushin, Mikhail V., 165
Alyushin, Victor M., 9
Anikin, Anton, 188
Artamonov, Alexey, 517
Atreides, Kyrtin, 15

B
Bal, Natalia A., 218
Baryshev, Gennady, 24, 32, 40
Berestov, Aleksandr, 32
Bernshtein, A., 466
Bobrysheva, Julia, 49
Boltuc, Piotr (Peter), 56
Bozhenko, A., 466
Bozhko, Yuri, 24
Bugaenko, Marina V., 391
Burnaev, E., 466

C
Chepin, Eugene V., 417, 558
Chervyakov, Vladimir, 40
Chubarov, Arthur A., 435
Connelly, Thomas P., 56

D
Davydov, Vasilii, 69
del Valle-Padilla, Juan Luis, 541
Demareva, Valeriia, 78
Devyatkina, Anastasia, 84
Dimitrichenko, Dmitriy, 91
Dmitriy, Kovtun, 253
Dolenko, Sergei A., 484
Dolenko, Sergey, 115
Dontsov, Dmitrii, 188
Dvoryankin, Sergey V., 1

E
Edeleva, Julia, 78
Enyagina, Irina, 380

F
Furnell, Steven, 270

G
Gavrilenko, Yuliya, 97
Golitsina, Olga L., 218
Gridnev, Alexander A., 417, 558
Grigoryan, Asmik, 156
Gurtueva, Irina, 327
Guseva, Anna, 106
Guseva, Anna I., 181

I
Ilyushin, Eugene, 97
Isaev, Igor, 115
Ismailova, Larisa, 565, 572, 578, 585
Ivlev, Vladimir I., 127

J
Jimenez, Juan P., 541

K
Kainova, Anna, 24
Kamennov, Yaroslav, 501
Kartashov, Sergey, 350
Kartashov, Sergey I., 138, 149, 591
Kazakov, Mukhamed, 143

A. V. Samsonovich et al. (Eds.): BICA 2020, AISC 1310, pp. 611–613, 2021.
https://doi.org/10.1007/978-3-030-65596-9

Kelley, David J., 15
Kholodny, Yuri I., 149, 591
Klimov, Valentin, 32
Klimova, Sofya, 156
Kolobashkina, Lyubov V., 165
Kosikov, Sergey, 565, 572, 578, 585
Kotov, Artemiy, 350
Kotov, Yuri, 453
Kozlov, Stanislav, 380
Kreinin, G. V., 309
Kreinin, German V., 320
Kudryavtsev, Konstantin Y., 558
Kulik, Sergey, 490
Kulik, Sergey D., 173

L
Lavrinova, Liubov I., 495
Lebedev, Alexander, 225
Lebedev, Mikhail, 365
Lebedeva, Anna V., 181
Levin, Evgeny O., 173
Litovkin, Dmitrii, 188
Liusko, Timofei, 69
Luzgina, K. S., 195
Lyutikova, Larisa A., 206, 212

M
Maksimov, Nikolay, 225
Maksimov, Nikolay V., 218
Malakhov, Denis G., 149, 591
Malanchuk, Irina, 233
Malynov, Andrey, 241
Manakhova, I. V., 195
Masi, Uplift, 15
Maslennikova, Aleksandra V., 138
Matrosov, Nikolay, 106, 253
Matrosova, Elena, 84, 106, 253
Medvedeva, Y. M., 265
Melnikova-Pitskhelauri, T., 466
Miloslavskaya, Natalia, 270, 283, 295
Misyurin, S. Yu., 309
Misyurin, Sergey Yu., 127, 320
Monankov, Kirill V., 218
Myklyuchenko, Natalia, 84

N
Nagoev, Zalimkhan, 327
Nelyubin, A. P., 309
Nelyubin, Andrey P., 127, 320
Neroznikova, Yulia M., 335, 342
Nikiforov, Andrey, 283
Nikishov, Kirill S., 165
Nosova, N. Yu., 309
Nosova, Natalia Yu., 320
Nosovets, Zakhar, 350

O
Obornev, Eugeny, 115
Obornev, Ivan, 115
Orlov, V., 466
Orlov, Vyacheslav, 350, 380
Orlov, Vyacheslav A., 138, 149, 591
Ossadtchi, Alexey, 365

P
Panov, Aleksandr I., 69
Perconti, Pietro, 359
Petrenko, Valeria, 32
Petrosyan, Artur, 365
Petrov, P., 466
Pimenova, O. Victoria, 371
Pitskhelauri, D., 466
Plaksiy, Kirill, 283
Plebe, Alessio, 359
Polevaia, Sofia, 78
Popova, G. I., 195
Poyda, Alexey, 380
Prokhorov, Igor, 241
Pronicheva, Larisa, 517
Pronin, I., 466
Putilov, Aleksandr, 40
Putilov, Aleksandr V., 391

R
Raghavachary, Saty, 400
Ramos, Félix, 541
Red'ko, Vladimir G., 410
Repkina, Olga B., 371
Rodionov, Eugeny, 115
Rudy, Michael A., 417
Rusakov, Victor A., 425
Rybka, Roman, 446

S
Saada, Daniel, 97
Samsonovich, Alexei V., 435, 484
Sboev, Alexander, 446
Semenova, Tatiana, 453
Serenko, Alexey, 446
Sharaev, M., 466
Sharaev, Maksim, 380
Shevchenko, Andrey, 97
Shibzukhov, Z. M., 472, 478
Shimelevich, Mikhail, 115
Shirokiy, Vladimir, 115
Shirokiy, Vladimir R., 484

Shmatova, Elena V., 212
Shtanko, Alexander, 490
Smirnov, A., 466
Smirnov, Dmitriy, 40
Sukhobokov, Artem A., 495
Suyuncheva, Alisa R., 547
Sychev, Oleg, 188, 501

T
Tikhomirova, Anna, 84, 106, 253
Tikhomirova, Daria V., 484
Timokhin, D. V., 508
Timokhin, Dmitriy V., 371, 391
Tokarev, Anton, 32
Tolokonskij, A. O., 553
Tolstoy, Alexander, 295
Tretyakov, Evheniy, 517
Treur, Jan, 527, 598
Tsyganov, Aleksandr, 24, 40
Tukumbetova, Rufina, 517

U
Ulizko, Mikhail, 517
Ushakov, Vadim, 350, 380
Ushakov, Vadim L., 138

V
van den Heuvel, Joey, 527
Vargas, Natividad, 541
Vartanov, Alexander V., 97, 335, 342, 547
Velichkovsky, Boris M., 350
Vladimirov, Roman D., 484
Vlasov, Danila, 446
Volodin, V. S., 553
Voznenko, Timofei I., 558

W
Wolfengagen, Viacheslav, 565, 572, 578, 585

Y
Yarkin, V., 466
Yudin, Igor, 24

Z
Zabotkina, Vera, 350
Zaidelman, Liudmila, 350
Zapechnikov, Sergey, 49
Zegerius, Lennart, 598